THE ELEMENTS

	IB	IIB	IIIA	IVA	VA	VIA	VIIA	O
							1 **H** 1.00797 ±0.00001	2 **He** 4.0026 ±0.00005
			5 **B** 10.811 ±0.003	6 **C** 12.01115 ±0.00005	7 **N** 14.0067 ±0.00005	8 **O** 15.9994 ±0.0001	9 **F** 18.9984 ±0.00005	10 **Ne** 20.183 ±0.0005
			13 **Al** 26.9815 ±0.00005	14 **Si** 28.086 ±0.001	15 **P** 30.9738 ±0.00005	16 **S** 32.064 ±0.003	17 **Cl** 35.453 ±0.001	18 **Ar** 39.948 ±0.0005
28 **Ni** 58.71 ±0.005	29 **Cu** 63.54 ±0.005	30 **Zn** 65.37 ±0.005	31 **Ga** 69.72 ±0.005	32 **Ge** 72.59 ±0.005	33 **As** 74.9216 ±0.00005	34 **Se** 78.96 ±0.005	35 **Br** 79.909 ±0.002	36 **Kr** 83.80 ±0.005
46 **Pd** 106.4 ±0.05	47 **Ag** 107.870 ±0.003	48 **Cd** 112.40 ±0.005	49 **In** 114.82 ±0.005	50 **Sn** 118.69 ±0.005	51 **Sb** 121.75 ±0.005	52 **Te** 127.60 ±0.005	53 **I** 126.9044 ±0.00005	54 **Xe** 131.30 ±0.005
78 **Pt** 195.09 ±0.005	79 **Au** 196.967 ±0.0005	80 **Hg** 200.59 ±0.005	81 **Tl** 204.37 ±0.005	82 **Pb** 207.19 ±0.005	83 **Bi** 208.980 ±0.0005	84 **Po** (210)	85 **At** (210)	86 **Rn** (222)

63 **Eu** 151.96 ±0.005	64 **Gd** 157.25 ±0.005	65 **Tb** 158.924 ±0.0005	66 **Dy** 162.50 ±0.005	67 **Ho** 164.930 ±0.0005	68 **Er** 167.26 ±0.005	69 **Tm** 168.934 ±0.0005	70 **Yb** 173.04 ±0.005	71 **Lu** 174.97 ±0.005

95 **Am** (243)	96 **Cm** (247)	97 **Bk** (247)	98 **Cf** (249)	99 **Es** (254)	100 **Fm** (253)	101 **Md** (256)	102 **No** (253)	103 **Lr** (257)

Atomic Weights are based on C^{12}—12.0000 and Conform to the 1961 Values

INTRODUCTION TO

Organic Laboratory Techniques

A CONTEMPORARY APPROACH

Third edition

INTRODUCTION TO

Organic Laboratory Techniques

A CONTEMPORARY APPROACH

Third edition

DONALD L. PAVIA   **GARY M. LAMPMAN**   **GEORGE S. KRIZ**

Western Washington University, Bellingham, Washington

SAUNDERS GOLDEN SUNBURST SERIES
SAUNDERS COLLEGE PUBLISHING
Philadelphia   New York   Chicago
San Francisco   Montreal   Toronto
London   Sydney   Tokyo

Text Typeface: Times Roman
Compositor: York Graphic Services, Inc.
Acquisitions Editor: John Vondeling
Copy Editor: Linda Kesselring
Art Director: Carol Bleistine
Art Assistant: Doris Roessner
Text Designer: Edward A. Butler
Cover Designer: Lawrence R. Didona
Production: York Production Services

Cover Credit: © COMSTOCK, Inc.

Printed in the United States of America

INTRODUCTION TO ORGANIC LABORATORY TECHNIQUES, Third Edition

ISBN 0-03-014813-8

Library of Congress Catalog Card Number: 87-28688

789 039 987654321

To All of Our Students in Chemistry 354 and 355

Preface

This new edition of *Introduction to Organic Laboratory Techniques* represents an evolution from our previous editions. Some new experiments have been added, a few have been deleted, and several experiments have been modified and improved. Carbon-13 nuclear magnetic resonance spectroscopy has been added to some of the experiments. The essential features of our original book have been retained, however. The topical essays, the thoroughly class-tested experiments, and the complete chapters on techniques have been kept. Our strong emphasis on laboratory safety remains an important feature of this edition.

New experiments in this edition include the following:

Experiment 3, *Acetaminophen*

Experiment 37, *Isolation of Carotenoid Pigments from Spinach*

Experiment 48, *The Aldol Condensation: Preparation of Benzalacetone, Benzalacetophenone, and Benzalpinacolone*

Experiment 49, *5,5-Dimethyl-1,3-cyclohexanedione (Dimedone)*

Experiment 50, *Preparation of meso-Tetraphenylporphyrin and Some Metalloporphyrins*

Experiment 54, *cis,trans-,* and *trans,trans-1,4-Diphenyl-1,3-butadiene*

Experiment 62, *Determination of the Optical Purity of ($\pm$)-α-Phenylethylamine Using Chiral Shift Reagents*

Experiments that have been modified extensively are listed below:

Experiment 26, *Chromic Acid Oxidation of Alcohols*

Experiment 29, *Markovnikov and anti-Markovnikov Hydration of Styrene* (Procedure 29A)

Experiment 30, *Triphenylmethanol and Benzoic Acid*

Experiment 33, *Friedel-Crafts Acylation*

We have also added a new essay, *Porphyrins,* and a new appendix chapter, *Carbon-13 Nuclear Magnetic Resonance Spectroscopy*. Many of the other essays and technique chapters have been updated.

The material dealing with laboratory safety and chemical hazards is current as of the time at which we prepared our latest manuscript. The nature of the study of chemical hazards is in such rapid flux, however, that new information may supersede what we have included. The instructor is cautioned, therefore, to try to stay abreast of the latest developments in this field.

We would not have been able to bring this writing project to a successful conclusion without the help of many persons. Several students worked on the development of new or modified experiments. In particular, we would like to thank Peter Bajema, Svea Bjorkstam, Caroline Chamblin, Lisa Hammond, Roger Koops, David Lyon, and Larry Wienkers. Chris Vandebrooke helped with photocopying and miscellaneous clerical chores. DeeDee Lombard and Denice Hougen provided important help in a number of areas. We must also acknowledge the valuable contribution made by succeeding classes of organic chemistry students on whom we sprung our half-baked ideas for new experiments and who had to suffer through our running modifications. We would like to thank the following reviewers for their suggestions: Lars Hellberg (San Diego State University), J. H. Looker (University of Nebraska), Eugene Losey (Elmhurst College), Robert Pinnell (Claremont Colleges), Michael Richmond (North Texas State University), George Thyvelikakath (Oral Roberts University).

We must again take special pains to thank Robert D. LaRiviere, who added new illustrations to the fine collection of drawings that he has created for our textbooks. Finally, we must thank our families and friends for their support and patience.

Donald L. Pavia
Gary M. Lampman
George S. Kriz

Contents

FOREWORD TO THE STUDENT AND WORDS OF ADVICE

Welcome to organic chemistry! Organic chemistry can be fun, and we hope to prove it to you. That word *fun* may be misleading; it doesn't mean "less work." On the contrary, the work in this laboratory course will teach you a lot, and you will have to hustle to keep abreast of it all. For you to get the most out of the laboratory course, you should strive to do several things. First, you need to understand the organization of this laboratory manual and how to use the manual effectively. It is your guide to learning. Second, you must try to understand both the purpose and the principles behind each experiment you do. Third, you must try to organize your time effectively **before** each laboratory period.

ORGANIZATION OF THE TEXTBOOK

Let us consider briefly how this textbook is organized. There are four introductory sections, of which this Foreword is the first; a section on laboratory safety is second; advance preparation and laboratory records make up the third; and laboratory glassware is the fourth. Beyond these introductory sections, the textbook is divided into two parts. Part One consists of about 60 experiments, which may be assigned as part of your laboratory course. Interspersed within Part One are various covering essays that provide background information related to the experiments. Part Two comprises a series of detailed instructions and explanations dealing with the techniques of organic chemistry. While the experiments themselves may not seem to deal explicitly with instruction in the ordinary techniques of organic chemistry, those techniques are extensively developed and used. You become familiar with them in the context of the experiments. Within each experiment, you will find a section, entitled "Special Instructions," that indicates which techniques should be studied to do that experiment. That section also lists special safety precautions and specific instructions to the student. The Appendix to this textbook contains sections dealing with infrared spectroscopy and nuclear magnetic resonance spectroscopy. While only a few experiments specifically using these spectroscopic techniques are included in Part One, the techniques can easily be incorporated in the experiments by your instructor.

ADVANCE PREPARATION

Earlier you were told that you would have to hustle to keep abreast of the material you might expect to learn in your organic chemistry laboratory course. This means that you should not treat these experiments as a novice cook would treat a Fanny Farmer cookbook. You should come to the laboratory with a plan for the use of your time and some understanding of what you are about to do. A really good cook does not follow the recipe line by line with a finger, nor does a good mechanic fix your car with the instruction manual in one hand and a wrench in the other. In addition, it is unlikely that you will learn much if you try to follow the instructions blindly, without understanding them. It can't be emphasized strongly enough that you should come to the lab **prepared.**

If there are items or techniques you don't understand, you should not hesitate to ask questions. Since other students are not always reliable sources, you should ask an instructor or a laboratory assistant. You will also learn more if you figure things out on your own. Don't rely on others to do your thinking for you.

You should read the section entitled ''Advance Preparation and Laboratory Records'' right away. Although your instructor will undoubtedly have a preferred format, much of the material here will help you in learning to think constructively about laboratory experiments in advance. You will also have to learn to keep complete records **for your own use,** no matter what the results may be. It would also save time if, as soon as possible, you read at least the first four technique sections in Part Two. These techniques are basic to all the experiments in this textbook. The laboratory class will begin with experiments almost immediately, and a thorough familiarity with this particular material will save you much valuable laboratory time. You should also read the section on safety. Knowing what to do and what not to do in the laboratory is of paramount importance, since the laboratory has many potential hazards associated with it.

BUDGETING TIME

As mentioned in the Advance Preparation section, you should have read several chapters of this book even before your first laboratory class meeting. You should also read the assigned experiment carefully before every class meeting. Having read the experiment, you should schedule your time wisely. Often you will be doing more than one experiment at a time. Experiments like the fermentation of sugar require a few minutes of advance preparation **one week** ahead of the actual experiment. At other times you will have to catch up on some unfinished details of a previous experiment. For instance, usually it is not possible to determine accurately a yield or a melting point of a product immediately after you first get the product. Products must be free of solvent to give an accurate weight or melting-point range; they have to be ''dried.'' This drying is done usually by leaving the product in an open container in your desk. Then when you have a pause in your schedule during the subsequent experiment, you can determine these missing data using a sample that is dry.

THE PURPOSE

The main purpose of an organic chemistry laboratory course is to teach you the techniques necessary for a person dealing with organic chemicals. You will learn how to handle equipment that is basic to every research laboratory. You will also learn the techniques needed for separating and purifying organic compounds. If the appropriate experiments are included in your course, you may also learn how to identify unknown compounds. The experiments themselves are only the vehicle for learning these techniques. The technique chapters in Part Two are the heart of this textbook, and you should learn them thoroughly. Your instructor may provide laboratory lectures and demonstrations explaining the techniques, but the burden is on you to master them fully by familiarizing yourself with the material in the technique chapters. Your instructor simply will not be able to explain or demonstrate everything.

In choosing experiments, we have tried whenever possible to make them relevant and, more important, interesting. To that end, we have tried to make them a learning experience of a different kind. Most experiments are prefaced by a background essay to place things in context and provide you with some new information. We hope to show you that organic chemistry pervades your lives (drugs, foods, plastics, perfumes, and so on). We may not have succeeded totally with this book, but we have made a good beginning. We hope that organic chemistry itself is the dominant aspect of the text and that the level of expected performance has not been downgraded one iota. You should leave your course well trained in organic laboratory techniques. We are enthusiastic about our subject and hope you will receive it with the same spirit.

Donald L. Pavia
Gary M. Lampman
George S. Kriz

LABORATORY SAFETY

In any laboratory course, familiarity with the fundamentals of laboratory safety is critical. Any chemistry laboratory, particularly an organic chemistry laboratory, can be a dangerous place in which to work. Understanding potential hazards will serve well in minimizing that danger for you. Remember that if you have a serious accident, it will not be reversible. You won't get a second chance!

EYE SAFETY

**Always Wear Approved
Safety Glasses or Goggles**

First and foremost, ALWAYS WEAR APPROVED SAFETY GLASSES OR GOGGLES. This sort of eye protection must be worn whenever you are in the laboratory. Even if you are not actually carrying out an experiment, a person near you might have an accident that could endanger your eyes, so eye protection is essential. Even dishwashing may be hazardous. Cases are known in which a person has been cleaning glassware only to have an undetected piece of reactive material explode, sending fragments into the person's eyes. To avoid this sort of accident, it is necessary to wear your safety glasses at all times.

**Learn Location of
Eyewash Facilities**

If there are eyewash fountains in your laboratory, you should determine which one is nearest to you. In case any chemical enters your eyes, go immediately to the eyewash fountain and flush your eyes and face with large amounts of water. If an eyewash fountain is not available, the laboratory will usually have at least one sink fitted with a piece of flexible hose. When the water is turned on, this hose can be aimed upward and directly into the face, thus working much like an eyewash fountain. Care should be taken not to set the water flow rate too high, or damage to the eyes can result.

FIRES

**Use Care With Open Flames
in the Laboratory**

No Smoking

Equally important is the need to stress caution about fire. Because an organic chemistry laboratory course deals with flammable organic solvents at all times, the

4

danger of fire is always present. Because of this danger, DO NOT SMOKE IN THE LABORATORY. Furthermore, exercise supreme caution when you light matches or use any open flame. Always check to see whether your neighbors on either side, across the bench, and behind you are using flammable solvents. If so, either delay your use of a flame or move to a safe location, such as a fume hood, to use your open flame. Many flammable organic substances are the source of dense vapors that can travel for some distance down a bench. These vapors present a fire danger, and you should be careful, since the source of those vapors may be far away from you. Do not use the bench sinks to dispose of flammable solvents. If your bench has a trough running along it, only **water** (no flammable solvents!) should be poured into it. The trough is designed to carry the water from the condenser hoses and aspirators—not flammable materials.

> **Learn Location of Fire Extinguishers,**
> **Fire Showers, and Fire Blankets**

For your own protection in case of a fire, you should learn immediately where the nearest fire extinguisher, fire shower, and fire blanket are. You should learn how these safety devices are operated, particularly the fire extinguisher. Your instructor can demonstrate how to operate it.

If you have a fire, the best advice is to get away from it and let the instructor or laboratory assistant take care of it. DON'T PANIC! Time spent in thought before action is never wasted. If it is a small fire in a flask, it usually can be extinguished quickly by placing a wire gauze screen with a ceramic fiber center or, possibly, a watch glass, over the mouth of the flask. It is good practice to have a wire screen or watch glass handy whenever you are using a flame. If this method does not take care of the fire and if help from an experienced person is not readily available, then extinguish the fire yourself with a fire extinguisher.

Should your clothing catch on fire, DO NOT RUN. Walk *purposefully* toward the nearest fire blanket or fire shower station. Running will fan the flames and intensify them. Wrapping yourself in the fire blanket will smother the flames quickly.

ORGANIC SOLVENTS: THEIR HAZARDS

> **Avoid Contact With**
> **Organic Solvents**

It is essential to remember that most organic solvents are **flammable** and will burn if they are exposed to an open flame or a match. Remember also that many are toxic or carcinogenic, or both. For example, many chlorocarbon solvents, when accumulated in the body, cause liver deterioration similar to the cirrhosis caused by the excessive use of ethanol. The body does not rid itself easily of chlorocarbons nor does it **detoxify** them; thus, they build up over time and may cause illness in the future. Some chlorocarbons are also suspected to be cancer-causing agents (carcinogens). MINIMIZE YOUR EXPOSURE. Constant and excessive exposure to benzene may cause a form of leukemia. Don't sniff benzene, and avoid spilling it on yourself. Many other solvents, such as chloroform and ether, are good anesthetics and will put you to sleep if you breathe too much of them. They subsequently cause nausea. Many of these solvents have a synergistic effect with ethanol, meaning that they enhance its effect. Pyridine causes temporary impotence. In other words, organic solvents are just as dangerous as corrosive chemicals, such as sulfuric acid, but manifest their hazardous nature in other, more subtle ways. Minimize any direct exposure to solvents, treating them with respect. The laboratory room should be well ventilated. Normal cautious handling of solvents should not cause any health problem. If you are trying to evaporate a solution in an open flask, you should do the evaporation in the hood. Excess solvents should be discarded in a container specifically intended for waste solvents, rather than down the drain at the laboratory bench.

> **Do Not Breathe
> Solvent Vapors**

If you want to check the odor of a substance, you should be careful not to inhale very much of the material. The technique for smelling flowers is **not** advisable here; you could inhale dangerous amounts of the compound. Rather, a technique for smelling minute amounts of a substance is used. You should pass a stopper moistened with the substance (if it is a liquid) under your nose. Alternatively, you may hold the substance away from you and waft the vapors toward you with your hand. But you should never hold your nose over the container and inhale deeply!

> **Solvent Hazards**
> _____
> **Learn These!**

A list of common organic solvents follows, with a discussion of toxicity, possible carcinogenic properties, and precautions that should be used when these solvents are being handled. A tabulation of the compounds currently suspected of being carcinogens can be found at the end of this chapter.

Acetic acid: Glacial acetic acid is corrosive enough to cause serious acid burns on the skin. Its vapors can irritate the eyes and nasal passages. Care should be exercised not to breathe the vapors and not to allow them to escape into the laboratory.

Acetone: Relative to other organic solvents, acetone is not very toxic. It is flammable, however. It should not be used near open flames.

Benzene: Benzene can cause damage to bone marrow, it is a cause of various blood disorders, and its effects may lead to leukemia. Benzene is considered a serious carcinogenic hazard. Benzene is absorbed rapidly through the skin. It also poisons the liver and kidneys. In addition, benzene is flammable. Because of its toxicity and its carcinogenic properties, benzene should not be used in the laboratory; some less dangerous solvent should be used instead. In this textbook, there are no experiments that call for benzene. Toluene is considered a safer alternative solvent in procedures that specify benzene.

Carbon tetrachloride: Carbon tetrachloride can cause serious liver and kidney damage as well as skin irritation and other problems. It is absorbed rapidly through the skin. In high concentrations it can cause death, owing to respiratory failure. Moreover, carbon tetrachloride is suspected of being a carcinogenic material. Although this solvent has the advantage of being nonflammable (in the past, it was used on occasion as a fire extinguisher), it should not be used routinely in the laboratory since it causes health problems. If no reasonable substitute exists, however, it must be used in small quantities, as in preparing samples for infrared (ir) and nuclear magnetic resonance (nmr) spectroscopy. In such cases, it must be used in a hood.

Chloroform: Chloroform is like carbon tetrachloride in its toxicity. It has been used as an anesthetic. However, chloroform is currently on the list of suspect carcinogens. Because of this, chloroform should not be used routinely as a solvent in the laboratory. Occasionally it may be necessary to use chloroform as a solvent for special samples. Then, it must be used in a hood. Methylene chloride is usually found to be a safer substitute in procedures that specify chloroform as a solvent. Deuterochloroform, $CDCl_3$, is a common solvent for nmr spectroscopy. Caution dictates that it should be treated with the same respect as chloroform.

1,2-Dimethoxyethane (Ethylene Glycol Dimethyl Ether or Monoglyme):   This is a relatively nontoxic solvent. Because it is miscible with water, it is a useful alternative to solvents such as dioxane and tetrahydrofuran, which are more hazardous. 1,2-Dimethoxyethane is flammable and should not be handled near open flames. On long exposure to light and oxygen, explosive peroxides may form.

Dioxane: Dioxane has been used widely in the past because it is a convenient, water-miscible solvent. It is now suspected, however, of being carcinogenic. Additionally, it is toxic, affecting the central nervous system, liver, kidneys, skin, lungs, and mucous membranes. Dioxane is also flammable and tends to form explosive peroxides when it is exposed to light and air. Because of its carcinogenic properties, it is no longer used in the laboratory unless absolutely necessary. Either 1,2-dimethoxyethane (also known as ethylene glycol dimethyl ether or Monoglyme) or tetrahydrofuran is a suitable, water-miscible alternative solvent.

Ethanol: Ethanol has well-known properties as an intoxicant. In the laboratory, the principal danger arises from fires, since ethanol is a flammable solvent. When using ethanol, take care to work where there are no open flames.

Diethyl ether: The principal hazard associated with diethyl ether is fire or explosion. Ether is probably the most flammable solvent one is likely to find in the laboratory. Because the vapors are much more dense than air, they may travel along a laboratory bench for a considerable distance from their source before being ignited. Before you begin to use ether, it is very important to be certain that no one is working with matches or any open flame. Ether is not a particularly toxic solvent, although in high enough concentrations it can cause drowsiness and perhaps nausea. It is used as a general anesthetic. Ether can form highly explosive peroxides when exposed to air. Consequently, it should never be distilled to dryness.

Hexane: Hexane may be irritating to the respiratory tract. It can also act as an intoxicant and a depressant of the central nervous system. It can cause skin irritation, since it is an excellent solvent for skin oils. The most serious hazard, however, comes from its flammable nature. The precautions recommended for the use of diethyl ether in the presence of open flames apply equally to hexane.

Ligroin: See Hexane.

Methanol: Much of the material outlining the hazards of ethanol apply to methanol. Methanol is more toxic than ethanol; ingestion can cause blindness and even death. Because methanol is more volatile, the danger of fires is more acute.

Methylene chloride (Dichloromethane):   Methylene chloride is not flammable. Unlike other members of the class of chlorocarbons, it is not currently considered as a serious carcinogenic hazard. Recently, however, it has been the subject of much serious investigation and there have been proposals to regulate it in industrial situations where workers have high levels of exposure on a day-to-day basis. Methylene chloride is less toxic than chloroform and carbon tetrachloride. It can cause liver damage when ingested, however, and its vapors may cause drowsiness or nausea.

Pentane: See Hexane.

Petroleum ether: See Hexane.

Pyridine: There is some fire hazard associated with pyridine. The most serious hazard arises from its toxicity, however. Pyridine may cause depression of the central nervous system; irritation of the skin and respiratory tract; damage to the liver, kidneys, and gastrointestinal system; and even temporary sterility. Pyridine should be treated as a highly toxic solvent. It should be handled only in the fume hood.

Tetrahydrofuran: Tetrahydrofuran may cause irritation of the skin, eyes, and respiratory tract. It should never be distilled to dryness, since it tends to form potentially explosive peroxides on exposure to air. Tetrahydrofuran does present a fire hazard.

Toluene: Unlike benzene, toluene is not considered a carcinogen. However, it is at least as toxic as benzene. It can act as an anesthetic and also cause damage to the central nervous system. If benzene is present as an impurity in toluene, then one must expect the hazards associated with benzene to manifest themselves. Toluene is also a flammable solvent, and the usual precautions about working near open flames should be applied.

Certain solvents, because of their carcinogenic properties, should not be used in the laboratory. Benzene, carbon tetrachloride, chloroform, and dioxane are among these solvents. For certain applications, however, notably as solvents for ir or nmr spectroscopy, there may be no suitable alternative solvent. When it is necessary to use one of these solvents, safety precautions will be recommended, or you will be referred to the discussion in Technique 17.

Because relatively large amounts of solvents may be used in a large organic laboratory class, one must consider safe means of storing these substances. Only the amount of solvent that is needed for a particular experiment being conducted should be kept in the laboratory room. The preferred location for bottles of solvents being used during a class period is in a hood. When the solvents are not being used, they should be stored in a fireproof storage cabinet for solvents. If possible, this cabinet should be ventilated into the fume hood system.

WASTE-SOLVENT DISPOSAL

> **Do Not Pour Flammable Solvents Into Troughs or Sinks**
>
> ---
>
> **Use Waste Containers**

Because of the toxicity and flammability hazards, it is not acceptable to dispose of solvents by pouring them down the sink. Municipal sewage-treatment plants are not equipped to remove these materials from sewage. Furthermore, with volatile and flammable materials, a spark or an open flame can cause an explosion in the sink or further down the drains.

The appropriate disposal method for waste solvents is to pour them into appropriately labeled waste-solvent containers. These containers should be placed in the hoods in the laboratory. When these containers are filled, they should be disposed of safely either by incineration or by burial in a designated hazardous-waste dump.

DISPENSING REAGENTS

Careless dispensing of reagents can cause additional safety hazards in the laboratory, unnecessary waste of expensive chemicals, and destruction of balance pans, laboratory benches, and clothing. Generally speaking, it is not a good idea to pour small amounts of chemicals from large bottles. The following discussion is intended to provide some better methods of dispensing reagents.

One should never pour from the large bottles of concentrated acids or bases into a small container. It is much safer to store these concentrated reagents in smaller, labeled bottles, from which pouring can be accomplished more easily. Alternatively, a pipet or a dropper can be supplied with each acid or base, so that small amounts of the chemical can be obtained conveniently. A practical method is to tape a test tube to the side of each large reagent bottle. This test tube is used to store a pipet or dropper to be used with that particular reagent. In this way, it is not necessary to pick up that large reagent bottle, and problems of cross-contamination are minimized. It is advisable to wear protective gloves when dispensing concentrated acids or bases.

One should never pour from large solvent or reagent bottles into small flasks. A better method is to pour from the large bottle into an intermediate-sized beaker, and then to pour from the beaker into the small flask. In some cases, a funnel can be helpful. Any remaining solvent should never be returned to the original bottle; it should be poured into the appropriate waste-solvent container. Therefore you should be frugal in your estimation of amounts needed.

It is very poor technique to pour chemicals directly from the reagent bottle into a container resting on a balance pan. This is especially true when the quantity desired weighs less than about 5 grams (g). With liquids, the chemical should be added to the container on the balance by a dropper or pipet. With solids, a scoopula or a spatula should be used.

In using a pipet to dispense liquids, you should never attempt to fill the pipet by mouth suction. A much safer practice is to use a rubber pipet bulb to fill the pipet. If mouth suction is used, there is a danger of filling the mouth with toxic or corrosive liquids or harmful vapors.

USE OF FLAMES

Even though organic solvents are frequently flammable (for example, hexane, ether, methanol, acetone, petroleum ether), there are certain laboratory procedures for which a flame may be used. Most often these procedures involve an aqueous solution. In fact, as a general rule a flame should be used to heat only aqueous solutions. Most organic solvents boil substantially below the boiling point of water (100 °C), and a steam bath may be used to heat these solvents. A listing of common organic solvents is given in Table 1–1, p. 505, of Technique 1. Solvents marked in that table with boldface type will burn. Ether, pentane, and hexane are especially dangerous, since in combination with the correct amount of air, they explode.

Some common sense rules apply to using a flame in the presence of flammable solvents. Again, we stress that you should check to see whether anyone in your vicinity is using flammable solvents before you ignite any open flame. If someone is using such a solvent, move to a safer location before you light your flame. Remember, IF THE SOLVENT BOILS BELOW 80 TO 85 °C, USE A STEAM BATH FOR HEATING. The drainage troughs or sinks should never be used in disposing of flammable organic solvents. They will vaporize if they are low boiling and may encounter a flame further down the bench on their way to the sink. If it is not prudent to use a flame at your bench, move to a safer location for your operations. Methods for heating solvents safely are discussed in detail in Technique 1.

INADVERTENTLY MIXED CHEMICALS

To avoid unnecessary hazards of fire and explosion, **never pour any reagent back** into a stock bottle. There is always the chance that you may accidentally pour back some foreign substance that will react explosively with the chemical in the stock bottle. Of course, by pouring reagents back into stock bottles you may introduce impurities that could spoil the experiment for the person using the stock reagent after you. Pouring things back into bottles is thus not only a dangerous practice; it is also inconsiderate.

UNAUTHORIZED EXPERIMENTS

You should never undertake any unauthorized experiments. The chances of an accident are high, particularly with an experiment that has not been completely checked to reduce the hazard. You should never work alone in the laboratory. Minimal safety considerations require that another person should be present also.

FOOD IN THE LABORATORY

Because all chemicals are toxic to some extent, you should avoid accidentally ingesting toxic substances; therefore, never eat or drink any food in the laboratory. There is always the possibility that whatever you are eating or drinking may become contaminated with a potentially hazardous material.

SHOES

You should always wear shoes in the laboratory. Even open-toed shoes or sandals offer inadequate protection against spilled chemicals or broken glass.

FIRST AID: CUTS, MINOR BURNS, AND ACID OR BASE BURNS

> **Any Injury, No Matter How Small,
> Must Be Reported to Your
> Instructor as Soon as Practicable**

If any chemical enters your eyes, immediately irrigate the eyes with copious quantities of water. Slightly warm water, if it is available, is preferable. Be sure that the eyelids are kept open. Continue flushing the eyes in this way for 15 minutes.

In case of a cut, wash the wound well with water, unless you are specifically instructed to do otherwise. If necessary, apply pressure to the wound to stop the flow of blood.

Minor burns caused by flames or contact with hot objects may be soothed by immediately immersing the burned area in cold water or cracked ice for about 5 minutes. Applying salves to burns is discouraged. Severe burns must be examined and treated by a physician.

For chemical acid or base burns, the burned area should be rinsed first with copious quantities of water. After thorough rinsing, as an option, one of the following neutralizing solutions may be applied. For acid burns, apply a dilute solution of sodium bicarbonate (no substitutes). For burns by strong bases, a dilute solution of a weak acid, such as 2% acetic acid solution, a 25% vinegar solution, or a 1% boric acid solution, may be applied. Follow the use of an optional neutralizing solution by more rinsing with water. Most well-equipped laboratories always have these safety solutions available in containers clearly marked ''FOR ACID BURNS'' and ''FOR BASE BURNS.'' You should learn where these safety solutions are. After applying either sodium bicarbonate or dilute acid solution, flush the affected area with water for 10 to

15 minutes. Do not use the safety solutions for chemical purposes. If you do, they may not be available when they are really needed.

If you accidentally ingest a chemical, immediately begin drinking large quantities of water while proceeding immediately to the nearest medical assistance. It is important that the examining physician be informed of the exact nature of the substance ingested.

CARCINOGENIC SUBSTANCES

The accompanying table is taken from a list published by the Occupational Safety and Health Administration (OSHA) of substances suspected of being carcinogens. The table is not complete but merely lists substances likely to be found in an organic chemistry laboratory. Complete listings can be found in *Chemical and Engineering News,* July 31, 1978, p. 20 and *NIH Guidelines for the Laboratory Use of Chemical Carcinogens,* May 1981, p. 11. An additional list of suspected carcinogens has been compiled by the U.S. Department of Labor. Some substances from this list are also included in the table.

Acetamide	Hydrazine and its salts
Acrylonitrile	Lead(II) acetate
4-Aminobiphenyl	Methyl chloromethyl ether
Asbestos	Methyl methanesulfonate
Aziridine (Ethyleneimine)	*N*-Methyl-*N*-nitrosourea
Benzene	4-Methyl-2-oxetanone
Benzidine	(β-Butyrolactone)
Bis(2-chloroethyl) sulfide	1-Naphthylamine
Bis(chloromethyl) ether	2-Naphthylamine
Carbon tetrachloride	4-Nitrobiphenyl
Chloroform	N-Nitrosodimethylamine
Chromic oxide	2-Oxetanone (β-Propiolactone)
Coumarin	Phenylhydrazine and its salts
Diazomethane	Polychlorinated biphenyl
1,2-Dibromo-3-chloropropane	Progesterone
1,2-Dibromoethane	Tannic acid
Dimethyl sulfate	Tannins
p-Dioxane	Testosterone
Ethyl carbamate	Thioacetamide
Ethyl diazoacetate	Thiourea
Ethyl methanesulfonate	*o*-Toluidine
1,2,3,4,5,6-Hexachloro-	Trichloroethylene
cyclohexane	Vinyl chloride

Besides the specific substances listed, there are some general classes of compounds cited by OSHA that tend to have carcinogenic behavior. They are listed below.

Alkylating reagents
Androgens
Arsenic and its compounds

Azo and diazo compounds
Beryllium and its compounds
Cadmium and its compounds
Chromium and its compounds
Estrogens
Hydrazine derivatives
Lead(II) compounds
Nickel and its compounds
Nitrogen mustards (β-halo amines)
N-Nitroso compounds
Polyhalogenated compounds
Polycyclic aromatic amines and hydrocarbons
Sulfur mustards (β-halo sulfides)

REFERENCES

Green, M. E., and Turk, A. *Safety in Working with Chemicals*. New York: Macmillan, 1978.

McKusick, B. C. ''Prudent Practices for Handling Hazardous Chemicals in Laboratories.'' *Science*, 211 (February 20, 1981): 777.

Merck Index. 10th ed. Rahway, N.J.: Merck and Co., 1983.

''OSHA Issues Tentative Carcinogen List.'' *Chemical and Engineering News*, 56 (July 31, 1978): 20.

Prudent Practices for Handling Hazardous Chemicals in Laboratories. Washington, D.C.: Committee on Hazardous Substances in the Laboratory, National Research Council, National Academy Press, 1981.

Prudent Practices for Disposal of Chemicals from Laboratories. Washington, D.C.: Committee on Hazardous Substances in the Laboratory, National Research Council, National Academy Press, 1983.

Safety in Academic Chemistry Laboratories. 3rd ed. Washington, D.C.: Committee on Chemical Safety, American Chemical Society, 1979.

Sax, N. I. *Dangerous Properties of Industrial Materials*. New York: Van Nostrand Reinhold, 1975.

Searle, C.E., ed. *Chemical Carcinogens*. ACS Monograph No. 173. Washington, D.C.: American Chemical Society, 1976.

Lenga. R. E., ed. *The Sigma-Aldrich Library of Chemical Safety Data*. Milwaukee, WI: Sigma-Aldrich Corp., 1985.

Renfrew, M. M., ed. *Safety in the Chemical Laboratory*. Easton, PA: Division of Chemical Education, American Chemical Society, 1967–1981.

NIH Guidelines for the Laboratory Use of Chemical Carcinogens. Washington, D.C.: National Institutes of Health, U.S. Department of Health and Human Services, 1981.

ADVANCE PREPARATION AND LABORATORY RECORDS

In the Foreword, mention was made of the importance of advance preparation for laboratory experimentation. Presented here are some suggestions about what specific information you should try to get in your advance studying. Much of this information must be obtained while preparing your notebook, and so the two subjects, advance study and notebook preparation, will be developed simultaneously.

An important part of any laboratory experience is learning to maintain very complete records of every experiment undertaken and every item of data obtained. Far too often, careless recording of data has resulted in mistakes, frustration, and lost time due to needless repetition of experiments. If reports are required, you will find that proper collection and recording of data can make your report writing much easier. Besides good laboratory technique and the methods of carrying out basic laboratory procedures, other things you should learn from this laboratory course are

1. How to take data carefully
2. How to record relevant observations
3. How to use your time effectively
4. How to assess the efficiency of your experimental method
5. How to plan for the isolation and purification of the substance you prepare
6. How to work safely

Because organic reactions are seldom quantitative, special problems result. Frequently reagents have to be used in large excess to increase the amount of product. Some reagents are expensive, and therefore care in the amounts of these substances used is called for. Very often, many more reactions take place than you desire. These extra reactions, or **side reactions,** may form other products besides the desired product. These are called **side products.** For these reasons, you must plan your experimental procedure carefully before undertaking the actual experiment.

THE NOTEBOOK

For recording data obtained during experiments, A BOUND NOTEBOOK SHOULD BE USED. The notebook should have consecutively numbered pages. If it does not, you should number the pages immediately. A spiral-bound notebook or any other type from which the pages can be removed easily is not acceptable, since the possibility of losing the pages is great.

The notebook is the place where all primary data must be recorded. Paper towels, napkins, toilet tissue, or scratch paper have a tendency to become lost or destroyed. It is bad laboratory practice to record primary data on such random and perishable pieces of paper. Data must be recorded in **permanent** ink. It can be frustrating to have important data disappear from the notebook because they were recorded in washable ink and could not survive a flood caused by the student at the next position on the bench. Because you will be using the notebook in the laboratory during your experiments, it is quite likely that the book will become soiled or stained by chemicals, filled with scratched-out entries, or even slightly burned. This damage is to be expected.

Your instructor may wish to check your notebook at any time, so you should always have it up to date. If your instructor requires reports, you can prepare them expeditiously from the material recorded in the laboratory notebook.

NOTEBOOK FORMAT: ADVANCE PREPARATION

It is reasonable that individual instructors vary greatly in the type of notebook format they prefer; such variation stems from differences in philosophies and experience. You will have to get specific directions in preparing a notebook from your own instructor. Certain features, however, are common to most notebook formats. The following discussion presents what might be included in a typical notebook.

You can save much time in the laboratory if you understand fully the procedure of the experiment and the theory underlying it. Knowing, before you come to the laboratory, the main reactions, the potential side reactions, the mechanism, the stoichiometry, and the procedure is very helpful. Understanding the procedure by which the desired product is to be separated from undesired materials is also very important. If you have examined each of these topics before coming to class, you will be prepared to do the experiment efficiently. You will have your equipment and reagents prepared ahead of when they are to be used. Your reference material will be at hand when you need it. Finally, with your time efficiently organized, you will be able to take advantage of long reaction or reflux periods to perform other tasks, such as doing shorter experiments or finishing previous ones.

For experiments in which a compound is synthesized from other reagents, that is, a **preparative** experiment, it is essential to know the main reaction. For easier subsequent stoichiometric calculations, the equation for the main reaction should be balanced. Therefore, before you begin the experiment, your notebook should contain the balanced equation for the pertinent reaction. Using the preparation of aspirin (Experiment 1) as an example, you would write

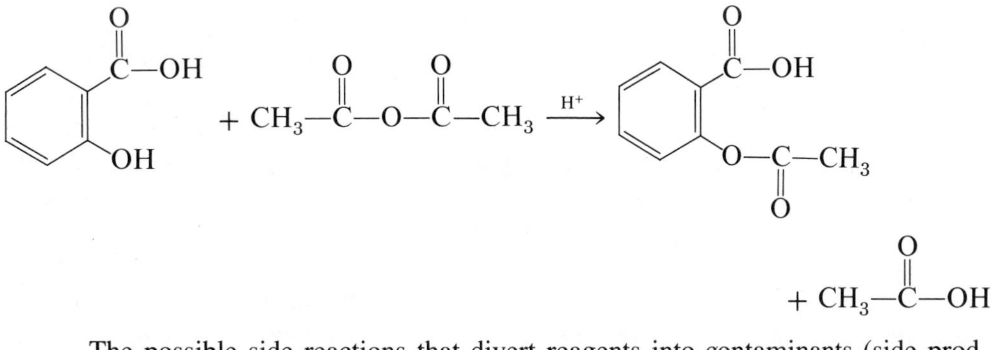

The possible side reactions that divert reagents into contaminants (side products) should, if they are known, also be entered in the notebook before the experiment is begun. These side products will have to be separated from the major product during purification. In the preparation of aspirin, the principal side reaction is the reaction of salicylic acid with itself to form a polymer:

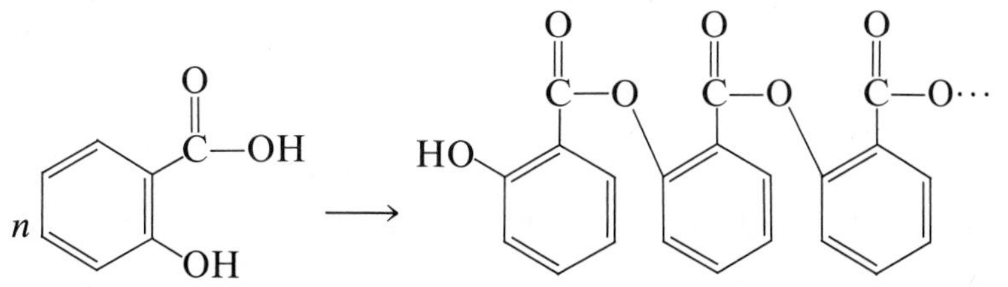

Other forms of advance preparation should also be included in the notebook. The stoichiometry of the reaction and the actual experimental procedure should be examined to determine the **limiting reagent.** The limiting reagent is the reagent that is not present in excess and on which the overall yield of product depends. You will need this information subsequently as a basis for calculating yield. In the aspirin experiment, the limiting reagent is **salicylic acid.** Such useful data as melting points, boiling points, molecular weights, and densities may also be recorded. These data are located in such sources as the *Handbook of Chemistry and Physics,* the *Merck Index,* or Lange's *Handbook of Chemistry.*

Advance preparation may also include examination of some subjects, information not necessarily recorded in the notebook, that would prove useful in understanding the experiment. Included among such subjects are an understanding of the mechanism of the reaction, an examination of other methods by which the same compound might be prepared, and a detailed study of the experimental procedure. Many students find that an outline of the procedure, prepared **before** they come to class, helps them use their time more efficiently once they begin the experiment. Such an outline could very well be prepared on some loose sheet of paper, rather than in the notebook itself.

Once the reaction has been completed, the desired product does not magically appear as purified material; it must be isolated from a frequently complex mixture of side products, unreacted starting materials, solvents, and catalysts. You should try to outline a **separation scheme** in your notebook for isolating the product from its contaminants. At each stage you should try to understand the reason for the particular instruction given in the experimental procedure. This not only will familiarize you with the basic separation and purification techniques used in organic chemistry but also will help you understand when to use these techniques. Such an outline might take the form of a flowchart; an example, again taken from the aspirin experiment, is the flowchart on the facing page. Careful attention to understanding the separation may, besides familiarizing you with the procedure by which the desired product is separated from impurities in your particular experiments, also prepare you for original research, where no experimental procedure exists.

In designing a separation scheme, you should note that such a scheme outlines those steps undertaken once the reaction period has been concluded. For this reason, the represented scheme did not include such steps as the treatment of salicylic acid with acetic anhydride, the addition of sulfuric acid, or the heating of the reaction mixture.

For experiments in which a compound is isolated from a particular source and not prepared from other reagents, much of the information described in this section will not be appropriate. Such experiments are called **isolation experiments.** A typical isolation experiment involves isolating a pure compound from a natural source. Some examples include the isolation of caffeine from coffee or the isolation of nicotine from tobacco. Although isolation experiments require somewhat different advance preparation, this advance study may include determining physical constants for the compound isolated and outlining the isolation procedure. A detailed examination of the separation scheme is very important here, since this is the heart of such an experiment.

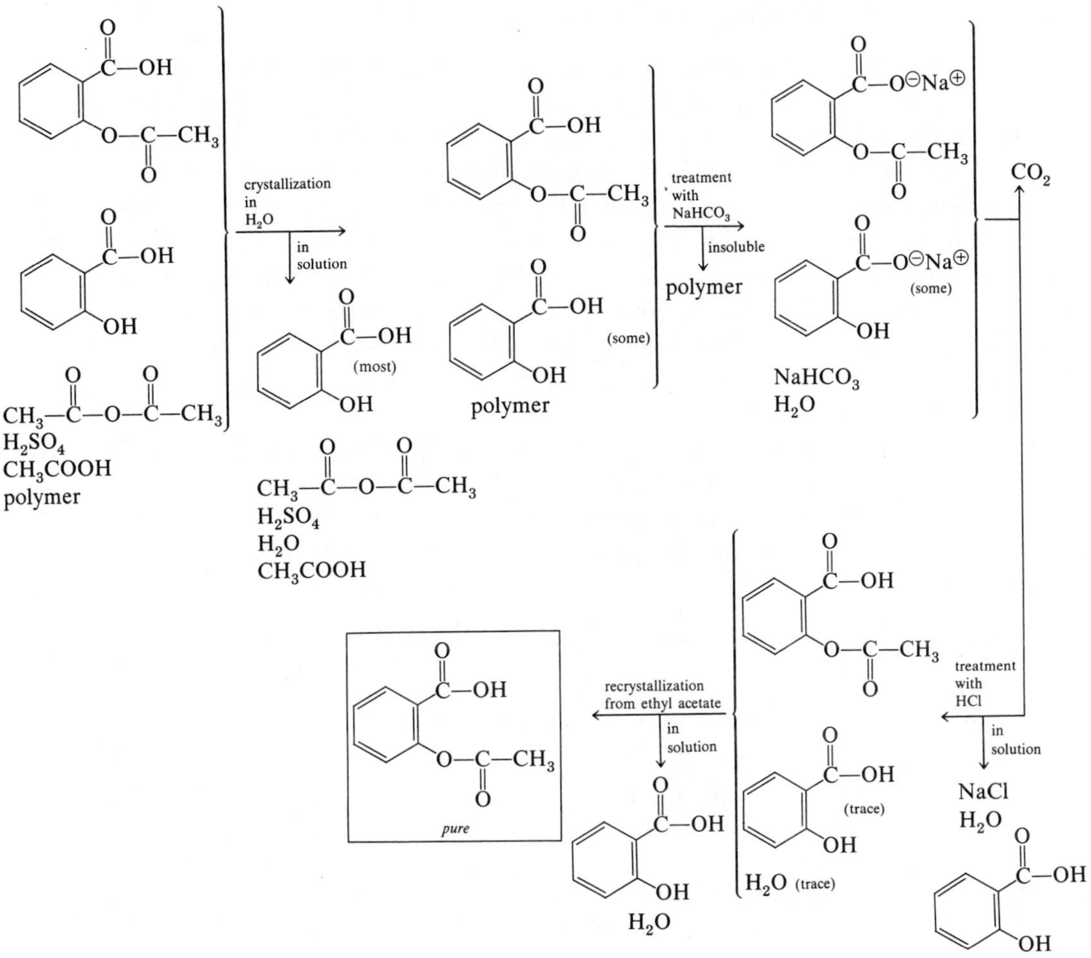

NOTEBOOK FORMAT: LABORATORY RECORDS

When you begin the actual experiment, your notebook should be kept nearby so that you will be able to record in it those operations you perform. When you are working in the laboratory, the notebook serves as a place where a rough transcript of your experimental method is recorded. In it, data from actual weighings, volume measurements, and determinations of physical constants, and the like are noted. This section of your notebook should **not** be prepared in advance. The purpose here is not to write a recipe; it is to provide a record of what you did and what you **observed.** These observations will help you to write reports without having to resort to memory. They will also help you or other workers to repeat the experiment **exactly.**

When your product has been prepared and purified, or isolated if this is an isolation experiment, you should record such pertinent data as the melting point or boiling point of the substance, its density, its index of refraction, and the conditions under which spectra were determined.

NOTEBOOK FORMAT: CALCULATIONS

A chemical equation for the overall conversion of the starting materials to products is written on the assumption of simple ideal stoichiometry. Actually the ideal assumption is seldom realized. Side reactions or competing reactions will also occur, giving other products. For some synthetic reactions, an equilibrium state will be reached in which an appreciable amount of starting material still is present and can be recovered. Some of the reactant may also remain if it is present in excess or if the reaction was incomplete. A reaction involving an expensive reagent illustrates another need for knowing how far a particular type of reaction converts reactants to products. In such a case it is preferable to use the most efficient method for this conversion. Thus, information about the efficiency of conversion for various reactions is of interest to the person contemplating the use of these reactions.

The quantitative expression for the efficiency of a reaction is given by a calculation of the **yield** for the reaction. The **theoretical yield** is the number of grams of product expected from the reaction on the basis of ideal stoichiometry, with side reactions, reversibility, losses and the like ignored. The theoretical yield is calculated from the expression

Theoretical yield = (moles of limiting reagent)(ratio)(molecular weight of product)

The ratio here is the stoichiometric ratio of product to limiting reagent. In aspirin preparation, that ratio is **1.** One mole of salicylic acid, under ideal circumstances, would yield one mole of aspirin.

The **actual yield** is simply the number of grams of desired product obtained. The **percentage yield** describes the efficiency of the reaction and is determined by

$$\text{Percentage yield} = \frac{\text{actual yield}}{\text{theoretical yield}} \times 100$$

A sample calculation, with a hypothetical value for the actual yield, can be provided; aspirin preparation is again the example:

Theoretical yield =

$$(0.015 \; \text{mole salicylic acid})\left(\frac{1 \; \text{mole aspirin}}{1 \; \text{mole salicylic acid}}\right)\left(\frac{180 \; \text{g aspirin}}{1 \; \text{mole aspirin}}\right)$$

$$= 2.7 \; \text{g aspirin}$$

Actual yield = 1.7 g aspirin

$$\text{Percentage yield} = \frac{1.7 \; \text{g}}{2.7 \; \text{g}} \times 100 = 63\%$$

For experiments that have the principal objective of isolating a substance such as a natural product, rather than preparing and purifying some reaction product, the **weight percentage recovery** and not the percentage yield is of interest. This value is determined by

$$\text{Weight percentage recovery} = \frac{\text{weight of substance isolated}}{\text{weight of original material}} \times 100$$

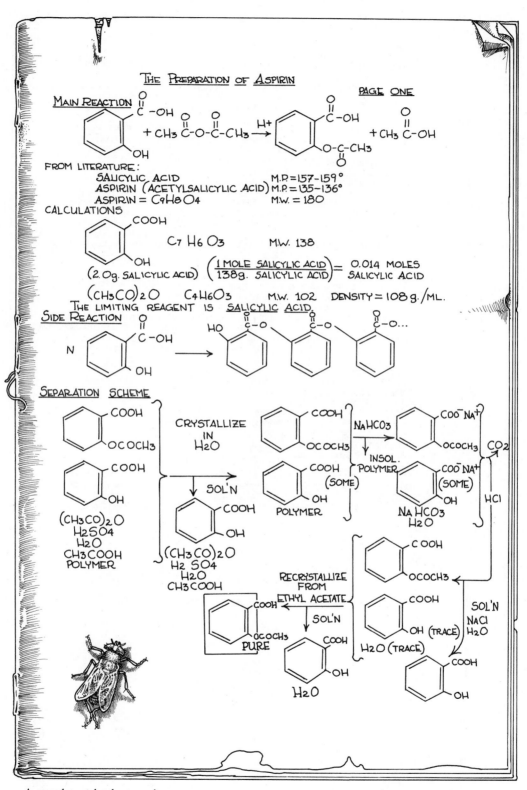

A sample notebook, page 1

PROCEDURE :   PAGE TWO

SALICYLIC ACID: SAMPLE + PAPER 3.12g.   2.00g. = 0.014 MOLES
                         PAPER 1.12g.   138g./MOLE
                         SAMPLE 2.00g. SALICYLIC ACID

THE SALICYLIC ACID WAS PLACED IN A 125ML. ERLENMEYER FLASK. ACETIC ANHYDRIDE (5ML.) WAS ADDED ALONG WITH 5 DROPS OF CONC. H_2SO_4. THE FLASK WAS SWIRLED UNTIL THE SALICYLIC ACID WAS DISSOLVED. THE SOLUTION WAS HEATED ON THE STEAMBATH FOR 10 MINUTES. THE FLASK WAS ALLOWED TO COOL TO ROOM TEMP, AND SOME CRYSTALS APPEARED. WATER (50ML.) WAS ADDED AND THE MIXTURE WAS COOLED IN AN ICE BATH. THE CRYSTALS WERE COLLECTED BY SUCTION FILTRATION, RINSED THREE TIMES WITH COLD H_2O, AND DRIED.

CRUDE YIELD: PRODUCT + PAPER 3.07g
                            PAPER 1.15g
                            1.92g. ASPIRIN

THEORETICAL YIELD = (0.014 MOLES) (180g. ASPIRIN/MOLE) = 2.52g.
ACTUAL YIELD   = 1.92g.
PERCENTAGE YIELD = 76%

THE CRUDE PRODUCT GAVE A FAINT COLOR WITH $FeCl_3$. PHENOL AND SALICYLIC ACID GAVE STRONG POSITIVE TESTS.

THE CRUDE SOLID WAS PLACED IN A 150ML. BEAKER, AND 25 ML. OF SATURATED $NaHCO_3$ WAS ADDED. WHEN THE REACTION HAD CEASED, THE SOLUTION WAS FILTERED BY SUCTION. THE BEAKER AND FUNNEL WERE WASHED WITH CA. 2 ML. OF H_2O. DILUTE HCl WAS PREPARED BY MIXING 3.5 ML. CONC. HCl, AND 10 ML. H_2O IN A 150 ML. BEAKER. THE FILTRATE WAS POURED INTO THE DILUTE ACID, AND A PRECIPITATE FORMED IMMEDIATLY. THE MIXTURE WAS COOLED IN AN ICE BATH. THE SOLID WAS COLLECTED BY SUCTION FILTRATION, WASHED THREE TIMES WITH COLD H_2O, AND PLACED ON A WATCH GLASS TO DRY OVERNIGHT. YIELD:

              PAPER AND PRODUCT: 2.78g.
                        PAPER : 1.08g.
                      PRODUCT : 1.70g.

THEORETICAL YIELD = 2.52g.
ACTUAL YIELD   = 1.70g.
PERCENTAGE YIELD = 67%   M.P. 133-135°C

THE SOLID DID NOT GIVE A POSITIVE $FeCl_3$ TEST. THE FINAL PRODUCT (CA. 0.75g.) WAS DISSOLVED IN A MINIMUM AMOUNT OF HOT ETHYL ACETATE. CRYSTALS APPEARED. THE CRYSTALS WERE COLLECTED BY SUCTION AND DRIED. M.P. 135-136°C.

A sample notebook, page 2

Thus, for instance, if 0.08 g of caffeine was obtained from 35.0 g of coffee, the weight percentage recovery of caffeine would be

$$\text{Weight percentage recovery} = \frac{0.08 \text{ g caffeine}}{35.0 \text{ g coffee}} \times 100 = 0.23\%$$

LABORATORY REPORTS

Various formats for reporting the results of laboratory experiments may be used. When original research is performed, these reports should include a detailed description of all the experimental steps undertaken. Frequently the style used in scientific periodicals such as the *Journal of the American Chemical Society* is applied to writing laboratory reports. Your instructor is likely to have his or her own requirements for laboratory reports and should detail the requirements to you.

SUBMISSION OF SAMPLES

In all preparative experiments, and in some of the nonpreparative ones, you will be required to submit the sample of the substance you prepared to your instructor. How this sample is labeled is very important. Again, learning a correct method of labeling bottles and flasks can save much time in the laboratory because fewer mistakes will be made. More important, learning to label properly can decrease greatly the danger inherent in any chemical laboratory.

Solid materials should be stored and submitted in containers that permit the substance to be removed easily. For this reason, narrow-mouthed bottles are not used for solid substances. Liquids should be stored in containers that will not let them escape through leakage. You should be careful not to store volatile liquids in containers that have plastic caps. The vapors from the liquid are likely to come in contact with the plastic and dissolve some of it, thus contaminating the substance being stored.

On the label, print the name of the substance, its melting or boiling point, the actual and percentage yields, and your name. An illustration of a properly prepared label follows:

> **Aspirin**
> **MP 135–136 °C**
> **Yield 1.7 g (63%)**
> **Joe Schmedlock**

LABORATORY GLASSWARE

Since your glassware is expensive and since you are responsible for it, you will want to give it proper care and respect. Needless maltreatment of your equipment may cost you money; so if you read this section carefully and follow the procedures presented here, you may be able to avoid some unnecessary expense. Mistreating equipment can also cause lost time in the laboratory. Cleaning problems and replacing destroyed glassware are time-consuming.

For those of you who are unfamiliar with the equipment typically found in an organic laboratory, or who are uncertain about how such equipment should be treated, this section should provide useful information. Included are such topics as cleaning glassware, greasing ground-glass joints, and caring for glassware when using corrosive or caustic reagents. At the end of this section are illustrations of most of the pieces of apparatus you are likely to find in your drawer or locker. Names are given to help you identify them.

CLEANING GLASSWARE

Glassware can be cleaned more easily if it is cleaned immediately after use. It is good practice to do your ''dishwashing'' right away. With time, the organic tarry materials left in a flask begin to attack the surface of the glass. The longer you wait to clean glassware, the more extensively this interaction of the residue with the glass will have progressed. It then becomes more difficult to wash the glassware, because water will no longer wet the surface of the glass as effectively as it did when the glass was clean.

Various washing powders and cleansers can be used as aids in dishwashing. Synthetic detergents can also be used, although they do not have the fine abrasives contained in cleansing powders, which help in the cleaning. Organic solvents can also be used in cleaning, since the residue is likely to be soluble in some organic solvent. After the solvent has been used, the flask probably will have to be washed with detergent (or cleanser) and water to remove the residual solvent. The need to wash the glassware after using an organic solvent is particularly great with a chlorocarbon solvent, such as methylene chloride. When you use solvents in cleaning glassware, use caution also, since the solvents are hazardous (see the section entitled Laboratory Safety). You should try to use fairly small amounts of a solvent for cleaning purposes. Acetone is a common solvent in cleaning glassware but is expensive. For this operation, either **waste acetone** (recycled) or **wet acetone** should be used. Do not use reagent-grade acetone for cleaning. Your **wash acetone** can be used effectively several times before it is ''spent.''

For troublesome stains and residues that insist on adhering to the glass in spite of your best efforts, a cleaning solution is effective. The most common cleaning solution consists of 35 mL of saturated aqueous sodium dichromate or potassium dichromate solution dissolved in 1 liter of concentrated sulfuric acid.[1] If you prepare the cleaning solution yourself, remember to ADD THE SULFURIC ACID TO THE SATURATED DICHROMATE SOLUTION! It is much more hazardous to add the aqueous solution to the concentrated acid, since spattering may result. The preparation of cleaning solution is hazardous; you should not undertake this operation without careful

---

[1]In preparing and using this solution, use caution, since chromium compounds are suspected carcinogens.

advance precautions. SAFETY GLASSES MUST BE WORN WHEN YOU ARE PREPARING OR USING CLEANING SOLUTION. Do not allow the cleaning solution to come into contact with your flesh or your clothing. When you are using the cleaning solution, use only small quantities. Swirl the cleaning solution in the glassware for a few minutes, and pour the remaining cleaning solution into a bottle designated for **heavy metal wastes** only. Be certain that the flask is rinsed with copious amounts of water. For most common organic chemistry applications, any stain that survives this treatment is not likely to cause difficulty in subsequent laboratory procedures.

If the glassware is contaminated with stopcock grease, the best way to remove the grease is by rinsing the glassware with a small amount of methylene chloride. Once the grease is removed, wash the glassware as described above.

The easiest way to dry glassware is to allow it to stand overnight. Flasks and beakers should be stored upside down to permit the water to drain from them. Drying ovens can be used to dry glassware if they are available and if they are not being used for other purposes. Rapid drying can be achieved by rinsing the glassware with acetone and air-drying it. For the acetone method, the glassware is thoroughly drained of water. It is then rinsed with one or two **small** portions (about 10 mL) of "wash acetone," which is a cheaper grade of acetone, suitable for cleaning glassware. Reagent-grade acetone should not be used. You should not use any more acetone than is suggested here. Using great quantities of acetone does not improve the drying but does waste expensive acetone. The used acetone is either poured down the drain or returned to a waste acetone container. Waste acetone can be used for cleaning, as mentioned before. It can also be recycled for further use in cleaning glassware.

After you rinse the flask with acetone, then dry it by blowing a **gentle** stream of compressed air into the flask. Before drying the flask with air, you should make sure that the air line is not filled with oil. Otherwise the oil will be blown into the flask, and you will have to clean it all over again. It is not necessary to blast the acetone out of the flask with a wide-open stream of air; a gentle stream of air is just as effective and will not offend the ears of other people in the room. The acetone can also be removed by aspirator suction. With either procedure, as the acetone evaporates, it carries the residual water away with it.

LUBRICATING GROUND-GLASS JOINTS

Standard-taper glassware has many advantages. It can be assembled easily into a completed apparatus without the need for laborious cork-boring, and parts can be made to fit together tightly for use under reduced pressure or with inert atmospheres. The apparatus can be disassembled and cleaned easily. One disadvantage of apparatus with ground-glass joints, however, besides the expense you may incur if you break a piece, is that the joints have some tendency to stick together.

To prevent this sticking, you can use a lubricant. Typical lubricants include a variety of hydrocarbon-based stopcock greases, such as Lubriseal and Apiezon. Silicone greases are also used. To apply the grease, you coat the inner (male) joint with a

very thin film of lubricant. You should be careful not to apply too much grease, since excess grease may contaminate the materials contained within the apparatus. A properly lubricated joint appears clear, with no striations. Contamination due to excess grease is a particular problem with silicone grease, which is much harder to remove than hydrocarbon-based greases. Because of this possible contamination, many chemists prefer to use grease only when necessary. The joints usually seal well without grease. In some situations grease is necessary, including distillations under reduced pressure, when joints must be airtight, and in reactions involving strong caustic solutions (see below). Whenever there is doubt, it never hurts to use grease. Before lubricating the joints, always make sure that they are free of any adhering liquid or solid.

ETCHING GLASSWARE

Glassware that has been used for reactions involving strong bases such as sodium hydroxide or sodium alkoxides must be cleaned thoroughly **immediately** after use. If these caustic materials are allowed to remain in contact with the glass, they will etch the glass permanently. The etching makes later cleaning attempts more difficult, since dirt particles may become trapped within the microscopic surface irregularities of the etched glass. Furthermore, the etched glass is weakened, so the lifetime of the glassware is decreased. If caustic materials are allowed to come into contact with ground-glass joints without being removed promptly, the joints will become fused. It is extremely difficult to separate fused joints without breaking them.

FROZEN JOINTS

Occasionally, ground-glass joints may become "frozen," or stuck together so strongly that it is difficult to separate them. The same is true of glass stoppers in bottles. If a joint becomes frozen, it can sometimes be loosened by tapping it gently with the wooden handle of a spatula or by tapping it gently on the edge of the bench-top. If this procedure fails, you may try heating the joint in hot water or a steam bath. A heated joint will often free more easily. If this heating fails, the instructor may be able to advise you. As a last resort, you may try heating the joint in a flame. You should not try this heating procedure unless the apparatus is hopelessly stuck. Heating by flame often causes the joint to expand rapidly and to crack or break. If you do heat, the joint should be clean and dry. The outer (female) part of the joint is heated slowly, in the yellow portion of a low flame, until it expands and breaks away from the inner section. The joint should be heated very slowly and carefully, or it may break due to rapid expansion.

ASSEMBLING THE APPARATUS

One final note of advice concerns assembling the glass components into the desired apparatus. You should always remember that Newtonian physics applies to assemblies

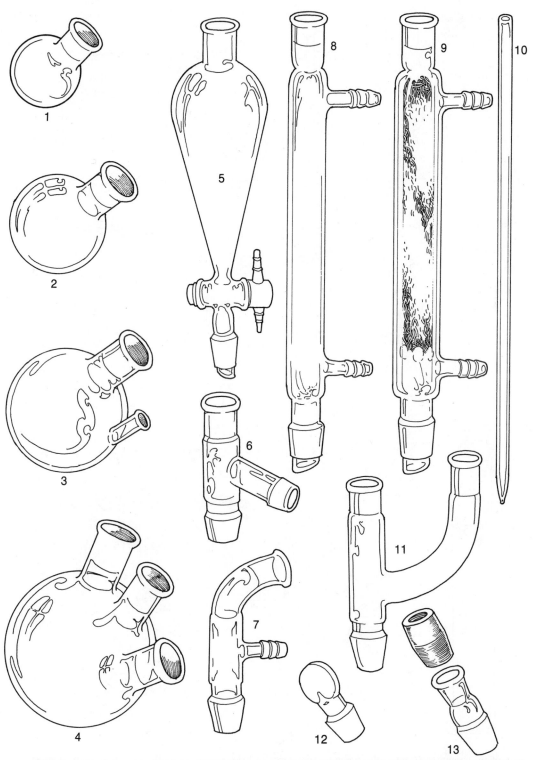

Components of the organic kit: 1) 50-mL round-bottomed boiling flask; 2) 100-mL round-bottomed boiling flask; 3) 250-mL round-bottomed boiling flask; 4) three-necked flask; 5) 125-mL separatory funnel; 6) distillation head; 7) vacuum adapter; 8) condenser (West); 9) fractionating column; 10) ebulliator tube; 11) Claisen head; 12) stopper; 13) thermometer adapter.

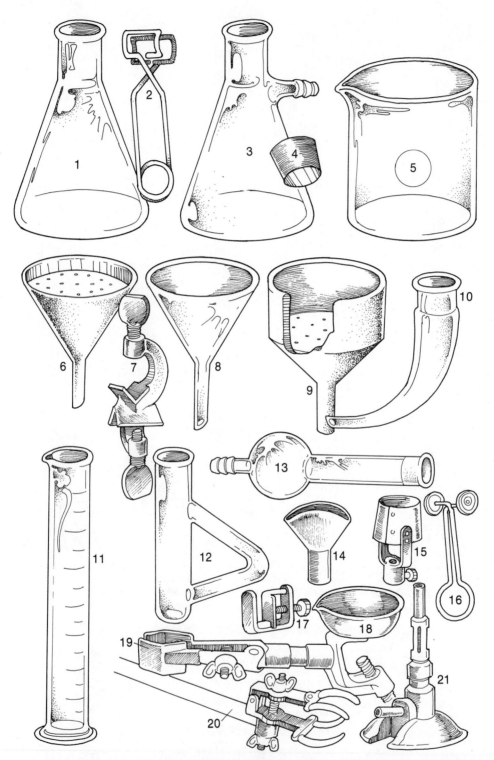

Equipment commonly used in the organic laboratory: 1) Erlenmeyer flask; 2) test tube clamp; 3) filter flask; 4) neoprene adapter; 5) beaker; 6) Hirsch funnel; 7) clamp holder; 8) conical funnel; 9) Büchner funnel; 10) bent adapter; 11) graduated cylinder; 12) melting point tube (Thiele); 13) drying tube; 14) wing top; 15) burner chimney; 16) pinch clamp; 17) screw clamp; 18) crystallizing dish (evaporating dish); 19) utility clamp; 20) condenser clamp; 21) micro burner.

of chemical apparatus and that unsecured pieces of glassware are certain to respond to gravity. You should be careful to clamp the glassware securely to a ring stand. Do not rely on friction to hold the apparatus together. This is particularly true of the vacuum adapter, which has a suicidal tendency to fall if it is not clamped. Throughout this textbook, the illustrations of the various glassware arrangements include the clamps that attach the apparatus to a ring stand. You should assemble your apparatus using the clamps as shown in the illustrations.

In assembling ground-glass equipment, you should make sure that no solid or liquid material is on the joint surfaces. Such materials will lessen the efficiency of the seal, and the joints may leak. Also, if the apparatus is to be heated, material caught between the joint surfaces will increase the tendency for the joints to stick. If the joint surfaces are coated with liquid or adhering solid, they should be wiped with a paper towel or a cloth before assembly.

When you are attaching rubber tubing to the glass apparatus or when you are inserting glass tubing into rubber stoppers, it is good practice to lubricate the rubber tubing or the rubber stopper with either water or glycerin beforehand. Without such lubrication, it can be difficult to attach rubber tubing to the sidearms of items of glassware such as condensers and suction flasks. Furthermore, glass tubing may break when it is being inserted into rubber stoppers. Water is a good lubricant for most purposes; glycerin should be used, however, when friction between glass and rubber becomes too high. Water should not be used as a lubricant when it might contaminate the reaction. If glycerin is the lubricant, be careful not to use too much.

Part One

The Experiments

Essay
ASPIRIN

Aspirin is one of the most popular cure-alls of modern life. Even though its curious history began over 200 years ago, we still have much to learn about this enigmatic remedy. No one yet knows exactly how or why it works, yet more than 20 million pounds of aspirin are consumed each year in the United States.

The history of aspirin began on June 2, 1763, when Edward Stone, a clergyman, read a paper to the Royal Society of London entitled ''An Account of the Success of the Bark of the Willow in the Cure of Agues.'' By ague, Stone was referring to what we now call malaria, but his use of the word *cure* was optimistic; what his extract of willow bark actually did was to reduce the feverish symptoms of the disease. Almost a century later, a Scottish physician was to find that extracts of willow bark would also alleviate the symptoms of acute rheumatism. This extract was ultimately found to be a powerful **analgesic** (pain reliever), **antipyretic** (fever reducer), and **anti-inflammatory** (reduces swelling) drug.

Soon thereafter, organic chemists working with willow bark extract and flowers of the meadowsweet plant (which gave a similar principle) isolated and identified the active ingredient as salicylic acid (from *salix,* the Latin name for the willow tree). The substance could then be chemically produced in large quantity for medical use. It soon became apparent that using salicylic acid as a remedy was severely limited by its acidic properties. The substance caused severe irritation of the mucous membranes lining the mouth, gullet, and stomach. The first attempts at circumventing this problem by using the less acidic sodium salt (sodium salicylate) were only partially successful. This substance gave less irritation but had such an objectionable sweetish taste that most

Salicylic acid 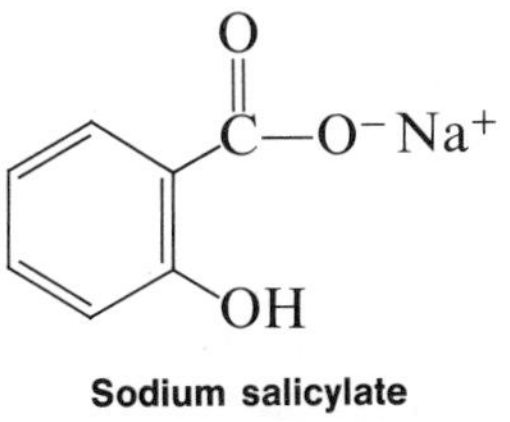   **Sodium salicylate** 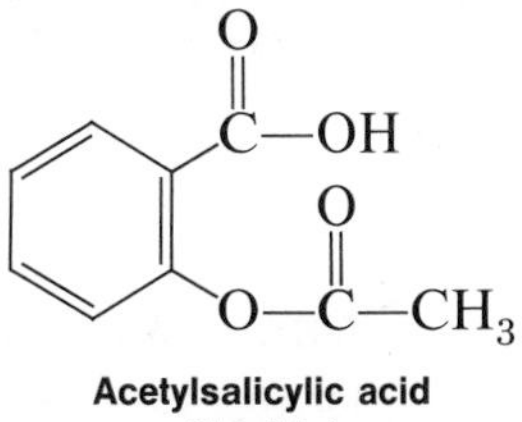   **Acetylsalicylic acid (Aspirin)**

people could not be induced to take it. The breakthrough came at the turn of the century (1893) when Felix Hofmann, a chemist for the German firm of Bayer, devised a practical route for synthesizing acetylsalicylic acid, which was found to have all the same medicinal properties without the highly objectionable taste or the high degree of mucosal-membrane irritation. Bayer called its new product ''aspirin,'' a name derived from *a* for acetyl, and the root *-spir,* from the Latin name for the meadowsweet plant, *spirea.*

31

The history of aspirin is typical of many of the medicinal substances in current use. Many began as a crude plant extract or folk remedy whose active ingredient was isolated and structure was determined by chemists, who then improved on the original.

In the last few years the mode of action of aspirin has just begun to unfold. A whole new class of compounds, called **prostaglandins,** has been found to be involved in the body's immune responses. Their synthesis is provoked by interference with the body's normal functioning by foreign substances or unaccustomed stimuli.

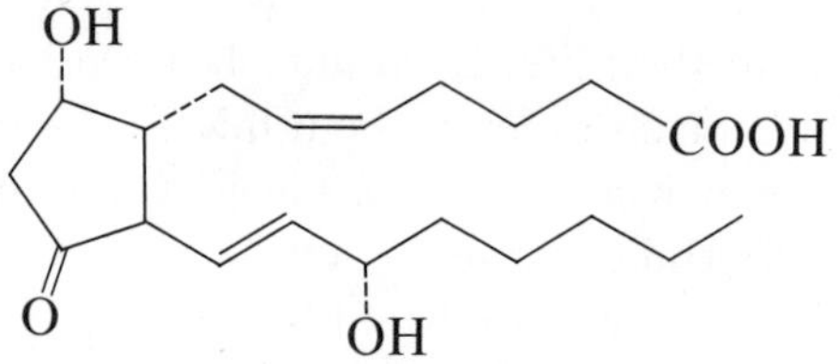

Prostaglandin E₂

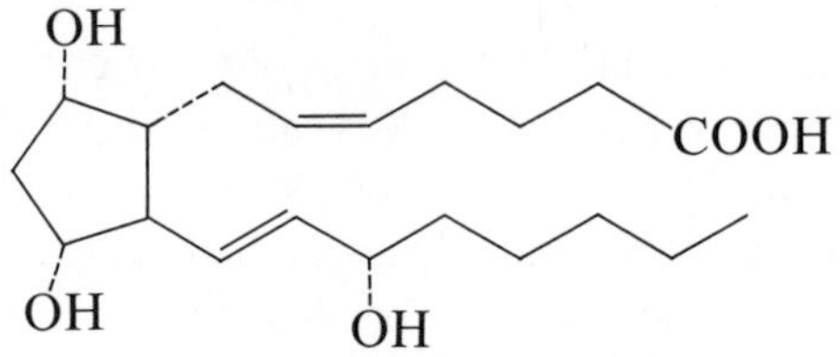

Prostaglandin F₂ₐ

These substances are involved in a wide variety of physiological processes and are thought to be responsible for evoking pain, fever, and local inflammation. Aspirin has recently been shown to prevent bodily synthesis of prostaglandins and thus to alleviate the symptomatic portion (fever, pain, inflammation, menstrual cramps) of the body's immune responses (that is, the ones that let you know something is wrong). A recent report suggests that the role of aspirin may be to inactivate one of the enzymes responsible for the synthesis of prostaglandins. The natural precursor for prostaglandin synthesis is **arachidonic acid.** This substance is converted to a peroxide intermediate by

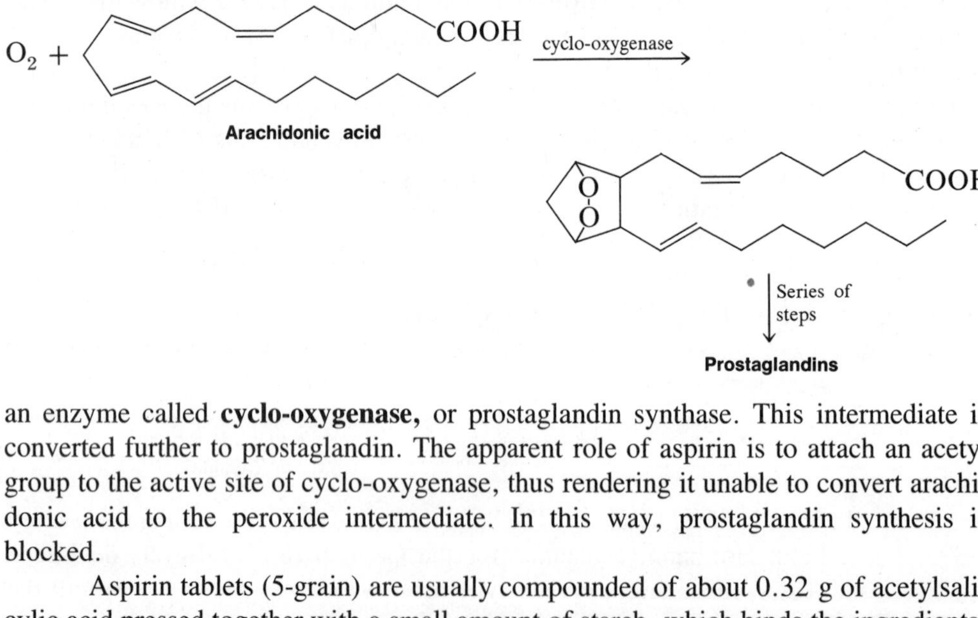

O_2 +

Arachidonic acid

cyclo-oxygenase

Series of steps

Prostaglandins

an enzyme called **cyclo-oxygenase,** or prostaglandin synthase. This intermediate is converted further to prostaglandin. The apparent role of aspirin is to attach an acetyl group to the active site of cyclo-oxygenase, thus rendering it unable to convert arachidonic acid to the peroxide intermediate. In this way, prostaglandin synthesis is blocked.

Aspirin tablets (5-grain) are usually compounded of about 0.32 g of acetylsalicylic acid pressed together with a small amount of starch, which binds the ingredients. Buffered aspirin usually contains a basic buffering agent to reduce the acidic irritation of mucous membranes in the stomach, since the acetylated product is not totally free of this irritating effect. Bufferin contains 5 grains of aspirin (1 grain = 0.0648 g), 0.75

grain of aluminum dihydroxyaminoacetate, and 1.5 grains of magnesium carbonate. Combination pain relievers usually contain aspirin, acetaminophen, and caffeine. Excedrin, for instance, contains 0.750 g aspirin, 0.250 g acetaminophen, and 0.065 g caffeine.

REFERENCES

"Aspirin Action Linked to Prostaglandins." *Chemical and Engineering News* (August 7, 1972): 4.
"Aspirin—Good News, Bad News." *Saturday Review* (November 25, 1972): 60.
Collier, H. O. J. "Aspirin." *Scientific American, 209* (November 1963): 96.
Collier, H. O. J. "Prostaglandins and Aspirin." *Nature, 232* (July 2, 1971): 17.
Pike, J. E. "Prostaglandins." *Scientific American, 225* (November 1971): 84.
Roth, G. J., Stanford, N., and Majerus, P. W. "Acetylation of Prostaglandin Synthase by Aspirin." *Proceedings of the National Academy of Science of the U.S.A., 72* (1975): 3073.
"The Uses of Aspirin." *The Medicine Show*. Mt. Vernon, N.Y.: Consumers Union, 1980.
Vane, J. R. "Inhibition of Prostaglandin Synthesis as a Mechanism of Action for Aspirin-like Drugs." *Nature-New Biology, 231* (June 23, 1971): 232.

Experiment 1
Acetylsalicylic Acid

Crystallization
Vacuum Filtration
Esterification
 (Acetylation)

Salicylic acid (*o*-hydroxybenzoic acid) is a bifunctional compound. It is a phenol (hydroxybenzene) and a carboxylic acid. Hence it can undergo two different types of esterification reactions, acting as either the alcohol or the acid partner in the reaction. In the presence of acetic anhydride, acetylsalicylic acid (aspirin) is formed, whereas in the presence of excess methanol, the product is methyl salicylate (oil of wintergreen). In this experiment we shall use the former reaction to prepare aspirin. (The synthesis of oil of wintergreen is described in another experiment.) The reaction is complicated because salicylic acid has a carboxyl as well as a phenolic hydroxyl group, and a small amount of polymeric by-product is also formed. Acetylsalicylic acid will react with

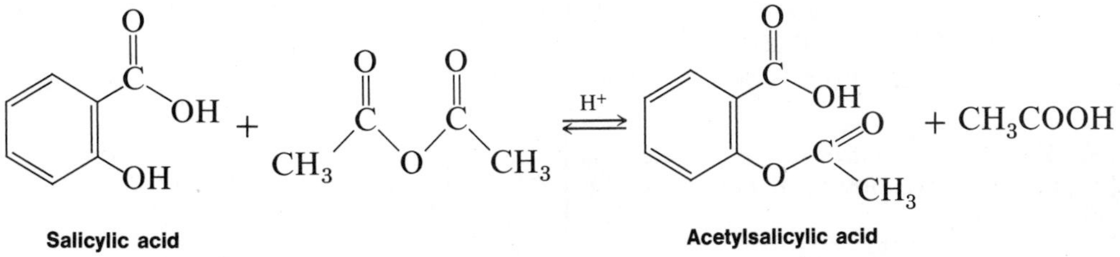

Salicylic acid   **Acetylsalicylic acid**

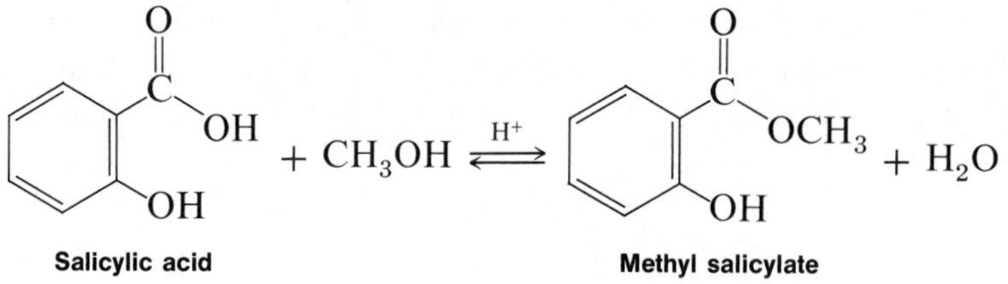

Salicylic acid   **Methyl salicylate**

sodium bicarbonate to form a water-soluble sodium salt, whereas the polymeric by-product is insoluble in bicarbonate. This difference in behavior will be used to purify the product aspirin:

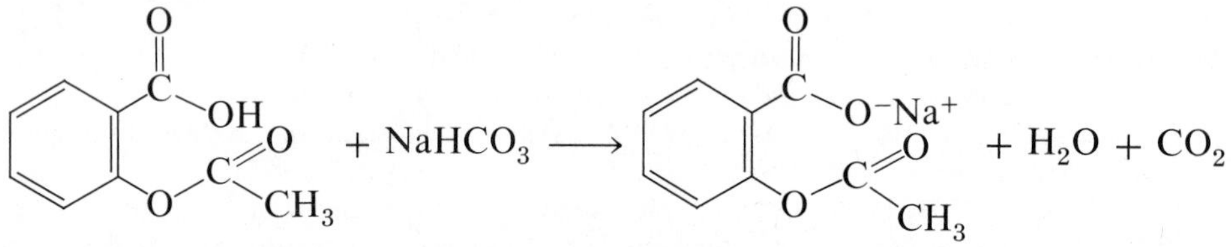

The most likely impurity in the final product is salicylic acid itself, which can arise from incomplete acetylation or from hydrolysis of the product during the isolation steps. This material is removed during the various stages of the purification and in the final crystallization of the product. Salicylic acid, like most phenols, forms a highly colored complex with ferric chloride (Fe^{3+} ion). Aspirin, which has this group acetylated, will not give the color reaction. Thus, the presence of this impurity in the final product is easily detected.

SPECIAL INSTRUCTIONS

Before beginning this experiment, you should read Techniques 1, 2, 3, and 4. There are many convenient stopping places in this experiment, but you should not leave the product in contact with sodium bicarbonate for prolonged periods. This experiment involves concentrated sulfuric acid, which is highly corrosive. Care should be exercised in handling it.

PROCEDURE

Weigh 2.0 g (0.015 mole) of salicylic acid crystals and place them in a 125-mL Erlenmeyer flask. Add 5 mL (0.05 mole) of acetic anhydride, followed by 5 drops of concentrated sulfuric acid from a dropper, and swirl the flask gently until the salicylic acid

> **CAUTION: Concentrated sulfuric acid is highly corrosive. You must handle it with great care.**

dissolves. Heat the flask gently on the steam bath for at least 10 minutes. Allow the flask to cool to room temperature, during which time the acetylsalicylic acid should begin to crystallize from the reaction mixture. If it does not, scratch the walls of the flask with a glass rod and cool the mixture slightly in an ice bath until crystallization has occurred. After crystals form, add 50 mL of water and cool the mixture in an ice bath. Do not add the water until crystal formation is complete. Usually the product will appear as a solid mass when crystallization has become complete. Collect the product by vacuum filtration on a Büchner funnel (see Technique 2, Section 2.3, p. 518, and Figure 2–4, p. 519). The filtrate can be used to rinse the Erlenmeyer flask repeatedly until all crystals have been collected. Rinse the crystals several times with small portions of *cold* water. Continue drawing air through the crystals on the Büchner funnel by suction until the crystals are free of solvent. Remove the crystals for air drying. Weigh the crude product, which may contain some unreacted salicylic acid, and calculate the percentage yield of crude acetylsalicylic acid.

PURIFICATION

In three test tubes, each containing 5 mL of water, separately dissolve a few crystals of phenol (first tube), salicylic acid (second tube), and your crude product (third tube). Add about 10 drops of 1% ferric chloride solution to each tube and note the color. Formation of an iron-phenol complex with Fe(III) gives a definite color ranging from red to violet, depending on the particular phenol present. At the end of the purification procedure you will be asked to repeat this test and note the difference between crude and purified materials.

Transfer the crude solid to a 150-mL beaker and add 25 mL of a saturated aqueous sodium bicarbonate solution. Stir until all signs (listen) of reaction have ceased. Filter the solution through a Büchner funnel. Any polymer should be left behind at this point. Wash the beaker and funnel with 5 to 10 mL of water. Prepare a mixture of 3.5 mL of concentrated hydrochloric acid and 10 mL of water in a 150-mL beaker. Carefully pour the filtrate, a small amount at a time, into this mixture while stirring. The aspirin should precipitate. If it does not, check to see whether the solution is acidic, using blue litmus paper. Enough hydrochloric acid must be added to ensure that the solution is definitely acidic. Cool the mixture in an ice bath, filter the solid by suction, using a Büchner funnel (see Technique 2, Section 2.3, p. 518), press the liquid from the crystals with a clean stopper or cork, and wash the crystals well with cold water. It is essential that the water used for this step be ice-cold. Place the crystals on a watch glass to dry. Weigh the product, determine its melting point (135–136 °C), and calculate the percentage yield. Test for the presence of unreacted salicylic acid, using the ferric chloride solution as described above.

RECRYSTALLIZATION

The product prepared according to the above instructions was isolated by *precipitation;* it should now be prepared as a pure *crystalline* substance. Water is not a suitable solvent for crystallization because aspirin will partially decompose when heated in water. Following the general instructions given in Technique 3 (pp. 526–529), dissolve the final product in a **minimum** amount of hot ethyl acetate (no more than 2–3 mL) in a 25-mL Erlenmeyer flask, while gently and continuously heating the mixture on a steam bath.[1]

When the mixture cools to room temperature, the aspirin should crystallize. If it does not, evaporate some of the ethyl acetate solvent to concentrate the solution and cool the solution in ice water while scratching the inside of the flask with a glass rod (not a fire-polished one). Collect the product by vacuum filtration, using a Büchner funnel. Any remaining material can be rinsed out of the flask with a few milliliters of cold petroleum ether. Dispose of the residual solvents in an appropriate waste container. Submit the crystalline sample in a small vial to your instructor, along with a record of its melting point. Do not forget to test the crystals with $FeCl_3$.

ASPIRIN TABLETS

Aspirin tablets are acetylsalicylic acid pressed together with a small amount of inert binding material (usually starch). One can test for the presence of starch by boiling approximately one-fourth of an aspirin tablet with 2 mL of water. The liquid is cooled, and a drop of iodine solution is added. If starch is present, it will form a complex with the iodine. The starch-iodine complex is deep blue violet. Repeat this test with a commercial aspirin tablet and with the acetylsalicylic acid prepared in this experiment.

---

[1]It will usually not be necessary to filter the hot mixture. If an appreciable amount of solid material remains, add 5 mL of additional ethyl acetate, heat the solution to boiling, and filter the hot solution by gravity into an Erlenmeyer flask through a fluted filter. Be sure to preheat the short-stemmed funnel by pouring hot ethyl acetate through it (see Technique 2, Section 2.1, p. 514, and Technique 3, Section 3.4, p. 526). Reduce the volume until crystals appear. Add a minimum additional amount of hot ethyl acetate until the crystals dissolve. Let the filtered solution stand.

QUESTIONS

1. What is the purpose of the concentrated sulfuric acid used in the acetylation reaction?

2. Give a possible structure for the polymeric by-product obtained in the reaction.

3. Why is the polymeric by-product not soluble in sodium bicarbonate solution, while salicylic acid itself is soluble?

4. If one were to use 5.0 g of salicylic acid and excess acetic anhydride in the above synthesis of aspirin, what would be the theoretical yield of acetylsalicylic acid in moles? in grams?

5. When aspirin (acetylsalicylic acid) is heated in boiling water, it decomposes. The resulting solution gives a positive $FeCl_3$ test. Why is this test positive? What is the reaction? Give an equation for the reaction.

Essay
ANALGESICS

Acylated aromatic amines (those having an acyl group, $R-\overset{\overset{\displaystyle O}{\|}}{C}-$, substituted on nitrogen) are important in over-the-counter headache remedies. Over-the-counter drugs are those you may buy without a prescription. Acetanilide, phenacetin, and acetaminophen are mild analgesics (relieve pain) and antipyretics (reduce fever) and are important, along with aspirin, in many nonprescription drugs.

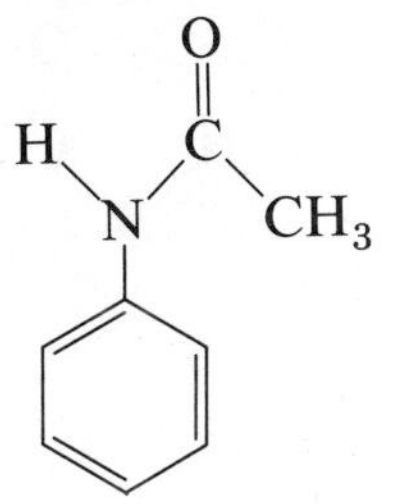

Acetanilide 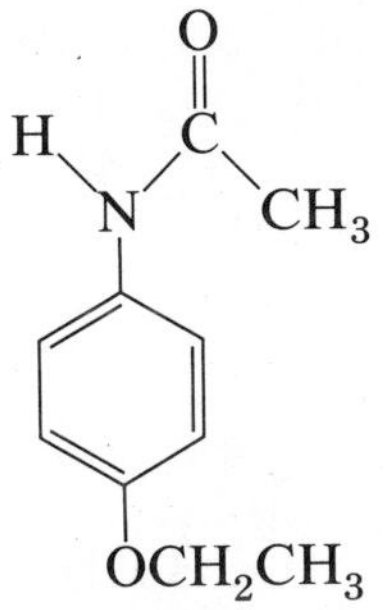   **Phenacetin** 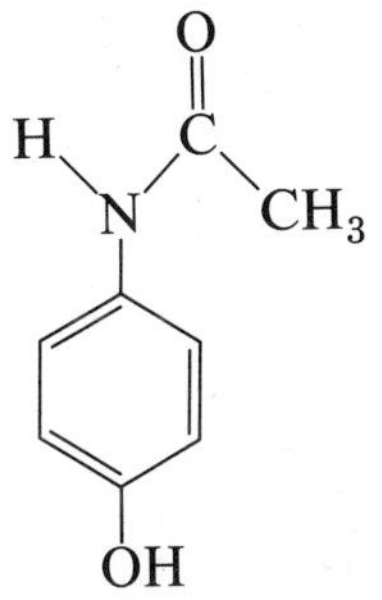   **Acetaminophen**

The discovery that acetanilide was an effective antipyretic came about by accident in 1886. Two doctors, Cahn and Hepp, had been testing naphthalene as a possible vermifuge (an agent that expels worms). Their early results on simple worm cases were very discouraging, so Dr. Hepp decided to test the compound on a patient with a larger variety of complaints, including worms—a sort of "shotgun" approach. A short time later, Dr. Hepp excitedly reported to his colleague, Dr. Cahn, that naphthalene had miraculous fever-reducing properties.

In trying to verify this observation, the doctors discovered that the bottle they thought contained naphthalene had apparently been mislabeled. In fact, the bottle brought to them by their assistant had a label so faint as to be illegible. They were sure that the sample was not naphthalene since it had no odor. Naphthalene has a strong odor reminiscent of mothballs. So close to an important discovery, the doctors were nevertheless stymied; they appealed to a cousin of Hepp, who was a chemist in a nearby dye factory, to help them identify the unknown compound. This compound turned out to be acetanilide, a compound with a structure not at all like that of naphtha-

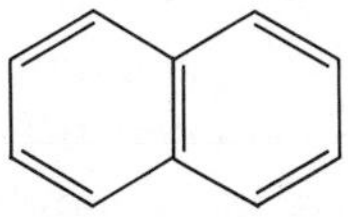

Naphthalene

lene. Certainly, Hepp's unscientific and risky approach would be frowned on by doctors today; and to be sure, the Food and Drug Administration (FDA) would never allow human testing before extensive animal testing (consumer protection has progressed). Nevertheless, Cahn and Hepp made an important discovery.

In another instance of serendipity, the publication of Cahn and Hepp, describing their experiments with acetanilide, caught the attention of Carl Duisberg, director of research at the Bayer Company in Germany. Duisberg was confronted with the problem of profitably getting rid of nearly 50 tons of p-aminophenol, a by-product of the synthesis of one of Bayer's other commercial products. He immediately saw the possibility of converting p-aminophenol to a compound similar in structure to acetanilide, by putting an acyl group on the nitrogen. It was then believed, however, that all compounds having a hydroxyl group on a benzene ring (that is, phenols) were toxic. Duisberg devised a scheme of structural modification of p-aminophenol to get the compound phenacetin. The reaction scheme is shown here.

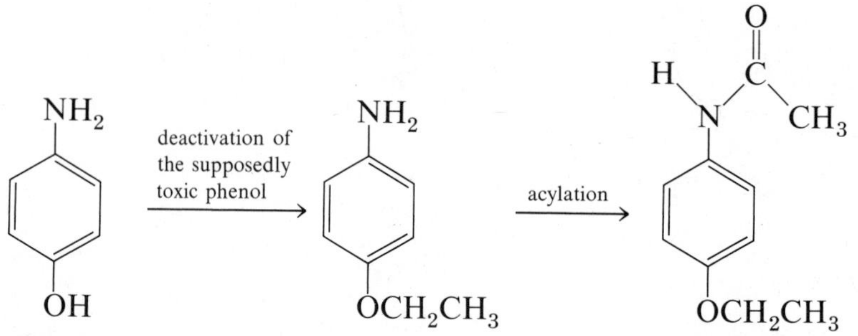

p-Aminophenol   deactivation of the supposedly toxic phenol   acylation   Phenacetin

Phenacetin turned out to be a highly effective analgesic and antipyretic. A common form of combination pain reliever, called an APC tablet, was once available. An APC tablet contained **A**spirin, **P**henacetin, and **C**affeine (hence, **APC**). Phenacetin is no longer used in commercial pain-relief preparations. It was later found that not all aromatic hydroxyl groups lead to toxic compounds, and today the compound acetaminophen is very widely used as an analgesic in place of phenacetin.

Another analgesic, structurally similar to aspirin, that has found some application is **salicylamide.** Salicylamide is found as an ingredient in some pain-relief preparations, although its use is declining.

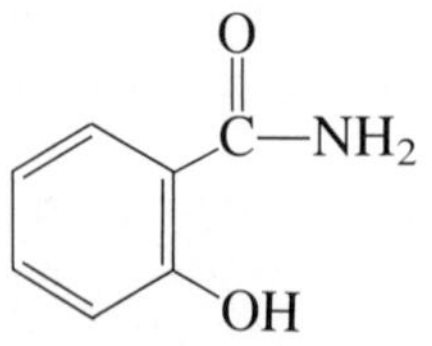

Salicylamide

On continued or excessive use, acetanilide can cause a serious blood disorder called **methemoglobinemia.** In this disorder, the central iron atom in hemoglobin is converted from Fe(II) to Fe(III) to give methemoglobin. Methemoglobin will not function as an oxygen carrier in the bloodstream. The result is a type of anemia (deficiency of hemoglobin or lack of red blood cells). Phenacetin and acetaminophen cause the same disorder, but to a much lesser degree. Since they are also more effective as antipyretic and analgesic drugs than acetanilide, they are preferred remedies. Acetaminophen is marketed under the trade names Tylenol and Datril and is often successfully used by persons who are allergic to aspirin.

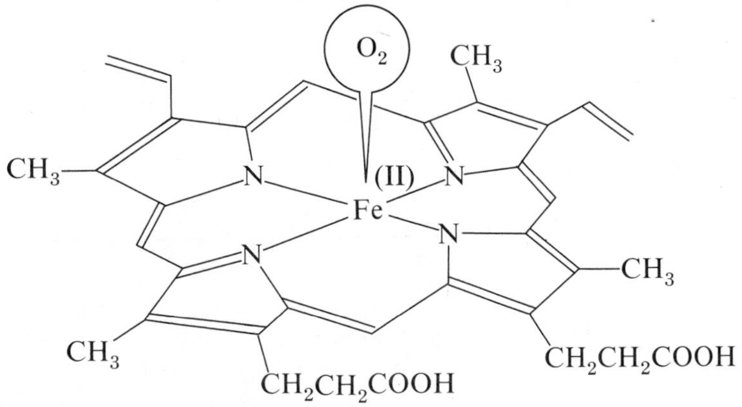

Heme portion of blood-oxygen carrier, hemoglobin

Recently a new drug has appeared in over-the-counter preparations. This drug is **ibuprofen,** which is marketed as a prescription drug in the United States under the name Motrin. Ibuprofen was first developed in England in 1964. United States marketing rights were obtained in 1974. Ibuprofen is now sold without prescription under brand names, which include Advil and Nuprin. Ibuprofen is principally an anti-inflammatory drug, but it is also effective as an analgesic and an antipyretic. It is particularly effective in the treatment of the symptoms of rheumatoid arthritis and menstrual cramps. Ibuprofen appears to control the production of prostaglandins, which parallels the mode of action of aspirin. An important advantage of ibuprofen is that it is a very powerful pain reliever. One 200-mg tablet is as effective as two tablets (650 mg) of aspirin. Furthermore, ibuprofen has a more advantageous dose-response curve, which means that taking two tablets of this drug is approximately twice as effective as one tablet for certain types of pain. Aspirin and acetaminophen reach their maximum effective dose at two tablets. Little additional relief is gained at doses above that level. Ibuprofen, however, continues to increase its effectiveness up to the 400-mg level (the equivalent of four tablets of aspirin or acetaminophen). Ibuprofen is a relatively safe drug, but its use should be avoided in cases of aspirin allergy, kidney problems, ulcers, asthma, hypertension, or heart disease.

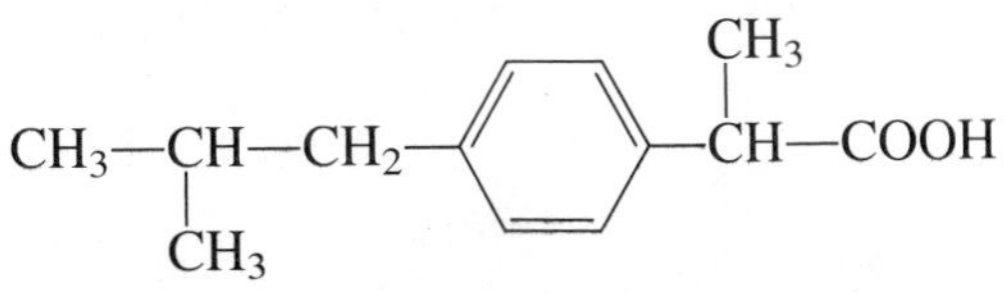

Ibuprofen

Analgesics and Caffeine in Some Common Preparations

	ASPIRIN	ACETAMINOPHEN	CAFFEINE	SALICYLAMIDE
Aspirin*	0.325 g	—	—	—
Anacin	0.400 g	—	0.032 g	—
Bufferin	0.324 g	—	—	—
Cope	0.421 g	—	0.032 g	—
Excedrin	0.250 g	0.250 g	0.065 g	—
(Extra-Strength)				
Tylenol	—	0.325 g	—	—
B. C. Powder	0.650 g	—	0.032 g	0.195 g

NOTE: Nonanalgesic ingredients, e.g., buffers, are not listed.
*5-grain tablet (1 grain = 0.0648 g).

REFERENCES

Barr, W. H., and Penna, R. P. "O-T-C Internal Analgesics." In *Handbook of Non-Prescription Drugs,* 7th ed., edited by G. B. Griffenhagen. American Pharmaceutical Association, 1982.

Hansch, C. "Drug Research or the Luck of the Draw." *Journal of Chemical Education, 51* (1974): 360.

Ray, O. S. "Internal Analgesics." *Drugs, Society, and Human Behavior,* 2nd ed. St. Louis: C. V. Mosby, 1978.

Consumers Union of the United States. "Aspirin and its Competitors." *The Medicine Show.* New York: Pantheon Books, 1980. Pp. 15–33.

Flower, R. J., Moncada, S., and Vane, J. R. "Analgesic-Antipyretics and Anti-inflammatory Agents; Drugs Employed in the Treatment of Gout." In A. G. Gilman, L. S. Goodman, T. W. Rall, and F. Murad, *The Pharmacological Basis of Therapeutics.* 7th ed. New York: Macmillan, 1985.

"The New Pain Relievers." *Consumer Reports, 49* (November 1984): 636–638.

Senozan, N. M. "Methemoglobinemia: An Illness Caused by the Ferric State." *Journal of Chemical Education, 62* (March 1985): 181.

Experiment 2
Acetanilide

Crystallization
Vacuum Filtration
Acylation
 (Preparation of an Amide)

An amine can be treated with an acid anhydride to form an amide. In this example, aniline, the amine, is treated with acetic anhydride to form acetanilide, the amide. The

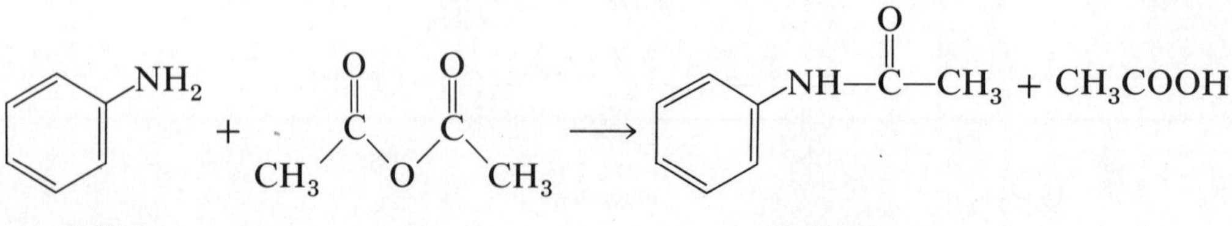

Aniline   Acetanilide

impure liquid amine, aniline, is converted to a solid product, acetanilide, which is readily purified by crystallization. The acylated amine precipitates as a crude solid, which is dissolved in hot water, decolorized with charcoal, filtered, and recrystallized.

THE REACTION

Amines can be acetylated in several ways. Among these are the use of acetic anhydride, acetyl chloride, or glacial acetic acid (with removal of the water formed in the reaction). The procedure with glacial acetic acid is of commercial interest since it is economical. It requires long heating, however. Acetyl chloride is unsatisfactory for several reasons. Principally, it reacts vigorously, liberating HCl; this converts half the amine to its hydrochloride salt, rendering it incapable of participating in the reaction.

Acetic anhydride is preferred for a laboratory synthesis. Its rate of hydrolysis is low enough to allow the acetylation of amines to be carried out in aqueous solutions. The procedure gives a product of high purity and in good yield, but it is not suitable for use with deactivated amines (weak bases) such as ortho- and para-nitroanilines.

Acetylation is often used to "protect" a primary or a secondary amine functional group. Acylated amines are less susceptible to oxidation, less reactive in aromatic substitution reactions, and less prone to participate in many of the typical reactions of free amines, since they are less basic. The amino group can be regenerated readily by hydrolysis in acid or base.

SPECIAL INSTRUCTIONS

Before beginning this experiment, you should read Techniques 1, 2, 3, and 4. You should be careful not to stop the experiment at any place where solid has not been filtered from solution. The starting material for this experiment is aniline. Aniline is a toxic substance, and it can be absorbed through the skin. Care should be exercised in handling it.

PROCEDURE

Measure 2.0 g of aniline by weighing it into a 125-mL Erlenmeyer flask. To avoid spills or contact with the skin, use a disposable pipet to transfer the aniline to the flask. Add 15 mL of water to the flask. Next, swirl the flask gently while adding 2.5 mL of acetic anhydride (density 1.08 g/mL) slowly. Note in the laboratory notebook any changes that might occur during this addition.

Once the crude acetanilide precipitates during this reaction, it must be recrystallized. This recrystallization will be accomplished in the same flask as the original reaction. Add 50 mL of water, along with a boiling stone, to the flask. Heat the mixture with a

Bunsen burner or on a hot plate until all the solid and oily materials have dissolved. After removing the heat, pour about 1 mL of the hot solution into a small beaker and allow this material to cool. The Erlenmeyer flask will be hot, so it is advisable to handle it with a paper towel. Add a small amount of activated charcoal, or Norite, to the Erlenmeyer flask, and allow the mixture to boil again. About a spatulaful should be enough charcoal. It is important not to add the charcoal to the solution while it is boiling vigorously, or violent frothing is likely. Swirl the mixture and allow it to boil gently for a few minutes. While the solution is boiling, assemble an apparatus for gravity filtration, equipped with a stemless funnel and fluted filter paper in a 125- or a 250-mL Erlenmeyer flask (see Technique 2, Section 2.1, p. 514). Also prepare about 25 mL of boiling water, which will be needed as a washing solvent in the steps that follow.

Warm the funnel by pouring about 10 mL of the boiling water through it. Discard this warming solution, and then filter the acetanilide-charcoal mixture, a little at a time, through the fluted filter. Keep the solution warm on a steam bath until it is poured into the funnel. At the same time, keep the collection flask warmed on a steam bath. The purpose of keeping these flasks on steam baths is to allow the solvent vapors to warm the funnel stem and to reduce the likelihood that crystals will form in the funnel, clogging it. If crystallization should begin in the funnel, add some hot water to the funnel to dissolve the crystals. Rinse the flask and the solid materials in the funnel with a little hot water and set the flask containing the filtrate aside to cool. To complete the crystallization, place the flask in an ice bath for about 15 minutes. Using the instructions given in Technique 2, Section 2.3, p. 518, prepare a Büchner funnel with filter paper. Collect the crystals by vacuum filtration and dry them as completely as possible by allowing air to be drawn through them while they remain on the Büchner funnel. Complete the drying by spreading the crystals on paper and allowing them to remain overnight.

Collect the crystals that were not treated with charcoal in a Hirsch funnel (equipped with filter paper) and compare the color of these crystals with the color of the crystals that were treated with charcoal. Record the results of the comparison in the laboratory notebook.

Record the weight of the product, percentage yield, and melting point of both the pure and the impure samples of dried acetanilide. Place each sample of crystals in a separate, labeled vial and submit both samples to the instructor.

QUESTIONS

1. Aniline is basic and acetanilide is not basic. Explain this difference.
2. Calculate the theoretical yield of acetanilide expected if 10 g of aniline were allowed to react with excess acetic anhydride.
3. Give equations for the reactions of aniline with acetyl chloride and with acetic acid, to give acetanilide.
4. In the introduction to this experiment, the hydrolysis of acetic anhydride is mentioned as a competing reaction. Write an equation for this reaction.

Experiment 3
Acetaminophen

Crystallization
Vacuum Filtration
Acylation
 (Preparation of an Amide)

Preparation of acetaminophen involves treating an amine with an acid anhydride to form an amide. In this case, *p*-aminophenol, the amine, is treated with acetic anhydride to form acetaminophen (*p*-acetamidophenol), the amide. Much of the material presented in the introduction to Experiment 2 also applies to the preparation of acetaminophen, so that section should be read. The acylated amine precipitates as a crude solid, which is then crystallized from hot water.

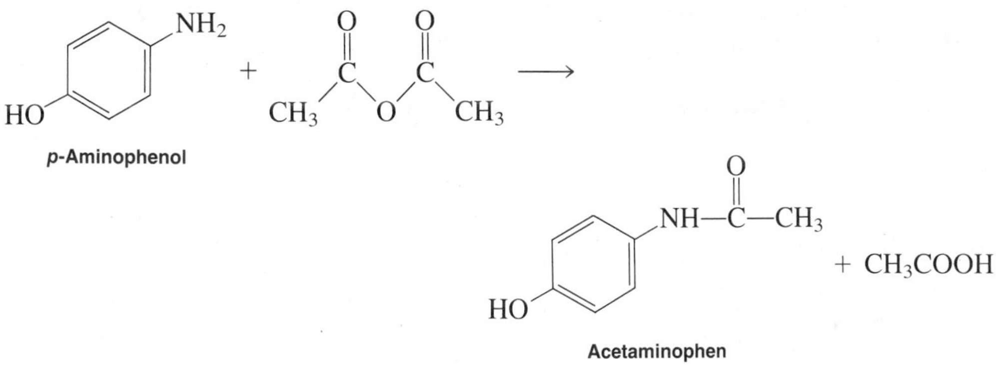

SPECIAL INSTRUCTIONS

Before beginning this experiment, read Techniques 1, 2, 3, and 4. Be careful not to stop the experiment at any place where solid has not been filtered·from solution.

PROCEDURE

Place 2.1 g of *p*-aminophenol and 35 mL of water in a 125-mL Erlenmeyer flask. To avoid spills or contact with the skin, use a disposable pipet to transfer the *p*-aminophenol to the flask. Add 1.5 mL of concentrated hydrochloric acid, and swirl the flask for a few minutes to ensure that the amine dissolves completely. If some amine remains undissolved, add a few more drops of concentrated hydrochloric acid and continue swirling. Add a spatulaful of decolorizing charcoal (Norite) to the solution, swirl the solution on a steam bath for a few minutes, and remove the charcoal by gravity filtration, using a fluted filter (see Technique 2, Section 2.1, p. 514).

Prepare a solution for use as a buffer by dissolving 2.5 g of sodium acetate ($CH_3COONa \cdot 3H_2O$) in 7.5 mL of water and clarify the solution by gravity filtration.

Transfer the filtered *p*-aminophenol hydrochloride solution to a 125-mL Erlenmeyer flask and warm it on a steam cone. Add the sodium acetate buffering solution in one portion. Immediately add 2 mL of acetic anhydride while swirling the solution. Continue swirling the solution vigorously for about 10 minutes to ensure mixing.

Cool the reaction mixture by immersing the flask in an ice-water bath, and stir the mixture vigorously until the crude acetaminophen crystallizes. It may be necessary to scratch the walls of the flask vigorously in order to initiate crystallization (see Technique 3, Section 3.4, p. 526). Allow the flask to remain in the ice bath for about an hour. Collect the crystals in a Büchner funnel by vacuum filtration (see Technique 2, Section 2.3, p. 518). Wash the crystals with a portion of cold water. Weigh the crude crystals and calculate the yield.

Using the crystallization procedures described in Technique 3, Section 3.4, p. 526, dissolve about 1 g of the crude acetaminophen in the minimum amount of *boiling* water. Allow the solution to cool slowly. When the first crystals appear, immerse the flask in an ice bath for 15 to 20 minutes. Collect the crystals by vacuum filtration, using a Büchner funnel (see Technique 2, Section 2.3, p. 518), and dry them. Determine the melting point of the dried crystals. Record the weight of the acetaminophen obtained in the experiment and, finally, the melting point of the recrystallized sample. Submit the sample of recrystallized material in a labeled vial to the instructor.

QUESTIONS

1. *p*-Aminophenol is purified by adding concentrated hydrochloric acid and decolorizing the resulting solution with Norite. Write the chemical equation for the reaction of *p*-aminophenol with hydrochloric acid. Explain why the *p*-aminophenol dissolves.

2. *p*-Aminophenol acts as a base, while acetaminophen does not. Explain this difference.

3. Phenacetin has the structure shown. How might it be prepared?

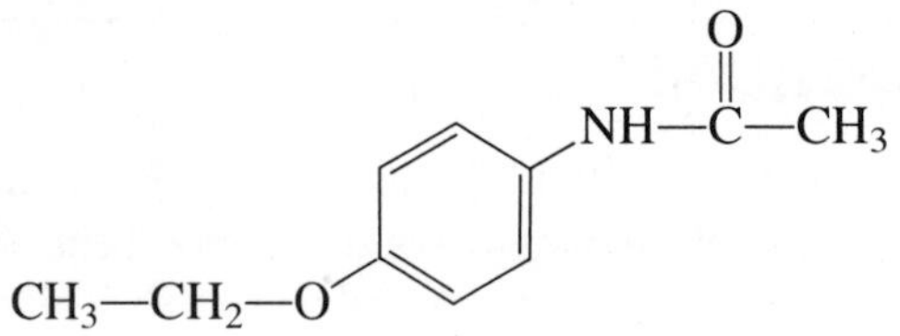

4. Calculate the theoretical yield of acetaminophen if 10 g of *p*-aminophenol are allowed to react with excess acetic anhydride.

Essay
IDENTIFICATION OF DRUGS

Frequently a chemist is called on to identify a particular unknown substance. If there is no prior information to work from, this can be a formidable task. There are several million known compounds, both inorganic and organic. For a completely unknown substance, the chemist must often use every available method. If the unknown substance is a mixture, then the mixture must be separated into its components and each component identified separately. A pure compound can often be identified from its physical properties (melting point, boiling point, density, refractive index, and so on) and a knowledge of its functional groups. These can be identified by the reactions that the compound is observed to undergo or by spectroscopy (ir, ultraviolet, nmr, and mass spectroscopy). The techniques necessary for this type of identification are introduced in a later section.

A somewhat simpler situation often arises in drug identification. The scope of drug identification is more limited, and the chemist working in a hospital trying to identify the source of a drug overdose, or the law enforcement officer trying to identify a suspected illicit drug or a poison, usually has some prior clues to work from. So does the medicinal chemist working for a pharmaceutical manufacturer who might be trying to discover why a competitor's product is better than his.

Consider a drug overdose case as an example. The patient is brought into the emergency ward of a hospital. This person may be in a coma or in a hyperexcited state, have an allergic rash, or clearly be hallucinating. These physiological symptoms are themselves a clue to the nature of the drug. Samples of the drug may be found in the patient's possession. Correct medical treatment may require a rapid and accurate identification of a drug powder or capsule. If the patient is conscious, the necessary information can be elicited orally; if not, the drug must be examined. If the drug is a tablet or a capsule, the process is often simple, since many drugs are coded by a manufacturer's trademark or logo, by shape (round, oval, bullet shape), by formulation (tablet, gelatin capsule, time-release microcapsules), and by color.

It is more difficult to identify a powder, but under some circumstances such identification may be easy. Plant drugs are often easily identified since they contain microscopic bits and pieces of the plant from which they are obtained. This cellular debris is often characteristic for certain types of drugs, and they can be identified on this basis alone. A microscope is all that is needed. Sometimes chemical color tests can be used as a confirmation. Certain drugs give rise to characteristic colors when treated with special reagents. Other drugs form crystalline precipitates of characteristic color and crystal structure when treated with appropriate reagents.

If the drug itself is not available and the patient is unconscious (or dead), identification may be more difficult. It may be necessary to pump the stomach or bladder contents of the patient (or corpse), or to obtain a blood sample, and work on these. These samples of stomach fluid, urine, or blood would be extracted with an appropriate organic solvent, and the extract would be analyzed.

Often the final identification of a drug, as an extract of urine, serum, or stomach fluid, hinges on some type of **chromatography.** Thin-layer chromatography (tlc) is often used. Under specified conditions, many drug substances can be identified by their R_f values and by the colors that their tlc spots turn when treated with various reagents or when they are observed under certain visualization methods. In the experiment that follows, tlc is applied to the analysis of an unknown analgesic drug.

REFERENCES

Keller, E. "Forensic Toxicology: Poison Detection and Homicide." *Chemistry, 43* (1970): 14.
Keller, E. "Origin of Modern Criminology." *Chemistry, 42* (1969): 8.
Lieu, V. T. "Analysis of APC Tablets." *Journal of Chemical Education, 48* (1971): 478.
Neman, R. L. "Thin Layer Chromatography of Drugs." *Journal of Chemical Education, 49* (1972): 834.
Rodgers, S. S. "Some Analytical Methods Used in Crime Laboratories." *Chemistry, 42* (1969): 29.
Tietz, N. W. *Fundamentals of Clinical Chemistry.* Philadelphia: W. B. Saunders, 1970.
Walls, H. J. *Forensic Science.* New York: Praeger, 1968.
A collection of articles on forensic chemistry can be found in:
Berry, K., and Outlaw, H. E., eds. "Forensic Chemistry—A Symposium Collection." *Journal of Chemical Education, 62* (December 1985): 1043–1065.

Experiment 4
TLC Analysis of Analgesic Drugs

Thin-Layer Chromatography

In this experiment, thin-layer chromatography (tlc) will be used to determine the composition of various over-the-counter analgesics. If the instructor chooses, you may also be required to identify the components and actual identity (trade name) of an unknown analgesic. You will be given two commercially prepared tlc plates with a flexible backing and a silica gel coating with a fluorescent indicator. On one tlc plate, you will spot four standard compounds often used in analgesic formulations. In addition, a standard reference mixture containing these same compounds will also be spotted. The reference substances are

        Acetaminophen (Ac)
        Aspirin   (Asp)
        Caffeine   (Cf)
        Salicylamide (Sal)

They will all be available as solutions of 1 g of each dissolved in 20 mL of a 50:50 mixture of methylene chloride and ethanol. The purpose of the first plate is to determine the order of elution (R_f values) of the known substances and to index the standard reference mixture. On the second plate, the standard reference mixture will be spotted along with several solutions prepared from commercial analgesic tablets. The crushed tablets will also be dissolved in a 50:50 methylene chloride–ethanol mixture. At your instructor's option, one of the analgesics to be spotted on the second plate may be an unknown.

Two methods of visualization will be used to observe the positions of the spots on the developed tlc plates. First, the plates will be observed while under illumination from a short-wavelength ultraviolet (uv) lamp. This is done best in a darkened room or in a fume hood that has been darkened by taping butcher paper or aluminum foil over the lowered glass cover. Under these conditions, some of the spots will appear as dark areas on the plate, while others will fluoresce brightly. This difference in appearance under uv illumination will help to distinguish the substances from one another. You will find it convenient to outline very lightly in pencil the spots observed and to place a small x inside those spots that fluoresce. For a second means of visualization, iodine vapor will be used. Not all the spots will become visible when treated with iodine, but at least two will develop a deep brown color. The differences in the behaviors of the various spots with iodine can be used to further differentiate among them.

It is possible to use several developing solvents for this experiment. Ethyl acetate is preferred. Acetone and cyclohexanone also give fairly good results. Cyclohexanone, however, is a fairly viscous liquid and it takes a good while longer than the other solvents to diffuse up the plate (>1 hour). It is also more difficult to evaporate this solvent from the slide. Ethyl acetate requires only about 20 minutes to ascend the plate and evaporates quickly. A more complicated developing solvent of 120:60:18:1—benzene, ether, glacial acetic acid, and methanol—has been recommended in the pharmaceutical literature.

In some analgesics you may find other ingredients besides the four mentioned above. Some include an antihistamine and some a mild sedative. For instance, Midol contains *N*-cinnamylephedrine (cinnamedrine), an antihistamine, while Excedrin PM contains the sedative methapyrilene hydrochloride. Cope contains the related sedative methapyrilene fumarate.

SPECIAL INSTRUCTIONS

To understand this experiment fully, read the two essays ''Aspirin'' and ''Analgesics.'' These precede Experiments 1 and 2. It is also necessary to read Technique 11, which explains the principles and methods of thin-layer chromatography. Note that the developed tlc plates must be examined under ultraviolet light before the iodine visualization. The iodine permanently affects some of the spots. Aspirin presents some special problems since it is present in large amount in many of the analgesics and since it hydrolyzes easily. For these reasons, the aspirin spots often show excessive tailing.

PROCEDURE

To begin, you will need at least 12 capillary micropipets to spot the plates. The preparation of these pipets is described and illustrated in Technique 11, Section 11.3, p. 620.

After preparing the micropipets, get two 10-cm × 6.6-cm tlc plates (Eastman Chromagram Sheet, No. 13181) from your instructor. These plates have a flexible backing but should not be bent excessively. They should be handled carefully, or the adsorbent may flake off them. Also, they should only be handled by the edges; the surface should not be touched. Using a lead pencil (not a pen) **lightly** draw a line across the plates (short dimension) about 1 cm from the bottom. Using a centimeter ruler, move its index about 0.6 cm in from the edge of the plate and lightly mark off five 1-cm intervals on the line (see figure). These are the points at which the samples will be spotted. On the first plate, starting from left to right, spot first acetaminophen, then aspirin, caffeine, and salicylamide. This order is alphabetic and will avoid any further memory problems or confusion. Solutions of these compounds will be found in small bottles on the side shelf. The standard reference mixture, also found on the side shelf, is spotted in the last position. The correct method of spotting a tlc slide is described in Technique 11, Section 11.3, p. 620. It is important that the spots be made as small as possible and that the plates not be overloaded. If these cautions are disregarded, the spots will tail and will overlap one another after development. The applied spot should be about 1 or 2 mm (¹⁄₁₆ in.) in diameter. If scrap pieces of the tlc plates are available, it would be a good idea to practice spotting on these before preparing the actual sample plates.

When the first plate has been spotted, obtain a 16-oz wide-mouthed screw-cap jar for use as a development chamber. The preparation of a development chamber is described in Technique 11, Section 11.4, p. 621. For this experiment, a slight modification of the development chamber must be made. Since the backing on the tlc slides is very thin, if they touch the filter paper liner of the development chamber **at any point,** solvent will begin to diffuse onto the absorbent surface at that point. To avoid this, a very narrow liner strip of filter paper (approximately 5 cm wide) should be used. It should be folded into an L shape and should be long enough to traverse the bottom of the jar and extend up the side to the top of the jar. Slides placed in the jar for development should straddle this liner strip but not touch it.

When the development chamber has been prepared, saturate the filter paper liner with ethyl acetate and fill the jar to a depth of about 0.5 cm with the same solvent. Recall that the solvent level must not be above the spots on the plate. Place the spotted plate in the development chamber (straddling the liner) and allow the plate to develop. When the solvent has risen on the plate to a level about 0.5 cm from the top, remove the plate from the chamber, and using a lead pencil, mark the position of the solvent front. Allow the plate to dry and then observe it under a short-wavelength uv lamp. Lightly mark the observed spots with a pencil. Next, place the plate in a jar containing a few iodine crystals, cap the jar, and warm it **gently** on a steam bath until spots begin to appear. Notice which spots become visible and record that information in your notebook. Also, note which spot or spots fluoresce under the uv lamp. Remove the plate from the iodine jar, and with a ruler marked in millimeters, measure the distance that each spot has traveled relative to the solvent front. Calculate its R_f value (see Technique 11, Section 11.8, p. 625).

Next, obtain half a tablet of each of the analgesics to be analyzed. If you were issued an unknown, you may analyze four other analgesics; if not, you may analyze five. Five choices that lead to a particularly wide spectrum of results are Anacin, Bufferin, B.C. Powder, Excedrin, and Tylenol. However, if you have a favorite analgesic, you may prefer to analyze it. Take each analgesic half-tablet, place it on a smooth piece of notebook paper, and crush it well with a spatula. Transfer each crushed half-tablet to a small, labeled test tube or an Erlenmeyer flask. Using a graduated cylinder, mix together 15 mL of absolute ethanol and 15 mL of methylene chloride. Mix the solution well. Add about 5 mL of this solvent mixture to each of the crushed half-tablets and then heat each of them **gently** for a few minutes on a steam bath. Not all the tablet will dissolve, since the analgesics usually contain starch as a binder for the tablet. In addition, many of them contain inorganic buffering agents that are insoluble in this solvent mixture. After heating the samples, allow them to settle and then spot the clear liquid extracts at five of the positions on the second plate. At the sixth position, spot the standard reference solution. Develop the plate in ethyl acetate as before. Observe the plate under uv illumination and mark the visible spots. Repeat the visualization using iodine. Record your conclusions about the contents of the various analgesics. If you were issued an unknown, try to determine its identity (trade name).

Essay
NICOTINE

Nicotine

Tobacco that outlandish weede
It spends the braine and spoiles the seede
It dulls the spirite, it dims the sight
It robs a woman of her right
WILLIAM VAUGHN (1617)

Tobacco is one of the few contributions to civilization that the New World can claim. Before this continent was discovered, tobacco was unknown to Europeans. In fact, the plants *Nicotiana tabacum* and *Nicotiana rustica* are indigenous to America, the former to South and Central America and the latter to eastern North America and the West Indies.

Nicotiana was used by Indians in both the northern and southern hemispheres in religious, curing, and intoxicating practices. The literature from 1492 to the present leaves little doubt that the native peoples of nearly all parts of the New World accorded supreme ritualistic and mythological status to these plants. There is also evidence that in many areas tobacco has been used to trigger ecstatic states much like those induced by the true hallucinogens. Even today the shamans (''witch doctors'') of the Warao Indians of Venezuela use a psychotomimetic snuff, composed mostly of tobacco, to achieve the trance states that are part of the purification, initiation, and supernatural curing rituals. The American Indian peace pipe ritual is also well known.

Tobacco is not generally considered a hallucinogen. In fact, it is not clear that tobacco is the only component that the Indian shamans use in their secret preparations. For the cases in which the preparation is smoked, oxygen starvation in the lungs may well be the cause, at least partly, of the trancelike state.

The Indians held the tobacco plant in high esteem, and its use was limited to ritualistic and religious ceremony. Not so with civilized culture! Its acceptance into civilized society was both rapid and thorough. Tobacco smoking was introduced into Europe by Rodrigo de Jerez. He was the first European to land in Cuba, where he picked up the smoking habit. When he returned to Portugal and continued the practice, people who saw the smoke coming out of his nose and mouth thought he was possessed by the devil. Rodrigo spent the next several years in jail. During this time, the spread of tobacco use was so rapid that when Rodrigo was released from jail he was amazed to see everyone indulging in the very offense for which he had been confined.

The active ingredient in tobacco, (−)-nicotine, was first isolated in 1828. Nicotine is a colorless liquid **alkaloid,** and the natural material is levorotatory due to a single chiral (asymmetric) center. Alkaloids are a class of naturally occurring compounds containing nitrogen and having the properties of an amine base—alkaline,

50

hence, "alkaloid." Like almost all alkaloids, nicotine has dramatic effects on the human system. One quickly develops tolerance to the effects of small amounts of nicotine (tobacco), along with a dependency. Mark Twain remarked how easy it is to stop smoking—"I've done it a thousand times!" Even in the face of a highly condemning report by the surgeon general of the United States, millions of Americans continue to smoke.

Nicotine itself is one of the most toxic drugs known to humans. A dose of 60 mg is lethal, with death following intake by only a few minutes. A cigar contains enough nicotine for two lethal doses. However, it is estimated that only 10% of that amount is absorbed on inhaling the smoke and that this dose is absorbed over a relatively long period. The typical filter cigarette contains 20 to 30 mg of nicotine. That smoking can be practiced at all is explained by the body's ability to degrade or metabolize nicotine rapidly and eliminate it, thus preventing its accumulation. In acute poisoning, nicotine causes tremors, which turn into convulsions, and frequently causes death. Death comes about by paralysis of the muscles used in respiration. This results from a blocking effect on the motor nerve system that normally activates these muscles. Curiously enough, with lower doses there is actually an increase in respiratory rate because the body tries to counteract the effects of the nicotine. The oxygen need of the carotid artery is stimulated, and an increased heart rate and blood flow and constriction of the arteries are observed. In smoking, this effect is enhanced because some of the combustion products, namely carbon monoxide and hydrogen cyanide, irreversibly bind to hemoglobin in the blood, so that it can no longer carry oxygen. In a regular smoker, up to 10% of all hemoglobin can be inactivated in this way. This is the basis of the "shortness of breath" phenomenon familar to smokers. At higher doses, nicotine interferes with the transmission of nerve impulses across the nerve junctions (synapses) of the cholinergic system. That is, nicotine competes with the body's natural **motor nerve** transmission agent, acetylcholine, for the stimulation of voluntary muscles (other types of nerves may use other chemical transmitter agents).

Nicotine
(which is 90%
protonated at pH 7)

Acetylcholine

Nerve fibers are connected to muscle by a special kind of junction called a synapse. Conduction of messages along nerve fibers, or axons, is primarily electrical. Transmission across the synapse is chemical, however. The chemical agent released from the nerve ending by the electrical pulse is acetylcholine. This molecule travels through the membrane, which is somehow opened to it, and across the synaptic gap to a **receptor site,** where it binds. When it binds, muscle contraction is initiated. Soon afterward an enzyme, acetylcholinesterase (AChE), destroys the acetylcholine by removing the acetyl group (hydrolysis) and producing choline:

$$H_2O + CH_3 \overset{\overset{\displaystyle O}{\displaystyle \|}}{C} O—CH_2CH_2—\overset{+}{N}(CH_3)_3 \xrightarrow{\text{AChE}}$$

Acetylcholine

$$HO—CH_2CH_2—\overset{+}{N}(CH_3)_3 + CH_3COOH$$

Choline

This enzyme clears the receptor sites, opening them up for another pulse. The choline finds its way back into the nerve fiber; there other enzymes resynthesize acetylcholine, which is stored for future use. (See diagram.)

Nicotine has the same ability as acetylcholine to bind to acetylcholine receptor sites and generate muscle stimulation. The enzyme AChE cannot destroy the nicotine, however. Hence in large enough quantity, nicotine can block all receptor sites and cause paralysis. One can thus observe, for instance, an initial stimulation of heart muscle but ultimately a paralysis of the heart, leading to death. Other cholinergic centers, such as those in the brain, are affected similarly.

The ability of nicotine to block cholinergic receptors is the basis of its use as a natural insecticide; the insect dies of nerve poisoning and respiratory failure. Extracts of tobacco plant were long used as a potent insecticide against certain types of insects. Today this extract has largely been replaced by synthetic insecticides. Many of these synthetics are also nerve poisons and block nerve transmissions in one way or another. For instance, Malathion, an organophosphate, inhibits or stops the action of AChE. The way DDT acts is, unfortunately, not currently well understood. It is, however, known to be a potent nerve poison.

REFERENCES

Axelrod, J. "Neurotransmitters." *Scientific American, 230* (June 1974): 59.

Emboden, W. "Tobaccos and Snuffs." *Narcotic Plants.* New York: Macmillan, 1972.

Fried, R. "Introduction to Neurochemistry." *Journal of Chemical Education, 45* (1968): 322.

Hammond, C. "The Effects of Smoking." *Scientific American, 207* (July 1962): 39.

Ray, O. S. "Nicotine." *Drugs, Society, and Human Behavior.* 2nd ed. St. Louis: C. V. Mosby, 1978.

Robinson, T. "Alkaloids." *Scientific American, 201* (July 1959): 113.

Taylor, N. "The Lively Image and Pattern of Hell—The Story of Tobacco." In *Narcotics—Nature's Dangerous Gifts.* New York: Dell, 1970. (Paper-bound revision of *Flight from Reality.*)

Volle, R. L., and Koelle, G. B. "Ganglionic Stimulating and Blocking Agents." In L. S. Goodman and A. Gilman, *The Pharmacological Basis of Therapeutics.* 7th ed. New York: Macmillan, 1985.

Wilbert, J. "Tobacco and Shamanistic Ecstasy among the Warao Indians." In *Flesh of the Gods,* edited by P. Furst. New York: Praeger, 1972.

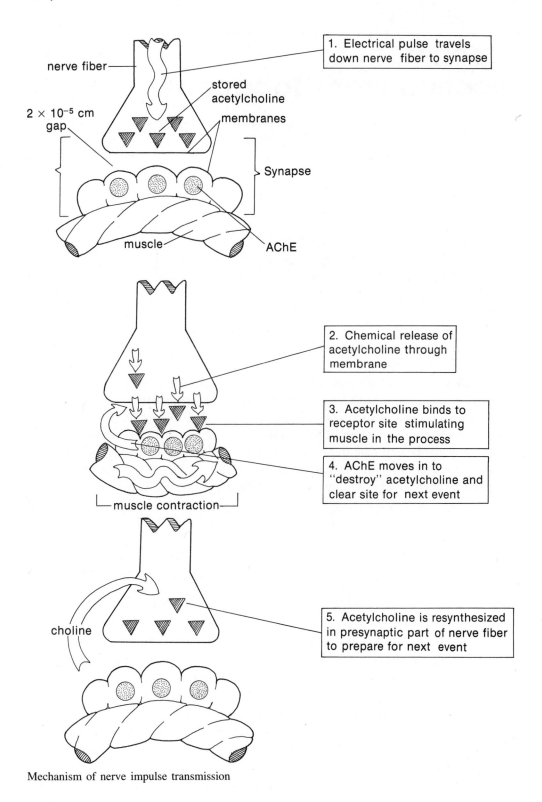

nerve fiber

1. Electrical pulse travels down nerve fiber to synapse

stored acetylcholine

2×10^{-5} cm gap

membranes

Synapse

muscle

AChE

2. Chemical release of acetylcholine through membrane

3. Acetylcholine binds to receptor site stimulating muscle in the process

4. AChE moves in to "destroy" acetylcholine and clear site for next event

muscle contraction

choline

5. Acetylcholine is resynthesized in presynaptic part of nerve fiber to prepare for next event

Mechanism of nerve impulse transmission

Experiment 5
Nicotine From Tobacco

Isolation of a Natural Product
Filtration
Extraction
Crystallization

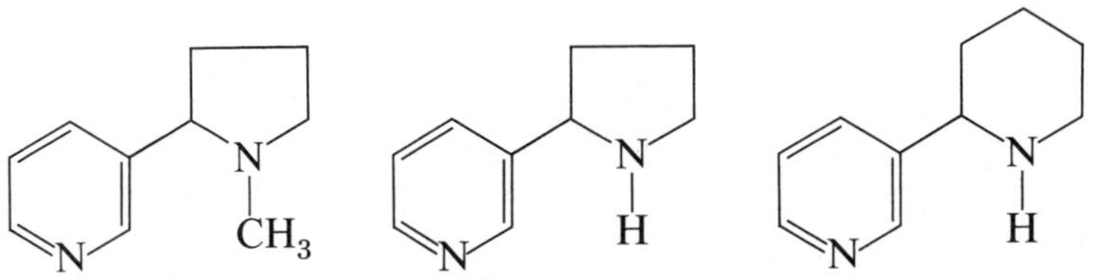

| Nicotine | Nornicotine | Anabasine |

The main alkaloidal component of the tobacco leaf is nicotine, along with smaller amounts of nornicotine, anabasine, and at least seven other lesser alkaloids. We shall isolate and characterize the pure chemical nicotine from its natural source. It is a nitrogen-containing base built from two heterocyclic rings (pyridine and pyrrolidine). Both amine functions are tertiary and can be protonated to form salts. In fact, the pyrrolidine ring has a pK_a of about 8 (pyridine has a pK_a of 3), which means that at pH 7 this ring would be expected to be about 90% protonated—one of the factors making nicotine a good mimic for the quaternary acetylcholine. This is also probably the chief reason for its ready solubility in water (ionic substance). We shall extract tobacco leaf with a strong base (5% NaOH) to generate the free base, which can then be extracted

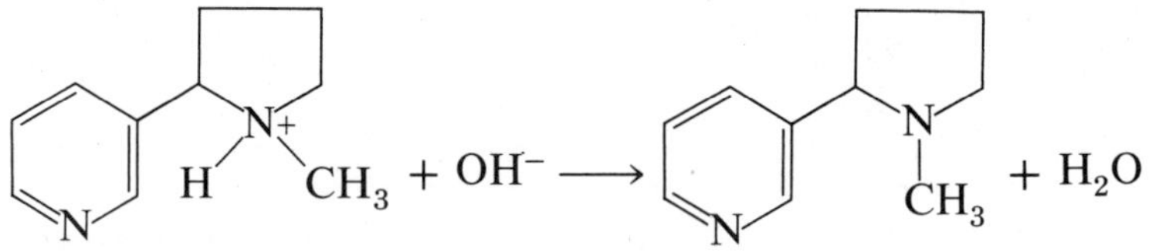

into the organic solvent, ether. Evaporation of the ether leaves nicotine as an oil (bp 246 °C) that in small amounts is not easily purified nor easily handled. Hence, we shall convert it to its salt with picric acid (a phenol). This will give a crystalline material that can be purified and characterized by its melting point. Nicotine forms a dipicrate salt by reaction at both basic centers; hence, the molecular weight of the product is more than three times the molecular weight of nicotine. This has the advantage of providing more material to work with.

54

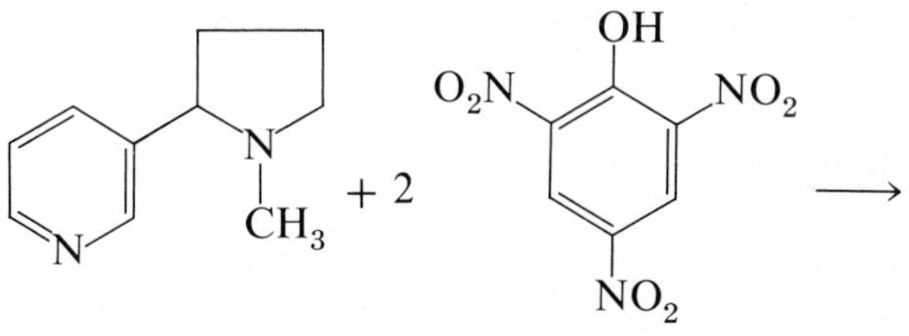

Nicotine   Picric acid

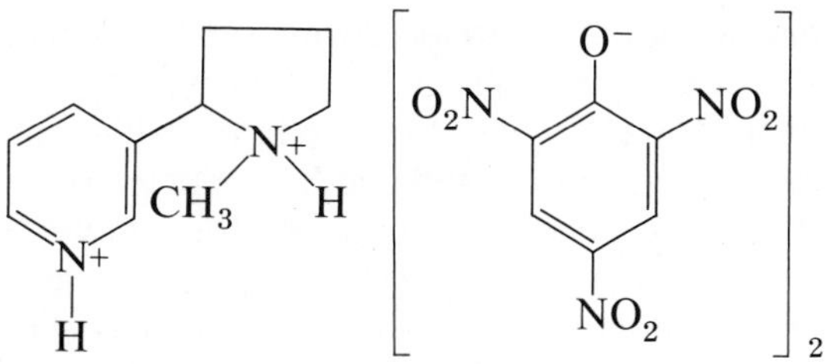

Nicotine dipicrate

Although cigarette tobacco could be used for our experiment, most manufactur-ers try to remove as much "nicotine and tar" as possible; hence, cigars or snuff provide a better source. Dried and crushed cigar tobacco is the more convenient source. Other components of the tobacco leaf are cellulose and tannic acid, which cannot be extracted into ether from a basic solution because it forms a salt, along with oxidation products of the once green chlorophyll content of the leaf. Chlorophyll oxidation products and tannic acid are primarily responsible for the brown color of the extract. A structure for tannic acids can be found in Experiment 6, Isolation of Caffeine from Tea.

SPECIAL INSTRUCTIONS

Read the essay on nicotine, which precedes this experiment. Also, you must read Techniques 1 through 5 before starting. Old cigars work well for a source of nicotine. They can be further dried by crushing them and exposing them to air for at least 1 week.

> **CAUTION:** Nicotine is extremely toxic. A dose of 60 mg is lethal. Handle it with ex-treme care.

PROCEDURE

ISOLATION OF NICOTINE DIPICRATE

Grind 8.5 g of cigar tobacco to a fine powder, using a mortar and pestle. Place the ground tobacco and 100 mL of 5% sodium hydroxide solution into a 400-mL beaker. Stir the mixture for 15 minutes. Assemble an apparatus for vacuum filtration (Technique 2, Section 2.3, p. 518) using a Büchner funnel fitted with fast filter paper (E&D 617, S&S 595, or Whatman 1) and a 250-mL filter flask. Turn on the aspirator, and filter the mixture. Wash the tobacco with a small volume of water to ensure that the basic solution has been removed from the tobacco completely. Press the tobacco with the bottom of a beaker or a large clean cork to remove the remaining base. Transfer the tobacco back to the original beaker and add 30 mL of water. Stir the mixture and refilter it with suction, using a new piece of filter paper. Discard the tobacco in a waste container (not a sink!). Remove any small particles that passed through the Büchner funnel by passing the basic extract through a thick layer of loosely compacted glass wool in a short-stemmed funnel. Wash the glass wool with a little water and add the washings to the aqueous extract. If solid particles remain in the filtrate (extract), repeat the filtration using a new layer of glass wool.

   Transfer the filtrate to a large (250-mL) separatory funnel and extract the aqueous phase with a 25-mL portion of ether (Technique 5, Section 5.4 p. 545). When extracting, separate the layers and save them both. Collect and save the lower aqueous phase in a beaker. When the upper, ether layer begins to approach the stopcock, some dark oily material (emulsion) may collect at the apex. Carefully decant the ether from the emulsion through the top of the separatory funnel into a dry beaker. Transfer the saved aqueous layer back into the funnel and add another 25-mL portion of ether. Shake and separate the layers. Again, decant the ether layer away from any remaining emulsion and into the beaker containing the first ether extract. Return the aqueous layer to the separatory funnel and again extract it with a third 25-mL portion of ether. Separate the phases and decant the ether layer away from the emulsion and into the storage vessel. Discard the aqueous layer. Carefully decant the combined ether extracts away from any remaining emulsion or water and into a dry container. Repeat if necessary until no emulsion or water remains.

   Remove the ether solvent under vacuum to prevent decomposition of the nicotine; this must be done in portions. Do not evaporate the solvent using heat. Using a 125-mL filter flask, assemble the apparatus shown in Technique 1, Figure 1–7, p. 513. Add about 25 mL of the clear ether extract to the 125-mL filter flask and place a wooden applicator stick in it; then cork the flask securely. Attach the flask to a trap and then to the aspirator. The trap should be cooled in an ice bath. Start the aspirator and remove the ether under vacuum. Gently warm the filter flask on the steam bath (with constant swirling) to keep the flask from becoming too cold. When the volume is reduced to about 5 mL, add another 25-mL portion of the ether extract from the storage vessel. Remove the old applicator stick and add a new one. Again reduce the volume under vacuum with the aspirator to about 5 mL. Finally, transfer the remaining ether extract to the filter flask and once again replace the applicator stick with a new one. Remove all the remaining ether.

   A small residue of oil and solid will remain after the ether is removed. Add 1 mL of water and swirl gently to dissolve the residue. Then add 4 mL of methanol and filter the solution into a small beaker through a **small** plug of glass wool placed loosely in the apex of a short-stemmed funnel. Rinse the funnel and glass wool with an additional 5 mL of methanol and combine the two methanol solutions. The solution must be **clear** (no suspended particles) at this point; if not, refiltering is necessary. Add 10 mL of saturated methanolic picric acid. Nicotine dipicrate will immediately appear as a pale yellow precipitate. Without delay, collect the precipitate by vacuum filtration, using a small Büchner funnel (use filter paper). Dry the precipitate in the funnel by drawing air through the filter. The melting point of the air-dried nicotine dipicrate is 216 to 217 °C. If the substance is allowed to dry overnight, the melting point is raised to 217 to 220 °C. Determine the melting point of your solid. Weigh the dry solid and calculate the percentage yield of actual nicotine from tobacco. The yield of nicotine dipicrate may be as high as 0.10 g. The precipitate that appears on evaporation of the methanol in the filter flask is mostly excess picric acid, and it is discarded.

RECRYSTALLIZATION OF NICOTINE DIPICRATE

Transfer the crude nicotine dipicrate to a 50-mL Erlenmeyer flask. Add hot 50% (by volume) ethanol–water until the solid just dissolves. Keep both containers hot during this process. Allow the contents of the flask to cool slowly. Since crystallization is slow, it is best to stopper the flask lightly, allowing it to stand until the next laboratory period. Collect the product by vacuum filtration, using a small Büchner funnel, and allow it to dry, preferably overnight. Obtain the melting point. The melting point of pure nicotine dipicrate is 222 to 223 °C. Submit the sample to the instructor in a labeled vial.

QUESTIONS

1. Outline a separation scheme for isolating nicotine from tobacco. Use a flowchart similar in format to that shown in the **Advance Preparation and Laboratory Records** section.

2. Explain why the pyrrolidine part of the nicotine molecule is more basic than the pyridine part.

3. A structure for tannic acid (pentadigalloyl derivative of glucose) is shown in Experiment 6. Which hydroxyl group or groups do you expect to be more acidic, and why?

4. Indicate how you might obtain gallic acid from tannic acid. A chemical reaction is involved.

5. Identify the chiral (asymmetric) center in nicotine.

Essay
CAFFEINE

The origins of coffee and tea as beverages are so old that they are lost in legend. Coffee is said to have been discovered by an Abyssinian goatherd who noticed an unusual friskiness in his goats when they consumed a certain little plant with red berries. He decided to try the berries himself and discovered coffee. The Arabs soon cultivated the coffee plant, and one of the earliest descriptions of its use is found in an Arabian medical book circa 900 A.D. The great systematic botanist, Linnaeus, named the tree *Coffea arabica.*

One legend of the discovery of tea—from the Orient, as one might expect—attributes the discovery to Daruma, the founder of Zen. Legend has it that he inadvertently fell asleep one day during his customary meditations. To be assured that this indiscretion would not recur, he cut off both eyelids. Where they fell to the ground, a new plant took root that had the power to keep a person awake. Although some experts assert that the medical use of tea was reported as early as 2737 B.C. in the pharmacopeia of Shen Nung, an emperor of China, the first indisputable reference is from the Chinese dictionary of Kuo P'o, which appeared in 350 A.D. The nonmedical, or popular, use of tea appears to have spread slowly. Not until about 700 A.D. was tea widely cultivated in China. Since tea is native to upper Indochina and upper India, it must have been cultivated in these places before its introduction to China. Linnaeus named the tea shrub *Thea sinensis;* however, tea is more properly a relative of the camellia, and botanists have renamed it *Camellia thea.*

The active ingredient that makes tea and coffee valuable to humans is **caffeine.** Caffeine is an **alkaloid,** a class of naturally occurring compounds containing nitrogen and having the properties of an organic amine base (alkaline, hence, **alkaloid**). Tea and coffee are not the only plant sources of caffeine. Others include kola nuts, maté leaves, guarana seeds, and in small amount, cocoa beans. The pure alkaloid was first isolated from coffee in 1821 by the French chemist Pierre Jean Robiquet.

XANTHINES
Xanthine R = R' = R'' = H
Caffeine R = R' = R'' = CH_3
Theophylline R = R'' = CH_3, R' = H
Theobromine R = H, R' = R'' = CH_3

Caffeine belongs to a family of naturally occurring compounds called **xanthines.** The xanthines, in the form of their plant progenitors, are possibly the oldest known stimulants. They all, to varying extents, stimulate the central nervous system and the skeletal muscles. This stimulation results in an increased alertness, the ability to put off sleep, and an increased capacity for thinking. Caffeine is the most powerful xanthine in this respect. It is the main ingredient of the popular No-Doz keep-alert tablets. While caffeine has a powerful effect on the central nervous system, not all xanthines are as effective. Thus theobromine, a xanthine found in cocoa, has fewer central nervous system effects. It is, however, a strong **diuretic** (induces urination) and is useful to doctors in treating patients with severe water-retention problems. Theophylline, a second xanthine found along with caffeine in tea, also has few central nervous system effects but is a strong myocardial (heart muscle) stimulant; it dilates (relaxes) the coronary artery that supplies blood to the heart. Theophylline, also called aminophylline, is often used in treating congestive heart failure. It is also used to alleviate and to reduce the frequency of attacks of angina pectoris (severe chest pains). In addition, it is a more powerful diuretic than theobromine. Since it is also a vasodilator (relaxes blood vessels), it is often used in treating hypertensive headaches and bronchial asthmas.

One can develop both a tolerance for the xanthines and a dependence on them, particularly caffeine. The dependence is real, and a heavy user (>5 cups of coffee per day) will experience lethargy, headache, and perhaps nausea after about 18 hours of abstinence. An excessive intake of caffeine may lead to restlessness, irritability, insomnia, and muscular tremor. Caffeine can be toxic; but to achieve a lethal dose of caffeine, one would have to drink about 100 cups of coffee over a relatively short period.

Caffeine is a natural constituent of coffee, tea, and kola nuts *(Kola nitida)*. Theophylline is found as a minor constituent of tea. The chief constituent of cocoa is theobromine. The amount of caffeine in tea varies from 2 to 5%. In one analysis of black tea, the following compounds were found: caffeine, 2.5%; theobromine, 0.17%; theophylline, 0.013%; adenine, 0.014%; and guanine and xanthine, traces. Coffee beans can contain up to 5% by weight of caffeine, and cocoa contains around 5% theobromine. Commercial cola is a beverage based on a kola nut extract. We cannot easily get kola nuts in this country, but we can get the ubiquitous commercial extract as a syrup. The syrup can be converted into "cola." The syrup contains caffeine, tannins, pigments, and sugar. Phosphoric acid is added, and caramel is also added to give the syrup a deep color. The final drink is prepared by adding water and carbon dioxide under pressure, to give the bubbly mixture. The Food and Drug Administration currently requires that a "cola" contain **some** caffeine but limits this amount to a maximum of 5 milligrams per ounce. To achieve a regulated level of caffeine, most manufacturers remove all caffeine from the kola extract and then re-add the correct amount to the syrup. The caffeine content of various beverages is listed in the accompanying table.

Amount of Caffeine (mg/oz) Found in Beverages

Brewed coffee	18–25	Tea	5–15
Instant coffee	12–16	Cocoa	1
		(but 20 mg/oz	
Decaffeinated	5–10	theobromine)	
coffee		Coca-Cola	3.5

NOTE: The average cup of coffee or tea contains about 5 oz of liquid. The average bottle of cola contains about 12 oz of liquid.

Because of the central nervous system effects from caffeine, many persons prefer to use **decaffeinated** coffee. The caffeine is removed from coffee by extracting the whole beans with an organic solvent. Then the solvent is drained off, and the beans are steamed to remove any residual solvent. The beans are dried and roasted to bring out the flavor. Decaffeination reduces the caffeine content of coffee to the range of 0.03 to 1.2% caffeine. The extracted caffeine is used in various pharmaceutical products, such as APC tablets.

Adenine   Guanine   Caffeine

Caffeine has always been a controversial compound. Some religions forbid the use of beverages containing caffeine, which they consider an addictive drug. In fact, many people consider caffeine an addictive drug. Recently there has been concern because caffeine is structurally similar to the purine bases adenine and guanine, which are two of the five principal bases organisms use to form the nucleic acids DNA and RNA. It is feared that the substitution of caffeine for adenine or guanine in either of these genetically important substances could lead to chromosome defects.

A portion of the structure of a DNA molecule is shown on the facing page. The typical mode of incorporation of both adenine and guanine is specifically shown. If caffeine were substituted for either of these, the hydrogen bonding necessary to link the two chains would be disturbed. Although caffeine is most similar to guanine, it could not form the central hydrogen bond, since it has a methyl group instead of a hydrogen in the necessary position. Hence, the genetic information would be garbled, and there would be a **break** in the chain. Fortunately, little evidence exists of chromosome breaks due to caffeine. Many cultures have been using tea and coffee for centuries without any apparent genetic problems.

Recent work by researchers at Ohio State University, however, suggests that consumption of caffeine and other xanthine compounds is related to the development of cystic breast disease, which is a nonmalignant but often painful condition characterized by a fibrous growth in the breast. When all coffee, tea, colas, and chocolate were eliminated from the diet of women suffering from cystic breast disease, most women experienced a resolution of this disease.

Another problem, not related to caffeine but rather to the beverage tea, is that in some cases persons who consume high quantities of tea may show symptoms of Vitamin B_1 (thiamine) deficiency. It is suggested that the tannins in the tea may complex with the thiamine, rendering it unavailable for use. An alternative suggestion is that caffeine may reduce the levels of the enzyme transketolase, which depends on the presence of thiamine for its activity. Lowered levels of transketolase would produce the same symptoms as lowered levels of thiamine.

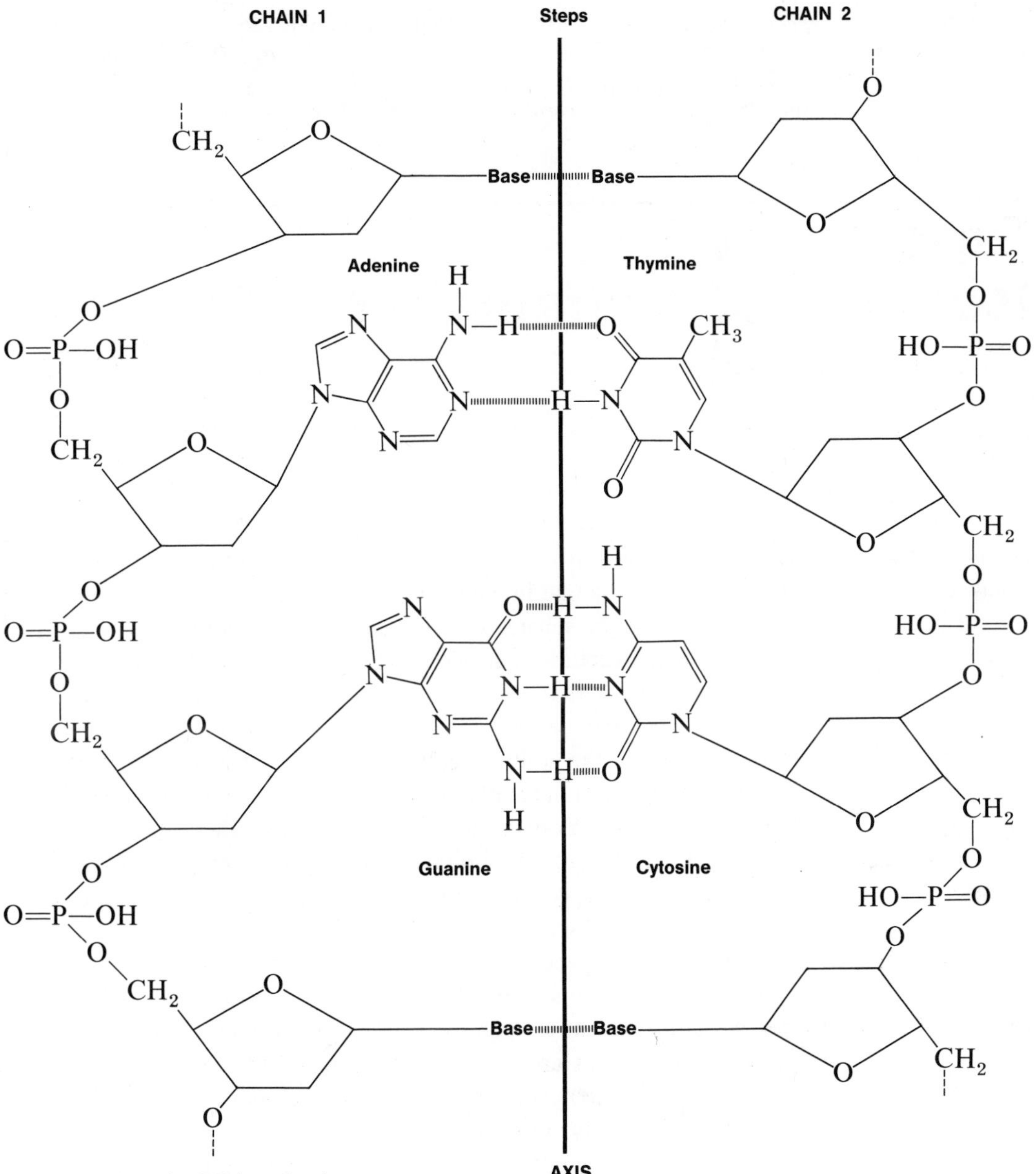

Portion of a DNA molecule

REFERENCES

"Caffeine May Damage Genes by Inhibiting DNA Polymerase." *Chemical and Engineering News, 45* (November 20, 1967): 19.

Emboden, W. "The Stimulants." *Narcotic Plants*. New York: Macmillan, 1972.

Minton, J. P., Foecking, M. K., Webster, D. J. T., and Matthews, R. H. "Caffeine, Cyclic Nucleotides, and Breast Disease." *Surgery, 86* (1979): 105.

Ray, O. S. "Caffeine." *Drugs, Society and Human Behavior*. 2nd ed. St. Louis: C. V. Mosby, 1978.

Ritchie, J. M. "Central Nervous System Stimulants. II: The Xanthines." In L. S. Goodman and A. Gilman, *The Pharmacological Basis of Therapeutics*. 7th ed. New York: Macmillan, 1985.

Taylor, N. *Plant Drugs that Changed the World*. New York: Dodd, Mead, 1965. Pp. 54–56.
Taylor, N. "Three Habit-Forming Nondangerous Beverages." In *Narcotics—Nature's Dangerous Gifts*. New York: Dell, 1970. (Paperbound revision of *Flight from Reality*.)
"Tea, Powdered Milk May Harm Health." *Chemical and Engineering News, 54* (May 3, 1976): 27.

Experiment 6
Isolation of Caffeine From Tea

Isolation of a Natural Product
Heating under Reflux
Filtration, Extraction
Crystallization
Sublimation

In this experiment, caffeine is isolated from tea leaves. The chief problem with the isolation is that caffeine does not exist alone in tea leaves but is accompanied by other natural substances from which it must be separated. The main component of tea leaves is cellulose, which is the principal structural material of all plant cells. Cellulose is a polymer of glucose. Since cellulose is virtually insoluble in water, it presents no problems in the isolation procedure. Caffeine, on the other hand, is water soluble and is one of the main substances extracted into the solution called tea. Caffeine constitutes as much as 5% by weight of the leaf material in tea plants. Tannins also dissolve in the hot water used to extract tea leaves. The term *tannin* does not refer to a single homogeneous compound, or even to substances that have similar chemical structure. It refers to a class of compounds that have certain properties in common. Tannins are phenolic compounds having molecular weights between 500 and 3000. They are widely used to "tan" leather. They precipitate alkaloids and proteins from aqueous solutions. Tannins are usually divided in two classes: those that can be hydrolyzed and those that cannot. Tannins of the first type that are found in tea generally yield glucose and gallic acid when they are hydrolyzed. These tannins are esters of gallic acid and glucose. They represent structures in which some of the hydroxyl groups in glucose have been esterified by digalloyl groups. The nonhydrolyzable tannins found in tea are condensation polymers of catechin. These polymers are not uniform in structure; catechin molecules are usually linked at ring positions 4 and 8.

When tannins are extracted into hot water, the hydrolyzable ones are partially hydrolyzed, meaning that free gallic acid is also found in tea. The tannins, because of their phenolic groups, and gallic acid, because of its carboxyl groups, are both acidic. If calcium carbonate, a base, is added to tea water, the calcium salts of these acids are formed. Caffeine can be extracted from the basic tea solution with methylene chloride, but the calcium salts of gallic acid and the tannins are not soluble in methylene chloride and remain behind as insoluble solids.

The brown color of a tea solution is due to flavonoid pigments and chlorophylls and to their respective oxidation products. Although chlorophylls are soluble in methylene chloride, most other substances in tea are not. Thus, the methylene chloride extraction of the basic tea solution removes nearly pure caffeine. The methylene chloride is

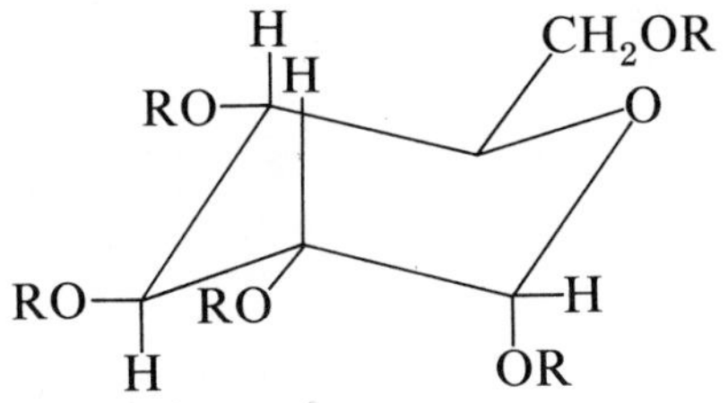

Glucose if R = H
A tannin if some R = Digalloyl

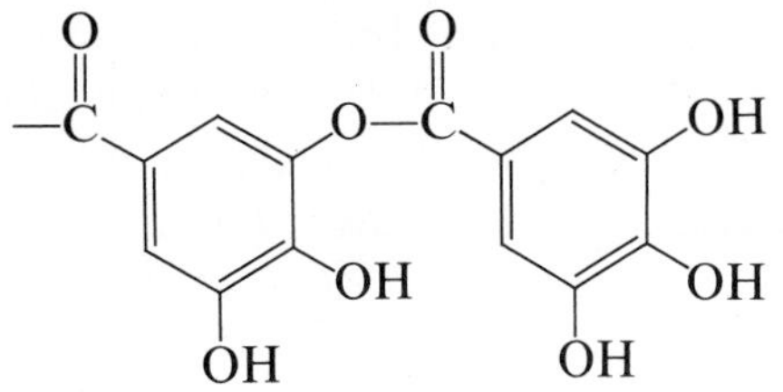

A digalloyl group

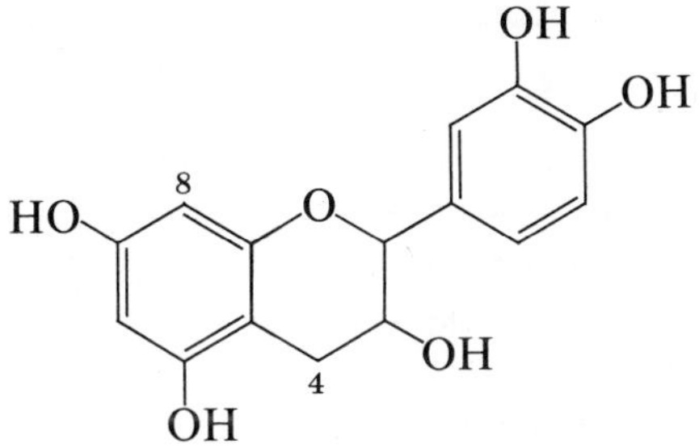

Catechin

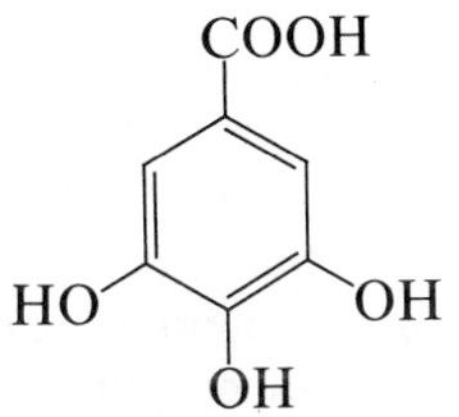

Gallic acid

easily removed by distillation (bp 40 °C) to leave the crude caffeine. The caffeine is then purified by recrystallization and sublimation.

In a second part of this experiment, caffeine may be converted to a **derivative.** A derivative of a compound is a second compound, of known melting point, formed from the original compound by a simple chemical reaction. In trying to make a positive identification of an organic compound, one must often convert it to a derivative. If the first compound, caffeine in this case, and its derivative have melting points that match those reported in the chemical literature (a handbook, for instance), it is assumed that there is no coincidence and that the identity of the first compound, caffeine, has been definitely established.

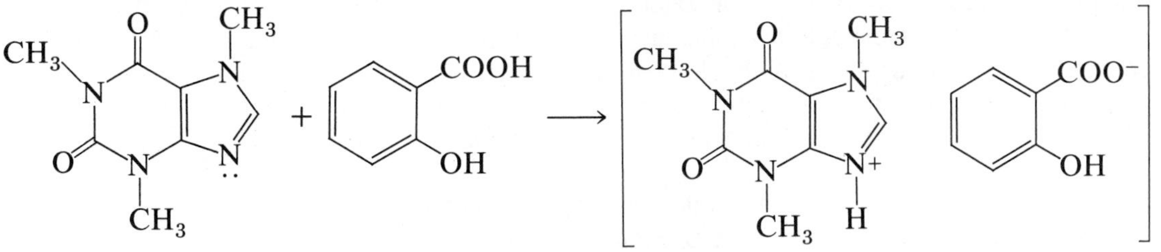

Caffeine   **Salicylic acid**   **Caffeine salicylate**

Caffeine is a base and will react with an acid to give a salt. With salicylic acid, a derivative **salt** of caffeine, caffeine salicylate, can be made to establish the identity of the caffeine isolated from tea leaves.

SPECIAL INSTRUCTIONS

You should read, as an introduction, the essay on caffeine. To perform this experiment, you should have read Techniques 1 through 5 and 13. Sections 6.3 and 6.4 of Technique 6 are also necessary. Be careful when handling methylene chloride. It is a toxic solvent, and you should not breathe it excessively or spill it on yourself. When you are discarding spent tea leaves, do not put them in the sink; they will clog the drain. Dispose of them in a waste container.

PROCEDURE

Place 25 g of dry tea leaves, 10 g of calcium carbonate powder, and 250 mL of water in a 500-mL three-necked round-bottomed flask equipped with a condenser for reflux (Technique 1, Figure 1–4, p. 510). Stopper the unused openings in the flask and heat the mixture under gentle reflux to prevent bumping (Technique 1, Section 1.7, p. 509) for about 20 minutes. Use a heating mantle or a Bunsen burner to heat. Shake the flask occasionally during this heating period. While the solution is still hot, vacuum-filter it through a fast filter paper such as E&D No. 617 or S&S No. 595 (Technique 2, Section 2.3, p. 518). A 500-mL filter flask is appropriate for this step.

Cool the filtrate (filtered liquid) to room temperature, and using a 500-mL separatory funnel, extract it (Technique 5, Section 5.4, p. 545) with a 50-mL portion of methylene chloride (dichloromethane). Shake the mixture vigorously for 1 minute. The layers should separate after standing for several minutes, although some emulsion will be present in the lower organic layer (Technique 5, Section 5.5, p. 548). The emulsion can be broken and the organic layer dried at the same time by passing the lower layer *slowly* through anhydrous magnesium sulfate as follows. Place a small piece of cotton (not glass wool) in the neck of a funnel and add a 1-cm layer of anhydrous magnesium sulfate on top of the cotton. Pass the organic layer directly from the separatory funnel into the drying agent and collect the filtrate in a dry flask. Rinse the magnesium sulfate with 1 to 2 mL of fresh methylene chloride solvent. Repeat the extraction with another 50-mL portion of methylene chloride on the aqueous layer remaining in the separatory funnel, and repeat the drying, as described above, with a *fresh* portion of anhydrous magnesium sulfate. Collect the organic layer in the flask containing the first methylene chloride extract. These extracts should now be clear, showing no visible signs of water contamination. If some water should pass through the filter, repeat the drying, as described above, with a fresh portion of anhydrous magnesium sulfate. Collect the extracts in a dry flask.

Pour the dry organic extracts into a round-bottomed flask. Assemble an apparatus for simple distillation (Technique 6, Figure 6–6, p. 557), add a boiling stone, and remove the methylene chloride by distillation on a steam bath. The residue in the distillation flask contains the caffeine and is purified as described in the next section on crystallization. Save the methylene chloride that was distilled. You may use some of it in the next step. The remainder should be placed in a waste collection container and **not** discarded in the sink.

CRYSTALLIZATION (PURIFICATION)

Dissolve the residue from the methylene chloride extraction of the tea solution in about 10 mL of the methylene chloride that you saved from the distillation. You may have to heat the mixture on a steam bath to dissolve the solid. Transfer the solution to a 50-mL Erlenmeyer flask. Rinse the distillation flask with an additional 5 mL of methylene chloride and combine this solution with the contents of the Erlenmeyer flask. Add a boiling stone and evaporate the now light green solution to dryness by heating it on a steam bath **in the hood.**

The residue obtained on evaporation of the methylene chloride is next crystallized by the mixed-solvent method (Technique 3, Section 3.7, p. 530). Using a steam bath, dissolve the residue in a small quantity (about 6 mL) of hot acetone[1] and add dropwise just enough low-boiling (30–60 °C) petroleum ether to turn the solution faintly cloudy. Cool the solution and collect the crystalline product by vacuum filtration, using a small Büchner funnel. A small amount of petroleum ether can be used to help in transferring the crystals to the Büchner funnel. A second crop of crystals can be obtained by concentrating the filtrate. Weigh the product (an analytical balance may be necessary). Calculate the weight percentage yield based on the 25 g of tea originally used and determine the melting point. The melting point of pure caffeine is 236 °C. Note the color of the solid for comparison with the material obtained after sublimation.

SUBLIMATION OF CAFFEINE

Caffeine can be finally purified by sublimation (Technique 13, p. 640). Assemble a sublimation apparatus as shown in Technique 13, Figure 13–2D, p. 644. Insert a 15 × 125-mm test tube into a No. 2 neoprene adapter, using a little water as lubricant, until the tube is fully inserted. Place chips of ice in the test tube and dry the outside surface. Check to see whether this assembly can be inserted into a 20 × 150-mm sidearm test tube.[2] Place all or part of your caffeine in the bottom of the sidearm test tube, which acts as the outer tube in the sublimation apparatus. Insert a trap (Technique 2, Figure 2–4, p. 519) between the aspirator and the sublimation apparatus. Turn on the aspirator and press the inner tube and the adapter into the outer tube until a good seal is obtained. At the point at which a good seal is achieved, one should hear or observe a change in the water velocity in the aspirator; at this time also, make sure that the central tube of the sublimation apparatus is firmly seated in the sidearm test tube. When a good vacuum seal has been obtained, heat the sample gently and carefully with a microburner to sublime the caffeine. Hold the burner in your hand (hold it at its base, **not** by the hot barrel) and apply heat by moving the flame back and forth under the outer tube and up the sides. If the sample begins to melt, remove the flame for a few seconds before you resume heating. When sublimation is complete, remove the burner and allow the apparatus to cool.

---

[1]If the residue does not dissolve in this quantity of acetone, magnesium sulfate may be present as an impurity (drying agent). Add additional acetone (up to about 20 mL), gravity-filter the mixture to remove the solid impurity, and reduce the volume of the filtrate to about 5 mL. Now add petroleum ether as indicated in the procedure.

[2]This sidearm test tube should be flared slightly. Although some come this way from the manufacturer, others may have to be altered by the instructor.

When the apparatus has cooled, stop the aspirator as you hold the inner tube down with your thumb and **carefully** remove the inner tube of the sublimation apparatus. If this operation is done carelessly, the sublimed crystals may be dislodged from the inner tube and fall back into the residue. Remove the remaining ice and water from the inner tube, and scrape the sublimed caffeine onto weighing paper, using a small spatula. Determine the melting point of this purified caffeine and compare it in melting point and color with the caffeine obtained following crystallization. Submit the sample to the instructor in a labeled vial, or if the instructor directs, prepare the caffeine salicylate derivative.

THE DERIVATIVE (OPTIONAL)

The amounts given in this part, including solvents, should be adjusted to fit the quantity of caffeine you obtained. Use an analytical balance. Dissolve 50 mg of caffeine and 37 mg of salicylic acid in 4 mL of toluene in a small Erlenmeyer flask by warming the mixture on a steam bath. Add about 1 mL (20 drops) of high-boiling (60–90 °C) petroleum ether or ligroin and allow the mixture to cool and crystallize. It may be necessary to cool the flask in an ice-water bath or to add a small amount of extra petroleum ether to induce crystallization. Collect the crystalline product by vacuum filtration, using a small Büchner funnel. Dry the product by allowing it to stand in the air and determine its melting point. Pure caffeine salicylate melts at 137 °C. Submit the sample to the instructor in a labeled vial.

QUESTIONS

1. Outline a separation scheme for isolating caffeine from tea. Use a flowchart similar in format to that shown in the **Advance Preparation and Laboratory Records** section.
2. Carefully inspect the structure of caffeine and caffeine salicylate. Why is the particular nitrogen that became protonated with salicylic acid more basic than the other three nitrogens?
3. Inspect the structures given for tannins. Which hydroxyl group or groups do you expect to be more acidic, and why?
4. Indicate how you might obtain gallic acid from tannic acid. A chemical reaction is involved.
5. In the crystallization of crude caffeine, petroleum ether (ligroin) is added to an acetone solution of caffeine. Why is the petroleum ether added? How does it work?
6. When methylene chloride first distills, it appears cloudy. Why?
7. Why does filtration through magnesium sulfate break the emulsion?
8. The crude caffeine isolated from tea has a green tinge. Why?

Experiment 7
Isolation of Caffeine From Coffee

Isolation of a Natural Product
Heating under Reflux
Filtration, Extraction
Crystallization
Sublimation

In this experiment, we shall isolate caffeine from coffee. The main problem of the isolation is that caffeine does not occur alone in coffee but is accompanied by other natural substances from which it must be separated. The green coffee bean (seed) contains caffeine (1–2%), tannins, glucose, fats, proteins, and cellulose. During roasting, the coffee beans swell and change their color to a dark brown. Roasting also develops the characteristic odor and flavor. The aroma is due to an oil called **caffeol.** This volatile oil contains mainly furfural, with traces of several hundred other compounds. The roasting also liberates the caffeine from the combination with chlorogenic acid in which it exists in the unroasted bean. Some caffeine sublimes from the beans during the roasting.

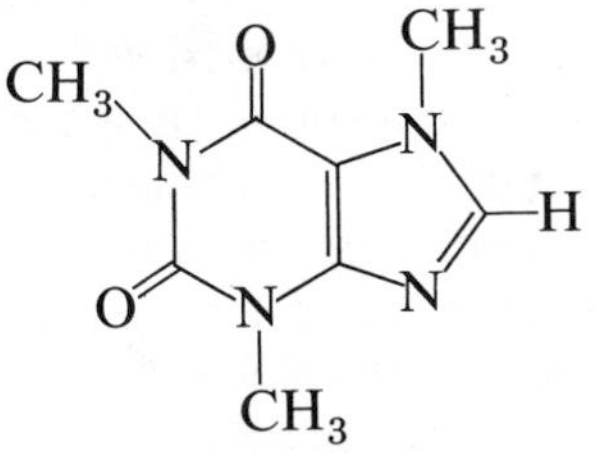

| Caffeine | Chlorogenic acid | Furfural |

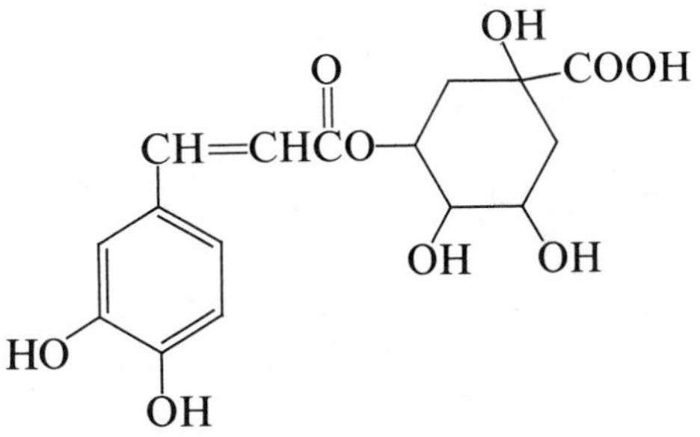

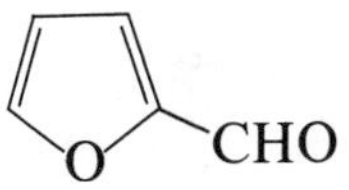

The separation depends on differences in the solubilities of the various constituents in coffee. The polymeric substances, protein and cellulose, are insoluble in water. Proteins are the polymers formed from amino acids, and cellulose is the polymer formed from glucose. Fats, which are esters formed from glycerol and three long-chain carboxylic acids, are also insoluble in water. On the other hand, caffeine, tannins, glucose, and chlorogenic acid are soluble in water.

Both the tannins (structures shown in Experiment 6) and chlorogenic acid are acidic. Adding calcium carbonate, a base, will precipitate these acidic substances as their calcium salts and will remove them from the aqueous solution:

$$2RCOOH + CaCO_3 \longrightarrow Ca^{2+}(RCOO^-)_2 + H_2O + CO_2$$

The calcium salts can then be removed along with the polymeric substances (protein and cellulose) by filtration. Caffeine and glucose remain behind in the aqueous solution.

Finally, the caffeine is removed from the aqueous solution by extracting with methylene chloride. Glucose is not soluble in methylene chloride. The methylene chloride is easily removed by distillation to yield crude caffeine. The caffeine is purified by recrystallization followed by sublimation. A derivative can be prepared as described in Experiment 6.

SPECIAL INSTRUCTIONS

Read the essay on caffeine, which precedes Experiment 6. To perform this experiment, you should have read Techniques 1 through 5 and 13. Sections 6.3 and 6.4 of Technique 6 are also necessary. Be careful when handling methylene chloride. It is a toxic solvent, and you should not breathe it excessively or spill it on yourself. When discarding spent coffee grounds, do not put them in the sink; they will clog the drain. Dispose of them in a waste container. Before doing the extractions, read *all* the instructions carefully. A bad emulsion will form if the instructions are not followed carefully.

PROCEDURE

Place 35 g of ground coffee (regular grind), 10 g of calcium carbonate powder, and 150 mL of water into a 500-mL three-necked round-bottomed flask equipped with a condenser for reflux (Technique 1, Figure 1–4, p. 510). Stopper the unused openings in the flask and heat the mixture under gentle reflux to prevent bumping (Technique 1, Section 1.7, p. 509) for about 20 minutes. Use a heating mantle or a Bunsen burner to heat. Shake the flask occasionally during this heating period. Also during the heating period, assemble a vacuum-filtration apparatus, using a 500-mL filter flask (Technique 2, Figure 2–4, p. 519). When the boiling action has stopped and the solids have settled, and while the solution is still hot, filter the solution through a Büchner funnel by vacuum filtration (Technique 2, Section 2.3, p. 518). Use a fast filter paper such as E&D No. 617 or S&S No. 595.

Cool the filtrate (filtered liquid) to room temperature, and using a 500-mL separatory funnel, extract it (Technique 5, Section 5.4, p. 545) with a 50-mL portion of methylene chloride (dichloromethane). In doing the extraction, if the separatory funnel is shaken too vigorously, a bad emulsion will form. To avoid an emulsion that can be broken only with difficulty, **gently** invert the separatory funnel as shown in Technique 5, Figure 5–4, p. 547. Vent the separatory funnel through the stopcock, and **gently** mix the liquids by moving the funnel repeatedly during a 5-minute period in a circular manner, as shown in the illustration on page 69. During that time, return the funnel occasionally and **carefully** to the upright position.

Allow the layers to separate so that a distinct interface is observed between the layers. The lower organic phase will contain a considerable amount of emulsion in spite

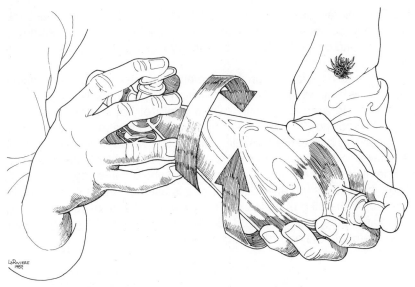

Correct manner of shaking and venting the separatory funnel

of the precautions you have taken. Drain the lower organic layer **and the emulsion** into a large Erlenmeyer flask until the interface just enters the stopcock. Place the organic layer aside and **carefully** reextract the aqueous layer remaining in the separatory funnel with another 50-mL portion of methylene chloride. Again, care should be taken to avoid bad emulsions by using the extraction procedure just described for a 5-minute period. Collect the lower organic layer with the emulsion in the flask containing the first methylene chloride extract. Discard the aqueous phase left in the separatory funnel and rinse the funnel with water. Add 10 g of anhydrous magnesium sulfate, or more if needed, to the combined methylene chloride extracts and swirl the mixture. At this point, distinct layers should form and the emulsion should break. Carefully decant the organic and aqueous phases away from the solids back into the separatory funnel. Separate the layers, taking care to save the organic layer. Transfer the organic phase to a **dry** flask, add a spatulaful of anhydrous magnesium sulfate to dry the solution (Technique 5, Section 5.6, p. 549, and swirl the mixture several times for about 5 minutes.

Remove the drying agent by gravity filtration of the solution through a fluted filter into a round-bottomed flask (Technique 2, Figure 2–3, p. 517). Assemble an apparatus for simple distillation (Technique 6, Figure 6–6, p. 557), add a boiling stone, and remove the methylene chloride by distillation on a steam bath. The light brown residue in the distillation flask contains the caffeine and is purified as described next (see "Crystallization"). Save the methylene chloride that was distilled. You will use some of it in the next step. The remainder should be placed in a waste collection container and **not** discarded in the sink.

CRYSTALLIZATION (PURIFICATION)

Dissolve the residue obtained from the methylene chloride extraction of the coffee solution in about 10 mL of the methylene chloride that you saved from the distillation. It may be necessary to heat the mixture on a steam bath to dissolve the solid. Transfer the solution to a 50-mL Erlenmeyer flask. Rinse the distillation flask with an additional 5 mL of methylene chloride and combine this solution with the contents of the Erlenmeyer flask. Add a boiling stone and evaporate the solution to dryness by heating it on a steam bath **in the hood.**

The residue obtained on evaporation of the methylene chloride is next crystallized by the mixed-solvent method (Technique 3, Section 3.7, p. 530). Using a steam bath, dissolve the residue in a small quantity (about 5 mL) of hot acetone[1] and add dropwise just enough low-boiling (30–60 °C) petroleum ether to turn the solution faintly cloudy. Cool the solution and collect the crystalline product by vacuum filtration, using a Büchner funnel. A small amount of petroleum ether can be used to aid in transferring the crystals to the Büchner funnel. A second crop of crystals can be obtained by concentrating the filtrate. Weigh the product (an analytical balance is useful). Calculate the weight percentage yield based on the 35 g of coffee originally used and determine the melting point. The melting point of pure caffeine is 236 °C. Record the color of the caffeine for comparison with the material obtained after sublimation.

SUBLIMATION OF CAFFEINE

The caffeine can be finally purified by the sublimation procedure in Experiment 6. A small amount of dark residue will remain unsublimed. Avoid splattering this residue into the crystals by subliming slowly. Following the sublimation, determine the melting point of this purified caffeine and compare in both melting point and color with the caffeine obtained after crystallization. Submit the sample to the instructor in a labeled vial, or at the option of the instructor, prepare the caffeine salicylate derivative by the procedure given in Experiment 6.

QUESTIONS

1. Outline a separation scheme for isolating caffeine from coffee. Use the flowchart format shown in the **Advance Preparation and Laboratory Records** section.

2. In the essay preceding Experiment 6, the decaffeination method is explained. Explain why decaffeination is conducted **before** roasting and not after. In answering this question, consider the differences between the chemical contents of the green bean and the roasted bean. Is something desirable extracted during decaffeination?

3. A structure of tannic acid (pentadigalloyl derivative of glucose) is shown in Experiment 6. Which hydroxyl group or groups do you expect to be more acidic and why?

4. Indicate how you might obtain gallic acid from tannic acid. A chemical reaction is involved.

5. In the crystallization of crude caffeine, petroleum ether is added to an acetone solution of caffeine. Why is the petroleum ether added? How does it work?

6. The crude caffeine is often brown. What do you suggest as a reason for the color? *Hint:* Consider the changes that might occur during roasting.

7. When methylene chloride first distills, it appears cloudy. Why?

8. Why did the magnesium sulfate break the emulsion in this experiment?

---

[1]If the residue does not dissolve in this quantity of acetone, magnesium sulfate may be present as an impurity (drying agent). Add additional acetone (up to about 20 mL), gravity-filter the mixture to remove the solid impurity, and reduce the volume of the filtrate to about 5 mL. Now add petroleum ether as indicated in the procedure.

Essay
STEROIDS

Steroids are an important class of natural products whose structures are related in that they all have the same basic four-ring carbon skeleton of perhydrocyclopentanophenanthrene. Steroids occur widely in both plants and animals; the most impor-

**THE BASIC STEROID
RING SKELETON**

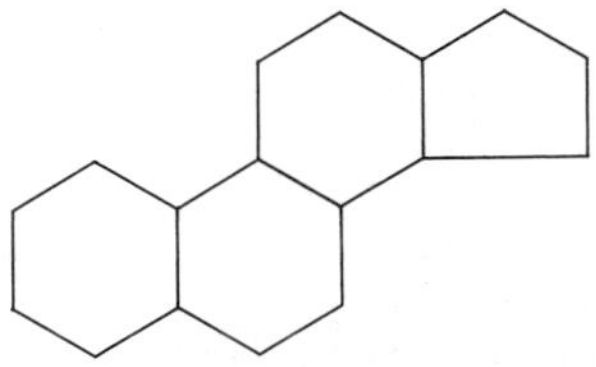

Perhydrocyclopentanophenanthrene

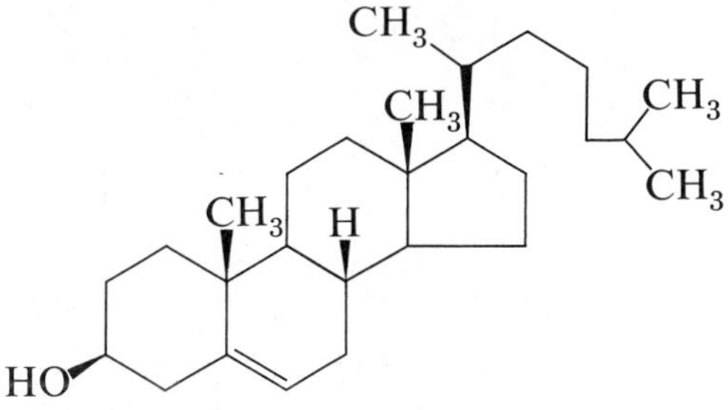

Cholesterol

tant steroids, however, are found in animals, where they have various essential biological functions. The male and female sex hormones in mammals are steroids, and so are the bile acids and adrenal hormones.

The most abundant steroid is cholesterol. Cholesterol exists in all tissues of the mammalian body. It is particularly abundant in the spinal cord, in the brain, and in gallstones, if they are present. The total cholesterol content of a 165-pound man averages about 250 g. Because of cholesterol's wide availability, it was the first steroid to be isolated and identified. Because of its complicated structure, however, more than 160 years intervened between its discovery in 1770 and the realization of its structure in 1932. Since cholesterol has eight asymmetric centers, and the possibility of 2^8, or 256, possible stereoisomers, another 23 years elapsed before its total three-dimensional structure was fully elucidated in 1955.

Mammals have the ability to absorb cholesterol from dietary sources. Early studies of cholesterol showed that since an animal often excretes more cholesterol than it consumes, animals must be able also to synthesize cholesterol. It actually turns out that all tissues in the body have the ability to synthesize cholesterol to some degree.

Biochemical studies have shown that cholesterol is important because it is one of the first intermediates the body synthesizes along the routes to the production of all the other steroid hormones. Cholesterol, for example, is the precursor of all the **bile acids.** Cholic acid and deoxycholic acid are the most important bile acids, but there are many different, though related, compounds found among the various species of animals. The bile acids are synthesized from cholesterol in the liver and converted to conjugates with simple peptides and amino acids. Glycocholic acid and deoxyglycocholic acid, formed by the conjugation (amide formation) of the above-mentioned bile acids with glycine, represent these conjugates. The bile acid conjugates are secreted as "liver bile" into the intestines as their sodium salts. They are natural, detergentlike substances, and their main purpose is to emulsify (carry into solution) fats and oils; the fats and oils can then be more easily digested and transported across the intestinal cell wall into the bloodstream. Because of this, cholesterol (the precursor of the bile acids) is inextricably involved in the metabolism of fats and oils.

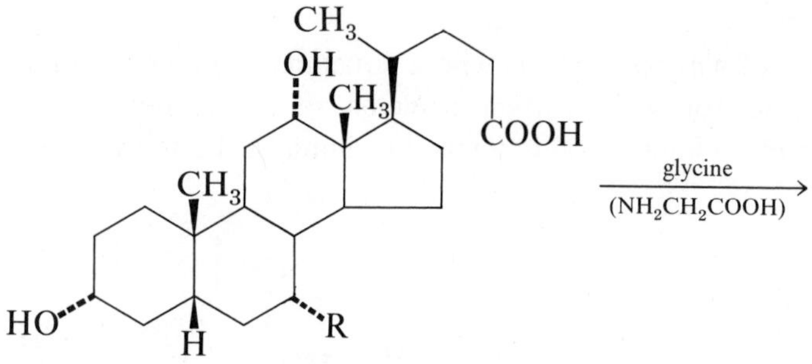

Cholic acid if R = OH
Deoxycholic acid if R = H

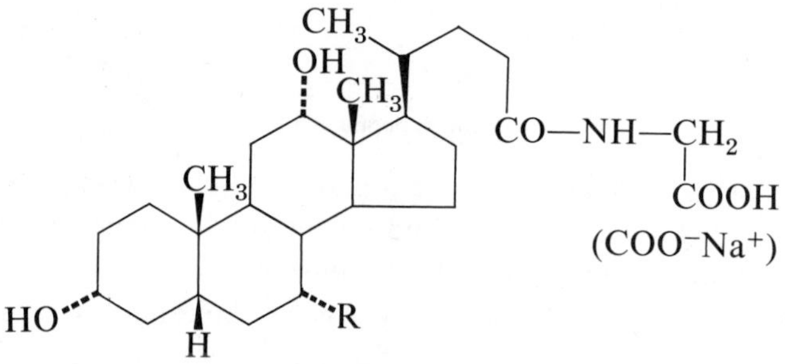

Glycocholic acid if R = OH
Deoxyglycocholic acid if R = H

Two diseases are commonly associated with an improper balance between cholesterol and the bile acids on the one hand and fats and oils in the diet on the other. Gallstones result from a condition in which the bile, which normally contains some cholesterol, becomes supersaturated with cholesterol. In the gallbladder, cholesterol crystallizes from the bile, producing large crystalline aggregates. The causes of gallstones are too complex for us to go into, but it can be stated that such factors as high levels of estrogens, multiple pregnancies, obesity, genetic factors, and certain drugs influence the degree to which the bile becomes supersaturated with cholesterol.

The second disease results from an elevated level of cholesterol in the blood. Many persons have difficulty in maintaining a proper cholesterol balance within their systems. Excesses of cholesterol in the blood lead to **atherosclerosis,** a hardening of the arteries due to cholesterol deposits. Atherosclerosis (a form of arteriosclerosis) is a dangerous condition. The loss of elasticity and the narrowing of the channel in the arteries lead to a strain on the heart and a likelihood of heart disease. Should some of the cholesterol deposits break loose, they could either block the flow of blood to the heart, causing a heart attack, or block the flow of blood to the brain, causing a stroke.

A far more likely possibility is that these deposits will simply grow in place until they occlude the artery altogether, thus stopping blood flow. For these reasons, determining the level of cholesterol in the blood has become routine in medical examinations.

The cholesterol level in the blood is often related to diet and physiology. Obese persons have generally higher serum cholesterol levels than lean persons do. Populations whose diets include a high proportion of animal fats (saturated fats) or foods containing cholesterol also tend to have higher levels. The role of cholesterol itself and the role of diet have both been studied with regard to atherosclerosis, but neither role is well defined. It is clear that persons with this disease have excess cholesterol and do not metabolize it correctly. The cholesterol, however, probably did not cause the condition, which more likely is a result of an abnormality in lipid (fat) metabolism. Persons who have this abnormality must avoid both cholesterol and saturated fats in their diets.

It has been widely assumed for several years that cholesterol, atherosclerosis, and heart disease are directly linked. However, a 1980 report of the Food and Nutrition Board of the National Academy of Sciences has suggested that the relation between cholesterol intake and heart disease may not be so clear as was once thought. This report suggests that there is no conclusive evidence for a direct connection between dietary cholesterol intake alone and the incidence of heart attacks.

A recent finding has shown that vitamin C will reduce the level of cholesterol in the blood and that it is also a required cofactor for the production of bile acids from cholesterol. Persons deficient in vitamin C usually have high cholesterol levels. This may result from the inability to produce bile acids, which would build up cholesterol, and this inability would lead to an incorrect metabolism of fats.

Estrone   Estradiol   Progesterone

A second class of important steroids the body produces from cholesterol includes the male and female sex hormones. The principal female hormones (estrogens) are estrone and estradiol, along with a third compound, progesterone. These compounds are responsible for the secondary female sex characteristics and for maintaining the female menstrual cycle. Progesterone, the pregnancy hormone, suppresses ovulation in the female. Synthetic compounds of similar structure are now used as birth control agents (for example, Norlutin and Norethynodrel). These two compounds mimic progesterone in their action and are more effective than progesterone when taken orally, since they are better absorbed from the stomach and intestines than progesterone itself is.

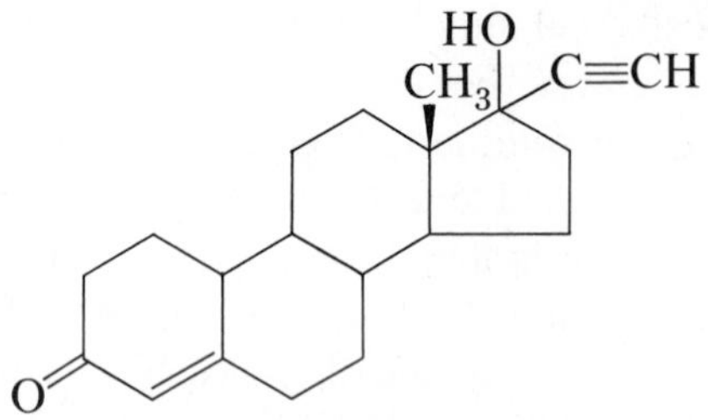

Norlutin

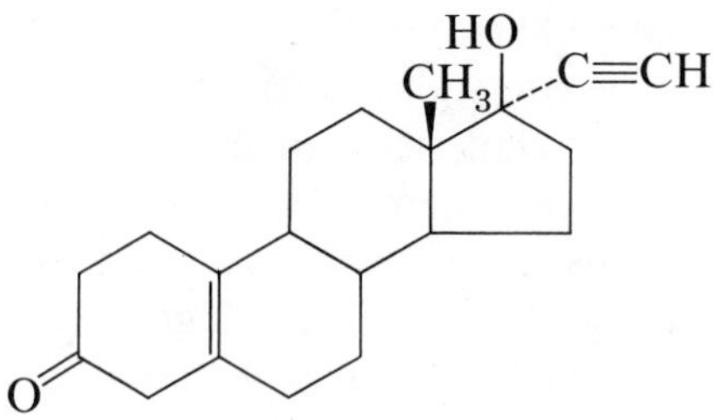

Norethynodrel

In spite of the widespread use of oral contraceptives, there is considerable controversy over their safety. There have been reports of a tendency for some women to experience formation of clots in their circulatory systems. There have also been some reports that these substances may form certain types of tumors in animals; but it has not been shown that these results apply to humans. To circumvent these problems, manufacturers of oral contraceptives have produced new formulations that include lower doses of the synthetic hormones.

The two principal male hormones are testosterone and androsterone. The structures of these two compounds are shown below.

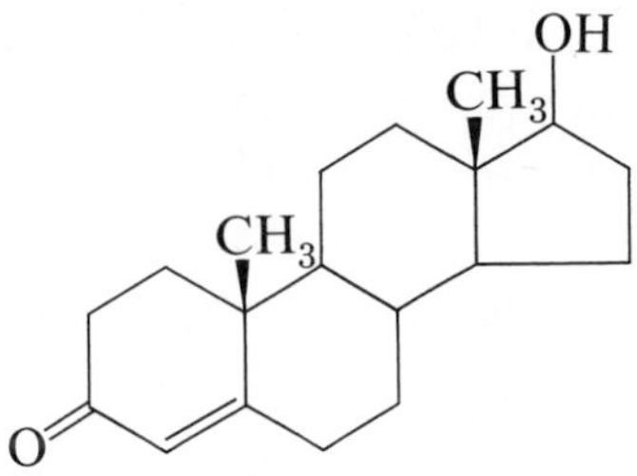

Testosterone

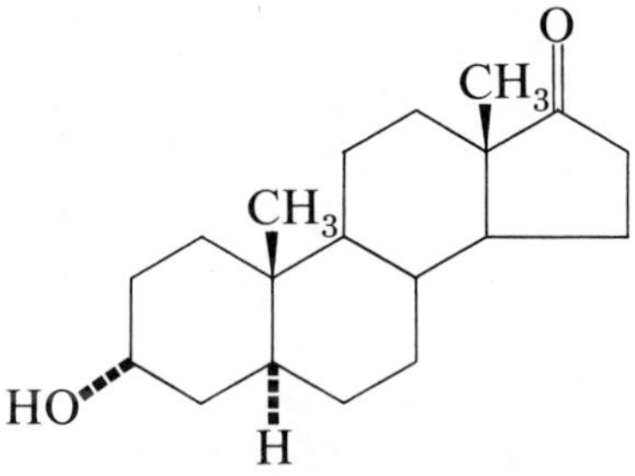

Androsterone

A third class of steroids synthesized biochemically from cholesterol contains the adrenal hormones. Two representative members of this class are corticosterone and cortisone. The adrenal hormones govern a wide variety of metabolic processes (too many to detail here). The principal use of cortisone is as an anti-inflammatory drug, to combat swelling and inflammation in the body. One of many possible examples of the medical use of cortisone is treating rheumatoid arthritis. Cortisone reduces swelling in inflamed joints.

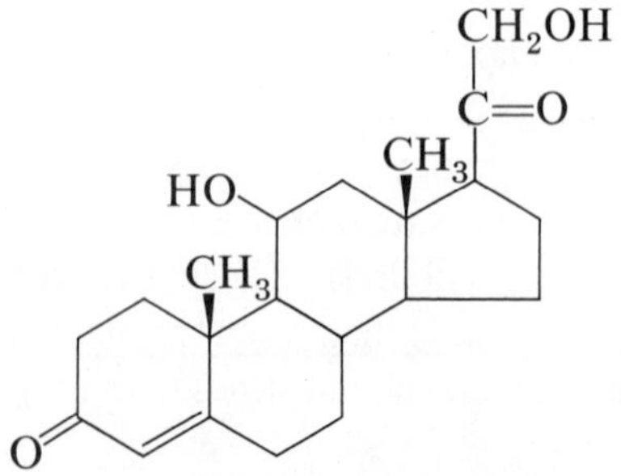

Corticosterone

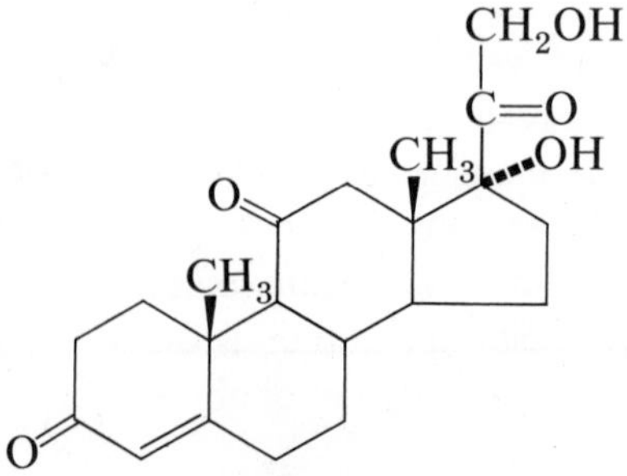

Cortisone

A variety of other important steroids are found in the plant world, where they serve many purposes.

REFERENCES

''Academy Says Curb on Cholesterol Not Needed.'' *Science, 208* (June 20, 1980): 1354.

Benditt, E. P. ''The Origin of Atherosclerosis.'' *Scientific American, 236* (February 1977): 74.

''Biochemistry of the Pill Largely Unknown.'' *Chemical and Engineering News, 45* (March 27, 1967): 44.

Bloch, K. ''The Biological Synthesis of Cholesterol.'' *Science, 150* (1965): 19.

Csapo, A. ''Progesterone.'' *Scientific American, 198* (April 1958): 48.

Fieser, L. F. ''Steroids.'' *Scientific American, 192* (January 1955): 52.

Ginter, E. ''Cholesterol: Vitamin C Controls Its Transformation to Bile Acids.'' *Science, 179* (1973): 704.

McKnight, J. D., and Julian, M. M. ''Chemistry of Gallstones.'' *Journal of Chemical Education, 60* (July 1983): 594.

O'Malley, B. W., and Schrader, W. T. ''The Receptors of Steroid Hormones.'' *Scientific American, 234* (February 1976): 32.

Petrow, V. ''Current Aspects of Fertility Control.'' *Chemistry in Britain, 6* (1970): 167.

Pincus, G. ''Control of Conception by Hormonal Steroids.'' *Science, 153* (1966): 493.

Sanders, H. J., ''Heart Disease and Drugs.'' *Chemical and Engineering News, 43.* Part I, p. 130 (March 8, 1965); Part II, p. 74 (March 22, 1965).

Zuckerman, S. ''Hormones.'' *Scientific American, 196* (March 1957): 76.

Zurer, P. S. ''Drugs in Sports.'' *Chemical and Engineering News, 62* (April 30, 1984): 69.

Experiment 8
Cholesterol From Gallstones

Isolation of a Natural Product
Bromination–Debromination
Filtration, Extraction

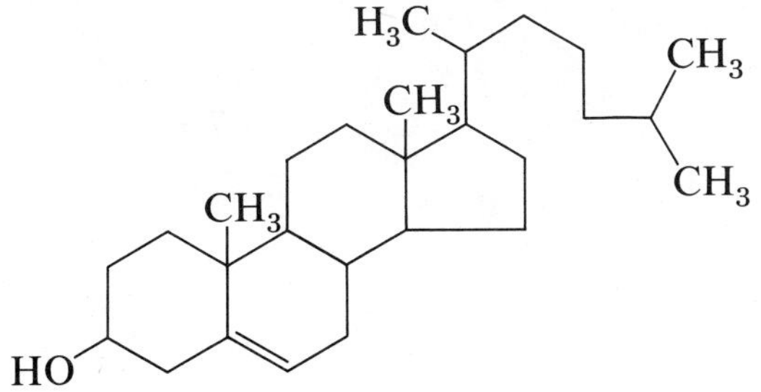

Cholesterol

In this experiment we isolate and purify cholesterol from human gallstones. The gallstones are pulverized in a blender, and the cholesterol is extracted with hot diethyl ether. Recrystallization is unfortunately not sufficient to purify cholesterol. This is because three other steroids, albeit in small amounts (0.1–3%), are found along with the cholesterol. These compounds are so much like cholesterol in their solubility characteristics that they are not separable by crystallization. The structure of one of the impurities, cholestanol, is shown in the accompanying diagram. The other impurities have exactly the same structure as cholestanol, except they contain double bonds. The substance 7-dehydrocholesterol has double bonds between carbons 5 and 6 and between carbons 7 and 8, and Δ^7-cholesten-3β-ol has one double bond between carbons 7 and 8.

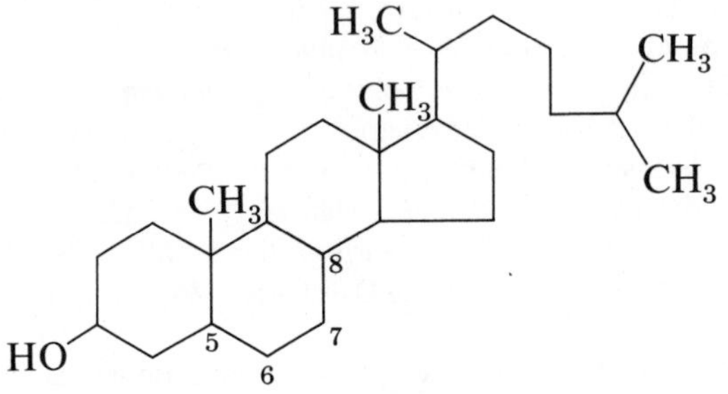

Cholestanol

The separation is done chemically. Cholesterol is converted to cholesterol dibromide by adding bromine across the double bond. The impurities have either no double bond (cholestanol) or less active ones, probably for steric reasons. If 7-dehydrocholesterol or Δ^7-cholesten-3β-ol should react with bromine, they would be dehydrobrominated to dienes and trienes, which would stay, along with cholestanol, in the mother liquors when cholesterol dibromide was recrystallized. The dibromide is different from any of the impurities in solubility properties.

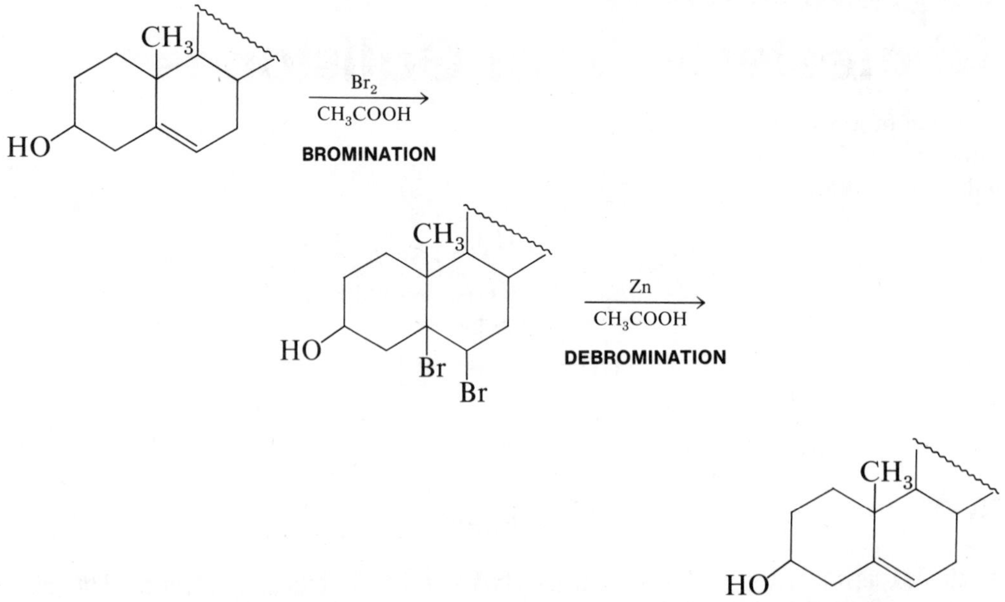

The cholesterol dibromide, now pure, can be converted back to cholesterol by debromination with zinc metal in acetic acid. Recrystallization yields pure cholesterol.

SPECIAL INSTRUCTIONS

Before beginning this experiment, you should read the essay on steroids and also Techniques 1, 2, 3, 4, and 5. The brominating solution can cause unpleasant burns if it touches the skin, so care should be exercised in handling it.

PROCEDURE

ISOLATION

Weigh 2 g of pulverized gallstones and place them in a 25-mL Erlenmeyer flask.[1] Add 10 mL of diethyl ether and heat the mixture, with swirling, on a steam bath until all the solid has disintegrated, and the cholesterol has dissolved (Technique 1, Section 1.7, p. 509, and Figure 1–5, p. 511). Filter the brownish yellow solution through a fluted filter while it is still hot (Technique 2, Section 2.1, p. 514). If it is necessary to replace evaporation loss, rinse the filter with a small amount (2–3 mL) of ether. The brown residue that collects is a metabolic oxidation product of hemoglobin—a bile pigment called **bilirubin.** Dilute the filtrate with 10 mL of methanol, add a little Norite to decolorize the solution, and heat the mixture on the steam bath. Preheat a funnel; then filter the hot solution through a fluted filter paper. Reheat the greenish yellow filtrate and add just enough water (dropwise) to make the solution cloudy (Technique 3, Section 3.7, p. 530). The solution is now saturated at the boiling point, and cholesterol will crystallize on cooling. Collect the crystals by vacuum filtration, using a small Büchner funnel (Technique 2, Section 2.3, p. 518). Wash them once with cold methanol and let them stand for a time in the open Büchner funnel to allow the solvent to dry. Weigh the product and determine the melting point. Approximately 1.5 g of impure cholesterol (mp ca 146–148 °C) should have been collected.

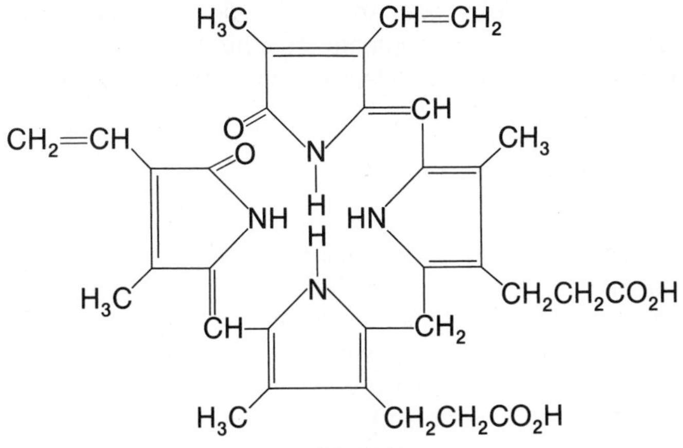

Bilirubin

[1]The gallstones should be prepared in advance by pulverizing them in a blender.

BROMINATION

Dissolve 1.0 g of the crude cholesterol obtained from the gallstones in 10 mL of **anhydrous** ether in a 25-mL Erlenmeyer flask. Gentle warming on the steam bath will probably be necessary. Using an eyedropper, slowly add 5 mL of a solution of bromine and sodium acetate in glacial acetic acid. The solution should be added until the yellowish

> **CAUTION: This solution causes bad burns if it comes in contact with the skin.**

color persists. The cholesterol dibromide should begin to crystallize in a minute or so. Cool the flask in an ice bath and stir the solution to ensure complete crystallization. Also cool a mixture of 3 mL of ether and 7 mL of glacial acetic acid in a separate flask in the ice bath. Collect the crystals by vacuum filtration, using a small Büchner or a Hirsch funnel and wash them with the cold ether–acetic acid solution. Wash the crystals a second time with 5 to 10 mL cold methanol and leave them in the funnel long enough for the passage of air to dry them.

DEBROMINATION

Transfer the dibromide crystals to a 50-mL Erlenmeyer flask and add 20 mL ether, 5 mL glacial acetic acid, and 0.2 g zinc dust. If the reaction is slow, add a little more zinc dust. Zinc deteriorates on exposure to air, so it is best to use a freshly opened bottle. Swirl the mixture. Gentle heating may be required. The dibromide should dissolve, and zinc acetate should separate as a white paste (5–10 minutes). Stir for an extra 5 minutes and then add enough water (dropwise) to dissolve any solid and give a clear solution. Decant the solution from the unreacted zinc into a separatory funnel (stopcock closed!) and extract the ether solution twice with water to remove the zinc acetate and zinc dibromide (Technique 5, Section 5.4, p. 545). Then wash the ether solution with 10% sodium hydroxide to remove acetic acid. Repeat until a drop of the ether solution no longer turns blue litmus red. Shake the solution with an equal volume of a saturated aqueous NaCl solution to reduce its water content; separate the ether layer and filter it through a filter paper containing anhydrous sodium sulfate to complete the drying.

RECRYSTALLIZATION

Add 10 mL of methanol to the dry ether solution and evaporate the solution on a steam bath until most of the ether is removed and the purified cholesterol begins to crystallize. Let the solution stand at room temperature for a while, then cool it in an ice bath to complete the crystallization. Collect the crystals by vacuum filtration, wash them with a little cold methanol, and leave them to dry. Weigh the product and determine the melting point. The melting point should be determined relatively quickly; otherwise, air oxidation may produce impurities that lower the melting point (pure cholesterol: mp 149–150 °C). Calculate the yield from gallstones, and submit the cholesterol to the instructor in a labeled vial.

QUESTIONS

1. Outline a separation scheme for isolating cholesterol from gallstones. Use a flowchart similar in format to that shown in the **Advance Preparation and Laboratory Records** section.

2. Bromine adds to the double bond in cholesterol with *trans* stereochemistry. Draw a three-dimensional representation of the mechanism of this process, showing the two rings involved in the reaction in their chair conformations.

3. Write a balanced equation for the formation of cholesterol from cholesterol dibromide and zinc metal. How is zinc acetate formed in the reaction?

4. Write a mechanism for the debromination of a dibromide by zinc metal. What stereochemistry is required for this reaction? Why?

5. Alkynes can be purified by forming the tetrabromide, crystallizing the product, and converting it back to the alkyne. Write equations for this process.

Essay
ETHANOL AND FERMENTATION CHEMISTRY

The fermentation processes involved in making bread, making wine, and brewing are among the oldest chemical arts. Even though fermentation had been known as an art for centuries, not until the nineteenth century did chemists begin to understand this process from the point of view of science. In 1810, Gay-Lussac discovered the general chemical equation for the breakdown of sugar into ethanol and carbon dioxide. The manner in which the process took place was the subject of much conjecture until Louis Pasteur began his thorough examination of fermentation. Pasteur demonstrated that yeast was required in the fermentation. He was also able to identify other factors that controlled the action of the yeast cells. His results were published in 1857 and 1866.

For many years, scientists believed that the transformation of sugar into ethanol and carbon dioxide by yeasts was inseparably connected with the life process of the yeast cell. This view was abandoned in 1897, when Büchner demonstrated that yeast extract will bring about alcoholic fermentation in the absence of any yeast cells. The fermenting activity of yeast is due to a remarkably active catalyst of biochemical origin, the enzyme zymase. It is now recognized that most of the chemical transformations that go on in living cells of plants and animals are brought about by enzymes. The enzymes are organic compounds, generally proteins, and establishment of structures and reaction mechanisms of these compounds is an active field of present-day research. Zymase is now known to be a complex of at least 22 separate enzymes, each of which catalyzes a specific step in the fermentation reaction sequence.

Enzymes show an extraordinary specificity—a given enzyme acts on a specific compound or a closely related group of compounds. Thus, zymase acts on only a few select sugars and not on all carbohydrates; the digestive enzymes of the alimentary tract are equally specific in their activity.

The chief sources of sugars for fermentation are the various starches and the molasses residue obtained from refining sugar. Corn (maize) is the chief source of starch in the United States, and ethyl alcohol made from corn is known commonly as **grain alcohol.** In preparing alcohol from corn, the grain, with or without the germ, is ground and cooked to give the **mash.** The enzyme diastase is added in the form of **malt** (sprouted barley that has been dried in air at 40 °C and ground to a powder) or of a mold such as *Aspergillus oryzae.* The mixture is kept at 40 °C until all the starch has been converted to the sugar **maltose** by hydrolysis of ether and acetal bonds. This solution is known as the **wort.**

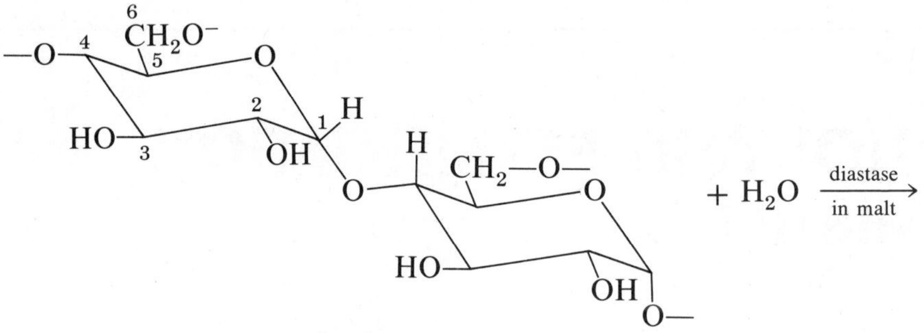

Starch

This is a glucose polymer with 1,4- and 1,6- glycosidic linkages. The linkages at C-1 are α.

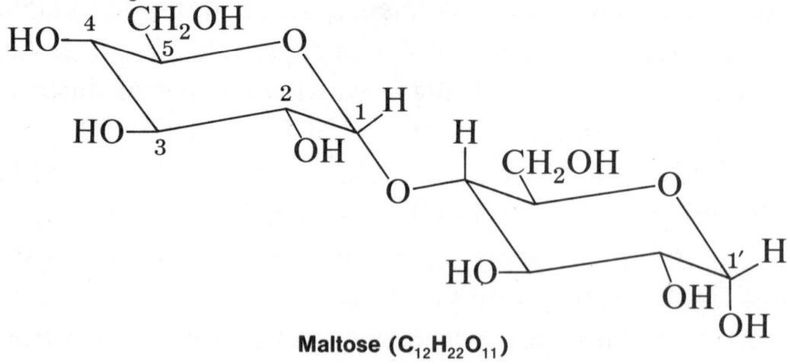

Maltose ($C_{12}H_{22}O_{11}$)

The α linkage still exists at C-1.
The —OH is shown α at the 1′ position (axial), but it can also be β (equatorial).

The wort is cooled to 20 °C and diluted with water to 10% maltose, and a pure yeast culture is added. The yeast culture is usually a strain of *Saccharomyces cerevisiae* (or *ellipsoidus*). The yeast cells secrete two enzyme systems: maltase, which converts the maltose into glucose, and zymase, which converts the glucose into carbon dioxide

$$\text{Maltose} + H_2O \xrightarrow{\text{maltase}} 2 \quad$$

β-**D**-(+)-**Glucose**

(*α*-D-(+)-Glucose, with an axial
—OH, is also produced.)

$$\text{Glucose} \xrightarrow{\text{zymase}} 2CO_2 + 2CH_3CH_2OH + 26 \text{ kcal}$$

C₆H₁₂O₆

and alcohol. Heat is liberated, and the temperature must be kept below 32 °C by cooling to prevent destruction of the enzymes. Oxygen in large amounts is initially necessary for the optimum reproduction of yeast cells, but the actual production of alcohol is anaerobic. During fermentation, the evolution of carbon dioxide soon establishes anaerobic conditions. If oxygen were freely available only carbon dioxide and water would be produced.

After 40 to 60 hours, fermentation is complete, and the product is distilled to remove the alcohol from solid matter. The distillate is fractionated by means of an efficient column. A small amount of acetaldehyde, bp 21 °C, distills first and is followed by 95% alcohol. Fusel oil is contained in the higher-boiling fractions. The fusel oil consists of a mixture of higher alcohols, chiefly 1-propanol, 2-methyl-1-propanol, 3-methyl-1-butanol, and 2-methyl-1-butanol. The exact composition of fusel oil varies considerably; it particularly depends on the type of raw material that is fermented. These higher alcohols are not formed by fermentation of glucose. They arise from certain amino acids derived from the proteins present in the raw material and in the yeast. These fusel oils cause the headaches familiarly associated with drinking alcoholic beverages.

Industrial alcohol is ethyl alcohol used for nonbeverage purposes. Most of the commercial alcohol is denatured, to avoid payment of taxes, the biggest cost in the price of liquor. The denaturants render the alcohol unfit for drinking. Methanol, aviation fuel, and other substances are used for this purpose. The difference in price between taxed and nontaxed alcohol is more than $20 a gallon. Before efficient synthetic processes were developed, the chief source of industrial alcohol was fermented blackstrap molasses, the noncrystallizable residue from refining cane sugar (sucrose). Most industrial ethanol in the United States is now manufactured from ethylene, a product of the ''cracking'' of petroleum hydrocarbons. By reaction with concentrated sulfuric acid, ethylene becomes ethyl hydrogen sulfate, which is hydrolyzed to ethanol by dilution with water. The alcohols 2-propanol, 2-butanol, 2-methyl-2-propanol, and higher secondary and tertiary alcohols also are produced on a large scale from alkenes derived from cracking.

Yeasts, molds, and bacteria are used commercially for the large-scale production of various organic compounds. An important example, in addition to ethanol production, is the anaerobic fermentation of starch by certain bacteria to yield 1-butanol, acetone, ethanol, carbon dioxide, and hydrogen.

REFERENCES

Amerine, M. A. "Wine." *Scientific American, 211* (August 1964): 46.
"Chemical Technology: Key to Better Wines." *Chemical and Engineering News, 51* (July 2, 1973): 14.
"Chemistry Concentrates on the Grape." *Chemical and Engineering News, 51* (June 25, 1973): 16.
Church, L. B. "The Chemistry of Winemaking." *Journal of Chemical Education, 49* (1972): 174.
Ough, C. S. "Chemicals Used in Making Wine." *Chemical and Engineering News, 65* (January 5, 1987): 19.
Ray, O. S. "Alcohol." *Drugs, Society, and Human Behavior*. St. Louis: C. V. Mosby, 1972.

Students wanting to investigate alcoholism and possible chemical explanations for alcohol addiction may consult the following references:

Cohen, G., and Collins, M. "Alkaloids from Catecholamines in Adrenal Tissue: Possible Role in Alcoholism." *Science, 167* (1970): 1749.
Davis, V. E., and Walsh, M. J. "Alcohol Addiction and Tetrahydropapaveroline." *Science, 169* (1970): 1105.
Davis, V. E., and Walsh, M. J. "Alcohols, Amines, and Alkaloids: A Possible Biochemical Basis for Alcohol Addiction." *Science, 167* (1970): 1005.
Labianca, D. A. "Acetaldehyde Syndrome and Alcoholism." *Chemistry, 47* (October 1974): 21.
Levitt, A. E. "Alcoholism: Costly Health Problem for U.S. Industry." *Chemical and Engineering News, 49* (September 13, 1971): 25.
Seevers, M. H., Davis, V. E., and Walsh, M. J. "Morphine and Ethanol Physical Dependence: A Critique of a Hypothesis." *Science, 170* (1970): 1113.
Yamanaka, Y., Walsh, M. J., and Davis, V. E. "Salsolinol, An Alkaloid Derivative of Dopamine Formed in Vitro during Alcohol Metabolism." *Nature, 227* (1970): 1143.

Experiment 9
Ethanol From Sucrose

Fermentation
Simple Distillation
Fractional Distillation
Azeotropes

Sucrose as well as maltose can be used as a starting material for making ethanol. Sucrose is a disaccharide with the formula $C_{12}H_{22}O_{11}$. It has one glucose molecule combined with fructose, unlike maltose, which has two glucose molecules. The enzyme **invertase** catalyzes the hydrolysis of sucrose, rather than maltase, which hydrolyzes maltose. The hydrolysis of maltose is discussed in the essay on ethanol and fermentation. Zymase is used to convert the sugars to alcohol and carbon dioxide.

Pasteur observed that growth and fermentation were promoted by adding small amounts of mineral salts to the nutrient medium. Later it was found that before fermentation, the hexose sugars combine with phosphoric acid, and the resulting hexose–phosphoric acid combination is then degraded into carbon dioxide and ethanol. The carbon dioxide is not wasted in the commercial process but is converted to Dry Ice.

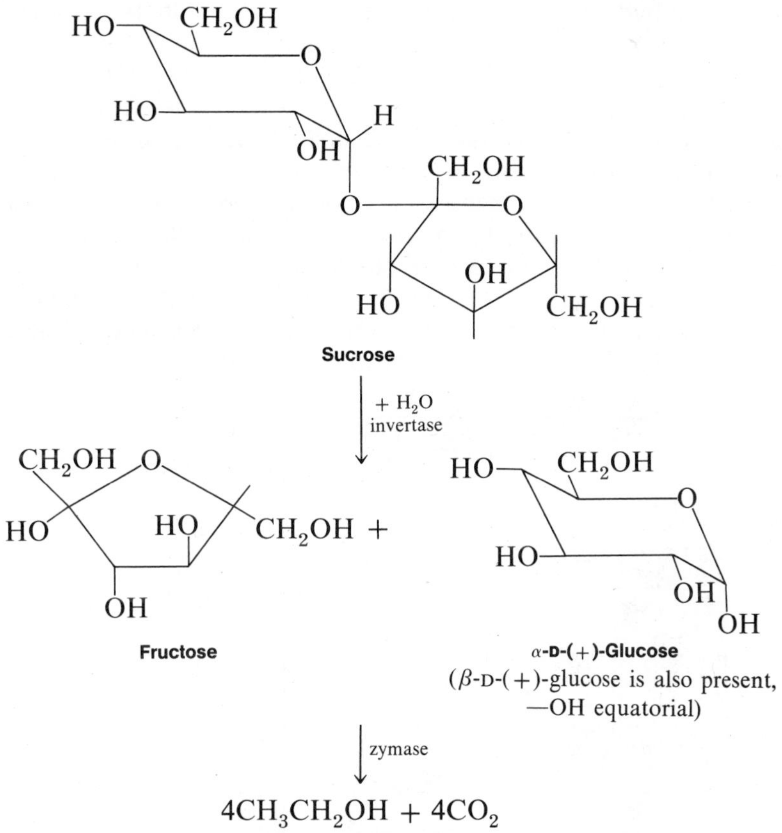

Sucrose

$+ H_2O$
invertase

Fructose   + α-**D-(+)-Glucose**
(β-D-(+)-glucose is also present,
—OH equatorial)

zymase

$$4CH_3CH_2OH + 4CO_2$$

The fermentation is inhibited by ethanol; it is not possible to prepare solutions containing more than 10 to 15% ethanol by this method. More concentrated ethanol can be isolated by fractional distillation. Ethanol and water form an azeotropic mixture consisting of 95% ethanol and 5% water by weight, which is the most concentrated ethanol that can be obtained by fractionation of dilute ethanol-water mixtures.

SPECIAL INSTRUCTIONS

Before you begin this experiment, you should read the essay on ethanol and Techniques 1, 2, 6, and 7. The fermentation must be started at least 1 week before ethanol is to be actually isolated. When the aqueous ethanol solution is to be separated from the yeast cells, it is important to siphon carefully as much of the clear, supernatant liquid as possible, without agitating the mixture.

PROCEDURE

Place 40 g of sucrose (common granulated sugar) in a 500-mL Erlenmeyer flask. Add 350 mL water, warmed to room temperature; 35 mL Pasteur's salts; and half a package of dried baker's yeast or 15 g of cake yeast.[1] Shake vigorously and fit the flask with a one-hole rubber stopper with a glass tube leading to a beaker or a test tube containing a solution of barium hydroxide. Protect the barium hydroxide from air by adding some mineral oil or xylene to form a layer above the barium hydroxide. The figure depicts the apparatus for this experiment. A precipitate of barium carbonate will form, indicating that CO_2 is being evolved. Alternatively, a balloon may be substituted for the barium hydroxide trap assembly. The gas will cause the balloon to expand as the fermentation continues. Oxygen from the atmosphere is excluded by these techniques. If oxygen were allowed to continue in contact with the fermenting solution, the ethanol could be further oxidized to acetic acid or even all the way to carbon dioxide and water. So long as carbon dioxide continues to be liberated, ethanol is being formed.

Allow the mixture to stand at about 25 °C until fermentation is complete, as indicated by the cessation of gas evolution. Usually about 1 week is required. After this time, carefully move the flask to a desk and remove the stopper. Siphon the liquid out of the

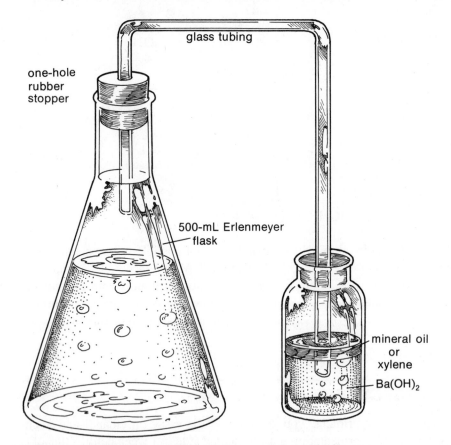

glass tubing

one-hole rubber stopper

500-mL Erlenmeyer flask

mineral oil or xylene

Ba(OH)$_2$

Apparatus for fermentation experiment

[1]A solution of Pasteur's salts consists of potassium phosphate, 2.0 g; calcium phosphate, 0.20 g; magnesium sulfate, 0.20 g; and ammonium tartrate, 10.0 g, dissolved in 860 mL water.

flask so that as little of the sediment as possible is removed. A siphon is easily started by filling a short section of rubber tubing with water, pinching one end closed, and placing the other end in the liquid in the flask. Release the end of the tubing that is not in the flask and hold it over the edge of the desk top. Allow the ethanol-water solution to run into a large beaker or flask. When the level of the liquid approaches the sediment, slow the siphoning by pinching the rubber tubing slightly. It is better to leave some liquid behind than to draw some sediment out of the flask.

If the siphoned liquid is not clear, clarify it as follows. Place about 2 tablespoons of Filter Aid (Johns-Manville Celite) in a beaker with about 200 mL water. Stir the mixture vigorously and then pour the contents into a Büchner funnel (with filter paper) while applying a vacuum, as in a vacuum filtration (Technique 2, Section 2.3, p. 518). This procedure will cause a thin layer of Filter Aid to be deposited on the filter paper (Technique 2, Section 2.4, p. 520). Discard the water that passes through this filter. The siphoned liquid containing the ethanol is then passed through this filter under gentle suction. The extremely tiny yeast particles are trapped in the pores of the Filter Aid. The liquid contains ethanol in water plus smaller amounts of dissolved metabolites (fusel oils) from the yeast.

Add about 46 g of anhydrous potassium carbonate to the filtered solution for each 100 mL of liquid. The solution, after becoming saturated with potassium carbonate, will be subjected to fractional distillation. Transfer the solution to a distillation apparatus equipped with a fractionating column packed with a metal sponge (Technique 7, Figure 7–2, p. 567, and Section 7.6, p. 575). Distill the liquid slowly through the fractionating column to get the best possible separation. Collect the fraction boiling between 78 and 88 °C and discard the residue in the distilling flask. The extent of purification of the ethanol is limited, since ethanol and water form a constant-boiling mixture, or an azeotrope, with a composition of 95% ethanol and 5% water (bp 78.1 °C). No amount of distillation will remove the last 5% of water (Technique 7, Section 7.7, p. 575).

Calculate the percentage yield of alcohol, assuming that the product is 85% alcohol and 15% water, and submit the ethanol to the instructor in a labeled vial.[2]

QUESTIONS

1. By doing some library research, see whether you can find out the method or methods commercially used to produce **absolute** ethanol.

2. Why is the air trap necessary in the late stages of fermentation?

3. The percentage of water in the alcohol can be estimated by measuring the density of the solution. Explain how this procedure works.

4. How does the acetaldehyde impurity arise in the fermentation?

5. Diethylacetal can be detected by gas chromatography. How does this impurity arise in fermentation?

6. Calculate how many milliliters of carbon dioxide would be produced theoretically from 40 g of sucrose at 25 °C and 1-atmosphere pressure.

---

[2]A careful analysis by flame-ionization gas chromatography on a typical student-prepared ethanol sample provided the following results:

Acetaldehyde	0.060%
Diethylacetal of acetaldehyde	0.005
Ethanol	88.3 (by hydrometer)
1-Propanol	0.031
2-Methyl-1-propanol	0.092
5-Carbon and higher alcohols	0.140
Methanol	0.040
Water	11.3 (by difference)

Essay
ESTERS—FLAVORS AND FRAGRANCES

Esters are a class of compounds widely distributed in nature. They have the general formula

$$R-\overset{\overset{\displaystyle O}{\|}}{C}-OR'$$

The simple esters tend to have pleasant odors. In many cases, although not exclusively so, the characteristic flavors and fragrances of flowers and fruits are due to compounds with the ester functional group. An exception is the case of the essential oils. The organoleptic qualities (odors and flavors) of fruits and flowers may often be due to a single ester, but more often, the flavor or the aroma is due to a complex mixture in which a single ester predominates. Some common flavor principles are listed in Table 1. Food and beverage manufacturers are thoroughly familiar with these esters and often use them as additives to spruce up the flavor or odor of a dessert or a beverage. Many times such flavors or odors do not even have a natural basis, as is the case with the "juicy fruit" principle, isopentenyl acetate. An instant pudding that has the flavor of rum may never have seen its alcoholic namesake—this flavor can be duplicated by the proper admixture, along with other minor components, of ethyl formate and iso-butyl propionate. The natural flavor and odor are not exactly duplicated, but most people can be fooled. Often only a trained person with a high degree of gustatory perception, a professional taster, can tell the difference.

A single compound is rarely used in good-quality imitation flavoring agents. A formula for an imitation pineapple flavor that might fool an expert is listed in Table 2. The formula includes 10 esters and carboxylic acids that can easily be synthesized in the laboratory. The remaining seven oils are isolated from natural sources.

Flavor is a combination of taste, sensation, and odor transmitted by receptors in the mouth (taste buds) and nose (olfactory receptors). The stereochemical theory of odor is discussed in the essay that precedes Experiment 21. The four basic tastes (sweet, sour, salty, and bitter) are perceived in specific areas of the tongue. The sides of the tongue perceive sour and salty tastes, the tip is most sensitive to sweet tastes, and the back of the tongue detects bitter tastes. The perception of flavor, however, is not so simple. If it were, it would require only the formulation of various combinations of four basic substances: a bitter substance (a base), a sour substance (an acid), a salty sub-stance (sodium chloride), and a sweet substance (sugar), to duplicate any flavor! In fact, we cannot duplicate flavors in this way. The human actually possesses 9000 taste buds. The combined response of these taste buds is what allows perception of a particu-lar flavor.

TABLE 1. Ester Flavors and Fragrances

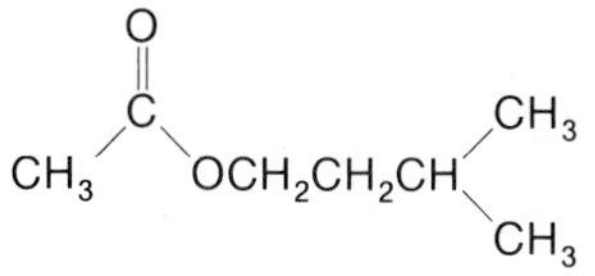 **Isoamyl acetate** Banana (Alarm pheromone of honeybee)	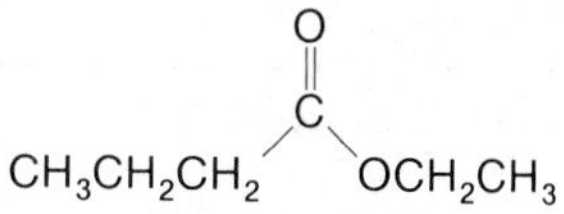 **Ethyl butyrate** Pineapple
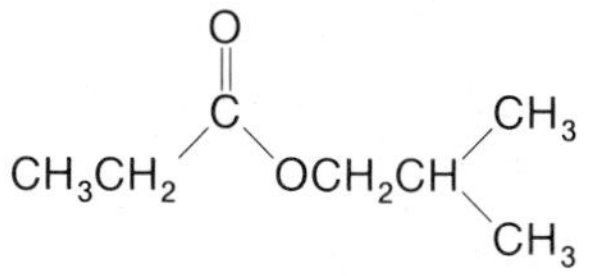 **Isobutyl propionate** Rum	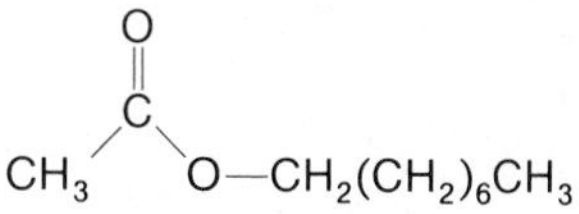 **Octyl acetate** Oranges
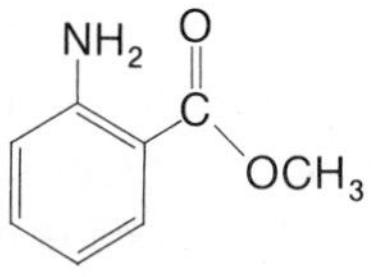 **Methyl anthranilate** Grape	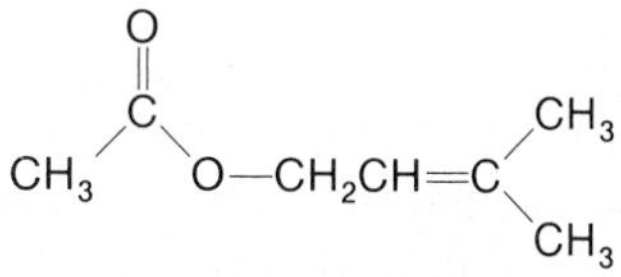 **Isopentenyl acetate** "Juicy fruit"
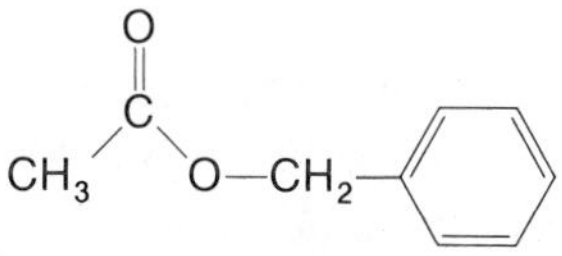 **Benzyl acetate** Peach	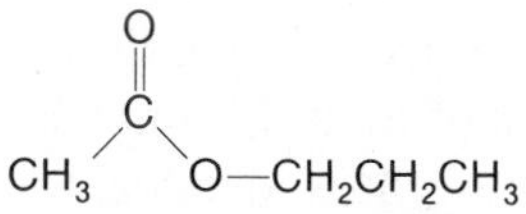 **n-Propyl acetate** Pear
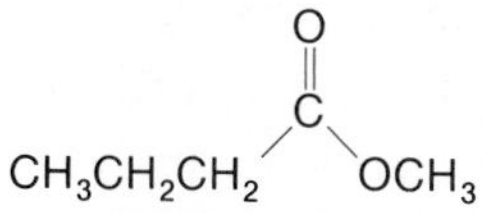 **Methyl butyrate** Apple	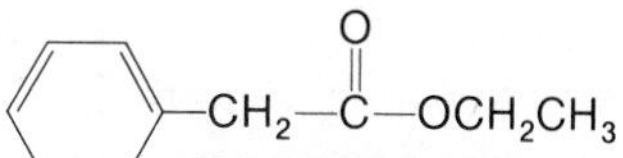 **Ethyl phenylacetate** Honey

Although the "fruity" tastes and odors of esters are pleasant, they are seldom used in perfumes or scents that are applied to the body. The reason for this is chemical. The ester group is not as stable to perspiration as the ingredients of the more expensive essential-oil perfumes. The latter are usually hydrocarbons (terpenes), ketones, and ethers extracted from natural sources. Esters, however, are used only for the cheapest toilet waters, since on contact with sweat, they undergo hydrolysis, giving organic acids. These acids, unlike their precursor esters, generally do not have a pleasant odor.

$$R-\overset{\displaystyle O}{\overset{\|}{C}}{}_{\displaystyle OR'} + H_2O \longrightarrow R-\overset{\displaystyle O}{\overset{\|}{C}}{}_{\displaystyle OH} + R'OH$$

TABLE 2. Artificial Pineapple Flavor

PURE COMPOUNDS	%	ESSENTIAL OILS	%
Allyl caproate	5	Oil of sweet birch	1
Isoamyl acetate	3	Oil of spruce	2
Isoamyl isovalerate	3	Balsam Peru	4
Ethyl acetate	15	Volatile mustard oil	1
Ethyl butyrate	22	Oil cognac	5
Terpinyl propionate	3	Concentrated orange oil	4
Ethyl crotonate	5	Distilled oil of lime	2
Caproic acid	8		19
Butyric acid	12		
Acetic acid	5		
	81		

Butyric acid, for instance, has a strong odor like that of rancid butter (of which it is an ingredient) and is a component of what we normally call body odor. It is this substance that makes foul-smelling humans so easy for an animal to detect when he is downwind of them. It is also of great help to the bloodhound, which is trained to follow small traces of this odor. Ethyl butyrate and methyl butyrate, however, which are the **esters** of butyric acid, smell like pineapple and apple, respectively.

A sweet fruity odor also has the disadvantage of possibly attracting fruit flies and other insects in search of food. Isoamyl acetate, the familiar solvent called banana oil, is particularly interesting. It is identical to the alarm **pheromone** of the honeybee. *Pheromone* is the name applied to a chemical secreted by an organism that evokes a specific response in another member of the same species. This kind of communication is common between insects who otherwise lack means of intercourse. When a honeybee worker stings an intruder, an alarm pheromone, composed partly of isoamyl acetate, is secreted along with the sting venom. This chemical causes aggressive attack on the intruder by other bees, who swarm after the intruder. Obviously it wouldn't be wise to wear a perfume compounded of isoamyl acetate near a beehive. Pheromones are discussed in more detail in the essay preceding Experiment 16.

REFERENCES

Benarde, M. A. *The Chemicals We Eat*. New York: American Heritage Press, 1971. Pp. 68–75.
The Givaudan Index. New York: Givaudan-Delawanna, 1949. (Gives specifications of synthetics and isolates for perfumery.)
Gould, R. F., ed. *Flavor Chemistry, Advances in Chemistry,* No. 56. Washington: American Chemical Society, 1966.
Pyke, M. *Synthetic Food*. London: John Murray, 1970.
Rasmussen, P. W. "Qualitative Analysis by Gas Chromatography—G. C. versus the Nose in Formulation of Artificial Fruit Flavors. *Journal of Chemical Education, 61* (January 1984): 62.
Shreve, R. N., and Brink, J. *Chemical Process Industries*. 4th ed. New York: McGraw-Hill, 1977.

Experiment 10
Isopentyl Acetate (Banana Oil)

Esterification
Heating under Reflux
Extraction
Simple Distillation

In this experiment, we prepare an ester, isopentyl acetate (isoamyl acetate). This ester is often referred to as banana oil, since it has the familiar odor of this fruit.

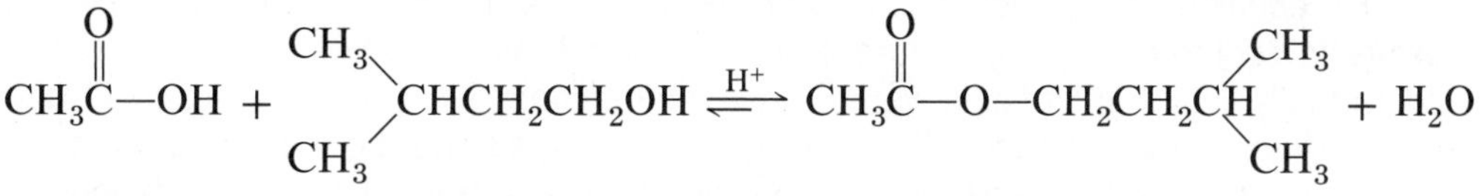

$$CH_3\overset{O}{\underset{}{C}}-OH \; + \; \underset{CH_3}{\overset{CH_3}{\diagdown}}CHCH_2CH_2OH \; \overset{H^+}{\rightleftharpoons} \; CH_3\overset{O}{\underset{}{C}}-O-CH_2CH_2\underset{CH_3}{\overset{CH_3}{CH}} \; + \; H_2O$$

Acetic acid   **Isopentyl alcohol**   **Isopentyl acetate**
(excess)

Isopentyl acetate is prepared by the direct esterification of acetic acid with isopentyl alcohol. Since the equilibrium does not favor the formation of the ester, it must be shifted to the right, in favor of the product, by using an excess of one of the starting materials. Acetic acid is used in excess because it is less expensive than isopentyl alcohol and more easily removed from the reaction mixture.

In the isolation procedure, much of the excess acetic acid and the remaining isopentyl alcohol are removed by extraction with water. Any remaining acid is removed by extraction with aqueous sodium bicarbonate. The ester is purified by distillation.

SPECIAL INSTRUCTIONS

Read the essay on esters, which precedes this experiment. To perform this experiment, you should read Techniques 1, 5, and 6. Since a 1-hour reflux is involved, the experiment should be started at the very beginning of the laboratory period. During the reflux period, other experimental work may be performed. Be careful in handling concentrated sulfuric acid. It will cause extreme burns if it is spilled on the skin.

PROCEDURE

Pour 15 mL (12.2 g, 0.138 mole) of isopentyl alcohol (also called isoamyl alcohol or 3-methyl-1-butanol) and 20 mL (21 g, 0.35 mole) of glacial acetic acid into a 100-mL round-bottomed flask. Carefully add 4 mL of concentrated sulfuric acid to the contents of the flask, with swirling. Add several boiling stones to the mixture.

CAUTION: Extreme care must be exercised to avoid contact with concentrated sulfuric acid. It will cause serious burns if it is spilled on the skin. If it comes in contact with the skin or clothes, it must be washed off immediately with excess water. In addition, sodium bicarbonate may be used to neutralize the acid. Clean up all spills immediately.

Assemble a reflux apparatus as shown in Technique 1, Figure 1–4, p. 510. Bring the mixture to a boil with a suitable heating source, such as a heating mantle or an oil bath (Technique 1, Sections 1.4 and 1.5, p. 507 to 509). Heat the mixture under reflux for 1 hour (Technique 1, Section 1.7, p. 509). Remove the heating source and allow the mixture to cool to room temperature. Pour the cooled mixture into a separatory funnel and carefully add 55 mL of cold water. Rinse the reaction flask with 10 mL of cold water and pour the rinsings into the separatory funnel. Use a stirring rod to mix the materials somewhat. Stopper the separatory funnel and shake it several times (Technique 5, Section 5.4, p. 545). Separate the lower aqueous layer from the upper organic layer (density 0.87 g/mL). Discard the aqueous layer after making certain that the correct layer has been saved.

The crude ester in the organic layer contains some acetic acid, which can be removed by extraction with 5% aqueous sodium bicarbonate solution. Carefully add 25 mL of the 5% base to the organic layer contained in the separatory funnel. Swirl the separatory funnel gently until carbon dioxide gas is no longer evolved. Stopper and gently shake the funnel once or twice, and then vent the vapors. Shake the funnel until no vapors are evolved when the separatory funnel is vented. Remove the lower layer, and repeat the above extraction with 25 mL of 5% sodium bicarbonate solution. Remove the lower layer and check to see whether it is basic to litmus. If it is not basic, repeat the procedure with additional 25-mL portions of 5% base until the aqueous layer is basic. Discard the basic washings and extract the organic layer with one 25-mL portion of water. Add 5 mL of saturated aqueous sodium chloride to aid in layer separation. Stir the mixture gently; do not shake it. Carefully separate the lower aqueous layer and discard it. When the water has been removed, pour the ester from the top of the separatory funnel into a flask. Add about 2 g of anhydrous magnesium sulfate to dry the ester (Technique 5, Section 5.6, p. 549). Stopper the flask and swirl it gently. Allow the crude ester to stand until the liquid is clear. About 15 minutes will be needed to complete drying. If the solution is still cloudy after this period, decant the solution, and add to it a fresh 0.5-g quantity of drying agent.

Assemble a simple distillation apparatus as shown in Technique 6, Figure 6–6, p. 557. Dry all glassware thoroughly before use. Carefully decant the ester into the distilling flask so that the drying agent is excluded. Add several boiling stones and distill the ester (Technique 6, Section 6.4, p. 556). The receiver must be cooled in an ice bath. Collect the fraction boiling between 134 and 143 °C in a dry flask. Weigh the product and calculate the percentage yield.

At your instructor's option, obtain an infrared spectrum (see Technique 17). Compare the spectrum with the one reproduced in this experiment. Include it with your report to the instructor. Submit the prepared sample to the instructor in a labeled vial.

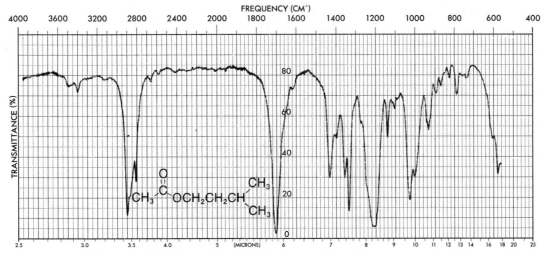

Ir spectrum of isopentyl acetate, neat

QUESTIONS

1. Give a mechanism for the formation of isopentyl acetate.
2. One method for favoring the formation of the ester is to add excess acetic acid. Suggest another method, involving the right-hand side of the equation, that will favor the formation of the ester.
3. Outline a separation scheme for isolating pure isopentyl acetate from the reaction mixture.
4. Why is it easier to remove excess acetic acid from the products than excess isopentyl alcohol?
5. Interpret the principal absorption bands in the infrared spectrum of isopentyl acetate.

Experiment 11
Methyl Salicylate (Oil of Wintergreen)

Synthesis of an Ester
Extraction
Vacuum Distillation

In this experiment you prepare a familiar-smelling organic ester—oil of wintergreen. Methyl salicylate was first isolated in 1843 by extraction from the wintergreen plant (*Gaultheria*). It was soon found that this compound had analgesic and antipyretic character almost identical to that of salicylic acid (see the essay "Aspirin") when taken internally. This medicinal character probably derives from the ease with which methyl salicylate is hydrolyzed to salicylic acid under the alkaline conditions found in the intestinal tract. Salicylic acid is known to have such analgesic and antipyretic properties. Methyl salicylate can be taken internally or absorbed through the skin; thus, it finds much use in liniment preparations. Applied to the skin, it produces a mild burning or soothing sensation, which probably comes from the action of its phenolic hydroxyl group. This ester also has a pleasant odor, and it is used to a small extent as a flavoring principle.

Salicylic acid   Methyl salicylate
                                                  (Oil of wintergreen)

Methyl salicylate will be prepared from salicylic acid, which is esterified at the carboxyl group with methanol. You should recall from your organic chemistry lecture course that esterification is an acid-catalyzed equilibrium reaction. The equilibrium does not lie far enough to the right to favor the formation of the ester in high yield. More product can be formed by increasing the concentrations of one of the reactants. In this experiment, a large excess of methanol will shift the equilibrium to favor more completely the formation of the ester.

This experiment also illustrates the use of distillation under reduced pressure for purifying high-boiling liquids. Distillation of high-boiling liquids at atmospheric pressure is often unsatisfactory. At the high temperatures required, the material being distilled (the ester, in this case) may partially or even completely decompose and cause loss of product and contamination of the distillate. When the total pressure inside the distillation apparatus is reduced, however, the boiling point of the substance is lowered. In this way the substance can be distilled without being decomposed.

SPECIAL INSTRUCTIONS

Before beginning this experiment, you should read the essay on esters, which precedes Experiment 10, and also Techniques 1, 5, 6, and 9. Handle concentrated sulfuric acid with caution, since it can cause severe burns. The experiment must be started at the beginning of the laboratory period, since a long reflux time is needed to esterify salicylic acid and obtain a respectable yield. Enough time should remain at the end of the reflux to perform the extractions, to place the product over the drying agent, and to assemble some of the distillation equipment. A supplementary experiment may be performed during this reaction period, or work from previous experiments that is pending may be completed. When a distillation is conducted under reduced pressure, it is important to guard against the dangers of an implosion. Inspect the glassware for flaws or cracks, and replace any glassware found to be defective. WEAR YOUR SAFETY GLASSES. Because the amount of crude methyl salicylate obtained in this reaction is small, your instructor may want two students to combine their products for the final vacuum distillation.

PROCEDURE

A reflux apparatus using a 100-mL round-bottomed flask will be needed (Technique 1, Section 1.7, p. 509). Place 9.7 g (0.07 mole) of salicylic acid and 25 mL (density 0.8 g/mL) of methanol in the flask. Swirl the flask and heat the contents slightly on a steam bath to help dissolve the salicylic acid. **Carefully** add 10 mL of concentrated sulfuric acid **in small portions** to the mixture in the flask. After each addition, swirl the flask.

> **CAUTION:** **Extreme care must be exercised to avoid contact with concentrated sulfuric acid. It will cause serious burns if it is spilled on the skin. Measure the acid in a dry graduated cylinder. Clean up all spills.**

A white precipitate may form, but it will dissolve after the mixture is heated. Add a boiling stone to the flask and assemble the apparatus by attaching the condenser. Heat the mixture on a steam bath for 1 hour, swirling it occasionally. The mixture will boil only slightly. During the heating period the mixture will turn cloudy and a layer will form near the top of the mixture.

Cool the mixture to room temperature and pour the liquid into a 125-mL separatory funnel. Extract the mixture with 25 mL of methylene chloride; remove the lower layer and save it. Reextract the liquid remaining in the funnel with another 25-mL portion of methylene chloride. Combine the lower layer with the other methylene chloride extract; return them to the separatory funnel and wash with 25 mL of water. Drain the lower organic layer, discard the water wash, return the organic layer to the funnel, and wash with a 25-mL portion of 5% sodium bicarbonate solution. After removing the lower organic layer, repeat the washing with another 25-mL portion of 5% sodium bicarbonate solution. Finally, drain the lower organic layer into a dry Erlenmeyer flask, add a spatulaful of anhydrous sodium sulfate to the flask, and swirl the mixture to dry the liquid. Decant the liquid away from the drying agent into a dry flask and evaporate the solvent on a steam bath **in a hood.**

Assemble an apparatus for distillation under reduced pressure (Technique 6, Figure 6–7, p. 559). Install a manometer in the system if one is available (Technique 9, Section 9.4, and Figure 9–6, pp. 592–593). Use the smallest round-bottomed flask as the distilling flask. Fit the distilling flask with a capillary ebulliator tube that reaches to the bottom of the flask (Technique 6, Figure 6–7, p. 559). The bleeder tube supplied with most organic kits will serve as an ebulliator if it is equipped with a piece of rubber tubing. The tubing is closed by a screw clamp until a gentle stream of bubbles is passed into the solution when the vacuum is applied to the system. A trap must be used between the aspirator and the distillation apparatus (Technique 9, Figure 9–6, p. 593).

The combined yields of crude methyl salicylate obtained by two students are placed in the distilling flask. It is important that all tubing connections and ground-glass joints fit tightly. Apply a vacuum to the apparatus by turning on the water aspirator to the maximum extent. Adjust the screw clamp on the ebulliator until a small stream of bubbles issues from the tube. Apply heat from a heating mantle or an oil bath to the distilling

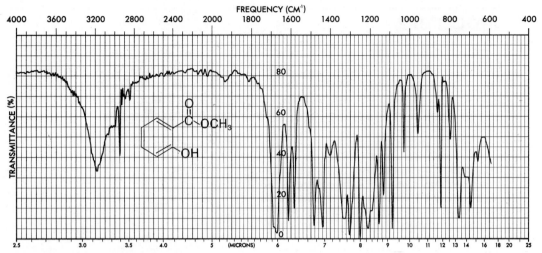

Ir spectrum of methyl salicylate, neat

flask. Collect all the liquid that distills at a temperature of 100 °C or higher.[1] If a manometer was used, be certain to record the pressure at which the liquid distills. When almost
all the liquid has been distilled, stop the heating, remove the hose from the aspirator or
open the trap valve, and turn off the water. Always disconnect the vacuum before turning
off the aspirator. Weigh the product and submit it to the instructor in a labeled vial. The

[1]The boiling point of methyl salicylate at atmospheric pressure is 222 °C, but the ester will distill at 105 °C at a pressure
of 14 mm. A pressure of 14 mm is the lowest that can be obtained with a good aspirator and with a water temperature of
16.5 °C. A manometer should be connected to the system to measure the pressure precisely (Technique 9, Section 9.4,
p. 592).

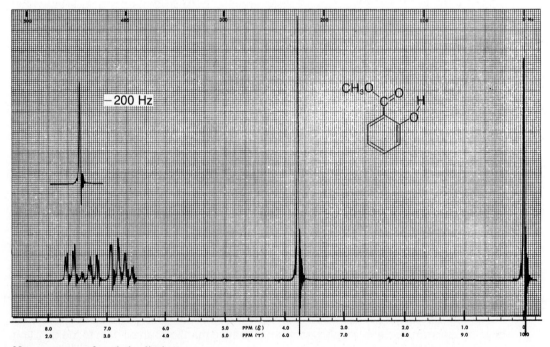

Nmr spectrum of methyl salicylate

report should include the names of the persons who combined their samples of crude ester for the vacuum distillation.

At your instructor's option, record the infrared spectrum of methyl salicylate (Technique 17). Compare the spectrum with the one reproduced in this experiment.

QUESTIONS

1. Write a mechanism for the acid-catalyzed esterification of salicylic acid with methanol.
2. What is the function of the sulfuric acid in this reaction? Is it consumed in the reaction?
3. In this experiment, excess methanol was used to shift the equilibrium toward the formation of more ester. Describe other methods for achieving the same result.
4. How are sulfuric acid and the excess methanol removed from the crude ester after the reaction has been completed?
5. Why was 5% $NaHCO_3$ used in the extraction? What would have happened if 5% NaOH had been used?
6. Interpret the principal absorption bands in the infrared and nmr spectra of methyl salicylate.

Experiment 12
Hydrolysis of Methyl Salicylate

Hydrolysis of an Ester
Heating under Reflux
Filtration, Crystallization
Melting-Point Determination

Esters can be hydrolyzed to their constituent carboxylic acid and alcohol parts under either acidic or basic conditions. In this experiment, **methyl salicylate,** an ester known as **oil of wintergreen** because of its natural source, is treated with aqueous base. The immediate product of this hydrolysis, besides methanol and water, is the sodium salt of salicylic acid. The reaction mixture is acidified with sulfuric acid, which converts the sodium salt to the free acid. The overall organic products of the reaction, therefore, are salicylic acid and methanol. The salicylic acid is a solid, which can be isolated and purified by crystallization. The chemical equations that describe this experiment are

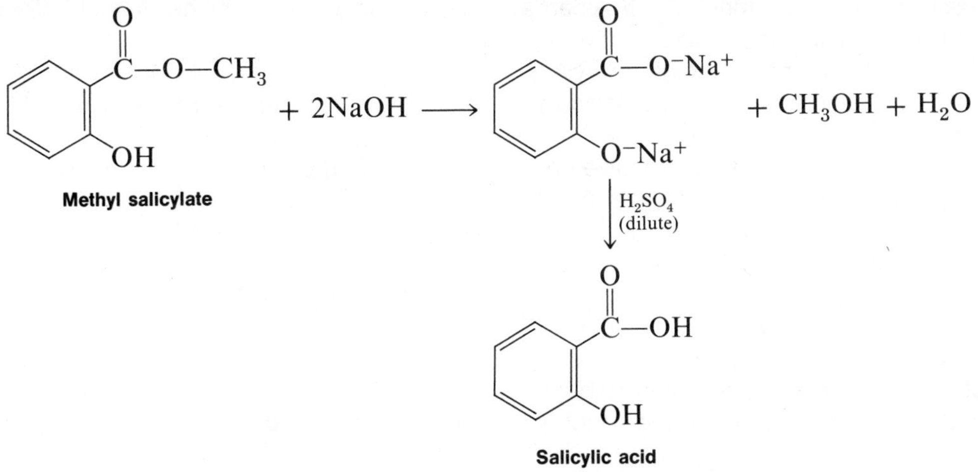

Methyl salicylate

$+ 2NaOH \longrightarrow$

$+ CH_3OH + H_2O$

H_2SO_4 (dilute)

Salicylic acid

Because the phenolic hydroxyl group is acidic, it is also converted to the corresponding sodium salt during the basic hydrolysis. In the subsequent acidification, this group becomes reprotonated.

SPECIAL INSTRUCTIONS

Before proceeding with this experiment, you should read Techniques 1, 2, 3, and 4. This procedure can be stopped at nearly any place.

PROCEDURE

Dissolve 10 g of sodium hydroxide in 50 mL of water. When the solution has cooled, place it, along with 5.0 g (0.033 mole) of methyl salicylate, in a 250-mL round-bottomed flask. A white solid may form at this point, but it will dissolve on heating. Attach a reflux condenser to the flask, according to the instructions given in Technique 1, Section 1.7, p. 509. The standard taper joints should be greased lightly.

Add one or two boiling stones to the reaction mixture to prevent bumping when the solution is heated. Heat the solution at its boiling point for about 20 minutes with a microburner. Use a wire gauze beneath the round-bottomed flask to disperse the flame. After heating the mixture, allow it to cool to room temperature. When the solution is cool, transfer it to a 250-mL beaker, and carefully add enough 1M sulfuric acid to make the solution acidic to litmus paper (blue litmus turns pink). As much as 150 mL of 1M sulfuric acid may have to be added at this stage. When the litmus turns pink, add an extra 15 mL of the sulfuric acid, thus causing the salicylic acid to precipitate from the solution. Cool the mixture in an ice-water bath to about 0 °C. Allow this cold mixture to settle. Collect the product by vacuum filtration, using a Büchner funnel with filter paper. See Technique 2, Section 2.3, p. 518, for details of this method. The filtration can be conducted most easily by decanting most of the supernatant liquid through the Büchner funnel before adding the mass of crystals.

Recrystallize the crude salicylic acid from water in a 125-mL Erlenmeyer flask. Add 100 mL of hot water and a boiling stone, and heat the mixture to boiling to dissolve the solid.

If the solid does not dissolve on boiling, add enough extra water to dissolve the solid. Filter the hot solution by gravity filtration through a fluted filter paper (Technique 2, Section 2.1, p. 514, and Figure 2–3, p. 517), using a fast filter paper, and set the solution aside to cool.

This gravity filtration must be carried out carefully. Filter the hot solution using only a small quantity at a time. Use a short-stemmed funnel for the filtration to reduce the probability that crystals might form in the stem, clogging the funnel. The filtration assembly should be placed on a steam bath. If salicylic acid begins to crystallize in the funnel, add to the filter the **minimum** amount of boiling water needed to redissolve the crystals.

After the filtered solution has cooled, place the flask in an ice-water bath to aid crystallization. When the crystals of salicylic acid have formed, collect them by vacuum filtration (Technique 2, Section 2.3, p. 518). Allow the crystals to dry overnight on a watch glass. When the crystals are thoroughly dry, weigh them and determine the percentage yield. Determine the melting point of the pure material (Technique 4, Sections 4.5 and 4.6, pp. 536–539). The melting point of pure salicyclic acid is 159 to 160 °C. Place the sample of product in a labeled vial and submit it to the instructor.

QUESTIONS

1. What is the white solid that forms when methyl salicylate is added to an aqueous solution of sodium hydroxide?

2. Give a mechanism for the hydrolysis of methyl salicylate.

3. Is sodium hydroxide consumed in the reaction or is it a catalyst? Explain.

Essay
FATS AND OILS

In the normal human diet, about 25 to 50% of the caloric intake consists of fats and oils. These substances are the most concentrated form of food energy in our diet. When metabolized, fats produce about 9.5 kcal of energy per gram. Carbohydrates and proteins produce less than half this amount. For this reason, animals tend to build up fat deposits as a reserve source of energy. They do this, of course, only when their food intake exceeds their energy requirements. In times of starvation, the body metabolizes these stored fats. Even so, some fats are required by animals for bodily insulation and as a protective sheath around some vital organs.

The constitution of fats and oils was first investigated by the French chemist Chevreul during the years 1810 to 1820. He found that when fats and oils were hydrolyzed, they gave rise to several "fatty acids" and the trihydroxylic alcohol, glycerol. Thus, fats and oils are **esters** of glycerol, called **glycerides** or **acylglycerols.** Since glycerol has three hydroxyl groups, it is possible to have mono-, di-, and triglycerides.

Fats and oils are predominantly triglycerides (triacylglycerols), constituted as follows:

3 Fatty acids + glycerol   =   A triglyceride

Thus, most fats and oils are esters of glycerol, and their differences result from the differences in the fatty acids with which glycerol may be combined. The most common fatty acids have 12, 14, 16, or 18 carbons, although acids with both lesser and greater numbers of carbons are found in several fats and oils. These common fatty acids are listed in the accompanying table along with their structures. As you see, these acids are both saturated and unsaturated. The saturated acids tend to be solids, while the unsaturated acids are usually liquids. This circumstance also extends to fats and oils. Fats are made up of fatty acids that are mostly saturated, while oils are primarily composed of fatty acid portions that have greater numbers of double bonds. In other words, unsaturation lowers the melting point. Fats (solids) are usually obtained from animal sources, while oils (liquids) are commonly obtained from vegetable sources. Therefore vegetable oils usually have a higher degree of unsaturation.

Common Fatty Acids

C_{12} Acids	Lauric	$CH_3(CH_2)_{10}COOH$
C_{14} Acids	Myristic	$CH_3(CH_2)_{12}COOH$
C_{16} Acids	Palmitic	$CH_3(CH_2)_{14}COOH$
	Palmitoleic	$CH_3(CH_2)_5CH{=}CH{-}CH_2(CH_2)_6COOH$
C_{18} Acids	Stearic	$CH_3(CH_2)_{16}COOH$
	Oleic	$CH_3(CH_2)_7CH{=}CH{-}CH_2(CH_2)_6COOH$
	Linoleic	$CH_3(CH_2)_4(CH{=}CH{-}CH_2)_2(CH_2)_6COOH$
	Linolenic	$CH_3CH_2(CH{=}CH{-}CH_2)_3(CH_2)_6COOH$
	Ricinoleic	$CH_3(CH_2)_5CH(OH)CH_2CH{=}CH(CH_2)_7COOH$

About 20 to 30 different fatty acids are found in fats and oils, and it is not uncommon for a given fat or oil to be composed of as many as 10 to 12 (or more) different fatty acids. Typically, these fatty acids are randomly distributed among the triglyceride molecules, and the chemist cannot identify anything more than an average composition for a given fat or oil. The average fatty acid composition of some selected fats and oils is given in the second table. As indicated, all the values in the table may vary in percentage, depending, for instance, on the locale in which the plant was grown or on the particular diet on which the animal subsisted. Thus, perhaps there is a basis for the claims that corn-fed hogs or cattle taste better than animals maintained on other diets.

Average Fatty Acid Composition (By Percentage) of Selected Fats and Oils

Column groupings: columns 1–6 = **SATURATED FATTY ACIDS (NO DOUBLE BONDS)**; columns 7–9 = **UNSATURATED (1 DOUBLE BOND)**; columns 10–12 = **UNSATURATED (>1 DOUBLE BOND)** with number of double bonds shown in parentheses; column 13 = **UNSATURATED**.

	C_{10} C_8 C_6 C_4	C_{12} LAURIC	C_{14} MYRISTIC	C_{16} PALMITIC	C_{18} STEARIC	C_{20} C_{22} C_{24}	C_{16} PALMITOLEIC	C_{18} OLEIC	C_{18} RICINOLEIC	C_{18} LINOLEIC (2)	C_{18} LINOLENIC (3)	C_{18} ELEOSTEARIC (3)	C_{20} C_{22} C_{24}
ANIMAL FATS													
Tallow			2–3	24–32	14–32		1–3	35–48		2–4			
Butter	7–10	2–3	7–9	23–26	10–13		5	30–40		4–5			2
Lard			1–2	28–30	12–18		1–3	41–48		6–7			2
ANIMAL OILS													
Neat's foot			4–5	17–18	2–3			74–77					
Whale			6–8	11–18	2–4		13–18	33–38		24–30 ↕			17–31
Sardine				10–16	1–2		6–15			24–30 ↕			12–19
VEGETABLE OILS													
Corn			0–2	7–11	3–4		0–2	43–49		34–42			
Olive			0–1	5–15	1–4		0–1	69–84		4–12			
Peanut				6–9	2–6	3–10	0–1	50–70		13–26			
Soybean			0–1	6–10	2–6			21–29		50–59	4–8		
Safflower				6–10	1–4			8–18		70–80	2–4		
Castor bean				0–1				0–9	80–92	3–7			
Cottonseed			0–2	19–24	1–2		0–2	23–33		40–48			
Linseed				4–7	2–5			9–38		3–43	25–58		
Coconut	10–22	45–51	17–20	4–10	1–5	←2–6→		2–10		0–2			
Palm			1–3	34–43	3–6			38–40		5–11			
Tung								4–16		0–1		74–91	

Vegetable fats and oils are usually found in fruits and seeds, and are recovered by three principal methods. In the first method, **cold pressing,** the appropriate part of the dried plant is pressed in a hydraulic press to squeeze out the oil. The second method is **hot pressing,** which is the same as the first method but done at a higher temperature. Of the two methods, cold pressing usually gives a better grade of product (more bland); the hot pressing method gives a higher yield, but with more undesirable constituents (stronger odor and flavor). The third method is **solvent extraction.** Solvent extraction gives the highest recovery of all and can now be regulated to give bland, high-grade food oils.

Animal fats are usually recovered by **rendering,** which involves cooking the fat out of the tissue by heating it to a high temperature. An alternative method involves placing the fatty tissue in boiling water. The fat floats to the surface and is easily recovered. The most common animal fats, lard (from hogs) and tallow (from cattle), can be prepared in either way.

Many triglyceride fats and oils are used for cooking. We use them to fry meats and other foods and to make sandwich spreads. Almost all commercial cooking fats and oils, except lard, are prepared from vegetable sources. Vegetable oils are liquids at room temperature. If the double bonds in a vegetable oil are hydrogenated, the resultant product becomes solid. Manufacturers, in making commercial cooking fats (Crisco, Spry, Fluffo, etc.), hydrogenate a liquid vegetable oil until the desired degree of consistency is achieved. This makes a product that still has a high degree of unsaturation (double bonds) left. The same technique is used for margarine. ''Polyunsaturated'' oleomargarine is produced by the partial hydrogenation of oils from corn, cottonseed, peanut, and soybean sources. The final product has a yellow dye (β-carotene) added to make it look like butter; milk, about 15% by volume, is mixed into it to form the final emulsion. Vitamins A and D are also commonly added. Since the final product is tasteless (try Crisco), salt, acetoin, and biacetyl are often added. The latter two additives mimic the characteristic flavor of butter.

$$CH_3-\underset{\underset{H}{|}}{\overset{\overset{HO}{|}}{C}}-\overset{\overset{O}{\|}}{C}-CH_3 \qquad CH_3-\overset{\overset{O}{\|}}{C}-\overset{\overset{O}{\|}}{C}-CH_3$$

Acetoin   **Biacetyl**

Many producers of margarine claim it to be more beneficial to health because it is ''high in polyunsaturates.'' Animal fats are low in unsaturated fatty acid content and are generally excluded from the diets of persons who have a high cholesterol level in the blood (see the essay on steroids before Experiment 8). Such people have difficulty in metabolizing saturated fats correctly and should avoid them since they encourage cholesterol deposits to form in the arteries. This ultimately leads to high blood pressure and heart trouble. Persons with normal metabolism, however, have no real need to avoid saturated fats.

Butter, when unrefrigerated and left exposed to air, turns rancid, giving an unpleasant odor and taste. This is due to the hydrolysis of the triglycerides by the moisture in air and to oxidation at the double bonds in the fatty acid components. Oxygen adds to allylic positions by an abstraction-addition reaction to give hydroperoxides. Rancid butter smells bad compared with a partially hydrolyzed margarine or cooking fat, because it contains, along with the fatty acids listed in the table, triglycer-

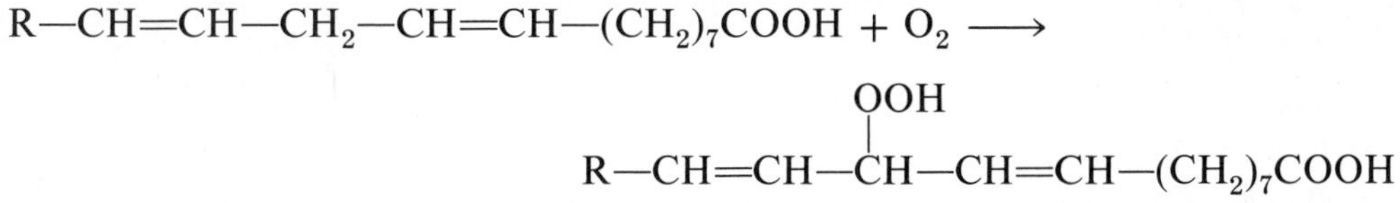

$$R-CH{=}CH-CH_2-CH{=}CH-(CH_2)_7COOH + O_2 \longrightarrow$$

$$\overset{\displaystyle OOH}{\underset{\displaystyle |}{R-CH{=}CH-CH-CH{=}CH-(CH_2)_7COOH}}$$

ides, which are composed of butyric (C_4), caproic (C_6), caprylic (C_8), and capric (C_{10}) acid moieties as well. These low-molecular-weight carboxylic acids are the source of the well-known objectionable odor. The hydroperoxides also decompose to low-molecular-weight aldehydes, which also have objectionable odors and tastes. The same reaction takes place when fats are burned, as in an oven fire. On combustion, an unsaturated fat produces large amounts of acrolein, a potent **lachrymator** (tear inducer), and other aldehydes that also irritate the eyes.

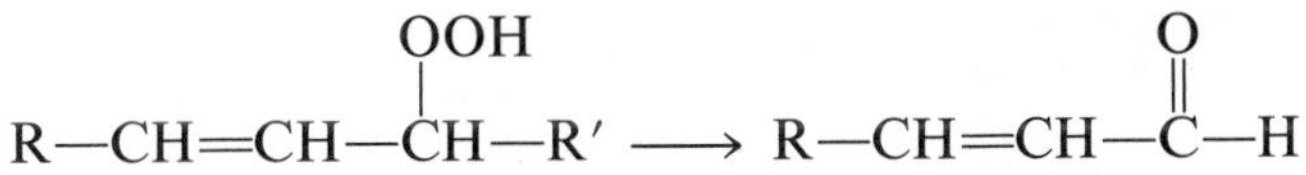

$$\overset{\displaystyle OOH}{\underset{\displaystyle |}{R-CH{=}CH-CH-R'}} \longrightarrow R-CH{=}CH-\overset{\displaystyle O}{\overset{\displaystyle \|}{C}}-H$$

Acrolein if R = H

Some oils, mainly those that are highly unsaturated (for example, linseed oil), thicken on exposure to air and eventually harden to give a smooth, clear resin. Such oils are called **drying oils,** and they are widely used in the manufacture of shellac, varnish, and paint. Apparently the double bonds in these compounds undergo both partial oxidation and polymerization on exposure to light and air.

REFERENCES

Dawkins, M. J. R., and Hull, D. "The Production of Heat by Fat." *Scientific American, 213* (August 1965): 62.

Eckey, E. W., and Miller, L. P. *Vegetable Fats and Oils*. ACS Monograph No. 123. New York: Reinhold, 1954.

Green, D. E. "The Metabolism of Fats." *Scientific American, 190* (January 1954): 32.

Green, D. E. "The Synthesis of Fats." *Scientific American, 202* (February 1960): 46.

Nawar, W. W. "Chemical Changes in Lipids Produced by Thermal Processing." *Journal of Chemical Education, 61* (April 1984): 299.

Shreve, R. N., and Brink, J. "Oils, Fats, and Waxes." *The Chemical Process Industries*. 4th ed. New York: McGraw-Hill, 1977.

Experiment 13
Methyl Stearate From Methyl Oleate

Catalytic Hydrogenation
Use of Filter Aid
Unsaturation Tests

In this experiment you will convert the liquid methyl oleate, an "unsaturated" fatty acid ester, to solid methyl stearate, a "saturated" fatty acid ester, by catalytic hydrogenation.[1]

$$CH_3(CH_2)_7 \overset{10}{-}CH = \overset{9}{C}H-(CH_2)_7-\overset{\displaystyle O}{\overset{\|}{C}}-OCH_3 \xrightarrow[H_2]{Pd/C}$$

Methyl oleate
(Methyl *cis*-9-octadecenoate)

$$CH_3(CH_2)_7-\underset{\underset{\displaystyle H}{|}}{C}H-\underset{\underset{\displaystyle H}{|}}{C}H-(CH_2)_7-\overset{\displaystyle O}{\overset{\|}{C}}-OCH_3$$

Methyl stearate
(Methyl octadecanoate)

By commercial methods like those described in this experiment, the unsaturated fatty acids of vegetable oils are converted to margarine (read the essay, "Fats and Oils," preceding this experiment). However, rather than use the mixture of triglycerides that would be present in a cooking oil such as Mazola (corn oil), we use as a model the pure chemical methyl oleate.

For this procedure, a chemist would normally use a tank of hydrogen gas. Since many students will be following the procedure simultaneously, however, we shall use the simpler expedient of causing zinc metal to react with dilute sulfuric acid:

$$Zn + H_2SO_4 \xrightarrow{H_2O} H_2(g)\uparrow + ZnSO_4$$

The hydrogen so generated will be passed into a solution containing methyl oleate and the palladium on carbon catalyst (10% Pd/C).

SPECIAL INSTRUCTIONS

Read the essay on fats and oils (p. 97). Also review the appropriate sections in your lecture textbook on the catalytic hydrogenation of double bonds.

---

[1]Compounds with double bonds are said to be **unsaturated;** those without are said to be **saturated** (with hydrogen).

Since this experiment calls for generating hydrogen gas, no flames will be allowed in the laboratory during this period.

```
No Flames Allowed
```

Since a pressure buildup of hydrogen gas is possible within the apparatus, it is especially important to remember to wear your safety goggles; you can thus protect yourself against the possibility of minor "explosions" from joints popping open or from any flask accidentally cracking under pressure.

```
Wear Safety Goggles
```

When you operate the hydrogen generator, be sure to add sulfuric acid at a rate that does not cause hydrogen gas to evolve too rapidly. The hydrogen pressure in the flask should not be allowed to rise much above atmospheric pressure. Neither should the hydrogen evolution be allowed to stop. If this happens, your reaction mixture may be "sucked back" into your hydrogen generator.

If your instructor indicates that you should do the optional unsaturation tests on your starting material and product, you should read the descriptions of the Br_2/CH_2Cl_2 test on p. 439, Experiment 56C, and on p. 204, in Experiment 27.

Finally, it is often the case, depending on the supplier, that commercial methyl oleate is only about 70% pure, containing several impurities. Consult your instructor about your sample and whether you should base your percentage yield calculation on 70% of theoretical rather than the usual 100%.

APPARATUS

Assemble the apparatus as illustrated in the accompanying figure. This apparatus consists of basically three parts:

1. Hydrogen generator
2. Reaction flask
3. Mineral-oil bubbler trap

The mineral-oil bubbler trap has two functions. First, it allows one to keep a pressure of hydrogen within the system that is slightly above atmospheric. Second, it prevents back-diffusion of air into the system. The functions of the other two units are self-explanatory.

So that hydrogen leakage is prevented, the tubing used to connect the various subunits of the apparatus should be either relatively new rubber tubing, without cracks or breaks, or Tygon tubing. The tubing can be checked for cracks or breaks simply by stretching and bending it before use. It should be of such a size that it will fit onto all glass tubing **tightly.** Similarly, the rubber stoppers should be fitted with glass tubing of a size that fits firmly through the holes in their centers. If the seal is tight, it will not be

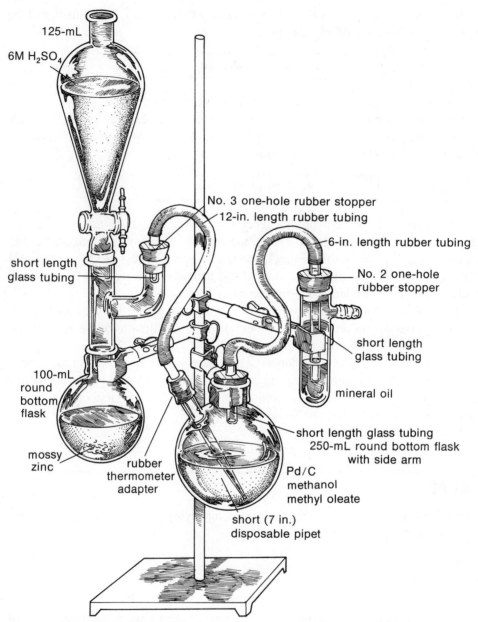

125-mL

6M H₂SO₄

No. 3 one-hole rubber stopper
12-in. length rubber tubing

6-in. length rubber tubing

short length
glass tubing

No. 2 one-hole
rubber stopper

short length
glass tubing

100-mL
round
bottom
flask

mineral oil

short length glass tubing
250-mL round bottom flask
with side arm

mossy
zinc

rubber
thermometer
adapter

Pd/C
methanol
methyl oleate

short (7 in.)
disposable pipet

Apparatus for catalytic hydrogenation

easy to slide the glass tubing up and down in the hole. The inlet tube in the reaction
flask should reach almost to the bottom of the flask. Hydrogen must bubble **through**
the solution.

Finally, all equipment should be assembled **as shown,** on a single ring stand, so
that the apparatus can be shaken from time to time to ensure thorough suspension of the
Pd/C catalyst in the solvent.

PROCEDURE

Fill the sidearm test tube (bubbler) about one-third full with mineral oil. The end of the glass tube should be submerged below the surface of the oil.

To charge the hydrogen generator, weigh out about 12 g of mossy zinc and place it in the 100-mL flask. Place 60 mL of 6M sulfuric acid in the addition funnel. Be sure to grease the ground-glass joint lightly.

Carefully weigh a 10-mL graduated cylinder. Pour 5 mL of methyl oleate into the cylinder and redetermine its weight to arrive at the amount of starting material used. Pour 70 mL of methanol into a second graduated cylinder (100-mL). Pour the contents of the small cylinder containing the methyl oleate into the reaction flask and rinse it several times, using the 70 mL of methanol in the large graduated cylinder. Add these rinses to the methyl oleate in the reaction flask. Finally, weigh 0.1 g of 10% Pd/C catalyst and also add it to the reaction flask (use smooth paper for this weighing).

Check all the seals to be sure they are gas-tight and then begin evolution of hydrogen by slowly adding the sulfuric acid, a little at a time, to the mossy zinc. The first addition should amount to only 2 to 3 mL of acid. Subsequent additions should be gauged by the rate of hydrogen bubbles passing through the solution (about one to two bubbles every second is a good rate). Continue passing hydrogen through the solution for about 1 hour (or a little longer to achieve the best yield). From time to time the apparatus should be shaken gently to maintain an even suspension of the catalyst in the methanol solution.

After the reaction period is complete, stop the reaction by disconnecting the reaction flask. Dilute the acid in the generator with water and decant the solution down a sink. Rinse the flask several times with water and then place any unreacted zinc in a waste container.

Prepare a large Büchner funnel with a layer of Celite (Filter Aid), depositing it from a methanol slurry (see Technique 2, Section 2.4, p. 520). Once the Celite has been deposited, disassemble the filter flask and rinse it well with water to remove any traces of Celite that may have passed through. Reassemble the filtering apparatus and rapidly filter the hydrogenation mixture with vacuum suction through the Celite bed to remove the finely divided Pd/C catalyst. The product may crystallize from the filtrate due to evaporation of the solvent. Don't worry about it. After filtering, warm the filtrate on a steam bath to dissolve any solids and transfer the solution to a 125-mL Erlenmeyer flask.[2] If necessary, use a small amount of warm methanol to rinse the filter flask. Cool the solution in an ice bath until the methyl stearate crystallizes.[3] Collect the product by vacuum filtration, using a small Büchner funnel. Be sure to save a portion of the filtrate as well as the crystals. If your instructor indicates that you should make the unsaturation tests (below), you will need some of the filtrate for these tests. After the crystals are dry, weigh them, determine their melting point (literature, 39 °C), and calculate the percentage yield.

---

[2]If any solid remains undissolved on heating, some Celite has seeped through the filter and must be removed by gravity filtration of the hot solution.

[3]If no crystals are obtained, reduce the volume of solution and cool again.

UNSATURATION TESTS (OPTIONAL)

Using a solution of bromine in methylene chloride, test for the number of drops of this solution decolorized by:

1. About 0.2 mL of methyl oleate dissolved in a small amount of methylene chloride
2. A small spatulaful of your methyl stearate product dissolved in a small amount of methylene chloride
3. About 0.2 mL of the filtrate that you saved as directed under "Procedure"

Use small test tubes and disposable pipets to make these tests. Include the results of the tests and your conclusions in your report.

QUESTIONS

1. Using the information in the essay on fats and oils, draw the structure of the triglyceride formed from one each of oleic acid, linoleic acid, and stearic acid. Give a balanced equation, showing how much hydrogen would be needed to reduce the triglyceride completely; show the product.
2. A 1.50-g sample of a pure compound subjected to catalytic hydrogenation takes up 250 mL of H_2 at 25 °C and 1-atm pressure. Calculate the molecular weight of the compound, assuming that it has only one double bond.
3. A compound with the formula C_5H_6 takes up 2 moles of H_2 on catalytic hydrogenation. Give one possible structure that would fit the information given.
4. A compound of formula C_6H_{10} takes up 1 mole of H_2 on reduction. Give one possible structure that would fit the information given.

Essay
SOAPS AND DETERGENTS

Soaps as we know them today were virtually unknown before the first century A.D. Clothes were cleaned primarily by the abrasive action of rubbing them on rocks in water. Somewhat later, it was discovered that certain types of leaves, roots, nuts, berries, and barks formed soapy lathers that solubilized and removed dirt from clothes. We now know these natural materials that lather as **saponins.** Many saponins contain pentacyclic triterpene carboxylic acids, such as oleanolic acid or ursolic acid, chemically combined with a sugar molecule. These acids also appear in the uncombined state. Saponins were probably the first known ''soaps.'' They may have also been an early source of pollution in that they are known to be toxic to fish. The pollution problem associated with the development of soap and detergents has been long and controversial.

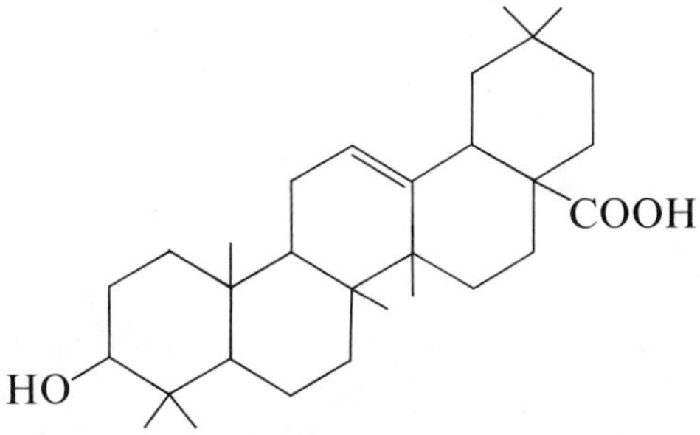

Oleanolic acid 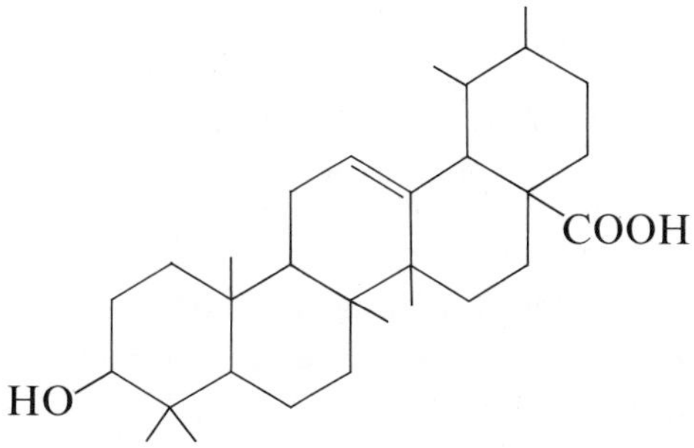   **Ursolic acid**

Soap as we know it today has probably evolved over many centuries from experimentation with crude mixtures of alkaline and fatty materials. Pliny the Elder described the manufacture of soap during the first century A.D. A modest soap factory was even built in Pompeii. During the Middle Ages, cleanliness of the body or clothing was not considered important. Those who could afford perfumes used them to hide their body odor. Perfumes, like fancy clothes, were a status symbol for the rich. An interest in cleanliness again emerged during the eighteenth century, when disease-causing microorganisms were discovered.

SOAPS

The process of making soap has remained practically unchanged for 2000 years. The procedure involves the basic hydrolysis (saponification) of a fat. Chemically, fats are usually referred to as **triglycerides.** They contain ester functional groups. Saponification involves heating fat with an alkaline solution. This alkaline solution was originally obtained by leaching wood ashes or from the evaporation of natural alkaline waters. Today, lye (sodium hydroxide) is used as the source of the alkali. The alkaline solution hydrolyzes the fat to its component parts, the salt of a long-chain carboxylic acid (soap) and an alcohol (glycerol). When common salt is added, the soap precipitates. The soap is washed free of unreacted sodium hydroxide and molded into bars. The carboxylic acid salts of soap usually contain 12 to 18 carbons arranged in a straight chain. The carboxylic acids containing even numbers of carbon atoms predominate, and the chains may contain unsaturation. The chemical structures of fats, and the related oils, are shown in the essay, ''Fats and Oils.'' The equation below shows how soap is produced

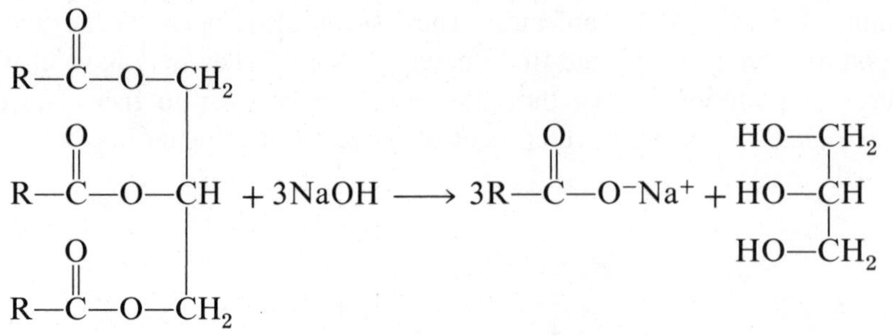

$$R-\overset{\overset{\displaystyle O}{\|}}{C}-O-CH_2$$

$$R-\overset{\overset{\displaystyle O}{\|}}{C}-O-CH \; + 3NaOH \longrightarrow 3R-\overset{\overset{\displaystyle O}{\|}}{C}-O^-Na^+ + \begin{array}{l} HO-CH_2 \\ HO-CH \\ HO-CH_2 \end{array}$$

$$R-\overset{\overset{\displaystyle O}{\|}}{C}-O-CH_2$$

| **Triglyceride (Fat)** | **Lye** | **Sodium salt of an acid (Soap)** | **Glycerol** |

from a fat. It is an idealized reaction, in which the R groups are the same. Usually, the R groups are **not** the same. Therefore, soap is actually a mixture of salts of carboxylic acids.

A disadvantage of soap is that it is an ineffective cleanser in hard water. Hard water contains salts of magnesium, calcium, and iron in solution. When soap is used in hard water, "calcium soap," the insoluble calcium salts of the fatty acids, and other precipitates are deposited as **curds.** This precipitate, or curd, is referred to as bathtub ring. Although soap is a poor cleanser in hard water, it is an excellent cleanser in soft water.

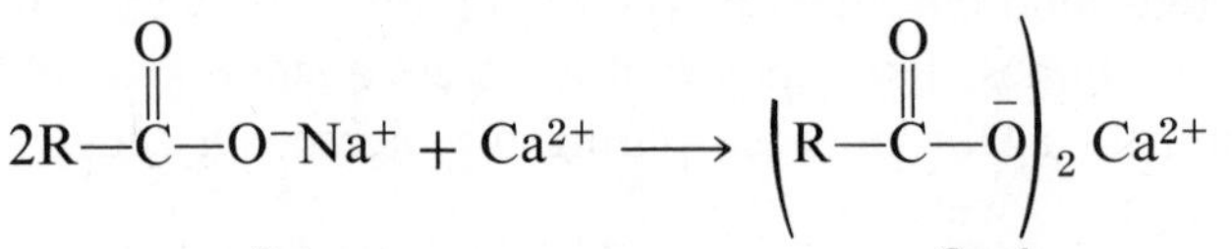

$$2R-\overset{\overset{\displaystyle O}{\|}}{C}-O^-Na^+ + Ca^{2+} \longrightarrow \left(R-\overset{\overset{\displaystyle O}{\|}}{C}-\overset{-}{O}\right)_2 Ca^{2+}$$

| **Soap** | **Curd** |

Water softeners are added to soaps to help remove the troublesome hard-water ions so that the soap will remain effective in hard water. Sodium carbonate or trisodium phosphate will precipitate the ions as the carbonate or phosphate. Unfortunately, the precipitate may become lodged in the fabric of items being laundered, causing a grayish or streaked appearance.

$$Ca^{2+} + CO_3^{2-} \longrightarrow CaCO_3 \;\; \downarrow$$

$$3Ca^{2+} + 2PO_4^{3-} \longrightarrow Ca_3(PO_4)_2 \;\; \downarrow$$

An important advantage of soap is that it is **biodegradable.** Microorganisms can consume the linear soap molecules and convert them to carbon dioxide and water. The soap is thus eliminated from the environment.

ACTION OF SOAP IN CLEANING

Dirty clothes, skin, or other surfaces have particles of dirt suspended in a layer of oil or grease. Polar water molecules cannot remove the dirt embedded in nonpolar oil or

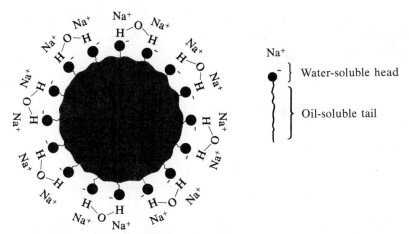

A soap micelle solvating a droplet of oil (from Linstromberg, Walter W., *Organic Chemistry: A Brief Course,* D. C. Heath, 1978).

grease. One can remove the dirt with soap, however, because of its dual nature. The soap molecule has a polar, **water-soluble** head (carboxylate salt) and a long, **oil-soluble** tail (the hydrocarbon chain). The hydrocarbon tail of soap dissolves in the oily substance but the ionic end remains outside the oily surface. When enough soap molecules have oriented themselves around an oil droplet with their hydrocarbon ends dissolved in the oil, the oil droplet, together with the suspended dirt particles, is removed from the surface of the cloth or skin. The oil droplet is removed because the heavily negatively charged oil droplet is now strongly attracted to water and solvated by the water. The solvated oil droplet is called a **micelle;** one is shown in the figure.

DETERGENTS

Detergents are synthetic cleaning compounds, often referred to as ''syndets.'' They were developed as an alternative to soaps because they are effective in **both** soft and hard water. No precipitates form when calcium, magnesium, or iron ions are present in a detergent solution. One of the earliest detergents developed was sodium lauryl sulfate. It is prepared by the action of sulfuric acid or chlorosulfonic acid on lauryl alcohol (1-dodecanol). This detergent is relatively expensive, however. The following reactions show one industrial method of preparation:

$$CH_3(CH_2)_{10}CH_2OH + H_2SO_4 \longrightarrow CH_3(CH_2)_{10}CH_2OSO_3H + H_2O$$
Lauryl alcohol

$$CH_3(CH_2)_{10}CH_2OSO_3H + Na_2CO_3 \longrightarrow$$

$$CH_3(CH_2)_{10}CH_2OSO_3{}^-Na^+ + NaHCO_3$$
Sodium lauryl sulfate

The first of the inexpensive detergents appeared about 1950. These detergents, called alkylbenzenesulfonates (ABS), can be prepared from inexpensive petroleum sources by the following set of reactions:

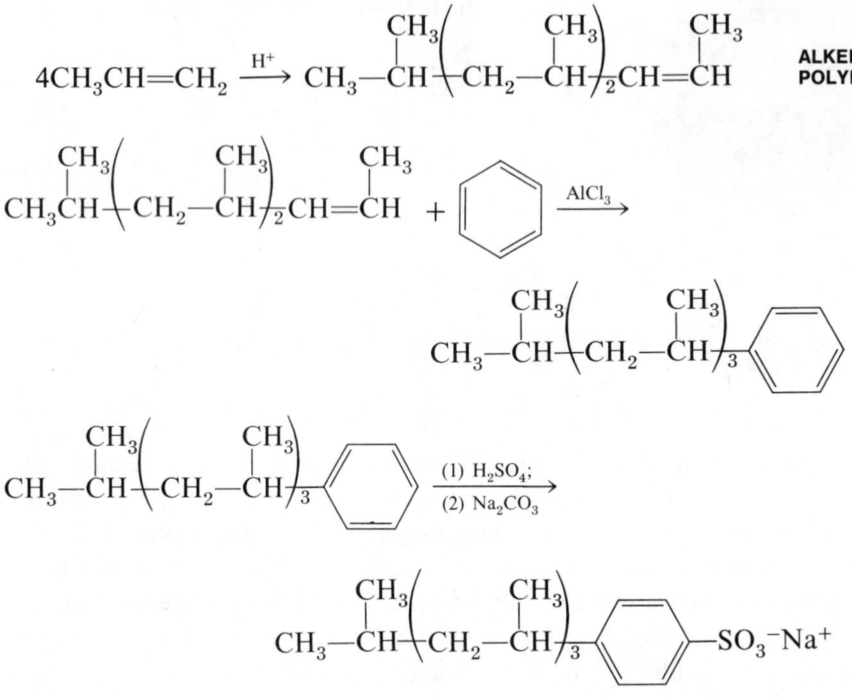

$$4CH_3CH{=}CH_2 \xrightarrow{H^+} CH_3{-}\underset{CH_3}{CH}\left({-}CH_2{-}\underset{CH_3}{CH}\right)_2 CH{=}\underset{CH_3}{CH}$$

ALKENE POLYMERIZATION

$$CH_3\underset{CH_3}{CH}\left({-}CH_2{-}\underset{CH_3}{CH}\right)_2 CH{=}\underset{CH_3}{CH} \;+\; \bigcirc \xrightarrow{AlCl_3}$$

$$CH_3{-}\underset{CH_3}{CH}\left({-}CH_2{-}\underset{CH_3}{CH}\right)_3 \bigcirc$$

FRIEDEL-CRAFTS AKLYLATION

$$CH_3{-}\underset{CH_3}{CH}\left({-}CH_2{-}\underset{CH_3}{CH}\right)_3 \bigcirc \xrightarrow[(2)\ Na_2CO_3]{(1)\ H_2SO_4;}$$

$$CH_3{-}\underset{CH_3}{CH}\left({-}CH_2{-}\underset{CH_3}{CH}\right)_3 \bigcirc{-}SO_3^-Na^+$$ **SULFONATION**

An alkylbenzenesulfonate (ABS)

Detergents became very popular because they could be used effectively in all types of water and were cheap. They rapidly displaced soap as the most popular cleaning agent. A problem with the detergents was that they passed through sewage-treatment plants without being degraded by the microorganisms present, a process necessary for the full sewage treatment. Rivers and streams in many sections of the country became polluted with detergent foam. The detergents even found their way into the drinking water supplies of numerous cities. The reason for the persistence of the detergent was that bacterial enzymes, which could degrade straight-chain soaps and sodium lauryl sulfate, could not destroy the highly branched detergents such as ABS.

It was soon found that the bacterial enzymes could degrade only a chain of carbons that contained, at the most, one branch. As numerous cities and states banned the sale of the nonbiodegradable detergents, by 1966 they were replaced by the new biodegradable detergents called linear alkylsulfonates (LAS). One example of an LAS detergent is shown below. Notice that there is one branch next to the aromatic ring.

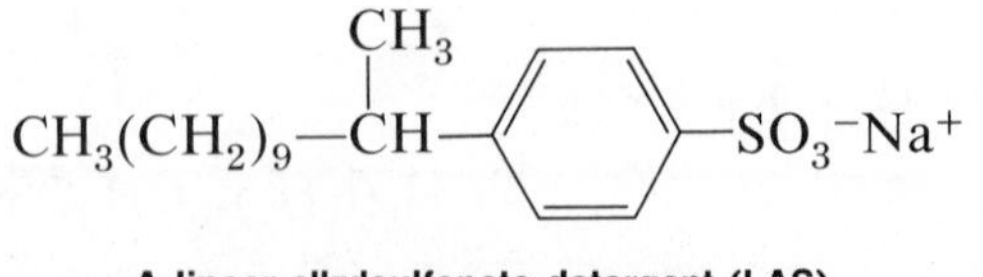

$$CH_3(CH_2)_9{-}\underset{CH_3}{CH}{-}\bigcirc{-}SO_3^-Na^+$$

A linear alkylsulfonate detergent (LAS)

NEW PROBLEMS WITH DETERGENTS

Detergents (also soaps) are not sold as pure compounds. A typical heavy-duty, controlled "sudser" may contain only 8 to 20% of the linear alkylsulfonate. A large quantity (30–50%) of a "builder" such as sodium tripolyphosphate, $Na_5P_3O_{10}$, may be present. Other additives include corrosion inhibitors, antideposition agents, and perfumes. Optical brighteners are also added. Brighteners absorb invisible ultraviolet light and reemit it as visible light, so laundry appears white and thus "clean." The phosphate builder is added to complex the hard-water ions, calcium and magnesium, and keep them in solution. Builders seem to enhance the washing ability of the LAS and also act as a cheap filler.

Unfortunately, phosphates speed the **eutrophication** of lakes and other bodies of water. The phosphates, along with other substances, are nutrients for algae. When algae begin to die and decompose, they consume so much dissolved oxygen from the water that no other life can exist in that water. The lake rapidly "dies." This is eutrophication.

Because phosphates have this undesirable effect, a search was initiated for a replacement for the phosphate builders. Some replacements have been made, but most also have problems associated with them. Two replacement builders, sodium metasilicate and sodium perborate, are highly basic substances, and they have caused injuries to children. In addition, they appear to destroy bacteria in sewage-treatment plants and may have other unknown environmental effects.

Many people have suggested a return to soap. The main problem is that we probably cannot produce enough soap to meet the demand because of the limited amount of animal fat available. Where do we go from here?

REFERENCES

Davidsohn, J., et al. *Soap Manufacture*. New York: Interscience, 1953.

Garner, W. *Detergents*. Amsterdam: Elsevier, 1966.

Kushner, L. M., and Hoffman, J. I. "Synthetic Detergents." *Scientific American, 185* (October 1951): 26.

"LAS Detergents End Stream Foam." *Chemical and Engineering News, 45* (1967): 29.

Levey, M. "The Early History of Detergent Substances." *Journal of Chemical Education, 31* (1954): 521.

"Makers Plan New Detergent Formulas." *Chemical and Engineering News, 48* (1970): 18.

Meloan, C. E., "Detergents—Soaps and Syndets." *Chemistry, 49* (September 1976): 6.

Puplett, P. A. R. *Synthetic Detergents*. London: Sidgwick and Jackson, 1957.

Snell, F. D. "Soap and Glycerol." *Journal of Chemical Education, 19* (1942): 172.

Snell, F. D., and Snell, C. T. "Syndets and Surfactants." *Journal of Chemical Education, 35* (1958): 271.

Stinson, S. C. "Consumer Preferences Spur Innovation in Detergents." *Chemical and Engineering News, 65* (January 26, 1987): 21.

Experiment 14
Preparation of Soap

Hydrolysis of a Fat (Ester)
Filtration

In this experiment we prepare soap from animal fat (lard). Animal fats and vegetable oils are esters of carboxylic acids; they have a high molecular weight and contain the alcohol glycerol. Chemically, these fats and oils are called **triglycerides.** The principal acids in animal fats and vegetable oils can be prepared from the natural triglycerides by alkaline hydrolysis (saponification). The natural acids are rarely of a single type in any

$$R_1\overset{\displaystyle O}{\overset{\|}{C}}\text{—O—CH}_2 \qquad R_1COO^-Na^+ \quad HO\text{—CH}_2$$

$$R_2\overset{\displaystyle O}{\overset{\|}{C}}\text{—O—CH} \xrightarrow[\substack{\text{saponification} \\ \text{or} \\ \text{hydrolysis}}]{\text{NaOH}} + R_2COO^-Na^+ + HO\text{—CH}$$

$$R_3\overset{\displaystyle O}{\overset{\|}{C}}\text{—O—CH}_2 \qquad R_3COO^-Na^+ \quad HO\text{—CH}_2$$

Triglycerides   **Carboxylic**   **Glycerol**
(Fat or oil)   **acid salts**
            (Soap)

given fat or oil. In fact, a single triglyceride molecule in a fat may contain three different acid residues (R_1COOH, R_2COOH, R_3COOH), and not every triglyceride in the substance will be identical. Each fat or oil, however, has a characteristic **statistical distribution** of the various types of acids possible. The composition of the common fats and oils is given in the essay on fats and oils.

     The fats and oils that are most common in soap preparations are lard and tallow from animal sources, and coconut, palm, and olive oils from vegetable sources. The length of the hydrocarbon chain and the number of double bonds in the carboxylic acid portion of the fat or oil determine the properties of the resulting soap. For example, a salt of a saturated long-chain acid makes a harder, more insoluble soap. Chain length also affects solubility.

     Tallow is the principal fatty material used in making soap. The solid fats of cattle are melted with steam, and the tallow layer formed at the top is removed. Soap-makers usually blend tallow with coconut oil and saponify this mixture. The resulting soap contains mainly the salts of palmitic, stearic, and oleic acids from the tallow, and the salts of lauric and myristic acids from the coconut oil. The coconut oil is added to produce a softer, more soluble soap. Lard (from hogs) differs from tallow (from cattle or sheep) in that lard contains more oleic acid.

112

TALLOW	$CH_3(CH_2)_{14}COOH$	$CH_3(CH_2)_{16}COOH$
	Palmitic acid	**Stearic acid**

$$CH_3(CH_2)_7CH{=}CH(CH_2)_7COOH$$
Oleic acid

COCONUT OIL	$CH_3(CH_2)_{10}COOH$	$CH_3(CH_2)_{12}COOH$
	Lauric acid	**Myristic acid**

Pure coconut oil yields a soap that is very soluble in water. The soap contains essentially the salt of lauric acid, with some myristic acid. It is so soft (soluble) that it will lather even in seawater. Palm oil contains mainly two acids, palmitic acid and oleic acid, in about equal amounts. Saponification of this oil yields a soap that is an important constituent of toilet soaps. Olive oil contains mainly oleic acid. It is used to prepare Castile soap, named after the region in Spain in which it was first made.

Toilet soaps generally have been carefully washed free of any alkali remaining from the saponification. As much glycerol as possible is usually left in the soap, and perfumes and medicinal agents are sometimes added. Floating soaps are produced by blowing air into the soap as it solidifies. Soft soaps are made by using potassium hydroxide, yielding potassium salts rather than the sodium salts of the acids. They are used in shaving creams and liquid soaps. Scouring soaps have abrasives added, such as fine sand or pumice.

SPECIAL INSTRUCTIONS

Read the essay on soaps and detergents, which precedes this experiment. In addition, it is necessary to read Techniques 1 and 2. This experiment is short and can easily be scheduled with another experiment.

PROCEDURE

Prepare a solution of 5 g of sodium hydroxide dissolved in a mixture of 20 mL of water and 20 mL of 95% ethanol. Place 10 g of cooking oil, fat, or lard (a commercial solid

> **CAUTION: This solution is caustic. Avoid skin contact.**

shortening works best) in a 250-mL beaker and add the solution to it. Heat the mixture on a steam bath for at least 45 minutes. Prepare another 40 mL of a 50:50 solution of ethanol-water and add it in small portions to the reaction over a 45-minute period. Stir the mixture constantly.

Prepare a solution of 50 g of sodium chloride in 150 mL of water in a 400-mL beaker. If the solution must be heated to dissolve the salt, it should be cooled before proceeding. Quickly pour the saponification mixture into the cooled salt solution. Stir the mixture thoroughly for several minutes and then cool it to room temperature in an ice bath. Collect the precipitated soap by vacuum filtration, using a Büchner funnel equipped with fast filter paper (Technique 2, Section 2.3, p. 518). Wash the soap with two portions of ice-cold water. Continue to draw air through the soap to dry the product partially. Allow the soap to dry overnight. Weigh the product.

If you are preparing the detergent in Experiment 15, save a small sample of the soap for the short comparison tests. Submit the remainder of the product to your laboratory instructor in a labeled vial.

QUESTIONS

1. Why should the potassium salts of fatty acids yield soft soaps?
2. Why is the soap derived from coconut oil so soluble?
3. Why does adding a salt solution cause soap to precipitate?
4. Why do you suppose a mixture of ethanol and water instead of simply water itself is used for saponification?
5. Sodium acetate and sodium propionate are poor soaps. Why?

Experiment 15
Preparation of a Detergent

$$CH_3(CH_2)_{10}CH_2O-\overset{\displaystyle O}{\underset{\displaystyle O}{\overset{\|}{\underset{\|}{S}}}}-O^-Na^+$$

We prepare a detergent, sodium lauryl sulfate, in this experiment. A detergent is usually defined as a synthetic cleaning agent, whereas a soap is derived from a natural source—a fat or an oil. The differences between the two basic types of cleaning agents are discussed in the essay on soaps and detergents, which precedes Experiment 14. In the second part of this experiment, the properties of soap are compared with the properties of the prepared detergent.

The reactions used to prepare the detergent are as follows:

$$CH_3(CH_2)_{10}CH_2OH + Cl-\overset{\displaystyle O}{\underset{\displaystyle O}{\overset{\|}{\underset{\|}{S}}}}-OH \longrightarrow CH_3(CH_2)_{10}CH_2O-\overset{\displaystyle O}{\underset{\displaystyle O}{\overset{\|}{\underset{\|}{S}}}}-OH + HCl$$

Lauryl alcohol	**Chlorosulfonic acid**		**Lauryl ester of sulfuric acid**

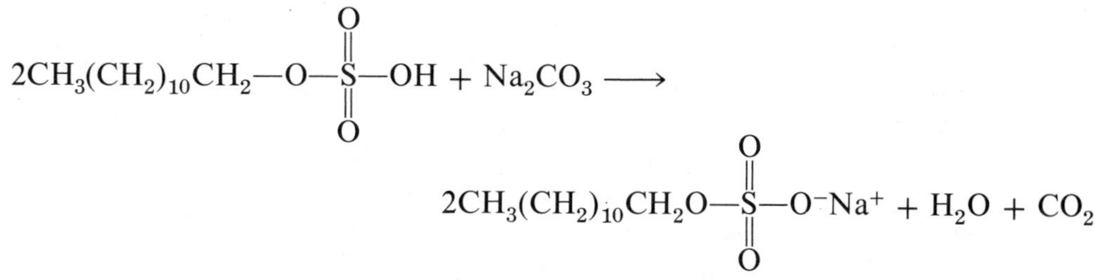

$$2CH_3(CH_2)_{10}CH_2-O-\overset{\overset{\displaystyle O}{\|}}{\underset{\underset{\displaystyle O}{\|}}{S}}-OH + Na_2CO_3 \longrightarrow$$

$$2CH_3(CH_2)_{10}CH_2O-\overset{\overset{\displaystyle O}{\|}}{\underset{\underset{\displaystyle O}{\|}}{S}}-O^-Na^+ + H_2O + CO_2$$

Sodium lauryl sulfate

In the first step, lauryl alcohol is allowed to react with chlorosulfonic acid to give the lauryl ester of sulfuric acid. Aqueous sodium carbonate is then added to produce the sodium salt (detergent). The aqueous mixture is saturated with solid sodium carbonate and extracted with 1-butanol. The sodium carbonate must be added to give phase separation; otherwise 1-butanol would be soluble in water. The sodium salt (detergent) is more soluble in the 1-butanol (organic layer) than in the aqueous layer because of the long hydrocarbon chain, which gives the salt considerable organic (nonpolar) character.

SPECIAL INSTRUCTIONS

Read the essay on soaps and detergents, which precedes Experiment 14, especially the sections dealing with their behavior in hard water. In addition, it is necessary to read Technique 5. Handle chlorosulfonic acid with extreme caution.

> **NOTE TO THE INSTRUCTOR: The lauryl alcohol is usually supplied in a narrow-mouthed bottle. Melt the alcohol (mp 24–27 °C) and pour the liquid into a wide-mouthed bottle. When crystallization is complete, the material is easily broken into small pieces with a spatula.**

PROCEDURE

Pour 9.5 mL of glacial acetic acid into a **dry** 250-mL beaker. Cool the acetic acid to about 5 °C in an ice bath. Slowly add 3.5 mL (density 1.77 g/mL) of chlorosulfonic acid from a buret (in the fume hood) directly to the acetic acid in the beaker. Allow the contents of the beaker to cool in the ice bath **in the hood.** Be careful not to allow any water to enter the beaker.

> **CAUTION: Use chlorosulfonic acid with extreme caution. It is an extremely strong acid similar to concentrated sulfuric acid. It will cause immediate burns on the skin. Use rubber gloves when dispensing the corrosive liquid.**

While stirring the mixture, slowly add 10 g of lauryl alcohol (1-dodecanol), in small pieces, over a 5-minute period. Stir the mixture until all the lauryl alcohol is dissolved and has reacted (about 30 minutes). Pour the reaction mixture into a 250-mL beaker containing 30 g of cracked ice.

Add 30 mL of 1-butanol to the beaker containing the reaction mixture and the cracked ice. Stir the mixture thoroughly for 3 minutes. While stirring, slowly add a saturated sodium carbonate solution in 3-mL portions until the solution is neutral or slightly basic to litmus. Add 10 g of solid sodium carbonate to the mixture to aid in the separation of the layers. Allow the layers to separate in the beaker and then pour the upper 1-butanol layer into a separatory funnel. Some of the aqueous layer may be transferred unavoidably to the separatory funnel with the organic phase. Add another 20-mL portion of 1-butanol to the aqueous layer remaining in the beaker. Stir the solution well for 3 minutes and allow the layers to separate. Again pour the upper layer into the separatory funnel containing the first extract. Discard the remaining aqueous layer.

Allow the phases to separate in the separatory funnel for about 5 minutes. Drain the lower aqueous phase until the upper layer begins to enter the stopcock and discard it (Technique 5, Section 5.4, p. 545). Pour the organic layer from the top of the separatory funnel into a large beaker (400 mL or larger). Evaporate the 1-butanol solvent (bp 117 °C) in a hood on a hot plate adjusted to a low temperature. The detergent precipitates as the solvent is removed. Stir the mixture occasionally to keep the product from decomposing. This is especially important near the end of the evaporation. When most of the 1-butanol has been removed, place the moist solid in an oven at 80 °C (or a steam cabinet) to dry the solid completely. The drying should be complete by the next laboratory period. It is important not to dry the sample at a high temperature (>80 °C) because of the danger of decomposition. When the sample has dried, remove the colorless product from the beaker. Weigh the detergent and calculate the percentage yield. If the detergent is not totally free of 1-butanol, the yield may exceed 100%. If necessary, continue to dry the sample. Part of the sample is used for the tests that follow. Submit the remaining detergent to the instructor in a labeled vial.

TESTS ON SOAPS AND DETERGENTS

Soap. Add 10 mL of a soap solution to a 125-mL Erlenmeyer flask.[1] Stopper the flask securely with a rubber stopper and shake it vigorously for about 15 seconds. Allow the solution to stand for 30 seconds and observe the level of the foam. Add 4 drops of 4% calcium chloride solution from an eye dropper. Shake the mixture for 15 seconds, and allow it to stand for about 30 seconds. Observe the effect of the calcium chloride on the foam. Do you observe anything else? Add 1 g of trisodium phosphate and shake the mixture again for about 15 seconds. Allow the solution to stand for 30 seconds. What do you observe? Explain the results of these tests in your laboratory report.

---

[1]The soap solution is prepared as follows. Add one bar of Ivory soap to 1 L of distilled or soft water. Stir the solution occasionally and allow the mixture to stand overnight. Remove the remainder of the bar. The mixture can be used directly. A large batch should be prepared by the laboratory instructor. Alternatively, a 0.15-g sample of soap from Experiment 14 can be added to 10 mL of distilled or soft water.

Detergent. Dissolve 0.75 g of your prepared detergent in 50 mL of water. Pour 10 mL of this solution into a 125-mL Erlenmeyer flask. Stopper the flask securely with a rubber stopper and shake it vigorously for 15 seconds. Allow the solution to stand for about 30 seconds and observe the level of the foam. Add 4 drops of 4% calcium chloride solution. Shake the mixture for about 15 seconds and allow it to stand for 30 seconds. What do you observe? Explain the results of these tests in your laboratory report.

QUESTIONS

1. Draw a mechanism for the reaction of lauryl alcohol with chlorosulfonic acid.

2. Why do you suppose sodium carbonate, instead of some other base, is used for neutralization?

3. Propose a model to explain how a cationic detergent works. A cationic detergent has its polar end positively charged.

4. Sodium methyl sulfate, $CH_3OSO_2^-Na^+$, is a poor detergent. Why?

5. Sodium lauryl sulfate can be prepared by replacing chlorosulfonic acid with another reagent. What could be used? Show the equations.

6. Suggest a method for synthesizing the linear alkyl sulfonate detergent shown on p. 110, starting with lauryl alcohol, benzene, and any needed inorganic compounds.

Essay
PHEROMONES: INSECT ATTRACTANTS AND REPELLENTS

It is difficult for humans, who are accustomed to heavy reliance on visual and verbal forms of communication, to imagine that there are forms of life that depend primarily on the release and perception of **odors** to communicate with one another. Among insects, however, this is perhaps the chief form of communication. Many species of insects have developed a virtual "language" based on the exchange of odors. These insects have well-developed scent glands, often of several different types, which have as their sole purpose the synthesis and release of chemical substances. When these chemical substances, known as **pheromones,** are secreted by insects and detected by other members of the same species, they induce a specific and characteristic response. Pheromones are usually of two distinct types: releaser pheromones and primer pheromones. **Releaser pheromones** produce an immediate **behavioral** response in the recipient insect; **primer pheromones** trigger a series of **physiological** changes in the recipient. Some pheromones, however, combine both releaser and primer effects.

SEX ATTRACTANTS

Among the most important types of releaser pheromones are the sex attractants. **Sex attractants** are pheromones secreted by either the female or, less commonly, the male of the species to attract the opposite member for the purpose of mating. Since in large concentrations, sex pheromones also induce a physiological response in the recipient (for example, the changes necessary to the mating act), they also have a primer effect and are therefore misnamed.

Anyone who has owned a female cat or dog will know that sex pheromones are not limited to insects. Female cats or dogs widely advertise, by odor, their sexual availability when they are "in heat." This type of pheromone is not uncommon to mammals. Some persons even believe that there are human pheromones that are responsible for attracting certain sensitive males and females to one another. This idea is, of course, responsible for many of the perfumes now widely available. Whether or not the idea is correct cannot yet be established, but there are proven sexual differences in the ability of humans to smell certain substances. For instance, Exaltolide, a synthetic lactone of 14-hydroxytetradecanoic acid, can be perceived only by females, or by males after they have been injected with an estrogen. Exaltolide is very similar in overall structure to civetone (civet cat) and muskone (musk deer), which are two naturally occurring compounds believed to be mammalian sex pheromones. Whether or not human males emit pheromones has never been established. Curiously, Exaltolide is used in perfumes intended for female as well as male use! But while the odor may lead a woman to believe that she smells pleasant, it cannot possibly have any effect on the male. The "musk oils," civetone and muskone, have also been long used in expensive perfumes.

One of the first identified insect attractants belonged to the gypsy moth, *Porthetria dispar*. This moth is a common agricultural pest, and it was hoped that the sex attractant that females emitted could be used to lure and trap males. Such a method of insect control would be preferable to inundating large areas with DDT and would be species-specific. Nearly 50 years of work were expended in identifying the chemical substance responsible. Early in this period, workers had found that an extract from the tail sections of female gypsy moths would attract males, even from a great distance. In experiments with the isolated gypsy moth pheromone, it was found that the male gypsy moth has an almost unbelievable ability to detect extremely small amounts of the substance. He can detect it in concentrations lower than a few hundred **molecules** per cubic centimeter (about 10^{-19} to 10^{-20} g/cc)! When a male moth encounters a small concentration of pheromone, he immediately turns into the wind and flies upward in search of higher concentrations and the female. In only a mild breeze, a continuously emitting female can activate a space 300 ft high, 700 ft wide, and almost 14,000 ft (nearly 3 miles) long!

In subsequent work, 20 mg of a pure chemical substance was isolated from solvent extracts of the two extreme tail segments collected from each of 500,000 female gypsy moths (about 0.1 μg/moth). This emphasizes that pheromones are effective in very minute amounts and that chemists must work with very small amounts to isolate them and prove their structures. It is not unusual to have to process thousands of insects to get even a very small sample of these substances. Very sophisticated analytical and instrumental methods, like spectroscopy, must be used to determine the structure of a pheromone.

In spite of these techniques, the original workers assigned an incorrect structure to the gypsy moth pheromone and proposed for it the name *gyplure*. Because of its great promise as a method of insect control, gyplure was soon synthesized. The synthetic material turned out to be totally inactive. After some controversy about why the synthetic material was incapable of luring male gypsy moths (see the references for the complete story), it was finally shown that the proposed structure for the pheromone (that is, the gyplure structure) was incorrect. The actual pheromone was found to be *cis*-7,8-epoxy-2-methyloctadecane, as shown in the accompanying table of structures. This material was soon synthesized, found to be active, and given the name *disparlure*. In recent years, disparlure traps have been found to be a convenient and economical method for controlling the gypsy moth.

A similar story of mistaken identity can be related for the structure of the pheromone of the pink bollworm, *Pectinophora gossypiella*. The originally proposed structure was called propylure. Synthetic propylure turned out to be inactive. Subsequently the pheromone was shown to be a mixture of two isomers of 7,11-hexadecadien-1-yl acetate, the *cis,cis* (7Z, 11Z) isomer and the *cis,trans* (7Z, 11E) isomer. It turned out to be quite easy to synthesize a 1:1 mixture of these two isomers, and the 1:1 mixture was named *gossyplure*. Curiously, adding as little as 10% of either of the other two possible isomers, either *trans,cis* (7E, 11Z) or *trans,trans* (7E, 11E), to the 1:1 mixture greatly diminishes its activity, apparently masking it. Geometric isomerism can be important! The details of the gossyplure story can also be found in the references.

Both these stories have been partly repeated here to point out the difficulties of research on pheromones. The usual method is to propose a structure determined by

INSECT SEX ATTRACTANTS

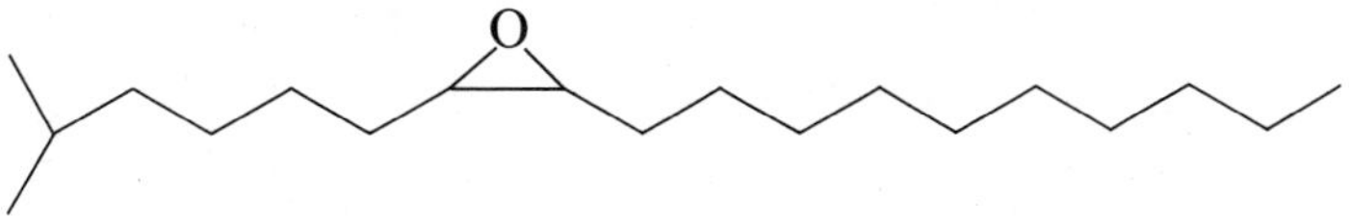

Disparlure
(Gypsy moth)

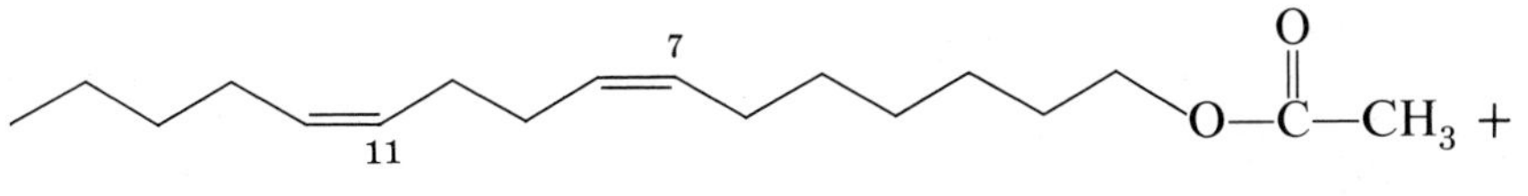

Gossyplure
(Pink bollworm)

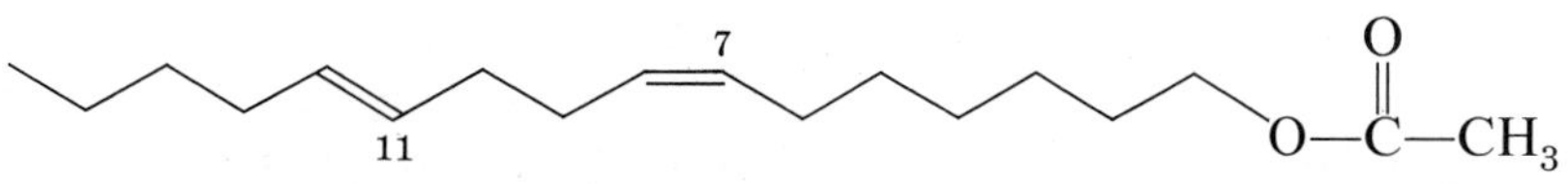

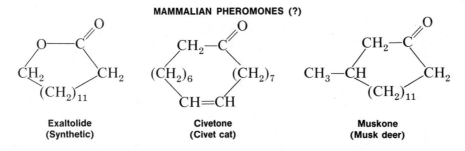

RECRUITING PHEROMONES

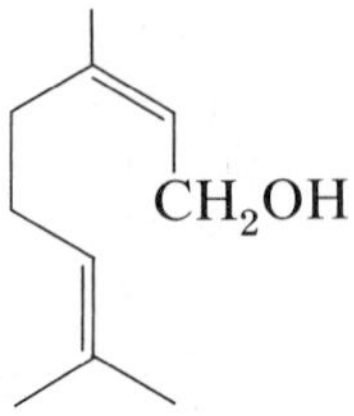

Geraniol
(Honeybee)

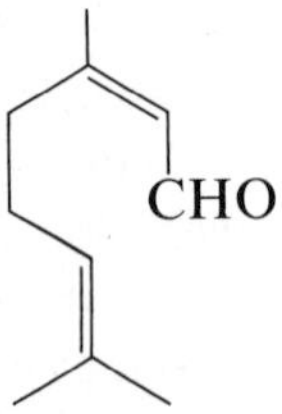

Citral
(Honeybee)

PRIMER PHEROMONE

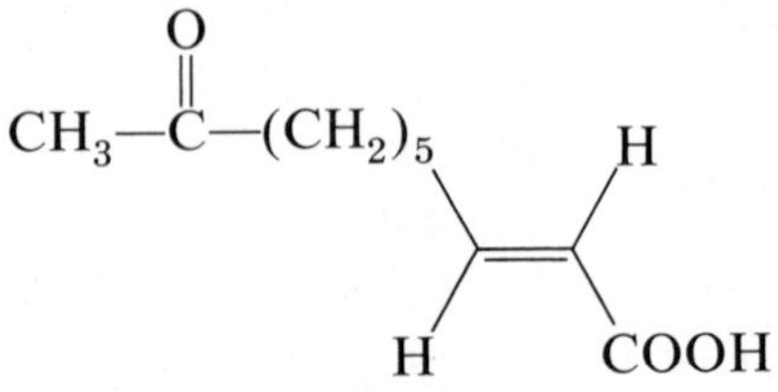

Queen substance
(Honeybee)

ALARM PHEROMONES

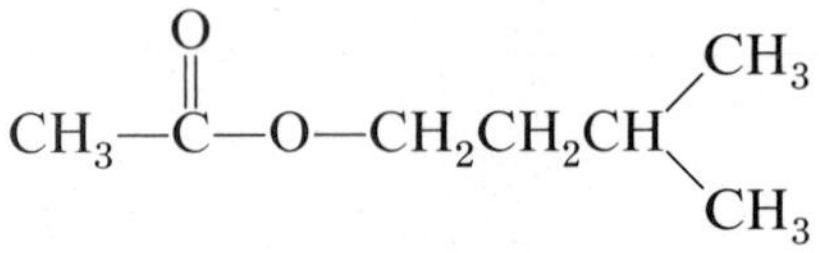

Isoamyl Acetate
(Honeybee)

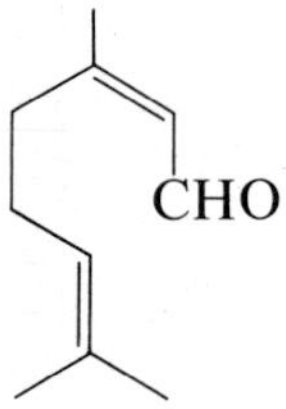

Citral 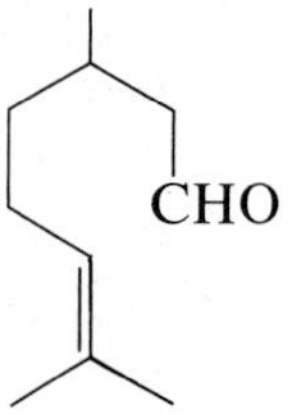   Citronellal

(Ant species)

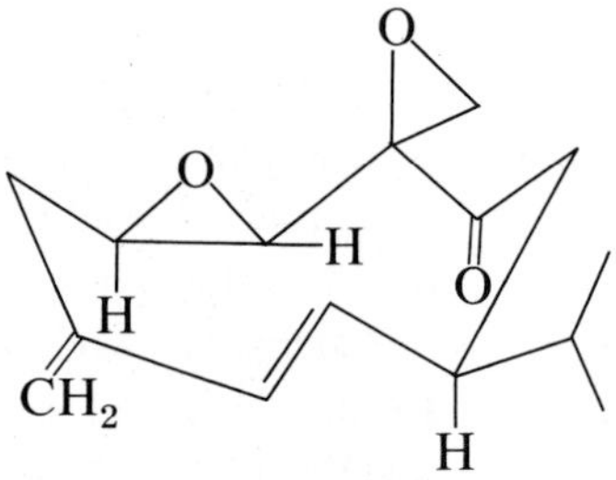

Periplanone B
(American cockroach)

MAMMALIAN PHEROMONES (?)

Exaltolide
(Synthetic)

Civetone
(Civet cat)

Muskone
(Musk deer)

work on **very tiny** amounts of the natural material. The margin for error is great. Such proposals are usually not considered "proved" until synthetic material is shown to be as biologically effective as the natural pheromone.

OTHER PHEROMONES

The most important example of a **primer pheromone** is found in honeybees. A bee colony consists of one queen bee, several hundred male drones, and thousands of worker bees, or undeveloped females. It has recently been found that the queen, the only female that has achieved full development and reproductive capacity, secretes a primer pheromone called the **queen substance.** The worker females, while tending the queen bee, continuously ingest quantities of this queen substance. This pheromone prevents the workers from rearing any competitive queens and also prevents the development of ovaries in all other females in the hive. The substance is also active as a sex attractant; it attracts drones to the queen during her "nuptial flight."

Honeybees also produce several other important types of pheromones. It has long been known that bees will swarm after an intruder. It has also been known that isoamyl acetate induces a similar type of behavior in bees. Isoamyl acetate (Experiment 10) is an **alarm pheromone.** When an angry worker bee stings an intruder, she discharges, along with the sting venom, a mixture of pheromones that incites the other bees to swarm upon and attack the intruder. Isoamyl acetate is an important component of the alarm pheromone mixture. Alarm pheromones have also been identified in many other insects. In less aggressive insects than bees or ants, the alarm pheromone may take the form of a **repellent,** which induces the insects to go into hiding or leave the immediate vicinity.

Honeybees also release **recruiting** or **trail pheromones.** These pheromones attract others to a source of food. Honeybees secrete recruiting pheromones when they locate flowers in which large amounts of sugar syrup are available. Although the recruiting pheromone is a complex mixture, both geraniol and citral have been identified as components. In a similar fashion, when ants locate a source of food, they drag their tails along the ground on their way back to the nest, continuously secreting a trail pheromone. Other ants follow this trail to the source of food.

In some species of insects, **recognition pheromones** have been identified. In carpenter ants, a caste-specific secretion has been found in the mandibular glands of the males of five different species. These secretions have several functions, one of which is to allow members of the same species to recognize one another. Insects not having the correct recognition odor are immediately attacked and expelled from the nest. In one species of carpenter ant, the recognition pheromone has been shown to have methyl anthranilate as an important component.

We do not yet know all the types of pheromones that any given species of insect may use, but it seems that as few as 10 or 12 pheromones could constitute a "language" that could adequately regulate the entire life cycle of a colony of social insects.

INSECT REPELLENTS

Currently, the most widely used **insect repellent** is the synthetic substance *N,N*-diethyl-*m*-toluamide (Experiment 16), also called Deet. It is effective against fleas, mosquitoes, chiggers, ticks, deerflies, sandflies, and biting gnats. A specific repellent is known for each of these types of insects, but none has the wide spectrum of activity that this repellent has. Exactly why these substances repel insects is not yet fully understood. The most extensive investigations have been carried out on the mosquito.

Originally, many investigators thought that repellents might simply be compounds that provided unpleasant or distasteful odors to a wide variety of insects. Others thought that they might be alarm pheromones for the species affected, or that they might be the alarm pheromones of a hostile species. Early research with the mosquito indicates that at least for several varieties of mosquitoes, none of these is the correct answer.

Mosquitoes seem to have hairs on their antennae that are receptors enabling them to find a warm-blooded host. These receptors detect the convection currents arising from a warm and moist living animal. When a mosquito encounters a warm and moist convection current, it moves steadily forward. If it passes out of the current into dry air, it turns until it finds the current again. Eventually it finds the host and lands. Repellents cause a mosquito to turn in flight and become confused. Even if it should land, it becomes confused and flies away again.

Researchers have found that the repellent prevents the moisture receptors of the mosquito from responding normally to the raised humidity of the subject. At least two sensors are involved, one responsive to carbon dioxide and the other responsive to water vapor. The carbon dioxide sensor is activated by the repellent, but if exposure to the chemical continues, adaptation occurs and the sensor returns to its usual low output of signal. The moisture sensor, on the other hand, simply seems to be deadened, or turned off, by the repellent. Therefore, mosquitoes have great difficulty in finding and interpreting a host when they are in an environment saturated with repellent. They fly right through warm and humid convection currents as if the currents did not exist. Only time will tell if other biting insects respond likewise.

REFERENCES

Batra, S. W. T. "Polyester-Making Bees and Other Innovative Insect Chemists." *Journal of Chemical Education, 62* (February 1985): 121.

Beroza, M., ed. *Chemicals Controlling Insect Behavior*. New York: Academic Press, 1970.

Beroza, M., "Insect Sex Attractants." *American Scientist, 59* (May–June 1971): 320.

Jacobson, M., and Beroza, M. "Insect Attractants." *Scientific American, 211* (August 1964): 20.

Katzenellenbogen, J. A. "Insect Pheromone Synthesis: New Methodology." *Science, 194* (October 8, 1976): 139.

Prestwick, G. D. "The Chemical Defenses of Termites." *Scientific American, 249* (August 1983): 78.

Silverstein, R. M. "Pheromones: Background and Potential Use for Insect Control." *Science, 213* (September 18, 1981): 1326.

Sondheimer, E., and Simeone, J. B. *Chemical Ecology*. New York: Academic Press, 1970.

Vollmer, J. J., and Gordon, S. A. "Chemical Communication." *Chemistry, 47* (November 1974):6; ibid., *48* (April 1975): 6; ibid., *48* (May 1975): 6.

Wilson, E. O. "Pheromones." *Scientific American, 208* (May 1963): 100.

Wood, W. F. "Chemical Ecology: Chemical Communication in Nature." *Journal of Chemical Education, 60* (July 1983): 531.

Wright, R. H. "Why Mosquito Repellents Repel." *Scientific American, 233* (July 1975): 105.

Gypsy moth

Beroza, M., and Knipling, E. F. "Gypsy Moth Control with the Sex Attractant Pheromone." *Science, 177* (1972): 19.

Bierl, B. A., Beroza, M., and Collier, C. W. "Potent Sex Attractant of the Gypsy Moth: Its Isolation, Identification, and Synthesis." *Science, 170* (1970): 87.

Pink bollworm

Anderson, R. J., and Henrick, C. A. "Preparation of the Pink Bollworm Sex Pheromone Mixture, Gossyplure." *Journal of the American Chemical Society, 97* (1975): 4327.

Hummel, H. E., Gaston, L. K., Shorey, H. H., Kaae, R. S., Byrne, K. J., and Silverstein, R. M. "Clarification of the Chemical Status of the Pink Bollworm Sex Pheromone." *Science, 181* (1973): 873.

American cockroach

Adams, M. A., Nakanishi, K., Still, W. C., Arnold, E. V., Clardy, J., and Persoon, C. J., *Journal of the American Chemical Society, 101* (1979): 2495.
Still, W. C., *Journal of the American Chemical Society, 101* (1979): 2493.
Stinson, S. C., *Chemical and Engineering News, 57* (April 30, 1979): 24.

Experiment 16
N,N-Diethyl-*m*-Toluamide:
The Insect Repellent "OFF"

Preparation of an Amide
Addition Funnel, Extraction
Column Chromatography or
Vacuum Distillation

In this experiment, we synthesize the active ingredient of the insect repellent "OFF," *N,N*-diethyl-*m*-toluamide. This substance belongs to the class of chemical compounds called **amides.** Amides have the generalized structure

$$\underset{\displaystyle R-\overset{\displaystyle \overset{O}{\|}}{C}-NH_2}{}$$

The amide to be prepared in this experiment is a **disubstituted** amide. That is, two of the hydrogens on the amide —NH_2 group have been replaced with ethyl groups.

The synthesis requires two reaction steps (see equations on page 124). First, *m*-toluic acid will be converted to its acid chloride derivative. Second, the acid chloride will be allowed to react with diethylamine. The acid chloride will not be isolated and purified but caused to react **in situ** (in the solution used to form it). This is a common way of preparing amides. Amides cannot be prepared directly from the admixture of an acid and an amine. If an acid and an amine are mixed, an **acid-base** reaction occurs, giving the conjugate base of the acid, which will not react further while in solution:

$$RCOOH + R_2NH \longrightarrow RCOO^-R_2NH_2^+$$

If the amine salt is isolated as a crystalline solid and strongly heated, the amide can be prepared:

$$RCOO^-R_2NH_2 + \xrightarrow{\text{heat}} RCONR_2 + H_2O$$

This is not a convenient laboratory method. Amines are usually prepared via the acid chloride, as in this experiment.

STEP 1

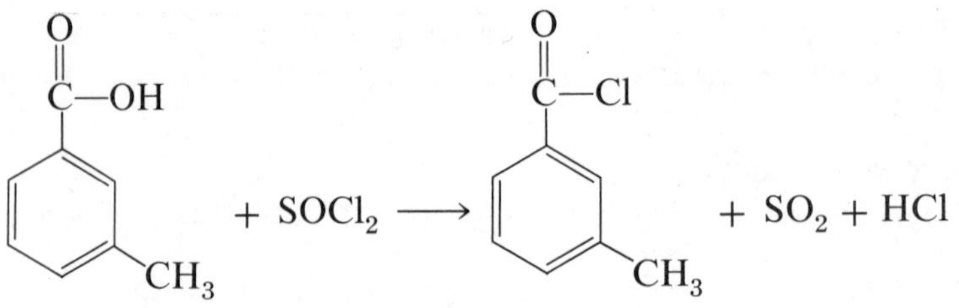

$$\text{m-Toluic acid} + SOCl_2 \longrightarrow [\text{acid chloride}] + SO_2 + HCl$$

m-Toluic acid

STEP 2

$$[\text{acid chloride}] + \underset{Et}{\overset{Et}{\text{NH}}} \longrightarrow N,N\text{-Diethyl-}m\text{-toluamide} + HCl$$

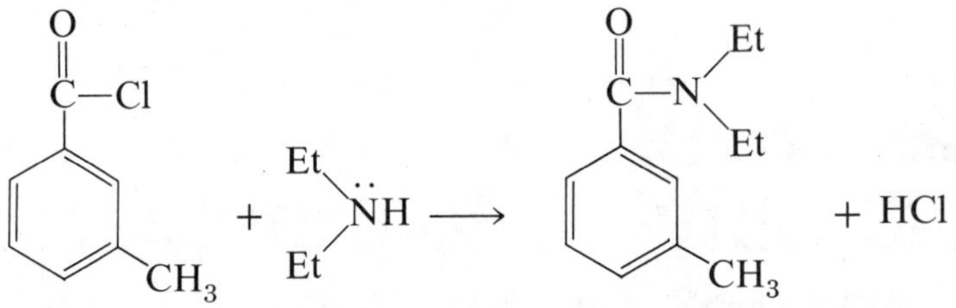

N,N-**Diethyl-***m***-toluamide**

SPECIAL INSTRUCTIONS

As an introduction, read the essay on pheromones that precedes this experiment. In measuring reagents, keep the following in mind. Since thionyl chloride reacts with water to liberate HCl and SO_2, all equipment should be dry, so that decomposition of the reagent is avoided. Likewise, in the second step of the reaction, **anhydrous** ether should be used, since water reacts with both thionyl chloride and the intermediate acid chloride. Thionyl chloride is a **noxious** and **corrosive** chemical and should be handled with care. If it is spilled on the skin, serious burns will result. Thionyl chloride also reacts with any moisture in the air and decomposes, giving hydrogen chloride gas. Therefore, the reagent should be dispensed **in the hood** from a bottle that should be kept tightly closed when not in use. The reagent should not be measured and left to stand on your desk until it is used, since it will hydrolyze. Diethylamine, used in the second step of the reaction, is also noxious and corrosive. In addition, it is quite volatile (bp 56 °C). It should be handled carefully, in the hood, using the same caution as for thionyl chloride.

PROCEDURE

Place 4.1 g of *m*-toluic acid (3-methylbenzoic acid) and 4.5 mL of thionyl chloride (density 1.65 g/mL) in a dry 500-mL three-necked round-bottomed flask. Refer to the accompanying figure, and equip the flask with a reflux condenser and a gas trap of the type shown. Hydrogen chloride is evolved in the reaction and the gas trap will prevent its escape into the room. Add a separatory funnel fitted with a drying tube (see the figure) and stopper the unused neck of the flask. Add a boiling stone, start the circulation of water in the reflux condenser, and heat the mixture **gently,** using either a steam bath or a heating mantle, until the evolution of gases stops (about 20 minutes).

Allow the flask to cool and add 50 mL of dry ether (bp 35 °C). Pour a solution of 10 mL of diethylamine (density 0.7108 g/mL) dissolved in 20 mL of anhydrous ether into the separatory funnel and replace the drying tube. Open the stopcock in the separatory funnel and add the diethylamine solution, a little at a time, over 20 to 25 minutes. Watch the boiling action and do not allow it to become too vigorous. When the diethylamine is added, a voluminous white cloud will form in the flask. Allow the cloud to settle each time before more diethylamine is added. Take care not to add the diethylamine in such large portions that the white material rises up in the arm of the flask and clogs the opening in the addition (separatory) funnel.

After adding the diethylamine, wash the contents of the flask into a separatory funnel, using about 20 mL of a 5% sodium hydroxide solution. If necessary, use a little water to wash into the flask any solid that has accumulated in the condenser; then also add this solution to the material in the separatory funnel. Shake the separatory funnel and allow the layers to separate (Technique 5, Section 5.4, p. 545). If there is not a distinct ether layer, much of the ether may have evaporated during the reaction and some additional ether should be added. Following the sodium hydroxide wash, wash the

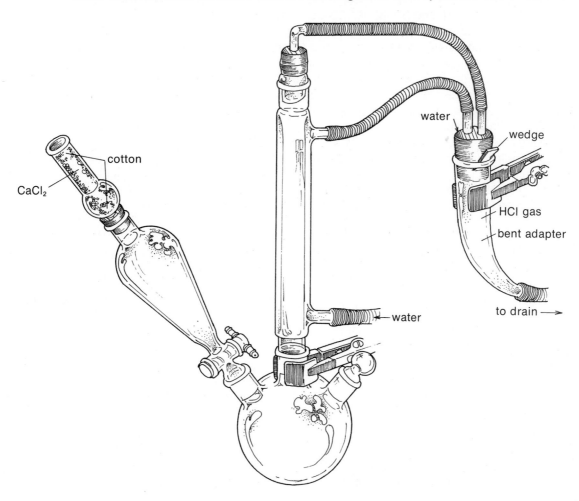

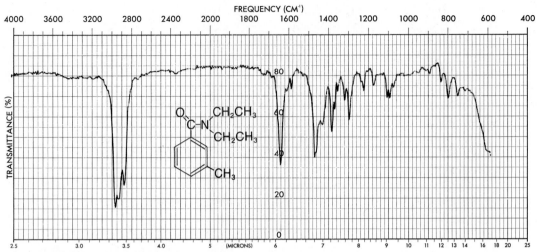

Ir absorption spectrum of *N,N*-diethyl-*m*-toluamide, neat

solution again with another 20-mL portion of 5% sodium hydroxide, then with 20 mL of 10% hydrochloric acid, and finally with 20 mL of water. Dry the ether layer over anhydrous sodium sulfate (Technique 5, Section 5.6, p. 549). Decant the solution from the sodium sulfate and evaporate the ether on a steam bath in the hood.

The product can be purified by a reduced-pressure distillation (Technique 6, Sections 6.6 and 6.7, pp. 558–563) or by column chromatography (Technique 10). Column chromatography is the easier procedure. If the product is distilled, several (three or four) students should combine their crude products and distill cooperatively. If it is purified by column chromatography, alumina should be used as the adsorbent (~25 g alumina in a 25-mm-diameter column), and the elution solvent should be petroleum ether or ligroin. The first compound to elute will be *N,N*-diethyl-*m*-toluamide. Evaporation of the petroleum ether should give the purified amide as a pure colorless or light tan liquid (bp 160–163 °C at 20 mm).

At your instructor's option, determine the infrared spectrum of your product.

REFERENCE

Wang, B. J-S. "An Interesting and Successful Organic Experiment." *Journal of Chemical Education, 51* (October 1974): 631. (The synthesis of *N,N*-diethyl-*m*-toluamide.)

QUESTIONS

1. Write an equation that describes the reaction of thionyl chloride with water.

2. What reaction would take place if the acid chloride of *m*-toluic acid were mixed with water?

3. What is the substance responsible for the voluminous white cloud that forms when diethylamine is added to the reaction mixture? Write an equation that describes its formation.

4. Why is the final reaction mixture extracted with 5% sodium hydroxide?

5. If *N,N*-diethyl-*m*-toluamide distills at 165 °C at 20 mm, what should its boiling point be at 760 mm?

6. Write a mechanism for each step in the preparation of *N,N*-diethyl-*m*-toluamide.

7. Interpret each of the principal peaks in the infrared spectrum of *N,N*-diethyl-*m*-toluamide.

Essay
PETROLEUM

Crude petroleum is a liquid that consists of hydrocarbons as well as some related sulfur, oxygen, and nitrogen compounds. Other elements, including metals, may be present in trace amounts. Crude oil is formed by the decay of marine animal and plant organisms that lived millions of years ago. Over many millions of years, under the influence of temperature, pressure, catalysts, radioactivity, and bacteria, the decayed matter has been converted into what we now know as crude oil. The crude oil is trapped in pools beneath the ground by various geological formations.

Most crude oils have a density between 0.78 and 1.00 g/mL. As a liquid, crude oil may be as thick and black as melted tar or as thin and colorless as water. Its characteristics depend on the particular oil field from which it comes. Pennsylvania crude oils are high in straight-chain alkane compounds (called **paraffins** in the petroleum industry); those crude oils are therefore useful in the manufacture of lubricating oils. Oil fields in California and Texas produce crude oil with a higher percentage of cycloalkanes (also called **naphthenes** by the petroleum industry). Some Middle East fields produce crude oil containing up to 90% cyclic hydrocarbons. Petroleum contains molecules in which the number of carbons ranges from 1 to 60.

When petroleum is refined to convert it into a variety of usable products, it is initially subjected to a fractional distillation. The accompanying table lists the various fractions obtained from fractional distillation. Each of these fractions has its own particular uses. Each fraction may be subjected to further purification, depending on the desired application.

GASOLINE

The gasoline obtained directly from fractional distillation of crude oil is called **straight-run gasoline.** An average barrel of crude oil will yield about 19% straight-run gasoline. This yield presents two immediate problems. First, there is not enough gasoline contained in crude oil to satisfy current needs for fuel to power automobile engines. Second, the straight-run gasoline obtained from crude oil is a poor fuel for modern engines. It must be "refined" at a chemical refinery.

The initial problem of the small quantity of gasoline available from crude oil can be solved by **cracking** and **polymerization.** Cracking is a refinery process by which large hydrocarbon molecules are broken down into smaller molecules. Heat and pressure are required for cracking, and a catalyst must also be used. Gaseous hydrogen may also be present during the cracking, in which case only saturated hydrocarbons are produced. Silica-alumina and silica-magnesia are among the most effective cracking

$$C_{16}H_{34} + H_2 \xrightarrow[\text{heat}]{\text{catalyst}} 2C_8H_{18} \qquad \textbf{CRACKING}$$

catalysts. The alkane mixtures produced tend to have a fairly high proportion of branched-chain isomers. These branched isomers improve the quality of the fuel.

In the polymerization, also carried out at a refinery, small molecules of alkenes are caused to react with each other to form larger molecules, which are also alkenes.

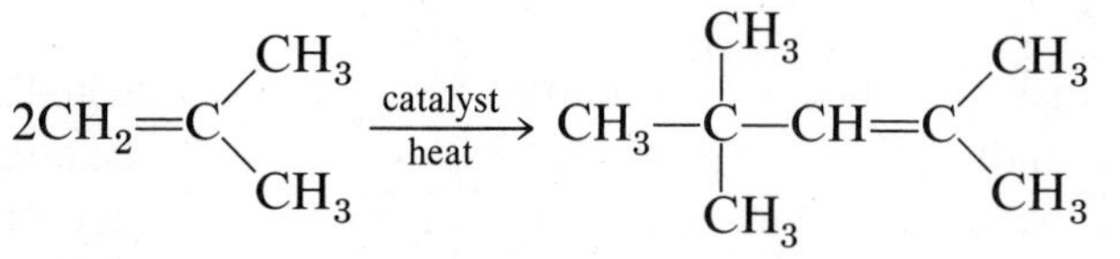

2-Methylpropene
(Isobutylene)   **2,4,4-Trimethyl-2-pentene**   **POLYMERIZATION**

The newly formed alkenes may be hydrogenated to form alkanes. The reaction sequence shown here is a very common and important one in petroleum refining, since

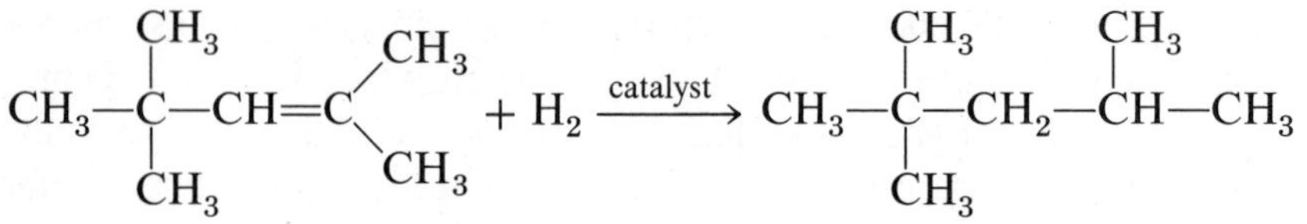

2,2,4-Trimethylpentane
("Iso-octane")

the product, 2,2,4-trimethylpentane (or "iso-octane") forms the basis for determining the quality of gasoline. By these refining methods, the percentage of gasoline that can be obtained from a barrel of crude oil may rise to as much as 45 or 50%.

PETROLEUM FRACTION	COMPOSITION	COMMERCIAL USE
Natural gas	C_1 to C_4	Fuel for heating
Gasoline	C_5 to C_{10}	Motor fuel
Kerosene	C_{11} and C_{12}	Jet fuel and heating
Light gas oil	C_{13} to C_{17}	Furnaces, diesel engines
Heavy gas oil	C_{18} to C_{25}	Motor oil, paraffin wax, petroleum jelly
Residuum	C_{26} to C_{60}	Asphalt, residual oils, waxes

KNOCKING

The internal combustion engine, as it is found in most automobiles, operates in four cycles, or strokes. The four strokes are intake, compression, power, and exhaust. They are illustrated in the figure. The power stroke is of greatest interest to us from the chemical point of view, since combustion occurs during this stroke.

When the fuel-air mixture is ignited, it does not explode. Rather, it burns at a controlled, uniform rate. The gases closest to the spark are ignited first; then they in

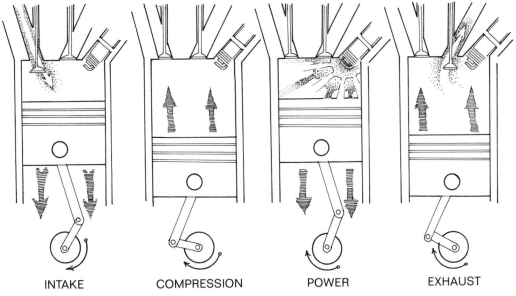

| INTAKE | COMPRESSION | POWER | EXHAUST |

Operation of a four-cycle engine

turn ignite the molecules farther from the spark; and so on. The combustion proceeds in a wave of flame, or the **flame front,** which starts at the spark plug and proceeds uniformly outward from that point until all the gases in the cylinder have ignited. Because a certain time is required for this burning, the initial spark is timed to ignite just before the piston has reached the top of its travel. In this way, the piston will be at the very top of its travel at the precise instant that the flame front, and the increased pressure that accompanies it, reach the piston. The result is a smoothly applied force to the piston, driving it downward.

If heat and compression should cause some of the fuel-air mixture to ignite before the flame front has reached it, the timing of the combustion sequence would be disturbed. Then, the gases would explode rather than burn smoothly. The result is **knocking,** or **detonation** (sometimes called pinging), when the fuel explodes in a violent and uncontrolled manner. The transfer of power to the piston under these conditions is much less efficient than in normal combustion. The wasted energy is merely transferred to the engine block. The explosive force may even damage the engine.

It has been found that the tendency of a fuel to knock is a function of the structures of the molecules composing the fuel. Normal hydrocarbons, those with straight carbon chains, have a greater tendency to lead to knocking than do those alkanes with highly branched chains. The quality of a gasoline, then, is a measure of its resistance to knocking, and this quality is improved by increasing the proportion of branched-chain alkanes in the mixture. Such chemical refining processes as **reforming** and **isomerization** are used to convert normal alkanes to branched-chain alkanes, thus improving the knock-resistance of gasoline.

$$CH_3(CH_2)_6CH_3 \xrightarrow{\text{catalyst}} \underset{\underset{CH_3}{|}}{\overset{\overset{CH_3}{|}}{CH_3-C-CH_2-}}CH-CH_3 \qquad \text{REFORMING}$$

$$CH_3(CH_2)_5CH_3 \xrightarrow{\text{catalyst}} \text{[benzene ring with } CH_3 \text{]} \qquad \text{REFORMING}$$

$$CH_3-CH_2-CH_2-CH_2-CH_3 \xrightarrow{\text{AlBr}_3} \underset{}{\overset{\overset{CH_3}{|}}{CH_3-CH-CH_2-CH_3}} \qquad \text{ISOMERIZATION}$$

None of these processes converts all the normal hydrocarbons into branched-chain isomers; consequently, additives are also put into gasoline to improve the knock-resistance of the fuel. Aromatic hydrocarbons can be considered additives that are effective in improving the knock-resistance of gasoline, and they are often used in unleaded as well as leaded gasolines. The most common additives used to reduce knocking are **tetraethyllead** and **tetramethyllead.** These substances are added in quantities that by law must not exceed 4.0 mL of tetraalkyllead per gallon of gasoline. The usual amount of tetraalkyllead added to gasoline is 3.0 mL per gallon. In recent

$$\underset{\underset{CH_2-CH_3}{|}}{\overset{\overset{CH_2-CH_3}{|}}{CH_3-CH_2-Pb-CH_2-CH_3}} \qquad \underset{\underset{CH_3}{|}}{\overset{\overset{CH_3}{|}}{CH_3-Pb-CH_3}}$$

Tetraethyllead   **Tetramethyllead**

years, because of concern over the possible health hazard associated with emission of lead into the atmosphere, the Environmental Protection Agency has issued regulations that will lead to the phaseout of tetraalkyllead in gasoline. Consequently, oil companies are testing various other additives that will improve the antiknock properties of gasoline without producing harmful emissions.

New cars are designed to operate on unleaded gasoline, which contains no tetraalkyllead additive. The quality of the gasoline is maintained by adding increased quantities of hydrocarbons that have very high antiknock properties themselves. Typical are the aromatic hydrocarbons, including benzene, toluene, and xylene. More expensive refining processes such as **hydrocracking** (cracking in the presence of hydrogen) and **reforming** produce mixtures of hydrocarbons that are more knock-resistant than typical gasoline components. Adding the products of the hydrocracking and reforming to gasoline improves its performance. However, increasing the proportion of aromatic hydrocarbons in gasoline brings with it certain hazards. These substances are toxic, and benzene is considered a serious carcinogenic hazard. Consequently, the risk that illness will be contracted by workers in refineries, and especially by persons who work in service stations, is increased. Research is also being directed toward developing nonhydrocarbon compounds that can improve the quality of unleaded gasoline. To this end, tests are being conducted on *t*-butyl alcohol and methyl *t*-butyl ether as potential antiknock additives.

A much publicized additive for gasoline is ethanol, to make the mixture known as "gasohol." The ethanol is added to ordinary unleaded gasoline to yield a solution that is about 10% ethanol. Adding ethanol improves the antiknock property of the fuel, and it extends the amount of fuel that can be produced from a barrel of crude oil. It is not clear, however, that the energy needed to produce the ethanol is significantly smaller than the amount of energy that could be produced when the ethanol is burned in an engine. It seems that more development is needed before gasohol can become a very important motor fuel nationwide.

OCTANE RATINGS

A fuel can be classified according to its antiknock characteristics. The most important rating system is the **octane rating** of gasoline. In this method of classification, the antiknock properties of a fuel are compared in a hot engine with the antiknock properties of a standard mixture of heptane and 2,2,4-trimethylpentane. This latter compound is called "iso-octane," hence the name *octane rating*. A fuel that has the same antiknock properties as a given mixture of heptane and "iso-octane" has an octane rating numerically equal to the percentage of "iso-octane" in that reference mixture. Today's 91 octane (research method) regular gasoline is a mixture of compounds that have, taken together, the same antiknock characteristics as a test fuel composed of 9% heptane and 91% "iso-octane." Other substances besides hydrocarbons may have high resistance to knocking. A table of various substances with their octane ratings is presented.

OCTANE RATINGS OF ORGANIC COMPOUNDS

COMPOUND	RESEARCH OCTANE NUMBER	COMPOUND	RESEARCH OCTANE NUMBER
Octane	−19	1-Pentene	91
Heptane	0	2-Hexene	93
Hexane	25	Butane	94
Pentane	62	Propane	97
Cyclohexane	83	1-Butene	97
2,2,4-Trimethylpentane	100	Methanol	106
Cyclopentane	101	*m*-Xylene	118
Benzene	106	Toluene	120

POLLUTION PROBLEMS

The number of grams of air required for the complete combustion of 1 mole of gasoline (assuming the formula C_8H_{18}) is 1735 g. This gives rise to a theoretical air-fuel ratio of 15.1:1 for complete combustion. For several reasons, however, it is neither easy nor advisable to supply each cylinder with a theoretically correct air-fuel mixture. The power and performance of an engine improve with a slightly richer mixture (lower air-fuel ratio). Maximum power is obtained from an engine when the air-fuel ratio is near 12.5 : 1, while maximum economy is obtained when the air-fuel ratio is near 16 : 1. Under conditions of idling or full load (that is, acceleration), the air-fuel ratio is lower than what would be theoretically correct. As a result, complete combustion does not take place in an internal combustion engine, and carbon monoxide, CO, is produced in the exhaust gases. Other types of nonideal combustion behavior give rise to the presence of unburned hydrocarbons in the exhaust. The high combustion temperatures cause the nitrogen and oxygen of the air to react, forming a variety of nitrogen oxides in the exhaust. Each of these materials contributes to air pollution. Under the influence of sunlight, which has enough energy to break covalent bonds, these materials may react with each other and with air to produce **smog.** Smog consists of **ozone,** which deteriorates rubber and damages plant life; **particulate matter,** which produces haze; **oxides of nitrogen,** which produce a brownish color in the atmosphere; and a variety of eye irritants, such as **peroxyacetyl nitrate** (PAN). Lead particles, from tetralkyllead, may also cause problems, since they are toxic. Sulfur compounds in the gasoline may lead to the production of noxious gases in the exhaust.

$$CH_3-\overset{\overset{\textstyle O}{\|}}{C}-O-O-NO_2 \quad \textbf{(PAN)}$$

Peroxyacetyl nitrate

Current efforts to reverse the trend of deteriorating air quality caused by automotive exhaust have taken many forms. Initial efforts at modifying the air-fuel mixture of engines produced some improvements in emissions of carbon monoxide, but at the cost of increased nitrogen oxide emissions. With the more stringent air-quality standards imposed by the Environmental Protection Agency, attention has been turned to alternative sources of power. Currently there is much interest in the **diesel engine** for powering passenger cars. The diesel engine has the advantage of producing only very small quantities of carbon monoxide and unburned hydrocarbons. It does, however, produce significant amounts of nitrogen oxides, soot (containing polynuclear aromatic hydrocarbons), and odor-causing compounds. At present, there are no legally established standards for the emission of soot or odor by motor vehicles. This does not mean that these substances are harmless; it means merely that there is no reliable method of analyzing exhaust gases quantitatively for these materials. It may well be that soot and odor may prove to be harmful, but today the emission of these substances remains unregulated. An additional advantage of diesel engines, important in these times of crude oil shortages, is that they tend to yield higher fuel mileage than gasoline engines of a similar size.

In the meantime, since the standard gasoline engine remains the most attractive power plant because of its great flexibility and reliability, efforts at improving its emissions continue. The advent of **catalytic converters,** which are mufflerlike devices containing catalysts that can convert carbon monoxide, unburned hydrocarbons, and nitrogen oxides into harmless gases, has resulted from such efforts. Unfortunately, the catalysts are rendered inactive by the lead additives in gasoline. Unleaded gasoline must be used, but it takes more crude oil to make a gallon of unleaded gasoline than it does to make leaded gasoline. Other hydrocarbons must be added as antiknock agents, formerly the role of tetraalkyllead. The active metals in the catalytic converters, principally platinum, palladium, and rhodium, are scarce and extremely expensive. Also, there has been concern that traces of other harmful substances may be produced in the exhaust gases by reactions catalyzed by these metals.

Some success in reducing exhaust emissions has been attained by modifying the design of the combustion chambers of internal-combustion engines. Additionally, the use of computerized control of ignition systems has shown promise. Efforts have also been directed at developing alternative fuels that would give greater mileage, lower emissions, better performance, and a lower demand on crude oil supplies. Methanol has been proposed as an alternative to gasoline as a fuel. Some preliminary tests have indicated that the amount of the principal air pollutants in automobile exhaust are greatly lowered when methanol is used instead of gasoline in a typical automobile. The mixture of ethanol and gasoline, **gasohol,** is also being examined carefully. Experiments are now under way toward developing such gases as methane and even hydrogen as future fuels. Although the technology for using these gases as fuels remains far away, these gases in theory should prove promising as automotive fuels. It will be interesting to see what further developments the dual problems of air quality and shortage of petroleum will bring to the design of automotive engines.

REFERENCES

"Facts about Oil." New York: American Petroleum Institute, 1970.

Greek, B. F. "Gasoline." *Chemical and Engineering News, 52* (November 9, 1970).

Kerr, J. A., Calvert, J. G., and Demerjian, K. L. "The Mechanism of Photochemical Smog Formation." *Chemistry in Britain, 8* (1972): 252.

Kolb, D., and Kolb, K. E. "Chemical Principles Revisited: Petroleum Chemistry." *Journal of Chemical Education, 56* (July 1979): 465.

Lane, J. C. "Gasoline and Other Motor Fuels." *Encyclopedia of Chemical Technology, 10* (1966). New York: John Wiley & Sons: 463.

Pierce, J. R. "The Fuel Consumption of Automobiles." *Scientific American, 232* (January 1975): 34.

Reed, T. B., and Lerner, R. M. "Methanol: A Versatile Fuel for Immediate Use." *Science, 182* (1973): 1299.

Wildeman, T. R. "The Automobile and Air Pollution." *Journal of Chemical Education, 51* (1974): 290.

Experiment 17
Gas Chromatographic Analysis of Gasolines

Gasoline
Gas Chromatography

It is possible to analyze samples of gasoline by gas chromatography. From your analysis, you should learn something about the composition of these fuels. Although all gasolines are compounded from the same basic hydrocarbon components, different companies blend these components in different proportions in order to obtain a gasoline with properties similar to those of other brands.

Sometimes the composition of the gasoline may vary depending on the composition of the crude petroleum from which the gasoline was derived. Frequently, refineries vary the composition of gasoline in response to differences in climate or seasonal changes. In the winter or in cold climates, the relative proportion of butane and pentane isomers is increased to increase the volatility of the fuel. This increased volatility permits easier starting. In the summer or in warm climates, the relative proportion of these volatile hydrocarbons is reduced. The decreased volatility thus achieved reduces the possibility of vapor-lock formation. Occasionally, differences in composition can be detected by examining the gas chromatograms of a particular gasoline over several months. In this experiment, we do not try to detect such differences.

There are different octane rating requirements for ''regular'' and ''premium'' gasolines. You will be able to observe differences in the composition of these two types of fuels. You should pay particular attention to increases in the proportions of those hydrocarbons that raise octane ratings in the premium fuels. If you analyze an unleaded gasoline, you should be able to observe differences in the composition of this type of gasoline compared with leaded fuels of a similar octane rating.

You will be asked to analyze a sample of regular leaded and one of regular unleaded gasoline, preferably from the same company. If a premium unleaded gasoline from that company is also available, you may be required to analyze it also. Other students in the class may be analyzing gasolines from other companies. If different brands are analyzed, equivalent grades from the different companies should be compared. If there is a supplier of gasohol in the region, a sample of that fuel should be examined as well.

Discount service stations usually buy their gasoline from one of the large petroleum-refining companies. If you analyze gasoline from a discount service station, you may find it interesting to compare that gasoline with an equivalent grade from a major supplier, noting particularly the similarities.

SPECIAL INSTRUCTIONS

Before performing this experiment, you should read the essay on petroleum, which precedes this experiment, and Technique 12. Your instructor may want each student in the class to collect samples of gasoline from different service stations. A list should be compiled of all the different gasoline companies represented in the nearby area, and each student should be assigned to collect a sample from a different company. You should collect the gasoline sample in a labeled screw-cap jar. It may be handy to take a funnel along with you also. An easy way to collect a gasoline sample for this experiment is to drain the excess gasoline from the nozzle and hose of the pump into the jar immediately after the gasoline tank of a car has been filled. The collection of gasoline in this manner must be done **immediately after** the gas pump has been used. If not, the volatile components of the gasoline may evaporate, thus changing the composition of the gasoline. Only a very small sample (a few **milli**liters) of gasoline is required, since the gas chromatographic analysis requires no more than a few **micro**liters (μL) of material. Be certain to close the cap of the sample jar tightly to prevent the selective evaporation of the most volatile components. The label on the jar should list the brand of gasoline and the grade.

If you live in a state in which collecting gasoline in glass containers is illegal, your instructor will supply you with a fireproof metal container in which to collect the gasoline sample. Alternatively, the instructor may provide you with gasoline samples.

Always remember that gasoline contains many highly volatile and flammable components. Do not breathe the vapors and do not use open flames near gasoline. Also recall that gasoline contains tetraethyllead and is therefore toxic

This experiment may be assigned along with another short one, since it requires only a few minutes of each student's time to carry out the actual gas chromatography. To work as efficiently as possible, it may be convenient to arrange an appointment schedule for using the gas chromatograph.

> **NOTE TO THE INSTRUCTOR:** The gas chromatograph should be prepared as follows: column temperature, 110–115 °C; injection port temperature, 110–115 °C; carrier gas flow rate, 40–50 mL/min. The column should be approximately 12 ft long and should contain a nonpolar stationary phase similar to silicone oil (SE-30) on Chromosorb W or some other stationary phase that separates components principally according to boiling point.

PROCEDURE

The instructor may require you to analyze a series of standard materials. Reference substances that should be used include pentane, hexane, cyclohexane, heptane, toluene, and *m*-xylene, although others may be included appropriately (ethanol, for example). If the samples are each analyzed individually, a sample size of about 0.5 microliters (μL) will be adequate for each reference substance. A better alternative is to analyze a reference mixture that contains all these standard substances and compare it with a similar gas chromatogram previously recorded by the instructor. The instructor may either post a copy of this chromatogram (with the peaks identified) or provide each student with a copy. An example of such a determination is provided at the end of this experiment. If a reference mixture is to be analyzed, a reasonable sample size is 1 μL. The sample is injected into the gas chromatograph, and the retention times of each of the components are measured and recorded (Technique 12, Sections 12.7 and 12.8, page 637).

The instructor may prefer to perform the sample injections or have a laboratory assistant perform them. The sample injection procedure requires careful technique, and the special microliter syringes that are required are very delicate and expensive. If students are to perform the sample injections themselves, they must have adequate instruction beforehand.

Inject the sample of regular leaded gasoline onto the gas chromatography column and wait for the gas chromatogram to be recorded. Next inject the sample of unleaded gasoline and obtain its gas chromatogram. Compare these chromatograms with each other and identify as many of the components as possible. For comparison, sample gas chromatograms of a regular leaded gasoline, a premium leaded gasoline, and gasohol are provided in the accompanying figure. Determine the difference between the two grades of gasoline. If a sample of premium unleaded gasoline was also obtained, record its gas chromatogram and compare it with the chromatograms of the regular and the premium gasoline. Do the same for a sample of gasohol if it is available. Be certain to compare very carefully the retention times of the components in each fuel sample with the standards in the reference mixture. Retention times of compounds vary with the conditions under which they are determined. It is best to analyze the reference mixture and each of the gasoline samples in succession to reduce the variations in retention times that may occur over time. If there are unidentified peaks in the gas chromatograms of the fuel samples, try to guess their probable identity. Compare the gas chromatograms with those of students who have analyzed gasolines from other dealers.

The report to the instructor should include the actual gas chromatograms as well as an identification of as many of the components in each grade of fuel as possible.

QUESTIONS

1. How do regular and premium grades of gasoline differ in this analysis?

2. Assuming one had a mixture of benzene, toluene, and *m*-xylene, what would be the expected order of retention times? Explain.

3. What do you expect the analysis of an unleaded gasoline to reveal in this experiment?

4. Is it possible to detect tetraethyllead in this experiment?

5. If you were a forensic chemist working for the police department, and the fire marshal brought you a sample of gasoline found at the scene of an arson attempt, could you identify the service station at which the arsonist purchased the gasoline? Explain.

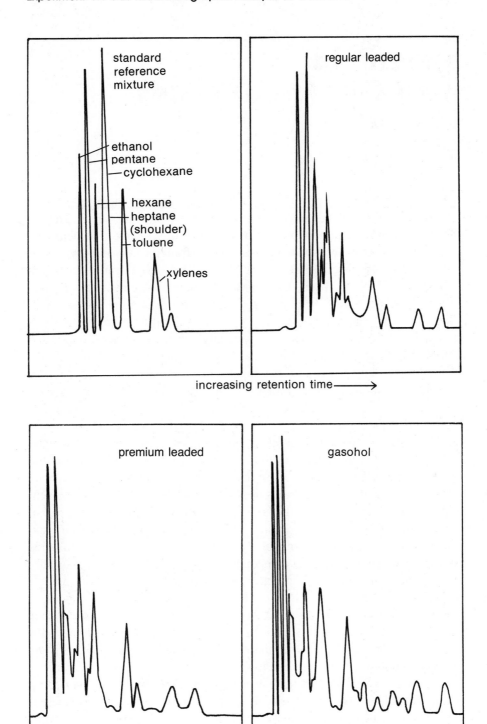

Gas chromatograms of a standard reference mixture, a regular leaded gasoline, a premium leaded gasoline, and gasohol

Experiment 18
Essential Oils From Spices

Isolation of a Natural Product
Steam Distillation
Derivative Formation

Anyone who has walked through a pine or a cedar forest, or anyone who loves flowers and spices, knows that many plants and trees have distinctly pleasant odors. The essences or aromas of plants are due to volatile or **essential oils,** many of which have been valued since antiquity for their characteristic odors (frankincense and myrrh, for example). A list of the commercially important essential oils would run to over 200 entries. Allspice, almond, anise, basil, bay, caraway, cinnamon, clove, cumin, dill, eucalyptus, garlic, jasmine, juniper, orange, peppermint, rose, sassafras, sandalwood, spearmint, thyme, violet, and wintergreen are but a few familiar examples of such valuable essential oils. Essential oils are used mainly for their pleasant odors and flavors in perfumes, incense, scents, and spices and as flavoring agents in foods. A few are also valued for their antibacterial and antifungal action. Some are used medicinally (camphor and eucalyptus), and others as insect repellents (citronella). Chaulmoogra oil represents one of the few known curative agents for leprosy. Turpentine is used as a solvent for many paint products.

Essential oil components are often found in the glands or intercellular spaces on plant tissue. They may exist in all parts of the plant but are often concentrated in the seeds or flowers. Many components of essential oils are steam-volatile and can be isolated by steam distillation. Other methods of isolating essential oils include solvent extraction and pressing (expression) methods. From the essay on esters, we saw that esters were frequently responsible for characteristic odors and flavors of fruits and flowers. Besides the esters, the ingredients of essential oils may be complex mixtures of hydrocarbons, alcohols, and carbonyl compounds. These other components usually belong to one of the two groups of natural products called **terpenes,** or **phenylpropanoids.**

The chief constituents of the essential oils from cloves, cinnamon, cumin, and allspice are aromatic and volatile with steam. In this experiment, we isolate the main component of the essential oil derived from these spices by steam distillation. At your instructor's option, you may also be required to convert these oils to a solid derivative. A derivative is a new compound to which the original compound is easily converted by a simple reaction. It may be difficult to purify the original compound or to characterize it accurately by its physical properties since it is a liquid or oil. A derivative is a solid crystalline compound of definite melting point. The derivative is easily characterized or identified. Its identity, once established, provides circumstantial evidence to help identify the material from which it was formed.

The technique of steam distillation permits separating volatile components from nonvolatile materials without the need for raising the temperature of the distillation above 100 °C. Steam distillation provides a means of isolating the essential oils without the risk of decomposing them thermally.

SPECIAL INSTRUCTIONS

Before beginning this experiment, read the sections in your lecture textbook on terpenes and phenylpropanoids; also read Techniques 3, 8, and 17 and Appendices 3 and 4. Your instructor may make the preparation of a derivative an optional part of the experiment. As the instructor indicates, you are to do only one of the following three experiments: 18A, 18B, or 18C.

Experiment 18A
Oil of Clove or Allspice

Both oil of cloves (from *Eugenia caryophyllata*) and oil of allspice (from *Pimenta officinalis*) are rich in **eugenol** (4-allyl-2-methoxyphenol). Caryophyllene is present in small amounts along with other terpenes. Eugenol (bp 250 °C) is a phenol, or an aromatic hydroxy compound.

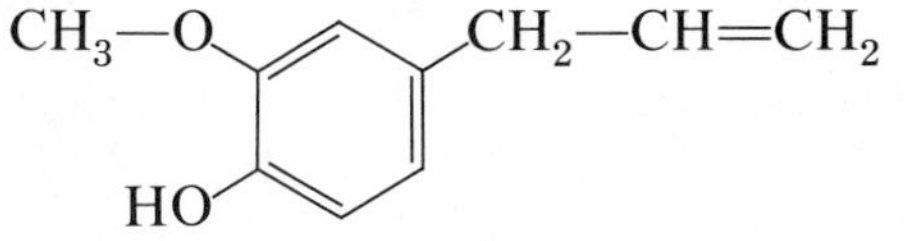

Eugenol 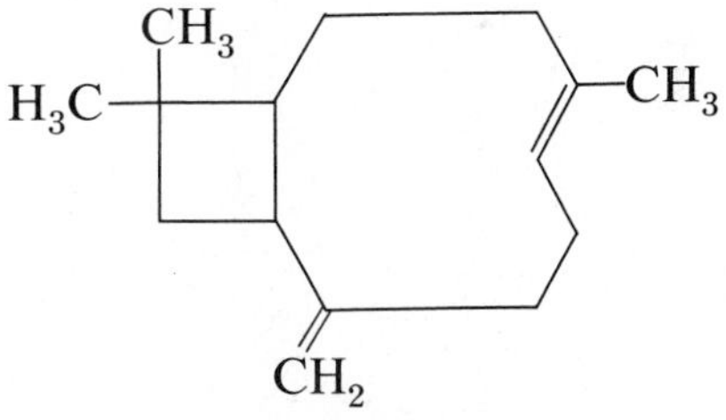   **Caryophyllene**

You are asked to isolate the eugenol and to characterize it by infrared and by nuclear magnetic resonance spectroscopy. At your instructor's option, you may also be asked to convert eugenol to its benzoate derivative by causing it to react with benzoyl chloride.

PROCEDURE

Assemble an apparatus for steam distillation, using a 500-mL three-necked round-bottomed flask (Technique 8, Section 8.3B, p. 587, Direct Method). The collection flask may be a 125-mL Erlenmeyer flask. The heat source will be a Bunsen burner flame or a heating mantle. Place 15 g of ground allspice or 5 g of clove buds or ground cloves into the flask and add 150 mL of water. Begin heating the liquid so as to provide a low but steady rate of distillation. During the distillation, continue to add water from the addition (separatory) funnel at a rate that will just maintain the original level of the liquid in the distilling flask. Continue the steam distillation until about 100 mL of distillate has been collected.

Empty the water from the addition funnel and place the distillate in it. Extract the distillate with two 10-mL portions of methylene chloride. Separate the layers and discard

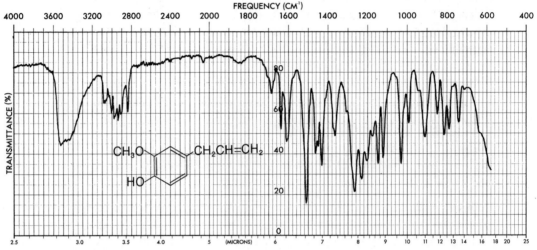

Ir spectrum of eugenol, neat

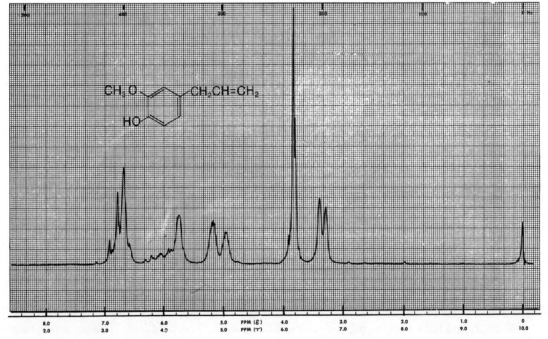

Nmr spectrum of eugenol, CDCl₃

the aqueous phase. If the methylene chloride layer is separated carefully, it will not need to be dried, although drying with a small amount of anhydrous sodium sulfate may be advisable. Decant the solution from any drying agent and evaporate most of the solvent on a steam bath in the hood. Transfer the remaining liquid to a previously weighed test tube. Using a wooden applicator stick to control bumping, concentrate the contents of the test tube by further careful heating on a steam bath until nothing but an oily residue remains. Dry the outside of the test tube and weigh it. Calculate the weight percentage recovery of the oil from the original amount of allspice or cloves used.

Determine the infrared spectrum and, at the instructor's option, the nuclear magnetic resonance spectrum of the oil. The report should include the spectra that were recorded. The major absorptions in the spectra should be interpreted.

PREPARATION OF A DERIVATIVE (OPTIONAL)

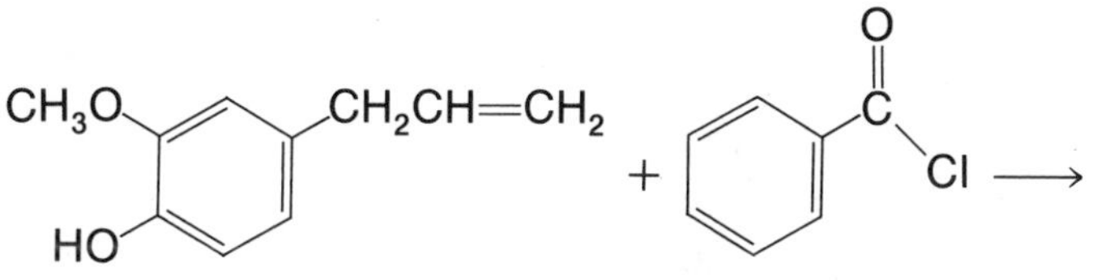

Eugenol   Benzoyl chloride

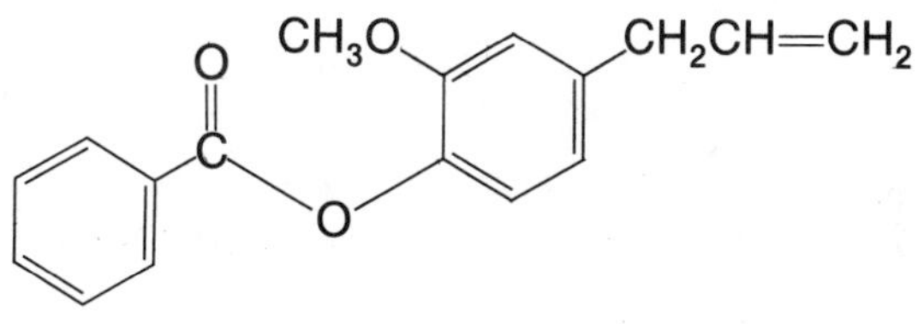

Eugenol benzoate

Carefully weigh out 0.25 g of the crude eugenol directly into a small test tube. Add 10 to 15 drops of a $1M$ solution of sodium hydroxide to the test tube until the oil just dissolves. Do not add any more sodium hydroxide than is needed to get the oil to dissolve. The final solution may be cloudy, but no large oil droplets should be apparent. Carefully weigh out 0.21 g of benzoyl chloride directly into the test tube. An excess of benzoyl chloride should be avoided since it will make achieving crystallization of the final product impossible. Heat the mixture on the steam bath for 3 to 5 minutes. Cool the mixture and scratch the inside of the test tube with a glass rod until the mixture solidifies. If it does not solidify, decant the aqueous layer, add a few drops of methanol, and continue scratching and cooling the oil.[1] Collect the solid on a small Büchner funnel and wash it carefully with a small volume of cold water. Recrystallize the solid from a minimum amount of boiling methanol. If an oil separates from the solution, reheat the solution and cool it

---

[1]If there is still no crystallization, dissolve the oil in a minimum amount of hot methanol. While the mixture is still hot, add water dropwise until precipitation of the derivative is complete.

more slowly, as you continue to scratch the inside of the test tube and add seed crystals. Collect the crystals on a small Büchner funnel. Dry the crystals and record their melting point. The melting point of eugenol benzoate is 70 °C. Submit the crystals in a labeled vial to the instructor.

Experiment 18B
Oil of Cumin

The major portion of the volatile oil from cumin seeds (*Cumin cymium*) is **cuminaldehyde** (*p*-isopropylbenzaldehyde). Cuminaldehyde has a boiling point of 235 to 236 °C.

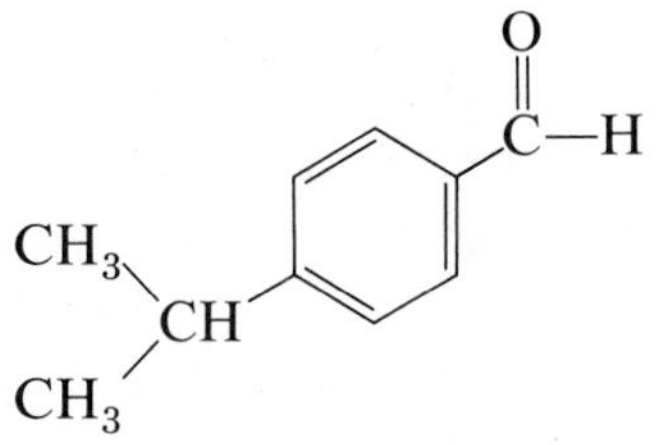

Cuminaldehyde

You are asked to isolate cuminaldehyde from cumin and to characterize it by infrared and nuclear magnetic resonance spectroscopy. At your instructor's option, you may also be asked to convert cuminaldehyde to its semicarbazone derivative by causing it to react with semicarbazide hydrochloride.

PROCEDURE

Assemble an apparatus for steam distillation using a 500-mL three-necked round-bottomed flask (Technique 8, Section 8.3B, Direct Method, p. 587). The collection flask may be a 125-mL Erlenmeyer flask. The heat source will be a Bunsen burner flame or a heating mantle. Place 15 g of ground cumin seeds in the flask and add 150 mL of water. Begin heating the liquid in the flask so as to provide a low but steady rate of distillation. During the distillation, continue to add water from the addition (separatory) funnel at a rate by which you just maintain the original level of the liquid in the distilling flask. Continue the steam distillation until about 100 mL of distillate has been collected.

Empty the water from the addition funnel and place the distillate in it. Extract the distillate with two 10-mL portions of methylene chloride. Separate the layers and discard the aqueous phase. If the methylene chloride layer is separated carefully, it will not need to be dried, although drying with a small amount of anhydrous sodium sulfate may be advisable. Decant the solution from any drying agent and evaporate most of the solvent on a steam bath in the hood. Transfer the remaining liquid to a previously weighed test

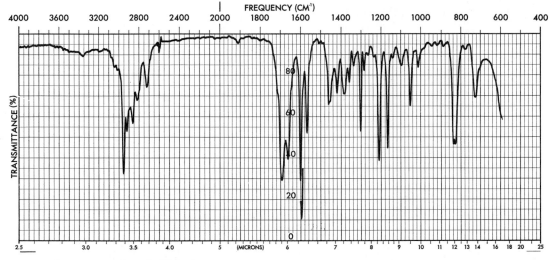

Ir spectrum of cuminaldehyde, neat

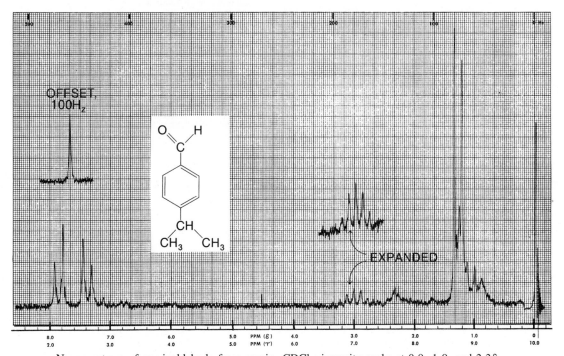

Nmr spectrum of cuminaldehyde from cumin, $CDCl_3$; impurity peaks at 0.9, 1.0, and 2.3δ

tube containing a boiling stone. Concentrate the contents of the test tube by further careful heating on a steam bath until nothing but an oily residue remains. Dry the outside of the test tube and weigh it. From the original amount of cumin used, calculate the weight percentage recovery of the oil.

Determine the infrared spectrum and, at the instructor's option, the nuclear magnetic resonance spectrum of the oil. The report should include the spectra recorded and an interpretation of the principal peaks in the spectra.

PREPARATION OF A DERIVATIVE (OPTIONAL)

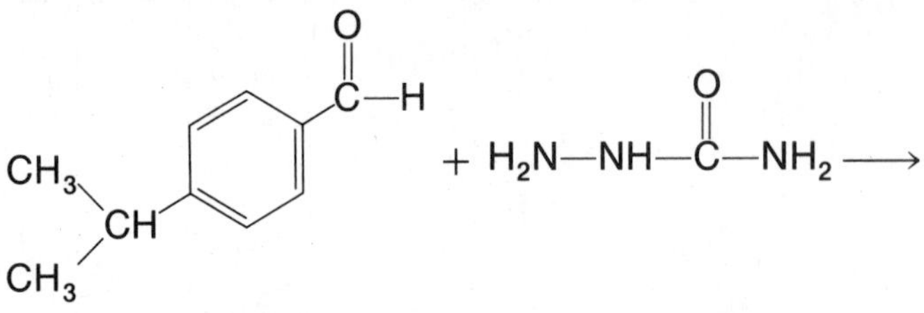

Cuminaldehyde   Semicarbazide

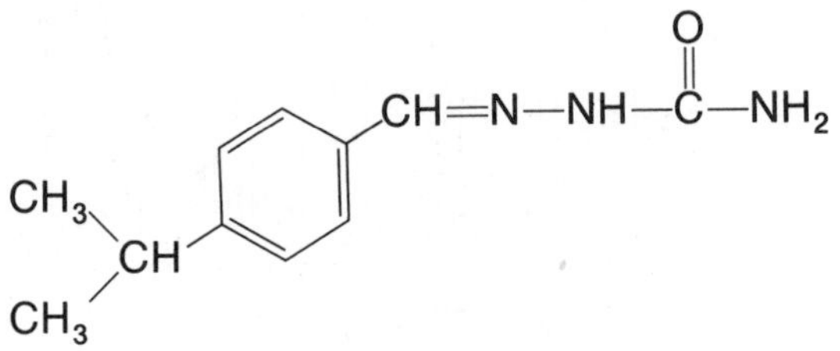

Cuminaldehyde semicarbazone

Dissolve 0.20 g of semicarbazide hydrochloride and 0.30 g of anhydrous sodium acetate, which will serve as a buffer, in 2 mL of water. To this mixture add 3 mL of absolute ethanol. Add this solution to the cumin oil and warm the mixture on the steam bath for 5 minutes. Cool the mixture and allow the cuminaldehyde semicarbazone to crystallize. Collect the crystals in a small Büchner funnel and recrystallize them from methanol. Allow the crystals to dry and determine their melting point. The melting point of cuminaldehyde semicarbazone is 216 °C. Submit the derivative in a labeled vial to the instructor.

Experiment 18C
Oil of Cinnamon

The principal component of cinnamon oil (from *Cinnamomum zeylanicum*) is cinnamaldehyde (*trans*-3-phenylpropenal). Cinnamaldehyde has a boiling point of 252 °C. You are asked to isolate cinnamaldehyde from cinnamon and to characterize it by

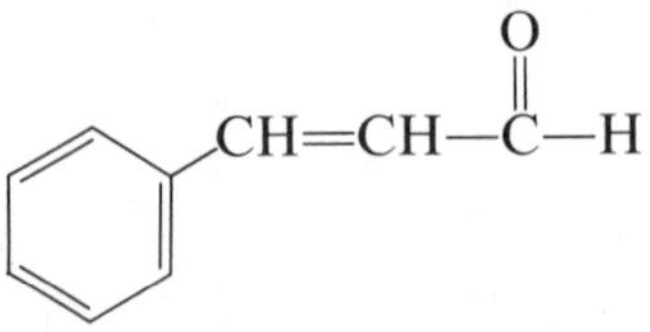

Cinnamaldehyde

means of infrared and nuclear magnetic resonance spectroscopy. At your instructor's option, you may also be asked to convert cinnamaldehyde to its semicarbazone derivative by causing it to react with semicarbazide hydrochloride.

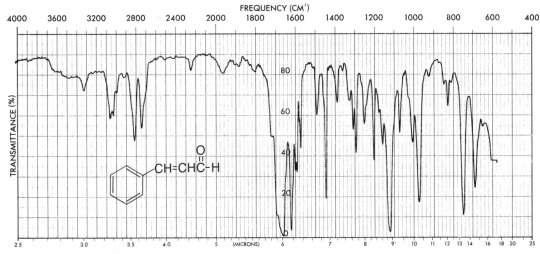

Ir spectrum of cinnamaldehyde, neat

PROCEDURE

Assemble an apparatus for steam distillation using a 500-mL three-necked round-bottomed flask (Technique 8, Section 8.3A, Live Steam Method, p. 585). The collection flask may be a 125-mL Erlenmeyer flask. Place 15 g of ground cinnamon bark in the flask and add 100 mL of hot water. Admit steam from the steam line so as to provide a low but steady rate of distillation. Continue the steam distillation until about 100 mL of distillate has been collected.

Place the distillate in a separatory funnel and extract it with two 10-mL portions of methylene chloride. Separate the layers and discard the aqueous phase. If the methylene chloride layer is separated carefully, it will not need to be dried, although drying with a small amount of anhydrous sodium sulfate may be advisable. Decant the solution from any drying agent and evaporate most of the solvent on a steam bath in the hood. Transfer the remaining liquid to a previously weighed test tube containing a boiling stone. Concentrate the contents of the test tube by further careful heating on a steam bath until nothing but an oily residue remains. Dry the outside of the test tube and weigh it. From the original amount of cinnamon bark used, calculate the weight percentage recovery of the oil.

Determine the infrared spectrum and, at the instructor's option, the nuclear magnetic resonance spectrum of the oil. The report should include the spectra recorded and an interpretation of the principal peaks in the spectra.

PREPARATION OF A DERIVATIVE (OPTIONAL)

Dissolve 0.20 g of semicarbazide hydrochloride and 0.30 g of anhydrous sodium acetate, which will serve as a buffer, in 2 mL of water. To this mixture add 3 mL of absolute ethanol. Add this solution to the cinnamon oil and warm the mixture on the steam bath

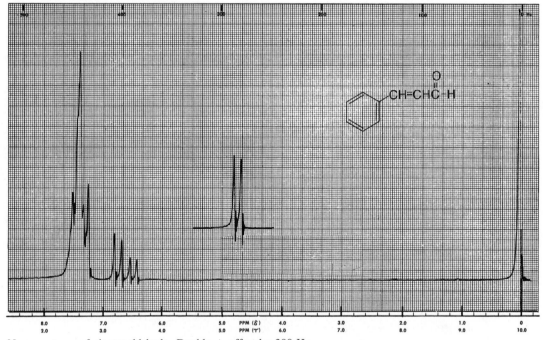

Cinnamaldehyde   **Semicarbazide**

Cinnamaldehyde semicarbazone

for 5 minutes. Cool the mixture and allow the cinnamaldehyde semicarbazone to crystallize. Collect the crystals in a small Büchner funnel and recrystallize them from methanol. Allow the crystals to dry and determine their melting point. The melting point of cinnamaldehyde semicarbazone is 215 °C. Submit the derivative in a labeled vial to the instructor.

Nmr spectrum of cinnamaldehyde. Double + offset by 300 Hz.

QUESTIONS

1. Why are eugenol, cuminaldehyde, and cinnamaldehyde steam-distilled rather than purified by simple distillation?

2. A natural product (MW 150) distills with steam at a boiling temperature of 99 °C at atmospheric pressure. The vapor pressure of water at 99 °C is 733 mm. (a) Calculate the weight of the natural product that co-distills with each gram of water at 99 °C. (b) How much water must be removed by steam distillation to recover this natural product when 100 g of a spice contains 10% of the desired substance?

3. In a steam distillation, the amount of water actually distilled is usually greater than the amount calculated, assuming that both water and the organic substance exert the same vapor pressure when they are mixed that they exert when each is pure. Why does one recover more water in the steam distillation than was calculated?

Experiment 19
Thin-Layer Chromatography for Monitoring the Oxidation of Borneol to Camphor

Chromic Acid Oxidation
Monitoring Reactions
Thin-Layer Chromatography

A typical reaction of secondary alcohols is their oxidation to ketones by such reagents as chromic acid, sodium dichromate, or potassium permanganate. A specific example of such an oxidation is shown for the oxidation of borneol to camphor. For a more

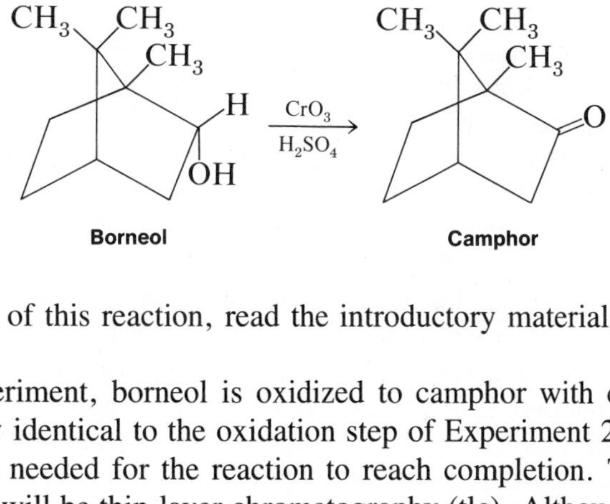

Borneol   Camphor

detailed discussion of this reaction, read the introductory material in Experiment 20, p. 149.

In this experiment, borneol is oxidized to camphor with chromic acid. This reaction is virtually identical to the oxidation step of Experiment 20. The object is to determine the time needed for the reaction to reach completion. The means used to follow the reaction will be thin-layer chromatography (tlc). Although both the starting material and the product are colorless, they can be visualized by iodine vapor.

The purpose of this experiment is to demonstrate the basic principles and techniques of thin-layer chromatography. This experiment also illustrates one method of monitoring the progress of a reaction, and it demonstrates that organic reactions often proceed slowly.

SPECIAL INSTRUCTIONS

Before beginning this experiment, read Technique 11 and the introductory section of Experiment 20. This experiment involves diethyl ether, which is a flammable solvent, sulfuric acid, which is corrosive, and chromic oxide, a possible carcinogen (see p. 12). Although the quantities of these materials used are small, safety precautions should be exercised.

PROCEDURE

Using a slurry of silica gel G in a methylene chloride–methanol (2:1) solvent, prepare about 10 hand-dipped microscope-slide tlc plates by the method described in Technique 11, Section 11.2A, p. 617. Prepare a developing chamber from a 4-oz wide-mouthed screw-cap jar, as described in Technique 11, Section 11.4, p. 621. Add some of the methylene chloride–methanol solvent to the developing chamber. Finally, prepare several capillary micropipets as described in Technique 11, Section 11.3, p. 620.

For this experiment, you use three solutions: Solution A is a 2% solution of borneol in diethyl ether; solution B is a solution of 10% chromic oxide and 5% sulfuric acid in water; and solution C is a 2% solution of camphor in diethyl ether. Solutions A and C provide reference spots on each tlc plate, and solution B provides the oxidizing medium to oxidize borneol to camphor.

Mix about 1 mL of solution A with 1 mL of solution B in a small test tube. Write down the time at which this mixture is prepared. Shake the mixture briefly. Spot a tlc plate with solution A, solution C, and the upper (ether) layer of the reaction mixture from the test tube. Place the plate in the developing chamber and develop it as described in Technique 11, Section 11.4, p. 621. After the reaction has been allowed to take place for 5 minutes, spot a second tlc plate with solution A, solution C, and the upper layer of the reaction mixture. Develop this plate as before. Continue spotting and developing plates at 5-minute intervals for a total reaction time of 40 minutes. During this reaction period, shake the test tube periodically.

When the tlc plates have all been developed, place each plate in a jar containing a few iodine crystals, cap the jar, and warm it gently on a steam bath until spots begin to appear (Technique 11, Section 11.6, p. 623). You will notice that the R_f values for borneol and camphor differ, with camphor more mobile than borneol. By comparing these R_f values with R_f values for the spots formed in the reaction mixture, you should be able to identify the presence of borneol and camphor in the mixture. Compare the intensities of the borneol and camphor spots in the reaction mixture on each of the tlc plates you have developed. Using this information, determine the time required for the oxidation reaction to reach completion. Report the reaction time and the R_f values for borneol and camphor to your instructor.

REFERENCE

Davis, M. ''Using TLC to Follow the Oxidation of a Secondary Alcohol to a Ketone.'' *Journal of Chemical Education, 45* (March 1968): 192.

QUESTIONS

1. What physical property differences between borneol and camphor make the separation of the two compounds on thin-layer chromatography plates possible?

2. Why is it necessary to shake the mixture periodically during the reaction?

3. Describe how you could use tlc for monitoring the progress of a reaction to follow the esterification of benzoic acid with methanol to form methyl benzoate. Indicate the relative order of R_f values expected for benzoic acid and methyl benzoate.

4. Are the structures given for borneol and camphor in this experiment different from those in Experiment 20? Explain the differences, if any.

5. A student spotted the tlc plates with the borneol–camphor reaction mixture as stated in the experiment, developed the plates, and placed them in a storage cabinet for a few days. When the student returned and placed the plates in the iodine chamber, no spots were observed. What happened?

Experiment 20
An Oxidation-Reduction Scheme: Borneol, Camphor, Isoborneol

Sodium Dichromate Oxidation
Sodium Borohydride Reduction
Stereochemistry
Sublimation
Spectroscopy (ir, nmr, cmr)

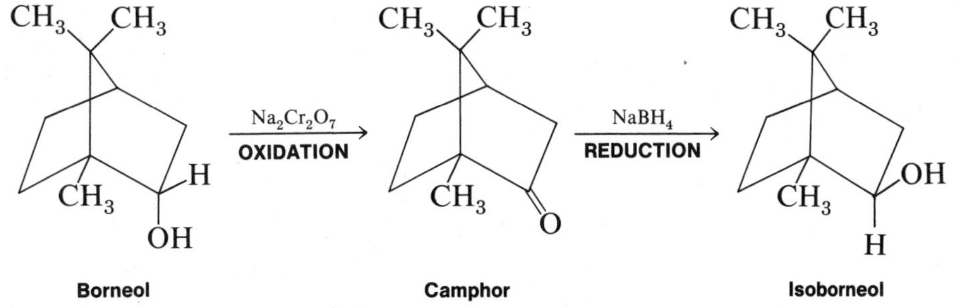

This experiment illustrates the use of an oxidizing agent (sodium dichromate) for converting a secondary alcohol (borneol) to a ketone (camphor). The camphor is purified by sublimation and then reduced by sodium borohydride to give the **isomeric** alcohol isoborneol. The isoborneol is also purified by sublimation. The spectra of borneol, camphor, and isoborneol are compared to detect structural differences and to determine the extent to which the final step produces a pure alcohol isomeric with the starting material.

OXIDATION OF BORNEOL WITH SODIUM DICHROMATE

Secondary alcohols are easily oxidized to ketones with sodium dichromate or other chromium(VI) compounds, such as chromium trioxide or sodium chromate.

The half-reactions and the overall reaction used in the oxidation of borneol with dichromate are as follows:

HALF-REACTION $\quad Cr_2O_7^{2-} + 14H^+ + 6e^- \longrightarrow 2Cr^{3+} + 7H_2O$

HALF-REACTION $\quad 3$ $\longrightarrow 3$ $+ 6H^+ + 6e^-$

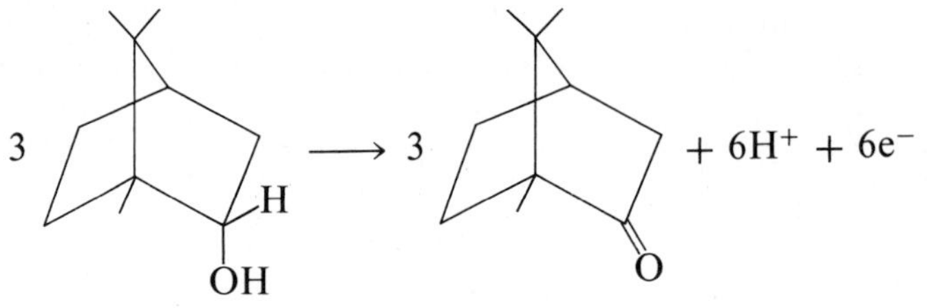

NET REACTION

3 $+ Cr_2O_7^{2-} + 8H^+ \longrightarrow 3$ $+ 2Cr^{3+} + 7H_2O$

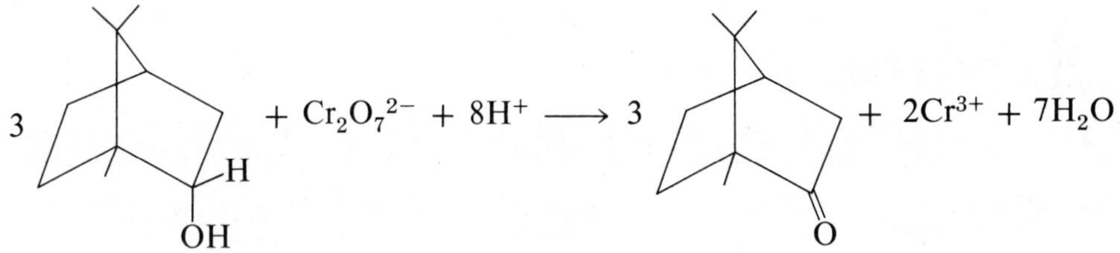

The mechanism for the oxidation is as follows:

$$Cr_2O_7^{2-} + H_2O \rightleftharpoons 2CrO_4^{2-} + 2H^+$$

Dichromate   **Chromate**

No net change
in oxidation
number of chromium

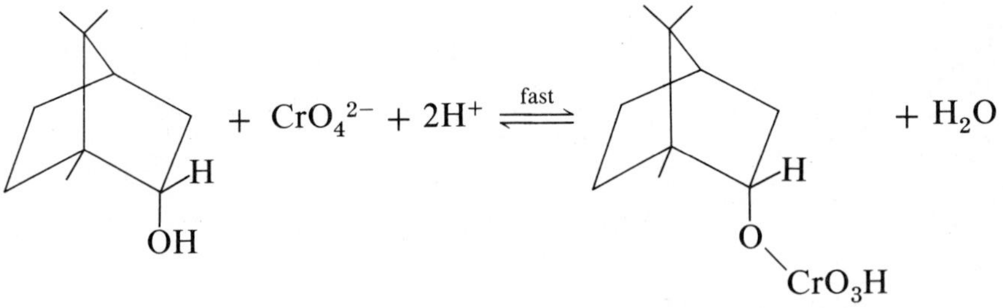

$+ CrO_4^{2-} + 2H^+ \xrightarrow{\text{fast}}$ $+ H_2O$

Borneol ester of chromic acid

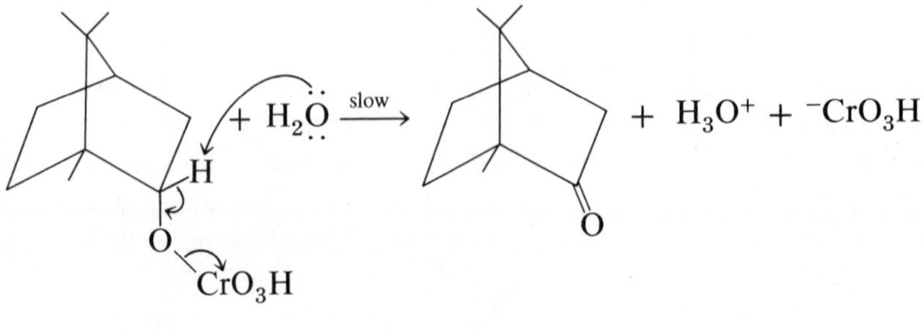

$+ H_2\ddot{O} \xrightarrow{\text{slow}}$ $+ H_3O^+ + {}^-CrO_3H$

REDUCTION OF CAMPHOR WITH SODIUM BOROHYDRIDE

Metal hydrides (sources of $H:^-$) of the Group III elements such as lithium aluminum hydride, $LiAlH_4$, and sodium borohydride, $NaBH_4$, are widely used in reducing carbonyl groups. Lithium aluminum hydride, for example, reduces many compounds containing carbonyl groups, such as aldehydes, ketones, carboxylic acids, esters, or amides, whereas sodium borohydride reduces only aldehydes and ketones. The reduced reactivity of borohydride allows it to be used even in alcohol and water solvents, whereas lithium aluminum hydride reacts violently with these solvents to produce hydrogen gas and thus must be used in nonhydroxylic solvents. In the present experiment, sodium borohydride is used because it is easily handled, and the results of reductions using either of the two reagents are essentially the same. The same care need not be taken in keeping sodium borohydride away from water as is required with lithium aluminum hydride.

The mechanism of action of sodium borohydride in reducing a ketone is as follows:

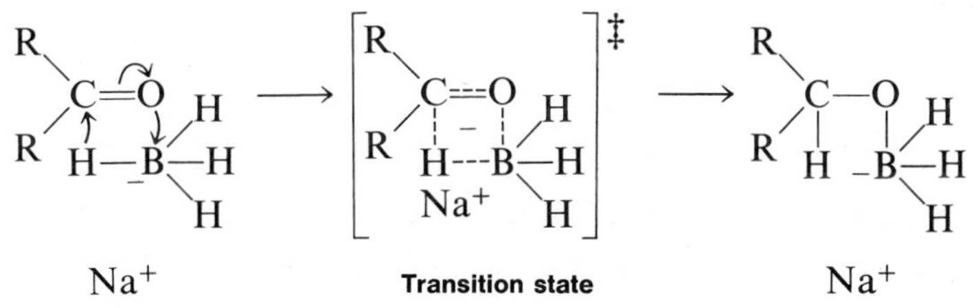

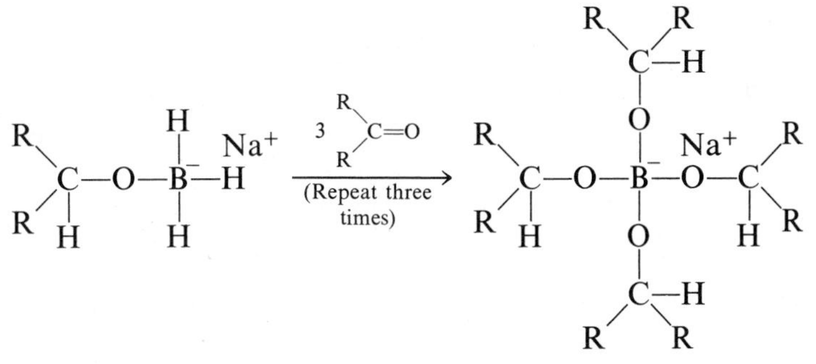

Note in this mechanism that all four hydrogen atoms are available as hydrides (H^-), and thus one mole of hydride can reduce four moles of ketone. All the steps are irreversible. Usually excess hydride is used because there is uncertainty regarding the purity of the material.

Once the final tetraalkoxyboron compound (I) is produced, it can be decomposed (along with excess hydride) at elevated temperatures as shown:

$$(R_2CH-O)_4B^-Na^+ + 4R'OH \longrightarrow 4R_2CHOH + (R'O)_4B^-Na^+$$

(I)

The stereochemistry of the reduction is very interesting. The hydride can approach the camphor molecule more easily from the bottom side (**endo** approach) than from the top side (**exo** approach). If attack occurs at the top, a large steric repulsion is created by one of the two **geminal** methyl groups. Attack at the bottom avoids this steric interaction.

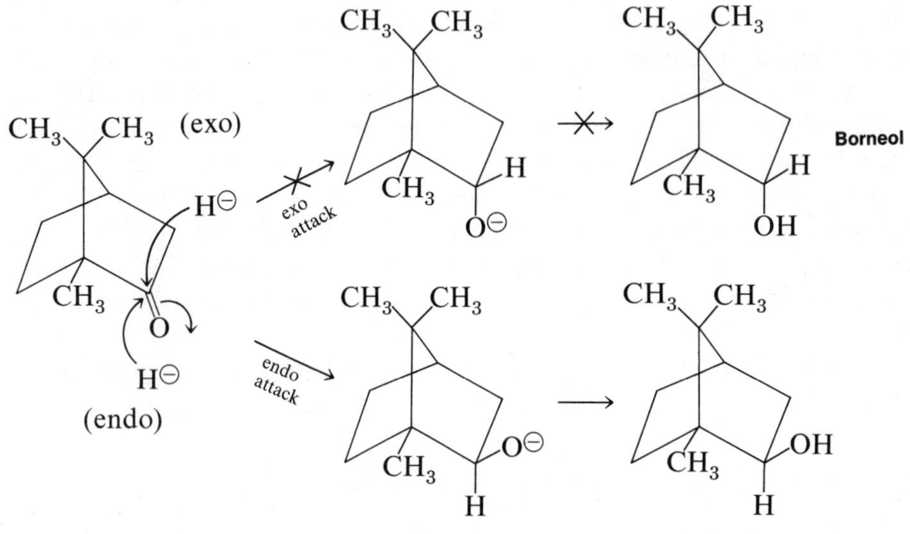

It is expected, therefore, that **isoborneol,** the alcohol produced from the attack at the **least**-hindered position, will **predominate but will not be the exclusive product** in the final reaction mixture. The percentage composition of the mixture can be determined by spectroscopy.

It is interesting to note that when the methyl groups are removed (as in 2-norbornanone), the top side (**exo** approach) is favored, and the opposite stereochemical result is obtained. Again, the reaction does not give exclusively one product.

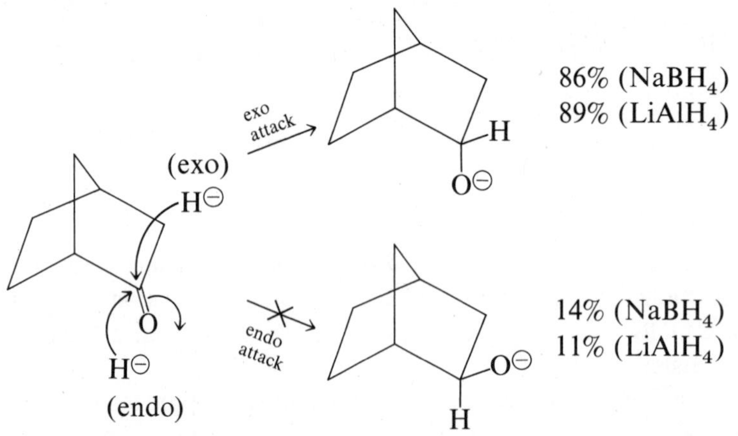

86% (NaBH$_4$)
89% (LiAlH$_4$)

14% (NaBH$_4$)
11% (LiAlH$_4$)

2-Norbornanone

Bicyclic systems such as camphor and 2-norbornanone react predictably according to steric influences. This effect has been termed **steric approach control.** In the reduction of simple acyclic and monocyclic ketones, however, the reaction seems to be influenced primarily by thermodynamic factors. This effect has been termed **product development control.** In the reduction of 4-*t*-butylcyclohexanone, the thermodynamically more stable product is produced by product development control.

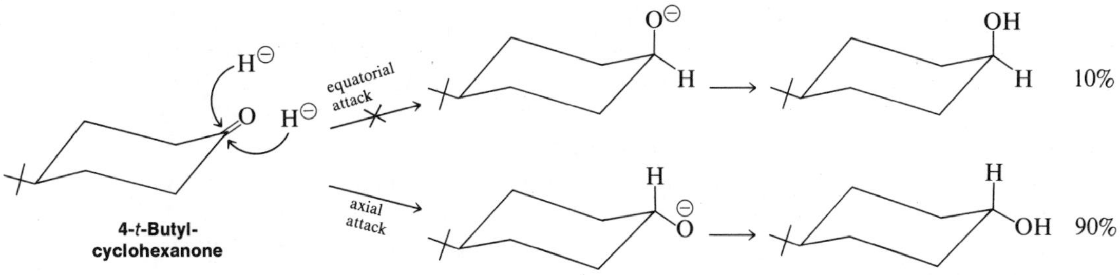

Equatorial product favored;
"product development control"

SPECIAL INSTRUCTIONS

It is necessary to read Techniques 1 through 5 and 13 before starting this experiment. You may also need to consult Technique 17 and Appendices 3, 4, and 5. The reactants and products are all highly volatile and must be stored in tightly closed containers. **Sodium dichromate and other chromium compounds are suspect carcinogens** (p. 12). Since volatile chromium compounds exist above acidic solutions of dichromate, you should work in a hood during the first part of the oxidation step (first paragraph of Procedure). If the instructor chooses, you may be asked to follow the progress of the oxidation of borneol to camphor by tlc (Experiment 19).

PROCEDURE

OXIDATION OF BORNEOL TO CAMPHOR

Dissolve 2.0 g of sodium dichromate dihydrate in 8 mL of water, and **carefully** add 1.6 mL of concentrated sulfuric acid with an eyedropper.[1] Place the oxidizing solution in

> **CAUTION: See p. 12.**

an ice bath. While this solution is cooling, dissolve 1.0 g of racemic borneol in 4 mL of ether in a 25-mL Erlenmeyer flask and cool it in an ice bath. Remove 6 mL of the sodium dichromate oxidizing mixture you have prepared, and **slowly** add it with an eyedropper

[1]You can monitor the progress of the oxidation by following the procedures outlined in Experiment 19.

to the **cold** ether solution over 10 minutes. **Swirl** the reaction mixture in the ice bath between additions and continue swirling for an additional 5 minutes following the final addition of oxidant. Pour the mixture into a separatory funnel and rinse the Erlenmeyer flask, first with a 10-mL portion of ether and then with 10 mL of water. **Add both** the rinsings to the separatory funnel.

The entire mixture will be very dark, and it may be difficult to see the interface between the aqueous and the organic phase. In this case, use a light source such as a lamp or flashlight to detect the interface. If there still is difficulty, begin to drain the lower aqueous phase until the interface is observed. When the lighter-colored ether phase passes into the narrow neck of the separatory funnel, it should become visible.

Complete the removal of the aqueous phase and pour the ether layer into a storage vessel. Return the aqueous layer to the separatory funnel and extract it with two successive 10-mL portions of ether. Each time add the ether phase to the storage vessel and return the aqueous layer to the separatory funnel.

Return the combined ether extracts to the separatory funnel and extract them with 10 mL of 5% sodium bicarbonate. A small amount of solid material may be produced at the interface. Carefully remove the lower aqueous layer and as much of this solid as possible without losing the ether layer. Finally, wash the ether layer with 10 mL of water and drain the lower aqueous layer. The ether layer contains the desired camphor. Pour it out of the top of the separatory funnel, decanting it away from any solids. Dry the ether layer thoroughly with a small amount (about 1 g) of anhydrous magnesium sulfate in a stoppered Erlenmeyer flask. Swirl the flask gently until the ether phase is **clear.**[2]

Decant the dry ether phase into a beaker and evaporate the solvent in the hood on a steam bath (using a boiling stone) or with a stream of **dry** air. When the ether has evaporated and a solid has appeared, remove the flask from the heat source **immediately,** otherwise the product may sublime prematurely and be lost. Weigh the product. At least 0.4 g of crude camphor should be obtained. Purify all the material by vacuum (aspirator) sublimation (Technique 13, Figure 13–2D, p. 644). Refer to the sublimation procedure given in Experiment 6, p. 65, and follow the details given there. A microburner is a convenient heating source, but great care must be taken to avoid fires. You must be certain that no one is using ether near your desk. You should sublime your camphor in portions. Be certain that the apparatus is **vacuum-tight** before the solid is sublimed. Scrape the purified material from the cold finger onto a piece of smooth paper with a spatula, weigh it, and calculate the percentage yield. Determine the melting point in a sealed capillary tube to prevent sublimation. The melting point of pure racemic camphor is 174 °C.[3] Save a small amount of purified camphor for an infrared spectrum determination. The remainder of the camphor is reduced in the next step to isoborneol. Store the compounds in tightly closed containers until needed. For the ir spectrum, dissolve the sample in carbon tetrachloride, place the solution between the salt plates, and mount the plates in a holder (see Technique 17, Method A, p. 663). An infrared spectrum for camphor is shown at the end of the experiment.

---

[2]Some solutions may not be clear. If you do not have a clear solution, gravity-filter the solution to remove as much solid as possible, then continue the procedure as indicated. The colored inorganic materials will be left behind when the camphor is sublimed.

[3]The observed melting point of camphor is often low. A small amount of impurity drastically reduces the melting point and increases the range.

REDUCTION OF CAMPHOR TO ISOBORNEOL

Use whatever amount of camphor you obtained in the first step unless it is less than 0.4 g. If it is less than 0.4 g, obtain some camphor from the supply shelf to supplement your yield. If it is more than 0.5 g, scale up the reagents appropriately from the amounts given below.

In a 25-mL Erlenmeyer flask dissolve 0.5 g of the racemic camphor obtained above in 2 mL of methanol. In portions, cautiously and intermittently add 0.3 g of sodium borohydride to the solution.[4] If necessary, cool the flask in an ice bath to keep the reaction mixture at room temperature. When all the borohydride is added, boil the contents of the flask on a steam bath for 2 minutes.

Pour the hot reaction mixture into about 15 g of chipped ice, using small portions of methanol to aid the transfer. When the ice melts, collect the white solid by suction, transfer the solid to a small flask, and add about 20 mL of ether to dissolve the product. Add a spatulaful of anhydrous magnesium sulfate to dry the solution, gravity-filter to remove the drying agent, and evaporate the solvent in a hood.

Purify the dry isoborneol by vacuum (aspirator) sublimation as before. Isoborneol rapidly sublimes under reduced pressure. Weigh the purified material and calculate the percentage yield. Determine the melting point (sealed tube); *pure* racemic isoborneol melts at 212 °C. Determine the infrared spectrum of the purified product by the method given above. Compare it with the spectra for borneol and isoborneol shown.

PERCENTAGE OF ISOBORNEOL AND BORNEOL OBTAINED FROM THE REDUCTION OF CAMPHOR

The percentage of each of the isomeric alcohols in the borohydride reduction mixture can be determined from the nmr spectrum.[5] (See Technique 17 Part B, p. 667, and Appendix 4.) The nmr spectra of the pure alcohols are shown on p. 158. The hydrogen on the carbon bearing the hydroxyl group appears at $\delta4.0$ (6.0τ) for borneol and $\delta3.6$ (6.4τ) for isoborneol. One can obtain the product ratio by integrating these peaks (using an expanded presentation) in the nmr spectrum of "isoborneol" obtained after the borohydride reduction. In the spectrum shown on p. 160, the isoborneol-borneol ratio 5:1 was obtained. The percentages obtained are 83% isoborneol and 17% borneol.

REFERENCES

Brown, H. C., and Muzzio, J. "Rates of Reaction of Sodium Borohydride with Bicyclic Ketones." *Journal of the American Chemical Society, 88* (1966): 2811.

Dauben, W. G., Fonken, G. J., and Noyce, D. S. "Stereochemistry of Hydride Reductions." *Journal of the American Chemical Society, 78* (1956): 2579.

---

[4]The sodium borohydride should be checked to see whether it is active. Place a small amount of powdered material in some methanol. Heat it on a steam bath. It should bubble vigorously if the hydride is active.

[5]The percentages can also be obtained by gas chromatography, using a Varian-Aerograph, Model A-90-P instrument. Use a 5-ft column of 5% Carbowax 4000 on acid-washed firebrick and operate the device at 110 °C with a 50 mL/min flow rate. The compounds are dissolved in a low-boiling solvent for analysis. The retention times for camphor, isoborneol, and borneol are 6, 10, and 12 minutes, respectively.

Flautt, T. J., and Erman, W. F. "The Nuclear Magnetic Resonance Spectra and Stereochemistry of Substituted Boranes." *Journal of the American Chemical Society, 85* (1963): 3212.

Markgraf, J. H. "Stereochemical Correlations in the Camphor Series." *Journal of Chemical Education, 44* (1967): 36.

QUESTIONS

1. Interpret the major absorption bands in the infrared spectra of camphor, borneol, and isoborneol.

2. Explain why the **gem**-dimethyl groups appear as separate peaks in the nmr spectrum of isoborneol although they are not resolved in borneol.

3. A sample of isoborneol prepared by reduction of camphor was analyzed by infrared spectroscopy and showed a strong band at 1760 cm^{-1} (5.7μ). This result was unexpected. Why?

4. Primary alcohols can be oxidized to carboxylic acids with dichromate in acid solution. Determine the balanced equation for this reaction:

$$RCH_2OH + Cr_2O_7^{2-} \longrightarrow RCOOH + Cr^{3+}$$

5. Are the structures for borneol and camphor given in this experiment different from the structures given in Experiment 19? Explain.

6. The observed melting point of camphor is often low. Look up the molal freezing-point-depression constant K for camphor and calculate the expected depression of the melting point of a quantity of camphor that contains 0.5 molal impurity. *Hint:* Look in a general chemistry book under freezing-point depression or colligative properties of solutions.

7. The peak assignments are shown on the carbon-13 magnetic resonance (cmr) spectrum of camphor. Using these assignments as a guide, assign all the peaks in the cmr spectra of borneol and isoborneol.

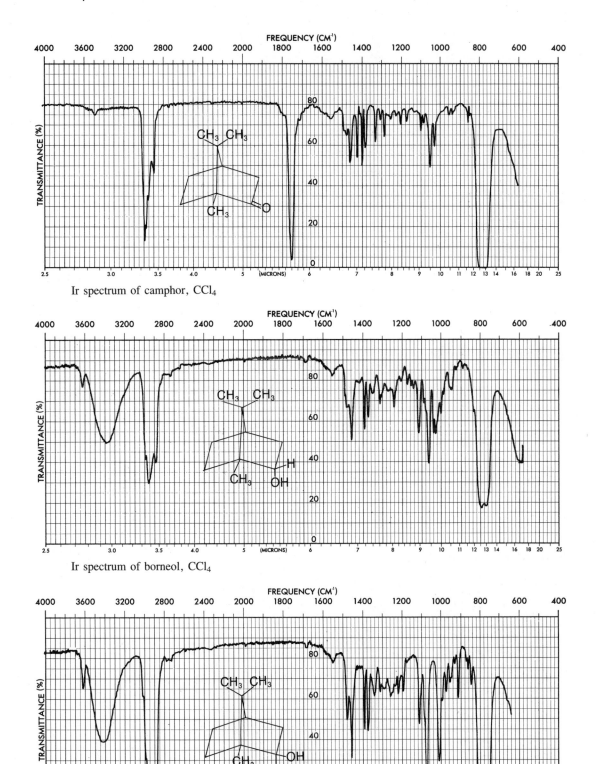

Ir spectrum of camphor, CCl₄

Ir spectrum of borneol, CCl₄

Ir spectrum of isoborneol, CCl₄

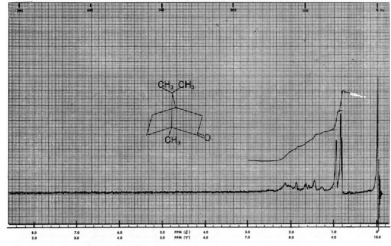

Nmr spectrum of camphor, CCl$_4$

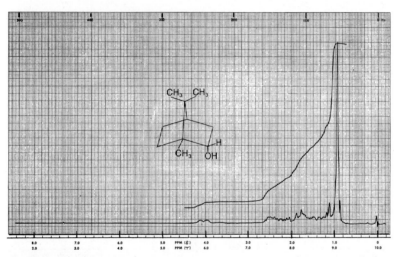

Nmr spectrum of borneol, CDCl$_3$

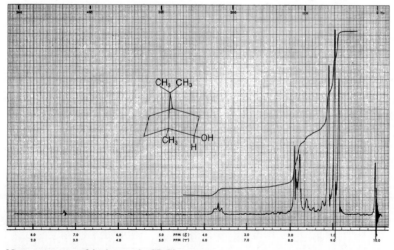

Nmr spectrum of isoborneol, CDCl$_3$

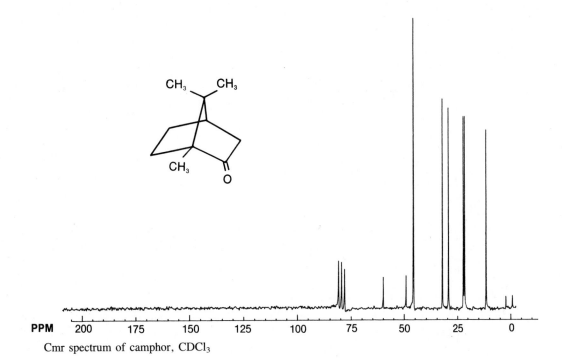

Cmr spectrum of camphor, CDCl₃

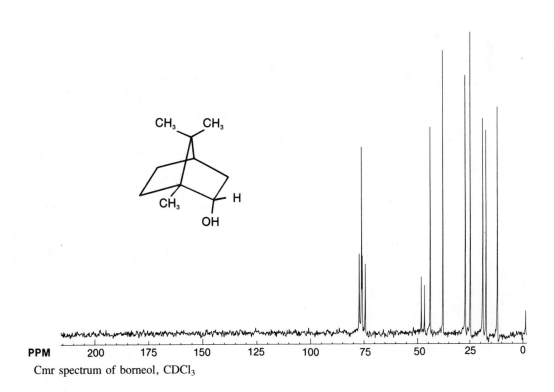

Cmr spectrum of borneol, CDCl₃

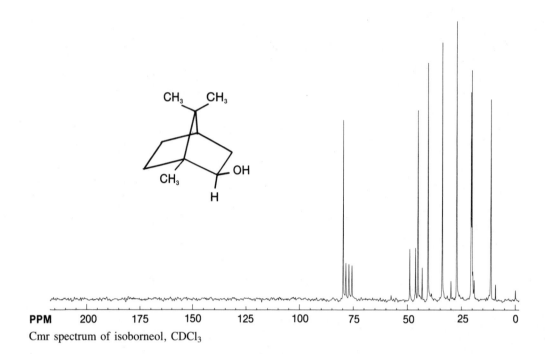

PPM   200   175   150   125   100   75   50   25   0

Cmr spectrum of isoborneol, CDCl$_3$

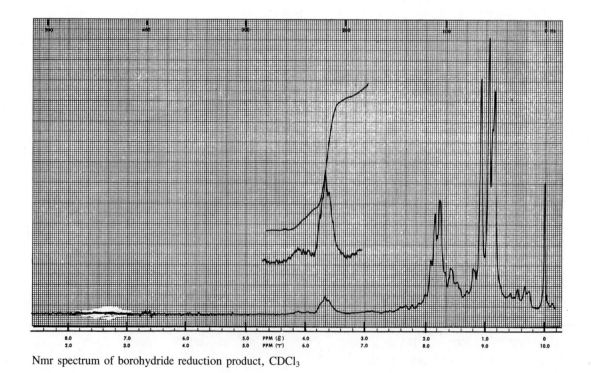

Nmr spectrum of borohydride reduction product, CDCl$_3$

Essay
STEREOCHEMICAL THEORY OF ODOR

The human nose has an almost unbelievable ability to distinguish odors. Just consider for a few moments the different substances you are able to recognize by odor alone. Your list should be a very long one. A person with a trained nose, a perfumer, for instance, can often recognize even individual components in a mixture. Who has not met at least one cook who could sniff almost any culinary dish and identify the seasonings and spices that were used? The olfactory centers in the nose can identify odorous substances even in very small amount. With some substances, studies have shown that as little as one ten-millionth of a gram (10^{-7} g) can be perceived. Many animals, for example, dogs and insects, have an even lower threshold of smell than humans (see the essay on pheromones that precedes Experiment 16).

There have been many theories of odor, but few have persisted very long. Strangely enough, one of the oldest theories, although in modern dress, is still the most current theory. Lucretius, one of the early Greek atomists, suggested that substances having odor gave off a vapor of tiny ''atoms,'' all of the same shape and size, and that they gave rise to the perception of odor when they entered pores in the nose. The pores would have to be of various shapes and the odor perceived would depend on which pores the atoms were able to enter. We now have many similar theories regarding the action of drugs (receptor-site theory) and the interaction of enzymes with their substrates (the lock-and-key hypothesis).

A substance must have certain physical characteristics to have the property of odor. First, it must be volatile enough to give off a vapor that can reach the nostrils. Second, once it reaches the nostrils, it must be somewhat water-soluble, even if only to a small degree, so that it can pass through the layer of moisture (mucus) that covers the nerve endings in the olfactory area. Third, it must also have lipid solubility to allow it to penetrate the lipid (fat) layers that form the surface membranes of the nerve cell endings.

Once we pass these criteria, we come to the heart of the question. Why do substances have different odors? In 1949, R. W. Moncrieff, a Scotsman, resurrected Lucretius' hypothesis. He proposed that in the olfactory area of the nose there is a system of receptor cells of several different types and shapes. He further suggested that each receptor site corresponded to a different type of primary odor. Molecules that would fit these receptor sites would display the characteristics of that primary odor. It would not be necessary for the entire molecule to fit into the receptor, so that for larger molecules, any portion might fit into the receptor and activate it. Molecules having complex odors would presumably be able to activate several different types of receptors.

Moncrieff's hypothesis has been strengthened substantially by the work of J. E. Amoore, who began studying the subject as an undergraduate at Oxford in 1952. After an extensive search of the chemical literature, Amoore concluded that there were only seven basic primary odors. By sorting molecules with similar odor types, he even formulated possible shapes for the seven necessary receptors. For instance, from the literature, he culled more than 100 compounds that were described as having a "camphoraceous" odor. Comparing the sizes and shapes of all these molecules, he postulated a three-dimensional shape for a camphoraceous receptor site. Similarly, he derived shapes for the other six receptor sites. The seven primary receptor sites he formulated are shown in the figure, along with a typical prototype molecule of the appropriate shape to fit the receptor. The shapes of the sites are shown in perspective. Pungent and putrid odors are not thought to require a particular shape in the odorous molecular but rather to need a particular type of charge distribution.

You can verify quickly that compounds with molecules of roughly similar shape have similar odors if you compare nitrobenzene and acetophenone with benzaldehyde or *d*-camphor and hexachloroethane with cyclooctane. Each group of substances has the same basic odor **type** (primary), but the individual molecules differ in the **quality** of the odor. Some of the odors are sharp, some pungent, others sweet, and so on. The second group of substances all have a camphoraceous odor and the molecules of these substances all have approximately the same shape.

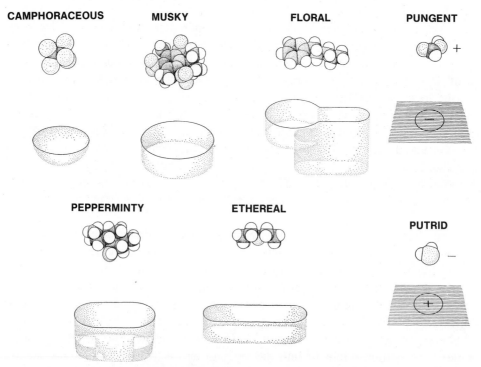

CAMPHORACEOUS   **MUSKY**   **FLORAL**   **PUNGENT**

PEPPERMINTY   **ETHEREAL**   **PUTRID**

An interesting corollary to the Amoore theory would be the postulate that if the receptor sites are chiral, then optical isomers (enantiomers) of a given substance might have **different** odors. This circumstance proves true in several cases. It is true for (+)- and (−)-carvone; we investigate the idea in Experiment 21 in this textbook.

Several workers have tested Amoore's hypothesis by experiment. The results of these studies are generally favorable to the hypothesis—so favorable that some chemists now elevate the hypothesis to the level of theory. In several cases, researchers have been able to ''synthesize'' odors almost indistinguishable from the real thing by properly blending primary odor substances. The primary odor substances used were unrelated to the chemical substances composing the natural odor. These experiments, and others, are described in the articles listed at the end of this essay.

REFERENCES

Amoore, J. E. *The Molecular Basis of Odor*. American Lecture Series Publication, No. 773. Springfield, Ill. Thomas, 1970.

Amoore, J. E., Johnston, J. W., Jr., and Rubin, M. ''The Stereochemical Theory of Odor.'' *Scientific American, 210* (February 1964): 1.

Amoore, J. E., Rubin, M., and Johnston, J. W., Jr. ''The Stereochemical Theory of Olfaction.'' *Proceedings of the Scientific Section of the Toilet Goods Association* (Special Supplement to No. 37) (October 1962): 1–47.

Burton, R. *The Language of Smell*. London: Routledge & Kegan Paul, 1976.

Moncrieff, R. W. *The Chemical Senses*. London: Leonard Hill, 1951.

Theimer, E. T., editor. *Fragrance Chemistry*. New York: Academic Press, 1982.

Experiment 21
Spearmint and Caraway Oil: (−)- and (+)-Carvones

Stereochemistry
Vacuum Fractional Distillation
Gas Chromatography, Spectroscopy
Optical Rotation, Refractometry

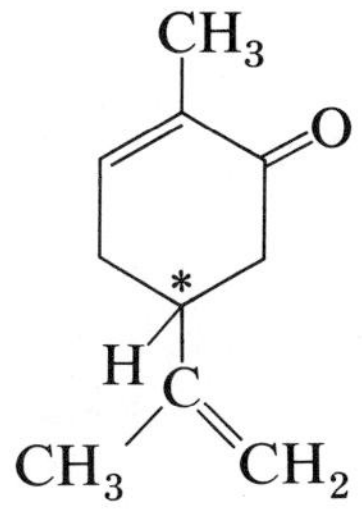

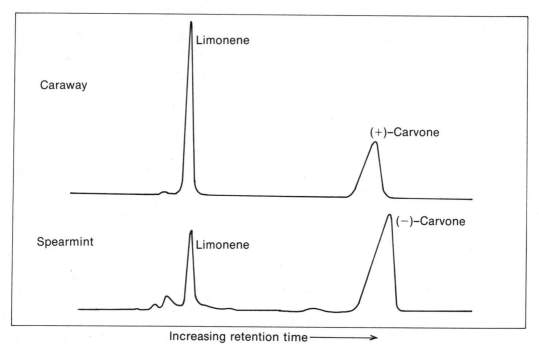

Gas chromatograms of caraway and spearmint oil

In this experiment, we isolate (+)-carvone from caraway-seed oil or (−)-carvone from spearmint oil by vacuum fractional distillation. The odors of these optical isomers are distinctly different from each other. The presence of one or the other of these isomers is responsible for the characteristic odors of each of the two oils. The difference in their odors is to be expected, since the odor receptors in the nose are chiral (see the essay that precedes this experiment). Although we should expect the optical rotations of the isomers (enantiomers) to be of opposite sign, the other physical properties should be identical. Thus, for both (+)- and (−)-carvone, we predict that the infrared and nuclear magnetic resonance spectra, the gas chromatographic retention times, and the refractive indices should all be identical, within experimental error. Hence, the only difference in properties one will observe for the two carvones are the odors and the signs of rotation in a polarimeter.

    Caraway-seed oil contains mainly limonene and (+)-carvone. The gas chromatogram for this oil is shown in the figure. Limonene (left-hand peak) has the lower retention time. The (+)-carvone (bp 230 °C) can easily be separated from the lower-boiling limonene (bp 177 °C) by vacuum fractional distillation. The separation is relatively easy due to the large boiling-point difference. **Spearmint oil** contains mainly (−)-carvone with a smaller amount of limonene and very small amounts of the lower-boiling terpenes α- and β-phellandrene. The gas chromatogram for this oil is also shown in the figure. The (−)-carvone can also easily be separated in this oil from the three lower-boiling components, and α- and β-phellandrenes and limonene, by vacuum fractional distillation. These three terpenes, however, are not easily separated because of their similar boiling points.

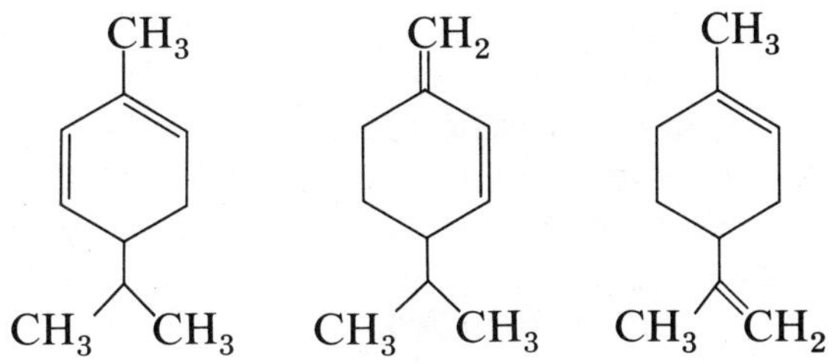

α-Phellandrene   β-Phellandrene   Limonene

SPECIAL INSTRUCTIONS

Read the essay on the stereochemical theory of odor, which precedes this experiment. In addition, it will be necessary to read Techniques 6, 7, 9, 12, 15, and 16. To conduct a successful fractional distillation under reduced pressure, one must be certain that the aspirator and hoses are in good condition. It is best to use a manometer in the system during the entire distillation.

> **NOTE TO THE INSTRUCTOR:** This experiment may be scheduled along with another experiment. Half the class should conduct this experiment while the other half conducts an experiment not requiring aspirators. In this way, the water pressure will be high enough to get adequate pressures at the aspirator. Alternatively, or in addition, the students may work in pairs. The column in the gas chromatograph must be heated and equilibrated well before the laboratory period.

PROCEDURE

VACUUM FRACTIONAL DISTILLATION

Place 30 mL of spearmint or caraway-seed oil in a 100-mL round-bottomed flask. Assemble the vacuum fractional distillation apparatus as shown in the accompanying figure. Use a trap arrangement and manometer as shown in Technique 9, Figure 9–6, p. 593. The distillation apparatus is like that in Technique 6, Figure 6–7, p. 559, except that a fractionating column, filled with stainless steel sponge (Technique 7, Figure 7–7C, p. 574), is inserted between the Claisen head and the distilling head. As explained in Technique 6, Section 6.6, p. 558, it is very important that the apparatus is assembled carefully. All the rubber tubing must be free of cracks. The ebulliator tube must fit securely in an adapter or rubber stopper. Wrap the fractionating column and the distilling head with glass wool for insulation.

When the apparatus has been assembled, conduct the distillation as outlined in Technique 6, Section 6.7, p. 561, and Technique 7, Section 7.6, p. 575. Use a heating mantle or an oil bath to heat the distilling flask. It is important to get a good vacuum, preferably about 20 mm pressure. If several students are using the aspirators on a given bench, the obtainable pressure may not be low enough to go ahead with the distillation. See Technique 6, Section 6.1, p. 551, for the effect of pressure on the boiling point when a pressure different from 20 mm is obtained.

With spearmint oil, the first fraction distills up to about 82 °C at 20 mm pressure. When the low-boiling fraction has been removed, the temperature will drop somewhat, and the rate of distillation takeoff in the receiver will slow considerably. The temperature of the heating source should then be raised, and the temperature at the thermometer should once again begin to rise. At that point, the distillation should be stopped and the heating source lowered. About 4 to 5 g of distillate may be collected in this first fraction.

To change receiving flasks, open the screw clamps on the trap assembly and ebulliator and turn off the aspirator. Remove the receiver and replace it with a clean one.

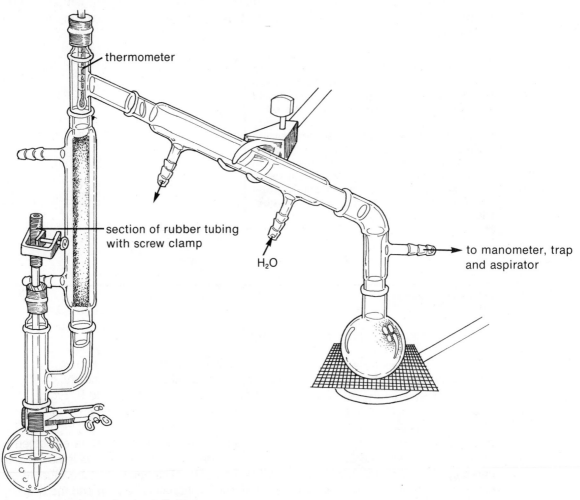

Apparatus for vacuum fractional distillation

Following this, continue the distillation (Technique 6, Section 6.7, p. 561). Place the heating source back under the distilling flask and raise its temperature. Collect the second fraction in the new receiver. This fraction, which contains the carvone, will boil from about 82 °C to about 113 °C at 20 mm. Most of this material will boil from about 105 to 113 °C at 20 mm of pressure. About 16 to 17 g of distillate may be obtained. This fraction, even though it may contain some low-boiling material, should be nearly pure carvone.

The caraway oil can be distilled in the same manner as the spearmint oil, and a larger first fraction (limonene) is obtained (10–11 g).[1] Again the temperature at the thermometer drops somewhat near the end of the first fraction. The temperature of the heating source is raised as before. When the temperature of the vapor again rises, a second fraction (9–10 g) of (+)-carvone is collected in another receiver.

ANALYSIS OF THE CARVONES

The samples obtained by distillation should be analyzed by the following methods; the instructor will indicate which methods to use. Compare your results with results obtained by someone who used a different oil. After the analyses are completed, submit the carvone to the instructor in a labeled **glass** vial.

Odor. About 8 to 10% of the population cannot detect the difference in the odors of the optical isomers. Most people, however, find the difference quite obvious.

Gas Chromatography. See Technique 12. With the help of the instructor or assistant, analyze each of the fractions on any relatively nonpolar column such as a 12 ft × ⅛ in. 30% GE SE-30 on a Chromosorb W column.[2] Determine the retention times of the components. Calculate the purity of the carvone sample by the method explained in Technique 12, Section 12.10, p. 638.

Polarimetry. See Technique 15. With the help of the instructor or assistant, obtain the observed optical rotation α of the (−)-carvone from spearmint and both the (+)-limonene and (+)-carvone from caraway oil. The sample or samples may all be analyzed directly, without dilution, in the smallest cell available, which may have a path length l of 0.5 dm. The specific rotation is calculated from the relation $[\alpha] = \alpha/(c)(l)$, given in Technique 15. The concentration c will equal the density of the substances analyzed at 20 °C. The values are 0.9608 g/mL for (+)-carvone, 0.8411 g/mL for (+)-limonene, and 0.9593 g/mL for (−)-carvone. The literature values for the specific rotations are as follows: $[\alpha]_D^{20} = +61.7°$ for (+)-carvone, $-62.5°$ for (−)-carvone, and $+125°$ for (+)-limonene. They are not identical because trace amounts of impurities are present.

Refractive Index. See Technique 16. Obtain the refractive index for the carvone sample. At 23 °C, the (+)- and (−)-carvones have the same refractive index, equal to 1.4950.

---

[1]It has been noted in the literature that caraway oil may foam in certain instances, especially at higher pressures. At pressures below 40 mm, foaming should not be expected. If foaming is experienced, it will usually stop once the limonene fraction (lower bp) has been removed.

[2]A 6 ft × ⅛ in. column of 15% QF-1 on Chromosorb W 80/100 support can also be used.

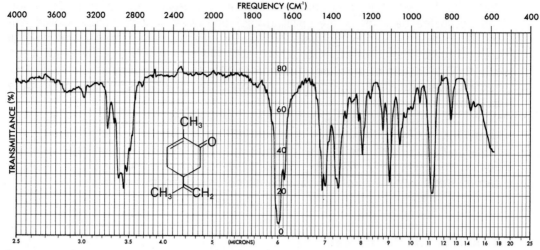

Ir spectrum of (+)-carvone from caraway oil, neat

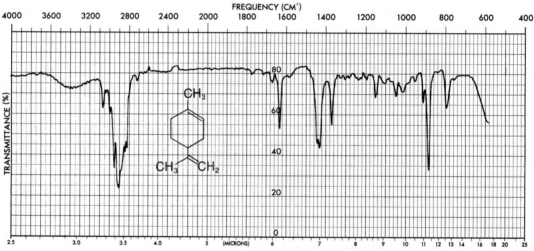

Ir spectrum of (+)-limonene, neat

Spectroscopy. See Technique 17. Obtain the infrared spectrum of the (−)-carvone sample from spearmint and of the (+)-carvone and (+)-limonene samples from caraway. Compare the carvone and limonene spectra with those shown. The nmr spectra for (−)-carvone and (+)-limonene are shown in this experiment. At the option of the instructor, obtain an nmr spectrum of the carvone for comparison.

REFERENCES

Friedman, L., and Miller, J. G. "Odor Incongruity and Chirality." *Science, 172* (1971): 1044.

Murov, S. L., and Pickering, M. "The Odor of Optical Isomers." *Journal of Chemical Education, 50* (1973): 74.

Russell, G. F., and Hills, J. I. "Odor Differences between Enantiomeric Isomers." *Science, 172* (1971): 1043.

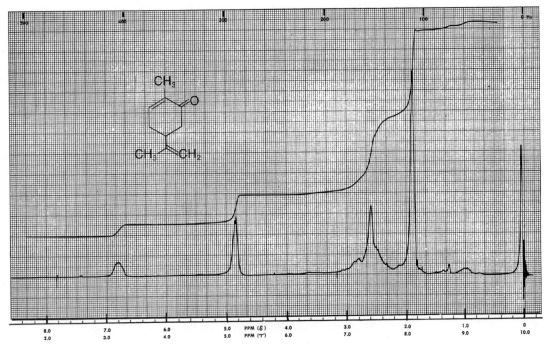

Nmr spectrum of (−)-carvone from spearmint oil

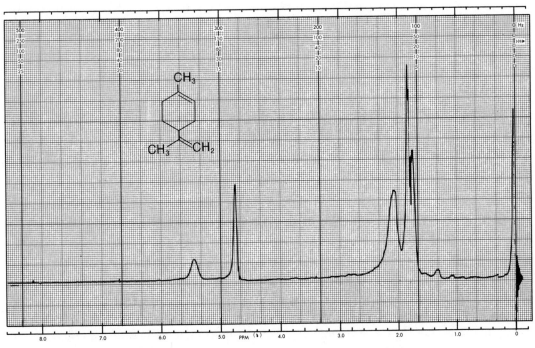

Nmr spectrum of (+)-limonene

QUESTIONS

1. Interpret the ir and the nmr spectra for carvone and the ir spectrum for limonene.
2. How could you establish that the difference in odor of the distilled carvones was not due to a trace component? Assume that the impurity is not abundant enough to appear in the spectra. In considering your answer, read Technique 12, Section 12.9.
3. Assign the chiral (asymmetric) center in α-phellandrene, β-phellandrene, and limonene.

Experiment 22
Reactivities of Some Alkyl Halides

S_N1/S_N2 Reactions

The reactivities of alkyl halides in nucleophilic substitution reactions depend on two important factors: reaction conditions and substrate structure.

SODIUM IODIDE OR POTASSIUM IODIDE IN ACETONE

A reagent composed of sodium iodide or potassium iodide dissolved in acetone is useful in classifying alkyl halides according to their reactivity in an S_N2 reaction. Iodide ion is an excellent nucleophile, and acetone is a nonpolar solvent. The tendency to form a precipitate increases the completeness of the reaction. Sodium iodide and potassium iodide are soluble in acetone, but the corresponding bromides and chlorides are not soluble. Consequently, as bromide ion or chloride ion is produced, it is precipitated from the solution. According to LeChâtelier's principle, the precipitation of a product from the reaction solution drives the equilibrium toward the right; such is the case in the reaction described here:

$$R\text{—}Cl + Na^+I^- \longrightarrow RI + NaCl\downarrow$$
$$R\text{—}Br + Na^+I^- \longrightarrow RI + NaBr\downarrow$$

SILVER NITRATE IN ETHANOL

A reagent composed of silver nitrate dissolved in ethanol is useful in classifying alkyl halides according to their reactivity in an S_N1 reaction. Nitrate ion is a poor nucleophile, and ethanol is a moderately powerful ionizing solvent. The silver ion, because of its ability to coordinate the leaving halide ion to form a silver halide precipitate, greatly assists the ionization of the alkyl halide. Again, a precipitate as one of the reaction products also enhances the reaction.

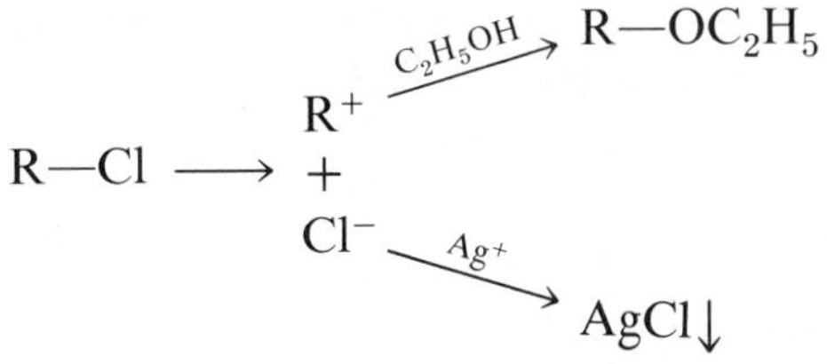

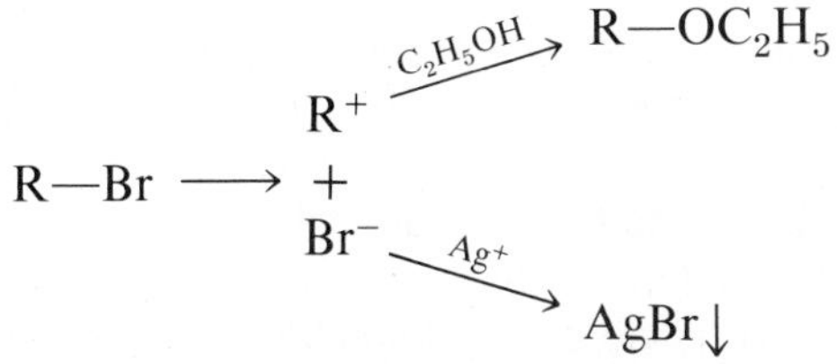

$$R-Cl \longrightarrow \begin{matrix} R^+ \\ + \\ Cl^- \end{matrix}$$

$$R^+ \xrightarrow{C_2H_5OH} R-OC_2H_5$$

$$Cl^- \xrightarrow{Ag^+} AgCl\downarrow$$

$$R-Br \longrightarrow \begin{matrix} R^+ \\ + \\ Br^- \end{matrix}$$

$$R^+ \xrightarrow{C_2H_5OH} R-OC_2H_5$$

$$Br^- \xrightarrow{Ag^+} AgBr\downarrow$$

SPECIAL INSTRUCTIONS

Before beginning this experiment, review the chapters dealing with nucleophilic substitution in your lecture textbook. The experiment requires very little time, and it may be performed along with another, longer experiment. Some compounds used in this experiment, particularly crotyl chloride, chloroacetone, and benzyl chloride, are powerful lachrymators, causing much eye irritation and the formation of tears. The tests with these substances should be done in the hood, and the compounds must be disposed of in a hood sink or appropriate waste container.

PROCEDURE

SODIUM IODIDE IN ACETONE

Label a series of 10 clean dry test tubes (10 × 75-mm test tubes may be used) from 1 to 10. In each test tube place 0.2 mL of one of the following halides: (1) 2-chlorobutane; (2) 2-bromobutane; (3) 2-chloro-2-methylpropane (*t*-butyl chloride); (4) 1-chlorobutane; (5) crotyl chloride, ($CH_3CH{=}CHCH_2Cl$) (Note: see Special Instructions, above); (6) chloroacetone, ($ClCH_2COCH_3$) (Note: see Special Instructions, above); (7) benzyl chloride, (α-chlorotoluene) (Note: see Special Instructions, above); (8) bromobenzene; (9) bromocyclohexane; and (10) bromocyclopentane.

    Add to the material in each test tube 2 mL of a 15% NaI-in-acetone solution, noting the time of each addition. After the addition, shake the test tube well to ensure adequate mixing of the alkyl halide and the solvent. Record the times needed for any precipitates to form. After about 5 minutes, place any test tubes that do not contain a precipitate in a 50 °C water bath. Be careful not to allow the temperature of the water bath to exceed 50 °C, since the acetone will evaporate or boil out of the test tube or both. At the end of 6 minutes, cool the test tubes to room temperature and note whether a reaction has occurred. Record the results and explain why each compound has the reactivity that you observed. Explain the reactivities in terms of structure.

Generally, reactive halides give a precipitate within 3 minutes, moderately reactive halides give a precipitate when heated, and unreactive halides do not give a precipitate even after being heated.

SILVER NITRATE IN ETHANOL

Label a series of 10 clean dry test tubes from 1 to 10, as described in the previous section. Place 0.2 mL of the appropriate halide in each test tube, as described for the sodium iodide test.

Add 2 mL of a 1% ethanolic silver nitrate solution to the material in each test tube, noting the time of each addition. After the addition, shake the test tubes well to ensure adequate mixing of the alkyl halide and the solvent. Record the times required for any precipitates to form.

After about 5 minutes, heat each solution that has not yielded any precipitate to boiling on the steam bath. Note whether a precipitate forms.

Record the results and explain why each compound has the reactivity observed. Explain the reactivities in terms of structure, as before.

Again, reactive halides will give a precipitate within 3 minutes, moderately reactive halides will give a precipitate when heated, and unreactive halides do not yield a precipitate even after being heated.

QUESTIONS

1. In the tests with sodium iodide in acetone and silver nitrate in ethanol, why should 2-bromobutane react faster than 2-chlorobutane?

2. In the test with silver nitrate in ethanol, why should the cyclopentyl compound react faster than the cyclohexyl compound?

3. When benzyl chloride is treated with sodium iodide in acetone, it reacts much faster than 1-chlorobutane, even though both compounds are primary alkyl chlorides. Explain this rate difference.

4. How do you expect the four compounds shown below to compare in behavior in the two tests?

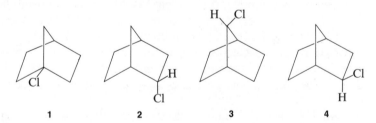

1   2   3   4

5. How do you predict that chlorocyclopropane will behave in each of these tests?

Experiment 23
Synthesis of *n*-Butyl Bromide and *t*-Pentyl Chloride

Synthesis of Alkyl Halides
Extraction
Simple Distillation

The synthesis of two alkyl halides from alcohols is the basis for this experiment. In the first procedure, a primary alkyl halide, *n*-butyl bromide, is prepared as shown in equation 1.

$$CH_3CH_2CH_2CH_2OH + NaBr + H_2SO_4 \longrightarrow$$

    _n_-Butyl alcohol

$$CH_3CH_2CH_2CH_2Br + NaHSO_4 + H_2O$$

      _n_-Butyl bromide   (1)

In the second procedure, a tertiary alkyl halide, *t*-pentyl chloride (*t*-amyl chloride), is prepared as shown in equation 2.

$$\underset{\underset{\textstyle OH}{|}}{CH_3CH_2\overset{\overset{\textstyle CH_3}{|}}{C}CH_3} + HCl \longrightarrow \underset{\underset{\textstyle Cl}{|}}{CH_3CH_2\overset{\overset{\textstyle CH_3}{|}}{C}CH_3} + H_2O$$

    _t_-Pentyl alcohol   **_t_-Pentyl chloride**   (2)

These reactions are an interesting contrast in mechanisms. The *n*-butyl bromide synthesis proceeds by an S_N2 mechanism, while *t*-pentyl chloride is prepared by an S_N1 reaction.

n-BUTYL BROMIDE

The alkyl halide *n*-butyl bromide can easily be prepared by allowing *n*-butyl alcohol to react with sodium bromide and sulfuric acid by equation 1. The sodium bromide reacts with sulfuric acid to produce hydrobromic acid. Excess sulfuric acid serves to shift the equilibrium, and thus to speed the reaction, by producing a higher concentration of hydrobromic acid. The sulfuric acid also protonates the hydroxyl group of *n*-butyl alcohol so that water is displaced rather than the hydroxide ion, OH^-. The acid also protonates the water as it is produced in the reaction and thus deactivates this nucleophile. This deactivation keeps the alkyl halide from being converted back to the alcohol by nucleophilic attack of water.

173

The reaction proceeds via an S_N2 mechanism as follows:

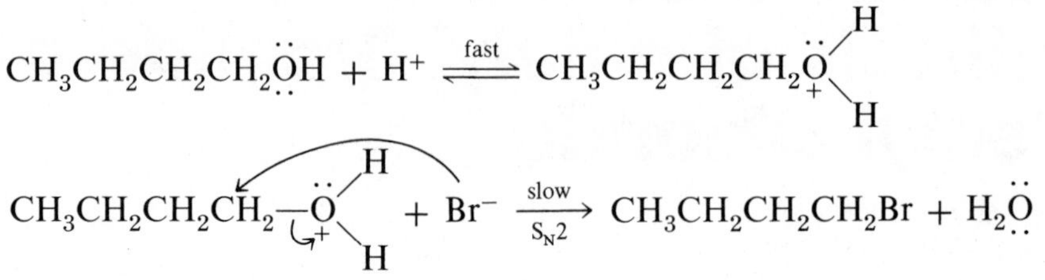

$$CH_3CH_2CH_2CH_2\overset{..}{\underset{..}{O}}H + H^+ \overset{fast}{\rightleftharpoons} CH_3CH_2CH_2CH_2\overset{H}{\underset{+}{\overset{..}{O}}}{\diagdown}H$$

$$CH_3CH_2CH_2CH_2{-}\overset{H}{\underset{+}{\overset{..}{O}}}{\diagdown}H + Br^- \overset{slow}{\underset{S_N2}{\longrightarrow}} CH_3CH_2CH_2CH_2Br + H_2\overset{..}{\underset{..}{O}}$$

Primary substrates such as *n*-butyl alcohol usually react by the S_N2 mechanism.

During the isolation of the *n*-butyl bromide, the crude product is washed with sulfuric acid to remove any remaining *n*-butyl alcohol. The sulfuric acid also can remove the by-products of the reaction, such as 1-butene and dibutyl ether, by making use of the basic character of these by-products. Alkyl halides are not basic, however.

t-PENTYL CHLORIDE

The alkyl halide *t*-pentyl chloride can easily be prepared by allowing *t*-pentyl alcohol to react with hydrochloric acid according to equation 2. The reaction is conducted in a separatory funnel. As the reaction proceeds, the insoluble alkyl halide forms as an upper phase.

The reaction proceeds via an S_N1 mechanism as follows:

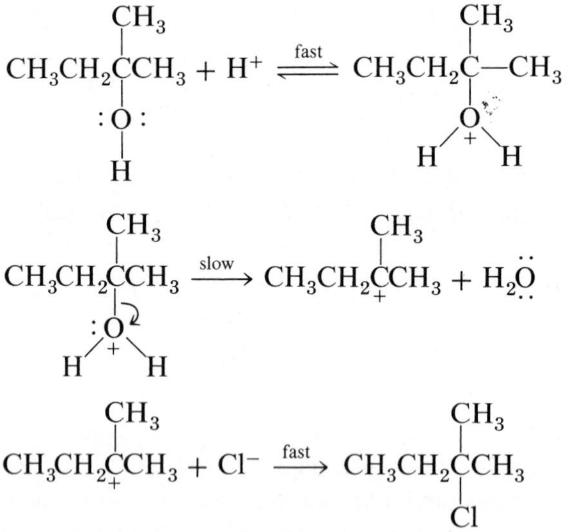

$$CH_3CH_2\underset{\overset{|}{\underset{H}{\overset{|}{:O:}}}}{\overset{CH_3}{\overset{|}{C}}}CH_3 + H^+ \overset{fast}{\rightleftharpoons} CH_3CH_2\underset{\underset{H}{\overset{+}{\overset{..}{O}}}H}{\overset{CH_3}{\overset{|}{C}}}{-}CH_3$$

$$CH_3CH_2\underset{\overset{|}{\underset{H}{\overset{+}{\overset{..}{O}}}H}}{\overset{CH_3}{\overset{|}{C}}}CH_3 \overset{slow}{\longrightarrow} CH_3CH_2\overset{CH_3}{\underset{+}{\overset{|}{C}}}CH_3 + H_2\overset{..}{\underset{..}{O}}$$

$$CH_3CH_2\overset{CH_3}{\underset{+}{\overset{|}{C}}}CH_3 + Cl^- \overset{fast}{\longrightarrow} CH_3CH_2\underset{\overset{|}{Cl}}{\overset{CH_3}{\overset{|}{C}}}CH_3$$

Tertiary substrates such as *t*-pentyl alcohol usually react by the S_N1 mechanism.

A small amount of an alkene, 2-methyl-2-butene, is produced as a by-product in this reaction. If sulfuric acid had been used, as it was for *n*-butyl bromide, a considerable amount of this alkene would have been produced.

SPECIAL INSTRUCTIONS

Before starting, review the appropriate chapters in your textbook. In addition, it is necessary to have read Techniques 1, 5, and 6 before starting this experiment. As your instructor indicates, perform either the *n*-butyl bromide or the *t*-pentyl chloride procedure.

Procedure 23A
n-Butyl Bromide

Place 24.0 g of sodium bromide in a 250-mL round-bottomed flask and add 25 mL of water and 17 mL (density 0.81 g/mL) of *n*-butyl alcohol (1-butanol). Cool the mixture in an ice bath and slowly add 20 mL (density 1.84 g/mL) of concentrated sulfuric acid with continuous swirling in the ice bath. Add several boiling stones to the mixture. Assemble the apparatus as shown in the figure. The inverted funnel in the beaker acts as a trap to absorb the hydrogen bromide gas evolved during the reaction period. An alternative trap arrangement is shown in Experiment 16. Place some water and sodium hydroxide pellets in the beaker. The funnel must be adjusted so that it projects only **slightly** below the surface of the liquid in the beaker. Heat the mixture with a heating mantle, an oil bath, or a flame until the mixture begins to reflux gently. Heat the mixture under reflux for 30 minutes. Two layers will form during this time.

At the end of this reflux period, remove the heating source and allow the mixture to cool. Remove the condenser and reassemble the apparatus for simple distillation (Technique 6, Figure 6–6, p. 557). Add several new boiling stones to the round-bottomed flask. Distill the mixture and collect the distillate in a receiver cooled in an ice bath. The alkyl halide co-distills with water and then separates into two phases in the receiver. Distill the mixture until the distillate appears clear. The temperature should reach 110 to 115 °C by that time. While the distillation is going on, remove the receiver and collect a few drops of distillate in a test tube containing some water. Check to see whether the distillate is completely soluble (miscible). If it is, no alkyl halide is present and the distillation may be stopped. If the distillate produces insoluble droplets in the water, then the distillation must be continued until the distillate becomes completely water-soluble. The distillate collected in the receiver contains principally *n*-butyl bromide and water, with smaller amounts of sulfuric acid and hydrogen bromide.

Transfer the distillate to a separatory funnel, add 25 mL of water to it, and shake the mixture (Technique 5, Section 5.4, p. 545). Drain the lower layer, which contains *n*-butyl bromide (density 1.27 g/mL), from the funnel. Discard the aqueous layer after making certain that the correct layer has been saved. Return the alkyl halide to the funnel and add to it 15 mL of **cold** concentrated sulfuric acid. Swirl the mixture (unstoppered) until it is thoroughly mixed. Then stopper the funnel and shake it thoroughly, but carefully, to ensure that sulfuric acid does not leak out of the stopper or stopcock. Allow several minutes for the phases to separate. Sulfuric acid has a density of 1.84 g/mL. Which phase contains *n*-butyl bromide? If there is some doubt about the identity of the

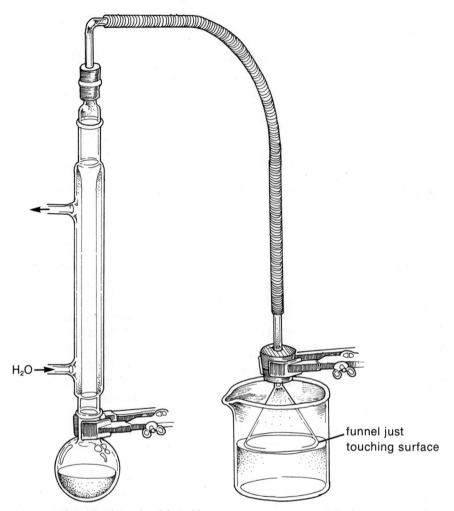

Apparatus for preparing *n*-butyl bromide

phases, read Technique 5, Section 5.4, p. 545. Drain the lower layer from the funnel. Allow several minutes for further layer separation and again drain the remaining lower layer. Wash the *n*-butyl bromide with 15 mL of 10% sodium hydroxide solution. Carefully separate the layers and be certain to save the organic layer. Dry the crude *n*-butyl bromide over 1.5 g of anhydrous calcium chloride in a small Erlenmeyer flask (Technique 5, Section 5.6, p. 549). Stopper the flask and swirl the contents until the liquid is **clear.** The drying process can be accelerated by **gently** warming the mixture on a steam bath.

　　Decant the **clear** liquid into a **dry** distilling flask. Add a boiling stone and distill the crude *n*-butyl bromide in a **dry** apparatus (Technique 6, Section 6.4, Figure 6–6, p. 557). Collect the material that boils between 98 and 102 °C. Weigh the product and calculate the percentage yield. Submit the sample in a labeled vial to the instructor.

Procedure 23B
t-Pentyl Chloride

In a 125-mL separatory funnel, place 22 mL (density 0.805 g/mL) of *t*-pentyl alcohol (*t*-amyl alcohol or 2-methyl-2-butanol) and 50 mL (density 1.18 g/mL; 37.3% HCl) of concentrated hydrochloric acid. Do not stopper the funnel. Gently swirl the mixture in the separatory funnel for about 1 minute. After this period of swirling, stopper the separatory funnel and carefully invert it. Without shaking the separatory funnel, immediately open the stopcock to release the pressure. Close the stopcock, shake the funnel several times, and again release the pressure through the stopcock (Technique 5, Section 5.4, p. 545). Shake the funnel for 2 to 3 minutes, with occasional venting. Allow the mixture to stand in the separatory funnel until the two layers have completely separated. The *t*-pentyl chloride has a density of 0.865 g/mL. Which layer contains the alkyl halide? Separate the layers.

The operations in this paragraph should be done as rapidly as possible since the *t*-pentyl chloride is unstable in water and sodium bicarbonate solution. Wash (swirl and shake) the organic layer with one 25-mL portion of water. Again, separate the layers and discard the aqueous phase after making certain that the proper layer has been saved (Technique 5, Section 5.4, p. 545). Wash the organic layer with a 25-mL portion of 5% aqueous sodium bicarbonate. Gently swirl the funnel (unstoppered) until the contents are thoroughly mixed. Stopper the funnel, and carefully invert it. Release the excess pressure through the stopcock. Gently shake the separatory funnel, with frequent release of pressure. Following this, vigorously shake the funnel, again with release of pressure, for about 1 minute. Allow the layers to separate, and drain the lower aqueous bicarbonate layer. Wash (swirl and shake) the organic layer with one 25-mL portion of water, and again drain the lower aqueous layer.

Transfer the organic layer to a small dry Erlenmeyer flask. Pour it from the top of the separatory funnel. Dry the crude *t*-pentyl chloride over anhydrous calcium chloride until it is clear (Technique 5, Section 5.6, p. 549). Swirl the alkyl halide with the drying agent to aid the drying. Decant the **clear** material into a small **dry** distilling flask. Add a boiling stone and distill the crude *t*-pentyl chloride in a **dry** apparatus (Technique 6, Section 6.4, Figure 6–6, p. 557), using a steam bath. Collect the pure *t*-pentyl chloride in a receiver cooled in ice. Collect the material that boils between 79 and 84 °C. Weigh the product and calculate the percentage yield. Submit the sample to the instructor in a labeled vial.

QUESTIONS

n-Butyl Bromide

1. Sulfuric acid was used to remove unreacted alcohol from the crude alkyl halide. Explain how it removes the alcohol. Write the equation.

2. Dibutyl ether and 1-butene can be formed as by-products in this reaction. Explain how they are removed by sulfuric acid. Give the reactions.

3. Look up the density of *n*-butyl chloride. Assume that this alkyl halide was prepared instead of the bromide. Decide whether the alkyl halide would appear as the upper or the lower phase in the separatory funnel at each stage of the isolation: after the reflux; after the codistillation; after the addition of water to the distillate; after the washing with sulfuric acid; after the washing with sodium hydroxide.

4. Why must the crude alkyl halide be dried carefully with calcium chloride before the final distillation? See Technique 7, Section 7.7.

t-Pentyl Chloride

1. Aqueous sodium bicarbonate was used to wash the crude *t*-pentyl chloride. Why would it be undesirable to wash the halide with aqueous sodium hydroxide?

2. Some 2-methyl-2-butene may be produced in the reaction as a by-product. Give the mechanism for its production. How can it be removed during the purification?

3. How is the unreacted *t*-pentyl alcohol removed in this experiment? Look up the solubility of the alcohol and alkyl halide in water.

4. Why must the crude alkyl halide be dried carefully with calcium chloride before the final distillation? See Technique 7, Section 7.7.

Experiment 24
Nucleophilic Substitution
Reactions: Competing
Nucleophiles

Nucleophilic Substitution
Refractometry
Gas Chromatography
Nmr Spectroscopy

The purpose of this experiment is to compare the relative nucleophilicities of chloride ions and bromide ions toward *n*-butyl alcohol (1-butanol) on one hand and toward *t*-butyl alcohol (2-methyl-2-propanol) on the other. The two nucleophiles will be present at the same time in each reaction, in equimolar concentrations, and they will be competing with each other for substrate.

In general, alcohols do not react readily in simple nucleophilic displacement reactions. If they are attacked by nucleophiles directly, hydroxide ion, a strong base, must be displaced. Such a displacement is not energetically favorable, and it cannot occur to any reasonable extent:

$$X^- + ROH \;\not\!\!\longrightarrow\; R\text{---}X + OH^-$$

To avoid this problem, one must carry out nucleophilic displacement reactions with alcohols as substrates in acidic media. In a rapid initial step, the alcohol is protonated; then water, a very stable molecule, is displaced. This displacement is energetically very favorable, and the reaction proceeds in high yield:

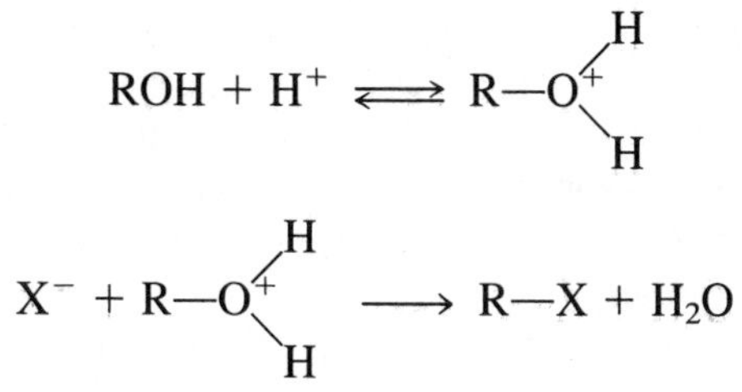

$$\text{ROH} + \text{H}^+ \rightleftharpoons \text{R---O}^+\!\!\begin{smallmatrix} \text{H} \\ \\ \text{H} \end{smallmatrix}$$

$$\text{X}^- + \text{R---O}^+\!\!\begin{smallmatrix} \text{H} \\ \\ \text{H} \end{smallmatrix} \longrightarrow \text{R---X} + \text{H}_2\text{O}$$

Once the alcohol is protonated, the substrate reacts by either the S_N1 or the S_N2 mechanism, depending on the structure of the alkyl group of the alcohol. For a brief review of these mechanisms, you should consult the chapters on nucleophilic substitution in your textbook.

You will analyze the products of the two reactions studied in this experiment by a variety of techniques to determine the relative amounts of alkyl chloride and alkyl bromide formed in each reaction. That is, using equimolar concentrations of chloride ions and bromide ions reacting against both *n*-butyl alcohol and *t*-butyl alcohol, you will try to determine which ion is the better nucleophile and for which of the two substrates (reactions) this difference is important. The ammonium halides (NH_4Cl and NH_4Br) are used as sources of halide ions in this experiment since they are more soluble at the concentrations required than are the corresponding sodium or potassium salts.

SPECIAL INSTRUCTIONS

Before beginning this experiment, review the appropriate chapters in your lecture textbook. Also read Techniques 1, 5, 12, 16, 17 and Appendix 4, Nuclear Magnetic Resonance. Concentrated sulfuric acid is very corrosive; be careful when handling it.

> **NOTE TO THE INSTRUCTOR:** Be certain that the *t*-butyl alcohol has been melted before the beginning of the laboratory period. Prepare the gas chromatograph as follows: column temperature, 85 °C; injection port temperature, 85 °C; carrier gas flow rate, 50 mL/min. The column should be approximately 12 ft long and should contain a stationary phase similar to silicone oil (SE-30) or 5% dinonyl phthalate on Chromosorb G.

PROCEDURE

Before preparing the solvent mixture and carrying out the reaction, assemble an apparatus for reflux using a 500-mL round-bottomed, three-necked flask and a condenser, as shown in the figure. Glass stoppers must be used to stopper the unused openings, as both corks and rubber stoppers will react with concentrated sulfuric acid. A trap for acidic gases should also be attached, and this too is shown in the illustration. A heating mantle should be used as a heating source. In addition, on a separate ring stand, a 125-mL separatory funnel should be ready and resting in a ring. A glass stopper must also be used in the separatory funnel.

First prepare the solvent medium containing equal concentrations of the two nucleophiles (described immediately below), and then proceed to **both** Procedure 24A and Procedure 24B.

THE SOLVENT-NUCLEOPHILE MEDIUM[1]

Place 50 g of ice in a 250-mL Erlenmeyer flask and carefully add 38 mL of concentrated sulfuric acid. Set this mixture aside to cool. Carefully weigh 9.5 g of ammonium chloride and 17.5 g of ammonium bromide into a 250-mL beaker. Crush any lumps of these reagents to powder, and then, using a powder funnel, transfer these halides to a 500-mL Erlenmeyer flask. Exercising caution, add the sulfuric acid mixture to the ammonium salts a little at a time. Swirl the mixture vigorously to induce the salts to dissolve. It will probably be necessary to heat the mixture on a steam bath to achieve total solution. If necessary, you may add as much as 5 mL of water at this stage. Do not worry if just a few small granules do not dissolve.

When solution has been achieved, allow the liquid to cool slightly and then pour 35 mL of it into the separatory funnel and the remainder into the reflux apparatus. Lightly replace the stoppers in the openings of both containers. A small portion of the salts in the separatory funnel or the boiling flask or both may precipitate as the solution cools. Do not worry about these at this point; they will redissolve during the reactions.

Procedure 24A
Competitive Nucleophiles With 1-Butanol

Add 5 mL of 1-butanol (*n*-butyl alcohol) to the solvent-nucleophile mixture contained in the reflux apparatus. To do this, remove the acidic gas trap and pour the 1-butanol carefully downward from the top of the condenser. Also add a boiling stone. Replace the trap fittings and start circulating the cooling water. Adjust the heat from the heating mantle so that this mixture maintains a gentle boiling action. Be very careful to adjust the

---

[1]NOTE TO THE INSTRUCTOR: As an alternative, an equimolar mixture of concentrated hydrochloric ($\sim$37%, 12M, density 1.18) and hydrobromic ($\sim$48%, 9M, density 1.50) acids can be used. To use these, exactly 36 mL of HCl should be mixed with exactly 48 mL of HBr, and then 35 mL of the mixture should be placed in the separatory funnel and the remainder in the reflux apparatus. We do not favor this method because the composition of the concentrated acids is somewhat variable, and it is difficult to achieve exact stoichiometric equivalence (especially with graduated cylinders). Also, this medium differs in ionic strength from the medium described, and the competitive ratio of the two nucleophiles (where it applies) is not so dramatic.

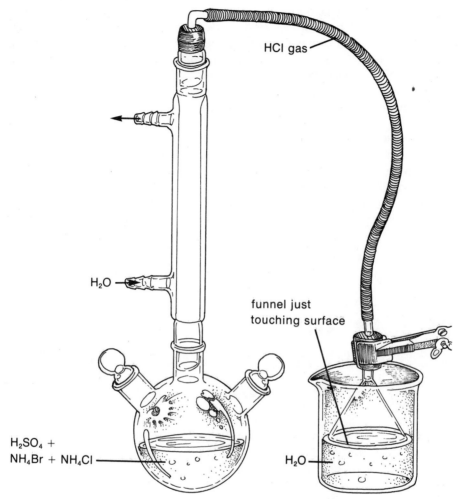

HCl gas

H$_2$O

funnel just
touching surface

H$_2$SO$_4$ +
NH$_4$Br + NH$_4$Cl

H$_2$O

Apparatus for S$_N$1/S$_N$2 experiment

reflux ring so that it remains in the lower fourth of the condenser. Violent boiling will cause loss of product. Continue heating the mixture for 75 minutes. During this heating period, go on to Procedure 24B and finish it before returning to this procedure.

REFLUX PERIOD

When the period of reflux has been completed, discontinue heating, remove the heating mantle, and allow the reaction mixture to cool. Be careful not to shake the hot solution as you remove the heating mantle, or a violent boiling and bubbling action will result; this could allow material to be lost out the top of the condenser. Prepare a bath of ice and water, and cool the reaction mixture by immersing the boiling flask in the bath. After a few minutes, it will be possible to begin to swirl the reaction mixture with safety, thereby increasing the efficiency of cooling.

Transfer the cooled solution to a 125-mL separatory funnel, taking care to leave behind any solids that may have precipitated. Allow the phases to separate and drain the aqueous layer. If the reaction was not yet complete, some 1-butanol may remain, which in some cases will form a **second organic layer;** that is, there will be **three** layers. Treat both these organic layers as if they were one, and add 10 mL of water to them. Shake this mixture in the separatory funnel, allow the layers to separate, and once again drain the lower organic layer. Extract the organic layer with 10 mL of saturated sodium bicarbonate solution, separate the layers, and drain the organic layer into a 50-mL beaker containing about 0.5 g of anhydrous sodium sulfate. When the solution is clear, decant the halide solution into a clean, dry, ground-glass stoppered flask (50-mL), taking care not to transfer any solid.[2] This sample can now be analyzed by as many of the methods given below as your instructor indicates. If this material is to be stored for any length of time, the flask should be sealed with stopcock grease. Do not store the liquid in a container with a cork or a rubber stopper, because these will absorb the halides.

Procedure 24B
Competitive Nucleophiles With 2-Methyl-2-Propanol

Carefully measure 5 mL of 2-methyl-2-propanol (*t*-butyl alcohol, mp 25 °C) using a **warm** graduated cylinder, and add it to the separatory funnel containing the other portion of the solvent-nucleophile medium, which should have cooled to room temperature by this time. Replace the stopper, carefully swirl the funnel a couple of times, and then invert it to release the mixing pressure. Repeat this until the pressures are substantially equalized; then invert the funnel and shake it vigorously, with occasional venting, for 2 minutes. Any solids that were originally present in the funnel should dissolve during this period. After shaking, place the funnel in a ring and allow the layer of alkyl halides to separate (1–2 minutes at most). A fairly distinct layer should have formed by this time. Slowly drain the lower layer into a beaker. After a wait of 10 to 15 seconds, drain another small portion of the material in the funnel, including a small amount of the organic layer, so as to be certain that it is not contaminated by any water. Then drain the remainder of the alkyl halide layer into a 50-mL beaker containing about 1 g of solid sodium bicarbonate. As soon as the bubbling stops and a clear liquid is obtained, decant it into a clean, dry, ground-glass-stoppered flask (25 mL), taking care not to transfer any solid. This sample can now be analyzed by as many of the methods given below as your instructor indicates. If this material is to be stored for any length of time, the flask should be sealed with stopcock grease. Do not store the liquid in a container with a cork or rubber stopper, since these will absorb the halides. When you have finished this procedure, return to Procedure 24A.

---

[2]If a very small amount of product is obtained, add about 1 mL of methylene chloride to help in the transfer. Methylene chloride will not interfere in the gas chromatographic analysis.

ANALYSIS PROCEDURES

The ratio of 1-chlorobutane to 1-bromobutane, or *t*-butyl chloride to *t*-butyl bromide, must be determined. At your instructor's option, you may do this by one of three methods: gas chromatography, refractive index, or nuclear magnetic resonance spectroscopy.

GAS CHROMATOGRAPHY

The instructor or laboratory assistant may either make the sample injections or allow the students in the class to make them. In the latter case, it is **essential** that instruction beforehand is adequate. A reasonable sample size is 1 μL. The sample is injected into the gas chromatograph, and the gas chromatogram is recorded. The alkyl chloride, because of its greater volatility, has a shorter retention time than the alkyl bromide.

Once the gas chromatogram has been obtained, determine the relative areas of the two peaks (Technique 12, Section 12.10, p. 638). While the peaks may be cut out and weighed on an analytical balance as a method of determining areas, triangulation is the preferred method. Record the percentages of alkyl chloride and alkyl bromide in the reaction mixture. (See Note on p. 185.)

REFRACTIVE INDEX

Measure the refractive index of the product mixture (Technique 16). To determine the composition of the mixture, assume a linear relation between the refractive index and the molar composition of the mixture. At 20 °C the refractive indices of the alkyl halides are

1-chlorobutane	1.4015	2-chloro-2-methylpropane	1.3877
1-bromobutane	1.4398	2-bromo-2-methylpropane	1.4280

If the temperature of the laboratory room is not 20 °C, the refractive index must be corrected. Add 0.0004 refractive index unit to the observed reading for each degree above 20 °C, and subtract the same amount for each degree below this temperature. Record the percentages of alkyl chloride and alkyl bromide in the reaction mixture.

NUCLEAR MAGNETIC RESONANCE

The instructor or a laboratory assistant will record the nmr spectrum of the reaction mixture. Submit a labeled sample vial containing the mixture for this spectral determination. The spectrum will also contain integration of the important peaks (Appendix 4, Nuclear Magnetic Resonance). If the substrate alcohol was 1-butanol, the resulting halide mixture will give rise to a complicated spectrum. Each alkyl halide will show a downfield triplet, caused by the CH_2 group nearest the halogen. This triplet will appear further downfield for the alkyl chloride than for the alkyl bromide. These triplets will overlap, but one branch of each triplet will be available for comparison. Compare the integral of the **downfield** branch of the triplet for 1-chlorobutane with the **upfield** branch of the triplet for 1-bromobutane. The spectrum shown on the next page provides an example. The relative heights of these integrals correspond to the relative amounts of each halide in the mixture.

If the substrate alcohol was 2-methyl-2-propanol, the resulting halide mixture will show two peaks in the nmr spectrum. Each halide will show a singlet, since all the CH$_3$ groups are equivalent and not coupled. In the reaction mixture, the upfield peak is due to t-butyl chloride, while the downfield peak is caused by t-butyl bromide. Compare the integrals of these peaks. The second spectrum shown provides an example. The relative heights of these integrals correspond to the relative amounts of each halide in the mixture.

REPORT

Record the percentages of alkyl chloride and alkyl bromide in the reaction mixture. The report must include the percentage of each alkyl halide determined by each method used in this experiment. The report should also include a discussion of the mechanism of the reaction studied. Any differences in product distribution and mechanism should

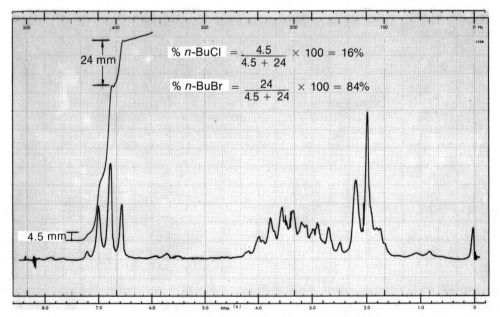

$$\% \; n\text{-BuCl} = \frac{4.5}{4.5 + 24} \times 100 = 16\%$$

$$\% \; n\text{-BuBr} = \frac{24}{4.5 + 24} \times 100 = 84\%$$

Nmr spectrum of 1-chlorobutane and 1-bromobutane, sweep width 250 Hz

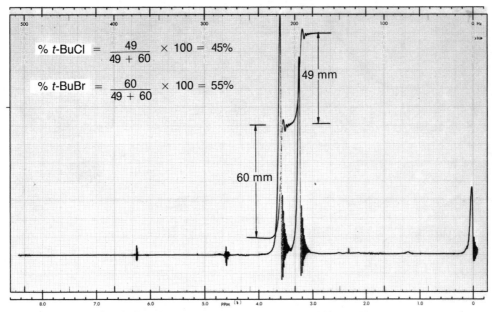

$$\% \text{ } t\text{-BuCl} = \frac{49}{49 + 60} \times 100 = 45\%$$

$$\% \text{ } t\text{-BuBr} = \frac{60}{49 + 60} \times 100 = 55\%$$

Nmr spectrum of *t*-butyl chloride and *t*-butyl bromide, sweep width 250 Hz

be discussed in the report. All gas chromatograms, refractive index data, and spectra should be attached to the report.

> **NOTE TO THE INSTRUCTOR:** If pure samples of each product are available, one can check the assumption (inherent here) that the gas chromatograph responds equally to each substance. Response factors (relative sensitivities) are easily determined by injecting an equimolar mixture of products and comparing the peak areas.

QUESTIONS

1. Which is the better nucleophile, chloride ion or bromide ion? Try to explain this.

2. What would be the expected product distribution if 2-butanol were the substrate in this experiment?

3. What is the principal organic by-product of these reactions?

4. A student left some alkyl halides (RCl and RBr) in an open container for several hours. What happened to the composition of the halide mixture during that time?

5. What would happen if a student decided not to worry about getting all the solids to dissolve in the nucleophile medium? How might this affect the outcome of the experiment?

6. Why does the alkyl chloride show nmr peaks further downfield than the corresponding peaks in the alkyl bromide for the primary products when the reverse is true for the tertiary products?

7. Draw a complete mechanism that shows why the resultant product distribution observed for the reaction of *t*-butyl alcohol is obtained.

8. What might have been the product ratios observed in this experiment if an aprotic solvent like dimethyl sulfoxide had been used instead of water?

9. Explain the order of elution you observed in doing the gas chromatography for this experiment. What property of the product molecules seems to be the most important in determining relative retention times?
10. Does it seem reasonable to you that the refractive index should be a temperature-dependent parameter? Try to explain.
11. When you calculate the percentage composition of the product mixture, exactly what kind of "percentage" are you dealing with?

Experiment 25
Hydrolysis of Some Alkyl Chlorides

Synthesis of an Alkyl Halide
Kinetics

Two chemical reactions are of interest in this experiment. The first is the preparation of the alkyl chlorides whose hydrolysis rates are to be measured. The chloride formation is a simple nucleophilic substitution reaction, carried out in a separatory funnel. Because the concentration of the initial alkyl chloride does not need to be determined for the kinetic experiment, isolation and purification of the alkyl chloride are not required.

$$ROH + HCl \longrightarrow R—Cl + H_2O$$

The second reaction is the actual hydrolysis, and the rate of this reaction will be measured. Under the conditions of this experiment, the reaction proceeds by an S_N1 pathway. The reaction rate is monitored by measuring the rate of appearance of hydrochloric acid. The concentration of hydrochloric acid is determined by titration with aqueous sodium hydroxide.

$$R—Cl + H_2O \xrightarrow{\text{solvent}} R—OH + HCl$$

The rate equation for the S_N1 hydrolysis of an alkyl chloride is

$$+ \frac{d[HCl]}{dt} = k[RCl]$$

Let c equal the initial concentration of RCl. At some time, t, x moles per liter of alkyl chloride will have decomposed and x moles per liter of HCl will have been produced. The remaining concentration of alkyl chloride at that value of time equals $c - x$. The rate equation becomes

$$+ \frac{dx}{dt} = k(c - x)$$

On integration, this becomes

$$\ln\left(\frac{c}{c-x}\right) = kt$$

which, converted to base 10 logarithms, is

$$2.303 \log\left(\frac{c}{c-x}\right) = kt$$

This equation is of the form appropriate for a straight line $y = mx + b$ with slope m and with intercept b equal to zero. If the reaction is indeed first-order, a plot of $\log(c/c-x)$ versus t will provide a straight line whose slope is $k/2.303$.

Evaluation of the term $c/c-x$ remains a problem, since it is experimentally difficult to determine the concentration of alkyl chloride. We can, however, determine the concentration of hydrochloric acid produced by titrating it with base. Because the stoichiometry of the reaction indicates that the number of moles of alkyl chloride consumed equals the number of moles of hydrochloric acid produced, c must also equal the number of moles of HCl produced when the reaction has gone to completion (the so-called infinity concentration of HCl), and x equals the number of moles of HCl produced at some particular value at time t. From these equalities, we can rewrite the integrated rate expression in terms of volume of base used in the titration. At the end-point of the titration,

$$\text{Number of moles HCl} = \text{number of moles of NaOH}$$

or

$$x = \text{number of moles NaOH at time } t$$

and

$$c = \text{number of moles NaOH at time } \infty$$

$$\text{Number of moles NaOH} = [\text{NaOH}]V$$

where V is the volume. Substituting and cancelling gives

$$\left(\frac{c}{c-x}\right) = \frac{[\cancel{\text{NaOH}}]V_\infty}{[\cancel{\text{NaOH}}](V_\infty - V_t)}$$

where V_∞ is the volume of NaOH used when the reaction is complete, and V_t is the volume of NaOH used at time t. This integrated rate equation becomes

$$2.303 \log\left(\frac{V_\infty}{V_\infty - V_t}\right) = kt$$

The concentration of the base used in the titration cancels out of this equation, so that it is necessary to know neither the concentration of base nor the amount of alkyl chloride used in the experiment.

A plot of $\log(V_\infty/V_\infty - V_t)$ versus t will provide a straight line whose slope equals $k/2.303$. The slope is determined according to the accompanying figure. If the

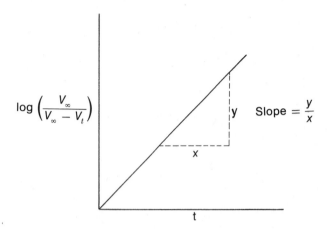

time is measured in minutes, the units of k are min^{-1}. The experimental points plotted on the graph may contain a certain amount of scatter, but the line drawn is the best **straight** line. The line should pass through the origin of the graph. With some reactions, competing processes may cause the line to contain a certain amount of curvature. In these cases, the slope of the initial portion of the line is used, before the curvature becomes too important.

One other value often cited in kinetic studies is the **half-life** of the reaction τ. The half-life is the time required for one half of the reactant to undergo conversion to products. During the first half-life, 50% of the available reactant is consumed. At the end of the second half-life, 75% of the reactant has been consumed. For a first-order reaction, the half-life is calculated by

$$\tau = \frac{\ln 2}{k} = \frac{0.69315}{k}$$

Two alkyl chlorides will be studied by the class in a variety of solvents. The class data will be compared, so that relative reactivities of the alkyl chlorides can be determined.

SPECIAL INSTRUCTIONS

Before beginning this experiment, you should read the material dealing with the methods of kinetics in your lecture textbook. Since concentrated hydrochloric acid is corrosive, care should be exercised in handling it. Avoid breathing the vapors. Some of the alkyl chlorides hydrolyze rapidly. Students must work in pairs in this experiment to make the measurements rapidly.

PROCEDURES

PREPARATION OF THE ALKYL CHLORIDES

The student should select an alcohol. The choices include *t*-butyl alcohol (2-methyl-2-propanol) and α-phenylethyl alcohol (1-phenylethanol). Place the alcohol (11 mL) in a separatory funnel along with 25 mL of cold, concentrated hydrochloric acid (specific gravity 1.18; 37.3% hydrogen chloride). Shake the separatory funnel vigorously, with frequent venting to relieve any excess pressure, over 30 minutes. Remove the aqueous layer. Wash the organic phase quickly with three 5-mL portions of cold water, followed by a washing with 5 mL of 5% sodium bicarbonate solution. Place the organic product in a small Erlenmeyer flask over 3 to 4 g of anhydrous calcium chloride. Shake the flask occasionally over 5 minutes. Carefully decant the alkyl chloride from the drying agent into a small Erlenmeyer flask, which can then be stoppered tightly. The alkyl chloride is used in this experiment without prior distillation. Because the true concentration of alkyl chloride introduced into the hydrolysis reaction is determined by titration, it is not necessary to purify the product prepared in this part of the experiment.

KINETIC STUDY OF THE HYDROLYSIS OF AN ALKYL CHLORIDE

Because the chlorides hydrolyze rapidly under the conditions used in this experiment, students must perform the kinetic studies working in pairs. One student will perform the titrations, while the other measures the time and records the data.

Prepare a stock solution of alkyl chloride by dissolving about 0.6 g of alkyl chloride in 50 mL of dry, reagent-grade acetone. Store this solution in a stoppered container to protect it from moisture. Use a 125-mL Erlenmeyer flask to carry out the hydrolysis. The flask should contain a magnetic stirring bar, 50 mL of solvent (see table for the appropriate solvent), and 2 to 3 drops of bromthymol blue indicator. Use absolute ethanol in preparing the aqueous ethanol solvent. Do not use denatured ethanol, as the denaturing agents may interfere with the reactions being studied. Bromthymol blue has a yellow color in acid solution and a blue color in alkaline solution.

Place a 50-mL buret filled with approximately 0.01N sodium hydroxide above the flask. The exact concentration of sodium hydroxide does not need to be known. Record the initial volume of sodium hydroxide at time t equal to 0.0 minutes. Add about 2 mL of sodium hydroxide from the buret to the Erlenmeyer flask, and precisely record the new volume in the buret. Start the stirrer. At time 0.0 minutes, **rapidly** add 1.0 mL of the acetone solution of the alkyl chloride from a pipet. Start the timer when the pipet is about half empty. The indicator will undergo a color change, passing from blue through green to yellow when enough hydrogen chloride has been formed in the reaction to neutralize the sodium hydroxide in the flask. Record the time at which the color changed. This color change may not be rapid. One should try to use the same color as the end point each time. Add another 2 mL of sodium hydroxide from the buret, precisely record the volume, and also record the time at which this second volume of sodium hydroxide is consumed. Repeat the sodium hydroxide addition twice more (four total). Finally, allow the reaction to go to completion for an hour without excess sodium hydroxide present. Stopper the Erlenmeyer flask during this period.

After the reaction has gone to completion, **accurately** titrate the amount of hydrogen chloride in solution to the end point. The end point is reached when the color of the solution remains constant for at least 0.5 minute. The time corresponding to this final volume is infinity ($t = \infty$). Repeat this process in the other two solvent mixtures indicated in the table below. These experiments can be carried out while you are waiting for the infinity titration of the previous experiments, provided that a separate buret is used for each run, so that the infinity concentrations of hydrogen chloride produced can be accurately determined.

The data should be plotted according to the method described in the introductory section of this experiment. The rate constant k and the half-life τ must be reported. The report to the instructor should include the plot of the data as well as a table of data. A sample table of data is shown. Explain your results, especially the effect of changing the water content of the solvent on the rate of the reaction. If the instructor so desires, the results from the entire class may be compared.

Experimental Conditions

COMPOUND	SOLVENT MIXTURES (volume percentage of organic phase in water)
t-Butyl chloride	40% Ethanol
	25% Acetone
	10% Acetone
α-Phenylethyl chloride	50% Ethanol
	40% Ethanol
	35% Ethanol

Hydrolysis of α-Phenylethyl Chloride in 50% Ethanol

TIME (min)	VOL. NaOH RECORDED	VOL. NaOH USED	$V_\infty - V_t$	$\dfrac{V_\infty}{V_\infty - V_t}$	$\ln\left(\dfrac{V_\infty}{V_\infty - V_t}\right)$
0.00	0.2	0.0	6.9	1.00	0.000
8.46	2.2	2.0	4.9	1.41	0.343
18.25	4.2	4.0	2.9	2.37	0.863
31.80	5.9	5.7	1.2	5.75	1.750
47.72	6.8	6.6	0.3	23.00	3.136
100 (∞)	7.1	6.9	0.0	. . .	. . .

QUESTIONS

1. Plot the data given in the table above. Determine the rate constant and the half-life for this example.

2. What are the principal by-products of these reactions? Give the rate equations for these competing reactions? Should the production of these by-products go on at the same rate as the hydrolysis reactions? Explain.

3. Compare the energy diagrams for an S_N1 reaction in solvents with two different percentages of water. Explain any differences in the diagrams and their effect on the reaction rate.

4. Compare the expected rates of hydrolysis of t-cumyl chloride (2-chloro-2-phenylpropane) and α-phenylethyl chloride (1-chloro-1-phenylethane) in the same solvent. Explain any differences that might be expected.

Essay
DETECTION OF ALCOHOL: THE BREATHALYZER

If one places organic compounds on a scale ranking their extent of oxidation, a general order such as

$$R\!-\!CH_3 < R\!-\!CH_2OH < R\!-\!CHO \text{ (or } R_2CO) < R\!-\!COOH < CO_2$$

is obtained. According to this scale, you can see that alcohols represent a relatively reduced form of organic compound, while carbonyl compounds and carboxylic acid derivatives represent highly oxidized structures. Using appropriate oxidizing agents, it should be possible to oxidize an alcohol to an aldehyde, a ketone, or a carboxylic acid, depending on the substrate and the oxidation conditions.

Primary alcohols can be oxidized to aldehydes by various oxidizing agents, including potassium permanganate, potassium dichromate, and nitric acid:

$$R\!-\!CH_2OH \xrightarrow{[O]} \left[R\!-\!\overset{\overset{\displaystyle O}{\|}}{C}\!-\!H \right] \xrightarrow{[O]} R\!-\!\overset{\overset{\displaystyle O}{\|}}{C}\!-\!OH$$

The aldehyde formed in this oxidation is unstable relative to further oxidation, and consequently the aldehyde is usually oxidized further to the corresponding carboxylic acid. The aldehyde is seldom isolated from such an oxidation, unless the oxidizing agent is relatively mild.

Chromium(VI) is a very useful oxidizing agent. It appears in various chemical forms, including chromium trioxide, CrO_3, chromate ion, CrO_4^{2-}, and dichromate ion, $Cr_2O_7^{2-}$. The chromium(VI) compounds are typically red to yellow. During the oxidation, they are reduced to Cr^{3+}, which is green. As a result, an oxidation reaction can be monitored by the color change. A typical chromium(VI) oxidation to illustrate the role of both the oxidizing and the reducing species is the dichromate oxidation of ethanol to acetaldehyde:

$$3CH_3CH_2OH + Cr_2O_7{}^{2-} + 8H^+ \longrightarrow 3CH_3\overset{\overset{\displaystyle O}{\|}}{C}-H + 2Cr^{3+} + 7H_2O$$

Because the aldehyde is also susceptible to oxidation, a second oxidation step of acetaldehyde to acetic acid can also take place:

$$3CH_3\overset{\overset{\displaystyle O}{\|}}{C}-H + Cr_2O_7{}^{2-} + 8H^+ \longrightarrow 3CH_3\overset{\overset{\displaystyle O}{\|}}{C}-OH + 2Cr^{3+} + 4H_2O$$

This oxidation reaction of alcohols by dichromate ion leads to a standard method of analysis for alcohols. The material to be tested is treated with acidic potassium dichromate solution, and the green chromic ion formed in the oxidation of the alcohol is measured spectrophotometrically by measuring the amount of light absorbed at 600 nm. By this method, it is possible indirectly to determine from 1 to 10 mg of ethanol per liter of blood with an accuracy of 5%. The alcohol content of beer can be determined within 1.4% accuracy.

THE BREATHALYZER

An interesting application of the oxidation of alcohols is in a quantitative method of determining the amount of ethanol in the blood of a person who has been drinking. The ethanol contained in alcoholic beverages may be oxidized by dichromate according to the equation shown above. During this oxidation, the color of the chromium-containing reagent changes from reddish orange ($Cr_2O_7{}^{2-}$) to green (Cr^{3+}). Law-enforcement officials use the color change in this reaction to estimate the alcohol content of the breath of suspected drunken drivers. This value can be converted to an alcohol content of the blood.

In most states, the usual legal definition of being under the influence of alcohol is based on a 0.10% alcohol content in the blood. Because the air deep within the lungs is in equilibrium with the blood passing through the pulmonary arteries, the amount of alcohol in the blood can be determined by measuring the alcohol content of the breath. The proper breath-blood ratio can be determined by simultaneous blood and breath tests. As a result of this equilibration, police officers do not need to be trained to administer blood tests. Instead, a simple instrument, a breath analyzer, which does not require any particular sophistication for its operation, can be used in the field.

In the simplest form, a breath analyzer contains a potassium dichromate–sulfuric acid reagent impregnated on particles of silica gel in a sealed glass ampoule. Before the instrument is to be used, the ends of the ampoule are broken off, and one end is fitted with a mouthpiece while the other is attached to the neck of an empty plastic bag. The person being tested blows into the tube to inflate the plastic bag. As air containing ethanol passes through the tube, a chemical reaction takes place, and the reddish-orange dichromate reagent is reduced to the green chromium sulfate, Cr^{3+}. When the green color extends beyond a certain point along the tube (the halfway point), it is determined that the motorist has a relatively high alcohol concentration in his breath, and he is usually taken to the police station for more precise tests. The device

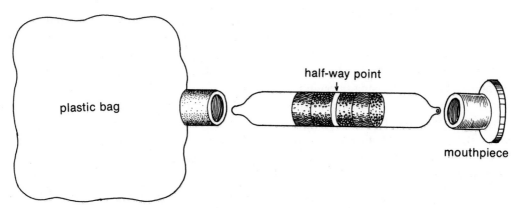

FIGURE 1. Breath alcohol screening device

described here is simple, and its precision is not high. It is used primarily as a **screening** device for suspected drunken drivers. An example of this simple device is shown in Figure 1.

A more precise instrument, the ''Breathalyzer,'' is shown in Figure 2. Air is blown into a cylinder A, whereupon a piston is raised. When the cylinder is full, the piston is allowed to fall and pump the measured volume of breath through a reaction ampoule B containing the potassium dichromate solution in sulfuric acid. As the alcohol-laden air is bubbled through this solution, the alcohol is oxidized to acetaldehyde and further to acetic acid, while the dichromate ion is reduced to Cr^{3+}. The instrument contains a light source C. Filters are used to select light in the blue region of the

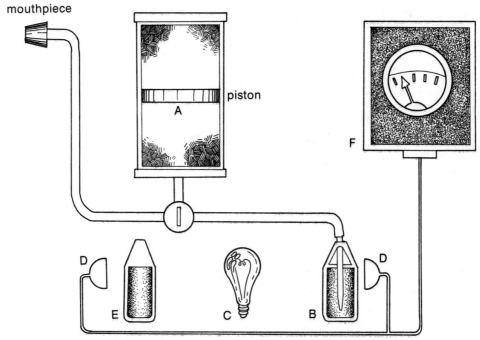

FIGURE 2. The Breathalyzer

spectrum. This blue light passes through the reaction ampoule and is detected by a photocell D. The light also passes through a sealed standard reference ampoule E, which contains exactly the same concentration of potassium dichromate in sulfuric acid as the reaction ampoule B had originally. No alcohol is allowed to enter this reference ampoule. The light passing through the reference ampoule is detected by another photocell. A meter F, calibrated in milligrams of ethanol per 100 mL of blood, or in percentage of blood alcohol, registers the difference in voltages between the two photocells. Before the test, both ampoules transmit blue light to the same extent, so the meter reads zero. After the test, the reaction ampoule transmits more blue light than the reference ampoule, and a voltage is registered on the meter.

Such an instrument, while more complicated and more delicate than the simple device shown in Figure 1, can be used in the field without a support laboratory. The instrument is portable, permitting it to be easily transported in the trunk of a patrol car.

A similar method is used in Experiment 26 to follow the rate of oxidation of several alcohols by dichromate ion. The color change that accompanies the oxidation is monitored by a spectrophotometer.

REFERENCES

Denney, R. C. "Analysing for Alcohol." *Chemistry in Britain, 6* (1970): 533.

Treptow, R. S. "Determination of Alcohol in Breath for Law Enforcement." *Journal of Chemical Education, 51* (1974): 651.

Lovell, W. S. "Breath Tests for Determining Alcohol in the Blood." *Science, 178* (1972): 264.

Symposium on Breath Alcohol Tests. *Journal of Forensic Science, 5* (1960): 395.

Timmer, W. C. "An Experiment in Forensic Chemistry—The Breathalyzer." *Journal of Chemical Education, 63* (October 1986): 897.

Experiment 26
Chromic Acid Oxidation
of Alcohols

Chromic Acid Oxidation of an Alcohol
Kinetics
Ultraviolet-visible Spectrophotometry

The chemical reaction of interest in this experiment is the oxidation of an alcohol to the corresponding aldehyde by an acidic solution of potassium dichromate:

$$3RCH_2OH + Cr_2O_7^{2-} + 8H^+ \longrightarrow 3R-\overset{\displaystyle O}{\overset{\|}{C}}-H + 2Cr^{3+} + 7H_2O$$

Normally, the aldehyde formed is also susceptible to oxidation by the dichromate ion, yielding the corresponding acid:

$$3R-\overset{\overset{\textstyle O}{\|}}{C}-H + Cr_2O_7{}^{2-} + 8H^+ \longrightarrow 3R-\overset{\overset{\textstyle O}{\|}}{C}-OH + 2Cr^{3+} + 4H_2O$$

In this experiment, however, the alcohol is present in large excess, and the likelihood that the second reaction will take place is thereby greatly reduced. A secondary alcohol is oxidized to a ketone by a similar process. A ketone is not easily oxidized further by the dichromate reagent.

$$3R-\overset{\overset{\textstyle}{|}}{\underset{\underset{\textstyle R'}{|}}{C}}-OH + Cr_2O_7{}^{2-} + 8H^+ \longrightarrow 3R-\overset{\overset{\textstyle O}{\|}}{C}-R' + 2Cr^{3+} + 7H_2O$$

Although various mechanisms have been proposed to explain how dichromate ion oxidizes alcohols, the most commonly accepted mechanism is the one F. H. Westheimer first proposed in 1949. In acid solution, dichromate ion forms two molecules of chromic acid, H_2CrO_4:

$$2H_3O^+ + Cr_2O_7{}^{2-} \overset{fast}{\rightleftharpoons} 2H_2CrO_4 + H_2O \qquad (1)$$

The chromic acid, in a rapid, reversible step, forms a chromate ester with the alcohol:

$$RCH_2OH + H_2CrO_4 \overset{fast}{\rightleftharpoons} RCH_2-O-\overset{\overset{\textstyle O}{\|}}{\underset{\underset{\textstyle O}{\|}}{Cr}}-OH + H_2O \qquad (2)$$

The chromate ester undergoes a rate-determining decomposition by a two-electron transfer with cleavage of the α-carbon-hydrogen bond, as seen in step 3.

$$R-\overset{\overset{\textstyle H}{|}}{\underset{\underset{\textstyle H}{|}}{C}}-O-\overset{\overset{\textstyle O}{\|}}{\underset{\underset{\textstyle O}{\|}}{Cr}}-OH \overset{slow}{\longrightarrow} R-\overset{}{\underset{\underset{\textstyle H}{|}}{C}}=O + H_2CrO_3 \qquad (3)$$

Cr in +6
oxidation state

Cr in +4
oxidation state

The H_2CrO_3 is further reduced to Cr^{3+} by interaction with chromium in various oxidation states and by further interaction with the alcohol. All these subsequent steps are very rapid relative to step 3. Consequently they are not involved in the rate-determining step of the mechanism and need not be considered further.

The rate-determining step, step 3, involves only one molecule of the chromate ester, which in turn arises from a prior equilibrium involving the combination of one molecule of alcohol and one molecule of chromic acid (step 2). As a result, this reaction, which is first-order in chromate ester, turns out to be *second-order* for the reacting alcohol and the dichromate reagent. The kinetic equation therefore is

$$-\frac{d[Cr_2O_7{}^{2-}]}{dt} = k[RCH_2OH][Cr_2O_7{}^{2-}]$$

The presence of the chromium atom strongly affects the distribution of electrons in the remainder of the chromate ester molecule. Electrons need to be transferred to the chromium atom during the cleavage step. If the R group includes an electron-withdrawing group, it diminishes the necessary electron density needed for reaction. Consequently, the reaction proceeds more slowly. An electron-releasing group would be expected to have the opposite effect.

THE EXPERIMENTAL METHOD

The rate of the reaction is measured by following the rate of disappearance of the dichromate ion as a function of time. The dichromate ion, $Cr_2O_7{}^{2-}$, is yellow orange, absorbing light at 350 and 440 nm. The chromium is reduced to the green Cr^{3+} during the reaction. The ion Cr^{3+} does not absorb light significantly at 350 or 440 nm, but rather at 406, 574, and 666 nm. Therefore, if we measure the amount of light absorbed at a single wavelength, such as 440 nm, we can follow the rate of disappearance of dichromate ion without any interfering absorption of light by the ion Cr^{3+}.

The instrument used to measure the amount of light absorbed at a particular wavelength, when that light lies within the visible region of the electromagnetic spectrum, is a **colorimeter.** It can be described simply. Ordinary visible light is passed through the sample and then through a prism, where the light of the particular wavelength being studied is selected. This selected light is directed against a photocell, where its intensity is measured. A meter provides a visible display of the intensity of the light of the desired wavelength.

The true rate equation for this reaction is second-order. However, because a large excess of alcohol will be used, its concentration will change imperceptibly during the reaction. The rate equation, under these conditions, will simplify to that of a pseudo first-order reaction. The mathematics involved will become much simpler as a result.

The rate equation for a first-order (or a pseudo first-order) reaction is

$$-\frac{d[A]}{dt} = k[A]$$

In this experiment, the rate equation becomes

$$-\frac{d[Cr_2O_7{}^{2-}]}{dt} = k[Cr_2O_7{}^{2-}]$$

Let a equal the initial concentration of dichromate ion. At some time t, an amount x moles/L of dichromate will have undergone reaction, and x moles/L of aldehyde will have been produced. The remaining concentration of dichromate at that value of time equals $a - x$. The rate equation becomes

$$+ \frac{dx}{dt} = k(a - x)$$

Integration provides

$$\ln\left(\frac{a}{a - x}\right) = kt$$

Converting to base 10 logarithms gives

$$2.303 \log\left(\frac{a}{a - x}\right) = kt$$

This equation is of the form appropriate for a straight line with intercept equal to zero. If the reaction is indeed first-order, a plot of log $(a/a - x)$ versus t will provide a straight line whose slope is $k/2.303$.

Since it is experimentally difficult to measure directly how much dichromate ion is consumed during this reaction, evaluating the term $a/a - x$ requires an indirect approach. What is needed is some measurable quantity from which the concentration of dichromate can be derived. Such a quantity is the amount of light absorbed by the solution at wavelength 440 nm.

The Beer-Lambert law relates the amount of light absorbed by a molecule or an ion to its concentration, according to the equation

$$A = \epsilon c l$$

where A is the absorbance of the solution, ϵ is the molar absorptivity (a measure of the efficiency with which the sample absorbs the light), c is the concentration of the solution, and l is the path length of the cell in which the solution is contained. The absorbance is read by the spectrophotometer.

At the initial concentration a of dichromate ion, we may write

$$A_0 = \epsilon a l \qquad \text{or} \qquad a = \frac{A_0}{\epsilon l}$$

The amount of dichromate ion remaining unreacted at any particular time t, which equals $a - x$, becomes

$$A_t = \epsilon(a - x)l \qquad \text{or} \qquad a - x = \frac{A_t}{\epsilon l}$$

Substituting absorbance values for concentrations and cancelling provides

$$\left(\frac{a}{a - x}\right) = \frac{A_0}{\epsilon l}\frac{\epsilon l}{A_t} \qquad \text{or} \qquad \left(\frac{a}{a - x}\right) = \frac{A_0}{A_t}$$

At this point a correction must be introduced. When the reaction reaches completion, at "infinite" time, a certain degree of absorption of 440-nm light remains. In other words, at time $t = \infty$, the value A_∞ does not equal zero. Therefore this residual absorbance must be subtracted from each of the absorbance terms written above. The difference $A_0 - A_\infty$ gives the actual amount of dichromate ion present initially, and the difference $A_t - A_\infty$ gives the actual amount of dichromate ion remaining unreacted at a value t of time. Introducing these corrections, we have

$$\left(\frac{a}{a - x}\right) = \frac{A_0 - A_\infty}{A_t - A_\infty}$$

The integrated rate equation becomes

$$2.303 \log \left(\frac{A_0 - A_\infty}{A_t - A_\infty}\right) = kt$$

Since the dimensions of the cell and the molar absorptivity cancel out of this equation, it is not necessary to have any particular knowledge of these parameters.

A plot of $\log \left(\dfrac{A_0 - A_\infty}{A_t - A_\infty}\right)$ versus time will provide a straight line whose slope equals $k/2.303$. The slope is determined according to the accompanying figure. If time is measured in minutes, the units of k are min^{-1}. The experimental points plotted on the graph may contain a certain amount of scatter, but the line drawn is the best **straight** line (use some mathematical method such as the method of averages or of least squares).

One other value often cited in kinetic studies is the *half-life, τ,* of the reaction. The half-life is the time required for one half the reactant to undergo conversion to products. During the first half-life, 50% of the available reactant is consumed. At the end of the second half-life, 75% of the reactant has been consumed. For a first-order reaction, the half-life is calculated by

$$\tau = \frac{\ln 2}{k} = \frac{0.69315}{k}$$

The class will study several alcohols in this experiment. The class data will be compared in determining the relative reactivities of the alcohols. Two particular alco-

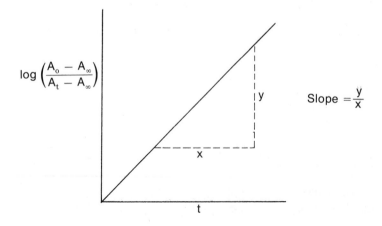

hols, 2-methoxyethanol and 2-chloroethanol, react more slowly than the other alcohols used in this experiment. In spite of this lower reactivity, the reactions will not be followed for more than a few minutes, because in these compounds, the second reaction—the oxidation of the aldehyde product to the corresponding carboxylic acid—becomes more important over longer periods. As a result of this second reaction, dichromate ion becomes consumed more rapidly than the calculations would suggest, and the graph of $\log \left(\dfrac{A_0 - A_\infty}{A_t - A_\infty} \right)$ versus time becomes curved. So that this complication is avoided, only the first few minutes of the reaction are used to calculate an initial reaction rate, which corresponds to the reaction being studied in this experiment. The other alcohols are sufficiently reactive that the second reaction does not introduce any significant error.

SPECIAL INSTRUCTIONS

Before beginning this experiment, read the material dealing with the methods of kinetics in your lecture textbook. You may also find the essay on the detection of alcohol, which precedes this experiment, to be interesting. Unlike the other alcohols that may be used in this experiment, 2-propanol and 1-phenylethanol are secondary alcohols. They are oxidized to form ketones, rather than aldehydes, but the experimental procedure required with the secondary alcohols is identical to the procedure used for the primary alcohols. The procedure described in this experiment is based on the controls found on a typical ultraviolet-visible spectrophotometer. Your instructor will have to show you how the specific controls must be adjusted on your instrument. This experiment can also be conducted using a colorimeter. Each kinetic run, including the temperature equilibration, requires from 1.5 to 2 hours, although most of that time is not involved in actually using the spectrophotometer.

This experiment involves using an acidic solution of potassium dichromate. Potassium dichromate solutions have been determined to be potential carcinogens. In this experiment, the dichromate solution will be prepared as a stock solution for the entire class to use. This stock solution should be stored in a hood. Students should wear gloves and use pipet bulbs when using this stock solution.

PROCEDURE

Select an alcohol from one of the following: ethanol, 1-propanol, 2-propanol, 2-methoxyethanol, 2-chloroethanol, ethylene glycol, and 1-phenylethanol (methylbenzyl alcohol). A stock solution of 3.9M sulfuric acid and a carefully prepared solution of 0.0196M potassium dichromate solution (prepared using distilled water in a volumetric flask) should be available for the entire class to use.

Turn on the instrument and allow it to warm up. Select the tungsten lamp as the light source. Select an operating mode that allows the instrument to operate at a fixed wavelength of **440 nm** and to record data as **absorbance.**

Using a small flask, prepare the test solution by transferring 1 mL of the stock dichromate solution and 10 mL of the stock sulfuric acid solution by pipet **(USE A PIPET BULB)** into the flask. Shake the solution well. Rinse a sample cuvette three times with this acidic dichromate solution and then fill the cell. Wipe the cuvette clean and dry. Place the cuvette into the sample cell compartment, and place a cuvette filled with distilled water into the reference cell compartment. Close the cell compartment lid and allow the chromic acid solution to reach temperature equilibrium by allowing it to remain in the instrument (with the instrument running) for 20 minutes. This preheating minimizes the problem of the solution being slowly heated during the experiment by the tungsten lamp, which is the light source in the spectrophotometer. Such heating would tend to accelerate the reaction as time passed.

At the end of the preheating, record the absorbance A_0 of the chromic acid solution, along with a time value of 0.0 minutes. Withdraw a 10.0-μL sample of the alcohol being studied into a hypodermic syringe and transfer it rapidly to the chromic acid solution. As the transfer is made, start the timer. Withdraw the sample cuvette from the cell compartment, shake it vigorously for 20 to 30 seconds, and return it to the cell compartment. Be sure to wipe off the cell again. Close the compartment lid. The measurements can now be started.

Take readings of the absorbance A_t and the corresponding time at 1-minute intervals over a 6-minute period (8 minutes for 2-propanol). At the end of this time, remove the cuvette from the cell compartment of the spectrophotometer and allow the solution in the cuvette to stand undisturbed for at least an hour. After this period, turn on the instrument as before and allow it to warm up. Return the cuvette to the cell compartment and read the absorbance value. This final value corresponds to "infinite" time (A_∞).

The instructor may require each student to perform a duplicate run. If so, repeat the experiment under precisely the same conditions used for the first run.

Plot the data according to the method described in the introductory section of the experiment. Report the value of each rate constant determined in this experiment (and the average of the rate constants, if duplicate determinations were made). Also, report the value of the half-life, τ. Include all data and graphs in the report. A table of sample data is shown. The results from the entire class may be compared, at the option of the instructor.

Oxidation of Ethanol

TIME (min)	ABSORBANCE (440 nm)	$A_t - A_\infty$	$\dfrac{A_0 - A_\infty}{A_t - A_\infty}$	$\log\left(\dfrac{A_0 - A_\infty}{A_t - A_\infty}\right)$
0.0	0.630	0.578	1.000	0.000
1.0	0.535	0.483	1.197	0.078
2.0	0.440	0.388	1.490	0.173
3.0	0.365	0.313	1.847	0.266
4.0	0.298	0.246	2.350	0.371
5.0	0.247	0.195	2.964	0.472
6.0	0.202	0.150	3.853	0.586
66.0 (∞)	0.052	0.000	. . .	. . .

REFERENCES

Lanes, R. M., and Lee, D. G. "Chromic Acid Oxidation of Alcohols." *Journal of Chemical Education*, *45* (1968): 269.

Westheimer, F. H. "The Mechanisms of Chromic Acid Oxidations." *Chemical Reviews, 45* (1949): 419.

Westheimer, F. H., and Nicolaides, N. "Kinetics of the Oxidation of 2-Deuterio-2-propanol by Chromic Acid." *Journal of the American Chemical Society, 71* (1949): 25.

Pavia, D. L., Lampman, G. M., and Kriz, G. S., Jr. *Introduction to Spectroscopy: A Guide for Students of Organic Chemistry*. Philadelphia: W. B. Saunders, 1979. Chap. 5.

QUESTIONS

1. Plot the data given in the table. Determine the rate constant and the half-life for this example.

2. Using data collected by the class, compare the relative rates of ethanol, 1-propanol, and 2-methoxyethanol. Explain the observed order of reactivities in terms of the mechanism of the oxidation reaction.

3. Using data collected by the class, compare the relative rates of 1-propanol and 2-propanol. Account for any differences that might be observed.

4. Balance the following oxidation-reduction reactions:

(a) $HO-CH_2CH_2-OH + K_2Cr_2O_7 \xrightarrow{H_2SO_4}$ H—C—C—H + Cr^{3+}

(b) 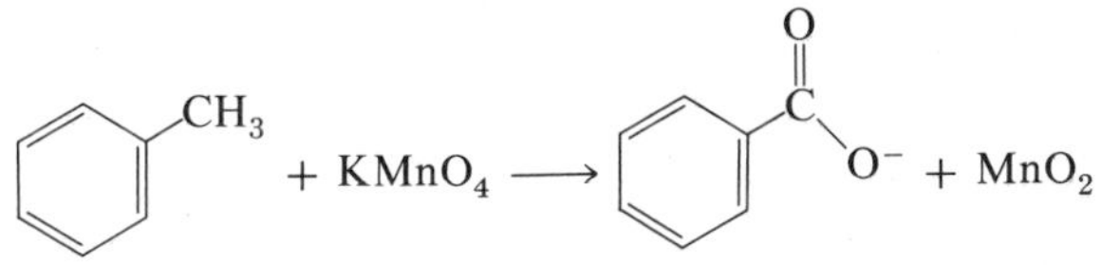 $+ KMnO_4 \longrightarrow$ $+ MnO_2$

(c) $HO-CH_2CH_2CH_2CH_2-OH + K_2Cr_2O_7 \xrightarrow{H_2SO_4}$ $HO-C-CH_2CH_2-C-H + Cr^{3+}$

(d) $CH_3CH_2CH_2CH{=}CH_2 + KMnO_4 \xrightarrow{KOH}$ $CH_3CH_2CH_2C-O^- + CO_2 + MnO_2$

(e) 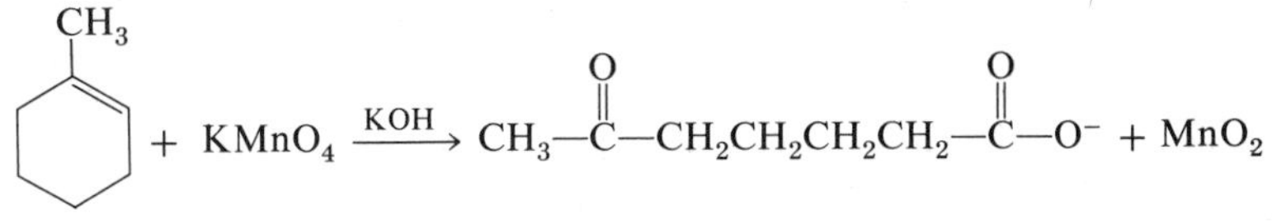 $+ KMnO_4 \xrightarrow{KOH}$ $CH_3-C-CH_2CH_2CH_2CH_2-C-O^- + MnO_2$

Experiment 27
Cyclohexene

Preparation of an Alkene
Dehydration of an Alcohol
Bromine and Permanganate Tests for Unsaturation

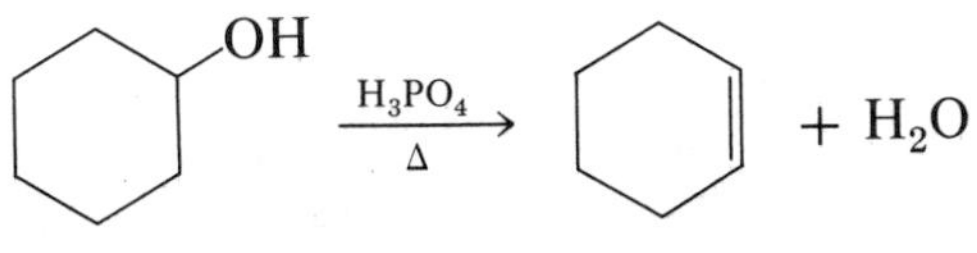

Cyclohexanol   Cyclohexene

Alcohol dehydration is an acid-catalyzed reaction performed by strong, concentrated mineral acids, such as sulfuric and phosphoric acids. The acids protonate the alcoholic hydroxyl group, permitting it to dissociate as water. Loss of a proton from the intermediate (elimination) brings about an alkene. Since sulfuric acid often causes extensive charring in this reaction, phosphoric acid, which is comparatively free of this problem, will be used.

The equilibrium that attends this reaction will be shifted in favor of the product, cyclohexene, by distilling it from the reaction mixture as it is formed. The cyclohexene (bp 83 °C) will co-distill with the water that is also formed. By continuously removing the products, one can obtain a high yield of cyclohexene. Since the starting material, cyclohexanol, is also rather low-boiling (bp 161 °C), the distillation must be done carefully, not allowing the temperature to rise much above 100 °C.

Unavoidably, a small amount of phosphoric acid co-distills with the products. It is removed by washing the distillate mixture with aqueous sodium carbonate. To remove the water that co-distills with cyclohexene, and any traces of water introduced in the base extraction, the product will be dried over anhydrous sodium sulfate.

Compounds containing double bonds react with a bromine solution (red) to decolorize it. Similarly, they react with a solution of potassium permanganate (purple)

$$
\underset{\substack{\text{Br} \quad \text{Br} \\ \text{(colorless)}}}{\diagup\!\!\diagdown} \xleftarrow[\text{(red)}]{\text{Br}_2} \;\diagup\!\!=\!\!\diagdown\; \xrightarrow[\text{(purple)}]{\text{KMnO}_4} \underset{\substack{\text{OH} \quad \text{OH} \\ \text{(colorless)}}}{\diagup\!\!\diagdown} + \underset{\text{(brown)}}{\text{MnO}_2}
$$

to discharge its color and produce a brown precipitate (MnO_2). These reactions are often used as qualitative tests to determine the presence of a double bond in an organic molecule (see Procedure 56C). Both tests will be performed on the cyclohexene formed in this experiment.

SPECIAL INSTRUCTIONS

Phosphoric acid is very corrosive. Do not allow any acid to touch your skin. Since a flame is used to carry out the distillation, keep in mind that cyclohexene will burn and is volatile. Be careful not to allow your flame to get near the collection flask. Also, do not leave your product in an open container on your desk top, because a neighbor's flame may ignite the vapors from your product. Finally, you should not flush your distillation residues down an open laboratory sink. This could fill the laboratory with dangerous vapors. Dispose of all residues in a suitable waste container. Before proceeding, you should read Techniques 1, 5, and 6.

PROCEDURE

Assemble a distillation apparatus as illustrated in Figure 6–6, Technique 6, p. 557. Use a 100-mL distilling flask and a 50-mL collection flask. The collection flask should be im-

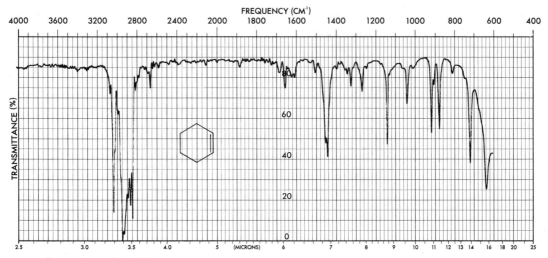

Ir spectrum of cyclohexene, neat

mersed to its neck in an ice-water bath to minimize the possibility that cyclohexene vapors will escape into the laboratory.

Place 20 mL of cyclohexanol (density 0.96) and 5 mL of 85% phosphoric acid in the distilling flask and mix the solution thoroughly. Add several boiling stones, start circulating the cooling water in the condenser, and heat the mixture until the product begins to distill. The temperature of the distilling vapor should be regulated so that it does not exceed 100 °C. This can best be done if the mixture is heated with a Bunsen burner. Hold the burner by its base and apply the heat so as to maintain a slow but steady rate of distillation. If the temperature rises above 100 °C, remove the burner for a few seconds before continuing to heat.

When only a few milliliters of residue remain in the distilling flask, stop the distillation. Saturate the distillate with solid sodium chloride. Add the salt, little by little, and shake the flask gently. When no more salt will dissolve, add enough 10% aqueous

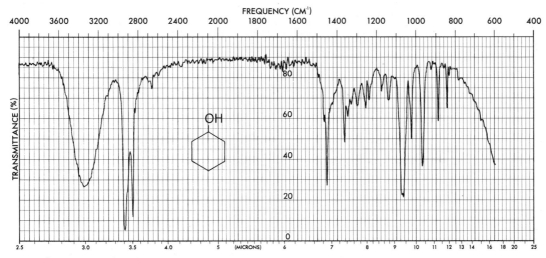

Ir spectrum of cyclohexanol, neat

sodium carbonate solution to make the distilled solution basic to litmus. Pour the neutralized mixture into a separatory funnel and separate the two layers. Drain the aqueous layer through the stopcock and then pour the upper layer (cyclohexene) through the neck of the separatory funnel into a 125-mL Erlenmeyer flask. Add about 2 or 3 g of anhydrous sodium sulfate to the flask and swirl occasionally until the solution is dry and clear (about 10–15 minutes). If there is not enough time to complete the experiment, you can stop it here. Store the cyclohexene over sodium sulfate in a lightly greased and stoppered standard-taper flask.

Reassemble a distillation apparatus as before, but this time use a 50-mL flask for the distilling flask. Again, cool the receiver in an ice-water bath. Decant the dry cyclohexene solution into the distilling flask and add a boiling stone. Distill the cyclohexene and collect the material that boils over the range of 80 to 85 °C. Weigh the product and calculate the yield. After performing the tests below, place the sample in a labeled glass screw-cap vial and submit it, along with the report, to the instructor.

UNSATURATION TESTS

Place 8 to 10 drops of cyclohexanol in each of two small test tubes. In each of another pair of small test tubes, place 8 to 10 drops of the cyclohexene you prepared. Do not confuse the test tubes. Take one test tube from each group and add to the contents of each a solution of bromine in carbon tetrachloride or methylene chloride (see Procedure 56C), drop by drop, until the red color is no longer discharged. Record the result in each case. Test the remaining two test tubes in a similar fashion with a solution of potassium permanganate. Since aqueous potassium permanganate is not miscible with organic compounds, you will have to add about 0.5 mL of 1,2-dimethoxyethane to each test tube before making the test. Record your results and explain them.

QUESTIONS

1. Draw a mechanism for the dehydration of cyclohexanol catalyzed by phosphoric acid.
2. What alkene would be produced on dehydration of each of the following alcohols?
    (a) 1-Methylcyclohexanol
    (b) 2-Methylcyclohexanol
    (c) 4-Methylcyclohexanol
    (d) 2,2-Dimethylcyclohexanol
    (e) 1,2-Cyclohexanediol
3. In the work-up procedure for cyclohexene, why is salt added before the layers are neutralized and separated?
4. What is the purpose of adding the sodium carbonate solution? Give an equation.
5. Compare and interpret the infrared spectra of cyclohexene and cyclohexanol (see above).

Experiment 28
Phase-Transfer Catalysis: Addition of Dichlorocarbene to Cyclohexene

Carbene Formation
Phase-Transfer Catalysis
Simple Distillation

It has long been known that a haloform, CHX_3, will react with a strong base to give a highly reactive carbene species, CX_2, by reactions 1 and 2. In the presence of an alkene, this carbene adds to the double bond to produce a cyclopropane ring (reaction 3).

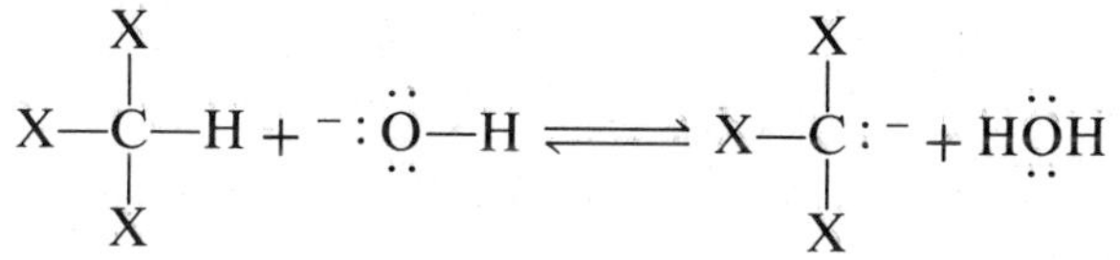

$$X-\underset{\underset{X}{|}}{\overset{\overset{X}{|}}{C}}-H + {}^-:\ddot{O}-H \rightleftharpoons X-\underset{\underset{X}{|}}{\overset{\overset{X}{|}}{C}}:{}^- + H\ddot{O}H \qquad (1)$$

$$X-\underset{\underset{X}{|}}{\overset{\overset{X\,\curvearrowright}{}}{C}}:{}^- \overset{slow}{\rightleftharpoons} X-\underset{\underset{X}{|}}{C}: + :\ddot{X}:{}^- \qquad (2)$$

$$X-\underset{\underset{X}{|}}{C}: + \overset{\diagup}{\underset{\diagup}{C}}{=}\overset{\diagdown}{\underset{\diagdown}{C}} \longrightarrow -\overset{|}{C}\overset{}{\underset{\underset{X}{\diagup}\underset{X}{\diagdown}}{\overset{|}{C}}}\overset{|}{C}- \qquad (3)$$

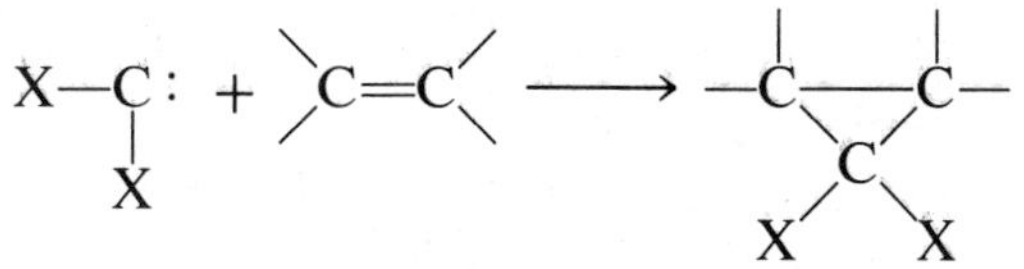

Traditionally, the reaction has been carried out in **one homogeneous phase** in anhydrous *t*-butyl alcohol solvent, using *t*-butoxide ion as the base [$t\text{-Bu} = C(CH_3)_3$]:

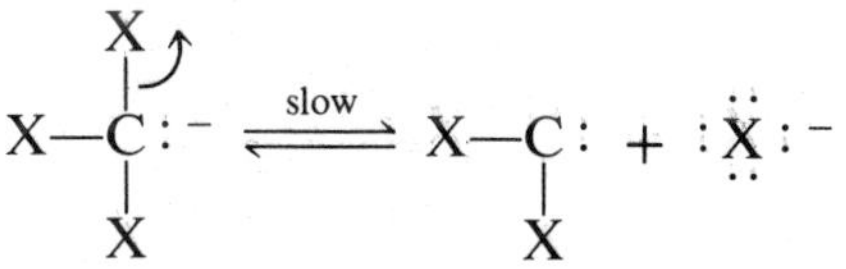

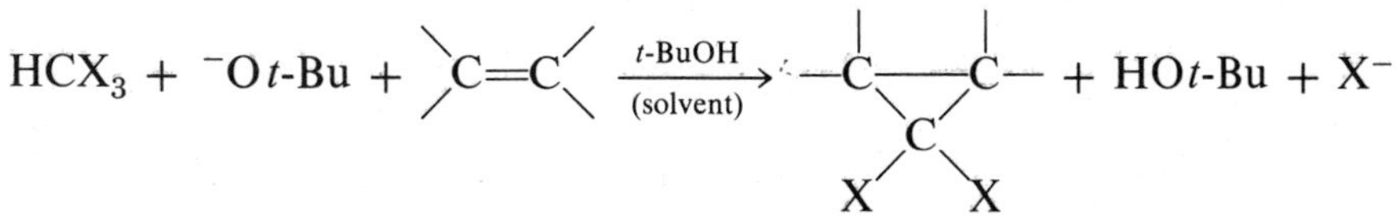

$$HCX_3 + {}^-Ot\text{-Bu} + \overset{\diagup}{\underset{\diagup}{C}}{=}\overset{\diagdown}{\underset{\diagdown}{C}} \xrightarrow[\text{(solvent)}]{t\text{-BuOH}} -\overset{|}{C}\overset{}{\underset{\underset{X}{\diagup}\underset{X}{\diagdown}}{\overset{|}{C}}}\overset{|}{C}- + HOt\text{-Bu} + X^-$$

Unfortunately, this technique requires time and effort to give good results. In addition, water must be avoided to prevent conversion of the haloform and carbene to formate ion and carbon monoxide by the undesirable base-catalyzed reactions, 4 and 5.

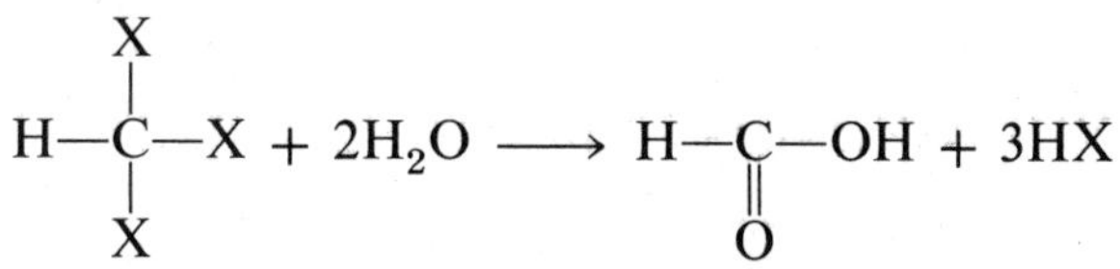

$$H-\underset{\underset{X}{|}}{\overset{\overset{X}{|}}{C}}-X + 2H_2O \longrightarrow H-\underset{\underset{O}{\|}}{C}-OH + 3HX \qquad (4)$$

$$H\underset{\displaystyle \underset{O}{\|}}{-C}-OH + {}^-OH \longrightarrow H\underset{\displaystyle \underset{O}{\|}}{-C}-O^- + H_2O$$

$$:CX_2 + H_2O \xrightarrow{\ {}^-OH\ } :C{\equiv}O + 2HX \qquad\qquad (5)$$

QUATERNARY AMMONIUM SALT CATALYSIS

As an alternative to a homogeneous reaction, a *two-phase* reaction can be considered when the organic phase contains the alkene and a haloform, CHX_3, and the aqueous phase contains the base, OH^-. Unfortunately, under these conditions the reaction will be very slow, since the two primary reactants, CHX_3 and OH^-, are in different phases. The reaction rate can be substantially increased, however, by adding a quaternary ammonium salt such as benzyltriethylammonium chloride as a **phase-transfer catalyst.**

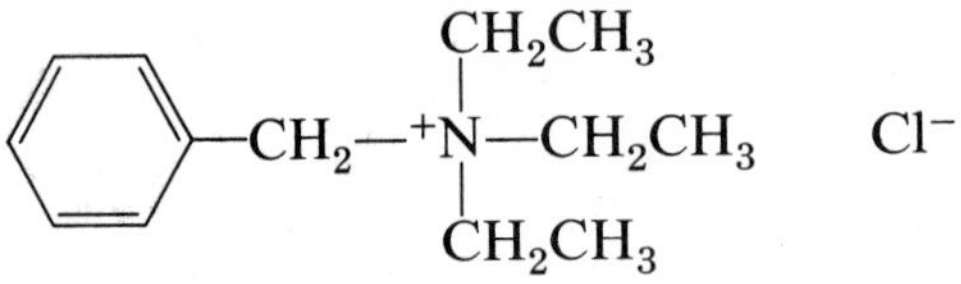

A phase-transfer catalyst: Benzyltriethylammonium chloride

Other common catalysts are tetrabutylammonium bisulfate, trioctylmethylammonium chloride and cetyltrimethylammonium chloride. All these catalysts, including benzyltriethylammonium chloride, have at least 13 carbon atoms. The numerous carbon atoms give the catalyst organic character (lipophilic) and allow it to be soluble in the organic phase. At the same time, the catalyst also has ionic character (hydrophilic) and can therefore be soluble in the aqueous phase.

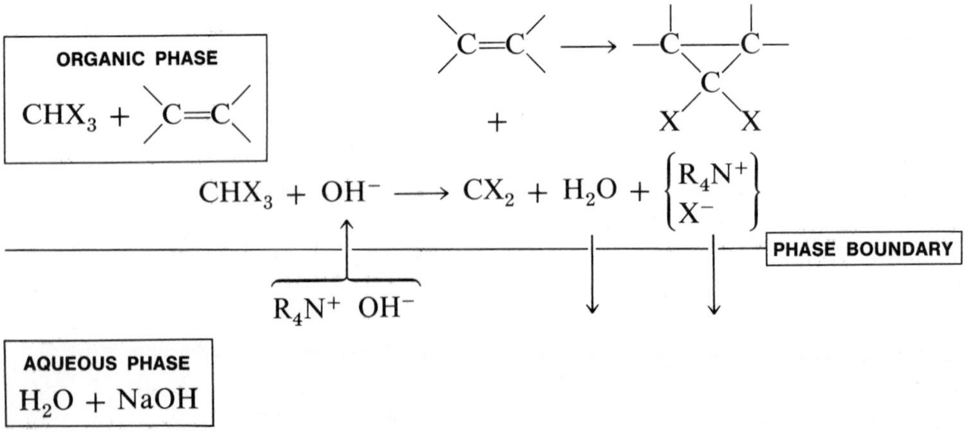

Because of this *dual* nature, the large cation can cross the phase boundary efficiently and transport a hydroxide ion from the aqueous phase to the organic phase (see figure). Once in the organic phase, the hydroxide ion will react with the haloform to give dihalocarbene by reactions 1 and 2. Water, a product of the reaction, will move from the organic phase to the aqueous phase, thus keeping the water concentration in the organic phase at a very low level. Because the water content in the organic phase is low, it will not interfere with the desirable reaction of the carbene with an alkene by reaction 3. Thus, the undesirable side reactions, 4 and 5, are minimized. Finally, the halide ion, which is also produced in reactions 1 and 2, is transported to the aqueous phase by the tetraalkylammonium cation. In this way, electrical neutrality is maintained and the phase-transfer catalyst, R_4N^+, is returned to the aqueous phase, to repeat the whole procedure. The accompanying figure summarizes the overall process. This process probably goes on at the interface rather than in the bulk, organic phase.

There are numerous examples of other reactions that might be effectively accelerated by a quaternary ammonium salt or other phase-transfer catalyst (see references). These reactions often involve simple experimental techniques, give shorter reaction times than noncatalyzed reactions, and avoid relatively expensive aprotic solvents that have been widely used to give one phase. Examples of reactions are shown.

Ether synthesis

$$ROH + R'Cl \xrightarrow[\text{aq. NaOH}]{\text{catalyst}} ROR'$$

Nitrile synthesis

$$R\text{—}Cl + CN^- \xrightarrow[\text{H}_2\text{O}]{\text{catalyst}} RCN$$

Oxidation reactions

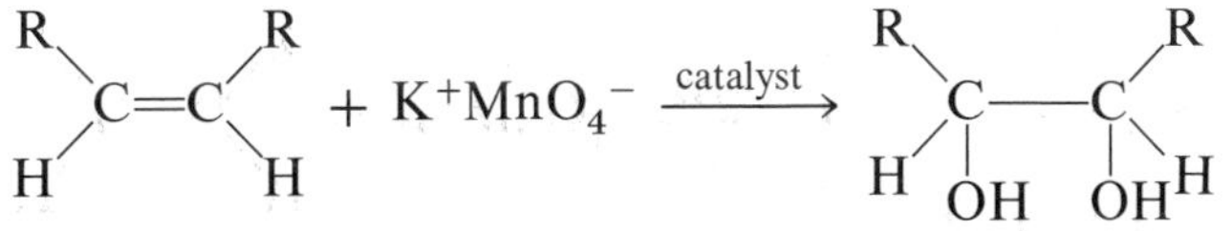

Alkylation reactions

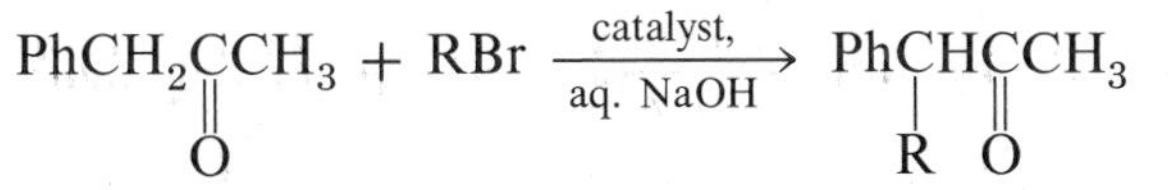

Wittig reactions

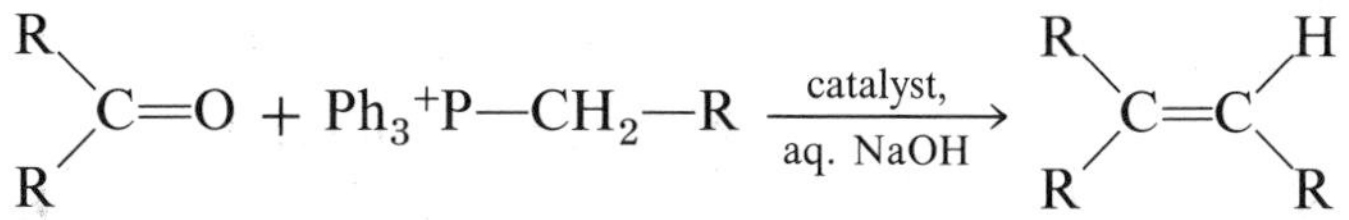

Phosphonium salts act as catalysts.

Increased nucleophilicity

Anions are heavily solvated in an aqueous solvent and are therefore poor nucleophiles in some S_N2 reactions. When they are transported into the organic phase with the catalyst, $R_4N^+\ X^-$, the anion, X^-, is no longer solvated with water and may have increased reactivity.

CROWN ETHER CATALYSIS

Another important class of phase-transfer catalysts includes the crown ethers (not used in this experiment). Crown ethers are used to dissolve organic and inorganic alkali metal salts in organic solvents. The crown ether complexes the cation and provides it with an organic exterior (lipophilic) so that it is soluble in organic solvents. The anion is carried along into solution as the counterion. One example of a crown ether is dicyclohexyl-18-crown-6. Potassium permanganate, $KMnO_4$, complexed to the crown ether is soluble in benzene and is known as "purple benzene." It is useful in various oxidation reactions.

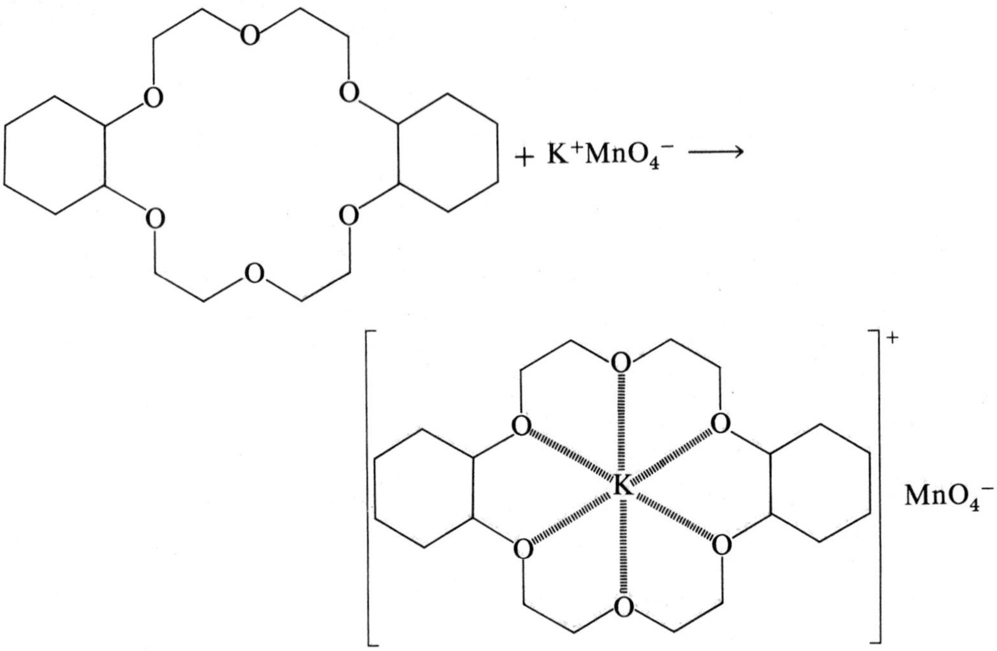

The crown ethers catalyze many of the same types of reactions listed in the preceding section on quaternary ammonium salt catalysis. Crown ethers are very expensive relative to the ammonium salts and are not used as widely for large-scale reactions. In some cases, however, these ethers may be necessary to obtain an efficient and high-yield reaction.

THE EXPERIMENT

We shall prepare 7,7-dichlorobicyclo[4.1.0]heptane, also known as 7,7-dichloronorcarane, by the reaction

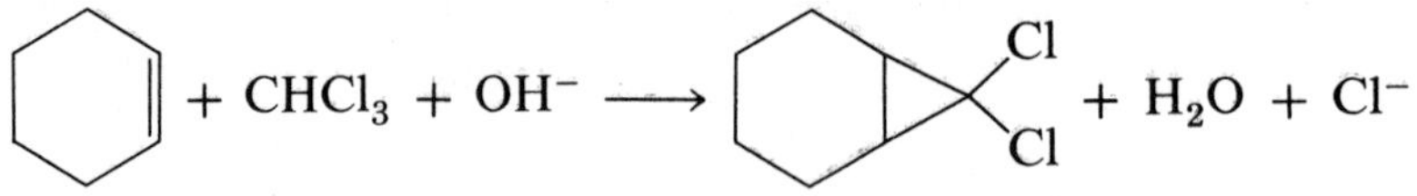

This reaction is exothermic and is complete after a short reaction period. Chloroform, $CHCl_3$, and base are used in excess in this reaction. Although most of the chloroform reacts to give the 7,7-dichloronorcarane via the carbene intermediate, a significant portion is hydrolyzed by the base to formate and carbon monoxide (equations 4 and 5, pp. 205–206). Bromoform, $CHBr_3$, can be used to prepare the corresponding 7,7-dibromonorcarane via the dibromocarbene. The reaction is not exothermic, however, and is much slower than the dichlorocarbene reactions.

SPECIAL INSTRUCTIONS

It is necessary to read Techniques 5 and 6 before starting this experiment. **Chloroform is a suspect carcinogen;** therefore, do not let it touch your skin, and avoid breathing the vapor. Laboratory operations must be conducted in a good hood. The distillation

> **CAUTION: Chloroform is a suspect carcinogen. Work in a hood during all operations involving this toxic substance. Do not breathe the vapor.**

can be done at the desk, since most of the chloroform will have been consumed in the reaction and little of the toxic substance will remain. Swirl the reaction mixture carefully to avoid splashing the caustic 50% sodium hydroxide on your skin; however, it is **absolutely essential** that the mixture be swirled continuously during the half-hour reaction period.

PROCEDURE

Dissolve 15 g of sodium hydroxide in 15 mL of water in a 500-mL Erlenmeyer flask to prepare a 50% aqueous sodium hydroxide solution. Swirl the mixture to help dissolve the solid. After the sodium hydroxide has dissolved, cool the solution to 20 °C with the aid of an ice bath. In another flask, place 10 mL of cyclohexene (density 0.80 g/mL) and 25 mL of chloroform (density 1.49 g/mL), and swirl the flask to mix the liquids.

> **CAUTION: All operations involving chloroform (and cyclohexene) should be conducted in a hood. Avoid contact with this toxic substance.**

Weigh out 1.0 g of the phase-transfer catalyst benzyltriethylammonium chloride on a smooth piece of paper and **reclose the bottle** immediately (it is hygroscopic!).[1] Avoid contact with skin. Transfer the catalyst to the 500-mL flask, and immediately add the cyclohexene-chloroform mixture. **Swirl the mixture as rapidly as possible** for 30 minutes to ensure adequate mixing, but also take care to avoid splashing the caustic mixture on your hand.[2] As the mixture is swirled, a thick emulsion will form, and the temperature will rise. Check the temperature periodically,[1] but do not leave the thermometer in the flask. Maintain a rapid swirling action during the **entire** 30-minute reaction period. Allow the mixture to return to room temperature, with occasional swirling, over an additional 30 to 45 minutes. Do not use an ice bath. Allow the solution to return to room temperature without external cooling.

---

[1]NOTE TO THE INSTRUCTOR: The benzyltriethylammonium chloride used in this experiment was obtained from Hexcel Specialty Chemicals, 215 N. Centennial St., Zeeland, Michigan 49464 (SUMQUAT 2355). If this catalyst is obtained from another supplier, it is advisable to try it in this reaction. A smaller amount may suffice (for example, 0.5 g), and the temperature may have to be moderated so that it does not exceed 50 °C.

[2]If a magnetic stirrer is available, stirring will be safer and more efficient.

Add 75 mL of water to the 500-mL Erlenmeyer flask and transfer the mixture to a 125-mL separatory funnel. Add 20 mL of methylene chloride to the funnel. The funnel will be nearly full, but it will accommodate all the liquids. Shake the funnel vigorously and drain the lower methylene chloride layer. Reextract the aqueous layer with another 20-mL portion of methylene chloride and combine the lower organic layer with the first extract. The small amount of emulsion that forms at the interface should be left behind with the aqueous layer. Discard the aqueous layer, avoiding contact with the basic solution, and rinse the funnel with water. Pour the combined organic layers back into the separatory funnel and extract the mixture with a 30-mL portion of a saturated sodium chloride solution. Drain the lower methylene chloride layer, avoiding any emulsion that might be present at the interface, into a **dry** Erlenmeyer flask containing 2 g of anhydrous magnesium sulfate. Swirl the mixture occasionally for at least 30 minutes to dry the solution or, alternatively, allow the mixture to stand over the drying agent (corked) until the next laboratory period.[3]

Gravity-filter the mixture into a dry flask, add a boiling stone, and evaporate the methylene chloride, together with some cyclohexene and chloroform, on a steam bath in a hood. If there are droplets of water in the residue, methylene chloride should be added and the drying procedure repeated. Pour the residue into a 25-mL round-bottomed flask with the aid of a small amount of methylene chloride, add a boiling stone, and distill the residual liquid by simple distillation, using a microburner as the heat source (Technique 6, Figure 6–6, p. 557). When the temperature reaches 150 °C, change receivers and collect the 7,7-dichloronorcarane in a preweighed dry receiver. Most of the liquid will boil in a narrow range, above 190 °C. You can distill to dryness. Record the observed boiling range. Weigh the 7,7-dichloronorcarane that you obtained and calculate the percentage yield as part of your laboratory report. At the instructor's option, determine the infrared spectrum or the nmr spectrum or both. Submit the product to the instructor in a labeled container.

REFERENCES

Dehmlow, E. V. ''Phase-Transfer Catalyzed Two-Phase Reactions in Preparative Organic Chemistry.'' *Angewandte Chemie, International Edition in English, 13* (1974): 170.

Dehmlow, E. V. ''Advances in Phase-Transfer Catalysis.'' Ibid., *16* (1977): 493.

Gokel, G. W., and Weber, W. P. ''Phase Transfer Catalysis.'' *Journal of Chemical Education, 55* (1978): 350, 429.

Makosza, M., and Wawrzyniewicz, M. ''Catalytic Method for Preparation of Dichlorocyclopropane Derivatives in Aqueous Medium.'' *Tetrahedron Letters* (1969): 4659.

Starks, C. M. ''Phase-Transfer Catalysis.'' *Journal of the American Chemical Society, 93* (1971): 195.

Weber, W. P., and Gokel, G. W. *Phase Transfer Catalysis in Organic Synthesis.* Berlin: Springer-Verlag, 1977.

QUESTIONS

1. Why did you swirl the mixture vigorously during the reaction?

2. Why did you wash the organic phase with saturated sodium chloride solution?

---

[3]If a lot of water seems to have formed a ''puddle,'' decant the liquid away from the drying agent into another dry flask and add some fresh drying agent.

3. What short chemical test could you make on the product to indicate whether cyclohexene is present or absent?

4. Would you expect 7,7-dichloronorcarane to give a positive sodium-iodide-in-acetone test?

5. If you determined the infrared spectrum, assign the C—H stretch for the cyclopropane hydrogens.

6. Suggest why it may be necessary to use a large excess of chloroform in this reaction.

7. A student obtained an nmr spectrum of 7,7-dichloronorcarane. The spectrum showed a peak at about 7.3δ and another peak at about 5.6δ. What do you think these peaks indicate? Are they part of the 7,7-dichloronorcarane spectrum?

8. Draw the structures of the products that you would expect from the reactions of *cis-* and *trans-*2-butene with dichlorocarbene.

9. Draw the structure of the expected dichlorocarbene adduct of methyl methacrylate (methyl 2-methylpropenoate). With compounds of this type, another product could have been obtained. It is the chloroform adduct to the double bond (Michael-type reaction). What would this structure look like?

10. Provide mechanisms for the following abnormal dichlorocarbene addition reactions. In both cases, the usual adduct is first obtained, and then a subsequent reaction occurs.

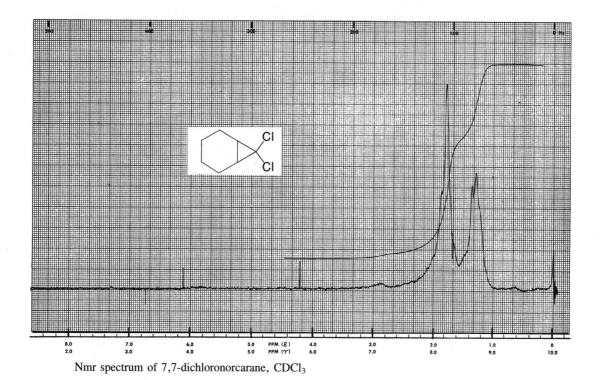

Nmr spectrum of 7,7-dichloronorcarane, CDCl$_3$

Experiment 29
Markovnikov and Anti-Markovnikov Hydration of Styrene

Addition to Double Bonds
Hydroboration-Oxidation
Oxymercuration
Gas Chromatography
Nuclear Magnetic Resonance
Regiospecificity-Regioselectivity

The most characteristic reaction of alkenes is the **addition reaction.** This reaction can be described by the general equation

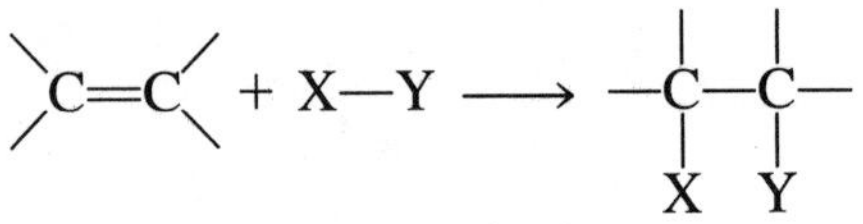

$$\text{\Large >C=C<} + \text{X—Y} \longrightarrow \text{—C—C—}\ _{\overset{|}{X}\ \overset{|}{Y}}$$

In this reaction, the π bond of the alkene is broken, and two σ bonds are formed.

Because the double bond is an electron-rich center, it behaves like a Lewis base, or a nucleophile, donating a pair of electrons to form a bond to a sufficiently electrophilic reagent. Some part of the attacking reagent XY must be electrophilic. As an example, in the reaction of an alkene with an acid, HA, the first step of the mechanism involves the reaction of the double bond with the electrophilic species, H^+. This step produces a cation:

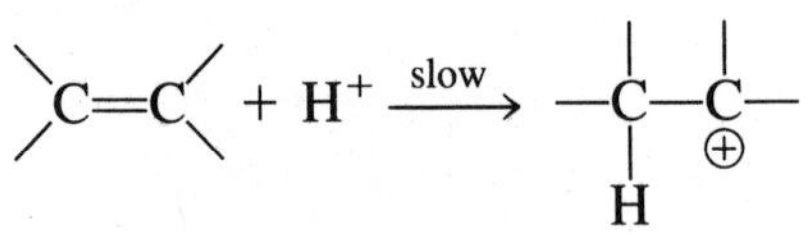

$$\text{\Large >C=C<} + H^+ \xrightarrow{\text{slow}} \text{—C—C—}\ _{\overset{|}{H}\ \oplus}$$

This intermediate cation can react rapidly with the anion, A^-, to yield the final product:

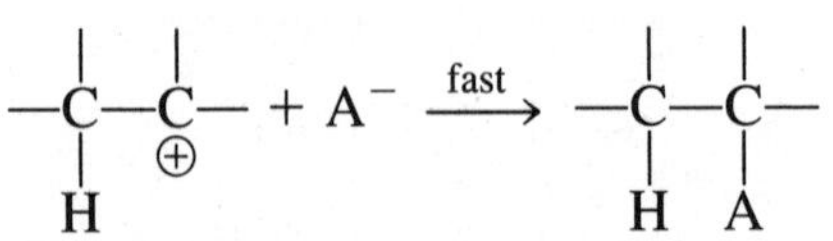

$$\text{—C—C—}\ _{\overset{|}{H}\ \oplus} + A^- \xrightarrow{\text{fast}} \text{—C—C—}\ _{\overset{|}{H}\ \overset{|}{A}}$$

A complication arises when an unsymmetrical reagent, HA, adds to an unsymmetrically substituted double bond. In this case, there are two possible products:

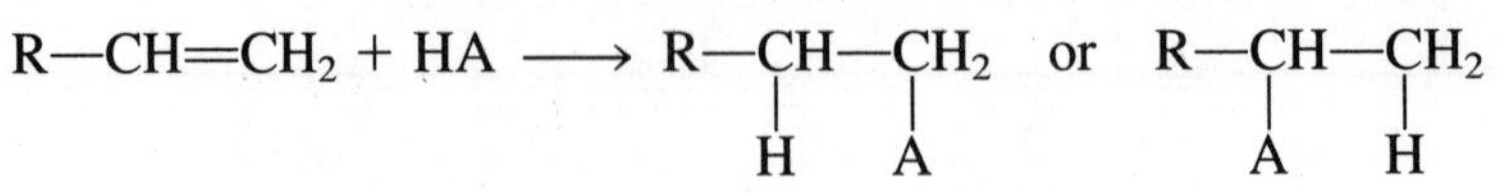

$$R—CH{=}CH_2 + HA \longrightarrow R—\underset{\overset{|}{H}}{CH}—\underset{\overset{|}{A}}{CH_2} \quad \text{or} \quad R—\underset{\overset{|}{A}}{CH}—\underset{\overset{|}{H}}{CH_2}$$

212

An empirical rule, known as *Markovnikov's rule,* was developed in 1868 by the Russian chemist V. V. Markovnikov to apply to such cases. Simply stated, the rule says

> In the ionic addition of an acid to the carbon-carbon double bond of an alkene, the hydrogen of the acid attaches itself to the carbon atom that already holds the **greater** number of hydrogens.

In recent years, chemists have determined that the reason Markovnikov's rule holds is that addition according to the rule always leads to the more highly substituted, and hence the more stable, of the two possible cationic intermediates.

The addition of water to alkenes under acidic conditions (hydration) follows Markovnikov's rule. The hydrogen of water attaches itself to the carbon atom that already carries the greater number of hydrogens:

$$R\!-\!CH\!=\!CH_2 + H_2O \longrightarrow R\!-\!\underset{\underset{OH}{|}}{CH}\!-\!\underset{\underset{H}{|}}{CH_2} \qquad \textbf{MARKOVNIKOV ADDITION}$$

If the addition of water to the alkene had proceeded contrary to Markovnikov's rule, the reaction would be said to have gone in an "anti-Markovnikov" fashion, and the product would be designated the anti-Markovnikov product:

$$R\!-\!CH\!=\!CH_2 + H_2O \longrightarrow R\!-\!\underset{\underset{H}{|}}{CH}\!-\!\underset{\underset{OH}{|}}{CH_2} \qquad \textbf{ANTI-MARKOVNIKOV ADDITION}$$

Our current experiment deals with two practical methods for carrying out the hydration of an alkene in both the Markovnikov and the anti-Markovnikov fashion. These methods are the **hydroboration-oxidation** and the **oxymercuration** reactions.

HYDROBORATION-OXIDATION

Diborane can react with alkenes by addition of the boron-hydrogen bond across the carbon-carbon double bond. This reaction is called **hydroboration.** Diborane is prepared by the reaction of sodium borohydride with boron trifluoride:

$$3NaBH_4 + 4BF_3 \xrightarrow{\text{ether}} 2B_2H_6 + 3NaBF_4$$

Diborane, B_2H_6, is an unusual substance. It has two three-center bonds. In these bonds, one pair of electrons bonds three atoms together—two boron atoms and one hydrogen atom. Without getting into the details of this bonding scheme, it seems a reasonable simplification to describe the reactions of diborane as if it were a dimer, composed of two units of borane, BH_3. Some studies even suggest that BH_3 may be in equilibrium with diborane.

**SIMPLIFIED
APPROXIMATION**

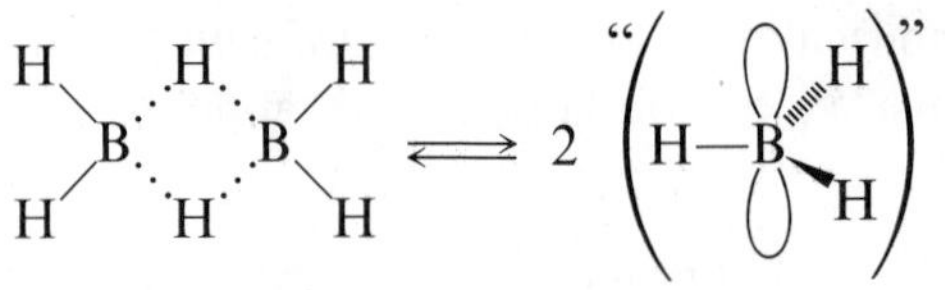

Diborane, because of its **three-center bonds,** is electron-deficient and is a good electrophile. Similarly, BH_3 is also a good electrophile, since it contains an incomplete octet of electrons. With BH_3 as the boron hydride, an alkene will donate electrons to boron as shown in the first step of the reaction below. This creates a boron atom that is electron-rich and a carbon atom that is electron-deficient. This situation is remedied in the next step by the transfer of a hydride ion ($H:^-$) to the carbon atom:

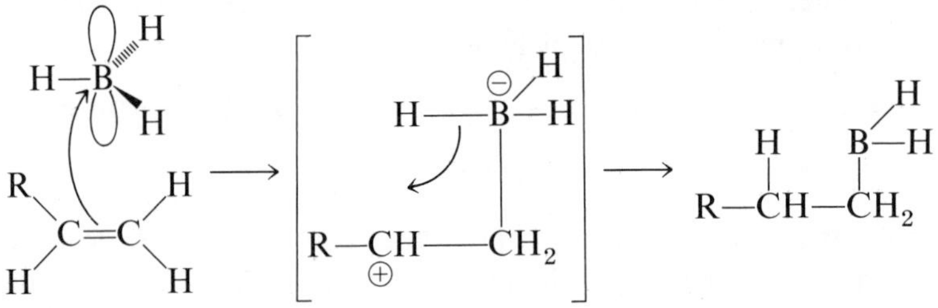

Experimental evidence shows that this entire process is **concerted** without the presence of intermediates and that the addition is stereospecific. The addition of borane to an alkene proceeds exclusively with **syn** stereochemistry. Both the boron and the hydrogen atoms add to the same side of the double bond:

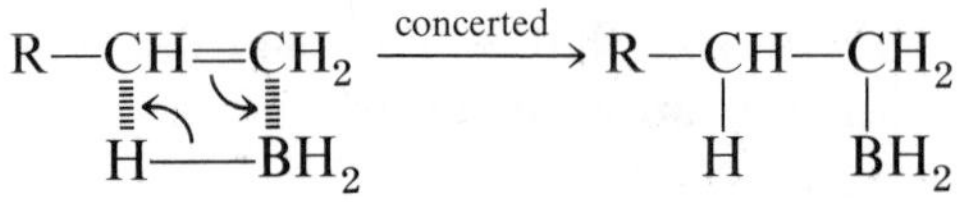

$$R-CH=CH_2 \xrightarrow{concerted} R-CH-CH_2$$

Notice that this addition is an **anti-Markovnikov** addition. In contrast to the situation that applies for the addition of an acid, HA, the hydrogen from diborane does not become attached to the carbon atom with the greater number of hydrogens. However, the electrophilic species that was added to the double bond in this case was not H^+ but the electron-deficient boron atom. The lowest-energy cationic intermediate would still predict the course of the reaction (even though it does not exist in hydroboration), and the hydrogen is transferred as $H:^-$ rather than as H^+. This addition occurs three times, because there are three B—H bonds, and causes a trialkylborane to form.

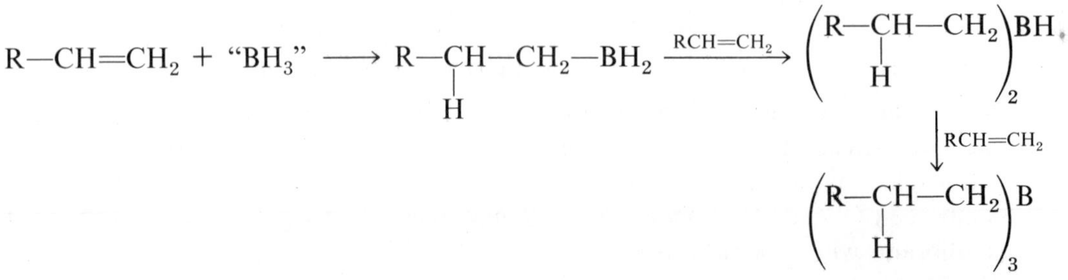

A trialkylborane

If the trialkylborane compound is treated with alkaline hydrogen peroxide, it is cleaved with oxidation to form three moles of an alcohol. As you can see from the mechanism shown, the replacement of a boron atom by an oxygen atom is an intramolecular process. As a result, this oxidation step is also stereospecific, proceeding entirely with **retention of configuration.** This oxidation proceeds via a migration of the alkyl group from boron to oxygen, generating a trialkoxyborane as an important intermediate:

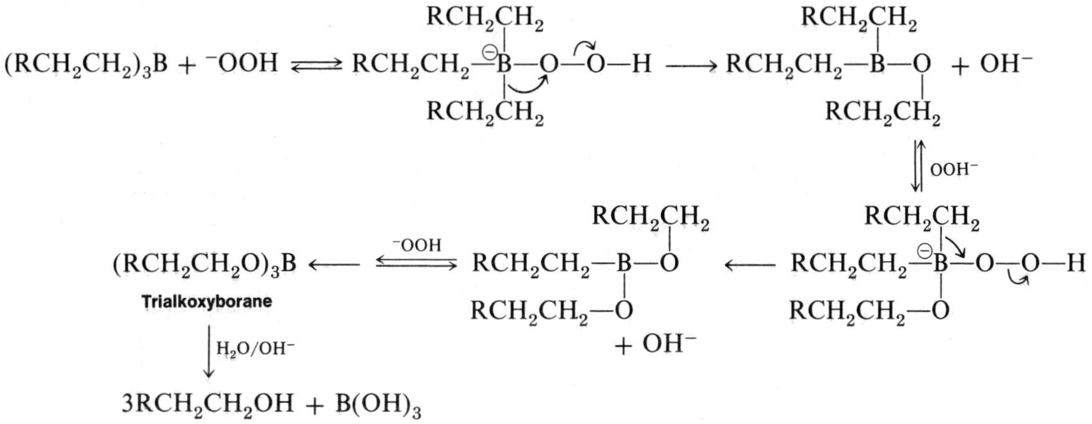

$$
(RCH_2CH_2)_3B + {}^-OOH \rightleftharpoons RCH_2CH_2\overset{\underset{RCH_2CH_2}{|}}{\overset{RCH_2CH_2}{|}}\!\overset{\ominus}{B}\!-\!O\!-\!O\!-\!H \longrightarrow RCH_2CH_2\overset{\underset{RCH_2CH_2}{|}}{\overset{RCH_2CH_2}{|}}\!B\!-\!O + OH^-
$$

$$\Updownarrow OOH^-$$

$$
(RCH_2CH_2O)_3B \longleftarrow \overset{{}^-OOH}{\rightleftharpoons} RCH_2CH_2\overset{\underset{RCH_2CH_2\!-\!O}{|}}{\overset{RCH_2CH_2}{|}}\!B\!-\!O + OH^- \longleftarrow RCH_2CH_2\overset{\underset{RCH_2CH_2\!-\!O}{|}}{\overset{RCH_2CH_2}{|}}\!\overset{\ominus}{B}\!-\!O\!-\!O\!-\!H
$$

Trialkoxyborane

$$\Big\downarrow H_2O/OH^-$$

$$3RCH_2CH_2OH + B(OH)_3$$

Hydroboration-oxidation of an alkene therefore leads to an alcohol that corresponds to the anti-Markovnikov addition of water across the double bond of an alkene. This seems to violate Markovnikov's rule. However, analysis of the mechanism of the hydroboration-oxidation reaction shows that this is not the case:

$$3RCH{=}CH_2 \xrightarrow{B_2H_6} (RCH_2CH_2)_3B \xrightarrow[OH^-]{H_2O_2} 3RCH_2CH_2OH$$

Finally, the anti-Markovnikov addition of water across the double bond proceeds with **syn** stereochemistry; this can best be illustrated using a cyclic alkene as an example. The reaction results from the intramolecular nature of the hydroboration and the oxidation steps, as already described.

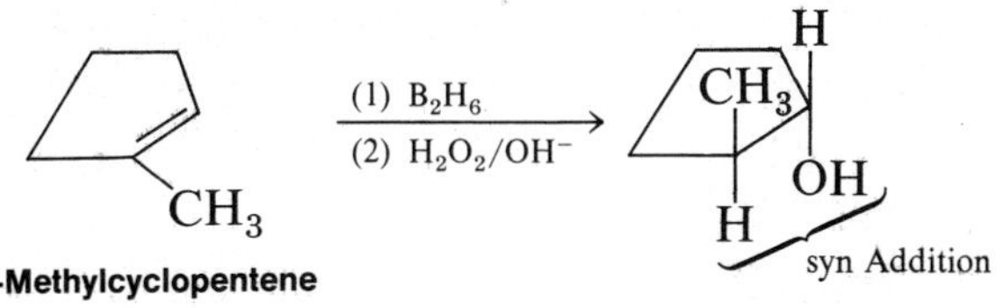

1-Methylcyclopentene → (1) B_2H_6 / (2) H_2O_2/OH^- → syn Addition → *trans*-2-Methylcyclopentanol

OXYMERCURATION

The second means of hydrating a double bond we consider is the **oxymercuration** reaction. In this reaction, mercuric acetate is added to an alkene to form an organomercury derivative:

$$RCH\!\!=\!\!CH_2 + Hg(OCOCH_3)_2 \xrightarrow{H_2O} \underset{\underset{OH}{|}}{RCH}\!\!-\!\!CH_2\!\!-\!\!HgOCOCH_3$$

The mechanism of this reaction involves initial ionization of mercuric acetate, followed by addition of the mercury atom across the double bond of the alkene to form a bridged cationic intermediate. This intermediate is then opened by the water, which is a nucleophile. This reaction is also stereospecific, proceeding with **anti** stereochemistry.

$$Hg(OCOCH_3)_2 \rightleftharpoons {}^+HgOCOCH_3 + CH_3COO^-$$

$$R\!\!-\!\!CH\!\!=\!\!CH_2 + {}^+HgOCOCH_3 \xrightarrow{\text{slow}} \begin{array}{c} H_2\ddot{O}: \\ R\!\!-\!\!CH\!\!-\!\!CH_2 \\ \underset{\oplus}{Hg\!\!-\!\!OCOCH_3} \end{array}$$

$$\downarrow$$

$$\underset{\underset{HgOCOCH_3}{|}}{R\!\!-\!\!\overset{\overset{OH}{|}}{CH}\!\!-\!\!CH_2} \qquad + H^+$$

The oxymercuration reaction is different from many addition reactions of alkenes since it involves no rearrangements.

     The organomercury intermediate is reduced with sodium borohydride, with the result that the mercury atom is replaced by hydrogen. The reduction of the organomercury intermediate is not stereospecific.

$$\underset{\underset{OH}{|}}{RCH}\!\!-\!\!CH_2\!\!-\!\!HgOCOCH_3 \xrightarrow[\text{NaOH}]{\text{NaBH}_4} \underset{\underset{OH}{|}}{RCH}\!\!-\!\!CH_3$$

     The product obtained from the oxymercuration of an alkene is the same as what would be expected for the hydration of an alkene according to Markovnikov's rule. In this particular reaction, the mercury atom is the original electrophile, so it adds to the end of the double bond that bears the greater number of hydrogens. When the mercury atom is replaced by hydrogen in the reduction step of the reaction, hydrogen is then attached to the carbon atom in the manner the rule predicts.

     In these two reactions, we see contrasting behavior as to the direction of addition of water. Although each of the reactions has been described as though it produced exclusively the product shown, in truth it must be mentioned that in some cases there may be a minor product with the opposite orientation. That is to say, while hydration of an alkene may produce the product predicted by Markovnikov's rule as the principal product, the anti-Markovnikov alcohol may also be produced as a minor product. Reactions that produce a product with only one of several possible orientations are called **regiospecific.** Reactions that produce one substance as the predominant product and small amounts of isomers with other orientations are called **regioselective.** One of the objects of this experiment is to contrast the hydroboration-oxidation and the oxy-mercuration of styrene to determine whether they are regiospecific or regioselective.

Procedure 29A Hydroboration-Oxidation of Styrene

The balanced equation for the hydroboration of styrene, using a borane-tetrahydrofuran solution, is

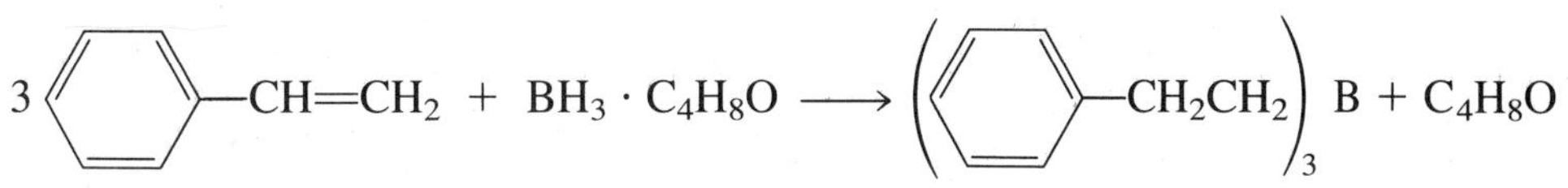

$$3 \text{ } \text{C}_6\text{H}_5-\text{CH}=\text{CH}_2 + \text{BH}_3 \cdot \text{C}_4\text{H}_8\text{O} \longrightarrow \left(\text{C}_6\text{H}_5-\text{CH}_2\text{CH}_2 \right)_3 \text{B} + \text{C}_4\text{H}_8\text{O}$$

Styrene

Once the organoborane intermediate is formed, it is oxidatively hydrolyzed with a basic solution of hydrogen peroxide:

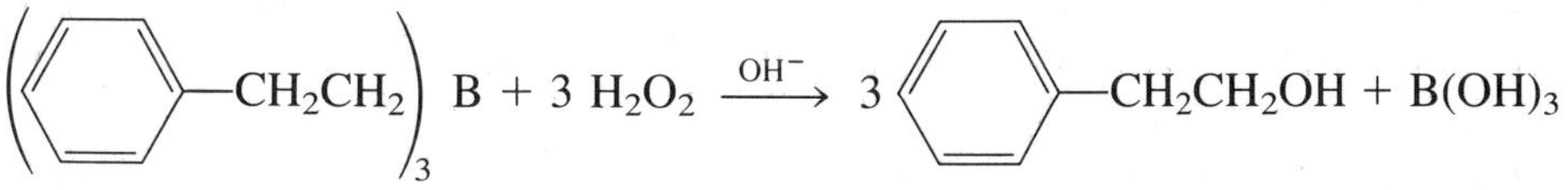

$$\left(\text{C}_6\text{H}_5-\text{CH}_2\text{CH}_2 \right)_3 \text{B} + 3 \text{ H}_2\text{O}_2 \xrightarrow{\text{OH}^-} 3 \text{ } \text{C}_6\text{H}_5-\text{CH}_2\text{CH}_2\text{OH} + \text{B(OH)}_3$$

2-Phenylethanol

In the method used in this experiment, a solution of diborane in tetrahydrofuran is added to the alkene styrene. The diborane is consumed as it is added. The solution is a commercially available product. The addition method avoids the need for handling the potentially hazardous (toxic and flammable) diborane.

SPECIAL INSTRUCTIONS

Because this reaction involves a somewhat lengthy period of preparation of apparatus and addition of reagents, it is important that it be started at the beginning of the laboratory period. The best place to stop is when the ether extracts are stored over magnesium sulfate, although the reaction can be stopped at any time after water has been added to the reaction, if necessary.

You should consult an organic chemistry lecture textbook for detailed informa-tion about the hydroboration-oxidation reaction.

This reaction involves the use of tetrahydrofuran, which is a potentially toxic and flammable solvent. Do not conduct this reaction in the presence of open flames. Avoid contact with 30% hydrogen peroxide, as it is a strong oxidant.

PROCEDURE

Assemble an apparatus consisting of a 50-mL round-bottomed flask equipped with a Claisen head. Attach a drying tube filled with calcium chloride to each of the openings of the Claisen head and heat all the glass surfaces *gently,* but thoroughly, with a Bunsen burner flame to drive out any water that may be absorbed on the surface of the glass. Be certain to conduct this flaming-out procedure in a place far from any flammable solvents. Allow the glassware to cool to room temperature.

  After the glass has cooled, remove the two drying tubes. Place a magnetic stirring bar in the flask and add 2.2 mL (density 0.9 g/mL) of styrene. Close the openings in the Claisen head with thermometer adapters sealed with rubber septa (see figure). Insert a hypodermic syringe needle through one of the septa. Using a syringe, draw 6.0 mL of a commercial borane–tetrahydrofuran solution from the stock bottle. Carefully insert the syringe through the other rubber septum. Place a magnetic stirring motor and an ice bath under the round-bottomed flask, and begin stirring the solution. Carefully add the

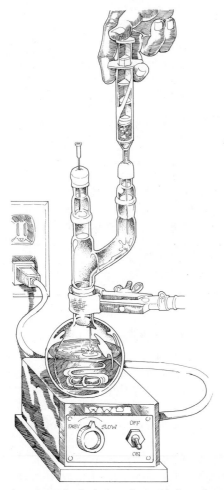

Apparatus for hydroboration-oxidation of styrene

borane–tetrahydrofuran solution dropwise to the reaction flask over a period of half an hour. After the borane–tetrahydrofuran solution has been added, allow the reaction mixture to stand in the ice bath, with stirring, for an hour.

At the end of this period, transfer the reaction mixture (including the magnetic stirring bar) to a 125-mL Erlenmeyer flask, and immerse this flask in an ice bath on a magnetic stirring motor. While the reaction is stirring, slowly add 2.1 mL of 30% hydrogen peroxide. During this addition, maintain the pH of the solution near 8 (use pH paper) by adding 3M sodium hydroxide as needed. The reaction mixture should bubble as the hydrogen peroxide is added.

Slowly add 20 mL of water to the reaction mixture, followed by 10 mL of diethyl ether (solvent grade). Allow the mixture in the flask to stir to mix the liquid and solid materials adequately; then decant the liquid into a 250-mL separatory funnel. Remove the ether layer, and reextract the aqueous phase with four successive 10-mL portions of ether. Combine the ether extracts, wash them with 20 mL of saturated sodium bicarbonate solution, and dry them over anhydrous magnesium sulfate.

Remove the ether solution from the drying agent and evaporate the ether from the reaction products, using a steam bath in the hood. When the ether has been evaporated, remove any residual tetrahydrofuran that may remain in the product by evaporation under reduced pressure (Technique 1, Figure 1–7, p. 513). Be sure to use a vacuum trap. Apply gentle heat from the steam bath during this evaporation. Do not overheat the solution, because some of the residual styrene may polymerize. When the bubbling in the flask ceases, remove the vacuum and weigh the residual alcohols to determine the percentage yield obtained from the reaction.[1] Analyze the mixture of alcohols using gas chromatography, nmr spectroscopy, or both to determine the relative percentages of Markovnikov and anti-Markovnikov products.[2] Compare your results with the results obtained by students who performed the oxymercuration of styrene (Procedure 29B). In your laboratory report, submit the gas chromatograms and nmr spectra you obtained. Calculate the percentage yield of alcohol isomers and the ratios of the two isomers from your chromatographic or spectral data. Comment on whether or not the reaction proceeded according to Markovnikov's rule and whether the reaction was regiospecific or regioselective.

---

[1]Some unreacted styrene may be present in the mixture of alcohols. If an appreciable amount of styrene is present, the percentage of styrene from the gas chromatogram can be used to subtract the contribution of the styrene from the total weight of products. This will give the actual weight of the isomeric alcohols.

[2]The correct conditions for the gas chromatographic analysis are

Column temperature: 150 °C
Injection port temperature: 150–155 °C
Column packing: SE-30 silicone oil on Chromosorb W
Relative order of retention times: styrene, 1-phenylethanol, 2-phenylethanol

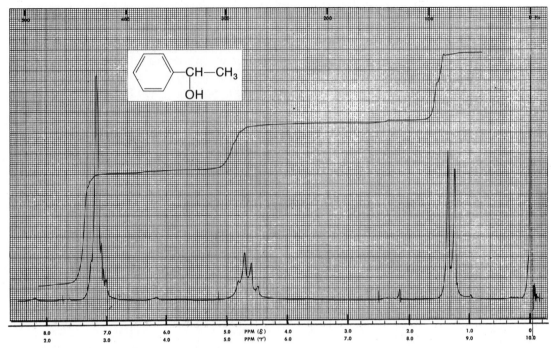

Nmr spectrum of 1-phenylethanol, neat

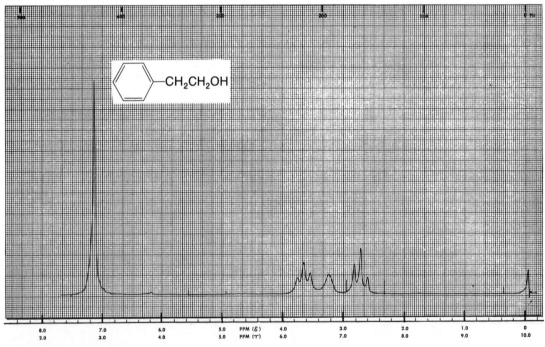

Nmr spectrum of 2-phenylethanol, neat

Procedure 29B
Oxymercuration of Styrene

The balanced equation for the oxymercuration of styrene is

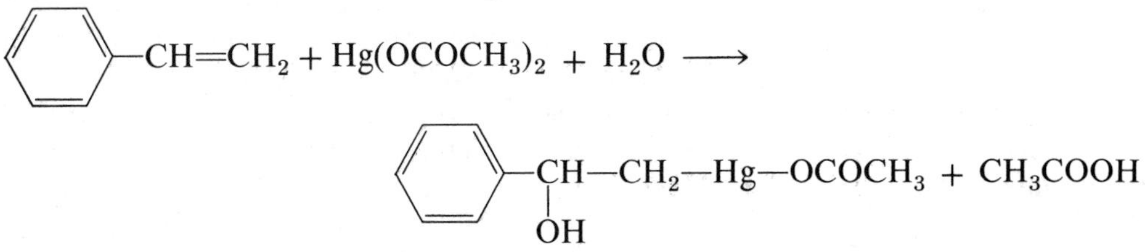

$$\text{C}_6\text{H}_5-\text{CH}=\text{CH}_2 + \text{Hg(OCOCH}_3)_2 + \text{H}_2\text{O} \longrightarrow$$

$$\text{C}_6\text{H}_5-\underset{\underset{\text{OH}}{|}}{\text{CH}}-\text{CH}_2-\text{Hg}-\text{OCOCH}_3 + \text{CH}_3\text{COOH}$$

Reduction of the organomercury derivative with sodium borohydride yields the alcohol:

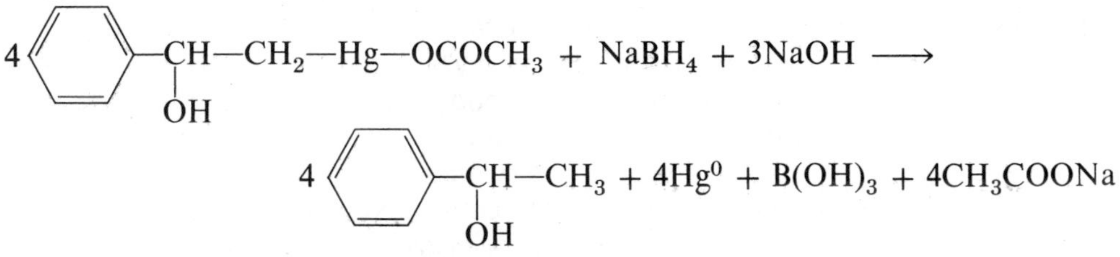

$$4\ \text{C}_6\text{H}_5-\underset{\underset{\text{OH}}{|}}{\text{CH}}-\text{CH}_2-\text{Hg}-\text{OCOCH}_3 + \text{NaBH}_4 + 3\text{NaOH} \longrightarrow$$

$$4\ \text{C}_6\text{H}_5-\underset{\underset{\text{OH}}{|}}{\text{CH}}-\text{CH}_3 + 4\text{Hg}^0 + \text{B(OH)}_3 + 4\text{CH}_3\text{COONa}$$

1-Phenylethanol

SPECIAL INSTRUCTIONS

Magnetic stirring motors and stirring bars are needed for this experiment. The experiment requires that the reaction mixture be allowed to stand overnight. As two lengthy periods of stirring are necessary, this experiment must be started at the beginning of the laboratory period.

You should consult an organic chemistry lecture textbook for detailed information about the oxymercuration reaction.

This experiment involves mercuric acetate. Like all mercury compounds, mercuric acetate is very toxic. Be careful to wash your hands thoroughly after handling this substance. Be careful not to get it on your hands or face.

PROCEDURE

Set up an apparatus consisting of a 250-mL two-necked, round-bottomed flask (see figure on p. 25, Number 3), equipped with a thermometer, a magnetic stirrer, and a

Claisen head. Equip the two necks of the Claisen head with an addition funnel and a condenser. Place 7.9 g of mercuric acetate and 25 mL of water in the flask. After the

> **CAUTION: Mercuric acetate is a highly poisonous substance.**

mercuric acetate has dissolved, add 25 mL of ether. While stirring the mixture vigorously, add 2.6 g of styrene and continue stirring at room temperature for 1 hour. After 1 hour, add 12.5 mL of 6M sodium hydroxide solution followed by a solution of 0.5 g of sodium borohydride in 25 mL of 3M sodium hydroxide. As this solution is added, the temperature of the reaction must be maintained at or below 25 °C, using an ice bath.

Stir this mixture for 1.5 to 2 hours, after which time most of the mercury should have settled to the bottom of the flask. Allow this mixture to settle until the next laboratory period. After the mixture has settled, carefully decant the supernatant liquid from the mercury, which has been deposited on the bottom of the flask. Discard the mercury in a suitable waste container. NEVER DISCARD THE MERCURY DOWN THE SINK DRAIN. Separate the ether layer from the aqueous layer and save it. Extract the aqueous layer with three successive 10-mL portions of ether. Combine the ether solutions and dry them over anhydrous magnesium sulfate.

Gravity-filter the ether solution from the drying agent and evaporate the ether solution to dryness on the steam bath in the hood. If the resulting product as a suspended fine precipitate, centrifuge the liquid before analyzing it by gas chromatography. Otherwise, the fine particles will likely plug the syringe used in the gas chromatographic analysis. Analyze the resulting liquid, using gas chromatography, to determine the relative percentages of Markovnikov and anti-Markovnikov products.[3] At the option of the instructor, obtain an nmr spectrum of the products, and analyze the spectrum to determine the relative percentages of the products. The nmr spectra of 2-phenylethanol and 1-phenylethanol are given in Procedure 29A for reference. In your report to the instructor, submit the gas chromatograms and nmr spectra you obtained. Calculate the ratios of the two alcohols, using the gas chromatograms and, if applicable, the nmr spectra. Comment on whether or not the reaction followed Markovnikov's rule and whether the reaction was regiospecific or regioselective. Compare your results with the results obtained by students who performed Procedure 29A.

REFERENCES

Brown, H. C. "Hydride Reductions: A 40-Year Revolution in Organic Chemistry." *Chemical and Engineering News, 57* (March 5, 1979): 24.

Brown, H. C. *Hydroboration.* Reading, MA: Benjamin/Cummings, 1980.

Brown, H. C., and Zweifel, G. "Hydroboration. VII. Directive Effects in the Hydroboration of Olefins." *Journal of the American Chemical Society, 82* (1960): 4708.

---

[3]Some unreacted styrene may be obtained in the products. The conditions for the gas chromatography are identical to those given in Procedure 29A, p. 219.

Jerkunica, J. M., and Traylor, T. G. "Oxymercuration-Reduction: Alcohols from Olefins: 1-Methylcyclohexanol." *Organic Syntheses, 53* (1973): 94.
Zweifel, G., and Brown, H. C. "Hydroboration of Olefins: (+)-Isopinocampheol." *Organic Syntheses, 52* (1972): 59.

QUESTIONS

1. Compare the relative percentages of 1-phenylethanol and 2-phenylethanol formed in each of the two reactions. Classify each reaction as regiospecific or regioselective.
2. Predict the products of the hydroboration-oxidation and the oxymercuration reactions for each of the following compounds. Include the correct stereochemistry, when appropriate.

(a)

(c)

(b)

(d)

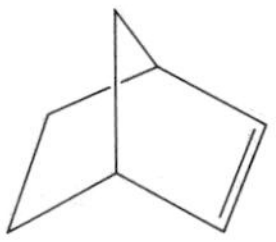

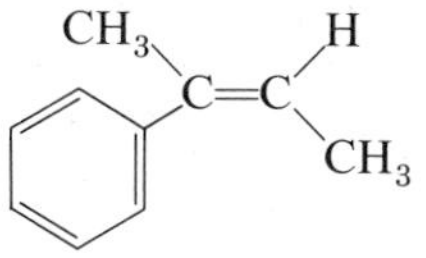

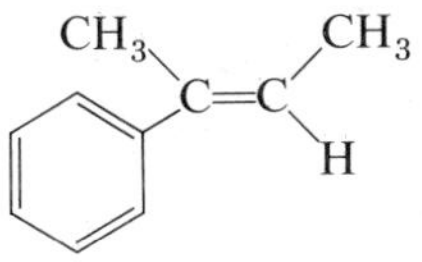

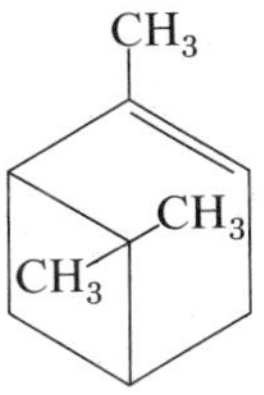

Experiment 30 Triphenylmethanol and Benzoic Acid

Grignard Reactions
Extraction

In this experiment a Grignard reagent, or organomagnesium reagent, is prepared. The reagent is phenylmagnesium bromide.

Bromobenzene **Phenylmagnesium bromide**

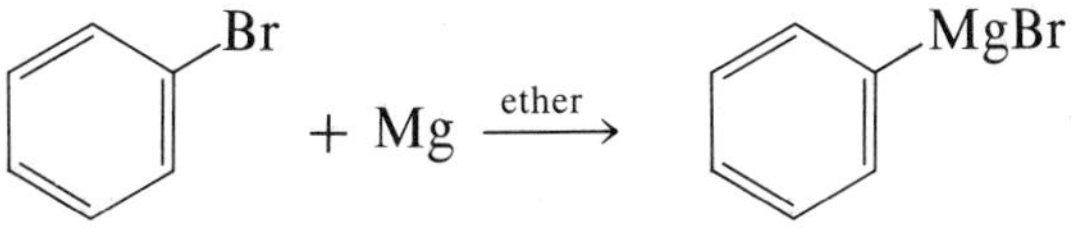

This reagent will be converted to a tertiary alcohol or a carboxylic acid, depending on the procedure selected.

PROCEDURE 30A

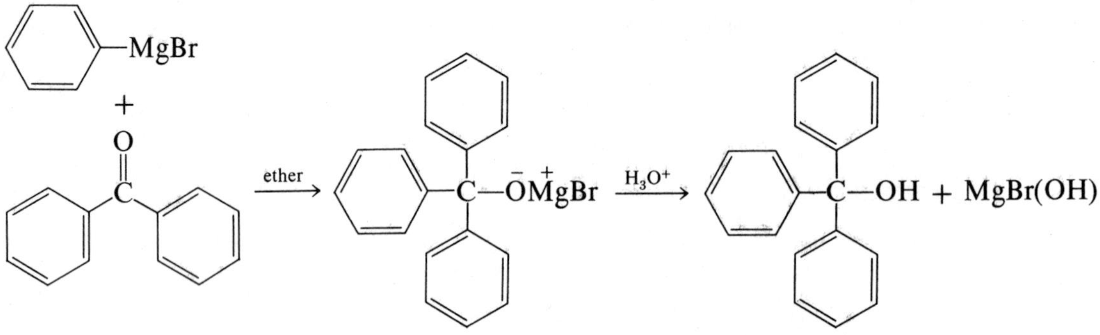

Benzophenone   Triphenylmethanol

PROCEDURE 30B

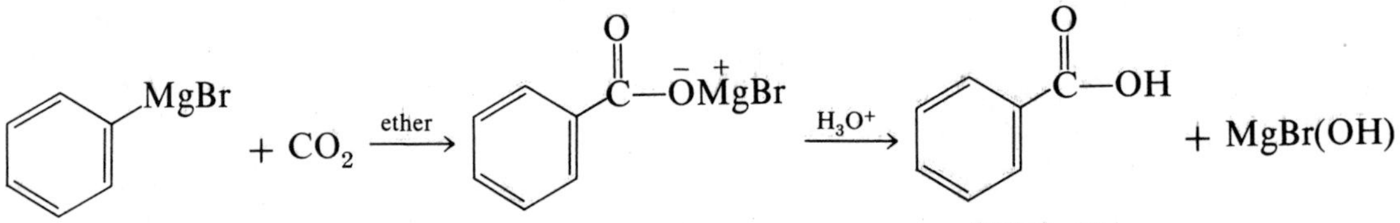

Benzoic acid

The alkyl portion of the Grignard reagent behaves as if it had the characteristics of a **carbanion.** We may write the structure of the reagent as a partially ionic compound: $\overset{\delta-}{R}\cdots\overset{\delta+}{MgX}$. This partially bonded carbanion is a Lewis base. It reacts with strong acids, as you would expect, to give an alkane:

$$\overset{\delta-}{R}\cdots\overset{\delta+}{MgX} + HX \longrightarrow R\!-\!H + MgX_2$$

Any compound with a suitably acidic hydrogen will donate a proton to destroy the reagent. Water, alcohols, terminal acetylenes, phenols, and carboxylic acids are all acidic enough to bring about this reaction.

The Grignard reagent also functions as a good nucleophile in nucleophilic addition reactions of the carbonyl group. The carbonyl group has electrophilic character at its carbon atom (due to resonance), and a good nucleophile seeks out this center for addition.

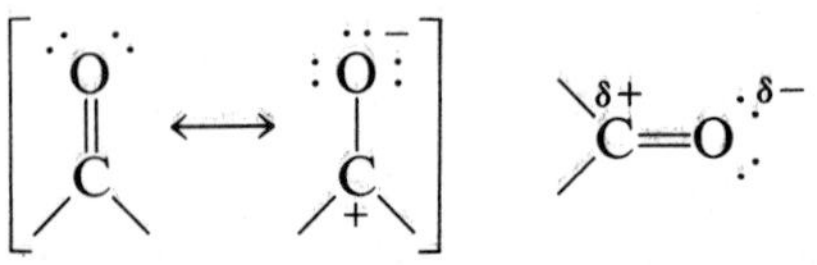

The magnesium salts produced form a complex with the addition product, an alkoxide salt, and in a second step of the reaction, these must be protonated by addition of dilute aqueous acid:

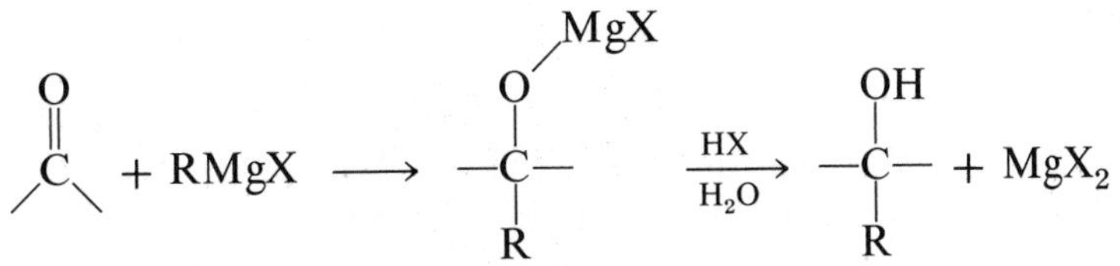

The Grignard reaction is used synthetically to prepare secondary alcohols from aldehydes and tertiary alcohols from ketones. The Grignard reagent will react with esters twice to give tertiary alcohols. Synthetically, it also can be allowed to react with carbon dioxide to give carboxylic acids

and with oxygen to give hydroperoxides

$$RMgX + O_2 \longrightarrow ROOMgX \xrightarrow[H_2O]{HX} ROOH$$

Because the Grignard reagent reacts with water, carbon dioxide, and oxygen, it must be protected from air and moisture when it is used. The apparatus in which the reaction is to be conducted must be scrupulously dry (recall that 18 mL of H_2O is 1 mole), and the solvent must be free of water, or anhydrous. During the reaction, the flask must be protected by a calcium chloride drying tube. Oxygen should also be excluded. In practice this can be done by allowing the solvent ether to reflux. This blanket of solvent vapor keeps air from the surface of the reaction mixture.

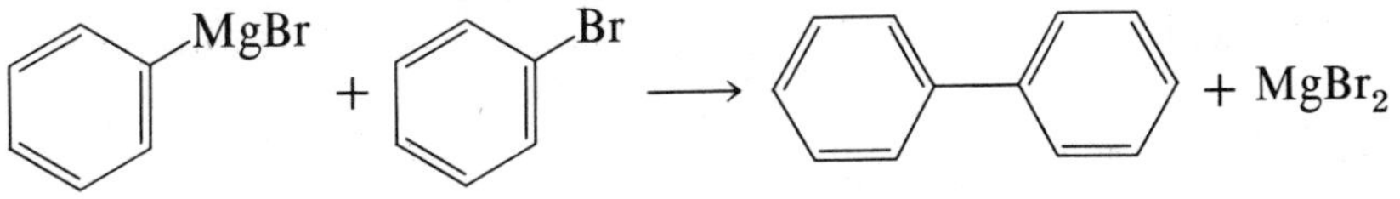

Biphenyl

In the experiment described here, the principal impurity is **biphenyl,** which is formed by a heat- or light-catalyzed coupling reaction of the Grignard reagent and unreacted bromobenzene. A high reaction temperature favors the formation of this product. Biphenyl is highly soluble in petroleum ether, and it is easily separated from triphenylmethanol by crystallization from that solvent and from benzoic acid by extraction.

SPECIAL INSTRUCTIONS

Before beginning this experiment, read Techniques 1 and 5. This reaction must be conducted in one laboratory period either to the point after which benzophenone is added (Procedure 30A) or to the point after which the Grignard reagent is poured over Dry Ice (Procedure 30B). The Grignard reagent cannot be stored.

FIRE SAFETY

This reaction involves large quantities of diethyl ether, which is extremely flammable. Be certain that no open flames of any sort are in your vicinity when you are using ether. We recommend that you reread the introductory chapter on laboratory safety.

APPARATUS

All glassware used in a Grignard reaction must be **scrupulously** dried. Surprisingly large amounts of water adhere to the walls of glassware, even glassware that is apparently dry. It is important to perform a flaming operation, since truly dry glassware is required in this reaction. The magnesium turnings must also be dried carefully. The glassware can be dried in a drying oven before use, but the best procedure is to "flame" the assembled apparatus. The apparatus is assembled as shown in the figure, with drying tubes attached in the proper locations. At this point, no hoses are attached, and no water is in the cooling jacket of the condenser. The glassware should appear dry, with no droplets of water visible on the inside of the apparatus. The magnesium to be used for the reaction is placed in the 500-mL flask, but no other reagents are included. The entire apparatus, including the 500-mL flask with the magnesium, is assembled on a ring stand, taken to a **hood** (ether is being used in the laboratory), and **gently** heated with a Bunsen burner flame for about 5 minutes. Do not heat the magnesium strongly. The apparatus is then allowed to cool. When it has cooled, return to your bench, add the remaining reagents, attach the water hoses to the condenser, turn on the cooling water, and begin the reaction.

In subsequent operations, all other equipment, including the graduated cylinders used for measurements, must be dry. Furthermore, **anhydrous** ether must be used as a solvent. It will absorb moisture from the air if ether containers are not kept tightly closed.

STARTING A GRIGNARD REACTION

Starting a Grignard reaction can often be a frustrating experience. Failure to form the Grignard reagent is the most common difficulty. This difficulty has several causes: (1) the halide may be too unreactive; (2) the magnesium may have an oxide coating; and (3) traces of water may be present in the reagents or the equipment.

Bromobenzene is fairly reactive, so the problem of an unreactive halide should not be important in this experiment. The oxide coating on the magnesium can be removed, at least partially, by rubbing or crushing the magnesium chips with a glass rod. If this is done in the flask, care should be taken not to punch a hole in the bottom of the flask. Crushing the magnesium exposes a fresh surface area at which the reaction may begin. Adding a small crystal of iodine sometimes helps to clean the surface of the magnesium. If traces of water are present, it is frequently helpful to prepare a small sample of the Grignard reagent in a test tube. When this reaction is started, it is added to the main reaction mixture. This "extra" reagent removes the trace of water quickly by reacting with it.

PROCEDURE

PHENYLMAGNESIUM BROMIDE

Have both a steam bath and an ice bath ready for use, since it may be necessary to control the rate of the reaction by alternately heating and cooling the reaction mixture. Fit a clean, dry 500-mL round-bottomed, three-necked flask with a dry reflux condenser and an addition (separatory) funnel. Place drying tubes on both the addition funnel and the condenser. Add 0.9 g of magnesium turnings and flame the apparatus according to the procedure described previously. Obtain 3.5 mL (density 1.49 g/mL) of bromobenzene. Add 5 to 10 mL of **anhydrous** ether (contained in metal cans) and 1 mL of the previously measured bromobenzene to the reaction flask. Attach the water hoses to the condenser and turn on the cooling water. Dissolve the rest of the bromobenzene in 20 mL of anhydrous ether, stir the solution, stopper it, and store it for later addition to the reaction. Use a stirring rod to crush two or three chips of the magnesium against the side of the reaction flask. Be careful not to punch a hole in the flask. Replace the stopper and note whether any reaction begins to take place.

If there is no reaction, heat the flask gently with a steam bath for a minute or two. Only a very small amount of steam is needed to heat the ether to its boiling point. Remember that the presence of water vapor in the reaction makes starting the reaction very difficult. Stop the heating, remove the flask from the steam bath, and check the reaction mixture to see whether any reaction is taking place. If not, try crushing some more of the magnesium chips with a stirring rod. Evidence of reaction will be the formation of a brownish gray cloudy material in the solution. Also, the ether will continue to reflux when no heat is being applied. The formation of small bubbles at the surface of certain pieces of magnesium will be noticeable. If the reaction does not commence within 2 to 3 minutes, repeat the heating procedure and continue crushing the magnesium with the stirring rod. If after two or three attempts the reaction still does not start, refer to "Starting a Grignard Reaction" in the Special Instructions, for additional advice.

While you are trying to start the reaction, place the remainder of the bromobenzene-ether solution in the addition (separatory) funnel. Replace the drying tube. When the reaction has started, as evidenced by the bubbling action in the absence of heat, add the bromobenzene solution to the magnesium suspension. This addition should be dropwise at a rate of 1 to 3 drops per second. If the condensation ring does not remain in the lower third of the condenser, slow the addition. Heating should not be necessary, since the reaction is sufficiently exothermic to allow the solvent to boil without additional heat. If the reaction begins to reflux too vigorously, moderate it with an ice bath. Do not

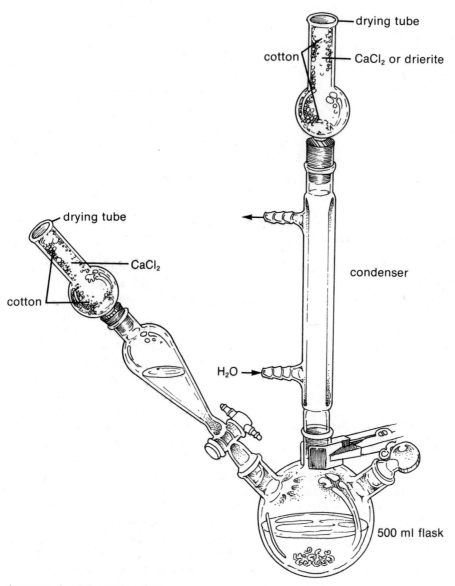

Apparatus for Grignard reactions

overcool the mixture, since this will slow the reaction excessively. When all the bromo-benzene has been added, heat the ether under reflux gently for at least 15 minutes on the steam bath. As your instructor designates, go on to either Procedure 30A or Procedure 30B.

Procedure 30A
Triphenylmethanol

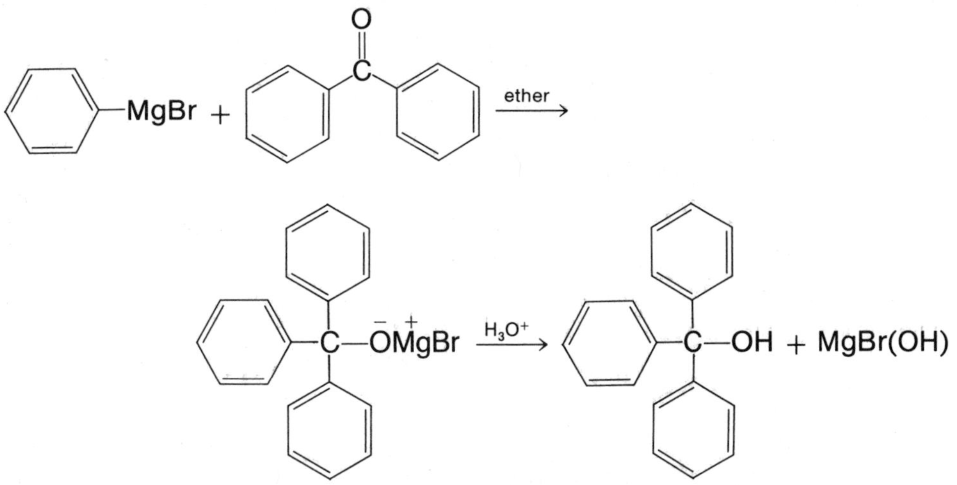

While the phenylmagnesium bromide solution is being heated under reflux, make a solution of 4.3 g of benzophenone in 15 mL of anhydrous ether. Place the solution in the addition (separatory) funnel. After the 15-minute reflux period, add the benzophenone solution to the phenylmagnesium bromide solution as rapidly as possible, again keeping the condensation ring in the lower third of the condenser. The ice bath can be used to moderate the reaction. Swirl the flask during the addition to ensure proper mixing. Cool the solution to room temperature, allow it to solidify, and add 6*M* hydrochloric acid **(dropwise at first)** until the aqueous solution is acidic to litmus. Use a stirring rod to take solution from the lower layer, since litmus does not work in nonaqueous solvents. It may be necessary to add more ether if some has evaporated. Separate the layers, using a separatory funnel. Wash the water layer with 15 mL of **solvent-grade** ether and combine the two ether fractions. Solvent-grade ether is cheaper than the anhydrous ether that was used initially. The principal difference is that solvent-grade ether contains some water. Water can no longer interfere at this point in the reaction. Dry the ether solution with 1 to 2 g of anhydrous powdered sodium sulfate and gravity-filter (through a fluted filter paper) the suspension to remove the drying agent. Evaporate the ether in a hood. Alternatively, the ether can be removed by distillation, using a steam bath as the heat source.

    The material that remains after all the ether has evaporated looks like a brown oil. The crude product contains the desired triphenylmethanol and the by-product biphenyl. Most of the biphenyl can be removed by adding 25 mL of low-boiling **petroleum** ether or ligroin. These solvents are mixtures of hydrocarbons that easily dissolve the hydrocarbon biphenyl and leave behind the alcohol triphenylmethanol. They are not to be confused with diethyl ether ("ether"). Swirl the crude product with the solvent for several minutes, and then suction-filter the mixture through a Büchner funnel. Dry the solid, weigh it, and calculate the percentage yield. Take 1.0 g of the remaining solid and recrystallize it from hot ligroin (65–90 °C). The solid dissolves slowly. If some solid

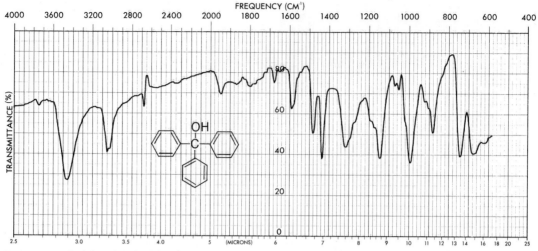

Ir spectrum of triphenylmethanol, KBr

remains undissolved after about 90 mL of hot ligroin has been added, filter it through a fluted filter (using fast filter paper, #617) and cool the solution in an ice bath. Scratch the contents of the flask vigorously to induce complete crystallization. Collect the solid on a small Büchner funnel and wash with some cold ligroin. Report the melting point of the purified material (literature value, 162 °C) and the yield in grams. Place both the crude and the recrystallized products in properly labeled vials and submit them to the instructor. At the option of the instructor, determine the infrared spectrum, as a KBr mull, of the purified triphenymethanol (Technique 17, Part A, p. 662). Your instructor may also assign certain tests on the product you prepared. These tests are described in the instructor's manual.

Procedure 30B
Benzoic Acid

When phenylmagnesium bromide has been prepared and the mixture has stopped refluxing, remove the condenser from the flask and pour the entire contents slowly but steadily over 10 g of crushed Dry Ice contained in a 250-mL beaker. The Dry Ice is best

> **CAUTION:** **Exercise caution in handling Dry Ice. Contact with the skin can cause severe frostbite. Always use the cotton gloves or tongs.**

crushed by wrapping one lump in a clean, dry towel and striking it on the bench top or the floor. Dry Ice condenses moisture from the atmosphere to form a coating of ice over its surface. Use the Dry Ice immediately after crushing it, to avoid side reactions caused by this condensed water. Cover the reaction mixture with a watch glass and allow it to stand until the excess Dry Ice has completely sublimed. The Grignard addition compound will appear as a viscous glassy mass. If the mass is too viscous to stir, add an additional 15 mL of ether.

Hydrolyze the Grignard addition product by adding a mixture of 15 g of crushed ice to which 5 mL of concentrated hydrochloric acid has been added. It may be necessary to add some extra acid to get any remaining solid material to dissolve completely. Stir the mixture until two layers appear; then transfer the mixture to a separatory funnel. Remove the lower aqueous layer and discard it. Wash the upper ether layer once with 15 mL of water. If the ether layer appears dark yellow or brown because iodine is present (if it has been used to help start the reaction), add 5 mL of a saturated solution of sodium bisulfate in water to the 15 mL of water used to wash the ether layer. Separate and discard the water layer. Extract the ether layer three times with successive 15-mL portions of 5% sodium hydroxide solution to remove the benzoic acid. Combine the basic extracts and discard the ether layer, which contains the by-product biphenyl. Place the combined basic extracts in a 250-mL beaker and remove the dissolved ether by heating on a steam bath in the hood. Stir the mixture until the ether dissolved in the alkaline solution has been completely removed. Ether is soluble in water to the extent of 7%. Unless the ether is removed before the benzoic acid is precipitated, the product may appear as a waxy solid instead of crystals. Cool the alkaline solution and precipitate the benzoic acid by adding 5 mL of concentrated hydrochloric acid. Collect the precipitated benzoic acid on a Büchner funnel by vacuum filtration. Wash the collected crystals with

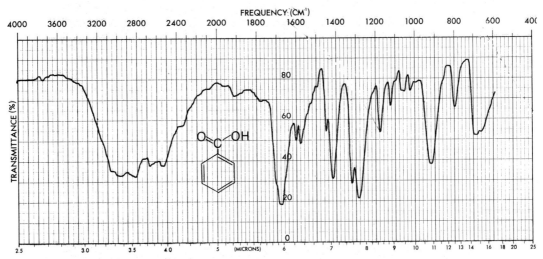

Ir spectrum of benzoic acid, KBr

several portions of cold water and allow the crystals to dry thoroughly at room tempera-ture.[1] When they are dry, weigh the product and calculate the percentage yield. Recrys-tallize a 1.0-g sample of benzoic acid from water. Determine the melting point of this recrystallized material (literature, 122 °C). At the option of the instructor, determine the infrared spectrum, as a KBr mull, of the purified benzoic acid (Technique 17, Part A, p. 661). Submit both the crude and the recrystallized products in properly labeled vials to the instructor. Your instructor may also assign certain tests on the product you prepared. These tests are described in the instructor's manual.

QUESTIONS

1. Benzene is often produced as a side product during Grignard reactions using phenylmagnesium bro-mide. How can its formation be explained? Give a balanced equation for its formation.

2. Why is ligroin or petroleum ether used, instead of a solvent like ethanol, to recrystallize triphenyl-methanol?

3. Write a balanced equation for the reaction of benzoic acid with hydroxide ion. Why is it necessary to extract the ether layer with sodium hydroxide?

4. Interpret the principal peaks in the infrared spectrum of either triphenylmethanol or benzoic acid, depending on the procedure used in this experiment.

5. Outline a separation scheme for isolating pure benzoic acid from the reaction mixture.

6. Provide methods for preparing the following compounds by the Grignard method:

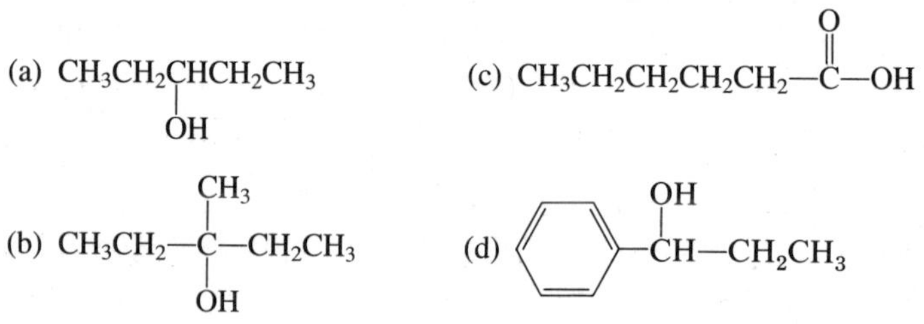

(a) $CH_3CH_2CHCH_2CH_3$
           $|$
           OH

(c) $CH_3CH_2CH_2CH_2CH_2-\overset{\overset{\displaystyle O}{\|}}{C}-OH$

(b) $CH_3CH_2-\overset{\overset{\displaystyle CH_3}{|}}{\underset{\underset{\displaystyle OH}{|}}{C}}-CH_2CH_3$

(d) a benzene ring with $\overset{\overset{\displaystyle OH}{|}}{CH}-CH_2CH_3$

Experiment 31
Nitration of Methyl Benzoate

Aromatic Substitution
Crystallization

The nitration of methyl benzoate to prepare methyl *m*-nitrobenzoate is an example of an electrophilic aromatic substitution reaction, in which a proton of the aromatic ring is replaced by a nitro group:

[1]In extreme cases, oven-drying may be required. Caution: Benzoic acid has a tendency to sublime.

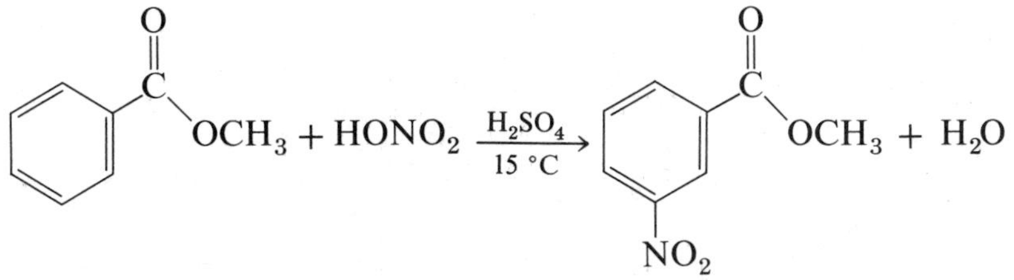

Methyl benzoate   **Methyl *m*-nitrobenzoate**

Many such aromatic substitution reactions are known to occur when an aromatic substrate is allowed to react with a suitable electrophilic reagent, and many other groups besides nitro may be introduced into the ring.

You may recall that alkenes (which are electron-rich due to an excess of electrons in the π system) can react with an electrophilic reagent. The intermediate formed is electron-deficient. It reacts with the nucleophile to complete the reaction. The overall sequence is called **electrophilic addition.** Addition of HX to cyclohexene is an example.

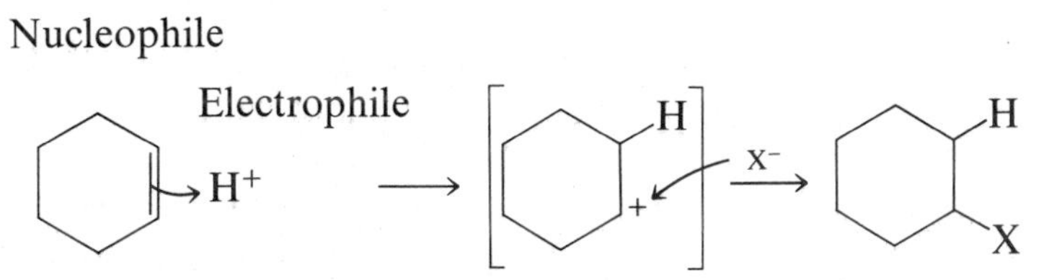

Nucleophile

Cyclohexene

ATTACK OF ALKENE   **CARBONIUM ION**   **NET ADDITION**
ON ELECTROPHILE (H$^+$)   **INTERMEDIATE**   **OF HX**

Aromatic compounds are not fundamentally different from cyclohexene. They can also react with electrophiles. However, due to resonance in the ring, the electrons of the π system are generally less available for addition reactions since an addition would mean the loss of the stabilization that resonance provides. In practice this means that aromatic compounds react only with **powerfully electrophilic reagents,** usually at somewhat elevated temperatures.

Benzene, for example, can be nitrated at 50 °C with a mixture of concentrated nitric and sulfuric acids; the electrophile is NO_2^+ (nitronium ion), whose formation is promoted by action of the concentrated sulfuric acid on nitric acid:

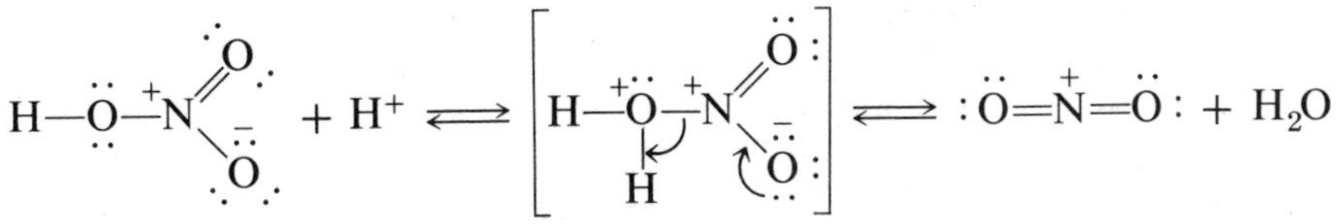

Nitric acid   **Nitronium ion**

The nitronium ion thus formed is sufficiently electrophilic to add to the benzene ring, **temporarily** interrupting ring resonance:

The intermediate first formed is somewhat stabilized by resonance and does not rapidly undergo reaction with a nucleophile; in this behavior, it is different from the unstabilized carbonium ion formed from cyclohexene plus an electrophile. In fact, aromaticity can be restored to the ring if **elimination** occurs instead. (Recall that elimination is often a reaction of carbonium ions.) Removal of a proton, probably by HSO_4^-, from the sp^3-ring carbon **restores the aromatic system** and yields a net **substitution** wherein a hydrogen has been replaced by a nitro group. Many similar reactions are known, and they are called **electrophilic aromatic substitution reactions.**

The substitution of a nitro group for a ring hydrogen occurs with methyl benzoate in the same way it does with benzene. In principle, one might expect that any hydrogen on the ring could be replaced by a nitro group. However, for reasons beyond our scope here (see your lecture textbook), the carbomethoxy group directs the aromatic substitution preferentially to those positions that are meta to it. As a result, methyl *m*-nitrobenzoate is the principal product formed. Additionally, one might expect the nitration to occur more than once on the ring. However, both the carbomethoxy group and the nitro group that has just been attached to the ring **deactivate** the ring against further substitution. Consequently, the formation of a methyl dinitrobenzoate product is much less favorable than the formation of the mononitration product.

While the products described above are the principal ones formed in the reaction, it is possible to obtain as impurities in the reaction small amounts of the ortho and para isomers of methyl *m*-nitrobenzoate and of the dinitration products. These side products are removed when the desired product is purified by crystallization.

Water has a retarding effect on the nitration since it interferes with the nitric acid–sulfuric acid equilibria that form the nitronium ions. The smaller the amount of water present, the more active the nitrating mixture. Also, the reactivity of the nitrating mixture can be controlled by varying the amount of sulfuric acid used. This acid must protonate nitric acid, which is a **weak** base, and the larger the amount of acid available, the more numerous the protonated species (and hence NO_2^+) in the solution. Water interferes since it is a stronger base than H_2SO_4 or HNO_3. Temperature is also a factor in determining the extent of nitration. The higher the temperature, the greater will be the amounts of dinitration products formed in the reaction.

SPECIAL INSTRUCTIONS

Before starting this experiment you should read Technique 3. It is important that the temperature of the reaction mixture be maintained at or below 15 °C. Nitric acid and sulfuric acid, especially when mixed, are very corrosive substances. Be careful not to get these acids on your skin. If you do get some of these acids on your skin, flush the affected area liberally with water.

PROCEDURE

In a 150-mL beaker cool 12 mL of concentrated sulfuric acid to about 0 °C and add 6.1 g of methyl benzoate. Using an ice-salt bath (see Technique 1, Section 1.10, p. 513), cool the mixture to 0 °C or below and add, VERY SLOWLY, using a Pasteur pipet, a cool mixture of 4 mL of concentrated sulfuric acid and 4 mL of concentrated nitric acid. During the addition of the acids, stir the mixture continuously and maintain the temperature of the reaction below 15 °C. If the mixture rises above this temperature, the formation of by-product increases rapidly, bringing about a decrease in the yield of the desired product.

   After all the acid has been added, warm the mixture to room temperature. After 15 minutes, pour the acid mixture over 50 g of crushed ice in a 250-mL beaker. After the ice has melted, isolate the product by vacuum filtration through a Büchner funnel and wash it with two 25-mL portions of cold water and then with two 10-mL portions of ice-cold methanol. Weigh the product and recrystallize it from an equal weight of methanol (Technique 3, Section 3.4, p. 526). The melting point of the recrystallized product should be 78 °C. Determine the infrared spectrum of the product as a Nujol mull (Technique 17, Section 17.4, p. 666). Submit the product to your instructor in a labeled vial, along with your infrared spectrum.

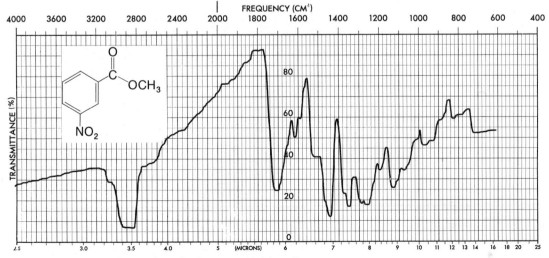

Ir spectrum of methyl *m*-nitrobenzoate, Nujol mull

QUESTIONS

1. Why is methyl *m*-nitrobenzoate formed in this reaction instead of the ortho or para isomers?
2. Why does the amount of the dinitration increase at high temperatures?
3. Interpret the infrared spectrum of methyl *m*-nitrobenzoate.
4. Indicate the product formed on nitration of each of the following compounds: benzene, toluene, chlorobenzene, and benzoic acid.

Experiment 32
p-Nitroaniline

Electrophilic Aromatic Substitution
Hydrolysis of an Amide
Protective Groups
Crystallization
Thin-Layer Chromatography

In this experiment, we convert acetanilide to *p*-nitroaniline. The sequence of reactions, beginning with aniline, is as shown. The conversion of aniline to acetanilide, the first step, was performed in Experiment 2.

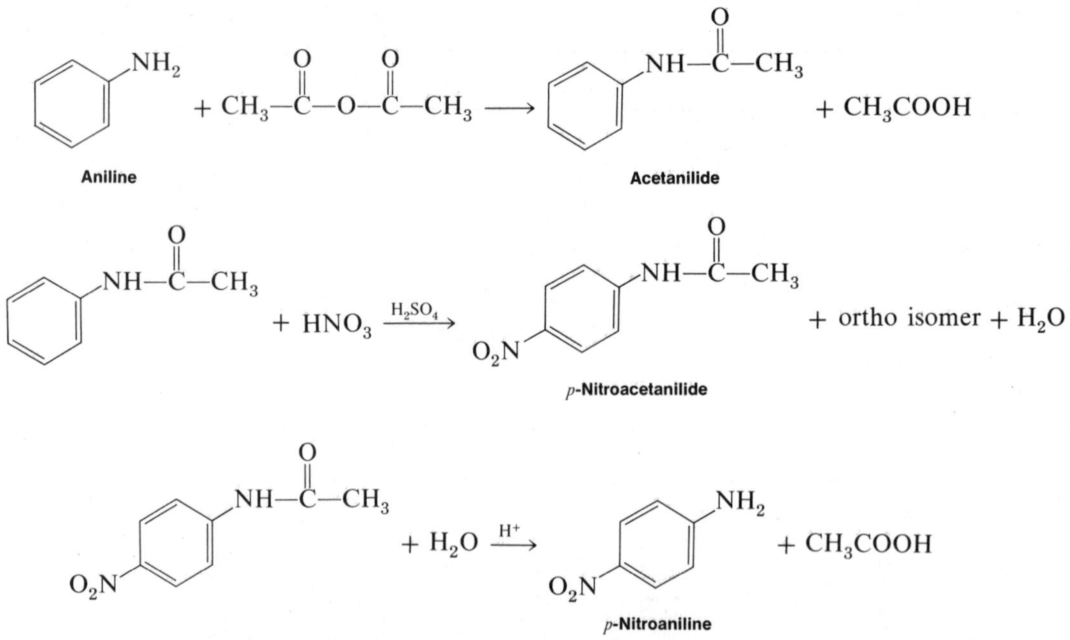

The mechanism for the nitration is essentially identical to that given for the nitration in Experiment 31. The nitronium ion is directed to the positions ortho and para to the acetamido (—NHCOCH$_3$) group. This occurs because the resonance electron-releasing effect of that group increases the electron density at those positions, helping to stabilize the intermediates that are formed. Substitution para to the acetamido group is favored over substitution ortho to that group, because the great bulk of the acetamido group shields the ortho positions from approach by reagents. This steric hindrance makes ortho substitution much less likely than para substitution, in which the bulk of the acetamido group has no influence. The ortho substitution product is formed in small quantities in this reaction, but we shall not try to isolate and purify it.

If one wished to convert aniline to *p*-nitroaniline, it seems reasonable, at first glance, to carry out the nitration directly on aniline, without passing through the amide intermediates. The amino group, due to its strong resonance electron-releasing effect, would theoretically direct substitution to the positions ortho and para to itself. In the usual reaction, the free amino group would also activate the benzene ring so greatly toward electrophilic aromatic substitution that substitution would occur at all three ortho and para positions. It would be difficult to get the monosubstituted aniline from this type of reaction. In the nitration of aniline there is the possibility that products with two or three nitro groups would be formed. Since most electrophilic aromatic substitution reactions occur in acidic media, however, the basic amino group is converted to the cationic ammonium group ($-NH_3^+$). This latter group is electron-withdrawing, meta-directing, and deactivating toward further substitution. Since a large proportion of the aniline molecules in acidic solution are protonated, there is a slow formation of the meta-substituted aniline, along with some ortho and para substitution. These competing reactions are illustrated for nitration:

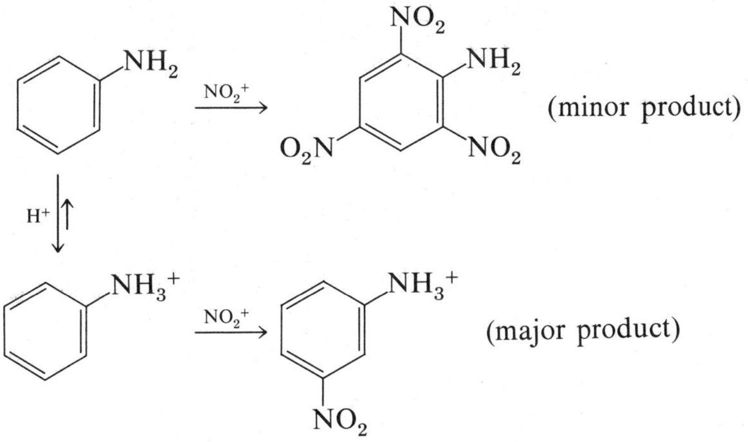

In the nitration of anilines, there is the additional complication that the anilines are susceptible to oxidation. This oxidation caused by the nitric acid, which is a powerful oxidizing agent, further reduces the yields of desired products in this reaction. You can see that it is very difficult to get reasonable yields of *p*-nitroaniline from the direct nitration of aniline.

For this reason, the amino group will be converted initially to the acetamido group with acetic anhydride. The acetyl group, which is thus attached to the amino group, serves as a **protective group.** The acetyl group reduces the reactivity of the amino group with acids, since the nitrogen, now part of an amide, is no longer basic.

The acetyl group also protects the amino function against oxidation. The acetyl group also reduces the electron-releasing resonance of the amino group. As a result, the substituent no longer activates the benzene ring toward multiple substitution as strongly as it did before the acetyl group was attached. Monosubstitution is now possible, so products with only one nitro group can now be isolated from the reaction mixture.

A protective group must fulfill three important requirements. First, it must be easily attachable to the molecule. The reaction to install the protective group must proceed in high yield, in order not to waste starting material, and it must not require reaction conditions that would cause decomposition of the molecule being protected. Second, the protective group must be stable under the conditions of the reaction in which it is expected to function protectively. Obviously, if the protective group were to come off the molecule during a reaction, it would no longer be able to protect the functional group of interest. Third, the protective group must be one that can be removed easily once it has fulfilled its protective function. This last requirement is very important. If one wished to restore the original functional group at the end of a reaction sequence, a protective group that could not be removed would not fulfill its role satisfactorily. In this experiment, the acetyl group is an example of a useful protective group. The original acetylation of the amino group proceeds in high yield and under mild reaction conditions. The acetyl group is stable under the nitration conditions. Finally, it can be removed easily by hydrolysis to regenerate the original amino group.

SPECIAL INSTRUCTIONS

Before you do this experiment, you should read the introductory material in Experiments 2 and 31 and also Techniques 3 and 11. Concentrated sulfuric and nitric acids, in combination, form a very hazardous and corrosive mixture. These acids should be poured together carefully, and this procedure should be carried out in a hood, since noxious vapors are produced. At the option of the instructor, you may begin with either a sample of acetanilide obtained from the stockroom or the sample of acetanilide you prepared in Experiment 2. If the acetanilide is obtained from Experiment 2, be certain that it has been thoroughly dried before you weigh it.

> **CAUTION:** The nitroanilines (ortho, meta, and para) are all toxic. They are absorbed readily through the skin. Avoid contact with skin, eyes, and clothing. Wash all contact areas with large quantities of water.

PROCEDURE

Place 3.0 g of acetanilide in a 125-mL Erlenmeyer flask. Add slowly about 5 mL of concentrated sulfuric acid to the acetanilide. Dissolve most of the solid by swirling and stirring the mixture. Do not be concerned if a small amount of undissolved solid remains. It will dissolve in later stages of this procedure. Place the flask in an ice bath. Place 1.8 mL of concentrated nitric acid in another small flask and add about 5 mL of concentrated sulfuric acid to it.

> **CAUTION: This is a hazardous mixture. This mixing should be done carefully in a hood. Mix the acids thoroughly.**

Using a disposable capillary pipet, add the mixed acids **dropwise** to the cooled sulfuric acid solution of acetanilide. After each addition of acids, swirl the mixture thoroughly in the ice bath. Do not allow the flask to become warm to the touch. After 20 minutes, including the time required for adding the nitric acid–sulfuric acid mixture, add 25 mL of an ice-water mixture to the reaction mixture. A suspension of nitroacetanilide isomers will result. Allow this mixture to stand for 5 minutes, with occasional stirring.

To hydrolyze the nitroacetanilides to the corresponding nitroanilines, heat the material in the flask, using the dilute sulfuric acid already present in the flask as the hydrolyzing medium. Add a boiling stone to the Erlenmeyer flask and heat the flask using a microburner. A wire gauze will disperse the burner flame.

> **NOTE TO THE INSTRUCTOR: Many types of boiling stones will dissolve in the acid medium.**

Heat the mixture gently until the solids dissolve, but do not overheat the mixture because the product may decompose. The solution may darken somewhat during this heating period. Cool the flask in an ice bath, and when it is cool, add 30 mL of concentrated aqueous ammonium hydroxide, in five or six portions, to the material in the flask. This addition must be conducted in a hood because noxious fumes are evolved during the addition. Swirl the flask in the ice bath after each portion of ammonium hydroxide is added. The nitroaniline isomers will precipitate during this addition.

Collect the precipitated nitroanilines on a Büchner funnel by vacuum filtration. Wash the solid **thoroughly** with small portions of water (total of about 50 mL). While continuing the vacuum, allow the solid to air-dry on the Büchner funnel for several minutes.

Scrape the solid material from the filter paper into a 50-mL Erlenmeyer flask and add enough hot ethanol to dissolve the solid when the ethanol is boiling. Use the steam bath to heat the solution. When enough ethanol has been added to just dissolve the solid while the ethanol is boiling, allow the solution to cool. When the first crystals appear, place the flask in an ice bath to complete the crystallization. Filter the crystals of *p*-nitroaniline by vacuum filtration, using a Büchner funnel and a small filter flask. **Save the filtrate for a later tlc analysis.** Wash the crystals with a minimum amount of **cold** ethanol and allow them to dry by drawing air through them on the filter for a few minutes. **Save and dry a small sample of the crystals** for a later determination of the melting point and a tlc analysis.

Dissolve the remaining crude *p*-nitroaniline in ethanol, using 15 mL of ethanol for each gram of *p*-nitroaniline. Warm the solution to dissolve the solid. Add about 0.5 g of activated charcoal to the solution and swirl it for a few minutes. Filter the charcoal from the solution by gravity, using a fluted filter paper. It probably will be necessary to repeat this gravity filtration.[1] Concentrate the filtrate to about one third of its original volume on the steam bath or hot plate. Allow the solution to cool. When the first crystals appear, place the flask in an ice bath. After crystallization is complete, collect the crystals on a Büchner funnel by vacuum filtration. Allow the crystals to air dry on the Büchner funnel by continuing to draw a vacuum on them. When the crystals are dry, weigh them. Determine the melting points of the two samples of crystals obtained before and after this final crystallization (literature mp 149 °C). **Save a small sample of this yellow purified *p*-nitroaniline for tlc analysis.** A labeled vial of the product should be submitted to the instructor.

For the thin layer chromatographic analysis, the three samples (filtrate from first crystallization, crude *p*-nitroaniline, and purified product) are each dissolved in a few drops of ethanol. Each sample is spotted on a single tlc plate that has been prepared from a microscope slide using silica gel G (Technique 11, Sections 11.2A, 11.3, and 11.4, pp. 617–623).[2] Using methylene chloride as the solvent, develop the plate and compare the pattern of spots obtained for each sample. The materials being studied are colored, so they should be visible. However, it may be necessary to intensify the colors of the spots by briefly exposing the tlc plate to iodine vapors (Technique 11, Section 11.6, p. 623).

The report should contain the melting points of the two samples of *p*-nitroaniline and a discussion of the differences observed in melting-point behavior. The percentage yield should also be reported. The report should also include a discussion of the tlc results obtained with the three samples analyzed, with particular emphasis on the differences between the pattern of spots for each sample.

QUESTIONS

1. Explain why the acetamido group is an ortho,para-directing group. Why should it be less effective in activating the aromatic ring toward further substitution than an amino group?

2. Outline the mechanism of the acid-catalyzed hydrolysis of *p*-nitroacetanilide to yield *p*-nitroaniline.

3. *o*-Nitroaniline is more soluble in ethanol than *p*-nitroaniline. Propose a scheme by which a pure sample of *o*-nitroaniline might be obtained from this reaction.

4. Explain how you could use column chromatography to separate *o*-nitroaniline from the filtrate. What adsorbent and solvent or solvents would you use? Which compound would elute first?

5. *N*-Methylbenzamide, an isomer of acetanilide, when allowed to react with HNO_3/H_2SO_4 gives a different product from what is obtained from acetanilide. What is the structure of the mononitrated product? Why is it produced?

---

[1] If continued problems exist in removing the decolorizing carbon, use Filter Aid (Technique 2, Section 2.4, p. 520).

[2] Commercially prepared plates, such as those described in Experiment 4, may also be used. Although methylene chloride gives an adequate separation in tlc, a better separation is obtained with a 5:1 (*v/v*) methylene chloride/pentane mixture.

Experiment 33
Friedel-Crafts Acylation

Aromatic Substitution
Directive Groups
Vacuum Distillation
Infrared Spectroscopy
Nmr Spectroscopy
  Proton/Carbon-13
Structure Proof

In this experiment a Friedel-Crafts acylation of an aromatic compound is undertaken, using acetyl chloride:

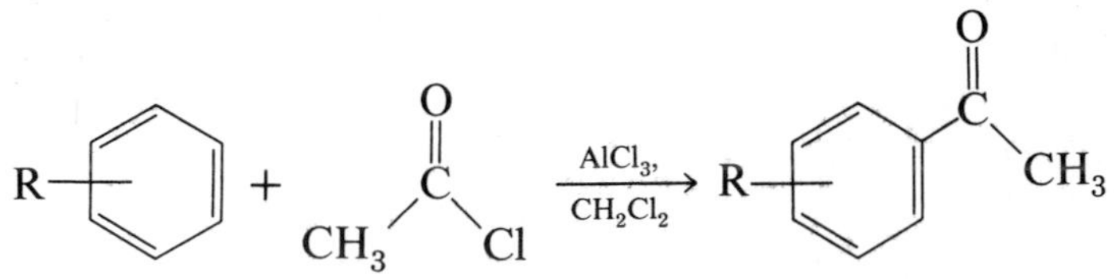

| Aromatic substrate | Acetyl chloride | An acetophenone derivative |

If benzene (R = H) were used as the substrate, the product would be a ketone, acetophenone. Instead of using benzene, however, you will perform the acylation on one of the following compounds:

Toluene   Ethylbenzene
o-Xylene ⎫   Mesitylene (1,3,5-trimethylbenzene)
m-Xylene ⎬ dimethylbenzenes   Cumene (isopropylbenzene)
p-Xylene ⎭   Anisole (methoxybenzene)

Each of these compounds will give a single product, a substituted acetophenone. You are to isolate this product by vacuum distillation and to determine its structure by infrared and nmr spectroscopy. That is, you are to determine at which position of the original compound the new acetyl group becomes attached.

This experiment is work of much the same kind that a professional chemist performs every day. A standard procedure, Friedel-Crafts acylation, is applied to a new compound for which the results are not known (at least not to you). A chemist who knows reaction theory well should be able to predict the result in each case. However, once the reaction is completed, it must be proved that the expected product has actually been obtained. If it has not, and sometimes surprises do occur, then the structure of the unexpected product must be determined.

To determine the position of substitution, several features of the product's spectra should be examined closely. These include the following.

INFRARED SPECTRUM

- The C–H out-of-plane bending modes found between 900 and 690 cm^{-1} (11.0–14.5 μ). The C–H out-of-plane absorptions (Figure IR–6A, p. 704) often allow one to determine the type of ring substitution by their numbers, intensities, and positions.
- The weak combination and overtone absorptions that occur between 2000 and 1667 cm^{-1} (5–6 μ). These combination bands (Figure IR–6B, p. 704), however, may not be as useful as those mentioned above, since the spectral sample must be very concentrated for them to be visible. They are often weak. In addition, a broad carbonyl absorption may overlap and obscure this region, rendering it useless.

PROTON NMR SPECTRUM

- The **integral ratio** of the downfield to the upfield peaks in the aromatic ring resonances found between 6 and 8 ppm. The acetyl group has a significant anisotropic effect, and those protons found ortho to this group on an aromatic ring usually have a greater chemical shift than the other ring protons (see Appendix 4, Section NMR.6, p. 720, and Section NMR.10, p. 727).
- A splitting analysis of the patterns found in the 6- to 8-ppm region of the nmr spectrum. The coupling constants for protons in an aromatic ring differ according to their positional relations:

ortho   J = 6–10 Hz
meta   J = 1–4 Hz
para   J = 0–2 Hz

If complex second-order splitting interaction does not occur, a simple splitting diagram will often suffice to determine the positions of substitution for the protons on the ring. For several of these products, however, such an analysis will be difficult. In other cases, an easily interpretable pattern like those described in Section NMR.10 (p. 727) will be found.

CARBON-13 NMR SPECTRUM

- In **completely decoupled** carbon-13 spectra, the number of resonances for the aromatic ring carbons (at about 120–130 ppm) will give some help in deciding the substitution patterns of the ring. Ring carbons that are equivalent by symmetry will give rise to a single peak, thereby causing the number of aromatic carbon peaks to fall below the maximum of six. A p-disubstituted ring, for instance, will only show four resonances. Carbons that bear a hydrogen will usually have a larger intensity than "quaternary" carbons. (See Appendix 5, Carbon-13 Nuclear Magnetic Resonance Spectroscopy, p. 733.)
- In **coupled** carbon-13 spectra, the ring carbons that bear hydrogen atoms will be split into doublets, allowing them to be easily recognized.

> **NOTE TO THE INSTRUCTOR:** For those not equipped to perform carbon-13 nmr spectroscopy, carbon-13 nmr spectra of all the products can be found reproduced in the instructor's manual.

As a final note, you should not eschew using the library. Technique 18 (p. 671) outlines how to find several important types of information. Once you think you know the identity of your compound, you might well try to find whether it has been reported previously in the literature and, if so, whether or not the reported data match your own findings. You may also wish to consult some spectroscopy books, such as Pavia, Lampman, and Kriz, *Introduction to Spectroscopy,* or one of the other textbooks listed at the end of either Appendix 3 (Infrared Spectroscopy) or Appendix 4 (NMR Spectroscopy), for additional help in interpreting your spectra.

SPECIAL INSTRUCTIONS

Before you begin this experiment, you should review the chapters in your lecture textbook that deal with electrophilic aromatic substitution. Pay special attention to Friedel-Crafts acylation and to the explanations of directing groups. You should also review what you have learned about the infrared and nmr spectra of aromatic compounds. To complete the vacuum distillation successfully, it is necessary to read Technique 6, especially Sections 6.5 to 6.7, pp. 558–563, and Technique 9, p. 588.

Both acetyl chloride and aluminum chloride are corrosive reagents. You should not allow them to come in contact with your skin, nor should you breathe them, since they generate HCl on hydrolysis. They may even react explosively on contact with water. Weighing and dispensing operations involving these reagents should be carried out in a hood. The work-up procedure wherein excess aluminum chloride is decomposed with ice water should also be done in the hood.

Your instructor will either assign you a compound or have you choose one yourself from the list given above. While you will acetylate only one of the above compounds, you should learn much more from this experiment by comparing results with other students.

Notice that the details of the vacuum distillation are left for you to figure out on your own. However, here are three hints. First, all the products boil between 100 and 150 °C at 20 mm pressure. Second, if your chosen substrate is anisole, the product will be a low-melting **solid.** It can be distilled, but you **should not** run cooling water through the condenser. In fact, you might attach steam to the condenser for use if the product should solidify before it reaches the collection flask. Third, you should look up or determine the boiling point of your substrate and of acetyl chloride at 20 mm so you will have an idea of where to cut fractions during distillation.

PROCEDURE

The entire reaction apparatus, excluding the traps, should be assembled on a single-ring stand so that it can be shaken gently from time to time. The apparatus is illustrated in the accompanying figure. Using a 500-mL round-bottomed three-necked flask, attach a reflux condenser to the central neck, a separatory funnel to one side neck, and a stopper to the unused third neck. Place a filled calcium chloride drying tube on top of the addition (separatory) funnel. Connect a gas trap to the top of the reflux condenser by attaching a length of flexible rubber tubing from its exit to your aspirator trap. Connect the other side of the aspirator trap to a funnel that is inverted and **just touching** the surface of a quantity of water in a 250-mL beaker. The funnel may be supported by placing a one-hole rubber stopper on its stem so as to allow it to be clamped to a ring stand. All the glassware must be **dry,** since both aluminum chloride and acetyl chloride react with water.

> **CAUTION: Both aluminum chloride and acetyl chloride are corrosive and noxious. Avoid contact and conduct all weighings in a hood. On contact with water, either compound may react violently.**

Measure 25 mL of dichloromethane (methylene chloride) in a graduated cylinder and have it at hand. Working quickly in a hood so as to avoid reaction with moisture in the air, weigh 14.0 g of anhydrous aluminum chloride into a 125-mL beaker. Using a powder funnel and a large spatula, transfer the aluminum chloride into the three-necked flask via the unused opening. Use the dichloromethane to transfer any last traces of powder into the flask and to rinse the neck of the flask. After adding all the dichloromethane, replace the stopper and start the cooling water in the condenser. Place an ice-water bath under the three-necked flask and support it with wooden blocks. Mix and cool the suspension of aluminum chloride in the reaction flask by carefully rotating the ring stand so as to induce the contents of the flask to swirl gently.

Again working in a hood, use a pipet to transfer 8.0 g of acetyl chloride into a 125-mL Erlenmeyer flask. Add 15 mL of dichloromethane to the flask and transfer this mixture to the addition funnel attached to the reaction apparatus. Taking a total of about 15 minutes, slowly add the acetyl chloride solution to the suspension of aluminum chloride in the lower flask. During this addition period, frequently rotate the ring stand so as to mix and cool the contents of the flask, which should still be immersed in the ice-water bath. After this addition is complete, dissolve 0.075 mole of your chosen aromatic compound in 10 mL of dichloromethane. Place this solution in the addition funnel and slowly add it to the cooled acylation mixture over about 30 minutes. Swirl the mixture occasionally, as you watch out for excessive bubbling from the liberation of hydrogen chloride.

After this second addition is complete, remove the ice bath and allow the mixture to stand at room temperature for an additional 30 minutes. Swirl the reaction mixture frequently during this period.

After moving to a hood, pour the reaction mixture into a mixture of 50 g of ice and 25 mL of concentrated hydrochloric acid placed in a 400-mL beaker. Stir this mixture thoroughly for 10 to 15 minutes. Using a separatory funnel, separate the organic layer

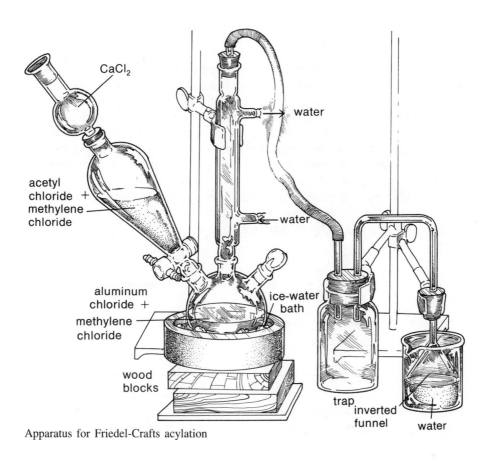

CaCl₂

water

acetyl
chloride +
methylene
chloride

water

aluminum
chloride +
methylene
chloride

ice-water
bath

wood
blocks

trap
inverted
funnel

water

Apparatus for Friedel-Crafts acylation

and save it. Extract the aqueous layer with 30 mL of dichloromethane and add it to the saved organic layers. Wash the combined organic layers with 50 mL of saturated sodium bicarbonate solution and dry them over anhydrous magnesium sulfate. Decant or filter to remove the drying agent.

Remove the dichloromethane by a simple distillation (Figure 6–6, p. 557) and place it in a container set aside for this purpose. Then, using a vacuum distillation, distill the residue at reduced pressure to yield the aromatic ketone. The apparatus for vacuum distillation is illustrated in Figure 6–7, p. 559; however, a manometer should be attached as shown in Figure 9–6, p. 593.

Weigh your product and calculate the percentage yield. Determine both the infrared and nmr spectra. The infrared spectra may be determined neat, using salt plates (Technique 17, Section 17.2, p. 660), except for the product from anisole, which is a solid. For this product, one of the solution spectrum techniques (Section 17.3, p. 661), either Method A or Method B, should be used. Again except for the anisole product, the nmr spectra can be determined neat as described in Technique 17 (Section 17.7, p. 667). If the samples are viscous, add a little carbon tetrachloride. The solid product from anisole will have to be dissolved in carbon tetrachloride or deuteriochloroform.

THE REPORT

In the usual fashion, your should report the boiling point (or melting point) of your product, calculate the percentage yield, and construct a separation scheme diagram. You should also give the actual structure of your product. Include the infrared and nmr spectra and discuss carefully what you learned from each spectrum. If they did not help your structure determination, explain why not. As many peaks as possible should be assigned on each spectrum and all important features explained, including the nmr splitting patterns, if possible. Discuss any literature you consulted and compare the reported results with your own.

You should also try to explain in terms of aromatic substitution theory why the substitution occurred at the position observed, and why a single product was obtained. Could you have predicted the result in advance?

REFERENCE

Schatz, Paul F. "Friedel-Crafts Acylation." *Journal of Chemical Education, 56* (July 1979): 480.

QUESTIONS

1. The following are all relatively inexpensive aromatic compounds that could have been used as substrates in this reaction. Predict the product or products, if any, that would be obtained on acylation of each of them using acetyl chloride.

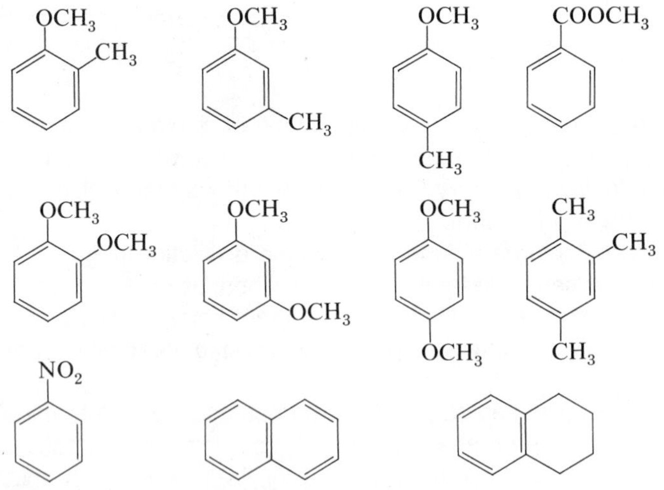

2. Why is it that only monosubstitution products are obtained in the acylation of the substrate compounds chosen for this experiment?

3. Give a full mechanism for the acylation of the compound you chose for this experiment. Include attention to any relevant directive effects.

4. Why do none of the substrates given as choices for this experiment include any with meta-directing groups?

5. Acylation of *n*-propylbenzene gives an unexpected (?) side product. Explain this occurrence and give a mechanism.

6. Give equations for what happens when aluminum chloride is hydrolyzed in water and do the same for acetyl chloride.

7. Explain carefully, with a drawing, why the protons substituted ortho to an acyl group normally have a greater chemical shift than the other protons on the ring.

8. The compounds shown are possible acylation products from 1,2,4-trimethylbenzene (pseudocumene). Explain the only way you could distinguish these two products by nmr spectroscopy.

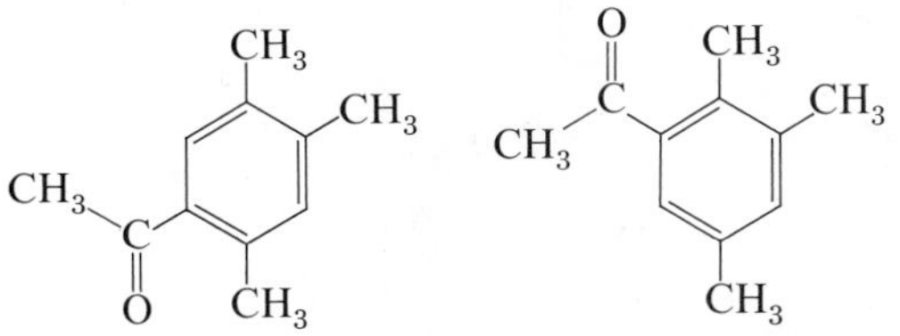

Essay
SYNTHETIC DYES

The practice of using dyes is an ancient art. Substantial evidence exists that plant dyestuffs were known long before humans began to keep written history. Before this century, practically all dyes were obtained from natural plant or animal sources. Dyeing was a complicated and secret art passed from one generation to the next. Dyes were extracted from plants mainly by macerating the roots, leaves, or berries in water. The extract was often boiled and then strained before use. In some cases, it was necessary to make the extraction mixture acidic or basic before the dye could be liberated from the plant tissues. Applying the dyes to cloth was also a complicated process. **Mordants** were used to fix the dye to the cloth or even to modify its color (see the essay on dyes and fabrics preceding Experiment 35).

Madder is one of the oldest known dyes. Alexander the Great was reputed to have used the dye to trick the Persians into overconfidence during a critical battle. Using madder, a root bearing a brilliant red dye, he simulated bloodstains on the tunics of his soldiers. The Persians, seeing the apparently incapacitated Greek army, became overconfident and much to their surprise were overwhelmingly defeated. Through modern chemical analysis, we now know the structure of the dye found in madder root. It is called **alizarin** (see structures) and is very similar in structure to another ancient dye, **henna,** which has been responsible for a long line of synthetic redheads. Madder is obtained from the plant *Rubia tinctorum*. Henna is a dye prepared from the leaves of the Indian henna plant (*Lawsonia alba*) and an extract of *Acacia catechu*.

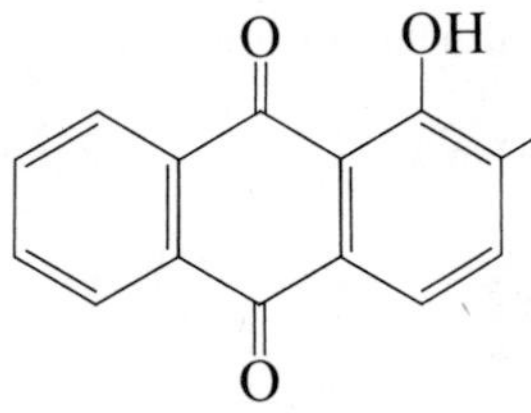

Alizarin

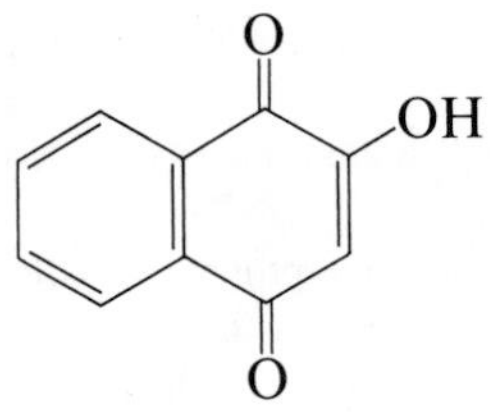

Henna

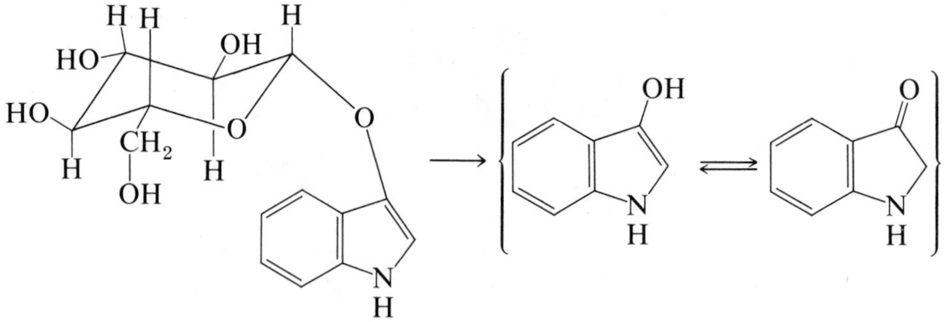

Indican

Indoxyl

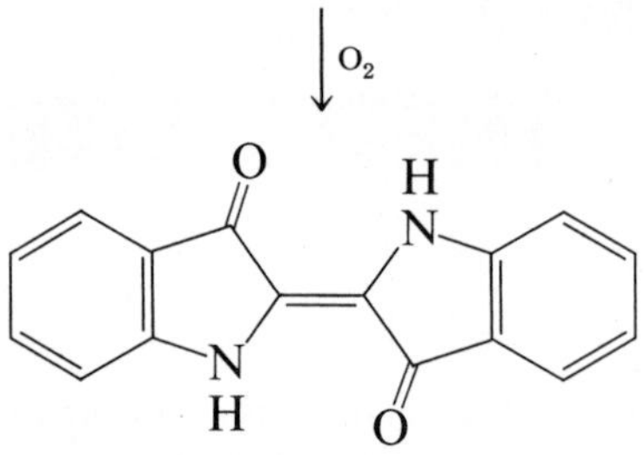

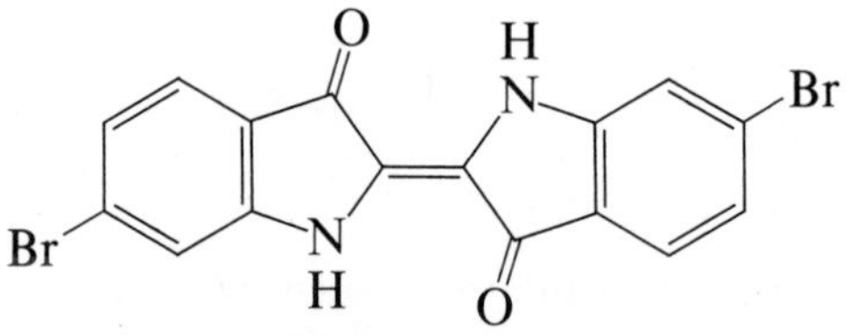

O_2

Tyrian purple

Indigo

Indigo is another plant dyestuff with a long history. This dye, obtained from the plant *Indigofera tinctoria,* has been known in Asia for more than 4000 years. By the ancient process for producing indigo, the leaves of the indigo plant are cut and allowed to ferment in water. During the fermentation, **indican** (see chart) is extracted into the solution, and the attached glucose molecule is split off to produce **indoxyl.** The fermented mixture is transferred to large open vats, in which the liquid is beaten with bamboo sticks. During this process, the indoxyl is air-oxidized to indigo. Indigo, a strong blue dye, is insoluble in water, and it precipitates. Today, indigo is made synthetically, and its principal use is in dyeing denim to produce "blue jeans" material.

Many plants yield dyestuffs that will dye wool or silk, but there are few of these that dye cotton well. Most do not dye synthetic fibers like polyester or rayon. In addition, the natural dyes, with a few exceptions, do not cover a wide range of colors, nor do they yield "brilliant" colors. Even though some people prefer the softness of the "homespun" colors from natural dyes, the **synthetic dyes,** which give rise to deep, brilliant colors, are much in demand today. Also, synthetic dyes that will dye the popular synthetic fibers can now be manufactured. Thus today we have available an almost infinite variety of colors as well as dyes to dye any type of fabric.

Before 1856, all dyes in use came from natural sources. However, an accidental discovery by W. H. Perkin, an English chemist, started the development of a huge synthetic dye industry, mostly in England and Germany. Perkin, then only aged 18, was trying to synthesize quinine. Structural organic chemistry was not very well developed at that time, and the chief guide to the structure of a compound was its molecular formula. Perkin thought, judging from the formulas, that it might be possible to synthesize quinine by the oxidation of allyltoluidine:

$$2C_{10}H_{13}N + 3[O] \longrightarrow C_{20}H_{24}N_2O_2 + H_2O$$

$$\text{Allyltoluidine} \qquad\qquad \text{Quinine}$$

He made allyltoluidine and oxidized it with potassium dichromate. The reaction was unsuccessful, because allyltoluidine bore no structural relation to quinine. He obtained no quinine, but he did recover a reddish brown precipitate with properties that interested him. He decided to try the reaction with a simpler base, aniline. On treating aniline sulfate with potassium dichromate, he obtained a black precipitate, which could be extracted with ethanol to give a beautiful purple solution. This purple solution subsequently proved to be a good dye for fabrics. After receiving favorable comments from dyers, Perkin resigned his post at the Royal College and went on to found the British coal tar dye industry. He became a very successful industrialist and retired at

AZO DYES

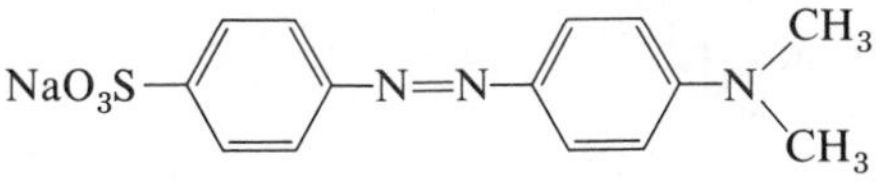

Methyl orange

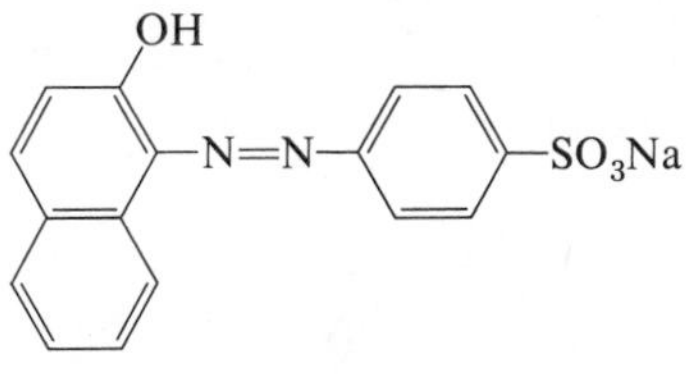

Orange II

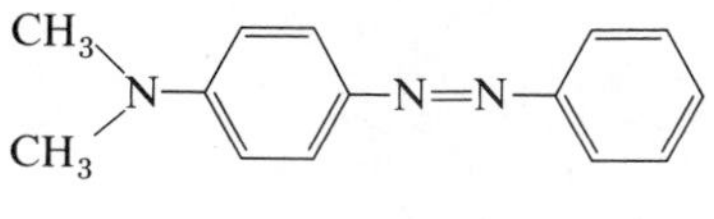

Butter yellow

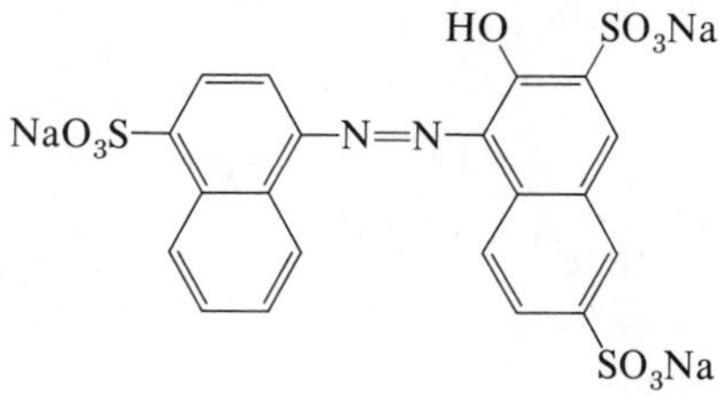

Amaranth red

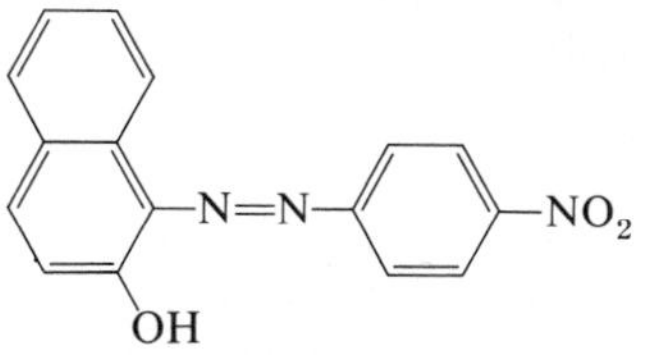

Para red

MAUVE

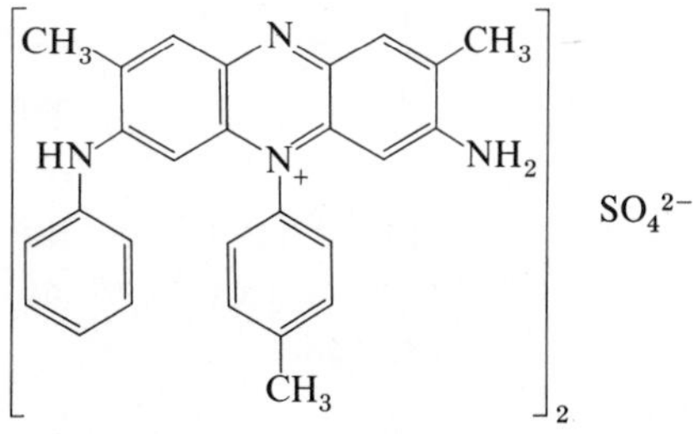

TRIPHENYLMETHANE DYES

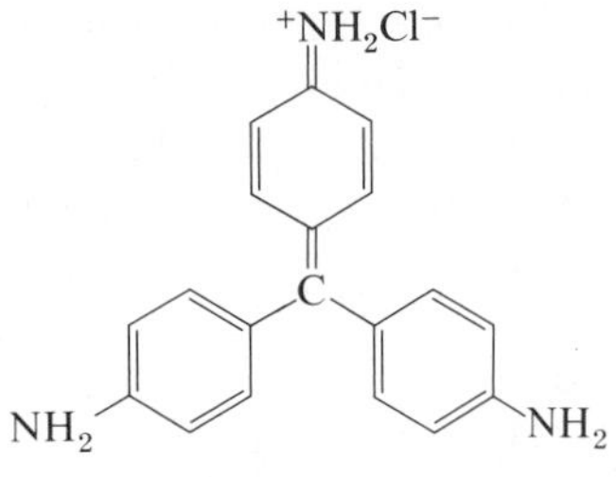

Pararosaniline 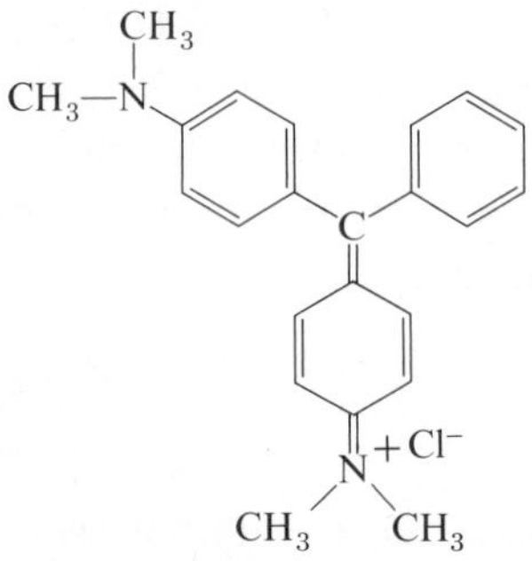   Malachite green 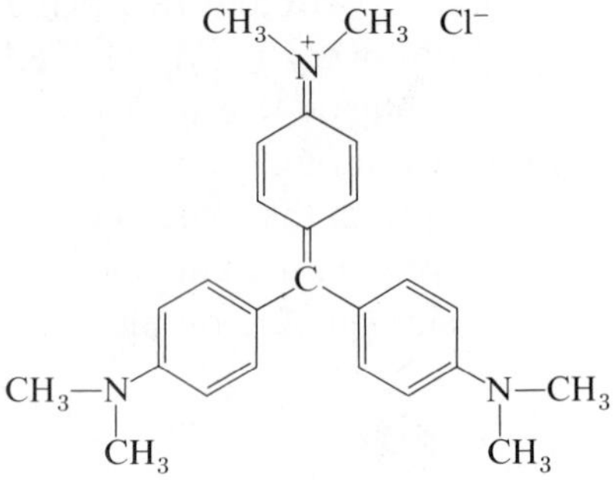   Crystal violet

age 36 (!) to devote full time to research. The dye he synthesized became known as **mauve.** The structure of mauve was not proved until much later. From the structure (see figures) it is clear that the aniline Perkin used was not pure and that it contained the *o-, m-,* and *p*-toluidines also.

Mauve was the first synthetic dye, but soon (1859) the triphenylmethyl dyes pararosaniline, malachite green, and crystal violet (see figures) were discovered in France. These dyes were produced by treating mixtures of aniline or of the toluidines, or of both, with nitrobenzene, an oxidizing agent, and in a second step, with concentrated hydrochloric acid. The triphenylmethyl dyes were soon joined by **synthetic** alizarin (Lieberman, 1868), **synthetic** indigo (Baeyer, 1879), and the azo dyes (Griess, 1862). The azo dyes, also manufactured from aromatic amines, revolutionized the dye industry.

The azo dyes are one of the most common types of dye still in use today. They are used as dyes for clothing (see essay preceding Experiment 35), as food dyes (see essay preceding Experiment 36), and as pigments in paints. In addition, they are used in printing inks and in certain color printing processes. Azo dyes have the basic structure

$$Ar—N{=}N—Ar'$$

Several of these dyes are illustrated in the accompanying figures. The unit containing the nitrogen-nitrogen bond is called an **azo** group, a strong chromophore that imparts a brilliant color to these compounds.

Producing an azo dye involves treating an aromatic amine with nitrous acid to give a **diazonium ion** intermediate. This process is called **diazotization:**

$$Ar—NH_2 + HNO_2 + HCl \longrightarrow Ar—\overset{+}{N}{\equiv}N{:} + Cl^- + 2H_2O$$

The diazonium ion is an electron-deficient (electrophilic) intermediate, and an aromatic compound, suitably rich in electrons (nucleophilic), will add to it. The most common species are aromatic amines and phenols. Both these types of compounds are usually more nucleophilic at a ring carbon than at either nitrogen or oxygen. This is due to resonance of the following types:

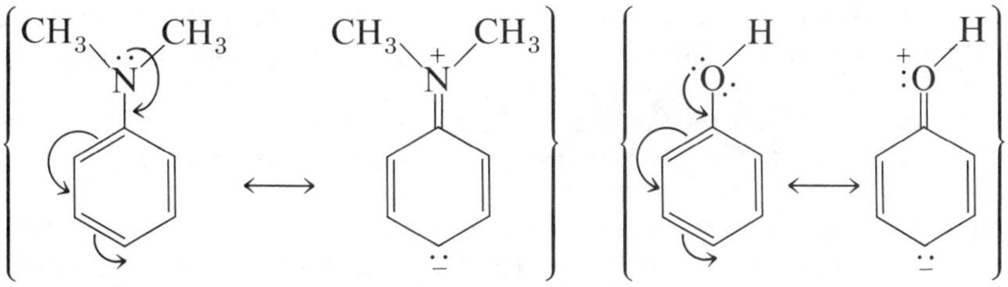

The addition of the amine or the phenol to the diazonium ion is called the **diazonium coupling** reaction, and it takes place as shown:

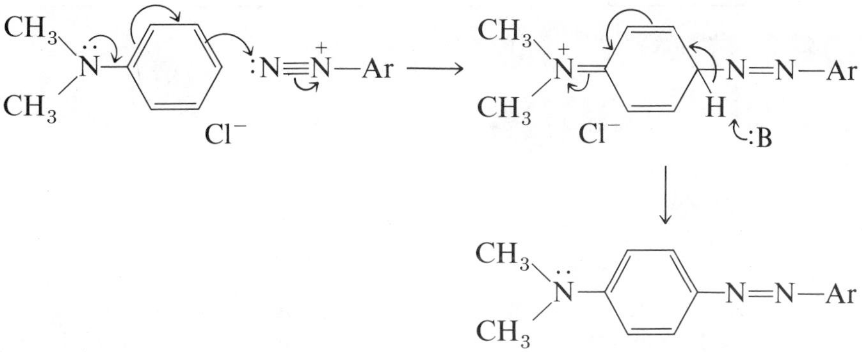

An azo dye

Azo dyes are both the largest and the most important group of synthetic dyes. In the formation of the azo linkage, many combinations of $ArNH_2$ and $Ar'NH_2$ (or $Ar'OH$) are possible. These combinations give rise to dyes with a broad range of colors, encompassing yellows, oranges, reds, browns, and blues. The preparation of an azo dye is given in Experiment 34.

The azo dyes, the triphenylmethyl dyes, and mauve are all synthesized from the anilines (aniline, o-, m-, and p-toluidine) and aromatic substances (benzene, naphthalene, anthracene). All these substances can be found in **coal tar,** a crude material that is obtained by distilling coal. Perkin's discovery led to a multimillion-dollar industry based on coal tar, a material that was once widely regarded as a foul-smelling nuisance. Today these same materials can be recovered from crude oil or from petroleum as by-products in the refining of gasoline. Although we no longer use coal tar, many of the dyes are still widely used.

REFERENCES

Abrahart, E. N. *Dyes and Their Intermediates*. London: Pergamon Press, 1968.
Allen, R. L. M. *Colour Chemistry*. New York: Appleton-Century-Crofts, 1971.

Decelles, C. "The Story of Dyes and Dyeing." *Journal of Chemical Education, 26* (1949): 583.

Jaffee, H. E., and Marshall, W. J. "The Origin of Color in Organic Compounds." *Chemistry, 37* (December 1964): 6.

Juster, N. J. "Color and Chemical Constitution." *Journal of Chemical Education, 39* (1962): 596.

Robinson, R. "Sir William Henry Perkin: Pioneer of Chemical Industry." *Journal of Chemical Education, 34* (1957): 54.

Sementsov, A. "The Medical Heritage from Dyes." *Chemistry, 39* (November 1966): 20.

Shaar, B. E. "Chance Favors the Prepared Mind. Part 1: Aniline Dyes." *Chemistry, 39* (January 1966): 12.

Experiment 34
Methyl Orange

Diazotization
Diazonium Coupling
Azo Dyes

In this experiment the azo dye **methyl orange** is prepared by the diazo coupling reaction. It is prepared from sulfanilic acid and *N,N*-dimethylaniline. The first product obtained from the coupling is the bright red acid form of methyl orange, called **helianthin.** In base, helianthin is converted to the orange sodium salt, called methyl orange.

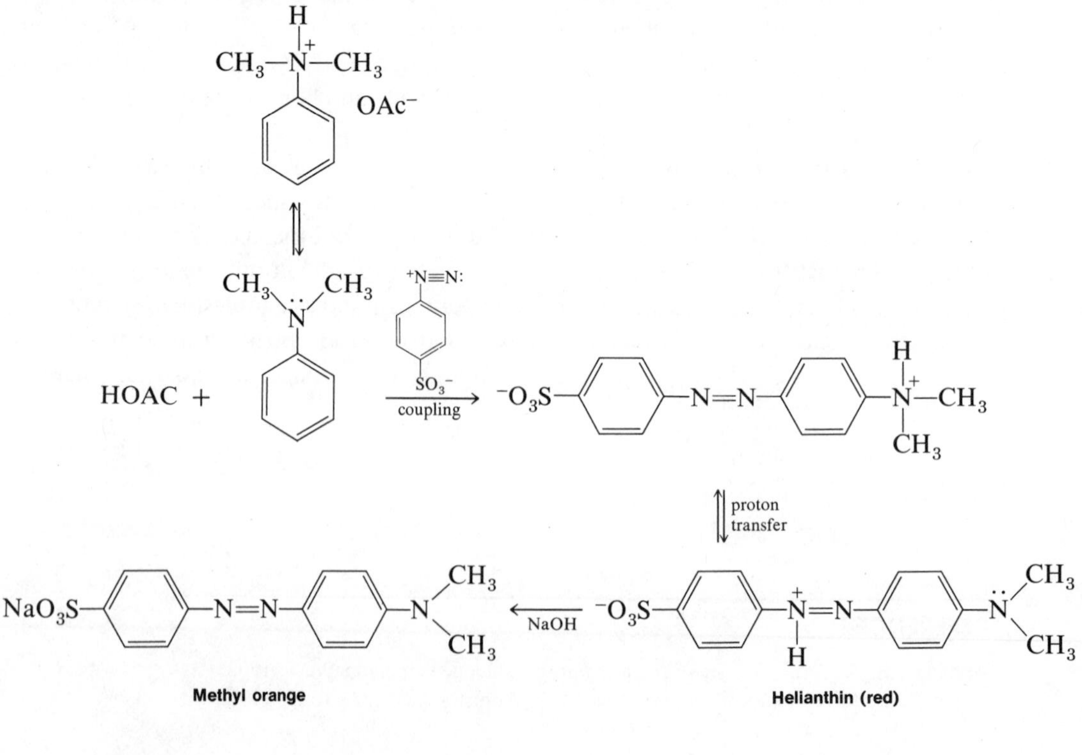

Methyl orange   **Helianthin (red)**

Although sulfanilic acid is insoluble in acid solutions, it is nevertheless necessary to carry out the diazotization reaction in an acid (HNO_2) solution. This problem can be avoided by precipitating sulfanilic acid from a solution in which it is initially soluble. The precipitate is a fine suspension and reacts instantly with nitrous acid. The first step is to dissolve sulfanilic acid in basic solution.

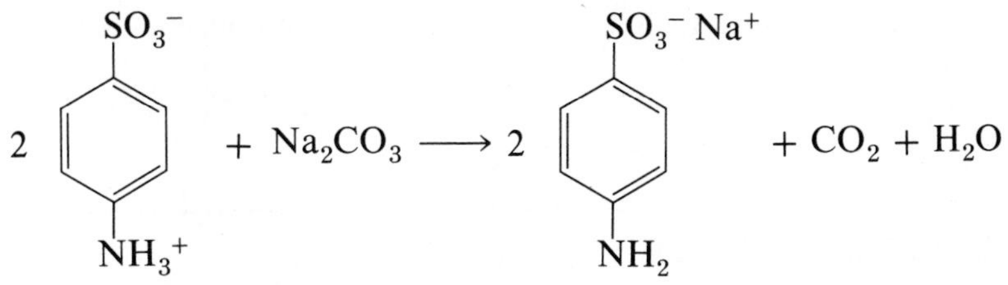

Sulfanilic acid

When the solution is acidified during the diazotization to form nitrous acid,

$$NaNO_2 + HCl \longrightarrow NaCl + HNO_2$$

the sulfanilic acid is precipitated out of solution as a finely divided solid, which is immediately diazotized. The finely divided diazonium salt is allowed to react immediately with dimethylaniline in the solution in which it was precipitated.

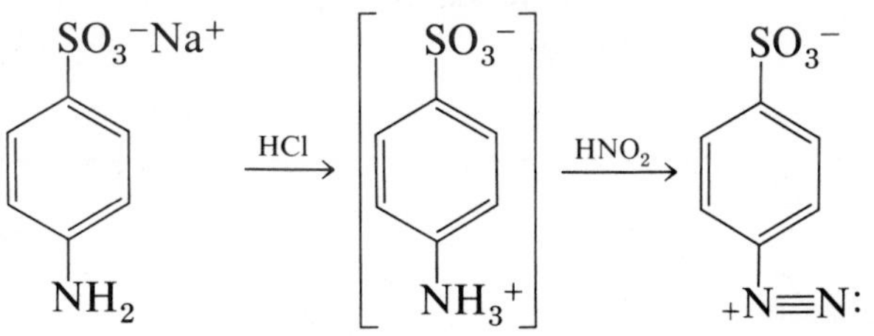

Methyl orange is often used as an acid-base indicator. In solutions that are more basic than pH 4.4, methyl orange exists almost entirely as the **yellow** negative ion. In solutions that are more acidic than pH 3.2, it is protonated to form a **red** dipolar ion.

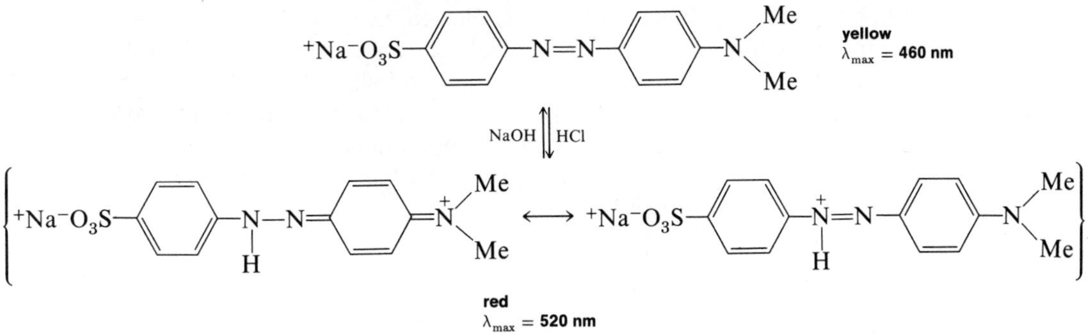

Thus, methyl orange can be used as an indicator for titrations that have their end points in the pH 3.2 to 4.4 region. The indicator is usually prepared as a 0.01% solution in water. In higher concentrations in basic solution, of course, methyl orange appears **orange.**

Azo compounds are easily reduced at the nitrogen–nitrogen double bond by reducing agents. Sodium hydrosulfite, $Na_2S_2O_4$, is often used to bleach azo compounds:

$$R'—N{=}N—R \xrightarrow{Na_2S_2O_4} R'—NH_2 + H_2N—R$$

Other good reducing agents, such as stannous chloride in concentrated hydrochloric acid, will also work.

SPECIAL INSTRUCTIONS

For an introduction to this experiment, you should read the essay on synthetic dyes, which precedes this experiment. You will also find of interest the essays preceding Experiments 35 and 36. To do the experiment, you must read Techniques 1 through 4.

Temperature is important in this reaction. You must keep the temperature of the reaction mixture below 10 °C when the diazonium salt is formed. If the diazonium solution is stored for even a few minutes, it should be kept in an ice bath. If the temperature of the aqueous diazonium salt solution is allowed to rise, the diazonium salt will be hydrolyzed to a phenol, reducing the yield of the desired product.

PROCEDURE

DIAZOTIZED SULFANILIC ACID

Dissolve 1.1 g of anhydrous sodium carbonate in 50 mL of water. Use a 125-mL Erlenmeyer flask. Add 4.0 g of sulfanilic acid monohydrate (3.6 g if anhydrous) to the solution and heat it on a steam bath until it dissolves. A small amount of suspended material may render the solution cloudy. As a remedy, a small portion of Norit can be added to the solution; the still hot solution should then be gravity-filtered, using fluted paper moistened with hot water. Rinse the filter paper with a little (2–5 mL) hot water. Cool the filtrate to **room temperature,** add 1.5 g of sodium nitrite, and stir until solution is complete. Pour this mixture, with stirring, into a 400-mL beaker containing 25 mL of ice water to which 5 mL of concentrated hydrochloric acid have been added. The diazonium salt of sulfanilic acid should soon separate as a finely divided white precipitate. Keep this suspension cooled in an ice bath until it is to be used.

METHYL ORANGE

In a test tube, mix together 2.7 mL of dimethylaniline and 2.0 mL of glacial acetic acid. Add this solution to the cooled suspension of diazotized sulfanilic acid in the 400-mL beaker. Stir the mixture vigorously with a stirring rod. In a few minutes, a red precipitate of helianthin should form. Keep the mixture cooled in an ice bath for about 15 minutes to ensure completion of the coupling reaction. Next, add 30 mL of a 10% aqueous sodium hydroxide solution. Do this slowly and with stirring, as you continue to cool the beaker in an ice bath. Check with litmus or pH paper to make sure the solution is basic. If it is not, add extra base. Heat the mixture to boiling with a Bunsen burner for 10 to 15 minutes to dissolve most of the newly formed methyl orange. When all (or almost all) the dye is dissolved, all 10 g of sodium chloride, and cool the mixture in an ice bath. The methyl orange should recrystallize. When the solution has cooled and the precipitation appears complete, collect the product by vacuum filtration, using a Büchner funnel. Rinse the beaker with two cold portions of a saturated aqueous sodium chloride solution (10 mL each) and wash the filter cake with these rinse solutions.

To purify the product, transfer the filter cake and paper to a large beaker containing about 150 mL of boiling water. Maintain the solution at a gentle boil for a few minutes, stirring it constantly with a glass stirring rod. Not all the dye will dissolve, but the salts with which it is contaminated will dissolve. Remove the filter paper and allow the solution to cool to room temperature. Cool the mixture in an ice bath, and when it is cold, collect the product by vacuum filtration, using a Büchner funnel. Allow the product to dry, weigh it, and calculate the percentage yield. The product is a salt. Since salts do not generally have well-defined melting points, the melting-point determination should not be attempted.

Your instructor may ask you to recrystallize a 1- or 2-g portion of the product isolated from the boiling water. Prepare a saturated solution of the dye in boiling water. It may not be possible to get all the sample to dissolve. Remove the insoluble material by filtering the hot solution by gravity through a fluted filter placed in a funnel that has been preheated on a steam bath. Set the filtered solution aside to cool and crystallize. Collect the product by vacuum filtration, using a Büchner funnel. Submit the purified sample, along with your report, to the instructor.

TESTS (OPTIONAL)

Obtain a square of Multifiber Fabric 10A from your instructor. This cloth contains alternate bands of acetate rayon, cotton, nylon, polyester, acrylic, and wool woven in sequence (see Experiment 35). Prepare a dye bath by dissolving 0.5 g of methyl orange (your crude material will suffice) in 300 mL of water to which 10 mL of a 15% aqueous sodium sulfate solution and 5 drops of concentrated sulfuric acid have been added. Heat the solution to just below its boiling point. Immerse the fabric in the bath for 5 to 10 minutes. Remove the fabric, rinse it well, and note the results.

Make the dye bath basic by adding sodium carbonate. Then add a solution of sodium dithionite (sodium hydrosulfite) until the color of the bath is discharged. Add a slight excess. Now place the very end portion of the dyed fabric in the bath for a few minutes. Note the result.

INDICATOR ACTION (OPTIONAL)

Dissolve a few crystals of methyl orange in a small amount of water in a test tube. Alternately add a few drops of dilute hydrochloric acid and a few drops of dilute sodium hydroxide solution until the color changes are apparent in each case.

QUESTIONS

1. Why does the dimethylaniline couple with the diazonium salt at the para position of the ring?

2. Give a mechanism for producing a phenol from the diazonium salt that was prepared from sulfanilic acid.

3. What would be the result if cuprous chloride were added to the diazonium salt prepared in this reaction?

4. The diazonium coupling reaction is an electrophilic aromatic substitution reaction. Give a mechanism that clearly indicates this fact.

5. In the essay on food colors that precedes Experiment 36, the structures of several azo food colors are given. Indicate how you would synthesize each of these dyes by the diazo coupling reaction.

6. Immediately after the coupling reaction in this experiment, a proton transfer occurs. The proton is transferred from the dimethylamino group to the azo linkage. Why is the latter protonated form lower in energy than the product formed initially?

Essay
DYES AND FABRICS

Not all dyes are suitable for every type of fabric. A dye must have the property of **fastness.** It must bond strongly to the fibers of the material and remain there even after repeated washings. It must not fade on exposure to light. Additionally, a dye must dye the fabric evenly if it is to be a commercially important dye. **Levelness** is the term used to refer to the uniformity of the dye on the fiber. A level dye, then, is one that colors the fabric evenly after its application. Just how a dye bonds to given fiber is a highly complex subject, since not all dyes and not all fibers are equivalent. A dye good for wool or silk may not dye cotton at all. Conversely, a dye that is fast and level on cotton may not be satisfactory for wool.

To understand the mechanism by which a dye bonds to a fiber, you must know both the structure of the dye and the structure of the fiber. The main types of fabric fiber are illustrated in Table 1.

The two natural fibers, wool and silk, are very similar in structure. They are both polypeptides, that is, polymers made of amino acid units. All amino acids have the same basic structure but differ in the nature of the substituent group, R. In naturally occurring amino acids, the various possible side-chain R groups consist of substituents that can be basic ($-NH_2$), neutral ($-CH_2OH$, $-CH_3$), or acidic ($-COOH$).

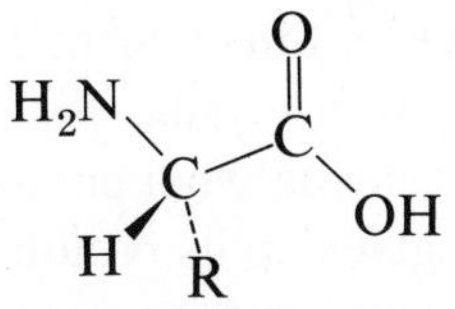

$$H_2N \diagdown \underset{\underset{\underset{R}{|}}{\overset{|}{C}}}{\overset{H}{}} \diagup \overset{\overset{O}{\|}}{C} \diagdown OH$$

**BASIC AMINO
ACID STRUCTURE**

      In wool and silk, there are numerous ways in which a dye can bind to the fiber. It may make chemical links either to the terminal —NH$_2$ or —COOH groups of the polypeptide chain or to the functional groups present in the side chains of the component amino acids.

TABLE 1. Fiber Types

NATURAL FIBERS

Wool (Polypeptide)
Silk

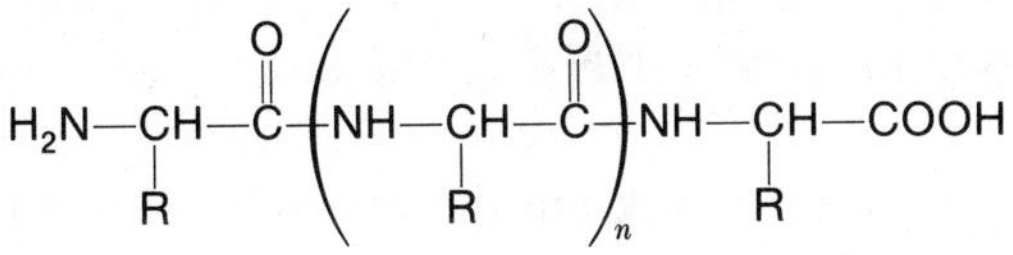

Cotton (Cellulose)

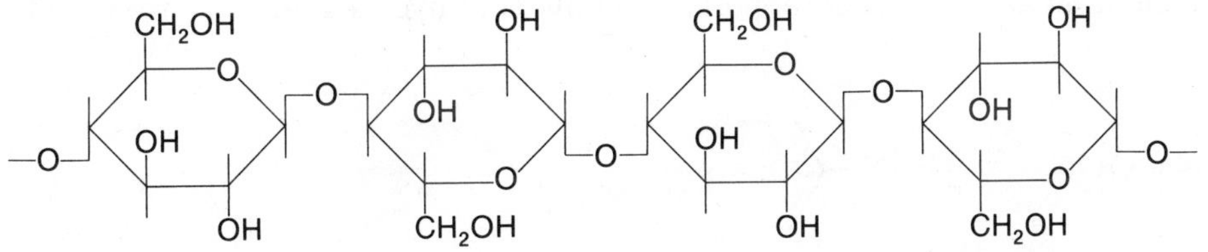

SYNTHETICS

Acetate (Cellulose acetate)

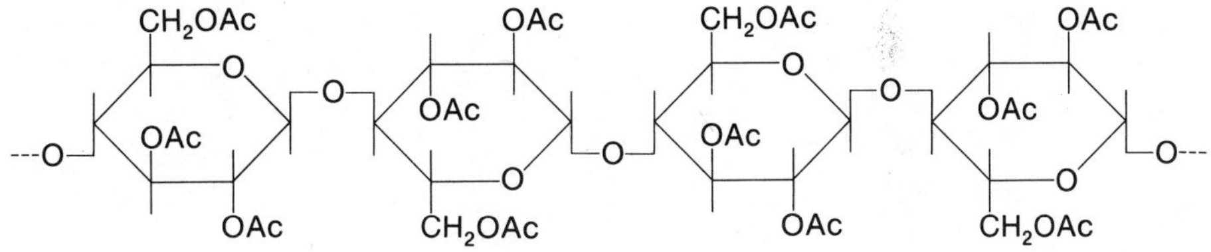

Nylon (Polyamide)

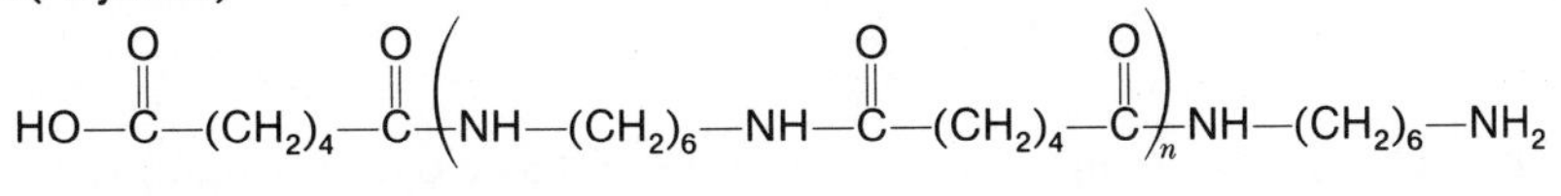

Dacron (Polyester)

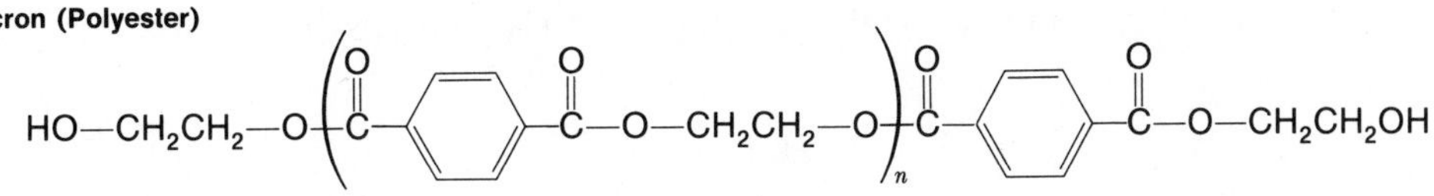

Orlon (Acrylic, Polyacrylonitrile)

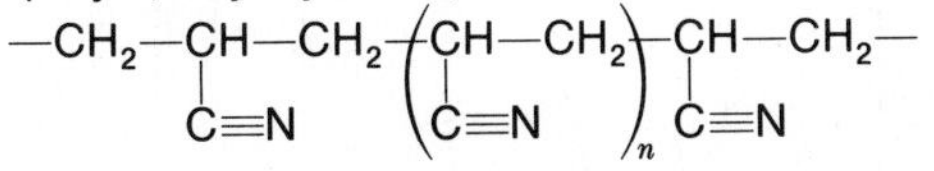

Dyes that attach themselves to a fiber by direct chemical interaction are called **direct dyes.** Picric acid is a direct dye for wool or silk. Since it is a strong acid, it interacts with the basic end groups or side chains in wool or silk to form a **salt linkage** between itself and the fiber. The picric acid gives up its proton to some basic group on the fiber and becomes an anion, which strongly bonds to a cationic group on the fiber by ionic interaction (salt formation). This interaction is illustrated in Figure 1.

At biological pH, most amino acids exist in their zwitterionic form:

$$H_3\overset{+}{N}-CH-COO^-$$
$$|$$
$$R$$

This means that in wool or silk there are $-NH_3^+$ and $-COO^-$ groups as well as $-NH_2$ and $-COOH$ groups. The $-NH_2$ and $-COO^-$ groups are basic, and the $-COOH$ and $-NH_3^+$ groups are acidic. Thus, in an alternative mechanism to the one given in Figure 1, picric acid might give up its proton to a $-COO^-$ group to form its anion. The picrate anion would then bond to some $-NH_3^+$ group already present on the chain.

Two other types of direct dyes for wool or silk are the **anionic** and **cationic** dyes. Anionic dyes are also called **acid dyes;** cationic dyes are also called **basic dyes.** Tartrazine, an azo dye, is an important yellow acid dye (anionic) for wool and silk.

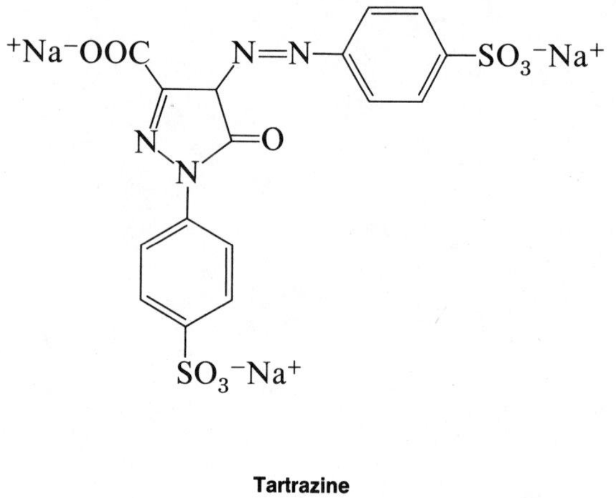

Tartrazine

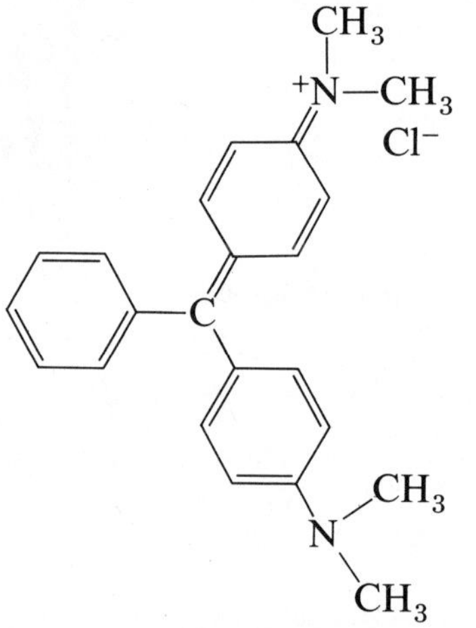

Malachite green

Malachite green, a triphenylmethane dye, is an important basic dye (cationic) for wool and silk. Anionic dyes interact with the acidic $-NH_3^+$ groups in the fiber.

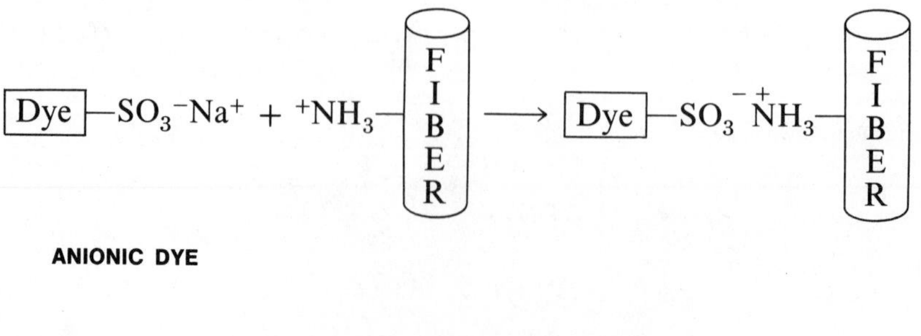

ANIONIC DYE

Cationic dyes interact with the —COO⁻ groups in the fiber.

$$\boxed{\text{Dye}}\text{—C}=\overset{+}{\text{N}}\diagdown \quad + \quad {}^{-}\text{OOC—}\begin{array}{|c|}\hline F \\ I \\ B \\ E \\ R \\ \hline\end{array} \quad \longrightarrow \quad \boxed{\text{Dye}}\text{—C}=\overset{+}{\text{N}}\diagdown \quad {}^{-}\text{OOC—}\begin{array}{|c|}\hline F \\ I \\ B \\ E \\ R \\ \hline\end{array}$$

CATIONIC DYE

Of the synthetic fibers, nylon is most like wool and silk. It is constructed with ''peptide'' or amide linkages,

$$\text{—NH—}\underset{\parallel}{\overset{}{\text{C}}}\text{—}$$
$$\text{O}$$

being formed from a diamine, $NH_2—(CH_2)_6—NH_2$, and a dicarboxylic acid, $HOOC—(CH_2)_4—COOH$. Nylon can be synthesized so that either $—NH_3^+$ or $—COO^-$ groups predominate at the ends of the chains. It is more usual for nylon fabrics to contain excess $—NH_3^+$ groups. Therefore, nylon is usually dyed with anionic (acid) dyes.

Dacron (a polyester fiber), Orlon (an acrylic fiber), and cotton (cellulose) do not contain many anionic or cationic groups in their structures and are not easily dyed by picric acid or by dyes of the anionic and cationic types. Orlon is sometimes synthesized with $—SO_3^-$ groups incorporated into the chain:

$$\begin{array}{ccccccc} & \text{C}\equiv\text{N} & & \text{C}\equiv\text{N} & & & \text{C}\equiv\text{N} \\ & | & & | & & & | \\ \text{—CH}_2\text{—} & \text{CH} & \text{—CH}_2\text{—} & \text{CH} & \text{—R—CH}_2\text{—} & \text{CH—} \\ & & & | & & \\ & & & \text{SO}_3^- & & \end{array}$$

When modified in this way, Orlon can be dyed with cationic dyes.

Azo dyes are important for dyeing cotton fibers. Two such dyes are Congo Red and Para Red. Congo Red is a direct dye for cotton. Para Red is a **developed dye** for cotton. We shall explain both types in detail.

Cotton, which has only hydroxylic groups in its structure, does not dye well with picric acid or with anionic and cationic dyes. Simple azo dyes do not bond well to cotton, but **disazo** dyes are direct dyes for cotton. Congo Red is a disazo dye. A disazo dye has two azo linkages (R—N=N∿N=N—R). When the two azo groups are separated by 10.2 to 10.8 Å, a disazo dye is a fast, direct dye for cotton. The mode of attachment for these dyes is not well understood. Models show that the repeating length between structurally similar hydroxyl groups in cellulose is about the same distance (10.3 Å) as that required for a disazo dye to be a direct dye. In addition, acetate fibers do not dye with disazo dyes. Since the hydroxyl groups are absent in the acetate fiber, it is thought that hydrogen bonding between the hydroxyl groups and the azo linkages accounts for the attachment of a disazo dye to cotton. This is shown in Figure 2.

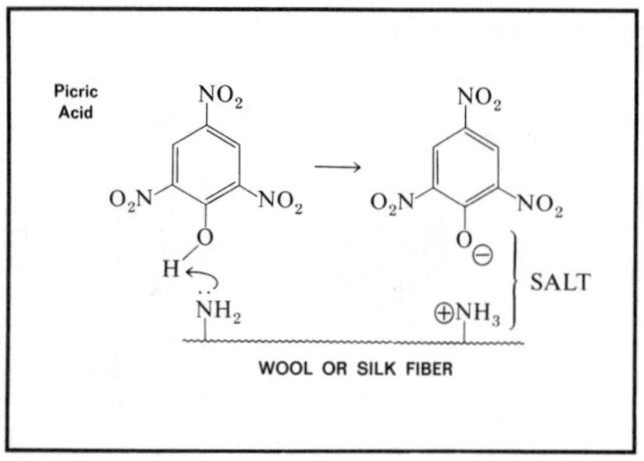

FIGURE 1. Direct dye

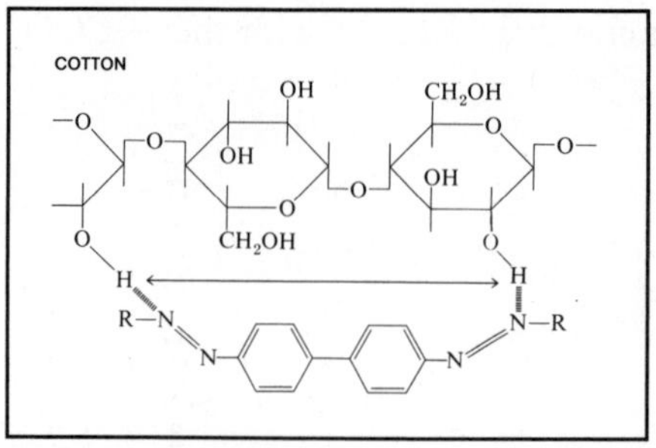

FIGURE 2. Disazo dye

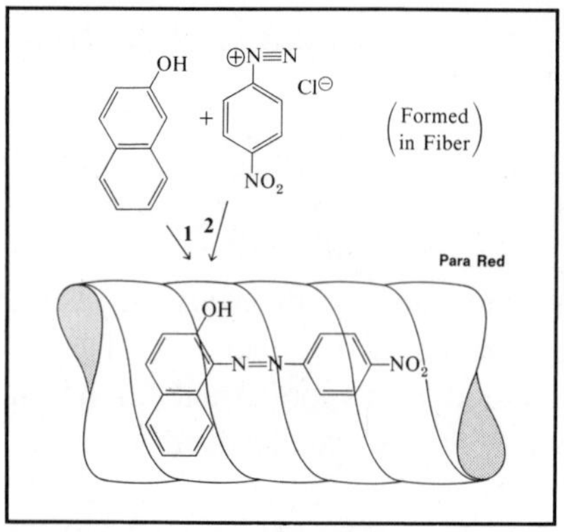

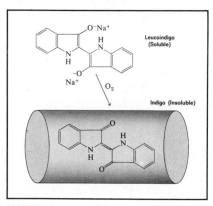

FIGURE 3. Ingrain dye

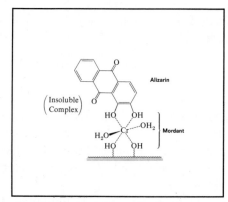

FIGURE 4. Mordant dye

FIGURE 5. Vat dye

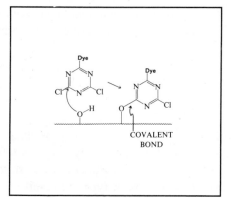

FIGURE 6. Fiber-reactive dye

Although simple azo dyes are not direct dyes for cotton, they can be used to dye cotton as **developed dyes,** also called **ingrain dyes.** Ingrain dyes are synthesized right inside the fiber. Individually, the two components used to synthesize the dye will diffuse into the pores and spaces between the fibers in the fabric. The fully formed dye would be too large a molecule to do this and could bond only to the surface of the fibers. When the individual components react to form the dye, it is trapped **inside,** or "in the grain," of the fibers. Azo dyes are used as ingrain dyes for cotton. Para Red is a typical ingrain azo dye. To form Para Red, the fabric is soaked first in a solution of

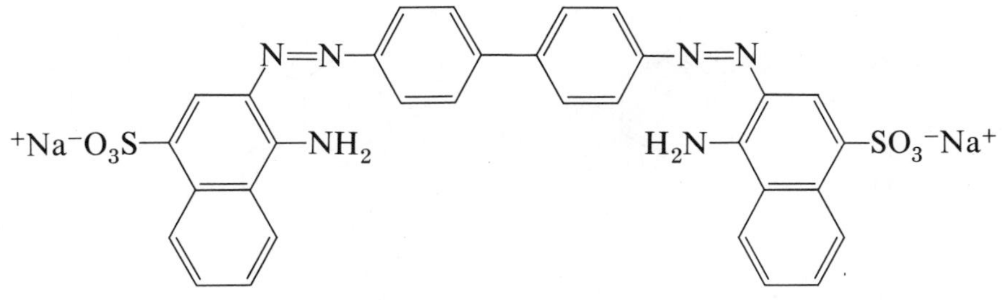

Congo Red (A disazo dye)

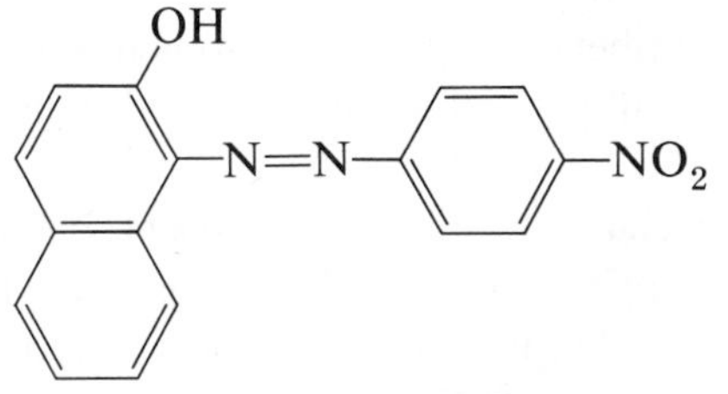

Para Red (An azo dye)

the coupling component β-naphthol, and then soaked in a solution of diazotized *p*-nitroaniline. This process is depicted in Figure 3.

A third type of dye for cotton is the **mordant dye.** Alizarin, shown in Figure 4, is a typical mordant dye. Mordant dyes usually have incorporated in their structures groups capable of forming chelate complexes with heavy metals such as copper, chromium, tin, iron, and aluminum. Cotton, which has many hydroxyl groups, can also coordinate with these metals. For a fabric to be dyed with a mordant dye, the fabric is first treated with a **mordant.** Mordants are heavy metal salts that can complex with the cotton fibers. Typical mordants are alum, copper sulfate, ferric chloride, stannous chloride, and potassium dichromate. After the fiber is treated with the mordant, it is dyed. The dye also complexes with the mordant, and the mordanting metal links the dye to the fabric. This process is depicted in Figure 4.

For a dye to be fast when a mordant is used, the complex formed among the dye, the fiber, and the mordanting metal must be very stable and **insoluble.** Different mordants (metals) often lead to different colors with the same dye.

When cationic dyes, such as malachite green, are used on cotton, the fabric must be treated with tannic acid as "mordant." In this case, the mordant (substance that binds the dye to the fiber) is not a metal. It is simply a substance that forms an insoluble precipitate when mixed with the dye. Tannic acid forms insoluble complexes with cationic dyes. In this process, the cotton is first impregnated with the mordant (tannic acid) and then treated with the dye.

A type of dye that can be used for all fibers, both natural and synthetic, is the **vat dye.** Vat dyes are normally water-soluble in their reduced form, but when oxidized, they become insoluble. Indigo is a vat dye. Indigo is an insoluble blue dye. It can be reduced by sodium dithionite (sodium hydrosulfite) to leucoindigo, a soluble form, which is colorless. In the vat process for indigo, the fabric is impregnated with the soluble leucoindigo in a hot dye bath or **vat.** Then the fabric is removed from the vat,

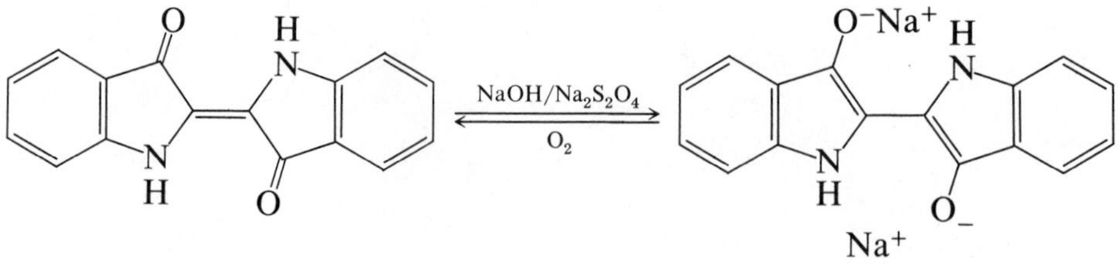

Indigo
(blue, insoluble)   **Leucoindigo**
                                                                  (colorless, soluble)

and the leucoindigo is allowed to air-oxidize to the insoluble indigo. The indigo precipitates both inside and on the surface of the fabric fibers. Since the indigo is insoluble in
aqueous solutions, the dye is fast. The process is illustrated in Figure 5. Indigo is used
for blue jeans.

On fibers like Dacron (polyester) and acetate, no groups are present that allow
any direct dye-fiber interactions. These fibers are hard to dye. **Disperse dyes,** however, can be used. In the disperse process, a dye that is water-insoluble is used. It is
allowed to penetrate the fiber with an organic solvent as "carrier." Often the dye is
suspended in an aqueous solution, and a small amount of the carrier solvent or substance is added. The carrier dissolves the dye and carries it into the fiber where it is
"dispersed." The dye becomes trapped within the fiber only because of water-insolu-
bility. It is not bonded to the fibers but merely dispersed among them.

The newest types of dyes are the **fiber-reactive dyes.** The main class of these
dyes, the Procion dyes, is based on cyanuric chloride. The chlorines in this compound
can be replaced by nucleophilic substitution reactions. If a dye with a good nucleophilic
group is allowed to react with cyanuric chloride in basic solution, the dye becomes
bound to the cyanuric chloride ring by a **covalent** bond and a Procion dye is produced.
Generally, the dye molecule must contain an —OH or an —NH$_2$ group to replace the
chlorine in cyanuric chloride. If the dye replaces only one of the chlorines, two chlorines still remain that can be replaced. If the fiber to be dyed has —OH or —NH$_2$

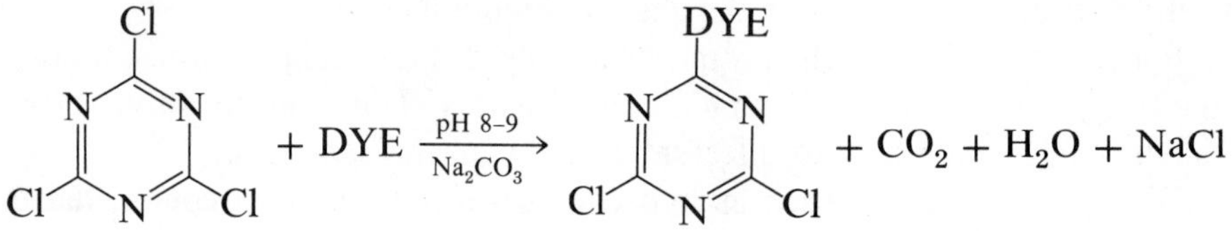

Cyanuric chloride   **Procion dye**

groups, these can displace the remaining chlorines, thus forming a **covalent** bond
between the cyanuric chloride and the fiber. The net result is that the dye is attached to
the fiber, through cyanuric chloride, by covalent bonds. This process is shown in
Figure 6.

REFERENCES

Abrahart, E. N. *Dyes and Their Intermediates*. London: Pergamon Press, 1968.
Allen, R. L. M. *Colour Chemistry*. New York: Appleton-Century-Crofts, 1971.

Davidson, M. F. *The Dye Pot.* Published by the author. Route 1, Gatlinburg, Tenn. 37738.

Editorial Committee of the Brooklyn Botanic Garden. *Dye Plants and Dyeing—A Handbook.* Published by the Brooklyn Botanic Garden, 100 Washington Avenue, Brooklyn, New York 11225. A special printing of *Plants and Gardens,* Vol. 20, No. 3.

Giles, C. H. *A Laboratory Course in Dyeing.* Society of Dyers and Colorists Publication, 1957.

Harper, R. S., Jr., and Reinhardt, R. M. "Chemical Treatments of Textiles." *Journal of Chemical Education, 61* (April 1984): 368.

Maille, A. *Tie and Dye.* New York: Ballantine Books, 1971.

Nea, S. *Tie-Dye.* New York: Van Nostrand, 1971.

Torimoto, N. "An Indigo Plant as a Teaching Material." *Journal of Chemical Education, 64* (April 1987): 332.

Experiment 35
Dyes, Fabrics, and Dyeing

Dyes and Fibers
Azo Dyes, Direct Dyes, Vat Dyes,
  Developed Dyes, Disperse Dyes,
  Substantive Dyes
Mordants. Wool, Cotton,
  Silk, Nylon, Polyester

In this experiment, we examine how dyes adhere to fibers in several fabrics—wool, cotton, polyester, and nylon (or silk). The various classes of dyes and their fastness are examined: direct (anionic and cationic), disazo, vat, ingrain (developed), disperse, and mordant. Finally, the action of mordants on several types of dyes is investigated. The theory of the experiment is discussed in the previous essay, "Dyes and Fabrics."

SPECIAL INSTRUCTIONS

This entire experiment can be completed in one lab period, but you must organize your time efficiently. It will be necessary to perform several experiments simultaneously. Since all the experiments involve waiting periods of as long as 15 to 20 minutes, you can use these waiting periods to begin the next experiment. To understand the experiment, it is necessary to read the preceding essay.

The experiment is written so as to include all the dyeing experiments. Your instructor may include only a few of these, however. In that case, you will be given specific instructions about which steps to perform and which materials to use.

Care should be taken not to immerse your hands in either the mordant or the dye baths. Many of these materials are toxic. In addition, all azo dyes should be suspected of possible carcinogenic activity. You will also find it socially uncomfortable to sport yellow, green, red, and blue digits.

PROCEDURES

Obtain from the instructor or laboratory assistant a kit that contains the following materials:

DYES (in waxed envelopes)[1]

Picric acid	(0.5 g)	Alizarin	(0.1 g)
Indigo	(0.1 g)	Methyl orange	(0.1 g)
Congo Red	(0.1 g)	Malachite green	(0.1 g)
Eosin	(0.1 g)		

FABRIC[2]

Six 4-cm × 10-cm pieces of Multifiber Fabric 10A
(Alternatively, small squares [2 in. × 2 in.] of wool, cotton, polyester, and silk [or nylon] may be used.)

YARNS

Thirty-six 10-in. pieces of wool yarn[3]

OTHER

Five 12-in. lengths of string
An index card
A 12-in. square of
   aluminum foil

Before beginning the experiments, tie the wool yarn pieces into 18 loops or hanks by the following method. Take two 10-in. strands of wool yarn, fold them in half, and tie an overhand knot in the center of the folded lengths. Prepare 18 such hanks. Tie the knot loosely, or the yarn will resist dyeing. See Figure 1, where this is illustrated for a *single* strand.

MORDANTING

Since mordanting takes some time, it should be done before starting the initial dyeing experiments in Part A. The mordanted yarns will be used in Part B. Five large (2-L) mordant baths will be found in the laboratory. They should be kept warm on steam baths or hot plates. The five mordants to be used are alum (potassium aluminum sulfate), copper (cupric sulfate), chrome (potassium dichromate), tin (stannous chloride), and tannic acid. A fixing bath of tartar emetic (potassium antimony tartrate) will also be found. It is to be used after the tannic acid mordant. All the solutions are 0.1 M in the mordants. The samples should be mordanted 15 to 20 minutes in the mordant baths. The tin mordant is especially hard on wool, so keep the time brief for this mordant. On the other hand, if possible, the tannic acid mordant should be left for a longer time.

It will be convenient to mordant several samples at once in the following way. Make five labels from an index card. Punch a hole in each label and tie a 12-in. length of string to it. Write your name and the name of one of each of the five mordants on these

---

[1]The amounts of dye will be very approximate. The instructor should weigh the amount once or twice and then approximate all other envelopes by eye.

[2]Available from Testfabrics, Inc., P.O. Box 118, 200 Blackford Ave., Middlesex, NJ 08846.

[3]Wind the yarn around a 10-in. board and then cut at both ends with scissors. This will prepare many lengths quickly.

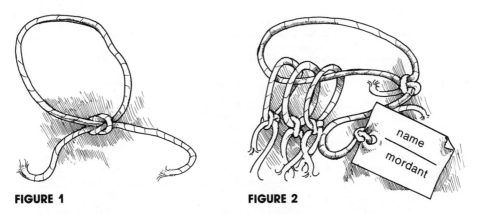

FIGURE 1   **FIGURE 2**

labels. The mordants are alum, copper, chrome, tin, and tannic acid. Then tie the following number of hanks of wool to the unfastened end of each label in the manner shown in Figure 2:

Alum	4
Copper	3
Chrome	3
Tin	3
Tannic acid	1

This will leave four unmordanted hanks of wool.

Place each set of samples in the appropriate bath. Use a long *wooden* stick (not your fingers) to make sure the samples are fully immersed and wetted. The label should not be immersed but should hang over the edge of the bath. Mordant the samples for 15 minutes in the hot baths. They should then be removed from the baths so as to allow excess mordant to drain back into the bath. In addition, squeeze out any excess mordant by pressing the sample against the side of the beaker with a spatula or stirring rod. Press each sample between a double thickness of paper toweling to remove any remaining mordant. After pressing, place all the samples, except that from the tannic acid bath, on a hot steam bath that has its top covered, first with a sheet of aluminum foil and then with a double thickness of paper towel. This will help to dry the samples somewhat. The last sample (from the tannic acid bath) should be fixed in a bath of tartar emetic for 5 minutes and then dried. Leave the samples to dry and start the next set of experiments. Experiments in Parts A and B may be done simultaneously since there are waiting periods in each set of procedures.

A. EXPERIMENTS WITH MULTIFIBER FABRIC (OR WITH MATERIAL PATCHES)

Table 1 outlines the experiments to be performed if patches of material are used. The entire set of experiments can be simplified if multifiber fabric is available. Multifiber Fabric 10A has six fibers—wool, acrylic, polyester, nylon, cotton, and acetate rayon—woven in sequence.[2] None of the experiments in this part involves mordants. All materials (fabrics) should have been laundered to remove sizing. If this has not been done, they should be washed in soapy water and thoroughly rinsed before dyeing. This is not necessary for the multifiber fabric.

TABLE 1. Experiments with Material Patches

DYE TYPE	DYE	WOOL	COTTON	POLYESTER	SILK OR NYLON
Direct	Picric acid	√	√	√	√
Vat	Indigo	√	√		√
Disazo (Anionic)	Congo Red	√	√		√
Azo	Para Red	√	√	√	
Ingrain	Para Red	√	√		
Disperse	Para Red			√	

1. Picric Acid (Direct to Wool, Silk, and Nylon)

Dissolve the picric acid (about 0.5 g) in about 60 mL of water to which 2 to 3 drops of concentrated sulfuric acid has been added. Use a small beaker and place it on a steam bath for warming. Immerse the multifiber fabric or one piece each of wool, cotton, polyester, and silk (or nylon) fabric. Be sure that each piece is thoroughly wetted in the dye solution. Add a little more water if necessary. After about 2 minutes, remove the patches with a forceps and transfer them to a beaker containing warm water. Rinse the samples well and then discard the water. Repeat the rinsing until the dye no longer runs (colors the rinse water). Note the results.

2. Indigo (Vat Dye)

Place the indigo powder (about 0.1 g) in a small Erlenmeyer flask containing a solution of 0.5 g of sodium dithionite (sodium hydrosulfite) and one sodium hydroxide pellet in 40 mL of warm water. Stopper the flask and shake it gently for several minutes until the indigo is dissolved and bleached. The leucoindigo solution will not be colorless but will appear somewhat green. Add more sodium dithionite if the indigo powder did not dissolve. Heat the solution on a steam bath. Unstopper the flask, and using a forceps or a glass rod, insert a patch of multifiber material into the leucoindigo solution. Stopper the flask and shake the solution to be sure that the material is thoroughly wetted. After about 30 seconds, remove the sample from the flask and squeeze it between a double thickness of paper toweling to remove the excess dye solution. Hang the sample to dry in the air and allow the leucoindigo to become oxidized. If a deeper blue is desired, repeat the dyeing. It may be necessary to add a bit more sodium dithionite to the indigo solution if it is still not bleached. If you are using separate material patches, repeat the dyeing, using wool, cotton, and silk (or nylon). Note and compare the results.

3. Congo Red (Disazo Dye for Cotton)

Dissolve the Congo Red (about 0.1 g) in a beaker containing 100 mL of hot water to which 1 mL of 10% aqueous sodium carbonate and 1 mL of 10% aqueous sodium sulfate have been added. Keep the solution warm on a steam bath. Introduce the multifiber fabric or pieces of wool, cotton, and silk (or nylon). After the samples have remained 10 minutes in the hot dye bath, remove them and wash them in warm water until no more dye is removed. Note the results.

4. Para Red (Developed or Ingrain Dye)

Solution 1—Diazotized _p_-Nitroaniline. Place 1.4 g _p_-nitroaniline, 25 mL of water, and 5 mL of 10% hydrochloric acid in a small beaker. Heat the solution until most of the _p_-nitroaniline dissolves (add a bit more acid if necessary). Cool the solution to about 5 °C in an ice bath. Add, all at once, a solution of 0.7 g of sodium nitrite dissolved in 10 mL of water. Stir the solution well, keeping it in the ice bath. If solid remains, the solution can be filtered by vacuum filtration using a Büchner funnel, but filtration is not necessary. Keep this solution in the ice bath, and prepare solution 2.

Solution 2—_β_-Naphthol. Place 0.5 g of _β_-naphthol in a small beaker containing 100 mL of hot water. While stirring, add a solution of 10% sodium hydroxide **by drops** until most of the _β_-naphthol dissolves. Not all the _β_-naphthol will dissolve, and it is important not to add a large amount of base. The cotton will disintegrate in a strongly basic solution; the wool will fare even worse.

Place a sample of multifiber fabric in solution 2 and allow it to soak for only 2 or 3 minutes. Remove the sample from the bath and pat it dry between paper towels. Dilute solution 1 with about 100 mL of cold water and place into it the sample removed from solution 2. After several minutes, remove the sample and rinse it well in water. Note the result.

If you are using separate material patches, place a sample each of cotton and wool in solution 2 for 2 or 3 minutes. Remove them and dry them between paper towels. After this, immerse the sample in solution 1 for several minutes, and then remove it and rinse it well. Note the result.

After completing the ingrain dyeings, mix the two ingrain solutions (1 and 2) together and stir them well. Split the resulting dye mixture into two parts and proceed to parts 5 and 6.

5. Para Red (Azo Dye)

Use half of the Para Red solution prepared by mixing the two ingrain solutions (part 4). Add enough sulfuric acid to make the solution acidic to litmus or pH paper (about 3 drops of concentrated sulfuric acid). Immerse the multifiber fabric or pieces of wool, cotton, and polyester in the hot dye bath for about 5 minutes. Remove the samples and rinse them well. Note the results.

6. Para Red (Disperse)

In its insoluble, finely dispersed form, Para Red may be used as a disperse dye. Use half of the suspension of Para Red produced above by mixing the two ingrain solutions (part 4). Add to this solution 0.1 g of biphenyl (the carrier) and several drops of a surfactant (liquid detergent). Immerse a piece of the multifiber fabric or of the polyester cloth in the solution and heat it on a steam bath for 15 to 20 minutes. Stir the solution frequently. Remove the sample and note the results.

B. EXPERIMENTS WITH YARN

Prepare four dye baths (use beakers) by dissolving each of the following dyes (0.1 g each) in 200 mL of hot water: eosin, alizarin, methyl orange, and malachite green. Add 0.1 g of sodium carbonate to the alizarin bath. Keep the dye baths hot on a steam bath. The experiments to be done are outlined in Table 2. Prepare a grid like Table 2 on a piece of notebook paper. You will be able to keep your samples straight by placing them on this sheet. Perform the dyeing operations in the sequence given. This is necessary as some of the mordants have a tendency to precipitate the dyes.

1. Immerse one hank of **unmordanted** wool in each bath. After about 20 minutes, remove the samples and rinse them thoroughly in a beaker of warm water. Change the rinse water for each sample. Blot the samples dry between paper towels. Note the results.
2. Place one hank of the wool mordanted in alum in each dye bath. After about 15 minutes, remove the samples from the dye baths and rinse them well in warm water. Blot them dry between paper towels. Note the result.
3. Place the sample mordanted in tannic acid and fixed in tartar emetic in the malachite green bath. After 10 to 15 minutes, remove the sample and wash it well in warm water. Blot it dry between paper towels. Note the results. The malachite green bath will not be used for any further experiments and can be discarded.
4. Place one hank of the wool mordanted in copper in each of the three remaining dye baths: eosin, alizarin, and methyl orange. After 10 to 15 minutes, remove and dry these samples, rinsing as before.
5. In each of the three dye baths (eosin, alizarin, and methyl orange) place one hank of the wool mordanted in chrome. After 10 to 15 minutes, remove and dry these samples.
6. In each of the three dye baths (eosin, alizarin, and methyl orange) place one hank of the wool mordanted in tin. After 10 to 15 minutes, remove and dry these samples.

TABLE 2. Experiments with Yarn

DYE TYPE	DYE	WOOL				
		UN	AL	CPR	CHR	TIN
Mordant	Eosin	√	√	√	√	√
	Alizarin	√	√	√	√	√
Azo (anionic)	Methyl orange	√	√	√	√	√
		UN	AL	TAR		
Triphenyl-methane (cationic)	Malachite green	√	√	√		

UN = unmordanted   AL = alum   CPR = copper   CHR = chrome
TIN = tin   TAR = tannic acid + tartar emetic

Essay
FOOD COLORS

Before 1850, most of the colors added to foods were derived from natural biological sources. Some of these natural colors are listed below.

Red	Alkanet root	Yellow	Annato seed (bixin)
	Beets (betanin)		Carrots (β-carotene)
	Cochineal insects (carminic acid)		Crocus stigmas (saffron)
	Sandalwood		Turmeric (rhizome)
Orange	Brazilwood	Green	Chlorophyll
Brown	Caramel (charred sugar)	Blue	Purple grape skins (oenin)

A wide variety of colors can be obtained from these natural sources, many of which are still used today, but they have been largely supplanted by synthetic dyes.

After 1856, when Perkin succeeded in synthesizing mauve, the first coal tar dye (see the essay on synthetic dyes preceding Experiment 34), and when chemists began to discover other new synthetic dyes, artificial colors began to find their way into food-stuffs with increasing regularity. Today, more than 90% of the coloring agents added to foods are synthetic.

The synthetic dyes have certain advantages over the natural coloring agents. Many natural dyes are sensitive to degradation by light and oxygen or by bacterial action; therefore, they are not stable or long-lasting. Synthetic colors can be devised that have a much longer shelf life. The synthetic dyes are also stronger and give more intense colors, and they can be used in smaller quantity to achieve a given color. Often the artificial coloring materials are cheaper than the natural colors. This fact of economics is often especially true when the smaller amounts that are required are taken into account.

Why should artificial colors be added to foods at all? It is easier to answer this question from the point of view of the manufacturer rather than of the consumer. The manufacturer knows that, to a certain extent, the eye appeal of a product will affect its sales. For example, a consumer is more likely to buy an orange that has a bright orange skin than to buy one with a mottled green and yellow skin. This is true, even though the flavor and nutritive value of the orange may not be affected at all by the color of the skin. Sometimes more than eye appeal is involved. The consumer is a creature of habit and is accustomed to having certain foodstuffs a particular color. How would you react to green margarine or blue steak? For obvious reasons, these products would not sell very well. Both butter and margarine are artificially colored yellow. Natural butter has a yellow color only in the summer; in the winter it is colorless, and manufacturers customarily add yellow coloring. Margarine must always be artificially colored yellow.

Thus, the colors that are added to foodstuffs are added for a different reason from what prompts the use of other types of food additives. Other additives may be added to foods for either nutritional or technological reasons. Some of these additives can be justified by good arguments. For instance, during the processing of many foods, valuable vitamins and minerals are lost. Many manufacturers replace these lost nutrients by ''enriching'' their product. In another instance, preservatives are sometimes added to food to forestall spoilage from oxidation or the growth of bacteria, yeasts, and molds. With modern marketing practices, which involve the shipping and warehousing of products over long distances and periods, preservatives are often a virtual necessity. Other additives, such as thickeners and emulsifiers, are often added for technological reasons, for example, to improve the texture of the foodstuff.

There is no nutritional or technological necessity for the use of food colors, however. In fact, in some cases, dyes have been used to deceive customers. For instance, yellow dyes have been used in both cake mixes and egg noodles to suggest a higher egg content than what is actually present. On the grounds that synthetic food dyes are unnecessary and perhaps dangerous, many persons have advocated that their use be abandoned.

Of all the food additives, dyes have come under the heaviest attack. As early as 1906, the government took steps to protect the consumer. At the turn of the century, more than 90 dyes were used in foods. There were no governmental regulations, and the same dyes that were used for dyeing clothes could be used to color foodstuffs. The first legislation governing dyes was passed in 1906, when food colors known to be harmful were removed from the market. At that time only seven colors were approved for use in food. In 1938 the law was extended, and any batch of dye destined for use in food had to be **certified** for chemical purity; previously, certification had been voluntary for the manufacturer. At that time there were 15 food colors in general use and each was given a color and a Food, Drug, and Cosmetic (F,D&C) number designation rather than a chemical name. In 1950, when the number of dyes in use had expanded to 19, an unfortunate incident led to the discontinuation of three of the dyes: two were F,D&C Oranges Number 1 and Number 2, and the other was F,D&C Red Number 32. These dyes were removed when several children became seriously ill after eating popcorn colored by them.

Since that time, research has revealed that many of these dyes are toxic, that they can cause birth defects, that they can cause heart trouble, or that they are carcinogenic (cancer-inducing). Because of experimental evidence, mainly with chick embryos, rats, and dogs, Reds Numbers 1 and 4 and Yellows Numbers 1, 2, 3, and 4 were also removed from the approved list in 1960. Subsequently, Reds Numbers 4 and 32 were reinstated but restricted to particular uses. In 1965, the ban on Red Number 4 was partly lifted to allow it to be used to color maraschino cherries. This use was allowed because there was no substitute dye available that would dye cherries, and it was thought that since maraschino cherries are mainly decorative, they are not properly a foodstuff. This use of Red Number 4 was considered to be a minor use. Similarly, Red Number 32, which may not be used to color food to be eaten, is now called Citrus Red Number 2, and is allowed only for dyeing the skins of oranges.

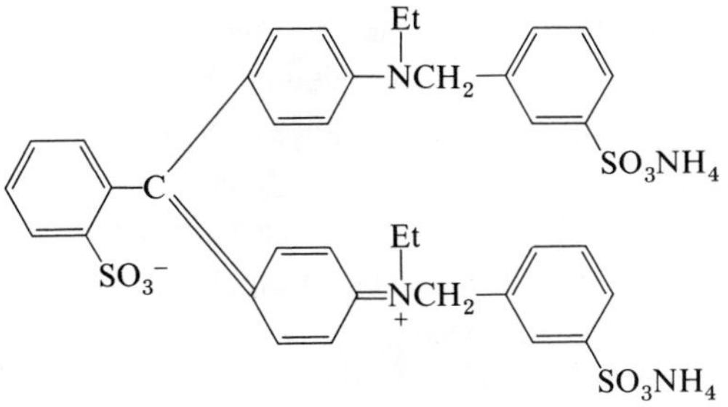

F,D&C Blue No. 1
(Brilliant Blue FCF)

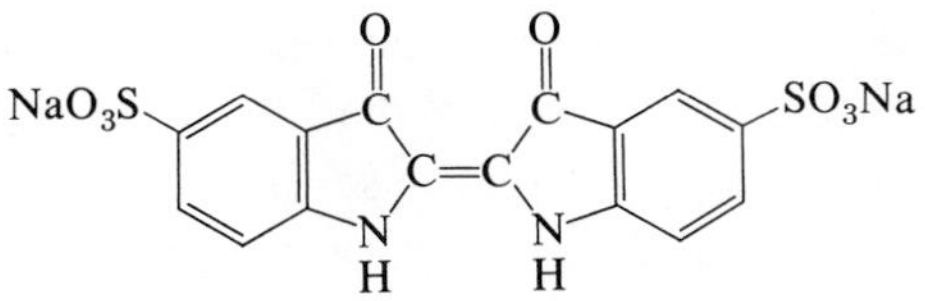

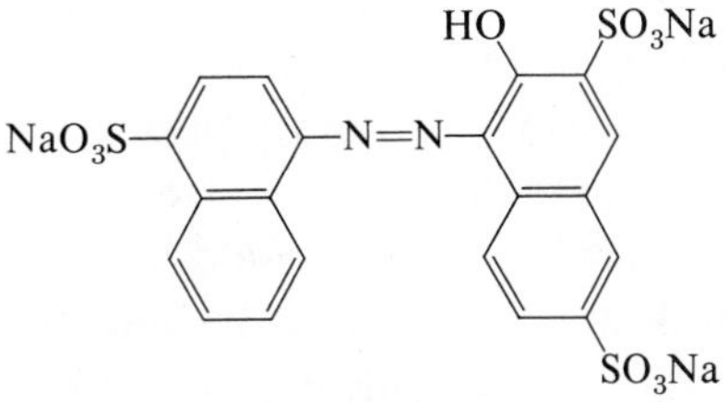

F,D&C Red No. 2
(Amaranth)

F,D&C Blue No. 2
(Indigo Carmine)

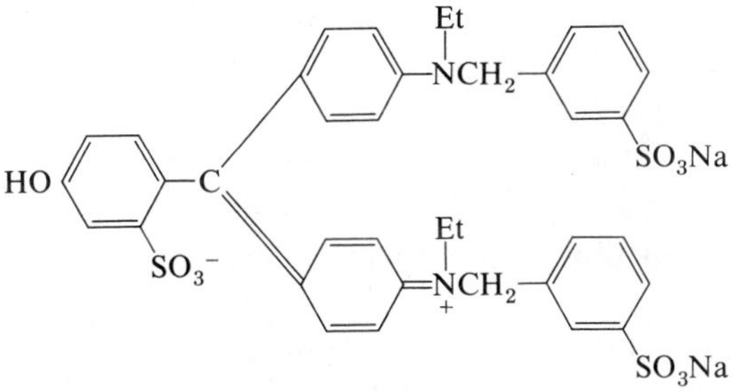

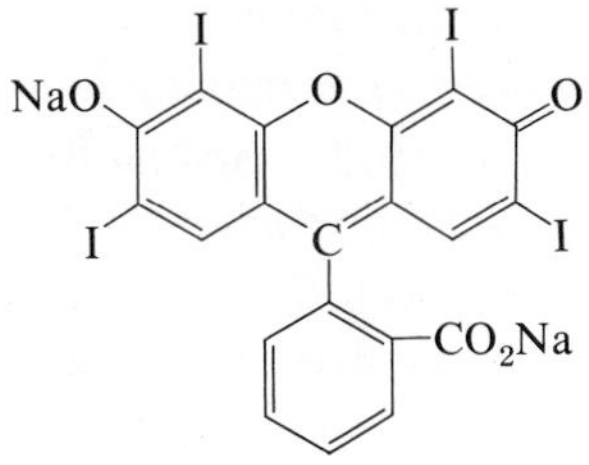

F,D&C Red No. 3
(Erythrosine)

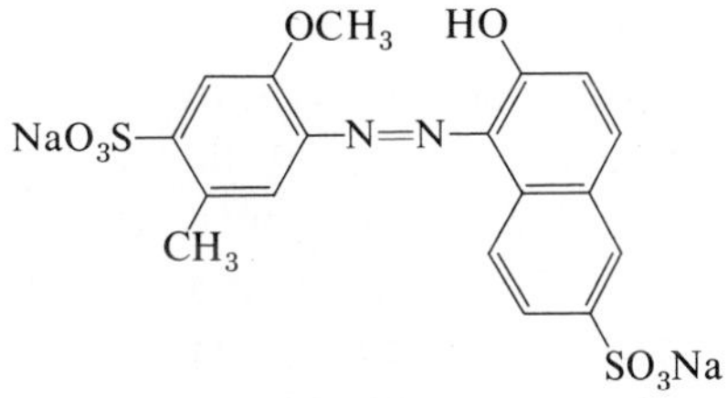

F,D&C Green No. 3
(Fast Green FCF)

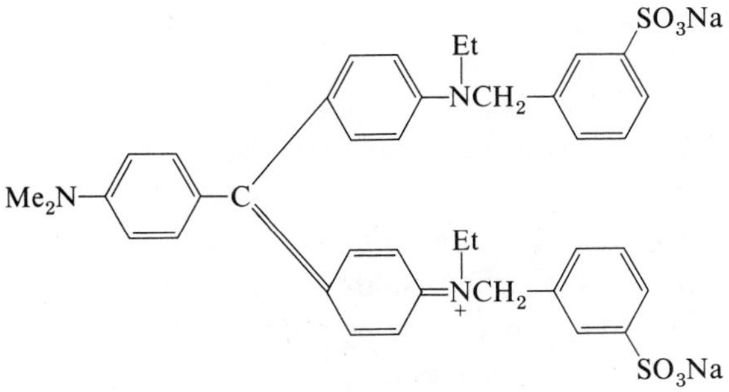

F,D&C Red No. 40
(Allura Red)

F,D&C Violet No. 1
(Benzyl Violet)

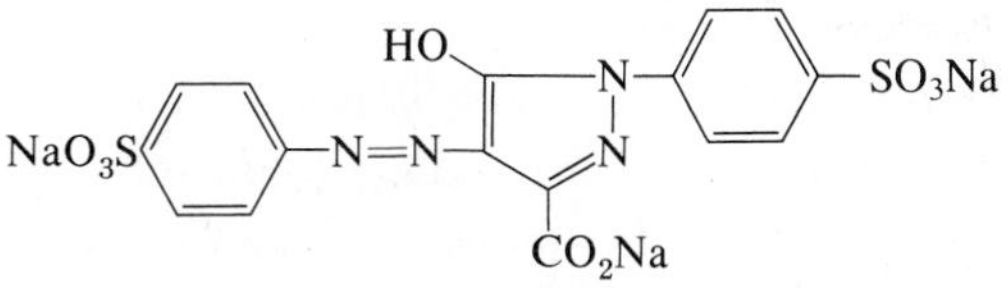

F,D&C Yellow No. 5
(Tartrazine)

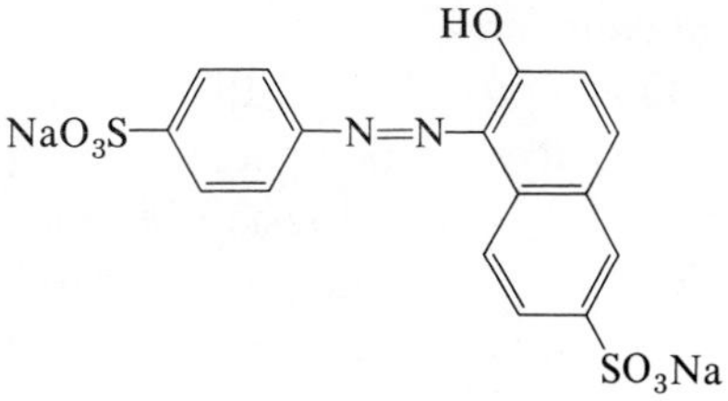

F,D&C Yellow No. 6
(Sunset Yellow)

Nine food colors approved by the Food and Drug Administration in 1975. All are still in use except for Red No. 2 (Amaranth), which was banned in 1976.

The structures of the main food dyes are shown in the figure. Note that a good many of them are azo dyes. Since many of the dyes with the azo linkage have been shown to be carcinogens, many persons suspect all such dyes. In 1960, the law was amended to require that any new dyes submitted for approval should undergo extensive scientific testing before they could be approved. They must be shown to be free from causing birth defects, organic dysfunction, and cancer. Old dyes may be subject to reconsideration if experimental evidence suggests that this is necessary.

Several recent studies have suggested that synthetic food dyes may be responsible, at least in part, for hyperkinetic activity in certain young children. It was shown that when these children were maintained on diets that excluded synthetic food dyes, many of them reverted to more normal behavior patterns. On the contrary, when they were administered a synthetic mixture of food dyes as a capsule along with this diet, the hyperkinetic syndrome would often manifest itself once again. Currently several groups of workers are involved in studying this apparent relationship.

Red Number 2 is the dye that has most recently been involved in a controversy concerning its safety. In many tests, some even performed by Food and Drug Administration (FDA) chemists, mounting evidence was found that this dye might be harmful, causing birth defects, spontaneous abortion of fetuses, and possibly cancer. However, the results of other workers were found to contradict these findings. Much controversy, involving the FDA, the opponents, the proponents, and the courts, ensued. Finally, in February 1976, this dye was banned for food use after the FDA and the courts decided that the bulk of the evidence argued for its discontinuation. More of this interesting story may be found in the references.

While Red Number 2 is proscribed in the United States, it is still approved for use in Canada and within the European Economic Community, and it may be found in products originating in those countries. Before the ban in the United States, Red Number 2 was the most widely used food dye in the industry, appearing in everything from ice cream to cherry soda. Fortunately, proscription of Red Number 2 has not been disastrous for the industry, since for most uses, either Red Number 3 or Red Number 40 are ready substitutes. This knowledge probably had much to do with the court decision finally to ban the dye.

Red Number 40, the most recently accepted food dye, was approved in 1971. Before gaining approval, the Allied Chemical Corporation, which holds exclusive patent rights to the dye, carried out the most thorough and expensive testing program ever given to a food dye. These tests even included a study of possible birth defects. Red Number 40, called Allura Red, seems destined to replace Red Number 2, since it has an extremely wide variety of applications, including the dyeing of maraschino cherries; in this it can replace the provisionally listed Red Number 4, which was also banned in 1976.

There are currently eight allowed dyes for food use. The structures of these eight approved food dyes are given in the accompanying chart. The use of these eight dyes is unrestricted. In addition, two other dyes are approved for restricted uses. Citrus Red Number 2 (old Red Number 32) may be used to color the skins of oranges, and Orange B may be used to color the skins of sausages.

REFERENCES

Augustine, G. J., Jr. and Levitan, H. ''Neurotransmitter Release from a Vertebrate Neuromuscular Synapse Affected by a Food Dye.'' *Science, 207* (March 28, 1980): 1489.

Boffey, P. M. ''Color Additives: Botched Experiment Leads to Banning of Red Dye No. 2.'' *Science, 191* (February 6, 1976): 450.

Boffey, P. M. ''Color Additives: Is Successor to Red Dye No. 2 Any Safer?'' *Science, 191* (February 22, 1976): 832.

Feingold, B. F. *Why Your Child Is Hyperactive*. New York: Random House, 1975.

Hopkins, H. ''Countdown on Color Additives.'' *FDA Consumer* (November 1976): 5.

Jacobson, M. F. *Eater's Digest: The Consumer's Factbook of Food Additives*. New York: Doubleday and Company (Anchor No. A 861), 1972.

Mebane, R. C., and Rybolt, T. R. ''Chemistry in the Dyeing of Eggs.'' *Journal of Chemical Education, 64* (April 1987): 291.

National Academy of Sciences, National Research Council. *Food Colors*. Printing and Publishing Office of the National Academy of Sciences, 2101 Constitution Avenue, NW, Washington, DC 20418 (1971).

''Red Food Coloring: How Safe Is It?'' *Consumer Reports* (February 1973): 130.

Sanders, H. J. ''Food Additives.'' *Chemical and Engineering News,* Part 1 (October 10, 1966): 100; Part 2 (October 17, 1966): 108.

Swanson, J. M., and Kinsbourne, M. ''Food Dyes Impair Performance of Hyperactive Children on a Laboratory Learning Test.'' *Science, 207* (March 28, 1980): 1485.

Turner, J. *The Chemical Feast*. New York: Grossman, 1970.

Weiss, B., *et. al.* ''Behavioral Responses to Artificial Food Colors.'' *Science, 207* (March 28, 1980): 1487.

Winter, R. *A Consumer's Dictionary of Food Additives*. New York: Crown Publishers, 1972.

Experiment 36
Chromatography of Some Dye Mixtures

Thin-Layer Chromatography
 Prepared Plates
 Hand-dipped Plates
Paper Chromatography
Column Chromatography

In this experiment, several different types of chromatography are used to separate mixtures of dyes. Three types of dye mixtures are involved. The first type of mixture will be that represented by the commercial food colors that can be bought in any grocery store (Parts I and II). These are usually available in small packages containing bottles of red, yellow, blue, and green food dye mixtures. As the experiment will show, each of these colors is rarely compounded of only a single dye. For instance, the blue dye usually has a small admixture of a red dye to make it more brilliant in color. A red dye is often added to the yellow food dye for a similar reason. The green dye will normally be a mixture of blue and yellow dyes.

The second type of dye mixture to be used is one compounded by the instructor from three F,D&C dyes approved for food use. This artificial mixture contains a red dye, a blue dye, and a yellow dye. The identities of these dyes will not be given; rather, you will be asked to identify them by thin-layer chromatography (Parts IV and V) using standard solutions of the individual pure dyes. Your instructor may also ask you to separate this mixture using column chromatography (Part VI).

The third type of mixture consists of the dyes obtained from a commercial powdered drink mix, such as Kool-Aid (Part II). In this case, you will be asked to try to identify the particular dyes used in its preparation.

For those students who are interested, the references listed at the end of this experiment give methods of extracting food dyes from various other types of foods. Also given is information on how to differentiate the various food dyes by their visible absorption spectra.

SPECIAL INSTRUCTIONS

To understand this experiment, you have to read the essay ''Food Colors,'' which precedes it. In addition, it is necessary to read Technique 10 (Column Chromatography), Sections 10.1–10.4, pp. 594–602, and Technique 11 (Thin-Layer Chromatography), p. 615.

The instructor may choose to do only selected portions of this experiment or perhaps all the parts. Several experiments may be done at the same time since much of the time involved is spent waiting for the solvent to ascend the chromatograms. To aid in your planning, an estimate of the amount of time required for development or separation is given at the beginning of each section. This time does not include preparation time for the solvents, development chambers, columns, or spotting procedures.

If you are instructed to complete Part VI (Separation of a Dye Mixture by Column Chromatography), it will be necessary to read **all** of Technique 10, rather than only the first four sections mentioned above.

PROCEDURES

I. PAPER CHROMATOGRAPHY OF FOOD COLORS (DEVELOPMENT TIME: 40 MINUTES)

At least 12 capillary micropipets will be required for the experiment. Prepare these according to the method described and illustrated in Technique 11, Section 11.3, p. 620.

Prepare about 90 mL of a development solvent consisting of

30 mL 2N NH_4OH (=4 mL conc. NH_4OH + 26 mL H_2O)
30 mL 1-Pentanol (=n-amyl or n-pentyl alcohol)
30 mL Absolute ethanol

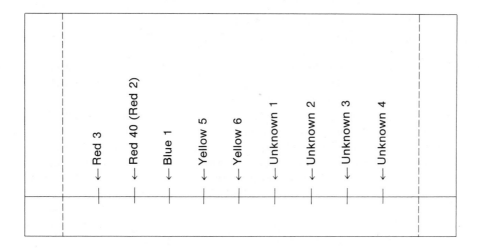

The entire mixture may be prepared in a 100-mL graduated cylinder. Mix the solvent well and pour it into the development chamber for storage. A 32-oz wide-mouthed screw-cap jar (or a Mason jar) is an appropriate development chamber. Cap the jar tightly to prevent losses of solvent from evaporation.

Next, obtain a 12-cm × 24-cm sheet of Whatman No. 1 paper. Using a pencil (not a pen), lightly draw a 24-cm-long line about 2 cm up from the long edge of the sheet. Using a centimeter ruler and the pencil, measure and mark off two dashed lines, each about 2 cm from each short end of the paper. Then make nine small marks at 2-cm intervals along the line on the long axis of the paper. These are the positions at which the samples will be spotted (see the illustration).

If they are available, starting from left to right, spot the reference dyes F,D&C Red No. 3 (Erythrosine), F,D&C Red No. 40 (Allura Red),[1] F,D&C Blue No. 1 (Erioglaucine), F,D&C Yellow No. 5 (Tartrazine), and F,D&C Yellow No. 6 (Sunset Yellow). These should be available in 2% aqueous solutions. It may be wise to practice the spotting technique on a small piece of Whatman No. 1 filter paper before trying to spot the actual chromatogram. The correct method of spotting is described in Technique 11, Section 11.3, p. 620. It is important that the spots be made as small as possible and that the paper not be overloaded. If either of these conditions is not met, the spots will tail and overlap after development. The applied spot should be 1 to 2 mm (1/16 in.) in diameter.

On the remaining four positions (nine if standards are not used) you may spot any dyes of your choice. Use of red, blue, green, and yellow dyes from a single manufacturer is suggested. If the dyes are supplied in screw-cap bottles, the pipets can be filled simply by dipping them into the bottle. If the dyes are supplied in squeeze bottles, however, it will be easiest to place a drop of the dye on a microscope slide and to insert the pipet into the drop. One microscope slide should suffice for all the samples.

When the samples have been spotted, hold the paper upright with the spots at the bottom and coil it into a cylinder. Overlap the areas indicated by the dashed lines and fasten the cylinder together (spots inside) with a paper clip or a staple. When the spots have dried, place the cylinder, spotted edge down, in the development chamber. The solvent level should be below the spots, or they will dissolve in the solvent. Cap the jar

---

[1] In Canada or the United Kingdom, substitute F,D&C Red No. 2 (Amaranth) for Red No. 40. Even in the United States, if the food color samples are old (predating 1977), they may contain Amaranth.

and wait until the solvent ascends to the top of the paper. This will take about 40 minutes, and the remaining parts of the experiment (if required) can be done while waiting.

When the solvent has ascended to within 1 cm from the top of the paper, remove the cylinder, open it quickly, and mark the level of the solvent with a pencil. This uppermost level is the solvent front. Allow the chromatogram to dry. Then, using a ruler marked in millimeters, measure the distance that each spot has traveled relative to the solvent front; calculate its R_f value (see Technique 11, Section 11.8, p. 625). Using the list of approved food dyes in the "Food Colors" essay and the reference dyes (if used), try to determine which particular dyes were used to formulate the food colors you tested. Be sure to examine the dye package (or the bottles) to see whether the desired information is given. What conclusions can you draw? Include your chromatogram along with your report.

II. PAPER CHROMATOGRAPHY OF THE DYES FROM A POWDERED DRINK MIX OR A GELATIN DESSERT (DEVELOPMENT TIME: 40 MINUTES)

Place a quantity of the powdered drink mix or gelatin dessert in a small test tube and add warm water dropwise until the sample just dissolves. Use this concentrated solution to spot the paper as described in the section above. Four drink mixes can be spotted on the same piece of Whatman No. 1 paper along with the five standards. Use a **pencil** to label each spot, and then develop the chromatogram in the solvent containing equal parts of $2N$ NH_4OH, pentanol, and ethanol, as previously described. Try to identify which dyes are used in samples of several drink mixes (for example, black cherry, cherry, grape, lemon-lime, lime, orange, punch, raspberry, or strawberry). Calculate and compare the R_f values of the standards as well as those of the dyes from the drink mixes. Methods of treating other types of foods to extract and identify the dyes that have been added are described in the references listed at the end of this experiment.

III. SEPARATION OF FOOD COLORS USING PREPARED TLC PLATES (DEVELOPMENT TIME: 90 MINUTES)

Obtain from the instructor a 5-cm × 10-cm sheet of a prepared silica gel tlc plate (Eastman Chromagram Sheet No. 13180 or No. 13181). These plates have a flexible backing, but they should not be bent excessively. They should be handled carefully, or the adsorbent may flake off of them. In addition, they should be handled only by the edges. The surface should not be touched.

Using a lead pencil (not a pen), **lightly** draw a line across the short dimension of the plate about 1 cm from the bottom. Using a centimeter ruler, mark off four 1-cm intervals on the line (see figure). These are the points at which the samples will be spotted.

Prepare at least four capillary micropipets as described and illustrated in Technique 11, Section 11.3, p. 620. Starting from left to right, spot first a red food dye, then a

blue dye, a green dye, and a yellow dye. The correct method of spotting a tlc slide is described in Technique 11, Section 11.3. It is important that the spots be made as small as possible and that the slide not be overloaded. If either of these cautions is disregarded, the spots will tail and will overlap after development. The applied spot should be about 1 to 2 mm (1/16 in.) in diameter. If small scrap pieces of the tlc plates are available, it would be a good idea to practice spotting on these before using the actual sample plate.

Prepare a development chamber from an 8-oz wide-mouthed screw-cap jar. It **should not** have the filter paper liner described in Technique 11, Section 11.4, p. 621. These plates are very thin, and if they touch a liner at any point, solvent will begin to diffuse onto the plate from that point. The development solvent, which can be prepared in a 10-mL graduated cylinder, should be a 4:1 mixture of isopropyl alcohol (2-propanol) and concentrated ammonium hydroxide.[2] Mix the solvent well and pour enough into the development chamber to give a solvent depth of about 0.5 cm (or less). If the solvent level is too high, it will cover the spotted substances, and they will dissolve into the solvent reservoir.

Place the spotted tlc slide in the development chamber, cap the jar tightly, and wait for the solvent to rise almost to the top of the slide. When the solvent is close to the top edge, remove the slide, and using a pencil (not a pen), quickly mark the position of the solvent front. Allow the plate to dry. Using a ruler marked in millimeters, measure the distance that each spot has travelled relative to the solvent front and calculate its R_f value (see Technique 11, Section 11.8, p. 625).

At your instructor's option, and if the dyes are available, you may be asked to spot a second plate with a set of reference dyes. The reference dyes will include F,D&C Red No. 40 (Allura Red),[1] F,D&C Blue No. 1 (Erioglaucine), F,D&C Yellow No. 5 (Tartrazine), and F,D&C Yellow No. 6 (Sunset Yellow). If this second set of dyes is analyzed, it should be possible (using the list of approved dyes in the "Food Colors" essay) to determine the identity of the dyes used to formulate the food colors tested on the first plate. Be sure to examine the package (or bottles) of the food dyes to determine if the desired information is given.

Since the plates are fragile, a sketch of the slide or slides instead of the plates should be included along with your report.

---

[2]An alternative solvent mixture, suggested by McKone and Nelson (see references), is a 50:25:25:10 mixture of 1-butanol, ethanol, water, and concentrated ammonia. In an earlier edition of this textbook, a 60:40 mixture of acetone and methanol was suggested. While this mixture does give separation, it also gives a fair amount of tailing.

IV. SEPARATION OF A DYE MIXTURE USING HAND-DIPPED TLC SLIDES (DEVELOPMENT TIME: 40 MINUTES)

Using a silica gel slurry, prepare three hand-dipped microscope-slide tlc plates by the method described in Technique 11, Section 11.2A, p. 617. Prepare developing chambers from 4-oz wide-mouthed screw-cap jars as described in Technique 11, Section 11.4, p. 621. Finally, prepare several capillary micropipets as described in Technique 11, Section 11.3, p. 620.

Next, obtain from the reagent shelf a bottle containing a mixture of three unknown dyes.[3] Using a capillary micropipet, spot the dye mixture twice on each plate. The correct method of spotting a tlc plate is described in Technique 11, Section 11.3. It is important that the spots be made as small as possible and that the plates not be overloaded. If either of these errors is made, the spots will tail and will overlap after development. The applied spot should be 1 to 2 mm ($\frac{1}{16}$ in.) in diameter.

Develop the first slide, using isopropyl alcohol (2-propanol) as the development solvent. The second slide should be developed in methanol, and the third in a 4:1 mixture of isopropyl alcohol and concentrated ammonium hydroxide.[2] Time will be saved if all three slides are run simultaneously. Two slides will easily fit in a single development chamber, and at least two students can develop slides simultaneously in the same jar.

When the solvents have risen to within 0.5 cm of the top of the slides, remove them, and with a pencil, quickly mark the position of the solvent fronts. Set the slides aside to dry. When they are dry, using a ruler calibrated in millimeters, measure the distance that each spot traveled relative to its solvent front and calculate an R_f value for that spot (Technique 11, Section 11.8, p. 625).

When you have determined which of the three solvents is the best development solvent (separates the dyes best), you can proceed in the same manner as before to try to identify which dyes are contained in the mixture. As an example of how to proceed, using another hand-dipped slide, spot two yellow dye standards and a spot of the dye mixture on the same slide. Three spots will easily fit on the same slide if they are small. If your mixture contains a yellow dye, you should readily be able to identify *which* yellow dye is in the mixture when you develop this slide. Proceed in a similar fashion to identify any red, blue, or green dyes present. Include sketches of your plates in your report, along with R_f values, and explain the results. By consulting the structures given for these dyes in the essay that precedes this experiment, you should be able to explain the relative R_f values of the three dyes. (*Hint:* Consider the substituent groups.) Also explain the function of the ammonia.

V. SEPARATION OF A DYE MIXTURE USING PREPARED TLC PLATES (DEVELOPMENT TIME: 90 MINUTES)

Obtain from the instructor three 5-cm × 10-cm sheets cut from a large prepared silica gel tlc plate (Eastman Chromagram Sheet No. 13180 or No. 13181). Although these plates have a flexible backing, they should not be bent excessively. They should be

---

[3]The composition of this three-component dye mixture will be found in the Instructor's Manual. The standard solutions have 0.1 g of each dye dissolved in 20 mL of methanol.

handled carefully, or the adsorbent may flake off them. They should be handled only by the edges; the surface should not be touched.

Using a lead pencil (not a pen), **lightly** draw a line across the short dimension of the plate about 1 cm from the bottom. Using a centimeter ruler, mark off four 1-cm intervals on the line (see figure shown in Section III). These are the points at which the samples will be spotted.

Prepare at least twelve capillary micropipets as described and illustrated in Technique 11, Section 11.3, p. 620. On each of the three plates, spot the unknown dye mixture in the lower left corner.[3] The remaining three positions on each slide may be used to spot standard solutions of known dyes. When spotting, it is important that the spots be made as small as possible and that the slide not be overloaded. If either of these errors is made, the spots will tail and will overlap one another after development. The applied spot should be about 1 or 2 mm (1/16 in.) in diameter. If small scrap pieces of the tlc plates are available, it would be a good idea to practice spotting on these prior to preparing the actual sample plate.

Prepare a development chamber from an 8-oz wide-mouth screw-cap jar. It *should not* have the filter paper liner that is described in Technique 11, Section 11.4, p. 621. These plates are very thin, and if they touch a liner at any point, solvent will begin to diffuse onto the plate from that point. The development solvent, which can be prepared in a 10-mL graduated cylinder, should be a 4:1 mixture of isopropyl alcohol and concentrated ammonium hydroxide.

Place the spotted tlc slides in the development chamber, cap the jar tightly, and wait for the solvent to rise almost to the top of the slide. When the solvent is close to the top edge, remove the slide, and using a pencil (not a pen), quickly mark the position of the solvent front. Allow the plate to dry. Using a ruler marked in millimeters, measure the distance that each spot has traveled relative to the solvent front and calculate its R_f value (see Technique 11, Section 11.8, p. 625).

If you have spotted the correct reference dye solutions along with the unknown mixture, you should now be able to identify which dyes are contained in the mixture. Since the plates are rather fragile, sketches of the slides instead of the plates should be included in your report. The spots should be labeled with their R_f values. Be sure to give your conclusions. By consulting the structures given for these dyes in the essay that precedes this experiment, you should be able to explain the relative R_f values of the dyes in the mixture. (*Hint:* Consider the substituent groups.) Also explain the function of the ammonia.

VI. SEPARATION OF A DYE MIXTURE BY COLUMN CHROMATOGRAPHY (ELUTION TIME FOR COLUMN: 2½–3 HOURS)[4]

Obtain a dry chromatography column made from 10-mm glass tubing and having a length of about 20 cm.[5] Place a small, loose plug of glass wool in the bottom of the

---

[4]To perform this separation, it is necessary to read **all** of Technique 10.

[5]Two columns can be made from a 42-cm-long piece of 10-mm glass tubing. Heat the tubing at the center while rotating it, and when it is hot, constrict the center by pulling the two ends about 4 cm apart. When the tubing is cool, score the constriction in the middle with a file; then with light pressure from the thumbs while pulling on the two ends, break the tubing into two equivalent columns. (See an illustration of the equivalent operation using capillary tubing to make micropipets in Figure 11–3, p. 620.)

column and, with a length of glass rod, gently tamp it level. Cover the glass wool with about 4 to 5 mm of white sand. With a utility clamp, attach the column vertically to a ring stand. Place a small piece of flexible tubing over the column's exit, loosely attach a screw clamp to the tubing, and close it.

Place a beaker under the exit of the column and fill the column ⅔ full with isopropyl alcohol (2-propanol). Start the column dripping **slowly** and add enough dry absorption alumina (about 5 g) to give a column of adsorbent 6 to 8 cm high.[6] Rinse the sides of the column with a little solvent. Allow any excess isopropyl alcohol to drain until the top of the column just becomes dry, and then close the screw clamp. Get the synthetic dye mixture from the supply shelf, and with a disposable pipet, carefully transfer enough of this mixture to the top of the column to make a layer of liquid 3 mm deep. Open the screw clamp, drain this mixture into the column, and then once again stop the flow. In the same way, place a similar amount of isopropyl alcohol on top of the column and then drain it also into the column, again stopping the flow when the surface of the alumina becomes dry. Finally, carefully add as much of 4:1 isopropyl alcohol–concentrated ammonia elution solvent on top of the column as the unfilled portion allows, and open the screw clamp to allow the chromatography to begin.

From time to time, refill the reservoir with more solvent, never allowing the column to run dry. Continue elution and carefully collect each of the dyes in a separate container. Once the first dye has been collected, you may switch to 4:1 ethanol–concentrated ammonia to elute the second dye. The third dye may be eluted by changing to 4:1 methanol–concentrated ammonia. If the last dye does not elute easily, the ratio of ammonia to methanol may be increased.

Each of the eluted dyes may be spotted on hand-dipped tlc slides (described in Section IV) and compared with standard solutions of the three dyes, using 4:1 isopropyl alcohol–concentrated ammonia for development. Give your conclusions in your report, and give the R_f values you found for each of the dyes when making the tlc comparisons. Finally, consult the structures given in the essay preceding this experiment and try to explain the order of elution (R_f values) observed for the three dyes. (*Hint:* Consider the substituent groups.) Also explain the function of the ammonia.

REFERENCES

McKone, H. T. "Identification of F,D&C Dyes by Visible Spectroscopy." *Journal of Chemical Education, 54* (June 1977): 376.

McKone, H. T., and Nelson, G. J. "Separation and Identification of Some F,D&C Dyes by TLC." *Journal of Chemical Education, 53* (November 1976): 722.

---

[6]We use Fisher No. A-540 Adsorption Alumina. Longer columns should be avoided since they lead to very long elution and separation times.

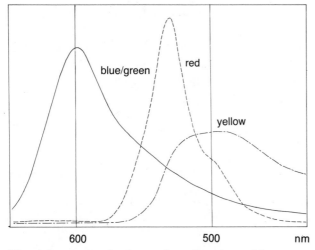

Ultraviolet spectra of unknown dyes, Experiment 36.

Essay
THE CHEMISTRY OF VISION

An interesting and challenging topic for chemists to investigate is how the eye functions. What chemistry is involved in detection of light and transmission of that information to the brain? The first definitive studies on how the eye functions were begun in 1877 by Franz Boll. Boll demonstrated that the red color of the retina of a frog's eye could be bleached yellow by strong light. If the frog was then kept in the dark, the red color of the retina slowly returned. Boll recognized that a bleachable substance had to be somehow connected with the ability of the frog to perceive light.

Most of what is now known about the chemistry of vision is the result of the elegant work of George Wald, of Harvard University; his studies, which began in 1933, ultimately brought about his receiving the Nobel Prize in biology. Wald identified the sequence of chemical events during which light is converted into some form of electrical information that can be transmitted to the brain. Here is a brief outline of that process.

The retina of the eye is made up of two types of photoreceptor cells: **rods** and **cones.** The rods are responsible for vision in dim light, and the cones are responsible for color vision in bright light. The same principles apply to the chemical functioning of the rods and of the cones; however, the details of functioning are less well understood for the cones than for the rods.

Each rod contains several million molecules of **rhodopsin.** Rhodopsin is a complex of a protein, **opsin,** and a molecule derived from Vitamin A, 11-*cis*-retinal (sometimes called **retinene**). Very little is known about the structure of opsin. The structure of 11-*cis*-retinal is shown below.

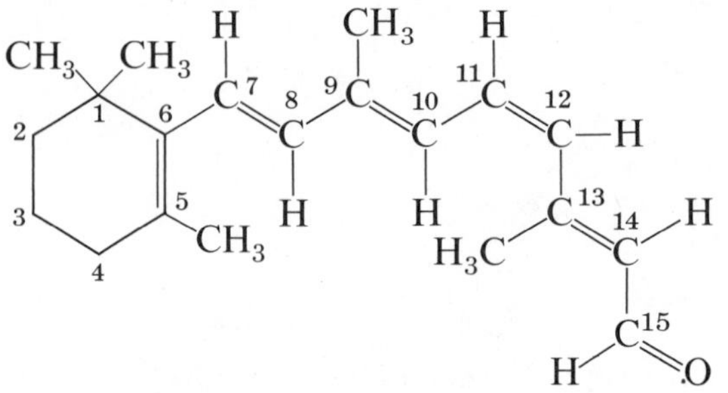

11-*cis*-retinal

The detection of light involves the initial conversion of 11-*cis*-retinal to its all-*trans* isomer. This is the only obvious role of light in this process. The high energy of a quantum of visible light promotes the fission of the π bond between carbons 11 and 12. When the π bond breaks, free rotation about the σ bond in the resulting radical is possible. When the π bond re-forms after such rotation, all-*trans*-retinal results. All-*trans*-retinal is more stable than 11-*cis*-retinal, which is why the isomerization proceeds spontaneously in the direction shown.

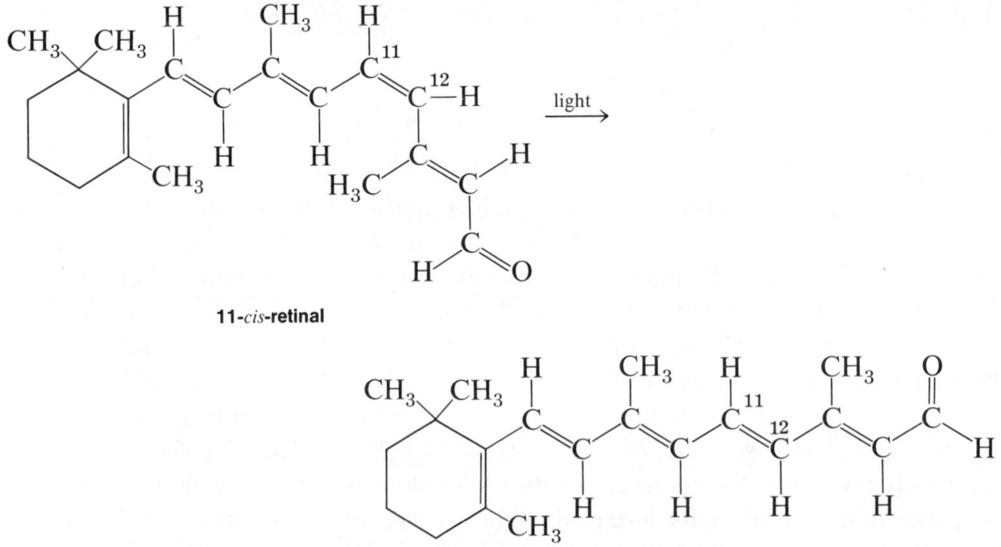

11-*cis*-retinal

All-*trans*-retinal

The two molecules have different shapes due to their different structures. The 11-*cis*-retinal has a fairly curved shape, and the parts of the molecule on either side of the *cis* double bond tend to lie in different planes. Because proteins have very complex and specific three-dimensional shapes (tertiary structures), 11-*cis*-retinal will associate

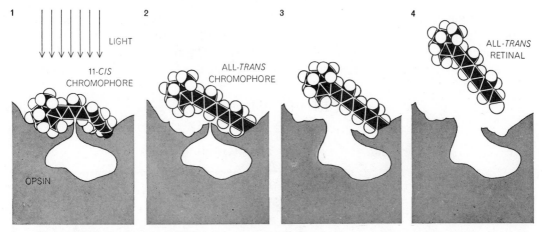

From "Molecular Isomers in Vision," by Ruth Hubbard and Allen Kropf. Copyright © 1967 by Scientific American Inc. All rights reserved.

with the protein opsin in a particular manner. All-*trans*-retinal has an elongated shape, and the entire molecule tends to lie in a single plane. This different shape for the molecule, compared with the 11-*cis* isomer, means that all-*trans*-retinal will have a different association with the protein opsin.

In fact, all-*trans*-retinal associates very weakly with opsin because its shape does not fit the protein. Consequently, the next step after the isomerization of retinal is the dissociation of all-*trans*-retinal from opsin. The opsin protein undergoes a simultaneous change in conformation as the all-*trans*-retinal dissociates.

At some time after the 11-*cis*-retinal–opsin complex receives a photon, a message is received by the brain. It was originally thought that either the isomerization of 11-*cis*-retinal to all-*trans*-retinal or the conformational change of the opsin protein was an event that generated the electrical message sent to the brain. Current research, however, indicates that both of these events occur too slowly relative to the speed with which the brain receives the message. Current hypotheses invoke involved quantum mechanical explanations, which hold it significant that the chromophores (light-absorbing groups) are arranged in a very precise geometrical pattern in the rods and cones, allowing the signal to be transmitted rapidly through space. The main physical and chemical events Wald discovered are illustrated in the figure for easy visualization. The question of how the electrical signal is transmitted still remains unsolved.

Wald was also able to explain the sequence of events by which the rhodopsin molecules are regenerated. After dissociation of all-*trans*-retinal from the protein, the following enzyme-mediated changes occur. All-*trans*-retinal is reduced to the alcohol all-*trans*-retinol, also called all-*trans*-Vitamin A:

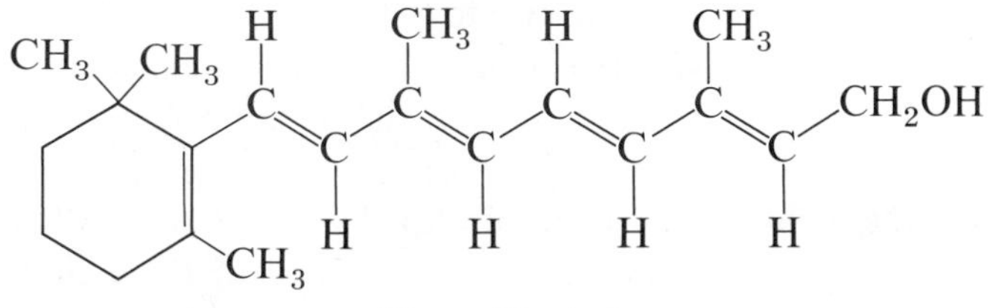

All-*trans*-Vitamin A

All-*trans*-Vitamin A is then isomerized to its 11-*cis*-Vitamin A isomer. Following the isomerization, the 11-*cis*-Vitamin A is oxidized back to 11-*cis*-retinal, which forthwith recombines with the opsin protein to form rhodopsin. The regenerated rhodopsin is then ready to begin the cycle anew, as illustrated in the accompanying diagram.

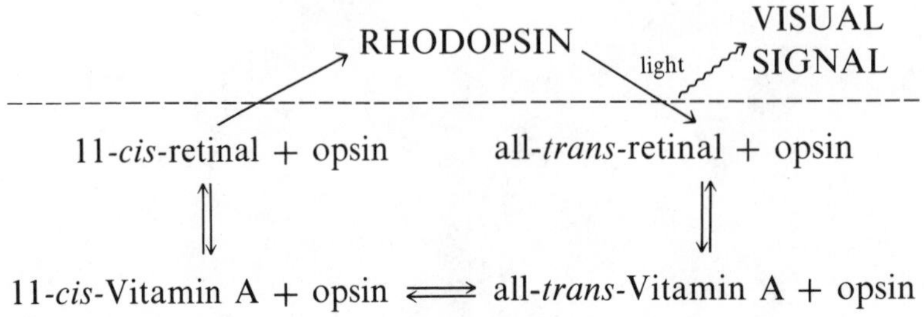

RHODOPSIN   light   VISUAL SIGNAL

11-*cis*-retinal + opsin   all-*trans*-retinal + opsin

11-*cis*-Vitamin A + opsin ⇌ all-*trans*-Vitamin A + opsin

By this process, as little light as 10^{-14} of the number of photons emitted from a typical flashlight bulb can be detected. The conversion of light into isomerized retinal exhibits an extraordinarily high quantum efficiency. Virtually every quantum of light absorbed by a molecule of rhodopsin causes the isomerization of 11-*cis*-retinal to all-*trans*-retinal.

As you can see from the reaction scheme, the retinal derives from Vitamin A, which merely requires the oxidation of a —CH_2OH group to a —CHO group to be converted to retinal. The precursor in the diet that is transformed to Vitamin A is β-carotene. The β-carotene is the yellow pigment of carrots and is an example of a family of long-chain polyenes called **carotenoids.**

In 1907 Wiltstätter established the structure of carotene, but it was not known until 1931 to 1933 that there were actually three isomers of carotene. The α-carotene differs from β-carotene in that the α isomer has a double bond between C_4 and C_5 rather than between C_5 and C_6, as in the β isomer. The γ isomer has only one ring, identical to the ring in the β isomer, while the other ring is opened in the γ form between $C_{1'}$ and $C_{6'}$. The β isomer is by far the most common of the three.

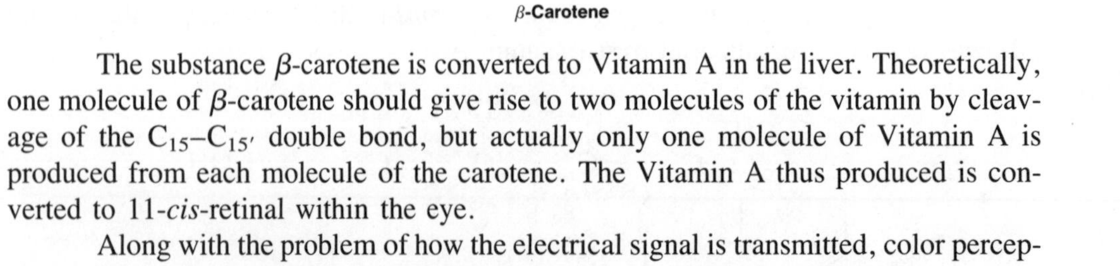

β-Carotene

The substance β-carotene is converted to Vitamin A in the liver. Theoretically, one molecule of β-carotene should give rise to two molecules of the vitamin by cleavage of the C_{15}–$C_{15'}$ double bond, but actually only one molecule of Vitamin A is produced from each molecule of the carotene. The Vitamin A thus produced is converted to 11-*cis*-retinal within the eye.

Along with the problem of how the electrical signal is transmitted, color perception is also currently under study. In the human eye there are three kinds of cone cells, which absorb light at 440, 535, and 575 nm, respectively. These cells discriminate among the primary colors. When combinations of them are stimulated, full color vision is the message received in the brain.

Since all these cone cells use 11-*cis*-retinal as a substrate-trigger, it has long been suspected that there must be three different opsin proteins. Recent work has begun to establish how the opsins vary the spectral sensitivity of the cone cells even though all of them have the same kind of light-absorbing chromophore.

Retinal is an aldehyde and it binds to the terminal amino group of a lysine residue in the opsin protein to form a Schiff base, or imine linkage ($\supset$C$=$N$-$). This imine linkage is believed to be protonated (with a plus charge) and to be stabilized by being located near a negatively charged amino acid residue of the protein chain. A second negatively charged group is thought to be located near the 11-*cis* double bond. Researchers have recently shown, from synthetic models that use a simpler protein than opsin itself, that forcing these negatively charged groups to be located at different distances from the imine linkage causes the absorption maximum of the 11-*cis*-retinal chromophore to be varied over a wide enough range to explain color vision.

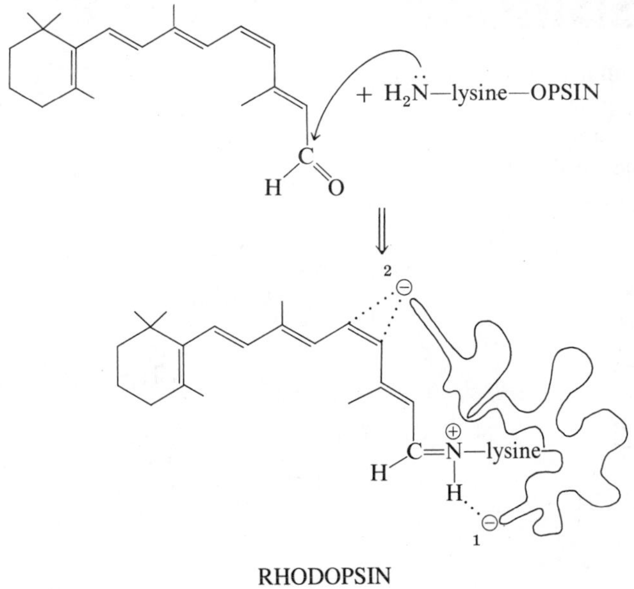

RHODOPSIN

Whether there are actually three different opsin proteins, or whether there are just three different conformations of the same protein in the three types of cone cells, will not be known until further work is completed on the structure of the opsin or opsins.

REFERENCES

Clayton, R. K. *Light and Living Matter*. Vol. 2: *The Biological Part*. New York: McGraw-Hill, 1971.
Fox, J. L. "Chemical Model for Color Vision Resolved." *Chemical and Engineering News, 57* (46) (November 12, 1979): 25.

A review of articles by Honig and Nakanishi in the *Journal of the American Chemical Society* (*101*, [1979]: 7082, 7084, 7086).

Hubbard, R., and Kropf, A. "Molecular Isomers in Vision." *Scientific American, 216* (June 1967): 64.

Hubbard, R., and Wald, G. "Pauling and Carotenoid Stereochemistry." In *Structural Chemistry and Molecular Biology*. Edited by A. Rich and N. Davidson. San Francisco: W. H. Freeman, 1968.

MacNichol, E. F., Jr. "Three Pigment Color Vision." *Scientific American, 211* (December 1964): 48.

Rushton, W. A. H. "Visual Pigments in Man." *Scientific American, 207* (November 1962): 120.

Wald, G. "Life and Light." *Scientific American, 201* (October 1959): 92.

Wolken, J. J. *Photobiology*. New York: Reinhold Book Co., 1968.

Zurer, P. S. "The Chemistry of Vision." *Chemical and Engineering News, 61* (November 28, 1983): 24.

Experiment 37
Isolation of Carotenoid Pigments From Spinach

Isolation of a Natural Product
Column Chromatography
Thin-Layer Chromatography
Ultraviolet-Visible Spectroscopy

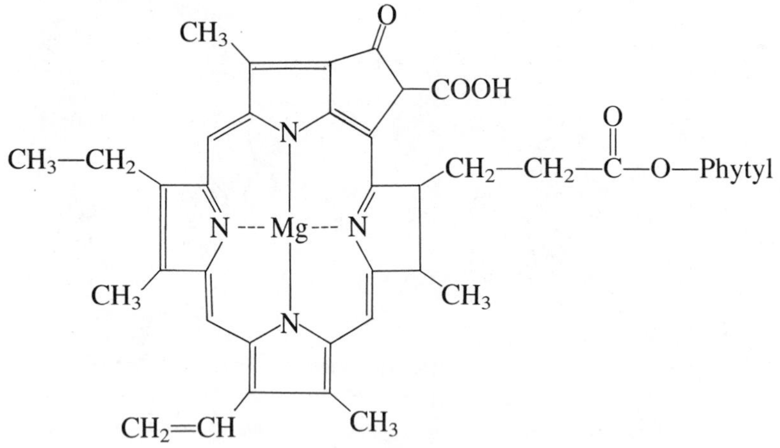

Chlorophyll a

Phytyl = $-CH_2-CH=\overset{\overset{\displaystyle CH_3}{|}}{C}-CH_2-(CH_2-CH_2-\overset{\overset{\displaystyle CH_3}{|}}{CH}-CH_2)_2-CH_2-CH_2-\overset{\overset{\displaystyle CH_3}{|}}{CH}-CH_3$

In this experiment you will isolate β-carotene (a yellow pigment) and chlorophyll (a green pigment) from spinach. β-Carotene is an example of a carotenoid, while chlorophyll is an example of a porphyrin. Each of these pigments can be characterized by

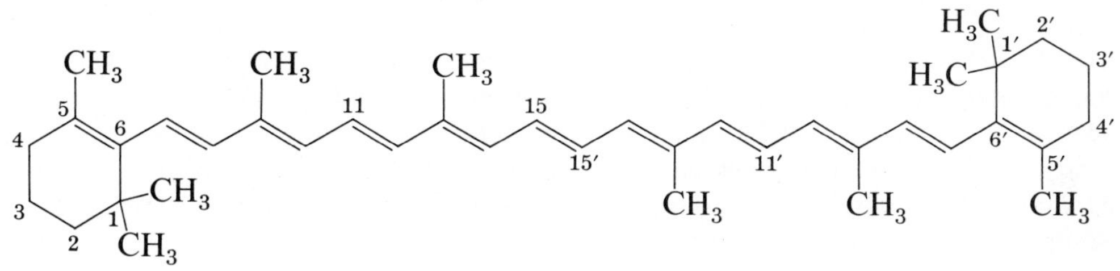

β-**Carotene**

determining its ultraviolet-visible spectrum and comparing it with a reference spectrum.

THE EXPERIMENT

A jar of strained spinach, sold as baby food, is a convenient source of β-carotene and chlorophyll. The procedure in this experiment calls for dehydrating the strained spinach with ethanol. This is followed by extraction with methylene chloride, an effective solvent for lipids. Methylene chloride, being immiscible with water, will not extract effectively the plant pigments from the tissue until the water is removed. The final purification by column chromatography illustrates an extremely important technique. This isolation procedure affords amounts of pigments that are unweighable except on a microbalance but are more than adequate for the tlc and spectroscopy experiments you will conduct after the isolation has been accomplished.

SPECIAL INSTRUCTIONS

Before beginning this experiment, you should read the preceding essay, which describes the chemistry of vision. Also read Techniques 1, 2, 10, and 11. Care should be exercised in handling toxic solvents such as methylene chloride, ethyl acetate, and ligroin. Time should be budgeted so that the entire column chromatography, including packing the column, can be completed without interruption. Allow 1 to 2 hours for the chromatography.

PROCEDURE

Weigh a 10-g sample of spinach paste in a 50-mL beaker. Add 12 mL of either absolute or 95% ethanol. Vigorously stir and macerate the mixture with a spatula until the stickiness of the paste disappears. After completing this dehydration of the paste, fit a small, short-stemmed funnel with a loose plug of glass wool. A very small amount of glass wool should be used, and it should not be forced into the stem of the funnel but should be made to lie snugly in the bottom, blocking the funnel exit completely. Filter the paste through this funnel into a clean 50-mL beaker. When the filtration is nearly complete, lightly press the pulp with your spatula to remove more liquid. Following this, clean the first beaker and transfer the pulp, glass wool and all, back into it. Discard the ethanol extract, which contains water and certain water-soluble salts, and clean that beaker for reuse. Next, add 10 mL of methylene chloride (dichloromethane) to the dehydrated pulp. Stir and macerate this mixture for about 2 minutes, then filter it as before, through a funnel fitted with a loose glass wool plug, into a clean 50-mL Erlenmeyer flask if the filtrate is to be stored or into a 50-mL beaker if you plan to continue ahead immediately. Stopper the Erlenmeyer flask and store it in a dark place until you are ready for the chromatography. Before the chromatography, the solution should be transferred to a 50-mL beaker and evaporated to dryness on a steam bath placed in a hood (use a boiling stone). The methylene chloride (bp 40 °C) should evaporate easily. Be careful not to overheat the collected pigments, or they will become oxidized and discolored. When the solvent has been evaporated, remove the beaker from the heat and allow the contents to cool. Add approximately 20 drops of fresh methylene chloride to the beaker and swirl the contents until the pigments dissolve in the solvent.

COLUMN CHROMATOGRAPHY

The crude pigment mixture is chromatographed on a 15-cm column of alumina (Fisher No. A-540) prepared with ligroin (63–75 °C) (see Technique 10, especially Sections 10.5 and 10.6, pp. 602–606, and refer to the illustration in this experiment). Place a short (3-cm) piece of rubber tubing on the bottom of the dry column and close it with a screw clamp. Place a piece of glass wool in the bottom of the column, gently tamp it down with a glass rod, and add a small amount of sand to form a layer about 3 mm thick at the bottom (Technique 10, Section 10.6A, p. 604). Add about 5 mL of the ligroin. Weigh 8 g of alumina into a beaker. Place 15 mL of ligroin in a 125-mL Erlenmeyer flask and slowly add the alumina powder, a little at a time, while swirling. Swirl the mixture vigorously to form a slurry and then add it quickly to the top of the column (see Technique 10, Section 10.6B, p. 605). Place the Erlenmeyer flask under the column, open the screw clamp, and allow the liquid to drain into it. During this operation it will be helpful to tap the side of the column **gently** with your fingers or a short length of pressure tubing. When a fair amount of liquid has passed through the column, change receiving flasks, pour the solvent that has been collected from the column into the flask containing the alumina, swirl the Erlenmeyer flask, and add more of the slurry to the column (with tapping) until a 15-cm column (approximately) of alumina is formed. Do not allow the column to run dry during the packing procedure. This slurry method eliminates air bubbles from the surface of the alumina. Drain the excess solvent until it just reaches the top level of the alumina. Close the screw clamp.

   If you have not already done so, evaporate the methylene chloride–pigment extract to dryness. Add 20 to 25 drops of methylene chloride and transfer **most** of the solution onto the column, using a small disposable pipet (Technique 10, Section 10.7, p. 606).[1] Save a few drops for later analysis by thin-layer chromatography. Open the screw clamp and allow the colored liquid to pass onto the column. When the top of the column just becomes dry, close the screw clamp. Using a capillary pipet, carefully add a few milliliters of ligroin to wash down the sides of the column and allow this solvent to

---

[1]If a larger volume of methylene chloride is used, some of the unchromatographed pigment mixture may be carried along with it through the column. This material will come off quite early in the first fractions. Fortunately, enough mixture usually adsorbs on the column, even in this event, for the chromatography to go on normally from there.

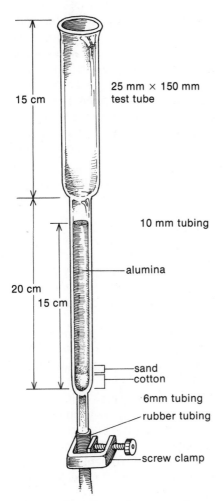

15 cm

25 mm × 150 mm
test tube

10 mm tubing

alumina

20 cm

15 cm

sand
cotton

6mm tubing

rubber tubing

screw clamp

Apparatus for column chromatography

pass onto the column by draining more of the solvent from the column until the surface again barely becomes exposed. All the colored material should now have been washed into the alumina. Allow the column to stand as is for about 5 minutes, while you prepare a solvent mixture composed of 70 volume percent ligroin and 30 volume percent ethyl acetate.

After the waiting period is complete, add a large amount of ligroin to the column reservoir, adding the first few milliliters carefully. Begin to elute the column (Technique 10, Section 10.8, p. 607), taking approximately 5-mL fractions in small test tubes. Any unadsorbed material that arises either from overloading the column or using too much methylene chloride will usually elute in the first fraction. From there on, the yellow β-carotene should move rapidly through the column, while the green chlorophyll moves more slowly. When the yellow band has been eluted from the column, replace the ligroin in the solvent reservoir with the ligroin–ethyl acetate solvent mixture. The green band will now begin to move down the column. Transfer any colored fractions to small beakers and evaporate the solvents (use a boiling stone) on a steam bath in the hood. Again, avoid overheating the pigment solutions.

Dissolve the samples in the smallest possible amount of methylene chloride (1 mL or less), hold the beakers in a slanting position for drainage into a small area, and, with disposable pipets, transfer the solutions into small test tubes labeled with fraction numbers. The few drops of material that were not column-chromatographed should also be transferred to a small, labeled test tube. These materials should be analyzed by thin-layer chromatography as soon as possible. If you have run out of time, you may have to store the samples until the next class period. If so, cork the test tubes tightly and store them in a dark, cool place. Carotenoids are sensitive to light and air; they decompose somewhat on standing. They should be analyzed in the very next laboratory period.

THIN-LAYER CHROMATOGRAPHY

Technique 11 describes the procedures needed for thin-layer chromatography (tlc). Silica gel G will be used. Use the dipping method to prepare the plates (Technique 11, Section 11.2A, p. 617). Analyze the materials that are extracted from spinach as described below.

If the solvent has evaporated from the pigment samples you have obtained, add a **few** drops of methylene chloride and shake the mixture to dissolve the material. The sample may not dissolve completely because of partial air oxidation of the carotenoids to form insoluble material, but there should be enough colored supernatant liquid for numerous experiments. Keep the solutions out of the light when they are not in use. The highly unsaturated hydrocarbons are subject to rapid photochemical autoxidation, but during chromatography they are protected by solvent vapor. When a plate is removed from the chamber, however, a spot may disappear rapidly, and it should be outlined in pencil immediately. Save a plate from which the outlined spots of pigment have disappeared and develop it in an iodine chamber, as described in Technique 11, Section 11.6, p. 623.

Up to three fractions can be spotted on a single plate (Technique 11, Section 11.3, p. 620). There should be at least three samples from the spinach chromatography experiment and from the unchromatographed material. It should not be necessary to analyze every column chromatography fraction by tlc, but those solutions that have the most intense colors and represent the centers of the eluted bands should be the minimum requirement for the interpretation of the results. In addition, you may choose to analyze a few other fractions that may appear interesting. Use the ligroin–ethyl acetate (70:30) solvent mixture to develop the plate. Remove the plates immediately after development; as soon as the plates have dried, outline the spots by marking them with a pencil. Calculate the R_f values for each of the spots.

Since spinach is known to contain both β-carotene and chlorophyll, it should have been possible to detect both pigments by analyzing the sample before column purification (at least two spots should be present) and comparing it with the two fractions (β-carotene and chlorophyll) obtained by column chromatography.

Compare the R_f values of each pigment both separately and in the crude (unchromatographed) mixture directly by measuring them on the **same** plate. Are the R_f values the same? Report the R_f values obtained in all experiments, along with a discussion of the results, to the instructor. At the instructor's option, determine the ultraviolet-visible spectrum of each pigment. Compare the spectra that you obtain with those illustrated below. Include the spectra with your laboratory report.

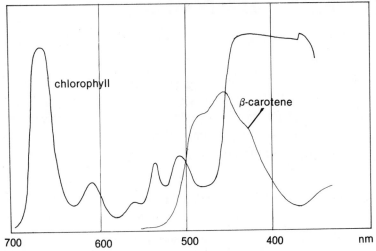

Visible-light absorption spectrum of isolated chlorophyll and β-carotene in petroleum ether

---

REFERENCE

McKone, H. T., "The Rapid Isolation of Carotenoids from Foods." *Journal of Chemical Education, 56:* 10 (October 1979): 676.

QUESTIONS

1. Why is chlorophyll less mobile on column chromatography and why does it have a lower R_f value than β-carotene?

2. Lycopene, an important carotenoid pigment in tomatoes, has the structure shown below. Would you expect it to be more or less mobile than β-carotene on column chromatography? Why?

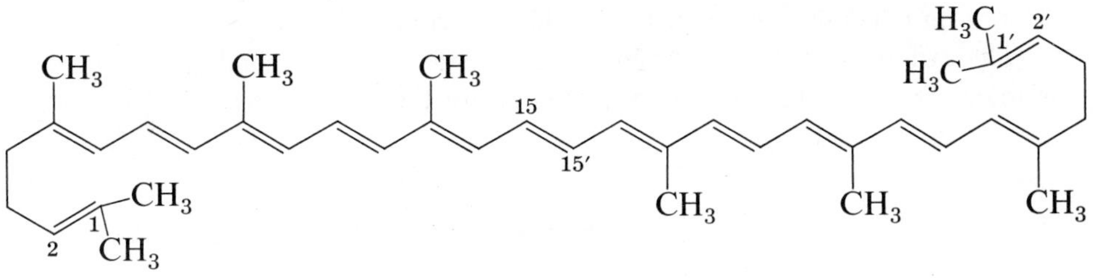

Lycopene

3. Why should lycopene be red, while β-carotene is yellow?

Essay
CYANOHYDRINS IN NATURE

The addition of hydrogen cyanide to aldehydes is a typical nucleophilic addition reaction of the aldehyde functional group. The adduct is called a **cyanohydrin.** Because cyanide ion is required to be the attacking nucleophile, the reaction is catalyzed by base. The reaction is usually carried out with sodium or potassium cyanide, by adding only that amount of acid needed to keep the pH near 7 or 8. In the presence of excess

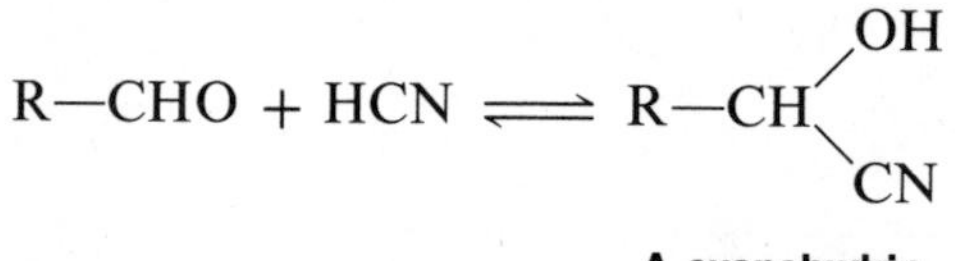

$$R-CHO + HCN \rightleftharpoons R-CH{\overset{\textstyle OH}{\underset{\textstyle CN}{}}}$$

A cyanohydrin

acid, the cyanide ion, which is a fairly strong base, is protonated and its effectiveness as a nucleophile is diminished.

In the cyanohydrin, the hydrogen derived from the original aldehyde is made acidic due to its new position alpha to the cyano group. The cyano group provides resonance stabilization.

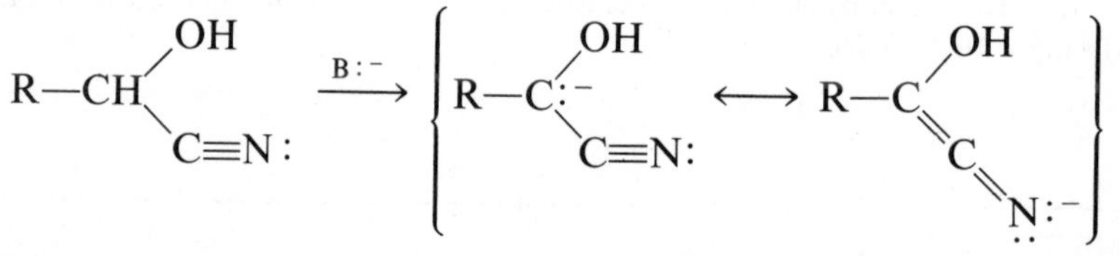

$$R-CH{\overset{\textstyle OH}{\underset{\textstyle C\equiv N:}{}}} \xrightarrow{B:^-} \left[R-\overset{\textstyle OH}{\underset{\textstyle C\equiv N:}{\overset{..}{C}:^-}} \longleftrightarrow R-\overset{\textstyle OH}{\underset{\textstyle \underset{..}{N}:^-}{C}} \right]$$

In a slightly basic KCN solution, the removal of this proton promotes the formation of the cyanohydrin by immediately converting it to its resonance-stabilized conjugate anion (LeChâtelier's principle). This anion can also behave as a nucleophile toward an unreacted molecule of benzaldehyde. An example of this type of behavior can be found in the benzoin condensation as described in Experiment 38.

The cyanohydrin formed from benzaldehyde is called mandelonitrile, since it yields mandelic acid when hydrolyzed in strongly **basic** solutions:

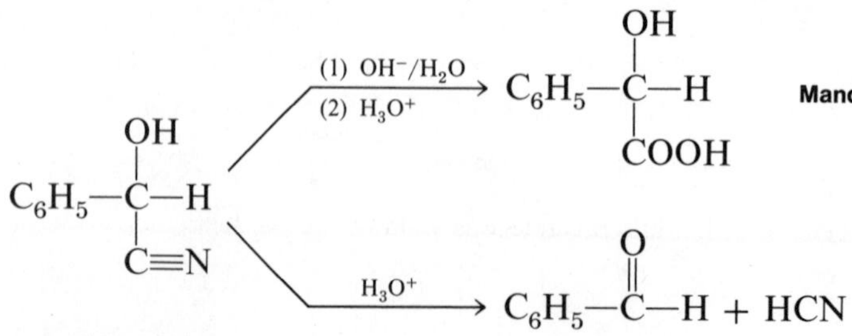

$$C_6H_5-\overset{\textstyle OH}{\underset{\textstyle C\equiv N}{C}}-H \quad \begin{array}{c} \xrightarrow[\text{(2) } H_3O^+]{\text{(1) } OH^-/H_2O} \quad C_6H_5-\overset{\textstyle OH}{\underset{\textstyle COOH}{C}}-H \quad \textbf{Mandelic acid} \\[30pt] \xrightarrow{H_3O^+} \quad C_6H_5-\overset{\textstyle O}{\overset{\|}{C}}-H + HCN \end{array}$$

Mandelonitrile

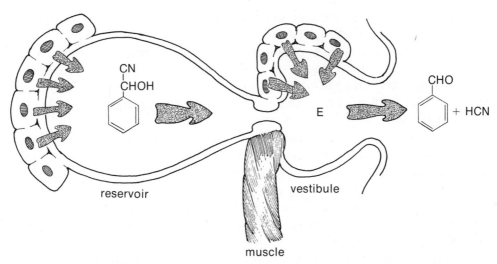

Reactor gland of *Apheloria corrugata*

In contrast, when mandelonitrile is hydrolyzed under **acidic** conditions, it decomposes to regenerate benzaldehyde and hydrogen cyanide.

Surprisingly, mandelonitrile is found in many forms in nature. Perhaps the most interesting example is in the millipede *Apheloria corrugata*. This millipede uses the compound as a part of its protective apparatus. It synthesizes and stores the cyanohydrin in a series of 22 glands, which are arranged in pairs on several of the body sections just above the legs. Each gland has two compartments. The inner compartment is a large saclike reservoir lined with cells that secrete mandelonitrile as an aqueous emulsion. The inner compartment is separated from the outer compartment by a muscular valve. The second compartment, the vestibule, is lined with cells that secrete an enzyme that decomposes mandelonitrile into benzaldehyde and hydrogen cyanide, a poisonous gas. When the millipede is alarmed by a predator, it opens the valve separating the two compartments and, by contraction of the storage reservoir, forces mandelonitrile into the vestibule, where it is mixed with the enzyme and forced outside. The products of the dissociation either kill or repel the predator. A single *Apheloria* can secrete enough hydrogen cyanide to kill a small bird or a small mouse. The benzaldehyde is also an effective repellent against many of the millipede's predators.

Mandelonitrile is also found in several plants of the Rosaceae family. It is found not in the free form but as a **glycoside,** that is, with its hydroxyl group attached through an acetal linkage to a sugar. The most common glycoside is **amygdalin,** which is found

Amygdalin

in the seeds (pits), leaves, and bark of many plants such as bitter almond (*Prunus amygdalus*), apricot, wild cherry, peach, and plum.

Two similar glycosides, **prunlaurasin** and **sambunigrin,** are found in cherry laurel and *Sambucus nigra,* respectively. These glycosides differ from amygdalin only in the stereochemistry. On hydrolysis, amygdalin yields two molecules of glucose and D-mandelonitrile. Sambunigrin yields L-mandelonitrile, and prunlaurasin yields racemic (D,L)-mandelonitrile.

The seeds of most of these plants also contain two enzymes, called **emulsin** and **prunase.** Emulsin hydrolyzes one glucose molecule from amygdalin to produce **prunasin,** or mandelonitrile monoglucoside. Prunasin is often found along with amygdalin. The second enzyme, prunase, cleaves the second glucose molecule from amygdalin (prunasin) to give mandelonitrile. The presence of the two enzymes, along with the glycoside, ensures that free mandelonitrile will be produced when the contents of the seed are digested in the stomach of a predator. The free mandelonitrile is then rapidly hydrolyzed to benzaldehyde and hydrogen cyanide in the acidic stomach medium. Thus, amygdalin, prunlaurasin, and sambunigrin probably constitute protective mechanisms for the plants in which they are found. Many of these plants produce an otherwise luscious and attractive fruit.

The controversial drug **Laetrile,** which is made from crushed apricot pits, consists mainly of amygdalin. Because it contains cyanide, it was originally considered too toxic for human use by its "discoverer," Dr. Ernst Krebs, Sr., a California physician. His son, Ernst Krebs, Jr., however, found an apparent way to purify it. Since that time they, along with others, have advocated its use as an effective treatment for cancer. They also developed an explanation for its action. Laetrile, it is maintained, goes directly to cancer cells, where an abundant enzyme releases cyanide, which in turn kills those cells. Normal cells are held to be low in this enzyme and therefore not affected. In addition, normal cells are held to contain a "protective" enzyme that detoxifies the Laetrile, whereas cancer cells lack the enzyme. Thus, normal cells live while cancer cells die, these workers conclude.

Unfortunately, other workers have shown that releasing enzymes are more abundant in normal cells than in cancer cells and that protective enzymes are distributed about equally. Further, the cyanide does not migrate preferentially to the cancerous cells but diffuses rapidly to all parts of the body.

In 1953, the Cancer Commission of the California Medical Association investigated Laetrile and found it ineffective. In one case study involving 44 patients treated with Laetrile, a follow-up showed that all but one of the patients either had died or still had cancer. Other studies gave similar results. Animal studies, some completed as recently as 1976, have corroborated clinical records. On the strength of this evidence, most states have banned Laetrile, and the Food and Drug Administration has banned its use in interstate commerce.

The proponents of Laetrile, however, vigorously maintain that it is effective, and that cancer patients should have free choice in deciding their treatment. In an apparent attempt to circumvent drug laws, Laetrile was promoted in 1970 as "Vitamin B-17," a vitamin needed to **prevent** rather than **cure** cancer. Vitamins are over-the-counter preparations and are exempt from the usual drug regulations. In 1974, the FDA won action in court stopping production of B-17. The claims for it as a vitamin could not be substantiated and were deemed fraudulent advertising.

Most scientific and medical authorities deem Laetrile ineffective and, in fact, dangerous, since it may keep some persons from receiving proper treatment until it is too late. Additionally, numerous deaths have resulted from overdoses, accidental ingestion by children, and misuse.

REFERENCES

Claus, E. P., Tyler, V. E., and Brady, L. R. *Pharmacognosy.* Philadelphia: Lea & Febiger, 1970. Pp. 114–118, "Cyanophore Glycosides."

Consumer's Union. "Laetrile: The Political Success of a Scientific Failure." *Consumer Reports* (August 1977): 444.

Culliton, B. J. "Sloan-Kettering: The Trials of an Apricot Pit." *Science, 182* (December 7, 1973): 1000.

Holden, C. "Laetrile: Quack Cancer Remedy Still Brings Hope to Sufferers." *Science, 193* (September 10, 1976): 982.

"Laetrile Poisoning." *Newsweek* (August 22, 1977): 80.

Sondheimer, E., and Simeone, J. B., eds. *Chemical Ecology.* New York: Academic Press, 1970. Chap. 8, "Chemical Defense against Predation in Arthropods."

Experiment 38
Benzoin Condensation

Condensation Reaction

Benzaldehyde does not possess alpha hydrogens and therefore does not undergo a self-aldol condensation. In fact, in strongly basic solution, benzaldehyde, like other aldehydes that lack alpha hydrogens, undergoes the Cannizzaro reaction, yielding benzyl alcohol and sodium benzoate:

$$2C_6H_5CHO + NaOH \longrightarrow C_6H_5CH_2OH + C_6H_5COO^-Na^+$$

In the presence of cyanide ion, however, benzaldehyde undergoes a unique self-condensation reaction, called the **benzoin condensation,** yielding an α-hydroxy ketone called **benzoin.** In this experiment, we use the benzoin condensation to synthesize benzoin:

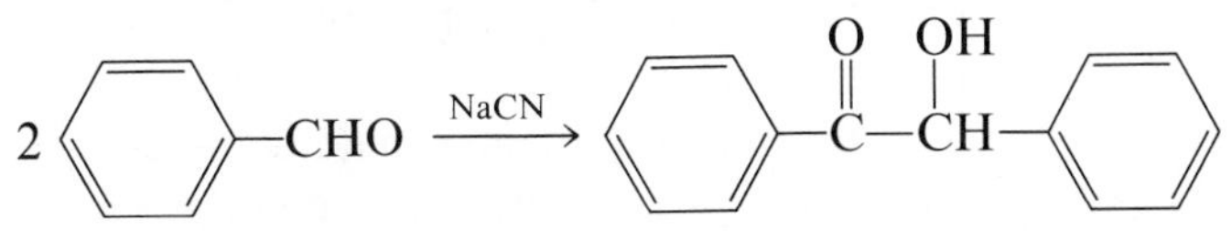

Benzaldehyde   **Benzoin**

The complete mechanism for this reaction is shown in the accompanying chart. The first step is the formation of the cyanohydrin **1,** which in the basic reaction medium immediately forms its conjugate anion **2.** The conjugate anion, **2,** is stabilized by resonance that involves both the cyano group and the aromatic ring. In a second step of the reaction, the cyanohydrin anion makes a nucleophilic addition to a second molecule of benzaldehyde, to give the adduct **3.** After a proton transfer to form the anion, **4,** cyanide is expelled, forming benzoin, an α-hydroxy ketone **5.**

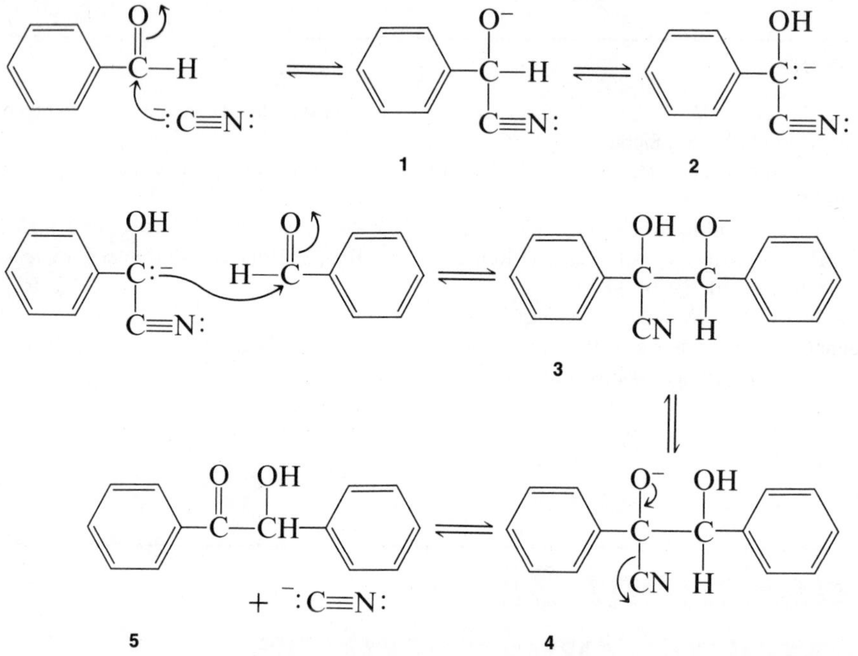

The reaction is carried out in 95% aqueous ethanol, and the product, which is sparingly soluble, crystallizes from the reaction mixture on cooling. The product is collected by vacuum filtration and recrystallized from 95% ethanol.

SPECIAL INSTRUCTIONS

A knowledge of Techniques 1 through 4 is needed for this experiment. In addition, the essay on cyanohydrins, which precedes the experiment, should be read.

Be sure to take all the cautionary measures mentioned in the note and to flush all cyanide residues down the sink with **large** volumes of water. If possible, the reaction should be carried out in a hood. If a hood is not available, it is wise to attach a trap like that shown in Experiment 23, to control any escaping vapors. The beaker should be filled with 15% aqueous sodium hydroxide.

The benzaldehyde used for this experiment should be pure. It is easiest to use a fresh bottle that has no benzoic acid, a solid white precipitate, evident in the bottom of the bottle. If benzoic acid is present, the benzaldehyde should be distilled before use.

> **CAUTION:** Sodium cyanide is extremely hazardous and toxic. When you are weighing this substance, be careful not to spill any of it, and do not allow it to come into contact with your skin. Do not breathe the dust. Any spilled material should be cleaned up immediately and disposed of down a drain in the hood. Any area of the skin that has come into contact with cyanide should be washed thoroughly with water. Do not allow sodium cyanide to come into contact with acid, since hydrogen cyanide, a toxic gas, will be generated.

PROCEDURE

Using a 100-mL round-bottomed flask, assemble an apparatus for heating under reflux (Technique 1, Figure 1–4, p. 510). Place 20 mL of 95% ethanol, 15.0 g of pure benzaldehyde, and a solution of 1.5 g of sodium cyanide in 15 mL of water in the flask. Heat the mixture gently under reflux for 30 minutes and then cool the flask in an ice bath (leave the condenser attached). The product should precipitate. Collect the crude benzoin by vacuum filtration using a Büchner funnel (Technique 2, Section 2.3, p. 518). Wash the product well with several portions of cold water to remove all the sodium cyanide (see the caution above). Set the benzoin aside to dry. A second batch of benzoin can be obtained by concentrating the filtrate. This should be done in a beaker in the hood. It will be necessary to use a hot plate to boil the solution. The crystals obtained should be colorless or pale yellow. Weigh the crude material and record the weight.

Recrystallize the crude benzoin from 95% ethanol (about 8 mL/g of crude crystals) to yield pure benzoin (mp 134–135 °C). Weigh the purified material, calculate the

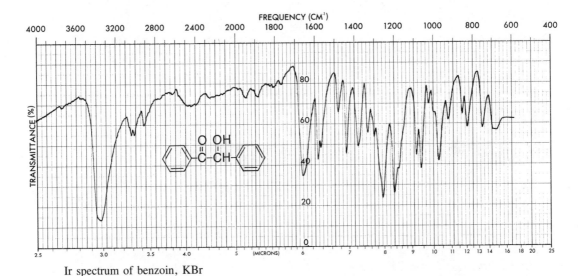

Ir spectrum of benzoin, KBr

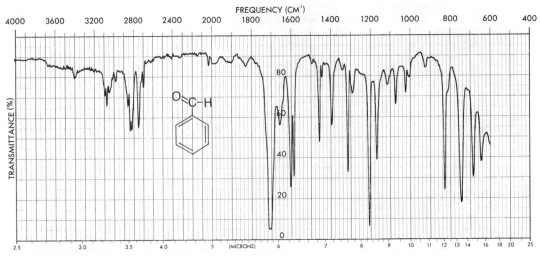

Ir spectrum of benzaldehyde, neat

yield, and determine the melting point. At your instructor's option, determine the infrared spectrum of the benzoin as a KBr mull. If the sample is to be used to prepare benzil (Experiment 40), go on to that experiment. If not, place the sample in a labeled vial and submit it with your report.

QUESTIONS

1. Give the structure of the product that would be formed by the action of cyanide ion on acetaldehyde. Will the reaction be the same as for benzaldehyde?

2. Interpret the principal peaks in the infrared spectra of benzoin and benzaldehyde.

3. Give all the possible resonance structures of the conjugate base of benzaldehyde cyanohydrin **(2).**

Essay
THIAMINE AS A COENZYME

Vitamin B_1, or thiamine, as its pyrophosphate derivative, thiamine pyrophosphate, is a coenzyme universally present in all living systems. It was originally discovered as a required nutritional factor (vitamin) in humans by its link with the disease beriberi.

Beriberi is a disease of the peripheral nervous system caused by a deficiency of Vitamin B_1 in the diet. Symptoms include pain and paralysis of the extremities, emaciation, or swelling of the body. The disease is most common in the Far East.

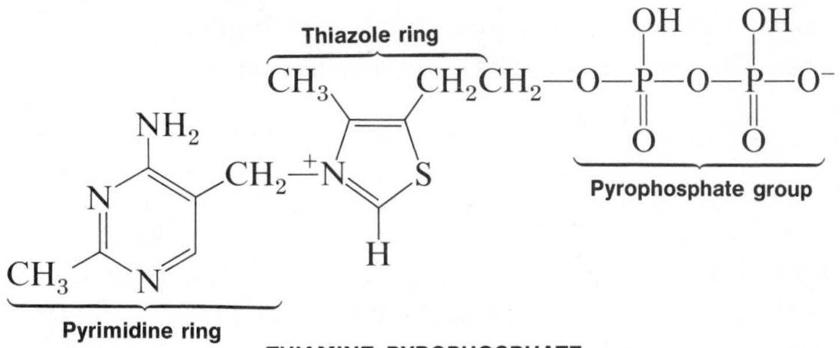

THIAMINE PYROPHOSPHATE

Thiamine serves as a coenzyme (defined later) for three important types of enzymatic reactions:

1. Nonoxidative decarboxylations of α-keto acids

$$R-\overset{\overset{O}{\|}}{C}-COOH \xrightarrow{B_1} R-\overset{\overset{O}{\|}}{C}-H + CO_2$$

2. Oxidative decarboxylations of α-keto acids

$$R-\overset{\overset{O}{\|}}{C}-COOH \xrightarrow{B_1, O_2} R-\overset{\overset{O}{\|}}{C}-OH + CO_2$$

3. Formation of acyloins (α-hydroxy ketones)

$$R-\overset{\overset{O}{\|}}{C}-COOH + R-\overset{\overset{O}{\|}}{C}-H \xrightarrow{B_1} R-\overset{\overset{O}{\|}}{C}-\overset{\overset{OH}{|}}{C}H-R + CO_2$$

or

$$2R-\overset{\overset{O}{\|}}{C}-COOH \xrightarrow{B_1} R-\overset{\overset{O}{\|}}{C}-\overset{\overset{OH}{|}}{C}H-R + 2CO_2$$

or

$$2R-\overset{\overset{O}{\|}}{C}-H \xrightarrow{B_1} R-\overset{\overset{O}{\|}}{C}-\overset{\overset{OH}{|}}{C}H-R$$

Most biochemical processes are no more than organic chemical reactions carried out under special conditions. It is easy to lose sight of this fact. Most of the steps of the ubiquitous metabolic pathways can, if they have been studied well enough, be explained mechanistically. Some simple organic reaction is a model for almost every biological process. Such reactions, however, are modified ingeniously through the intervention of a protein molecule (''enzyme'') to make them more efficient (have greater yield), more selective in choice of substrate (molecule being acted on), and more stereospecific in their result and to enable them to occur under milder conditions (pH) than would normally be possible.

Experiment 39 is designed to illustrate the last circumstance. As a biological reagent, the coenzyme thiamine is used to carry out an organic reaction **without** resorting to an enzyme. The reaction is an acyloin condensation (see above) of benzaldehyde:

$$2C_6H_5-CHO \longrightarrow C_6H_5-\overset{\displaystyle O}{\underset{\displaystyle \|}{C}}-\overset{\displaystyle OH}{\underset{\displaystyle |}{CH}}-C_6H_5$$

In Experiment 38, a similar condensation is described, in which sodium cyanide is used as the catalyst. Nature would clearly prefer to use a reactant that is milder and less toxic than cyanide ion to carry out the acyloin condensations necessary to everyday metabolism. Thiamine constitutes just such a reagent.

In the chemical view, the most important part of the entire thiamine molecule is the central ring—the thiazole ring—which contains nitrogen and sulfur. This ring constitutes the **reagent** portion of the coenzyme. The other portions of the molecule, although important in a biological sense, are not necessary to the chemistry that thiamine initiates. Undoubtedly the pyrimidine ring and the pyrophosphate group have important ancillary functions, such as enabling the coenzyme to make the correct attachment to its associated protein molecule (enzyme) or enabling it to achieve the correct degree of polarity and the correct solubility properties necessary to allow free passage of the coenzyme across the cell membrane boundary (that is, to allow it to get to its site of action). These properties of thiamine are no less important to its biological functioning than to its chemical reagent abilities; only the latter is our concern here, however.

Experiments with the model compound 3,4-dimethylthiazolium bromide have explained how thiamine-catalyzed reactions work. It was found that this model thiazolium compound rapidly exchanged the C-2 proton for deuterium in D_2O solution. At a pD of 7 (no pH here!), this proton was completely exchanged in seconds!

3,4-Dimethylthiazolium bromide   **ylide**

This indicates that the C-2 proton is more acidic than one would have expected. It is apparently easily removed because the conjugate base is a highly stabilized **ylide.** An ylide is a compound or intermediate with positive and negative formal charges on adjacent atoms.

The sulfur atom plays an important role in stabilizing this ylide. This was shown by comparing the rate of exchange of 1,3-dimethylimidazolium ion with the rate for the thiazolium ion shown. The dinitrogen compound exchanged its C-2 proton more slowly than the sulfur-containing ion. Sulfur, being in the third row of the periodic chart, has *d* orbitals available for bonding to adjacent atoms. Thus, it has fewer geometrical restrictions than carbon and nitrogen atoms do and can form carbon-sulfur multiple bonds in situations in which carbon and nitrogen normally would not.

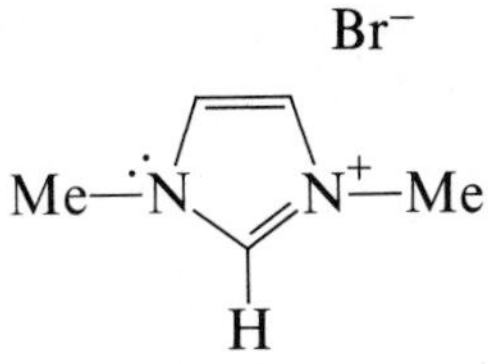

1,3-Dimethylimidazolium bromide

DECARBOXYLATION OF α-KETO ACIDS

From the knowledge described above, it is now thought that the active form of thiamine is its ylide. The system is interestingly constructed, as is seen in the decarboxylation of pyruvic acid by thiamine (see figure). Notice especially how the positively charged nitrogen provides a site to accommodate the electron pair that is released on decarboxylation. Thiamine is regenerated by use of this same pair of electrons that become protonated in vinylogous fashion on carbon. The other product is the protonated form of acetaldehyde, the decarboxylation product of pyruvic acid.

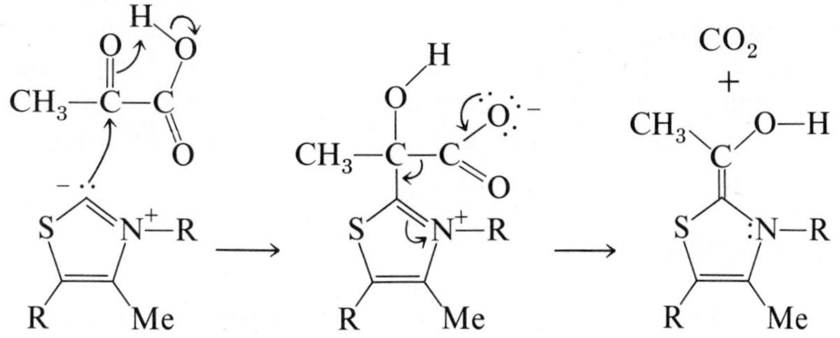

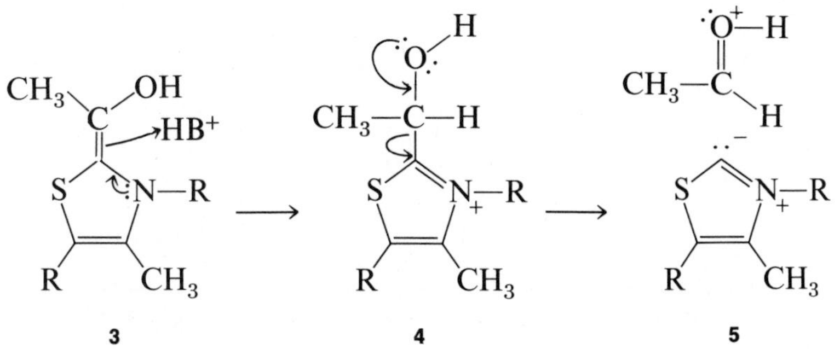

OXIDATIVE DECARBOXYLATION OF α-KETO ACIDS

In oxidative decarboxylations, two additional coenzymes—lipoic acid and coenzyme A—are involved. An example of this type of process, which characterizes all living organisms, is found in the metabolic process **glycolysis.** It is found in the steps that convert pyruvic acid to acetyl coenzyme A, which then enters the citric acid cycle (Krebs cycle, tricarboxylic acid cycle) to provide an energy source for the organism. In this process, the enamine intermediate, **3** (see diagram), is first oxidized by lipoic acid and then transesterified by coenzyme A.

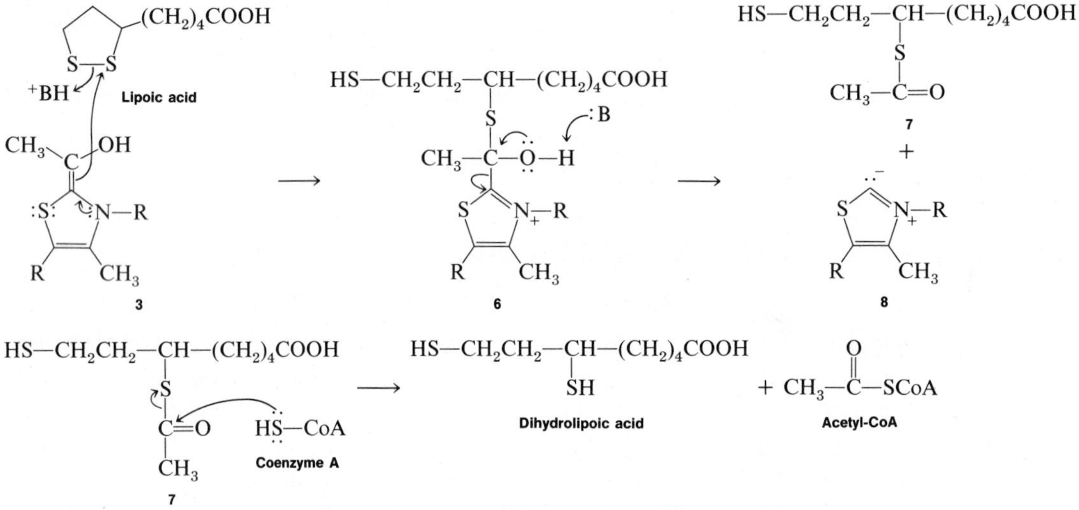

Following this sequence of events, the dihydrolipoic acid is oxidized (through a chain of events involving molecular oxygen) back to lipoic acid, and the acetyl coenzyme A is condensed with oxaloacetic acid to form citric acid. The formation of citric acid begins the citric acid cycle. Notice that acetyl coenzyme A is a thioester of acetic acid and could be hydrolyzed to give acetic acid, not an aldehyde. Thus, an oxidation has taken place in this sequence of events.

ACYLOIN CONDENSATIONS

The enamine intermediate **3** can also function much like the enolate partner in an acid-catalyzed aldol condensation. It can condense with a suitable carbonyl-containing

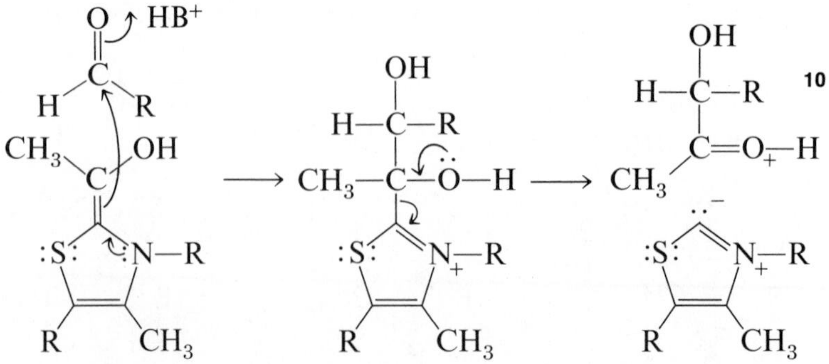

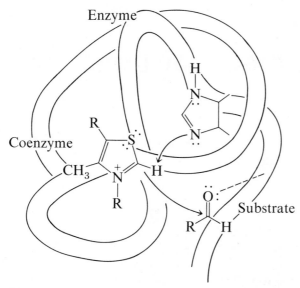

Thiamine (the coenzyme) and the substrate aldehyde are bound to the protein molecule, here called an enzyme. A possible catalytic group (imidazole) is also shown.

acceptor to form a new carbon-carbon bond. Decomposition of the adduct **9** to regenerate the thiamine ylide yields the protonated acyloin **10.**

FUNCTION OF A COENZYME

In biological terminology, thiamine is a **coenzyme.** It must bind to an enzyme before the enzyme is activated. The enzyme also binds the substrate. The coenzyme reacts with the substrate while they are both bound to the enzyme (a large protein). Without the coenzyme thiamine, no chemical reaction would occur. The coenzyme is the **chemical reagent.** The protein molecule (the enzyme) helps and mediates the reaction by controlling stereochemical, energetic, and entropic factors, but, in this case, it is nonessential to the overall result (see Experiment 39). A special name is given to coenzymes that are essential to the nutrition of an organism. They are called **vitamins.** Many biological reactions are of this type, in which a chemical reagent (coenzyme) and a substrate are bound to an enzyme for reaction and, after the reaction, are again released into the medium.

REFERENCES

Bernhard, S. *The Structure and Function of Enzymes*. New York: W. A. Benjamin, 1968. Chap. 7, ''Coenzymes and Cofactors.''

Bruice, T. C., and Benkovic, S. *Bioorganic Mechanisms*. Vol. 2. New York: W. A. Benjamin, 1966. Chap. 8, ''Thiamine Pyrophosphate and Pyridoxal-5′-Phosphate.''

Lowe, J. N., and Ingraham, L. L. *An Introduction to Biochemical Reaction Mechanisms*. Englewood Cliffs, N.J.: Prentice-Hall, 1974. Chap. 5, ''Coenzyme Function and Design.''

Experiment 39
Coenzyme Synthesis of Benzoin

Coenzyme Chemistry
Benzoin Condensation

In this experiment, a benzoin condensation of benzaldehyde is carried out with a biological coenzyme, thiamine hydrochloride, as the catalyst:

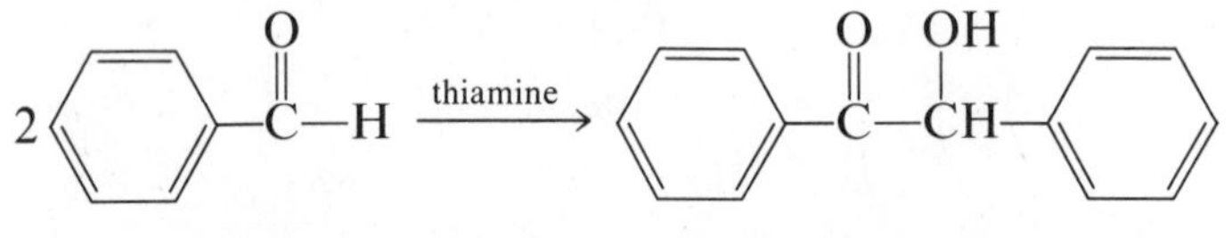

The same reaction can be accomplished with cyanide ion, an inorganic reagent, as the catalyst. A mechanism for the cyanide-catalyzed benzoin condensation is given in Experiment 38. The mechanistic information needed for understanding how thiamine accomplishes this same reaction is given in the essay that precedes this experiment.

SPECIAL INSTRUCTIONS

You have to be familiar with the material in Techniques 1 through 4 to perform this experiment. A careful reading of the essay "Thiamine as a Coenzyme" is also essential, because it explains the mechanism of action of thiamine. It will also be helpful to compare the mechanisms of the cyanide-catalyzed benzoin condensation (Experiment 38) and the reaction in this experiment.

The benzaldehyde used for this experiment **must** be free of benzoic acid. Benzaldehyde is oxidized easily in air, and crystals of benzoic acid are often visible in the bottom of the reagent bottle. If solid appears in the bottle of reagent, the benzaldehyde **must** be redistilled. If the lab assistant or instructor has not redistilled the benzaldehyde provided for this experiment, you will have to distill it. If a **newly opened** bottle of benzaldehyde is available, it will usually be unnecessary to redistill it. If in doubt, consult the instructor.

Thiamine hydrochloride is a heat-sensitive reagent. It should be stored in a refrigerator when not in use. Since it may decompose on being heated, you should take care not to heat the reaction mixture too vigorously. It is best to use a fresh bottle of this reagent.

PROCEDURE

Dissolve 3.5 g of thiamine hydrochloride in about 10 mL of water in a 100-mL round-bottomed flask equipped with a condenser for reflux (Technique 1, Figure 1–4, p. 510).

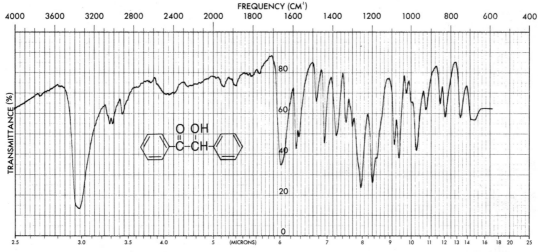

Ir spectrum of benzoin, KBr

Add 30 mL of 95% ethanol and cool the solution by swirling the flask in an ice-water bath. Meanwhile, place about 10 mL of 2*M* sodium hydroxide solution in a small Erlenmeyer flask. Cool this solution in the ice bath also. Then, over about 10 minutes, add the cold sodium hydroxide solution, through the condenser, to the thiamine solution. Measure 20 mL of benzaldehyde and add it, also through the condenser, to the reaction mixture. Add a boiling stone and heat the mixture gently on a steam bath for about 90 minutes. Do not heat the mixture under vigorous reflux. Allow the mixture to cool to room temperature, and then induce crystallization of the benzoin (it may already have begun) by cooling the mixture in an ice-water bath. If the product separates as an oil, reheat the mixture until it is once again homogeneous, and then allow it to cool more slowly than before. You may have to scratch the flask with a glass rod.

Collect the product by vacuum filtration, using a Büchner funnel. Wash the product with two 50-mL portions of **cold** water. Weigh the crude product and then recrystallize it from 95% ethanol. The solubility of benzoin in boiling 95% ethanol is about 12 to 14 g per 100 mL. Weigh the product, calculate the percentage yield, and determine its melting point (mp 134–136 °C).

At your instructor's option, determine the infrared spectrum of the benzoin as a KBr mull. A spectrum of benzoin is shown here for comparison.

The benzoin may be converted to benzil (Experiment 40). However, if you are not scheduled to do this experiment, submit the sample of benzoin, along with your report, to the instructor.

QUESTIONS

1. The infrared spectrum of benzoin is given in this experiment. Interpret the principal peaks in the spectrum.

2. Why is sodium hydroxide added to the solution of thiamine hydrochloride?

3. Using the information given in the essay that precedes this experiment, formulate a complete mechanism for the thiamine-catalyzed conversion of benzaldehyde to benzoin.

4. How do you think the appropriate enzyme would have affected the reaction (degree of completion, yield, stereochemistry)?

5. What modifications of conditions would be appropriate if the enzyme were to be used?

6. Refer to the essay that precedes this experiment. It gives a structure for thiamine pyrophosphate. Using this structure as a guide, draw a structure for thiamine hydrochloride. The pyrophosphate group is absent in this compound.

Experiment 40
Benzil

Oxidation
Crystallization

In this experiment an α-diketone, benzil, is prepared by the oxidation of an α-hydroxyketone, benzoin (Experiment 38 or 39).

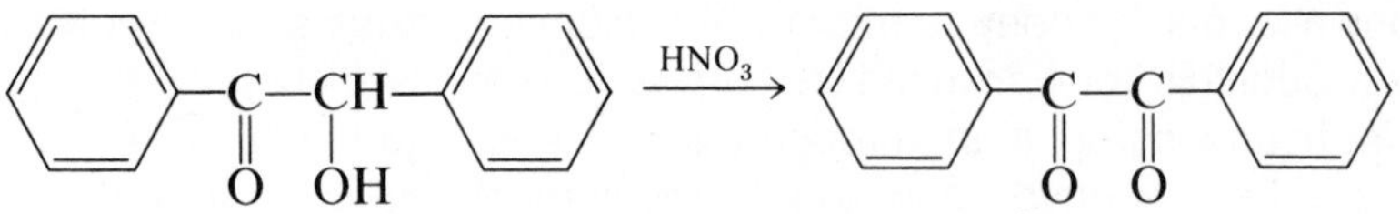

Benzoin   **Benzil**

This oxidation can easily be done with mild oxidizing agents such as Fehling's solution (alkaline cupric tartrate complex) or copper sulfate in pyridine. In this experiment, the oxidation is performed with nitric acid.

SPECIAL INSTRUCTIONS

To do this experiment, you should read Techniques 1 through 4. The benzoin prepared in Experiment 38 or 39 may be used in this experiment.

PROCEDURE

Place 10.0 g of benzoin (Experiment 38 or 39) in a 250-mL Erlenmeyer flask with 50 mL of concentrated nitric acid. Heat the mixture on a steam bath in the hood. Shake the mixture occasionally until nitrogen oxide gases (red) are no longer evolved (about 1 hour). Alternatively, the mixture may be heated at the desk if a trap arrangement such as that shown in Experiment 23 is used. Sodium hydroxide is added to the trap to react with the nitrogen oxide gases. Pour the reaction mixture into 150 mL of cool tap water and

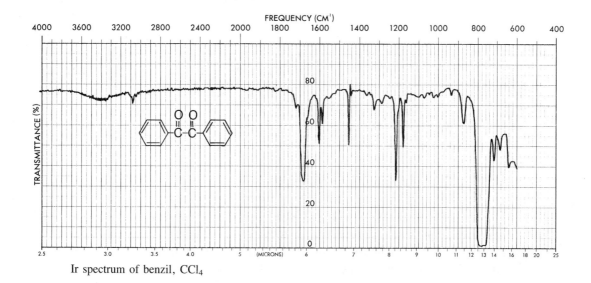

FREQUENCY (CM⁻¹)

Ir spectrum of benzil, CCl₄

stir it vigorously until the oil crystallizes completely as a yellow solid. Vacuum-filter the crude benzil and wash it well with cold water to remove the nitric acid.

Recrystallize the product from 95% ethanol (4 mL/g). Scratch the solution with a stirring rod as it cools. The solution will become supersaturated unless this is done. Yellow, needle-like crystals are formed. Cool the mixture in an ice bath to complete the crystallization. Vacuum-filter the crystals, using a Büchner funnel. Press the crystals with a clean stopper or cork to remove the excess solvent. Dry the product either in air overnight or in an oven at 75 °C for about 10 minutes. Weigh the benzil and calculate the percentage yield; then determine the melting point. The melting point of pure benzil is 95 °C. The value obtained is often lower than this, ranging from a low value of 84 °C to a high value of 92 °C. Material in this range of melting points is pure enough for conversion to benzilic acid (Experiment 41) or tetraphenylcyclopentadienone (Experiment 46). Submit the benzil to the instructor unless it is to be used to prepare benzilic acid or tetraphenylcyclopentadienone. At the instructor's option, obtain the infrared spectrum of benzil in carbon tetrachloride or chloroform. Compare it with the infrared spectrum of benzoin shown in Experiment 38.

Experiment 41
Benzilic Acid

Anionic Rearrangement

In this experiment, benzilic acid is prepared by causing the rearrangement of the α-diketone benzil. Preparation of benzil is described in Experiment 40. The reaction proceeds in the following way:

$$\text{Ph—}\overset{\overset{\displaystyle O}{\|}}{C}\text{—}\overset{\overset{\displaystyle O}{\|}}{C}\text{—Ph} \xrightleftharpoons{\text{(1) KOH}} \text{Ph—}\overset{\overset{\displaystyle O}{\|}}{C}\text{—}\underset{\underset{\displaystyle Ph}{|}}{C}\text{—OH} \longrightarrow \text{Ph—}\underset{\underset{\displaystyle Ph}{|}}{C}\text{—}\overset{\overset{\displaystyle O}{\|}}{C}\text{—OH} \longrightarrow$$

Benzil
(An α-diketone)

$$\text{Ph—}\underset{\underset{\displaystyle Ph}{|}}{\overset{\overset{\displaystyle OH}{|}}{C}}\text{——}\overset{\overset{\displaystyle O}{\|}}{C}\text{—O}^- \text{ K}^+ \xrightarrow{\text{(2) } H_3O^+} \text{Ph—}\underset{\underset{\displaystyle Ph}{|}}{\overset{\overset{\displaystyle OH}{|}}{C}}\text{——}\overset{\overset{\displaystyle O}{\|}}{C}\text{—OH}$$

Potassium   **Benzilic acid**
benzilate   **(An α-hydroxyacid)**

The driving force for the reaction is provided by the formation of a stable carboxylate salt (potassium benzilate). Once this salt is produced, acidification yields benzilic acid. The reaction can generally be used to convert aromatic α-diketones to aromatic α-hydroxyacids. Other compounds, however, also will undergo a benzilic acid type of rearrangement (see questions).

SPECIAL INSTRUCTIONS

This reaction involves a 15-minute reflux period, followed by an overnight crystallization period. The reaction may be scheduled with another experiment.

PROCEDURE

Dissolve 5.5 g of potassium hydroxide in 12 mL of water in an Erlenmeyer flask. The mixture may be heated to dissolve the base, then cooled again to room temperature. In a 100-mL round-bottomed flask, dissolve 5.5 g of benzil (Experiment 40) in 17 mL of 95% ethanol, heating slightly, if necessary, to dissolve the solid. Add the potassium hydroxide solution to the round-bottomed flask and swirl the contents of the flask. Attach a reflux condenser to the round-bottomed flask. Heat the mixture under reflux on a steam bath for 15 minutes. During this period, the initial blue-black coloration will be replaced by brown. Transfer the contents of the flask to a beaker or an evaporating dish and cover it with a watch glass. Allow the mixture to stand until the next laboratory period. The potassium salt of benzilic acid will crystallize during this period. Collect the crystals by vacuum filtration and wash the crystals with 2 mL of ice-cold 95% ethanol.

Dissolve the potassium benzilate in a minimum amount of hot water in a 250-mL Erlenmeyer flask. More than 100 mL of hot water will be needed to dissolve this solid. Add a small amount of decolorizing carbon and shake or stir the mixture for a few minutes. Filter the **hot** solution by gravity, using a fluted filter. Acidify the filtrate with concentrated hydrochloric acid to a pH of about 2. Allow the mixture to cool slowly to room temperature and then complete the cooling in an ice bath. Collect the benzilic acid by vacuum filtration, using a Büchner funnel. Wash the crystals thoroughly with water to remove salts, and remove the wash water by drawing air through the filter. Dry the product thoroughly by allowing it to stand until the next laboratory period.

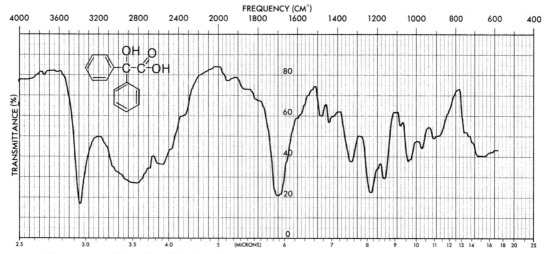

Ir spectrum of benzilic acid, KBr

Determine the melting point of the product. Pure benzilic acid melts at 150 °C. If necessary, recrystallize the product from water. Heat the mixture to boiling on a hot plate (some impurities will remain undissolved) and gravity-filter it through a fast fluted filter paper. Cool the filtrate and scratch the solution to induce crystallization. Allow the mixture to stand at room temperature until crystallization is complete (about 15 minutes). Cool the mixture in an ice bath and vacuum-filter the crystals. Wash them with a small amount of **cold** water and allow them to dry overnight. Determine the melting point of the crystallized product.

At the instructor's option, determine the infrared spectrum of the benzilic acid in potassium bromide. Calculate the percentage yield. Submit the sample or samples to your laboratory instructor in labeled vials.

QUESTIONS

1. Show how to prepare the following compounds, starting from the appropriate aldehyde (see Experiments 38 and 39).

(a) CH_3O—⌬—$\overset{\overset{\displaystyle OH}{|}}{\underset{\underset{\displaystyle OCH_3}{|}}{C}}$—$CO_2H$ (b) ⟨furan⟩—$\overset{\overset{\displaystyle OH}{|}}{C}$—$CO_2H$ ⟨furan⟩

2. Give the mechanisms for the following transformations:

(a) ⟨phenanthrenequinone structure⟩ $\xrightarrow[\text{(2) } H^+]{\substack{\text{(1) KOH,}\\ \text{alcohol}}}$ ⟨fluorenol structure with $OH\ CO_2H$⟩

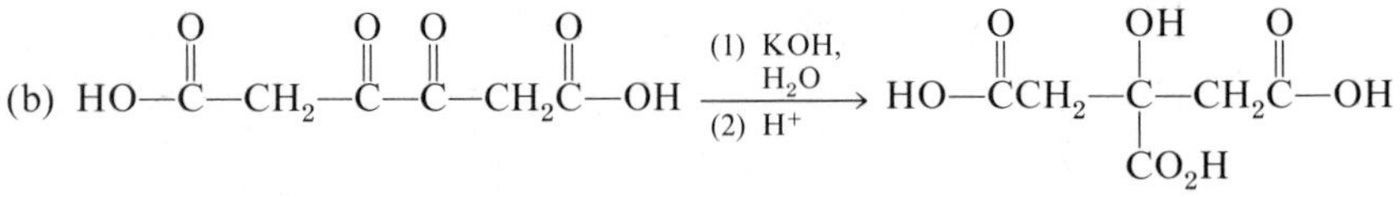

(b)
$$\text{HO—C(=O)—CH}_2\text{—C(=O)—C(=O)—CH}_2\text{C(=O)—OH} \xrightarrow[\text{(2) H}^+]{\text{(1) KOH, H}_2\text{O}} \text{HO—C(=O)CH}_2\text{—C(OH)(CO}_2\text{H)—CH}_2\text{C(=O)—OH}$$

Citric acid

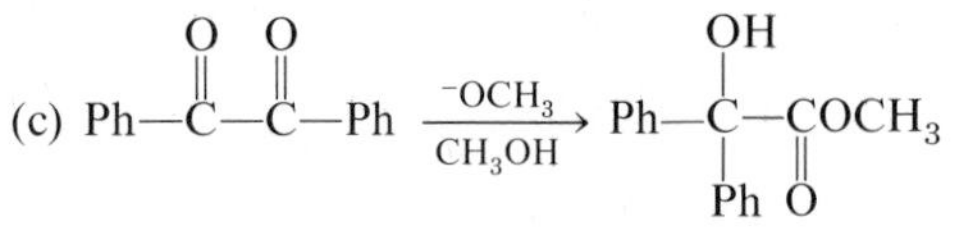

(c)
$$\text{Ph—C(=O)—C(=O)—Ph} \xrightarrow[\text{CH}_3\text{OH}]{^-\text{OCH}_3} \text{Ph—C(OH)(Ph)—C(=O)OCH}_3$$

3. Interpret the infrared spectrum of benzilic acid.

Essay
SULFA DRUGS

The history of chemotherapy extends back as far as 1909, when Paul Ehrlich first used the term. Although Ehrlich's original definition of chemotherapy was limited, he is recognized as one of the giants of medicinal chemistry. **Chemotherapy** might be defined as "the treatment of disease by chemical reagents." It is preferable that these chemical reagents exhibit a toxicity toward only the pathogenic organism, and not toward the organism and the host also. A chemotherapeutic agent would not be useful if it poisoned the patient at the same time that it cured the patient's disease!

In 1932, the German dye manufacturing firm I. G. Farbenindustrie patented a new drug, Prontosil. Prontosil is a red azo dye, and it was first prepared for its dye properties. Remarkably, it was discovered that Prontosil showed antibacterial action when it was used to dye wool. This discovery led to studies of Prontosil as a drug capable of inhibiting the growth of bacteria. The following year, Prontosil was successfully used against staphylococcal septicemia, a blood infection. In 1935, Gerhard Domagk published the results of his research, which indicated that Prontosil was capable of curing streptococcal infections of mice and rabbits. Prontosil was shown to be active against a wide variety of bacteria in later work. This important discovery, which paved the way for a tremendous amount of research on the chemotherapy of bacterial infections, earned for Domagk the 1939 Nobel Prize in Medicine, but an order from Hitler prevented Domagk from accepting this honor.

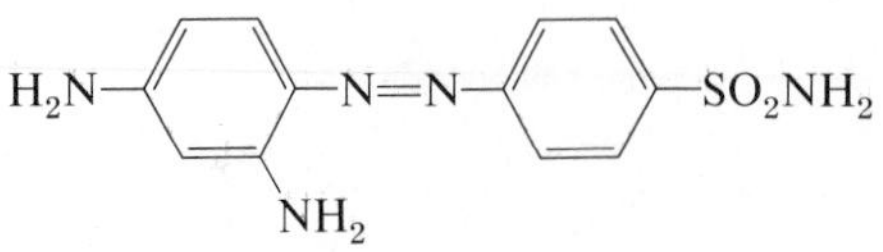

$$\text{H}_2\text{N—C}_6\text{H}_3(\text{NH}_2)\text{—N=N—C}_6\text{H}_4\text{—SO}_2\text{NH}_2 \qquad\qquad \text{H}_2\text{N—C}_6\text{H}_4\text{—SO}_2\text{NH}_2$$

Prontosil 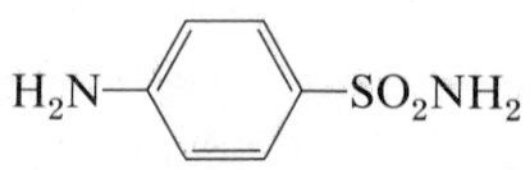   **Sulfanilamide**

Prontosil is an effective antibacterial substance **in vivo,** that is, when injected into a living animal. Prontosil is not medicinally active when the drug is tested **in vitro,** that is, on a bacterial culture grown in the laboratory. In 1935, the research group at the Pasteur Institute in Paris headed by J. Tréfouël learned that Prontosil is metabolized in animals to **sulfanilamide.** Sulfanilamide had been known since 1908. Experiments with sulfanilamide showed that it had the same action as Prontosil in vivo and that it was also active in vitro, where Prontosil was known to be inactive. It was concluded that the active portion of the Prontosil molecule was the sulfanilamide moiety. This discovery led to an explosion of interest in sulfonamide derivatives. Well over a thousand sulfonamide substances were prepared within a few years of these discoveries.

Although many sulfonamide compounds were prepared, only a relative few showed useful antibacterial properties. These few medicinally active sulfonamides, or **sulfa drugs,** as the first useful antibacterial drugs became the wonder drugs of their day. An antibacterial drug may be either **bacteriostatic** or **bactericidal.** A bacteriostatic drug suppresses the growth of bacteria; a bactericidal drug kills bacteria. Strictly speaking, the sulfa drugs are bacteriostatic. The structures of some of the most common sulfa drugs are presented in the accompanying table. These more complex sulfa drugs have various important applications. Although they do not have the simple structure characteristic of sulfanilamide, they tend to be less toxic than the simpler compound.

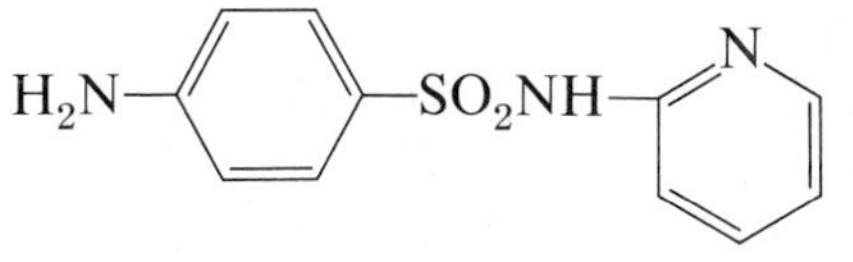

Sulfapyridine

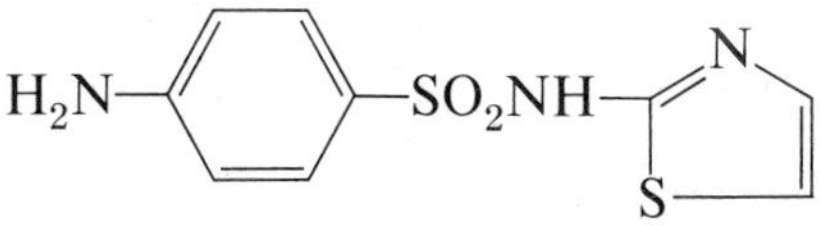

Sulfathiazole

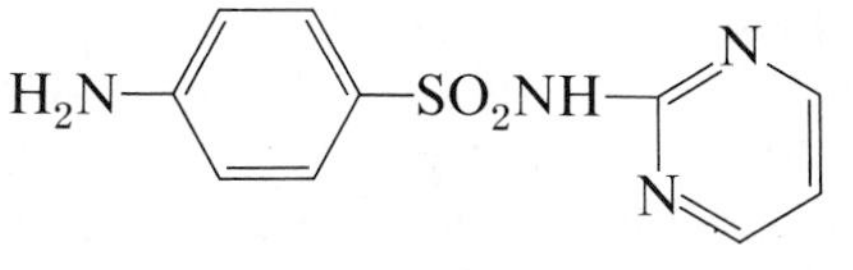

Sulfadiazine

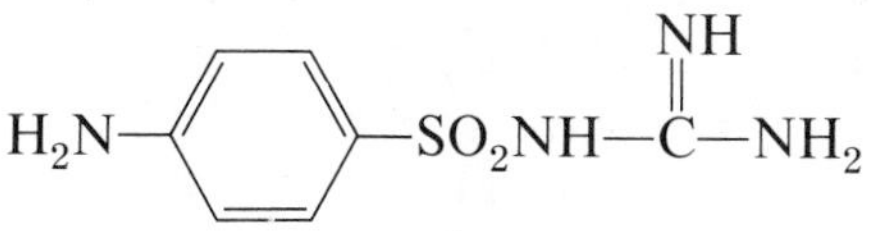

Sulfaguanidine

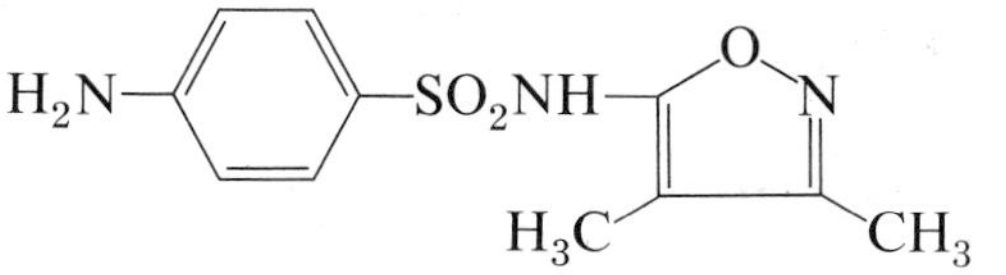

Sulfisoxazole

Sulfa drugs began to lose their importance as generalized antibacterial agents when production of antibiotics in large quantity began. In 1929, Sir Alexander Fleming made his famous discovery of **penicillin.** In 1941, penicillin was first used successfully on humans. Since that time, the study of antibiotics has spread to molecules that bear little or no structural similarity to the sulfonamides. Besides penicillin derivatives, antibiotics that are derivatives of **tetracycline,** including Aureomycin and Terramycin, were also discovered. These newer antibiotics have high activity against bacteria, and they do not usually have the severe unpleasant side effects of many of the sulfa drugs.

Nevertheless, the sulfa drugs are still widely used in treating malaria, tuberculosis, leprosy, meningitis, pneumonia, scarlet fever, plague, respiratory infections, and infections of the intestinal and urinary tracts.

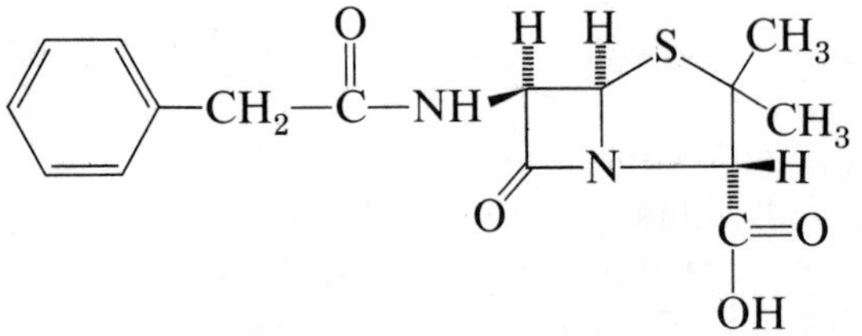

Penicillin G

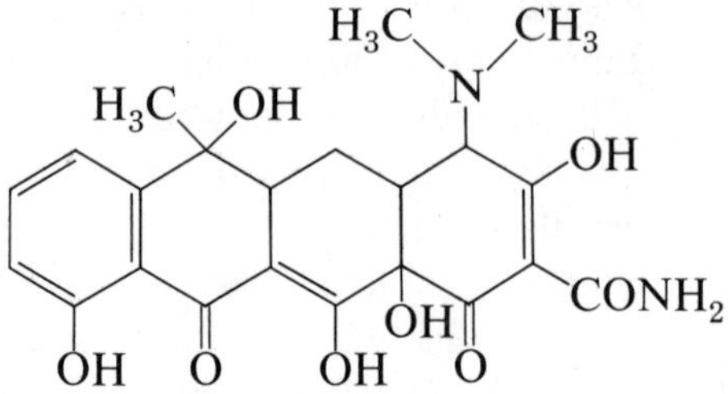

Tetracycline

Even though the importance of sulfa drugs has declined, studies of how these materials act provide a very interesting insight into how chemotherapeutic substances might behave. In 1940, Woods and Fildes discovered that *p*-aminobenzoic acid (PABA) inhibits the action of sulfanilamide. They concluded that sulfanilamide and PABA, because of their structural similarity, must compete with each other within the organism even though they cannot carry out the same chemical function. Further studies indicated that sulfanilamide does not kill bacteria but inhibits their growth. In order to grow, bacteria require an enzyme-catalyzed reaction that uses **folic acid** as a cofactor. Bacteria synthesize folic acid, using PABA as one of the components. When sulfanilamide is introduced into the bacterial cell, it competes with PABA for the active site of the enzyme that carries out the incorporation of PABA in the molecule of folic acid. Because sulfanilamide and PABA compete for an active site due to their structural similarity, and because sulfanilamide cannot carry out the chemical transformations characteristic of PABA once it has formed a complex with the enzyme, sulfanilamide is called a **competitive inhibitor** of the enzyme. The enzyme, once it has formed a complex with sulfanilamide, is incapable of catalyzing the reaction required for the synthesis of folic acid. Without folic acid, the bacteria cannot synthesize the nucleic acids required for growth. As a result, bacterial growth is arrested until the body's immune system can respond and kill the bacteria.

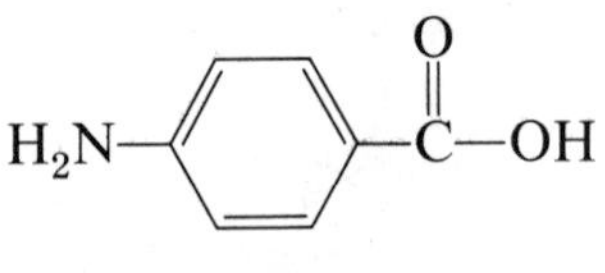

***p*-Aminobenzoic acid**
(PABA)

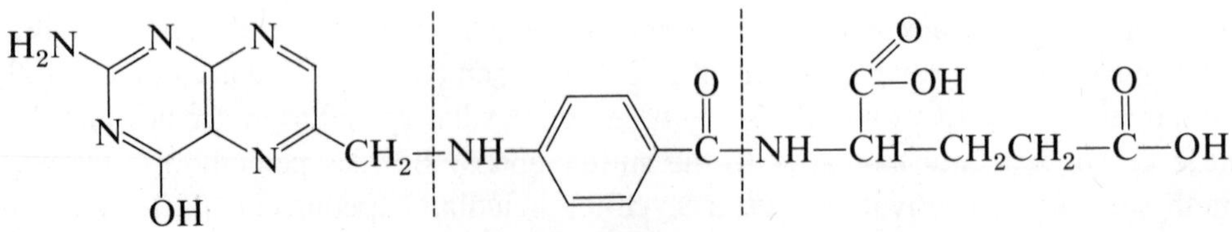

PABA residue

Folic acid

One might well ask the question, "Why, when someone takes sulfanilamide as a drug, doesn't it inhibit the growth of **all** cells, bacterial and human alike?" The answer is simple. Animal cells cannot synthesize folic acid. Folic acid must be a part of the diet of animals and is therefore an essential vitamin. Since animal cells receive their fully synthesized folic acid molecules through the diet, only the bacterial cells are affected by the sulfanilamide, and only their growth is inhibited.

For most drugs, a detailed picture of their mechanism of action is unavailable. The sulfa drugs, however, provide a rare example from which we can theorize how other therapeutic agents carry out their medicinal activity.

REFERENCES

Amundsen, L. H. "Sulfanilamide and Related Chemotherapeutic Agents." *Journal of Chemical Education, 19* (1942): 167.

Evans, R. M. *The Chemistry of Antibiotics Used in Medicine*. London: Pergamon Press, 1965.

Fieser, L. F., and Fieser, M. *Topics in Organic Chemistry*. New York: Reinhold, 1963. Chap. 7, "Chemotherapy."

Garrod, L. P., and O'Grady, F. *Antibiotic and Chemotherapy*. Edinburgh: E. and S. Livingstone, Ltd., 1968.

Goodman, L. S., and Gilman, A. *The Pharmacological Basis of Therapeutics*. 7th ed. New York: Macmillan, 1985. Chap. 56, "The Sulfonamides," by L. Weinstein.

Sementsov, A. "The Medical Heritage from Dyes." *Chemistry, 39* (November 1966): 20.

Zahner, H., and Maas, W. K. *Biology of Anitibiotics*. Berlin: Springer-Verlag, 1972.

Experiment 42
Sulfa Drugs: Sulfanilamide, Sulfapyridine, and Sulfathiazole

In this experiment you will prepare one of the sulfa drugs, sulfanilamide (Procedure 42A), sulfapyridine (Procedure 42B), or sulfathiazole (Procedure 42C), by one of the following synthetic schemes:

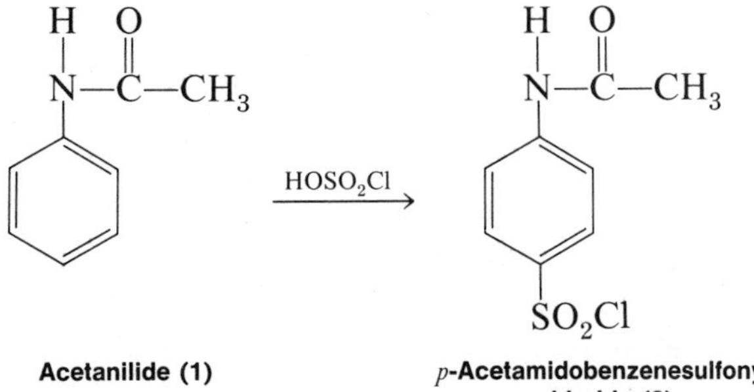

Acetanilide (1)   *p*-Acetamidobenzenesulfonyl chloride (2)

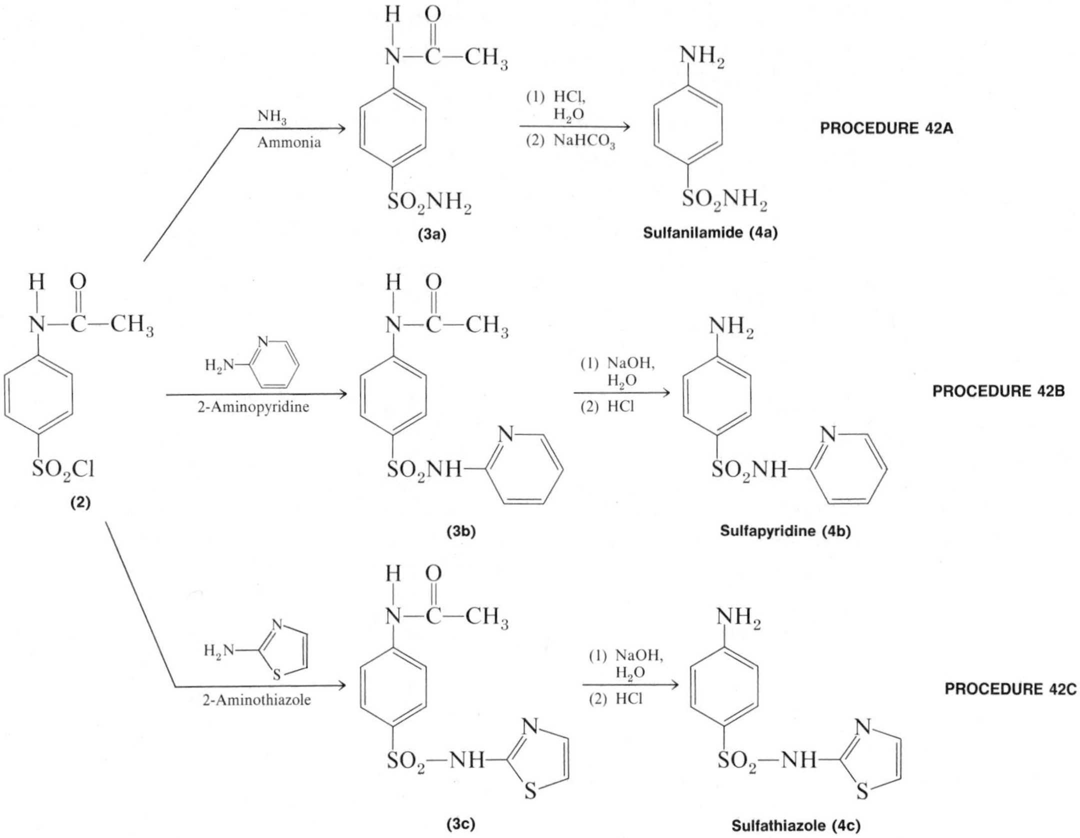
NH3
Ammonia
(1) HCl, H2O
(2) NaHCO3
PROCEDURE 42A
(3a)
Sulfanilamide (4a)
2-Aminopyridine
(1) NaOH, H2O
(2) HCl
PROCEDURE 42B
(2)
(3b)
Sulfapyridine (4b)
2-Aminothiazole
(1) NaOH, H2O
(2) HCl
PROCEDURE 42C
(3c)
Sulfathiazole (4c)

Each of the syntheses involves converting acetanilide (**1**) to *p*-acetamidobenzenesulfonyl chloride (**2**). This intermediate, which is prepared by everyone in the class, is then converted to one of three possible acetyl derivatives of sulfa drugs (**3a, 3b, 3c**). From these, you will then obtain the sulfa drugs (**4a, 4b,** or **4c**) on hydrolysis.

p-ACETAMIDOBENZENESULFONYL CHLORIDE (2)

Acetanilide (**1**), which can easily be prepared from aniline (see Experiment 2), is allowed to react with chlorosulfonic acid to yield *p*-acetamidobenzenesulfonyl chloride (**2**). The acetamido group directs substitution almost totally to the para position. The reaction is an example of a typical electrophilic aromatic substitution reaction. Two problems would result if aniline itself were used in the reaction. First, the amino group in aniline would be protonated in strong acid to become a meta director, and secondly, the chlorosulfonic acid would react with the amino group, rather than with the ring, to give C_6H_5—$NHSO_3H$. For these reasons, the amino group has been "protected" by acetylation. The acetyl group will be removed in the final step, after it is no longer needed, to regenerate the free amino group present in sulfa drugs. The product is isolated by pouring the reaction mixture into ice water, which decomposes the excess chlorosulfonic acid. The product, *p*-acetamidobenzenesulfonyl chloride (**2**), is fairly

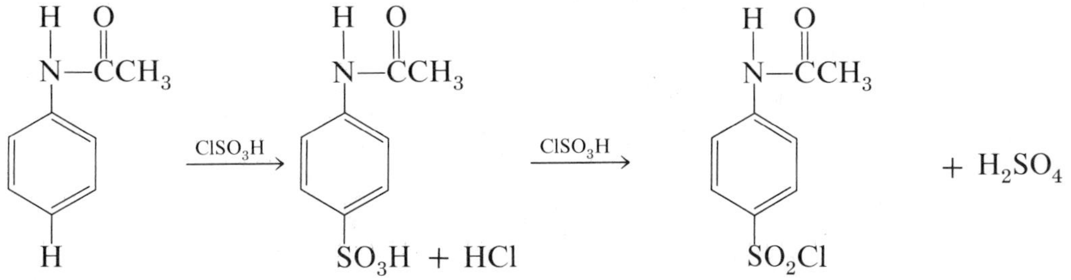

Acetanilide (1) → $ClSO_3H$ → **SO$_3$H + HCl** → $ClSO_3H$ → **SO$_2$Cl** **+ H$_2$SO$_4$**

p-Acetamidobenzenesulfonyl chloride (2)

stable in water; nevertheless, it is converted slowly to the corresponding sulfonic acid (Ar—SO$_3$H). Thus, water must be removed as soon as possible after this intermediate has been prepared. This is done by dissolving the sulfonyl chloride (2) in methylene chloride, which separates the immiscible water. Anhydrous material is obtained from the methylene chloride on crystallization.

SULFANILAMIDE (4a)

The intermediate sulfonyl chloride (2) is converted to the amide (3a) in a reaction with aqueous ammonia. Excess ammonia neutralizes the hydrogen chloride produced. The only side reaction is the hydrolysis of the sulfonyl chloride, in the presence of water, to the sulfonic acid. The next step is the acid-catalyzed hydrolysis of the protecting acetyl group to generate the protonated amino group. Note that of the two amide linkages present, only the carboxylic acid amide (acetamido group) is cleaved, not the sulfonic acid amide (sulfonamide). The salt of the sulfa drug that is formed is converted to sulfanilamide (4a) when the base, sodium bicarbonate, is added.

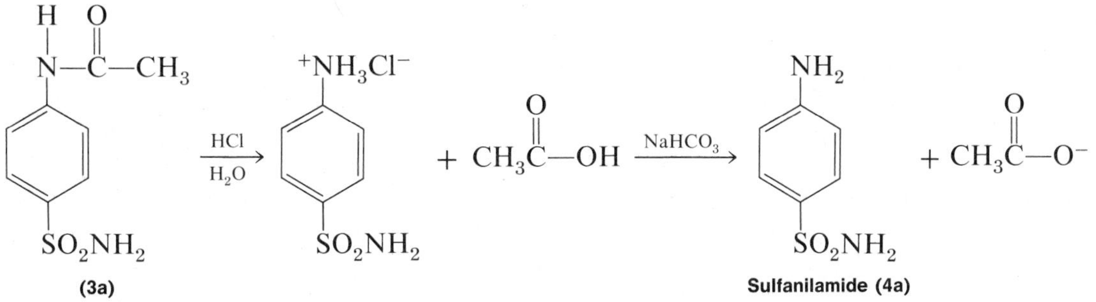

(3a) → $\frac{HCl}{H_2O}$ → $^+NH_3Cl^-$ ... **SO$_2$NH$_2$** $+ CH_3\overset{O}{\overset{\|}{C}}—OH$ → $\xrightarrow{NaHCO_3}$ → **NH$_2$** ... **SO$_2$NH$_2$** **Sulfanilamide (4a)** $+ CH_3\overset{O}{\overset{\|}{C}}—O^-$

SULFAPYRIDINE (4b)

The intermediate sulfonyl chloride (2) is converted to the amide (3b) in a reaction with 2-aminopyridine. The other product of the reaction, hydrogen chloride, reacts with the solvent pyridine to form a salt. If pyridine were not present, then the 2-aminopyridine would be inactivated as a nucleophile by salt formation, and this would reduce the yield of the reaction. The pyridine must be very dry, since sulfonyl chloride reacts rapidly with water in pyridine solution.

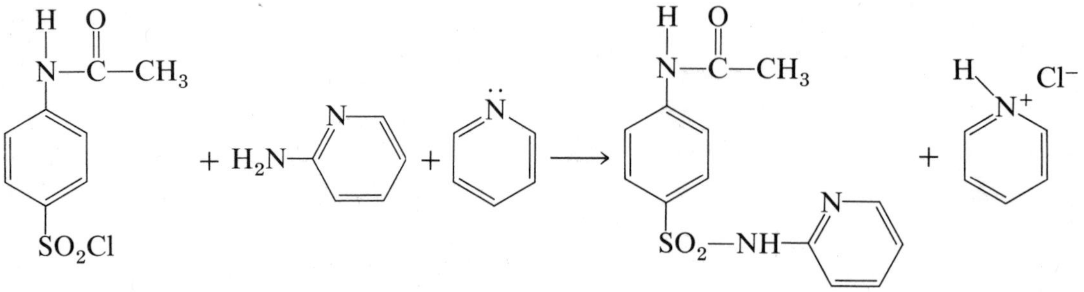

| (2) | 2-Aminopyridine   Pyridine | (3b) | Pyridine hydrochloride |

The second step (**3b** to **4b**) is a base-catalyzed hydrolysis that preferentially cleaves the acetamido group (see p. 314). The salt of the sulfa drug is converted to sulfapyridine (**4b**) when acid is added. The salt is formed because sulfonamides are reasonably acidic compounds.

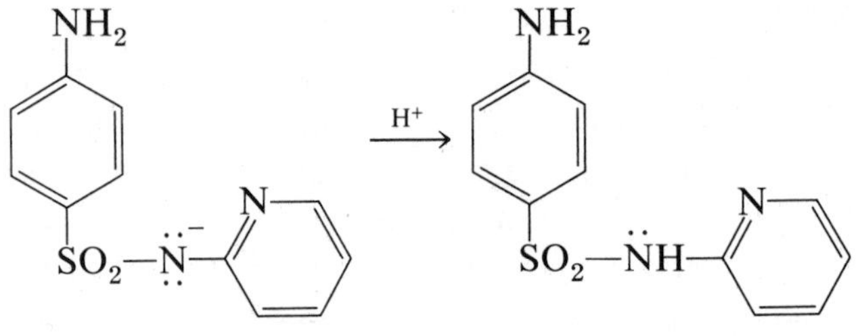

| Salt | Sulfapyridine (4b) |

SULFATHIAZOLE (4c)

The intermediate sulfonyl chloride (**2**) is converted to the amide (**3c**) in a reaction with 2-aminothiazole in the presence of pyridine. The reaction proceeds like the one for sulfapyridine. There may be a side reaction from possible tautomerism of the 2-amino-thiazole. The tautomer **B** can react with the sulfonyl chloride (**2**) to give an isomer of sulfapyridine (**C**). Fortunately, little of this isomer (**C**) is produced in this experiment. The second step (**3c** to **4c**) is a base-catalyzed hydrolysis (discussed previously for sulfapyridine).

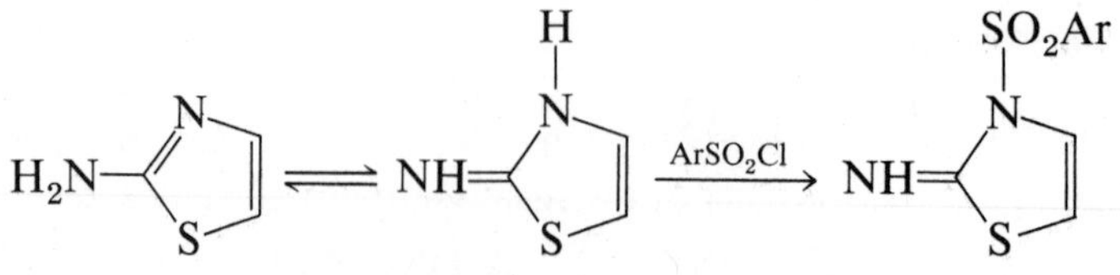

| A | B | C |

SPECIAL INSTRUCTIONS

Read the essay on sulfa drugs, which precedes this experiment. The chlorosulfonic acid must be handled with care since it is a corrosive liquid. The p-acetamidobenzenesulfonyl chloride (2) should be used during the same laboratory period in which it is prepared or during the very next period. It is unstable and will not survive long storage.

The instructor may divide the class into groups of three students. Each student in the group may then pick a different sulfa drug and synthesize it. The three drugs (sulfanilamide, sulfapyridine, and sulfathiazole) may then be tested on several kinds of bacteria (Instructor's Manual). If a 500-mL separatory funnel is not in your desk equipment, you will need to get one.

PROCEDURES

p-ACETAMIDOBENZENESULFONYL CHLORIDE (2)

Assemble an apparatus as shown in the figure to help trap the hydrogen chloride gas that forms as a by-product of this reaction. Insert a short section of glass tubing into a one-hole rubber stopper. Insert the stopper in the flask after the acetanilide and chlorosulfonic acid have been added as directed below. Use a piece of rubber tubing to lead the vapors away from the reaction flask to an inverted funnel placed **just below** the surface of a beaker of water containing about 1 g of sodium hydroxide. As the gas is evolved, it dissolves in the water and reacts with the sodium hydroxide.

Place 12 g of dry acetanilide (1) in a **dry** 250-mL Erlenmeyer flask. Melt the acetanilide by gentle heating with a flame. Swirl the heavy oil so that it is deposited uniformly on the lower wall and bottom of the flask. Cool the flask in an ice bath. To the solidified material, add 33 mL of chlorosulfonic acid, $ClSO_2OH$ (density 1.77 g/mL), in one portion, and then attach the trap.

> **CAUTION: Chlorosulfonic acid is an extremely noxious and corrosive chemical and should be handled with care. Use only dry glassware with this reagent. Should the chlorosulfonic acid be spilled on your skin, wash it off immediately with water. Wear safety glasses.**

Remove the flask from the ice bath and swirl it. Hydrogen chloride gas is evolved vigorously, so be certain that the rubber stopper is securely placed in the neck of the flask. The reaction mixture usually will not have to be cooled. If the reaction becomes too vigorous, however, slight cooling may be necessary. After 10 minutes, the reaction should have subsided and only a small amount of acetanilide should remain. Heat the flask for an additional 10 minutes on the steam bath to complete the reaction (continue to use the trap). After this time, remove the trap assembly and cool the flask in an ice bath in the hood.

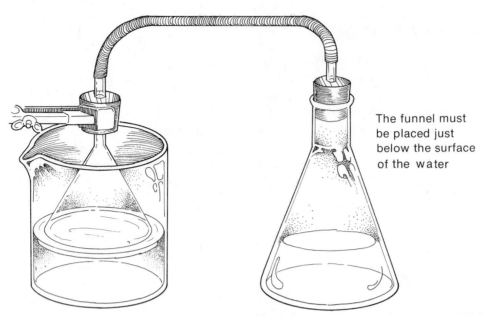

The funnel must be placed just below the surface of the water

Apparatus for making *p*-acetamidobenzenesulfonyl chloride

The operations in this paragraph should be conducted as rapidly as possible since the *p*-acetamidobenzenesulfonyl chloride reacts with water. **Slowly** pour the cooled mixture with vigorous stirring (it will splatter somewhat) into a beaker containing 200 mL of crushed ice. Rinse the flask with some cold water and transfer the contents to the beaker containing the ice. Stir the precipitate to break up the lumps and then vacuum-filter the mixture. Wash the crude *p*-acetamidobenzenesulfonyl chloride with a small amount of cold water. Dissolve the solids in 150 mL of boiling methylene chloride in the hood, taking care to avoid the toxic vapors. The solid may dissolve slowly. Replace the methylene chloride solvent as it evaporates. Transfer the mixture to a **warm** 500-mL separatory funnel. Drain the lower methylene chloride layer rapidly, but carefully, from the funnel and away from the upper water layer.

Cool the mixture in an ice bath and collect the crystalline *p*-acetamidobenzenesulfonyl chloride by vacuum filtration, using a Büchner funnel. Draw air through the funnel to dry the product. Weigh the material and calculate the percentage yield. Determine the melting point of the product. The anhydrous compound melts at 140 °C. The *p*-acetamidobenzenesulfonyl chloride should be used during the same laboratory period in which it is prepared or the very next one. If it must be stored, the dry product must be placed in a tightly stoppered bottle. Convert the *p*-acetamidobenzenesulfonyl chloride (**2**) into one of three possible sulfa drugs by one of the following procedures: 42A, 42B, or 42C.

Procedure 42A
Sulfanilamide (4a)

Place 5 g of *p*-acetamidobenzenesulfonyl chloride (**2**) in a 125-mL Erlenmeyer flask and add (in the hood) 15 mL of concentrated ammonium hydroxide. Stir the mixture well with

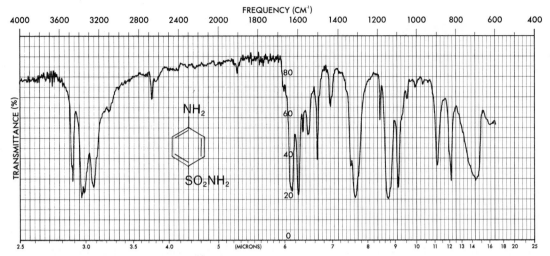

Ir spectrum of sulfanilamide, KBr

a stirring rod. A reaction usually begins immediately and the mixture becomes warm. Heat the mixture on a steam bath in the hood for 15 minutes, stirring frequently. During this time the material becomes a pasty suspension. Remove the flask and place it in an ice bath. When the mixture is well cooled, add 6N hydrochloric acid (5–10 mL) until the mixture is acidic to litmus paper. Continue cooling the mixture in the ice bath until it is thoroughly cold, and filter the product on the Büchner funnel with the vacuum. Wash the product (compound **3a**) with 50 mL of cold water. The material can be used directly in the next step or allowed to dry in air.

Transfer the crude product (**3a**) to a small round-bottomed flask and add 3 mL of concentrated hydrochloric acid, 6 mL of water, and a boiling stone. Attach a reflux condenser to the flask. Allow the mixture to reflux until the solid has dissolved (about 10 minutes) and then reflux for an additional 10 minutes. Cool the mixture to room temperature. If a solid (unreacted starting material) appears, bring the mixture to a boil again for several minutes. When it is cooled to room temperature, no further solids should appear. To this cooled solution, add 5 mL of water and a small amount of decolorizing carbon. Shake the mixture and then filter it by gravity into a 200-mL or larger beaker. Rinse the flask with 10 mL of water and pour the rinsings through the filter. To the filtrate, cautiously add a solution of 4 g of sodium bicarbonate (dissolved in a minimum amount of water) with stirring, until the solution is neutral to litmus. Foaming will occur after each addition of the bicarbonate solution because of carbon dioxide evolution. The sulfanilamide will precipitate during the neutralization. Cool the mixture thoroughly in an ice bath and filter the product by vacuum filtration through a Büchner funnel. Dry the sulfanilamide as much as possible by drawing air through the filter. Recrystallize the solid from water, using about 10 to 12 mL of water per gram of crude product. Allow the material to dry until the next laboratory period. Determine the melting point. Pure sulfanilamide melts from 163 to 164 °C. Weigh the product and calculate the percentage yield. Submit the sulfanilamide to the instructor in a labeled vial or save it for the tests with bacteria (Instructor's Manual). At the instructor's option, determine the infrared spectrum in potassium bromide.

Procedure 42B
Sulfapyridine (4b)

Dissolve 2.4 g of 2-aminopyridine in 10 mL of anhydrous pyridine (previously dried over KOH pellets) in a small round-bottomed flask. Add 6 g of dry *p*-acetamidobenzenesulfonyl chloride (**2**) to the mixture. Attach a reflux condenser to the flask. Allow the mixture to heat for 20 minutes on a steam bath. Cool the mixture slightly and pour it into 50 mL of water.[1] Add some water to the flask to aid in the transfer. Stir the mixture in an ice bath until the oil crystallizes. Filter the solid (**3b**) by vacuum filtration through a Büchner funnel.

      Transfer the crude product (**3b**) to a round-bottomed flask and dissolve it in 20 mL of 10% sodium hydroxide solution. Use stopcock grease on the joints before assembling the apparatus; otherwise the joints are likely to freeze. Heat the resulting solution under reflux for 40 minutes. After cooling the mixture, carefully neutralize it with 6*N* hydrochloric acid. Filter the precipitated sulfapyridine by vacuum filtration and recrystallize it from 95% ethyl alcohol. About 100 mL of alcohol will be needed to dissolve the sulfapyridine. Watch it carefully as it dissolves. When the point is reached at which the remaining solid no longer dissolves, remove the remaining solid by gravity filtration. Cool the filtrate slowly to room temperature. Filter the mixture by vacuum filtration on a Büchner funnel. Dry the sulfapyridine in air and determine its melting point. The recorded melting point is 190 to 193 °C. Weigh the product and calculate the percentage yield. Submit the sulfapyridine to the instructor in a labeled vial or save it for the tests with bacteria (Instructor's Manual). At the instructor's option, determine the infrared spectrum in potassium bromide.

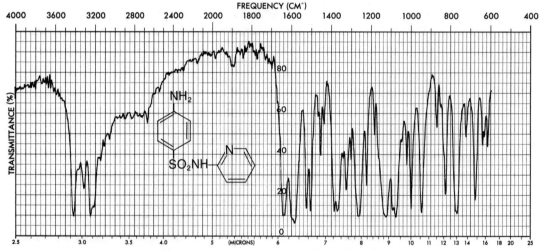

Ir spectrum of sulfapyridine, KBr

---

[1]Do not overcool the mixture at this point, or it will form a glassy mass at the bottom of the flask. If such a mass does form, it can be dissolved in warm water to yield a product.

320

Procedure 42C
Sulfathiazole (4c)

Dissolve 2.5 g of 2-aminothiazole in 10 mL of anhydrous pyridine (dried over KOH pellets) in an Erlenmeyer flask. To this mixture, add 6.5 g of dry *p*-acetamidobenzenesulfonyl chloride (**2**) in small portions with swirling. The rate of addition should be such that the temperature does not rise above 40 °C. Cool the mixture slightly if the temperature exceeds this value. After the addition, complete the reaction by heating it for 30 minutes on a steam bath. Cool the mixture and then pour it into 75 mL of water. Add some water to the flask to aid in the transfer. The oil (**3c**) should solidify when stirred with a glass rod. Isolate the crystals by vacuum filtration. Wash them with cold water and then press them dry.

Weigh the solid (**3c**) and dissolve it in 10% sodium hydroxide (10 mL per gram of solid) in a round-bottomed flask. Attach a reflux condenser and heat the mixture under reflux for 1 hour. Cool the solution and add concentrated hydrochloric acid until the mixture is at pH 6. If you add too much acid, adjust the pH to 6 by adding 10% sodium hydroxide. Then add solid sodium acetate until the solution is just basic to litmus. Heat the mixture to a boil and cool it in an ice bath. Filter the sulfathiazole by vacuum filtration and recrystallize it from water (the solubility of sulfathiazole in hot water is quite low). Cool the solution slowly to room temperature. Filter the mixture by vacuum filtration. Allow the sulfathiazole to dry until the next laboratory period. Determine the melting point. Pure sulfathiazole melts from 201 to 202 °C. Weigh the product and calculate the percentage yield. Submit the sulfathiazole in a labeled vial to the instructor or save it for the bacteria tests (Instructor's Manual). At the instructor's option, determine the infrared spectrum in potassium bromide.

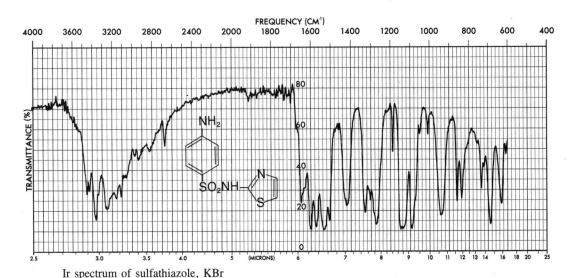

Ir spectrum of sulfathiazole, KBr

QUESTIONS

1. Give an equation showing how excess chlorosulfonic acid is decomposed.

2. In the preparation of sulfanilamide, why was aqueous sodium bicarbonate used, rather than aqueous sodium hydroxide, to neutralize the solution in the final step?

3. In the final steps of preparing sulfapyridine and sulfathiazole, the acidity must be carefully controlled. Why? Give equations to show the behavior of these sulfa drugs in both acidic and basic solutions.

4. At first glance, it might seem possible to prepare sulfanilamide from sulfanilic acid by the set of reactions shown below.

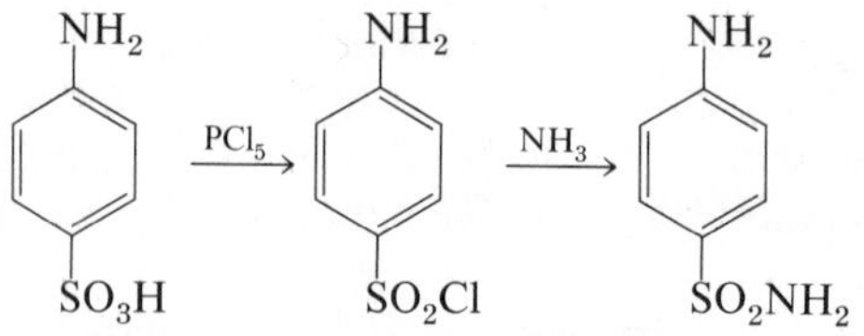

When the reaction is conducted in this way, however, a polymeric product is produced after the first step. What is the structure of the polymer? Why does *p*-acetamidobenzenesulfonyl chloride not produce a polymer?

5. In the preparation of sulfathiazole, an impurity is sometimes produced (shown below). How could you separate this impurity from sulfathiazole by using acids or bases or both?

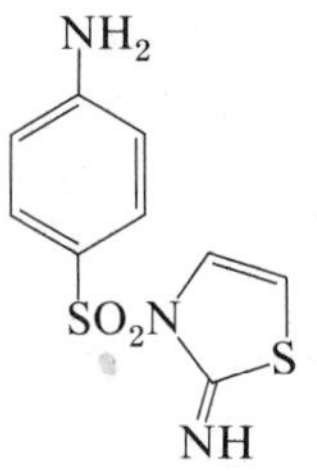

Experiment 43
p-Aminobenzoic Acid

Synthesis of a Vitamin
Amide Formation
Permanganate Oxidation
Amide Hydrolysis

The object of this set of chemical reactions is to prepare the vitamin (for bacteria) *p*-**aminobenzoic acid,** or PABA. For a description of the importance of PABA in biological processes, read the essay on the sulfa drugs immediately preceding Experiment 42. Because PABA can absorb the ultraviolet component of solar radiation, it also finds an important application in sunscreen preparations.

     The synthesis of *p*-aminobenzoic acid involves three reactions, which are outlined in the following paragraphs. The first of these reactions is the conversion of the commercially available *p*-**toluidine** into *N*-**acetyl-*p*-toluidine** (or *p*-acetotoluidide), the corresponding amide. The reaction is carried out by treating the amine, *p*-toluidine, with acetic anhydride. Such a procedure is a standard method for preparing amides. The reason for acetylating the amine in the initial step is to protect it during the second step, a permanganate oxidation. If the oxidation of the methyl group to the corresponding carboxyl group were carried out directly with *p*-toluidine, the highly reactive amino group would also be oxidized in the reaction. To prevent this undesired oxidation, a **protective group** is used. A protective group is a functional group that is added during a reaction sequence to shield a particular position on a molecule from undesired reactions. A good protective group is one that is easily added to the substrate molecule,

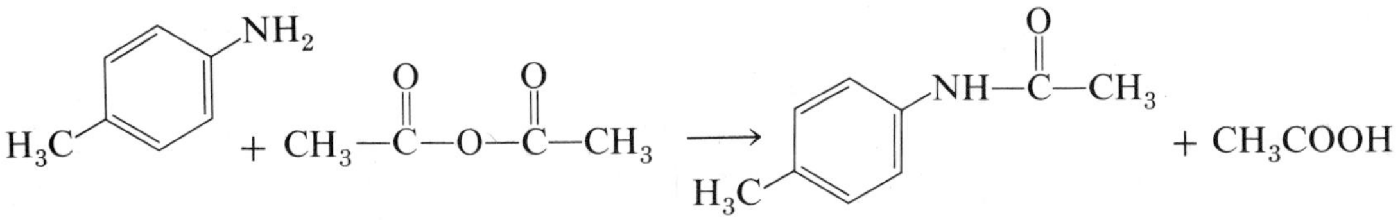

p-**Toluidine**   *N*-**Acetyl**-*p*-**toluidine**

does not permit the protected group to undergo reactions, and is easily removed after the protective role has been played. This last point is important, since if the protective group cannot be removed, the original functional group cannot be used. In the example used in this experiment, the protective group is the **acetyl group.** The amide formed in this initial reaction is stable toward the oxidation conditions used in the second step. As a result, the methyl group can be oxidized without destroying the amino function in the process.

The second step of this reaction sequence is the oxidation of the methyl group to the corresponding carboxyl group, with potassium permanganate serving as the oxidizing agent. Because of the great stability of substituted benzoic acids, alkyl groups

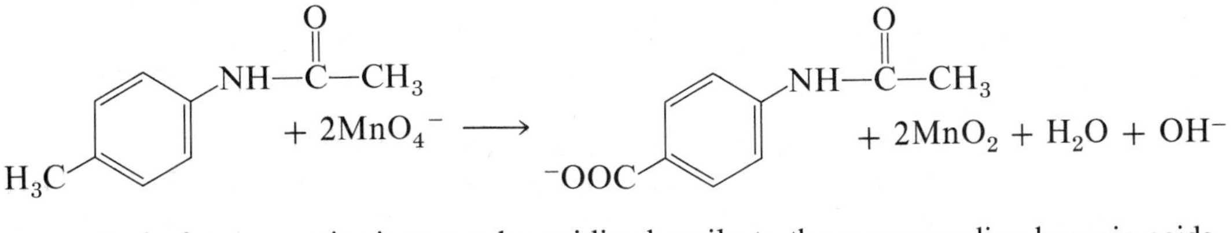

attached to aromatic rings can be oxidized easily to the corresponding benzoic acids. During the oxidation, the violet solution of permanganate ion is converted to a brown precipitate, manganese dioxide, as Mn(VII) is reduced to Mn(IV). A small amount of magnesium sulfate is added as a buffer, so that the solution does not become excessively basic as hydroxide ion is produced. The product of the reaction is not the carboxylic acid but rather its salt that is produced directly from the reaction. Acidification of the reaction mixture yields the carboxylic acid, which precipitates from solution.

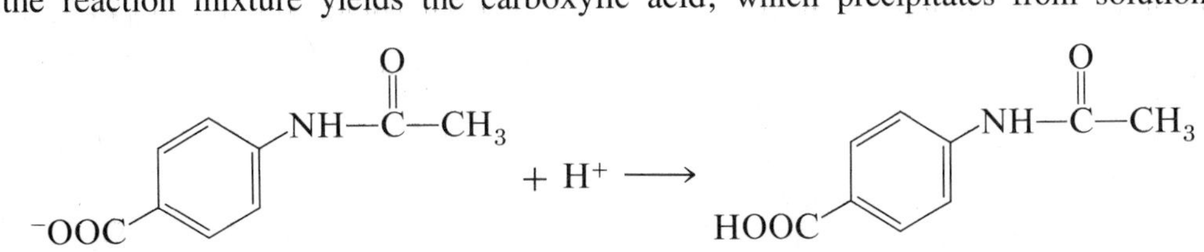

p-**Acetamidobenzoic acid**

The final step of this procedure is the hydrolysis of the amide functional group to remove the protecting acetyl group; *p*-aminobenzoic acid is thus produced. The reaction proceeds easily in dilute aqueous acid, and it is a characteristic reaction of amides in general. The product is crystallized from dilute aqueous acetic acid.

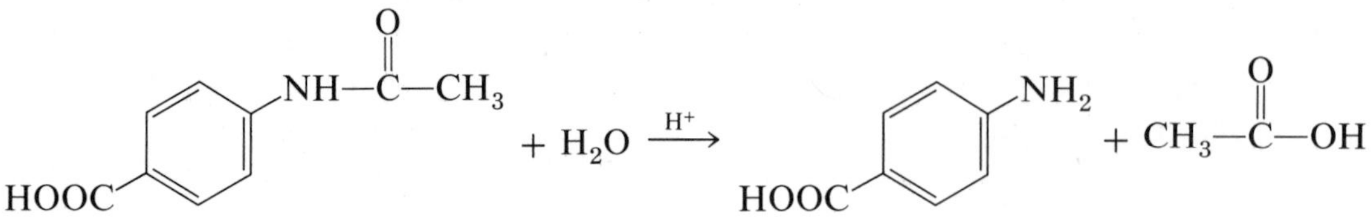

$$\text{HOOC}\text{—C}_6\text{H}_4\text{—NH—C(=O)—CH}_3 + \text{H}_2\text{O} \xrightarrow{\text{H}^+} \text{HOOC}\text{—C}_6\text{H}_4\text{—NH}_2 + \text{CH}_3\text{—C(=O)—OH}$$

p-**Aminobenzoic acid**

SPECIAL INSTRUCTIONS

Before performing this experiment, read Techniques 1, 2, 3, and 4. This experiment should be conducted in two parts. It is essential that the procedure through placing the crystals of *p*-acetamidobenzoic acid in the drying oven be carried out in the first laboratory period. Between 1 and 1.5 hours of reaction time is needed to reach this point, along with considerable time spent on other operations.

PROCEDURE

N-ACETYL-*p*-TOLUIDINE

Place 8 g of powdered *p*-toluidine in a 500-mL Erlenmeyer flask. Add 200 mL of water and 8 mL of concentrated hydrochloric acid. If necessary, warm the mixture on a steam bath, with stirring, to facilitate solution. If the solution is dark, add 0.5 to 1.0 g of decolorizing charcoal, stir it for several minutes, and filter it by gravity.

Prepare a solution of 12 g of sodium acetate trihydrate in 20 mL of water. If necessary, use a steam bath to warm the solution until all the solid has dissolved.

Warm the decolorized solution of *p*-toluidine hydrochloride to 50 °C. Add 8.4 mL of acetic anhydride (density 1.08 g/mL), stir rapidly, and immediately add the previously prepared sodium acetate solution. Mix the solution thoroughly, and cool the mixture in an ice bath. A white solid should appear at this point. Filter the mixture by vacuum, using a Büchner funnel, wash the crystals three times with cold water, and allow the crystals to stand in the filter to air-dry while the vacuum is maintained. These crystals will not be isolated and dried but used directly in the next step.

p-ACETAMIDOBENZOIC ACID

Place the previously prepared, wet *N*-acetyl-*p*-toluidine in a 1-L beaker, along with 25 g of magnesium sulfate hydrate and 350 mL of water. Place the flask on a steam bath, and adjust the steam flow to a gentle rate. Add a sludge made up of 30 g potassium permanganate and a small amount of water to the reaction in approximately teaspoon quantities. Increase the flow rate of the steam and allow the reaction to proceed for 1 hour. It is important to heat the reaction mixture thoroughly. It is necessary to place the beaker **down in** the steam cone to heat the beaker effectively. During the reaction period, the mixture must be stirred every few minutes. After 1 hour, the mixture should be quite brown. Vacuum-filter the **hot** solution through a bed of Celite. Wash the precipitated manganese dioxide with a small amount of hot water. If the filtrate shows the presence of excess permanganate by its purple color, add **not more than** 1 mL of ethanol, heat the solution **in** the steam bath for another 30 minutes, and filter the **hot** solution through fluted filter paper once more. Cool the colorless filtrate and acidify with excess 20% sulfuric acid solution. A white solid should form at this point. Filter the solid by vacuum and dry it in an oven. The yield based on *p*-toluidine and the melting point should be determined at this point. The melting point of the pure material is 250 to 252 °C. In some cases an inorganic salt may be produced. If a salt is produced, simply go to the next step of the procedure.

p-AMINOBENZOIC ACID

Prepare a dilute solution of hydrochloric acid by mixing 24 mL of concentrated hydrochloric acid and 24 mL of water. Place the previously prepared *p*-acetamidobenzoic acid in a 250-mL round-bottomed flask that has a reflux condenser attached to it. Add the hydrochloric acid solution. Heat the mixture gently under reflux for 30 minutes. Allow the reaction mixture to cool, transfer it to a 250-mL Erlenmeyer flask, add 48 mL of water, and make the reaction mixture just alkaline (pH 8–9; use pH paper) with dilute ammonia solution. For each 30 mL of the final solution, add 1 mL of glacial acetic acid, chill the solution in an ice bath, and initiate the crystallization, if necessary, by scratching the inside of the flask with a glass rod. Filter the crystals by vacuum and allow them to dry.

Determine the yield for this step, basing the overall yield on *p*-toluidine. Determine the melting point of the product. The melting point of the pure *p*-aminobenzoic acid

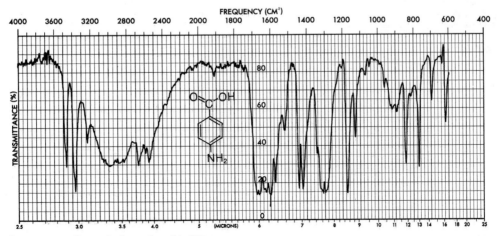

Ir spectrum of *p*-aminobenzoic acid, KBr

is 186 to 187 °C. Frequently the melting point of the product is somewhat lower. Attempts to recrystallize the product are not promising enough to be recommended. Recrystallization is not necessary before the *p*-aminobenzoic acid is used in Experiment 44. At the option of the instructor, an infrared spectrum of the *p*-aminobenzoic acid may be determined (Technique 17, Part A, p. 659, and Appendix 3, p. 695).

The *p*-aminobenzoic acid may be saved for use in Experiment 44, the preparation of benzocaine. If it is not to be used for the later experiment, submit the sample of *p*-aminobenzoic acid, in a labeled vial, to the instructor.

REFERENCE

Kremer, C. B. "The Laboratory Preparation of a Simple Vitamin: *p*-Aminobenzoic Acid." *Journal of Chemical Education, 33* (1956): 71.

QUESTIONS

1. Write a mechanism for the reaction of *p*-toluidine with acetic anhydride. Why is sodium acetate added to this reaction?

2. In the oxidation step, if excess permanganate remains after the reaction period, a small amount of ethanol is added to discharge the purple color. Write a chemical equation that describes the reaction of permanganate with ethanol.

3. Write a mechanism for the acid-catalyzed hydrolysis of *p*-acetamidobenzoic acid to form *p*-aminobenzoic acid.

4. Interpret the principal absorption bands in the infrared spectrum of *p*-aminobenzoic acid.

Essay
LOCAL ANESTHETICS

Local anesthetics, or "painkillers," are a well-studied class of compounds with which chemists have shown their ability to study the essential features of a naturally occurring drug and to improve on them by substituting totally new, synthetic surrogates. Often such substitutes are superior in desired medical effects and also in lack of unwanted side effects or hazards.

The coca shrub (*Erythroxylon coca*) grows wild in Peru, specifically in the Andes Mountains, at elevations of 1500 to 6000 ft above sea level. The natives of

Cocaine

Eucaine

South America have long chewed these leaves for their stimulant effects. Leaves of the coca shrub have even been found in pre-Inca Peruvian burial urns. The leaves bring about a definite sense of mental and physical well-being and have the power to increase endurance. For chewing, the Indians smear the coca leaves with lime and roll them. The lime, $Ca(OH)_2$, apparently releases the free alkaloid components; it is remarkable that the Indians learned this subtlety long ago by some empirical means. The pure alkaloid responsible for the properties of the coca leaves is **cocaine.**

Moderate indulgence, as practiced by the coca-chewing Indians, probably produces no more ill effects than moderate tobacco smoking does. The amounts of cocaine consumed in this way by the Indians are extremely small. Without such a crutch of central-nervous-system stimulation, the natives of the Andes would probably find it more difficult to perform the nearly Herculean tasks of their daily lives, such as carrying heavy loads over the rugged mountainous terrain. Unfortunately, overindulgence can lead to mental and physical deterioration and eventually an unpleasant death.

The pure alkaloid in large quantities is a common drug of addiction, which is psychological if not physical. Sigmund Freud first made a detailed study of cocaine in 1884. He was particularly impressed by the ability of the drug to stimulate the central nervous system, and he used it as a replacement drug to wean one of his addicted colleagues from morphine. This attempt was successful, but unhappily, the colleague became the world's first known cocaine addict.

An extract from coca leaves was one of the original ingredients in Coca-Cola. However, early in the present century, government officials, with much legal difficulty, forced the manufacturer to omit coca from its beverage. The company has managed to this day to maintain the *coca* in its trademarked title even though ''Coke'' contains none!

Our interest in cocaine lies in its anesthetic properties. The pure alkaloid was isolated in 1862 by Niemann, who noted that it had a bitter taste and produced a queer numbing sensation on the tongue, rendering it almost devoid of sensation. (Oh, those brave, but foolish chemists of yore who used to taste everything!) In 1880, Von Anrep found that the skin was made numb and insensitive to the prick of a pin when cocaine was injected subcutaneously. Freud and his assistant, Karl Koller, having failed at attempts to rehabilitate morphine addicts, turned to a study of the anesthetizing properties of cocaine. Eye surgery is made difficult by involuntary reflex movements of the eye in response to even the slightest touch. Koller found that a few drops of a solution of cocaine would overcome this problem. Not only can cocaine serve as a local anesthetic, but it can also be used to produce mydriasis (dilation of the pupil). The ability of cocaine to block signal conduction in nerves (particularly of pain) led to its rapid medical use in spite of its dangers. It soon found use as a "local" in both dentistry (1884) and in surgery (1885). In this type of application, it was injected directly into the particular nerves it was intended to deaden.

Soon after the structure of cocaine was established, chemists began to search for a substitute. Cocaine has several drawbacks for wide medical use as an anesthetic. In eye surgery, it also produces mydriasis. It can also become a drug of addiction. Finally, it has a dangerous effect on the central nervous system.

The first totally synthetic substitute was eucaine (see above). This was synthesized by Harries in 1918 and retains many of the essential skeletal features of the cocaine molecule. The development of this new anesthetic partly confirmed the portion of the cocaine structure essential for local anesthetic action. The advantage of eucaine over cocaine is that it does not produce mydriasis and is not habit-forming. Unfortunately, it is highly toxic.

A further attempt at simplification led to piperocaine. The molecular portion common to cocaine and eucaine is outlined by dotted lines in the structure shown. Piperocaine is only a third as toxic as cocaine itself.

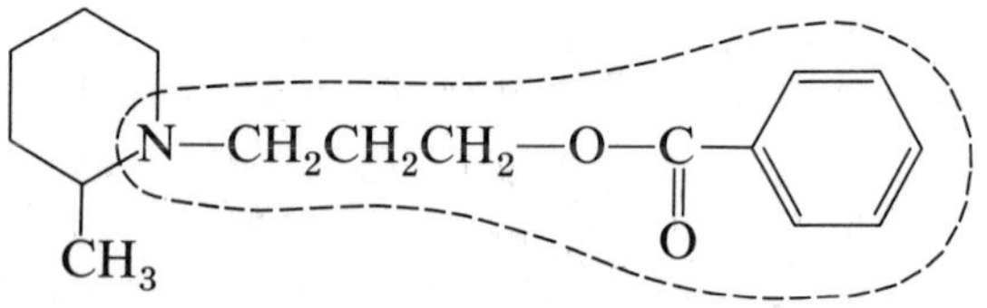

Piperocaine

The most successful synthetic for many years was the drug procaine, also known more commonly by its trade name Novocain (see table). Novocain is only a fourth as toxic as cocaine, giving a better margin of safety in its use. The toxic dose is almost 10 times the effective amount, and it is not a habit-forming drug.

Over the years, hundreds of new local anesthetics have been synthesized and tested. For one reason or another, most have not come into general use. The search for the perfect local anesthetic is still under way. All the drugs found to be active have certain structural features in common. At one end of the molecule is an aromatic ring. At the other is a secondary or tertiary amine. These two essential features are separated by a central chain of atoms usually one to four units long. The aromatic part is usually an ester of an aromatic acid. The ester group is important to the bodily detoxification of these compounds. The first step in deactivating them is a hydrolysis of this ester linkage, a process that occurs in the bloodstream. Compounds that do not have the ester link are both longer lasting in their effect and generally more toxic. An exception is lidocaine, which is an amide. The tertiary amino group is apparently necessary to enhance the solubility of the compounds in the injection solvent. Most of these com-

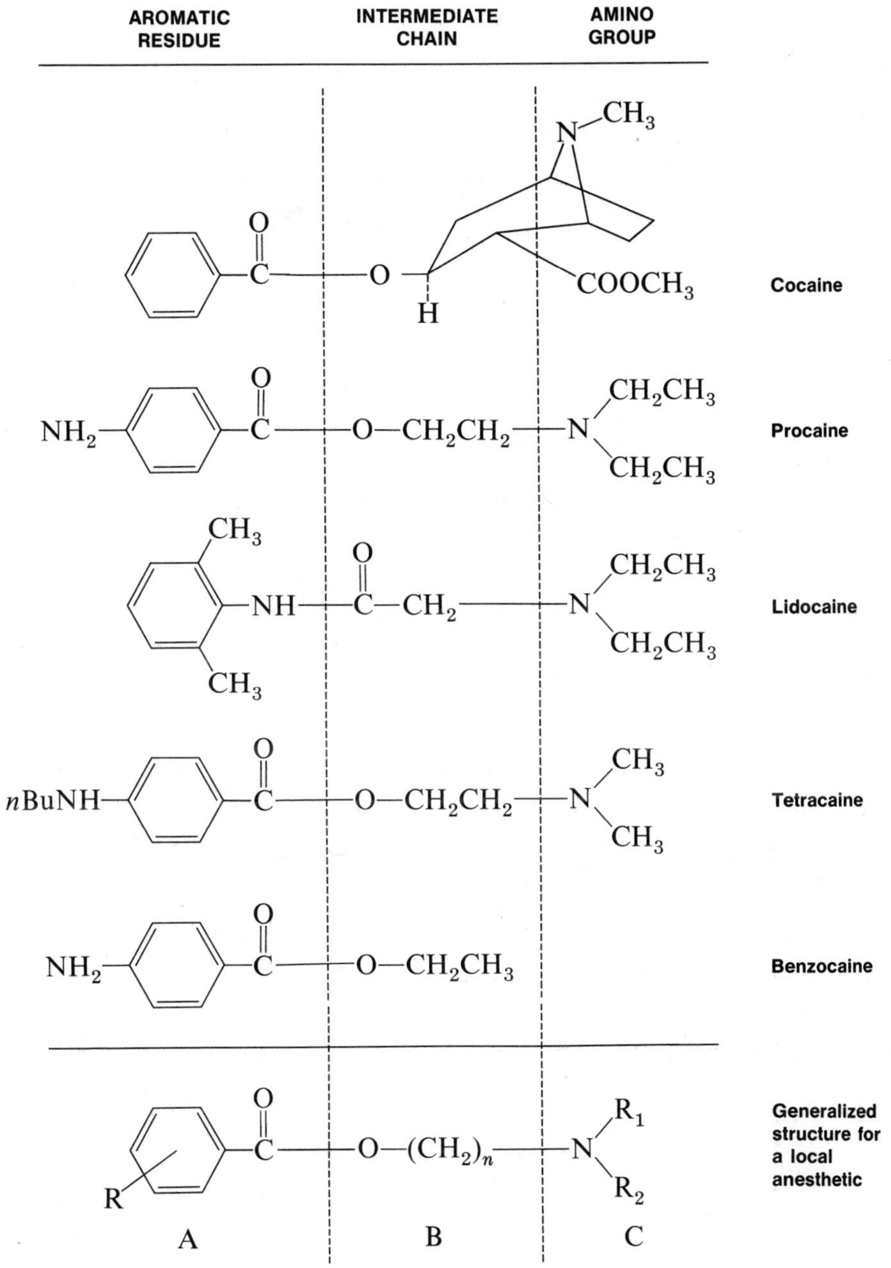

AROMATIC RESIDUE	INTERMEDIATE CHAIN	AMINO GROUP	

Cocaine

Procaine

Lidocaine

Tetracaine

Benzocaine

Generalized structure for a local anesthetic

A   B   C

Local Anesthetics

pounds are used in their hydrochloride salt forms, which can be dissolved in water for injection. Benzocaine, in contrast, is active as a local anesthetic but is not used for injection. It does not suffuse well into tissue and is not water-soluble. It is used primarily in skin preparations, in which it can be included in an ointment of salve for direct application. It is an ingredient of many sunburn preparations.

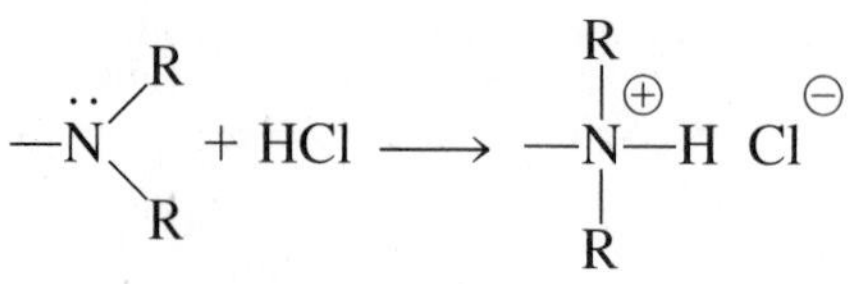

$$-\overset{..}{N}\overset{R}{\underset{R}{<}} + HCl \longrightarrow -\overset{R}{\underset{R}{\overset{|}{N}}}\overset{\oplus}{-}H \quad Cl^{\ominus}$$

How these drugs act to stop pain conduction is not well understood. Their main site of action is at the nerve membrane. They seem to compete with calcium at some receptor site, altering the permeability of the membrane and keeping the nerve slightly depolarized electrically.

REFERENCES

Foye, W. O. *Principles of Medicinal Chemistry*. Philadelphia: Lea & Febiger, 1974. Chap. 14, "Local Anesthetics."

Goodman, L. S., and Gilman, A. *The Pharmacological Basis of Therapeutics*. 7th ed. New York: Macmillan, 1985. Chap. 20, "Cocaine: Procaine and Other Synthetic Local Anesthetics," J. M. Ritchie, et al.

Ray, O. S. *Drugs, Society, and Human Behavior*. 3rd ed. St. Louis: C. V. Mosby, 1983. Chap. 11, "Stimulants and Depressants."

Snyder, S. H. "The Brain's Own Opiates." *Chemical and Engineering News* (November 28, 1977): 26–35.

Taylor, N. *Narcotics: Nature's Dangerous Gifts*. New York: Dell, 1970. Paperbound revision of *Flight from Reality*. Chap. 3, "The Divine Plant of the Incas."

Taylor, N. *Plant Drugs that Changed the World*. New York: Dodd, Mead, 1965. Pp. 14–18.

Wilson, C. O., Gisvold, O., and Doerge, R. F. *Textbook of Organic Medicinal and Pharmaceutical Chemistry*. 6th ed. Philadelphia: J. B. Lippincott, 1971. Chap. 22, "Local Anesthetic Agents," R. F. Doerge.

Experiment 44
Benzocaine

Esterification

In this experiment, a procedure is given for the preparation of a local anesthetic, benzocaine, by the direct esterification of *p*-aminobenzoic acid with ethanol. At the instructor's option, you may test the prepared anesthetic on a frog's leg muscle.

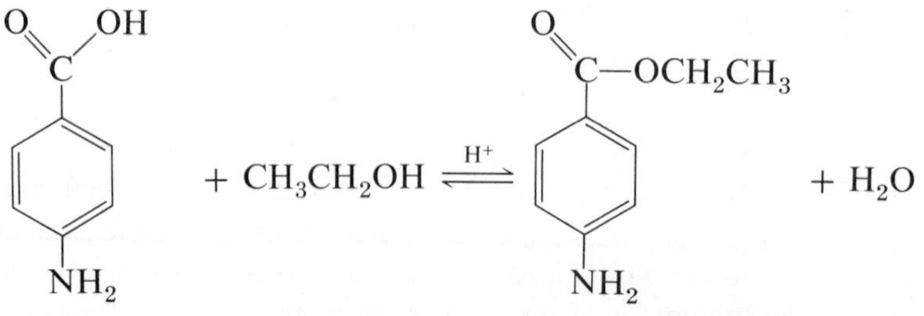

$$+ CH_3CH_2OH \underset{}{\overset{H^+}{\rightleftharpoons}} \qquad + H_2O$$

p-**Aminobenzoic acid**   **Ethyl *p*-aminobenzoate**
                                              (Benzocaine)

SPECIAL INSTRUCTIONS

Read the essay on local anesthetics, which precedes this experiment. Since this experiment involves a 2-hour reflux, the experiment must be started at the beginning of the laboratory period. It can then be completed in one period.

> **NOTE TO THE INSTRUCTOR.** If instruction for testing benzocaine on a frog's leg muscle is needed, please refer to the instructor's manual.

PROCEDURE

Place 5.0 g of *p*-aminobenzoic acid (Experiment 43) in a 250-mL round-bottomed flask and add 65 mL of 95% ethanol, swirling gently to help dissolve the solid (not all the solid will dissolve). Cool the mixture in an ice bath and slowly add 5 mL of concentrated sulfuric acid. A large amount of precipitate will form when the sulfuric acid is added, but

> **CAUTION: Be careful in handling concentrated sulfuric acid.**

this solid will slowly dissolve during the reflux that follows. Attach a reflux condenser and heat the mixture under reflux on a steam bath for 2 hours. The flask may have to be placed **down in** the steam cone to obtain the proper reflux rate. Swirl the contents of the flask at approximately 15-minute intervals during the first hour of the reflux.

Transfer the contents of the flask to a 400-mL beaker and add in small portions a 10% sodium carbonate solution (about 60 mL needed) to neutralize the mixture. After each addition of the sodium carbonate solution, extensive gas evolution (frothing) will be perceptible until the mixture is nearly neutralized. When gas no longer evolves as you

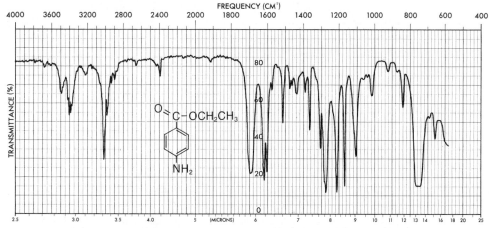

Ir spectrum of benzocaine, $CHCl_3$ ($CHCl_3$ solvent: 3030, 1220, and 750 cm^{-1})

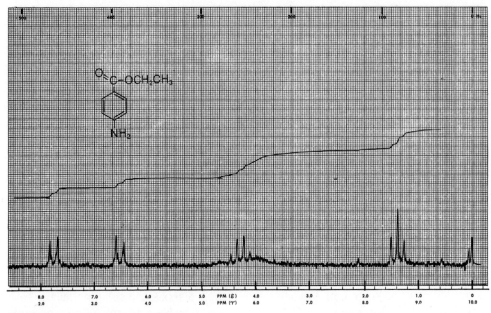

Nmr spectrum of benzocaine, CCl₄

add a portion of sodium carbonate, check the pH of the solution and add further portions of sodium carbonate until the pH is 9 or above. Decant the aqueous mixture away from any solids formed from neutralization, pouring it into a separatory funnel (250-mL or larger). Add a 100-mL portion of ether to the separatory funnel, and shake vigorously. Separate the mixture and save the upper ether layer. Dry it with magnesium sulfate (about 2 spatulafuls), gravity-filter the ether layer to remove the drying agent, and remove the ether and ethanol by evaporating them over a hot plate in a hood. When most of the solvent has been removed (about 5 mL remaining), an oil will be visible in the flask. Add hot 95% ethanol and heat the mixture on the hot plate until the oil dissolves. Add water to the alcohol solution until the oil again appears (extensive cloudiness) and then cool the mixture in an ice bath. The oil that may form initially will crystallize during the swirling in the ice bath. Collect the benzocaine by vacuum filtration, using a Büchner funnel. After the solid has dried overnight at room temperature on a piece of filter paper, weigh it, calculate the percentage yield, and determine its melting point. The melting point of pure benzocaine is 92 °C. At the option of the instructor, obtain the infrared spectrum in chloroform and the nmr spectrum in carbon tetrachloride or CDCl₃. Submit the sample in a labeled vial to the instructor.

QUESTIONS

1. Interpret the infrared and nmr spectra of benzocaine.
2. What is the structure of the precipitate that forms after the sulfuric acid has been added?
3. When 10% sodium carbonate solution is added, a gas evolves. What is the gas? Give a balanced equation for this reaction.
4. After the neutralization, the aqueous mixture is decanted from a solid. What is the structure of the inorganic solid? Benzocaine did not precipitate during the neutralization. Why not?

5. Refer to the structure of procaine in the table in the essay that precedes this experiment (''Local Anesthetics''). Using *p*-aminobenzoic acid, give equations showing how procaine and procaine *mono*hydrochloride should be prepared. Which of the two possible amino functional groups in procaine will be protonated first? Defend your choice. (*Hint:* Consider resonance.)

Essay
BARBITURATES

Barbituric acid results from the condensation of malonic acid and urea. It was first prepared in 1864 by Adolph von Baeyer, a young research assistant to Kekulé at the University of Ghent. One story has it that von Baeyer celebrated the advent of the new compound by visiting a nearby tavern. The tavern was one also frequented by artillery officers of the region; as it happened, it was the day of their patron saint, St. Barbara. Somehow, in ensuing festivities, the name of Barbara was joined to that of urea, and thereby the new compound was christened—barbituric acid.

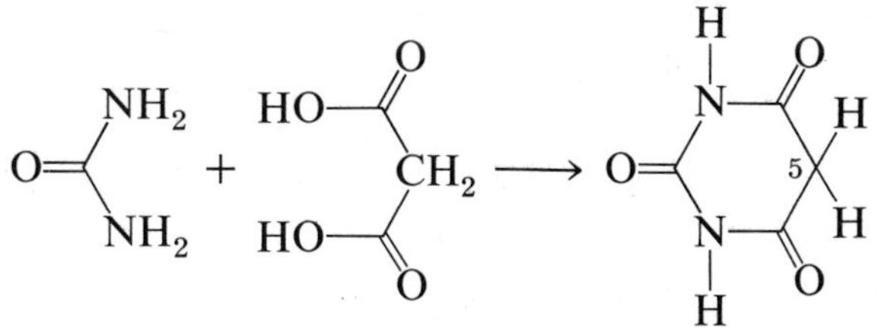

| Urea | Malonic acid | Barbituric acid |

The derivatives of barbituric acid are today called barbiturates, and many of them are among the most widely used of the sedative-hypnotic drugs. The first physiologically active drug, barbital, or Veronal, was introduced in 1903. The method of synthesis of this compound and its many later analogs has undergone little change. The usual method begins with diethyl malonate. This diester has two acidic hydrogens adjacent to the ester groups, and they can be removed with base. The resultant anion is a good nucleophile and can be **alkylated** with a suitable substrate.

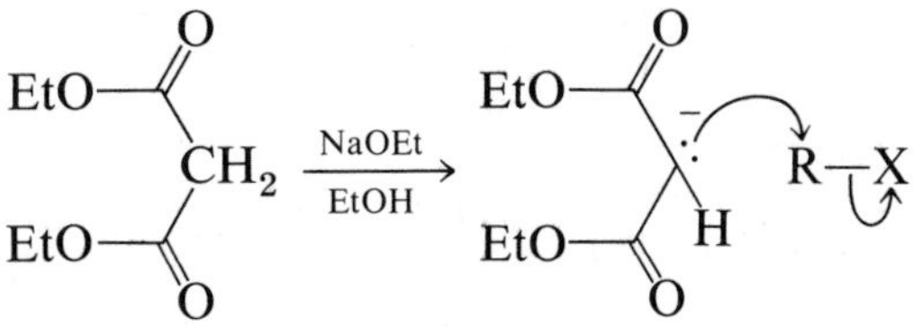

Diethyl malonate

Since there are two α-hydrogens, this process can be repeated to give a dialkyl diethyl malonate derivative. This product is condensed with urea to give a barbiturate, a 5,5-dialkylbarbituric acid:

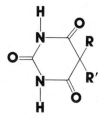

Both hydrogens must be replaced by alkyl groups for the compound to show sedative-hypnotic character as a drug. This is probably because the α-hydrogens are susceptible to rapid metabolic attack within the body, with consequent degradation of the compound.

Chemists have synthesized many of these drugs, and pharmacologists have tested them. Some of the common ones are listed in the accompanying table. The

Structures and Time Factors for Different Barbiturates

GENERIC NAME	BRAND NAME	R	R'
		Long-Acting	
Barbital	Veronal	—CH_2CH_3	—CH_2CH_3
Phenobarbital	Luminal	—CH_2CH_3	—phenyl
		Intermediate-Acting	
Amobarbital	Amytal	—CH_2CH_3	—CH_2CH_2—$CH\begin{smallmatrix}CH_3\\CH_3\end{smallmatrix}$
		Short-Acting	
Pentobarbital	Nembutal	—CH_2CH_3	—$\underset{\underset{CH_3}{\vert}}{CH}$—$CH_2CH_2CH_3$
Secobarbital	Seconal	—CH_2—$CH{=}CH_2$	—$\underset{\underset{CH_3}{\vert}}{CH}$—$CH_2CH_2CH_3$

	TIME TO TAKE EFFECT (hours)	DURATION OF ACTION (hours)
Long-acting	1	6–10
Intermediate-acting	0.5	5–6
Short-acting	0.25	2–3

barbiturates exhibit a wide variety of responses in the body, depending mainly on the identity of the substituted groups. Some generalizations can be made. Increasing the length of an alkyl chain up to five or six carbon atoms in the 5 position enhances the sedative action; beyond that, depressant action decreases, and the drugs become more effective as anticonvulsants for control of epileptic seizures. Branched or unsaturated chains in the 5 position generally produce a briefer duration of action. Barbiturates are classified into three categories, which are based on the time required for them to take effect and the duration of their activity (see table). Compounds with phenyl or ethyl groups in the 5 position seem to have the longest duration.

The medical and physiological use of barbiturates depends on the dose size in a manner typical of all these drugs. In small doses, the drugs are mild sedatives, acting to relieve tension and anxiety. In this use they have been replaced by the more modern tranquilizer drugs. At three to five times the sedative dose, sleep is produced. In large doses, barbiturates act as anesthetics. Sodium pentothal, the sodium salt of thiopental, is one of the most widely used anesthetics for surgery. It has an ultrashort action.

At lower dose levels, sodium pentothal was used during World War II as a "truth drug." With the right dose level, a kind of twilight sleep, or hypnosis, could be induced, with the patient only half-conscious. At this level of narcosis (sleep), a subject had little self-control or will power and was very susceptible to suggestion. He could not hide truthful answers, even if he so wished. Such drug treatment was also an integral part of "brainwashing," a practice of the Communist Chinese that gained notoriety during the Korean War. A combination of drugs and psychological warfare on the patient was used to change his perspectives totally. Sodium pentothal and another drug, scopolamine (not a barbiturate), were widely used for these purposes.

The barbiturates are widely used as prescription sleeping pills. Many persons find sleep brought about by these drugs as refreshing as natural sleep. Many, however, awake with a hangover, dizziness, drowsiness, and a headache. Tests have proved that whether or not a person experiences this hangover, his or her level of efficiency is reduced. Experimental subjects consistently score lower than average on mental and memory tests.

Sodium pentothal

High levels of barbiturates cause death. An overdose of sleeping pills is one of the most common forms of suicide. Barbiturate use can also lead to addiction and chronic intoxication. The addiction is real, and withdrawal is accompanied by symptoms of nausea, tremors, and vomiting. From the person trying to avoid anxiety to the chronic sleeping-pill user, all categories contribute to a legal barbiturate misuse second only to alcohol abuse; a large illegal market is also extant. Both alcohol and barbiturates are difficult drugs to use if one wishes to achieve the initial euphoria produced. With each, the user inevitably slips into sedation and sleep. In combination, these two central-nervous-system depressants, alcohol and barbiturates, can be fatal and nearly always are. Unfortunately, just how the barbiturates bring about narcosis, sedation, and anesthesia is not currently well understood.

REFERENCES

Adams, E. "Barbiturates." *Scientific American, 198* (January 1958): 60.

Foye, W. O. *Principles of Medicinal Chemistry*. Philadelphia: Lea & Febiger, 1974. Pp. 165–171.

Freedman, L. S. "'Truth' Drugs." *Scientific American, 202* (March 1960): 145.

Goodman, L. S., and Gilman, A. *The Pharmacological Basis of Therapeutics*. 7th ed. New York: Macmillan, 1985. Chap. 9, "Hypnotics and Sedatives. I. The Barbiturates." S. K. Sharpless.

Kauffman, G. B. "Adolf von Baeyer and the Naming of Barbituric Acid." *Journal of Chemical Education, 57* (1980): 222.

Ray, O. S. *Drugs, Society, and Human Behavior*. 3rd ed. St. Louis: C. V. Mosby, 1983. Chap. 11, "Stimulants and Depressants."

Experiment 45
5-*n*-Butylbarbituric Acid

Condensation Reactions
Amide Formation, Alkylation
Handling Sodium Metal

A derivative of barbituric acid that has little or no sedative-hypnotic potential and is relatively ineffective on humans is synthesized in this experiment. The product is toxic, however, and should be treated with respect. The synthetic target is 5-*n*-butylbarbituric acid, a monoalkyl barbiturate. The synthetic scheme is shown in the accompanying reactions.

The first step is the alkylation of diethyl malonate with *n*-butyl bromide. The high boiling point of diethyl butylmalonate dictates the use of reduced pressure distillation in its isolation. In the experiment, you will learn how to handle sodium metal and how to use a manometer in conjunction with a reduced pressure distillation. Note that potassium iodide is used to catalyze the alkylation reaction. Presumably *n*-butyl iodide is formed in situ (in the solution), and iodide ion is more easily displaced by malonate ion than bromide ion is:

$$I^- + R{-}Br \rightleftharpoons R{-}I + Br^-$$

Note also that, during the alkylation, ethanol is used as the solvent. This is dictated,

$$2C_2H_5OH + 2Na \xrightarrow{C_2H_5OH} 2NaOC_2H_5 + H_2$$

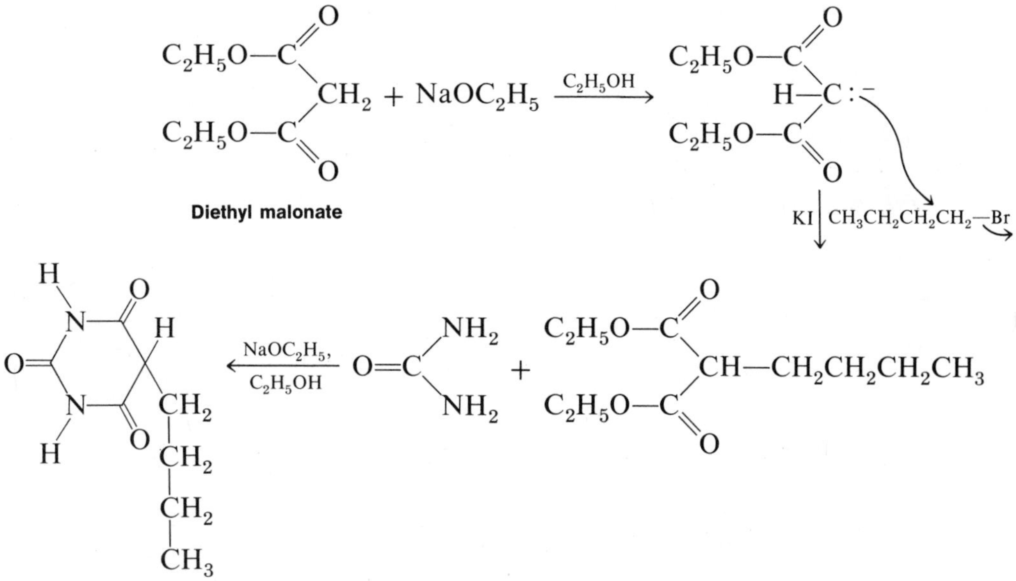

Diethyl malonate

5-*n*-Butylbarbituric acid   **Urea**   **Diethyl *n*-butylmalonate**

since the diethyl ester of malonic acid is being used. Another alcohol would lead to ester interchange and a more complex mixture of products.

The final step is a condensation reaction between urea and diethyl butylmalonate. The product, 5-*n*-butylbarbituric acid, is capable of two tautomeric forms, one having aromatic resonance.

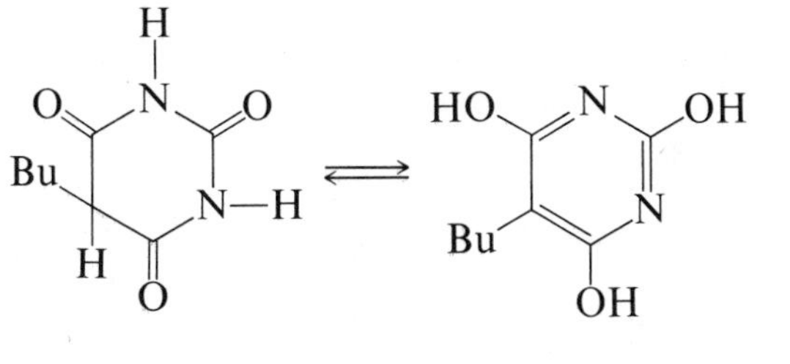

SPECIAL INSTRUCTIONS

Before beginning this experiment you should read the essay that precedes it. It is important to read Techniques 1 through 6 and particularly important to read Technique 14. When sodium reacts with ethanol, hydrogen gas is evolved in large quantities. Be certain that all flames in the laboratory are extinguished before dissolving sodium in ethanol. Alternatively, this operation may be conducted in a hood. The reaction must be carried out under strict anhydrous conditions; the apparatus and the alcohol must be dry. The formation of diethyl *n*-butylmalonate requires at least 45 minutes of heating under reflux. The formation of 5-*n*-butylbarbituric acid requires at least 2 hours of heating under reflux.

PROCEDURE

DIETHYL *n*-BUTYLMALONATE

Working in a hood, add 1.32 g of sodium metal (cut into small pieces; see Technique 14, Section 14.4, p. 647) cautiously and piece by piece to a 250-mL round-bottomed flask fitted with a reflux condenser and containing 25 mL of dry absolute ethanol.[1] Use forceps to handle the sodium metal. It is important that the amount of sodium used be neither more nor less than the amount specified. Be certain that any unused necks of the flask are stoppered and that all ground-glass joints are greased. The heat evolved when the sodium dissolves in the ethanol will cause the ethanol to boil. To keep the reaction from becoming too vigorous, cool the flask in an ice bath if necessary. After the sodium has dissolved and the boiling has subsided, attach a calcium chloride drying tube to the top of the reflux condenser. Heat the mixture on a steam cone if necessary to dissolve any particles of sodium that remain. Add 0.75 g of anhydrous **powdered** KI down the condenser and continue heating the mixture on the steam cone until all the solid has dissolved (or nearly so). Add 8.8 g of dry diethyl malonate. Loosen the clamp holding the apparatus and shake it gently or swirl it to mix the ingredients. Heat the mixture for an additional 5 to 10 minutes and then add 7 g of 1-bromobutane (*n*-butyl bromide) in three equal portions down the condenser; allow the reaction to subside after each portion is added. Heat the mixture for an additional 45 minutes on the steam cone. The product forms as a yellow oil on the top of the mixture.

   Cool the mixture to about room temperature, decant as much liquid as possible away from the inorganic salts that form as a white solid at the bottom of the flask, and save the solid material. Evaporate as much ethanol as possible from the decanted liquid on a steam bath in the hood, using a stream of air directed at the surface of the solution to accelerate the evaporation. Add 20 mL of water and 0.5 mL of concentrated hydrochloric acid to the resulting concentrated yellow liquid and pour the mixture back into the 250-mL flask containing the saved solids. After all the solid has dissolved and the mixture has cooled, add 20 mL of ether, swirl the solution, and transfer it to a separatory funnel. Shake the mixture and then separate the layers. Wash the organic layer once with 10 mL of water, once with 15 mL of 5% sodium bicarbonate solution, and finally with about 15 mL of water. When you are washing the organic layer with water, an emulsion may form; it can be broken with 2 or 3 mL of fresh ether. Dry the ether layer with 1 to 2 g of sodium sulfate until the solution appears clear. This should take about 2 minutes. Gravity-filter the solution and place the filtrate in a 50-mL flask. Remove the ether by distillation on a steam bath in the hood (use a boiling stone). Assemble an apparatus for distillation under reduced pressure, using a manometer if possible (Technique 6, Sections 6.6 and 6.7, pp. 558–563, and Technique 9, Section 9.4, p. 592). Be certain that all ground-glass joints are greased, but avoid excess grease. Make certain that the rubber

---

[1]The ethanol used in this experiment must be anhydrous. A freshly opened bottle of absolute (100%) ethanol works well. Do not use denatured or 95% ethanol.

tubing used has no cracks in it. Good rubber seals for the thermometer and air bleed tube are absolutely necessary to obtain a high vacuum. Collect the fraction boiling from 120 to 138 °C at 21 to 25 mmHg (115–130 °C at 17 mmHg).[2] Weigh the clear, colorless product and calculate the percentage yield. At the option of the instructor, obtain the infrared spectrum of diethyl *n*-butylmalonate as a neat liquid.

5-*n*-BUTYLBARBITURIC ACID

Prepare a sodium ethoxide solution in a 250-mL round-bottomed flask, as in the previous step. Use 1 mole of sodium metal per mole of diethyl *n*-butylmalonate. Dissolve the sodium in absolute ethanol at a ratio of 47 mL of ethanol per gram of sodium used. The last few pieces can be dissolved by placing the flask on a steam bath, taking care not to introduce water vapor in the reaction (use a calcium chloride drying tube). After all the sodium has dissolved, add the diethyl *n*-butylmalonate, followed by a warm (about 70 °C) solution of **dry** urea dissolved in **absolute** ethanol at a ratio of 18 mL of ethanol for each gram of urea.[3] Use 1 mole of urea for each mole of diethyl *n*-butylmalonate. Swirl the mixture well and heat it under reflux for 2 hours, using a heating mantle or an oil bath. This heating period may be conducted over more than one laboratory period. At the beginning of the reflux period, place a Claisen adapter (with the central joint stoppered) on top of the condenser. Attach a calcium chloride drying tube to the side arm of the Claisen adapter, using a thermometer adapter. All the pieces of the apparatus must be clamped firmly so that they will be secure in case of bumping. It is advisable to tape a wire gauze over the top of the drying tube to prevent spillage of the drying agent when the mixture bumps. It is a normal expectation that the reaction mixture will bump occasionally; these precautions avoid loss of product. Once the mixture begins to boil smoothly, careful monitoring of the reaction is no longer necessary.

After the reflux period, add 44 mL of warm (50 °C) water per gram of sodium used. Acidify the resulting solution with 4 mL of concentrated hydrochloric acid per gram of sodium used in the reaction. Concentrate the solution to about 50 mL, using a hot plate (add a boiling stone) and an air stream. Cool the solution in an ice bath for 15 minutes and collect the crude product by vacuum filtration. If a rubbery white solid smelling of diethyl *n*-butylmalonate is formed, wash it with 75 mL of petroleum ether while it is still in the Büchner funnel and set the solution aside. Dry the solid and recrystallize it from boiling water, using a hot plate to heat the water. The recrystallization is conducted using 20 mL of water per gram of diethyl *n*-butylmalonate used initially. Crystallize the product by placing it in an ice-salt bath for 30 minutes, stirring the solution occasionally after the first 20 minutes of cooling. Collect the product by vacuum filtration, using a large Büchner funnel, and air-dry the solid for 20 to 30 minutes. Occasionally turn the crystals with a spatula to expose the wet surfaces. Air-dry the product overnight or in an oven at 105 to 110 °C for 3 hours. Weigh the solid and determine the melting point of the pure 5-*n*-butylbarbituric acid (literature mp 209–210 °C). Calculate the percentage yield for the second step and for the overall reaction sequence.

---

[2]Some unreacted diethyl malonate may be obtained in this reaction. If an appreciable amount of this lower-boiling material is obtained, change receiver flasks when the correct boiling point is reached.

[3]The urea should be dried for about 45 minutes at 105 to 110 °C. The laboratory instructor may prepare a large quantity for use by the entire class.

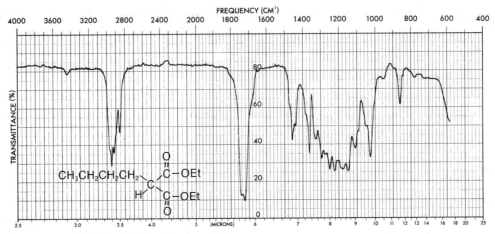

Ir spectrum of diethyl *n*-butylmalonate, neat

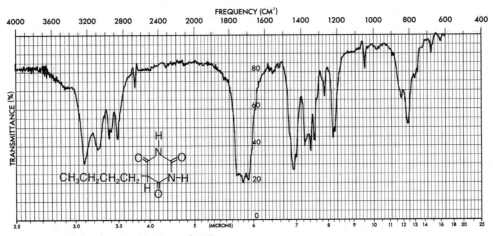

Ir spectrum of 5-*n*-butylbarbituric acid, KBr

Dry the petroleum ether extract obtained earlier, remove the solvent by distillation with a steam bath in the hood, and weigh the remaining liquid. This residue is mostly unreacted diethyl *n*-butylmalonate. Calculate the percentage of starting material recovered. Recalculate the percentage yield for the second step from the amount of reagent actually consumed. At the option of the instructor, determine the infrared spectrum of *n*-butylbarbituric acid as a KBr mull.

QUESTIONS

1. Diethyl malonate has two acidic hydrogens. With this in mind, what by-product would you expect in the first step of this reaction sequence?

2. The presence of water poses a serious problem in both steps of this reaction sequence. Write equations that explain this.

3. After the reaction of urea with diethyl *n*-butylmalonate, the reaction mixture must be neutralized with dilute hydrochloric acid. Why? Give the equations for the reaction.

4. Interpret the principal absorption bands in the infrared spectra of diethyl *n*-butylmalonate and *n*-butyl-barbituric acid.

Experiment 46
Tetraphenylcyclopentadienone

Aldol Condensation

In this experiment tetraphenylcyclopentadienone is prepared by the reaction of dibenzyl ketone (1,3-diphenyl-2-propanone) with benzil (Experiment 40) in the presence of base.

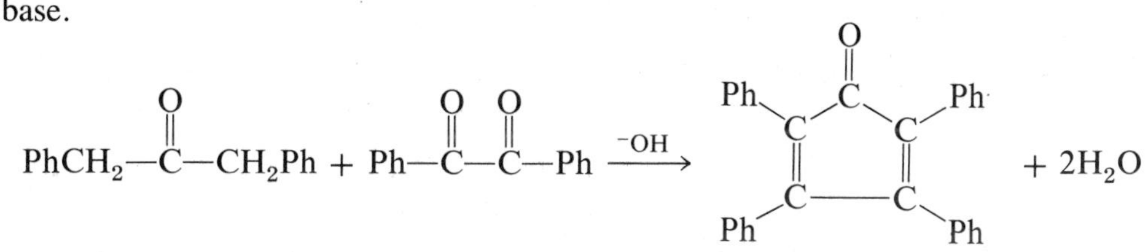

| Dibenzyl ketone | Benzil | Tetraphenylcyclopentadienone |

This reaction proceeds via an aldol condensation reaction, with dehydration giving the purple unsaturated cyclic ketone. A stepwise mechanism for the reaction may proceed as follows:

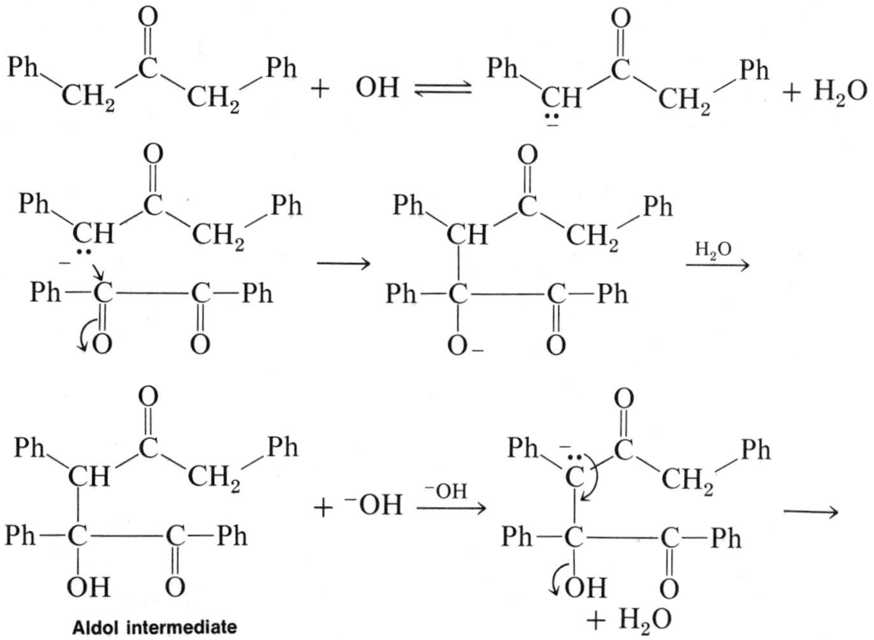

Aldol intermediate

(A)

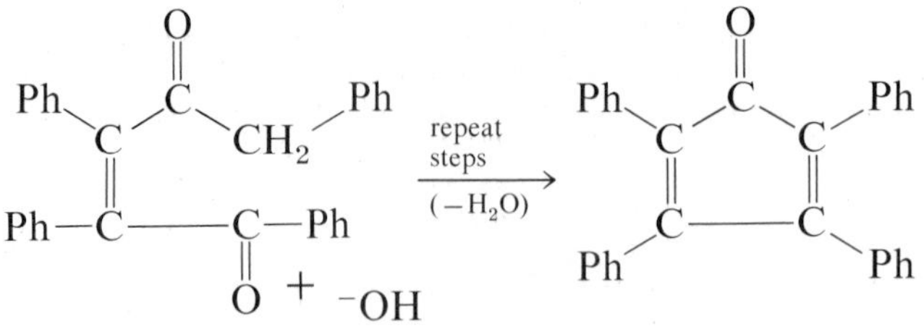

(B)

The aldol intermediate, **A,** readily loses water to give the highly conjugated system, **B,** which reacts further by intramolecular aldol condensation to give the dienone product following dehydration.

SPECIAL INSTRUCTIONS

This reaction can be completed in about 1 hour. The product can then be converted to 1,2,3,4-tetraphenylnaphthalene (Experiment 51).

PROCEDURE

Add 2.1 g of benzil (Experiment 40), 2.1 g of dibenzyl ketone (1,3-diphenyl-2-propanone, 1,3-diphenylacetone), and 15 mL of absolute ethanol to a 100-mL round-bottomed flask. Attach a reflux condenser to the flask. Heat the mixture on a steam bath until the solids dissolve. Also dissolve 0.3 g of potassium hydroxide in 3 mL of absolute ethanol in another container. As the solid dissolves, crush the pieces with a spatula to aid in the solution process.

Raise the temperature of the benzil and dibenzyl ketone solution until it is just below its boiling point and slowly add the solution of potassium hydroxide through the top of the condenser in two portions. The mixture will immediately turn deep purple.

> **CAUTION: Foaming may occur.**

Once the potassium hydroxide has been added, allow the mixture to reflux for 15 minutes. Shake the flask several times during this period. Cool the mixture below 5 °C in an ice bath. Vacuum-filter the deep purple crystals, wash them with three 5-mL portions of cold 95% ethanol, and dry them in an oven for 30 minutes or in air overnight.

The crude product is pure enough (mp 218–220 °C) for the preparation of 1,2,3,4-tetraphenylnaphthalene (Experiment 51). Weigh the product and calculate the percentage yield. Determine the melting point. A small portion may be recrystallized, if desired, from a 1:1 mixture of 95% ethanol and toluene (12 mL/0.5 g; mp 219–220 °C). At the instructor's option, determine the infrared spectrum of tetraphenylcyclopentadienone in potassium bromide. Submit the product to the instructor in a labeled vial or save it for Experiment 51.

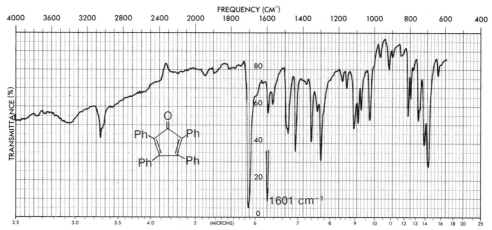

Ir spectrum of tetraphenylcyclopentadienone, KBr

QUESTIONS

1. Interpret the infrared spectrum of tetraphenylcyclopentadienone.
2. Draw the structure of the product you would expect from the reaction of a mixture of benzaldehyde and acetophenone with base.
3. Suggest several possible by-products of this reaction.

Experiment 47
Enamine Reactions:
2-Acetylcyclohexanone

Enamine Reaction
Azeotropic Distillation
Keto-Enol Tautomerism
Infrared, Ultraviolet, and Nmr Spectroscopy

Hydrogens on the α-carbon of ketones, aldehydes, and other carbonyl compounds are weakly acidic and are removed in a basic solution (Equation 1). Although resonance stabilizes the conjugate base (**A**) in such a reaction, the equilibrium is still unfavorable because of the high pK_a (about 20) of a carbonyl compound.

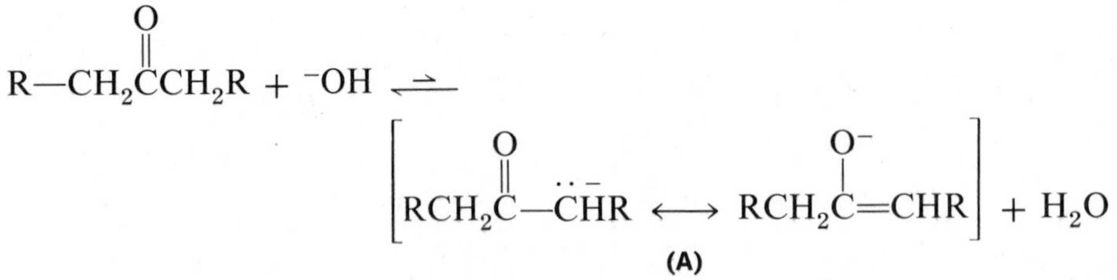

$$R-CH_2\overset{\overset{\displaystyle O}{\|}}{C}CH_2R + {}^-OH \rightleftharpoons$$

$$\left[RCH_2\overset{\overset{\displaystyle O}{\|}}{C}-\overset{..}{\overset{}{C}}\overset{-}{H}R \longleftrightarrow RCH_2\overset{\overset{\displaystyle O^-}{|}}{C}=CHR \right] + H_2O \qquad (1)$$

(A)

Typically, carbonyl compounds are alkylated (Equation 2) or acylated (Equation 4) only with difficulty in the presence of aqueous sodium hydroxide because of more important secondary side reactions (Equation 3, 5, and 6). In effect, the concentration of the nucleophilic conjugate base species (**A** in Equation 1) is low because of the unfavorable equilibrium (Equation 1), while the concentration of the competing nucleophile (OH$^-$) is very high. A significant side reaction occurs when hydroxide ion reacts with an alkyl halide by Equation 3 or acyl halide by Equation 5. In addition, the conjugate base can react with unreacted carbonyl compound by an aldol condensation reaction (Equation 6). Enamine reactions, described below, avoid many of the problems described here.

ALKYLATION

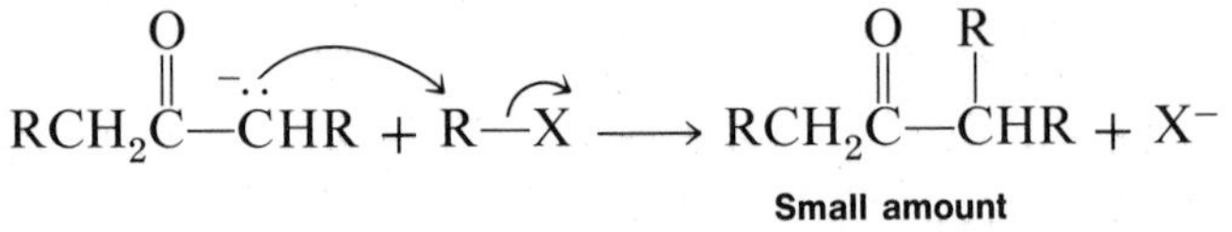

$$RCH_2\overset{\overset{\displaystyle O}{\|}}{C}-\overset{..}{\overset{}{C}}HR + R-X \longrightarrow RCH_2\overset{\overset{\displaystyle O}{\|}}{C}-\overset{\overset{\displaystyle R}{|}}{C}HR + X^- \qquad (2)$$

Small amount

$$HO^- + R-X \longrightarrow ROH + X^- \qquad \text{COMPETING REACTION} \qquad (3)$$

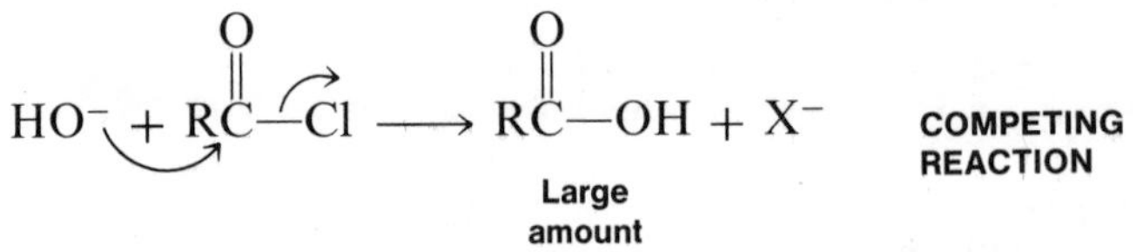

Large amount

ACYLATION

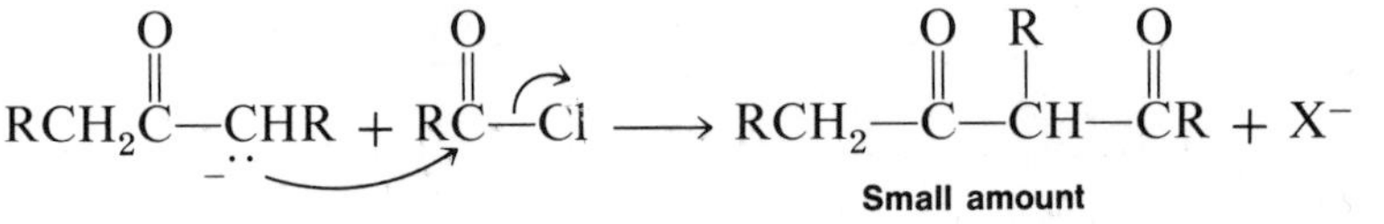

$$RCH_2\overset{\overset{\displaystyle O}{\|}}{C}-CHR + R\overset{\overset{\displaystyle O}{\|}}{C}-Cl \longrightarrow RCH_2-\overset{\overset{\displaystyle O}{\|}}{C}-\overset{\overset{\displaystyle R}{|}}{C}H-\overset{\overset{\displaystyle O}{\|}}{C}R + X^- \qquad (4)$$

Small amount

$$HO^- + R\overset{\overset{\displaystyle O}{\|}}{C}-Cl \longrightarrow R\overset{\overset{\displaystyle O}{\|}}{C}-OH + X^- \qquad \text{COMPETING REACTION} \qquad (5)$$

Large amount

ALDOL CONDENSATION

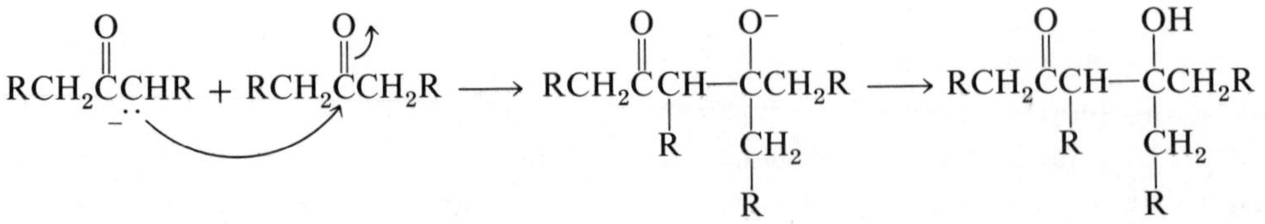

$$RCH_2\overset{\overset{\displaystyle O}{\|}}{C}CHR + RCH_2\overset{\overset{\displaystyle O}{\|}}{C}CH_2R \longrightarrow RCH_2\overset{\overset{\displaystyle O}{\|}}{C}CH-\underset{\underset{\displaystyle R}{|}}{\overset{\overset{\displaystyle O^-}{|}}{C}}CH_2R \longrightarrow RCH_2\overset{\overset{\displaystyle O}{\|}}{C}CH-\underset{\underset{\displaystyle R}{|}}{\overset{\overset{\displaystyle OH}{|}}{C}}CH_2R \qquad (6)$$

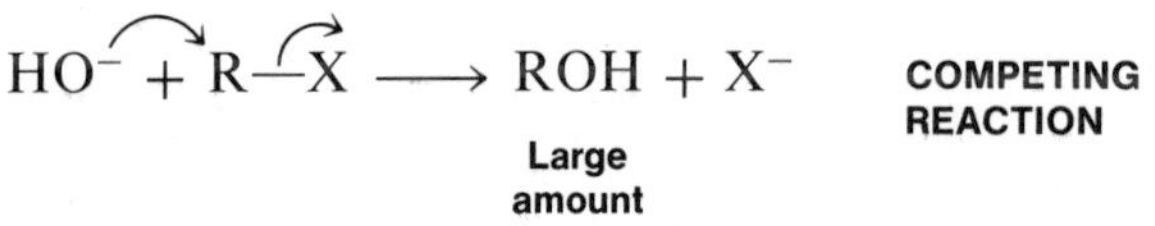

FORMATION AND REACTIVITY OF ENAMINES

Enamines are prepared easily from carbonyl compounds (for example, cyclohexanone) and a secondary amine (for example, pyrrolidine) by an acid-catalyzed addition-elimination reaction. Water, the other product of the reaction, is removed by azeotropic distillation with toluene (Technique 7, Section 7.8, p. 578), which drives the equilibrium to the right:

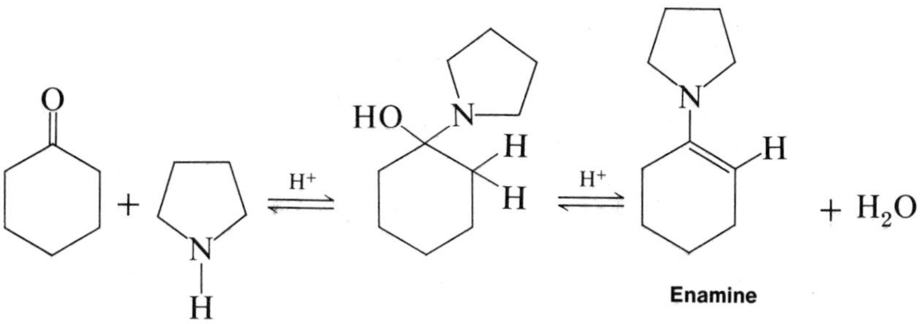

If the water were not removed, the equilibrium would be unfavorable, and only a small amount of enamine would be produced. Azeotropic distillation of water is an important "trick" used in organic chemistry to produce desired products in spite of an unfavorable equilibrium.

An enamine has the desirable property of being nucleophilic (carbon alkylation

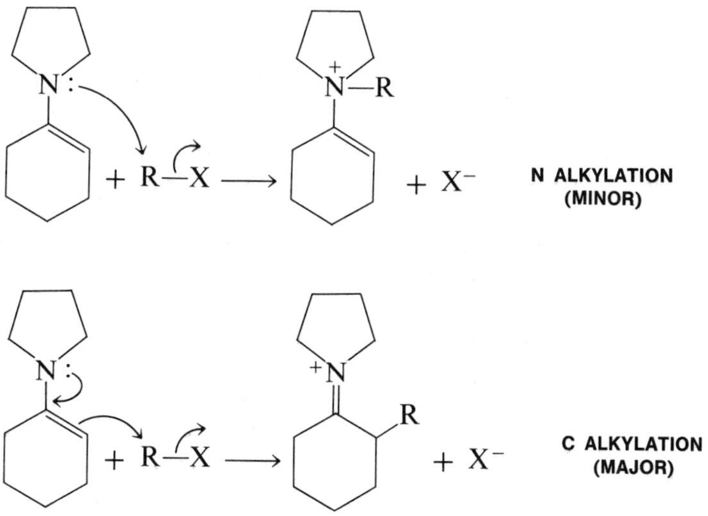

is more important than nitrogen alkylation) and is easily alkylated. The **key point** is that the resonance hybrid (**B**) is like the resonance hybrid (**A**) shown in Equation 1. However, **B** has been produced under nearly neutral conditions so that it is the *only* nucleophile present.

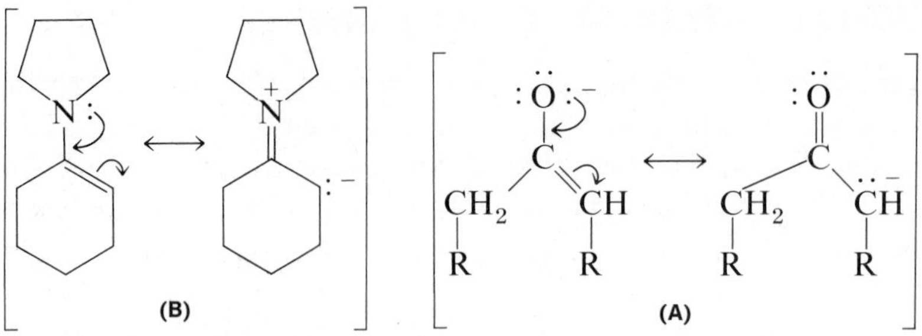

Contrast this situation to the one in Equation 1, where hydroxide ion, present in large amount, produces undesirable side reactions (Equations 3 and 5).

The alkylation step is followed by removal of the secondary amine by an acid-catalyzed hydrolysis:

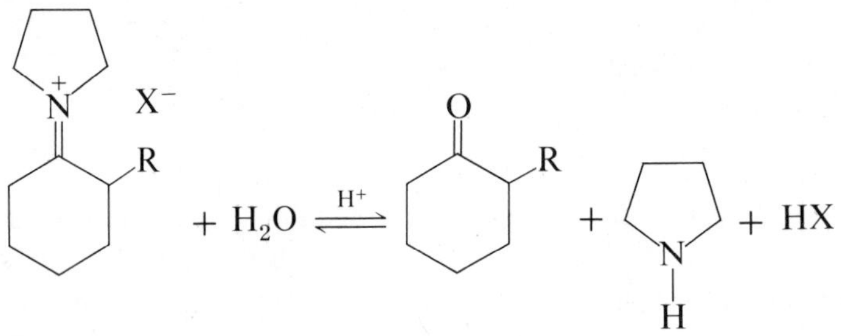

EXAMPLES OF ENAMINE REACTIONS

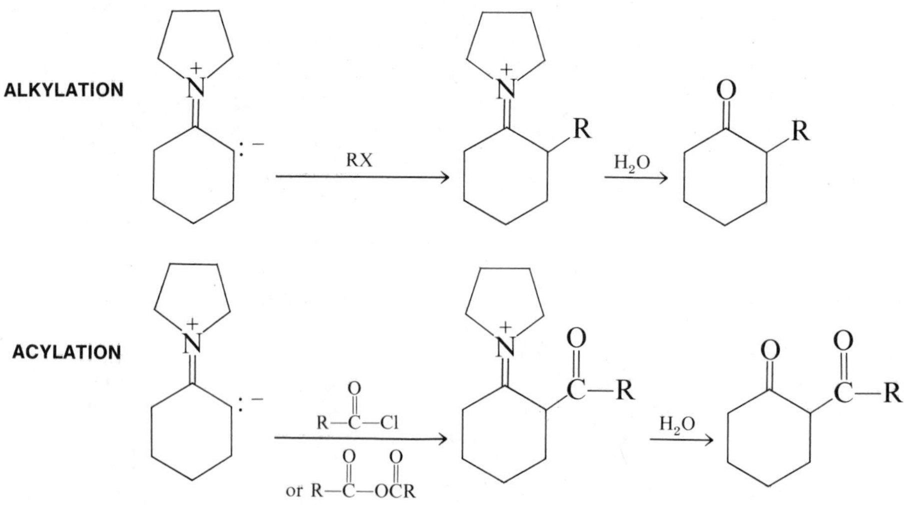

CONJUGATE ADDITION (MICHAEL REACTION)

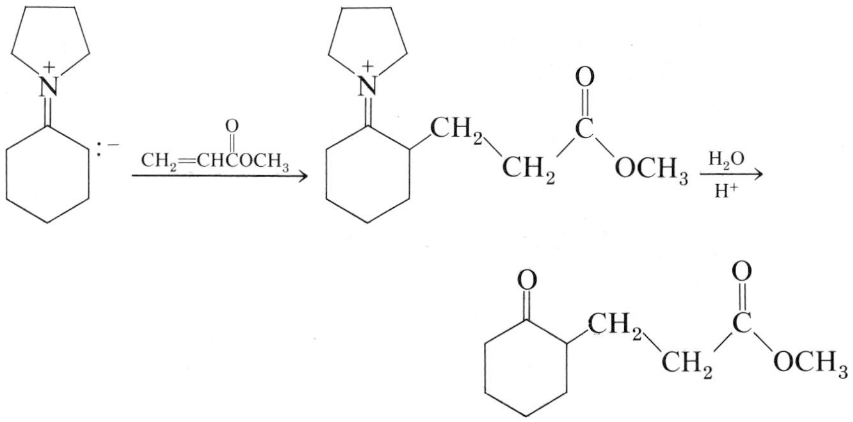

ROBINSON ANNELATION (RING-FORMATION) REACTION

Reactions that combine the Michael addition reaction and aldol condensation to form a six-membered ring fused on another ring are well known in the steroid field. These reactions are known as **Robinson annelation reactions.** An example is the formation of $\Delta^{1,9}$-2-octalone.

MICHAEL ADDITION (CONJUGATE ADDITION)

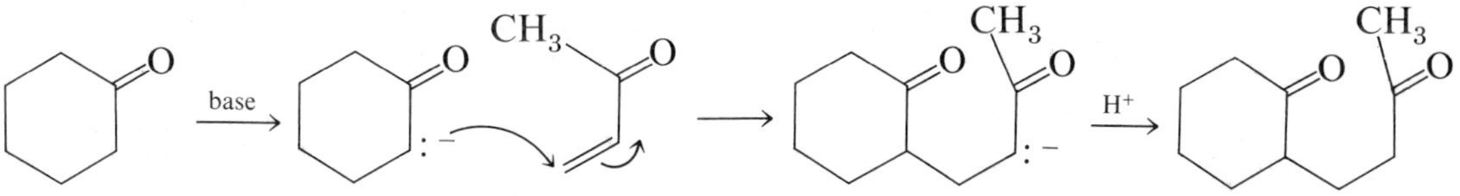

ALDOL CONDENSATION

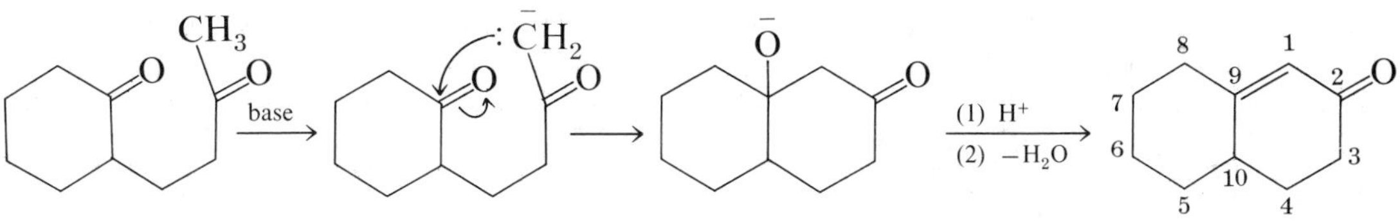

$\Delta^{1,9}$-2-Octalone

Robinson annelation reactions can also be conducted by enamine chemistry. One advantage of enamines is that the unsaturated ketones are not easily polymerized under the mild conditions of this reaction. Base-catalyzed reactions often give large amounts of polymer.

THE EXPERIMENT

In this experiment, pyrrolidine reacts with cyclohexanone to give the enamine. This enamine is used to prepare 2-acetylcyclohexanone.

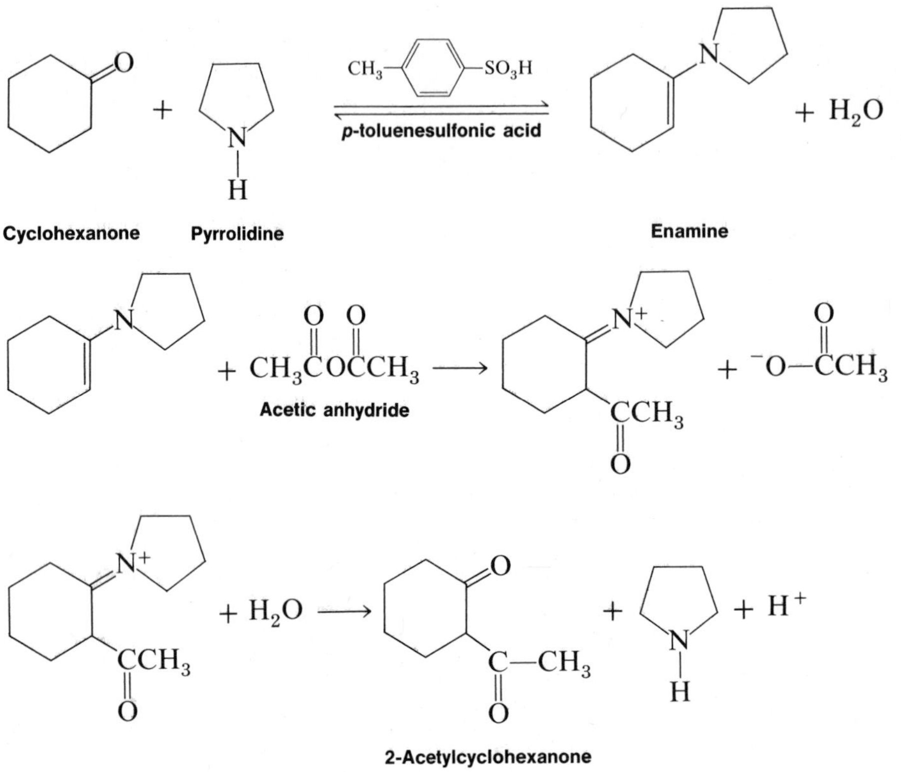

Cyclohexanone   **Pyrrolidine**   **Enamine**

Acetic anhydride

2-Acetylcyclohexanone

SPECIAL INSTRUCTIONS

Before beginning this experiment, you should read Techniques 6 and 9. Pyrrolidine and acetic anhydride are toxic and noxious. You should keep these substances in a good hood whenever possible. If you are not careful, the entire room will be filled with vapors of pyrrolidine, and it will not be pleasant to work in the laboratory.

The enamine should be made during the first part of the laboratory period and used as soon as possible. Once the acetic anhydride has been added, the reaction mixture must be allowed to stand in your locker for at least 24 hours. Enough time should be allowed for you to get to this point by the end of the first period. The second period is used for the work-up and vacuum distillation. The yields in these reactions are low (less than 20%), partly due to reduced reaction periods necessary to fit the experiment into convenient 3-hour laboratory periods.

PREPARATION OF ENAMINE

Place 12.8 g of cyclohexanone in a 250-mL round-bottomed flask. Add 11 mL (density 0.852 g/mL) of pyrrolidine **(in the hood)** and 0.2 g of *p*-toluenesulfonic acid monohy-

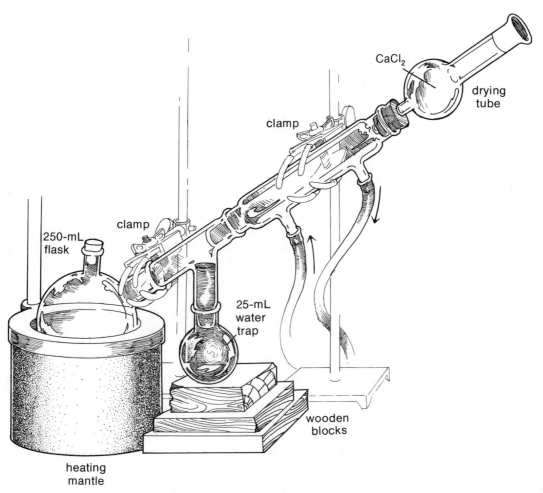

Apparatus for water separation in the enamine synthesis

drate to the flask. Add a boiling stone and assemble the apparatus as shown in the figure, being certain to support the water trap firmly to avoid leakage. Once the apparatus is assembled, pour 60 mL of toluene through the top of the condenser so that the water trap is filled completely and the excess passes into the 250-mL round-bottomed flask. Place a drying tube on the condenser. Heat the mixture under reflux for 30 minutes. The condensation ring should extend for a third to a half of the length of the condenser. Adjust the reflux rate in this way to aid in the rapid removal of the water-toluene azeotrope. Water will collect in the bottom of the water trap as the reaction proceeds.

   At the end of the 30-minute reflux period, allow the reaction mixture to cool somewhat, remove the water separator (trap), and reassemble the apparatus for simple distillation (Technique 6, Figure 6–6, p. 557). Do not allow the noxious vapors of pyrrolidine to be released into the room; cool the receiving flask in an ice bath. Distill the mixture until the temperature reaches 108 to 110 °C (boiling point of toluene). At this point most of the remaining pyrrolidine and water have been removed. Stop the distillation. The enamine and toluene solvent remain in the flask. The enamine is used without further purification and should be used without delay. Pour the toluene solution of the enamine into a 125-mL Erlenmeyer flask and allow the mixture to cool to room temperature. The enamine will be used to prepare 2-acetylcyclohexanone. Pour the distillate, containing toluene and water, into a suitable waste container.

2-ACETYLCYCLOHEXANONE

Add a solution of 13.3 g of acetic anhydride dissolved in 20 mL of toluene to the enamine solution contained in the 125-mL Erlenmeyer flask and swirl the mixture. Stopper the flask and allow the mixture to stand for at least 24 hours.

   Following this reaction period, transfer the mixture to a round-bottomed flask, add 15 mL of water, and heat the mixture under reflux for 30 minutes. When the solution has cooled to room temperature, pour it into a separatory funnel, and wash the organic layer with 25 mL of water. After draining the aqueous layer, reextract the organic layer with three successive 25-mL portions of 3M hydrochloric acid to remove any nitrogen-containing contaminants; finally, wash the organic layer with 25 mL of water. Dry the organic phase over 4 to 5 g of anhydrous magnesium sulfate. Set up a simple distillation apparatus and remove most of the toluene (bp 110 °C) by distillation at atmospheric pressure. When it appears that most of the toluene has been removed, transfer the crude 2-acetylcyclohexanone to a small flask with the aid of a small amount of toluene. Vacuum-distill the material according to the procedure in Technique 6, Sections 6.6 and 6.7, pp. 558–563, using the apparatus shown in Figure 6–7, p. 559. Use a trap and manometer as shown in Technique 9, Figure 9–6, p. 593. The first fraction contains mainly toluene, and the second fraction, distilling at 107 to 114 °C (14-mmHg pressure), contains the pure 2-acetylcyclohexanone. At the option of the instructor, obtain the infrared spectrum or the nmr spectrum or both. The nmr spectrum may be used to

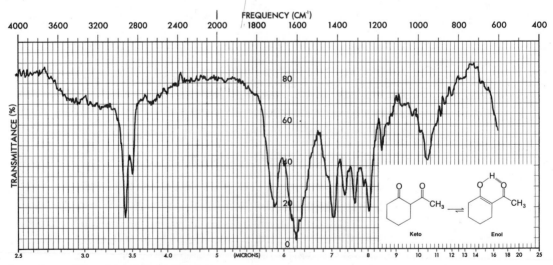

Ir spectrum of 2-acetylcyclohexanone, neat

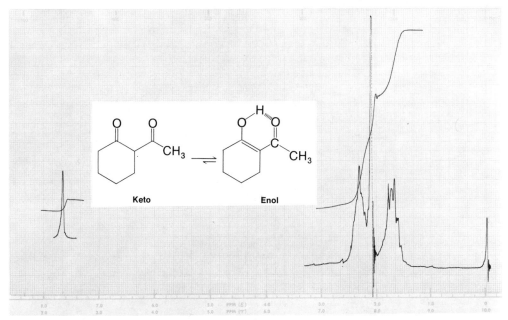

Nmr spectrum of 2-acetylcyclohexanone, CDCl₃, offset peak, 500 Hz

determine the percentage of enol content for 2-acetylcyclohexanone. This compound is **highly** enolic, giving a calculated value in excess of 70%.[1] Calculate the percentage yield and submit the product in a labeled vial with your laboratory report.

REFERENCES

Augustine, R. L., and Caputo, J. A. "$\Delta^{1,9}$-2-Octalone." *Organic Syntheses*, Coll. Vol. 5 (1973): 869.

Cook, A. G., ed. *Enamines: Synthesis, Structure, and Reactions*. New York: Marcel Dekker, 1969.

Dyke, S. F. *The Chemistry of Enamines*. London: Cambridge Univ. Press, 1973.

Mundy, B. P. "The Synthesis of Fused Cycloalkenones via Annelation Methods." *Journal of Chemical Education, 50* (1973): 110.

Pavia, D. L., Lampman, G. M., and Kriz, G. S. *Introduction to Spectroscopy*. Philadelphia: W. B. Saunders, 1979. Chap. 5, Pp. 201–205.

Stork, G., Brizzolara, A., Landesman, H., Szmuszkovicz, J., and Terrell, R. "The Enamine Alkylation and Acylation of Carbonyl Compounds." *Journal of the American Chemical Society, 85* (1963): 207.

---

[1]The percentage enol content can be calculated as follows, using the nmr spectrum reproduced above as an example. The offset peak is assigned to the enolic hydrogen (integral height, 6 mm). The remaining absorptions at 1.5 to 2.85 δ (integral height, 92 mm) are assigned to the 11 protons remaining in the enol structure and the 12 protons in the keto structure. Thus, 66 mm (6 × 11) of the 92 mm integral height is assigned to the enol hydrogens. Enol % = ⁶⁶/₉₂ = 72; keto % = ²⁶/₉₂ = 28. Further, the data can be refined by taking into account the fact that there are 11 hydrogens in the enol form and 12 in the keto form that give rise to the 1.5 to 2.8 peak. This correction gives 73% enol and 27% keto forms.

QUESTIONS

1. Draw a mechanism for the enamine synthesis of $\Delta^{1,9}$-2-octalone. Why is this octalone rather than the $\Delta^{9,10}$-2-octalone the main product in the reaction? On the other hand, why is there a relatively large amount of the $\Delta^{9,10}$-2-octalone produced in the reaction?

2. (a) The enamine formed from pyrrolidine and 2-methylcyclohexanone has the **A** structure. What reason can you give for the less substituted enamine being formed instead of the more substituted enamine, **B**? (*Hint:* Consider steric effects.)

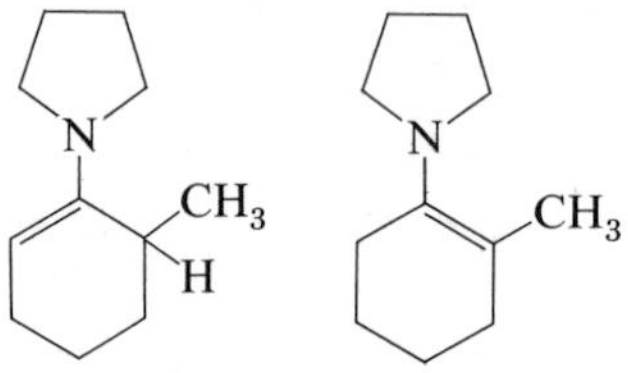

(A)   (B)

   (b) Draw the structure of the product that would result from the reaction of enamine **A** with methyl vinyl ketone. Compare its structure with the product obtained in Question 3.

3. (a) The enolate formed from 2-methylcyclohexanone has the structure shown below. What is the structure of the other possible enolate, and why is it not as stable as the one shown here?

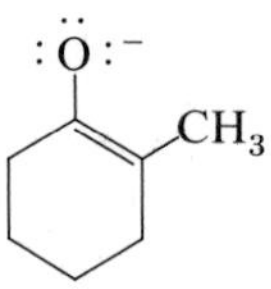

   (b) Draw the structure of the product that would result from the reaction with methyl vinyl ketone. Compare its structure with the product obtained in Question 2.

4. Draw the structures of the Robinson annelation products that would result from the following reactions.

(a) 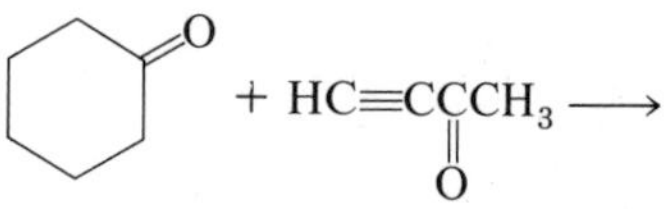 $+ \ HC\equiv CCCH_3 \longrightarrow$

(b) 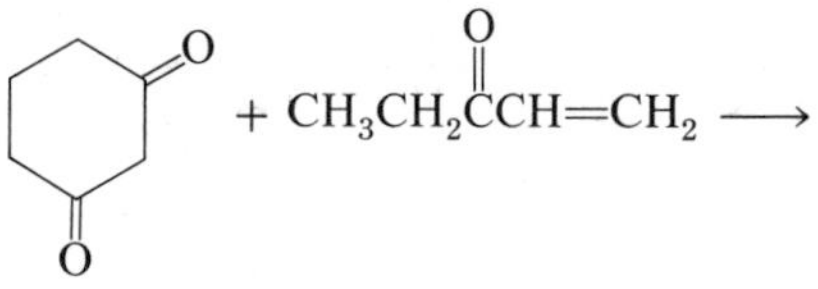 $+ \ CH_3CH_2CCH=CH_2 \longrightarrow$

5. Draw the structures of the products that would result from the following enamine reactions. Use pyrrolidine as the amine, and write equations for the reaction sequence.

(a) 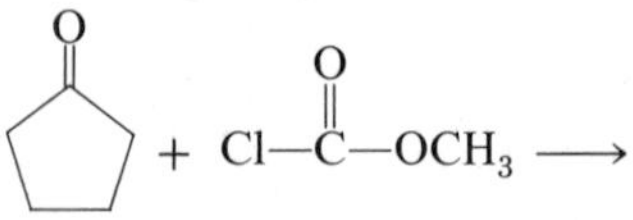 $+ \ Cl-C-OCH_3 \longrightarrow$

(b) 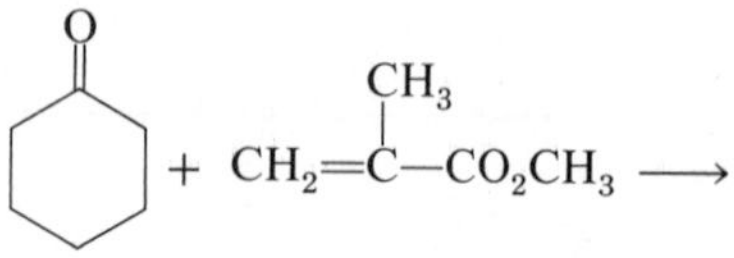 $+ \ CH_2=C-CO_2CH_3 \longrightarrow$

(c) 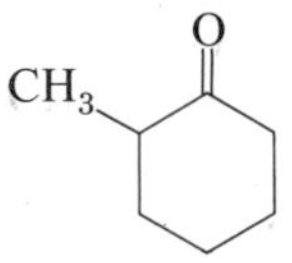 + $CH_3I \longrightarrow$ (see Question 2)

(d) 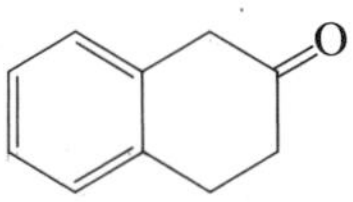 + $CH_2{=}CH{-}CH_2{-}Br \longrightarrow$

6. Interpret the infrared spectrum of 2-acetylcyclohexanone, especially in the O—H and C=O stretch regions of the spectrum.

7. Calculate the amount of water produced during the formation of the enamine in this experiment.

8. Write equations showing how one could carry out the following multistep transformation, starting from the indicated materials. One need not use an enamine synthesis.

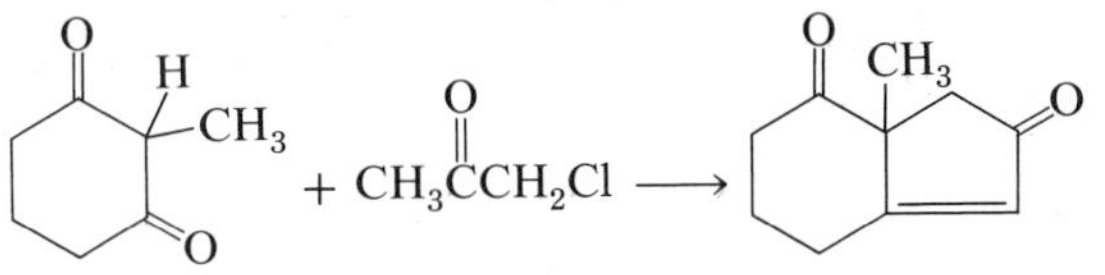

+ $CH_3\overset{O}{\overset{\|}{C}}CH_2Cl \longrightarrow$

Experiment 48
The Aldol Condensation Reaction: Preparation of Benzalacetone, Benzalacetophenone, and Benzalpinacolone

Aldol Condensation
Crystallization

Benzaldehyde reacts with a ketone in the presence of base to give α,β unsaturated ketones. This reaction is an example of a crossed aldol condensation in which the intermediate dehydrates to produce the resonance-stabilized unsaturated ketone.

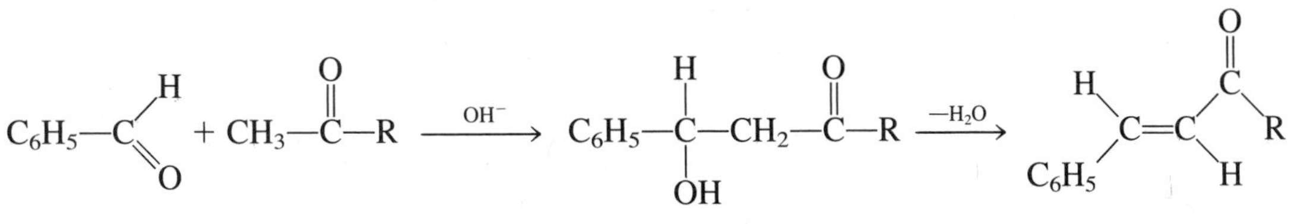

Intermediate

Crossed aldol condensations of this type proceed in high yield since benzaldehyde cannot react with itself by an aldol condensation reaction because it has no α hydrogen. Likewise, ketones do not react easily with themselves in aqueous base. Therefore, the only possibility is for a ketone to react with benzaldehyde.

In this experiment, procedures are given for the preparation of benzalacetones, benzalacetophenones, and benzalpinacolones. You should choose one of the substituted benzaldehydes and react it with one of the ketones: acetone, acetophenone, or pinacolone. All of the products are solids that can be easily recrystallized.

BENZALACETONES

Benzalacetones are prepared by the reaction of a substituted benzaldehyde with excess acetone in aqueous base. Piperonaldehyde and anisaldehyde are used.

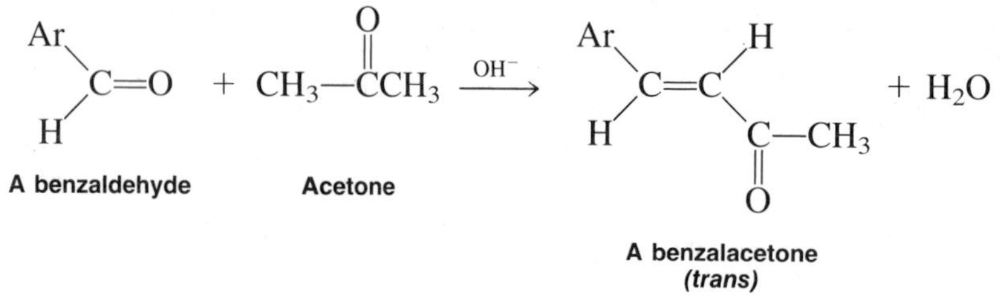

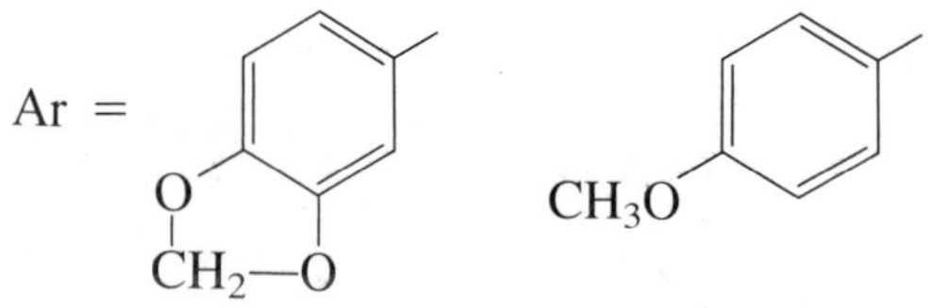

BENZALACETOPHENONES

Benzalacetophenones (chalcones) are prepared by the reaction of a substituted benzaldehyde with acetophenone in aqueous base. Piperonaldehyde, anisaldehyde, and 3-nitrobenzaldehyde are used.

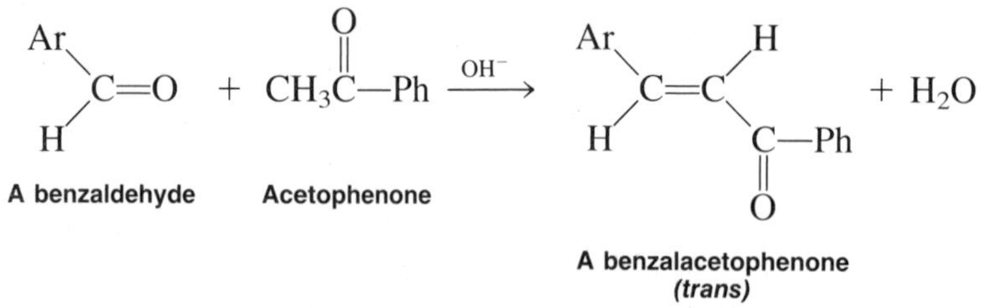

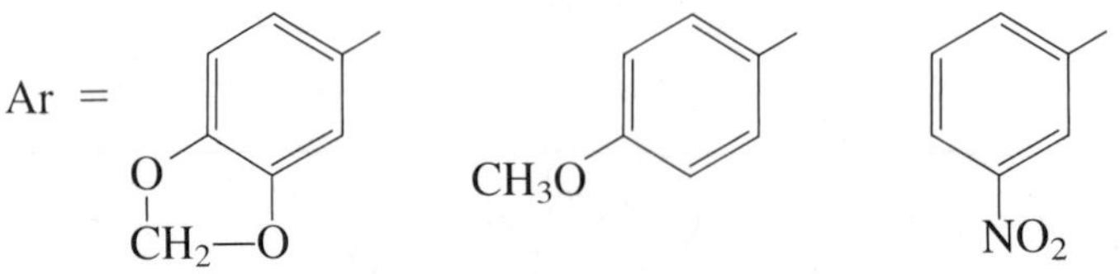

BENZALPINACOLONES

Benzalpinacolones are prepared by the reaction of a substituted benzaldehyde with pinacolone in aqueous base. Piperonaldehyde and 3-nitrobenzaldehyde are used.

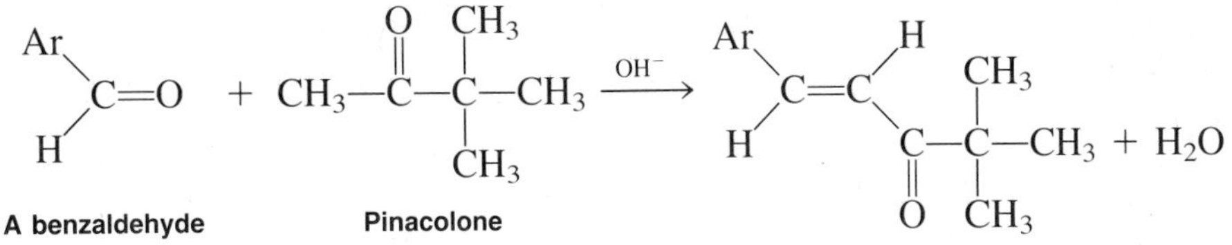

SPECIAL INSTRUCTIONS

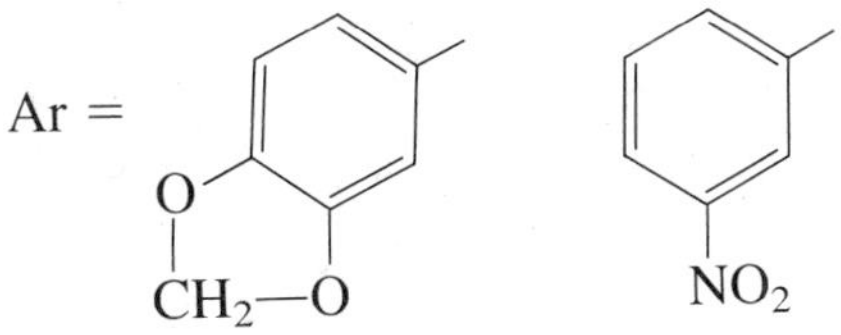

SPECIAL INSTRUCTIONS

Before beginning this experiment, you should select one of the three procedures and a substituted benzaldehyde. Alternatively, your instructor may assign a particular compound to you. In either case, you will need to calculate the molecular weights for the starting benzaldehyde and ketone.

PROCEDURES

BENZALACETONES

Dissolve 0.01 mole of piperonaldehyde (3,4-methylenedioxybenzaldehyde) or anisaldehyde (4-methoxybenzaldehyde) in 10 mL of acetone in a 50-mL Erlenmeyer flask. In a separate 50-mL Erlenmeyer flask, dissolve 0.6 g of sodium hydroxide in 1 mL of water. It may be necessary to warm these mixtures on a steam bath to dissolve the solids. If this is necessary, then the solutions should be cooled to room temperature before proceeding with the next step.

Combine the contents of the two flasks by adding the aldehyde/acetone mixture to the alkaline solution. After the addition, swirl the mixture for several minutes and allow the reaction to proceed for 1.5 hours. Swirl the mixture every 10 to 15 minutes during this period.

When the reaction is complete, place the flask in an ice bath for 3 to 5 minutes. Add 25 mL of ice cold water to the mixture and cool it for about 5 minutes. The precipitate should be broken into small pieces with a stirring rod. Collect the product on a small Büchner funnel and wash it with four 25-mL portions of water. Recrystallize the compound from a minimum amount of hot 95% ethanol in the usual way. Allow the solid to dry completely and weigh the compound.

Determine the percentage yield, the melting point, and the infrared spectrum as a potassium bromide pellet. The literature melting points for the benzalacetones formed from piperonaldehyde and anisaldehyde are 108 and 74 °C, respectively. Include a balanced equation for the reaction in your report and submit the sample to the instructor in a labeled vial. At the instructor's option, obtain the proton and/or carbon nmr spectrum.

BENZALACETOPHENONES

In a 50-mL Erlenmeyer flask, dissolve 0.01 mole of piperonaldehyde (3,4-methylenedioxybenzaldehyde) or anisaldehyde (4-methoxybenzaldehyde) or 3-nitrobenzaldehyde and an equimolar amount of acetophenone in 4 mL of 95% ethanol. To this solution, add 2 mL of 10% aqueous sodium hydroxide. Swirl this mixture for several minutes and allow the reaction to proceed for 30 minutes. Swirl the mixture every 5 minutes during this period.

When the reaction is complete, place the flask in an ice bath for 3 to 5 minutes. Add 25 mL of ice cold water to the mixture and cool it for about 5 minutes. The precipitate should be broken into small pieces with a stirring rod. Collect the product on a small Büchner funnel and wash it with four 25-mL portions of water. Recrystallize the compound from a minimum amount of hot 95% ethanol in the usual way. Allow the solid to dry completely and weigh the compound.

Determine the percentage yield, the melting point, and the infrared spectrum as a potassium bromide pellet. The literature melting points for the benzalacetophenones (chalcones) formed from piperonaldehyde, anisaldehyde, and 3-nitrobenzaldehyde are 122, 77, and 148 °C, respectively. Include a balanced equation for the reaction in your report and submit the sample to the instructor in a labeled vial. At the instructor's option, obtain the proton and/or carbon nmr spectrum.

BENZALPINACOLONES

In a 50-mL Erlenmeyer flask, dissolve 0.01 mole of piperonaldehyde (3,4-methylenedioxybenzaldehyde) or 3-nitrobenzaldehyde and an equimolar amount of pinacolone (3,3-dimethyl-2-butanone) in 8 mL of 95% ethanol. In a separate 50-mL Erlenmeyer flask, dissolve 0.6 g of sodium hydroxide in 1 mL of water. It may be necessary to warm these mixtures on a steam bath to dissolve the solids. If this is necessary, then the solutions should be cooled to room temperature before proceeding with the next step.

Combine the contents of the two flasks by adding the aldehyde/pinacolone mixture to the alkaline solution. After the addition, swirl the mixture for several minutes and allow the reaction to proceed for 1.5 hours. Swirl the mixture every 10 to 15 minutes during this period.

When the reaction is complete, place the flask in an ice bath for 3 to 5 minutes. Add 25 mL of ice cold water to the mixture and cool it for about 5 minutes. The precipitate should be broken into small pieces with a stirring rod. Collect the product on a small Büchner funnel and wash it with two 25-mL portions of water, followed by two 10-mL portions of cold 95% ethanol, and then two 25-mL portions of cold water. Recrystallize the compound from a minimum amount of hot 95% ethanol in the usual way. Allow the solid to dry completely and weigh the compound.

Determine the percentage yield, the melting point, and the infrared spectrum as a potassium bromide pellet. The literature melting points for the benzalpinacolones formed from piperonaldehyde and 3-nitrobenzaldehyde are 96 and 94 °C, respectively. Include a balanced equation for the reaction in your report and submit the sample to the instructor in a labeled vial. At the instructor's option, obtain the proton and/or carbon nmr spectrum.

QUESTIONS

1. Give a mechanism for the preparation of the appropriate benzalacetone, benzalacetophenone, or benzalpinacolone using the aldehyde and ketone that you selected in this experiment.

2. Draw the structure of the *cis* and *trans* isomers of the compound that you prepared. Why did you obtain the *trans* isomer?

3. Using proton nmr, how could you experimentally determine that you have the *trans* isomer rather than the *cis* one? (*Hint:* Consider the use of coupling constants for the vinyl hydrogens.)

4. When the amount of acetone is decreased significantly, the benzalacetone becomes contaminated with a side-product that has consumed 2 moles of aromatic aldehyde. What is its structure and why is it produced when the amount of acetone is decreased? Would you expect that the benzalacetophenone preparation should have this problem? Why?

5. Provide the starting materials needed to prepare the following compounds:

a) 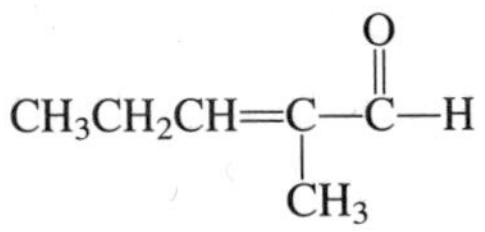 $CH_3CH_2CH{=}\overset{\underset{\textstyle CH_3}{|}}{C}{-}\overset{\overset{\textstyle O}{\|}}{C}{-}H$

b) $\overset{\textstyle CH_3}{\underset{\textstyle CH_3}{{>}}}C{=}CH\overset{\overset{\textstyle O}{\|}}{C}{-}CH_3$

c) $\overset{\textstyle Ph}{\underset{\textstyle CH_3}{{>}}}C{=}CH{-}\overset{\overset{\textstyle O}{\|}}{C}{-}Ph$

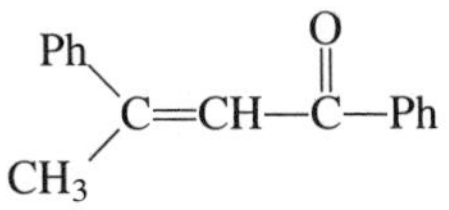

d) 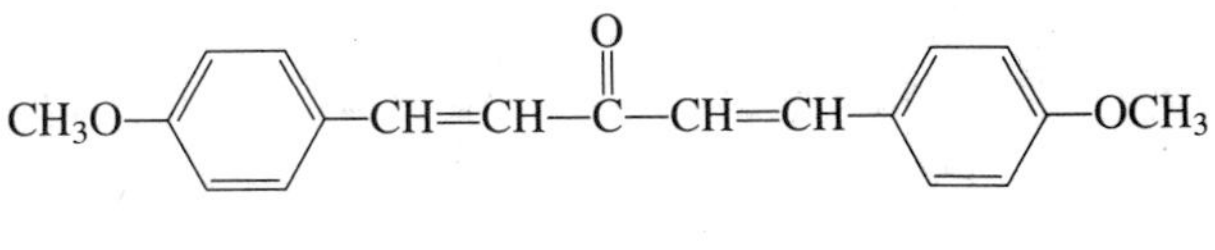 $CH_3O{-}\!\!\bigcirc\!\!{-}CH{=}CH{-}\overset{\overset{\textstyle O}{\|}}{C}{-}CH{=}CH{-}\!\!\bigcirc\!\!{-}OCH_3$

e) 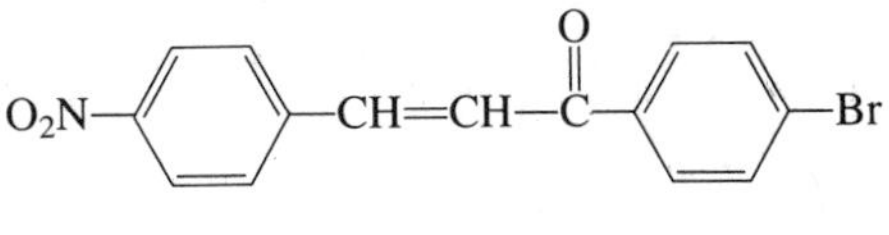 $O_2N{-}\!\!\bigcirc\!\!{-}CH{=}CH{-}\overset{\overset{\textstyle O}{\|}}{C}{-}\!\!\bigcirc\!\!{-}Br$

f) 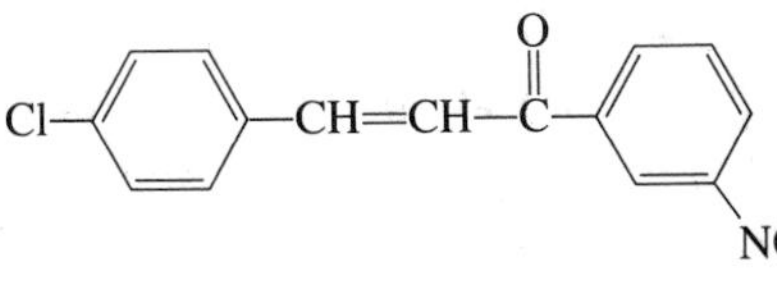 $Cl{-}\!\!\bigcirc\!\!{-}CH{=}CH{-}\overset{\overset{\textstyle O}{\|}}{C}{-}\!\!\underset{\underset{\textstyle NO_2}{|}}{\bigcirc}$

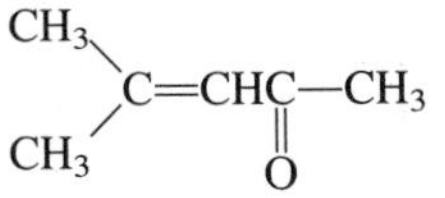

6. Prepare the following compounds starting from benzaldehyde and the appropriate ketone. Provide reactions for preparing the ketones starting from aromatic hydrocarbon compounds (see Experiment 33).

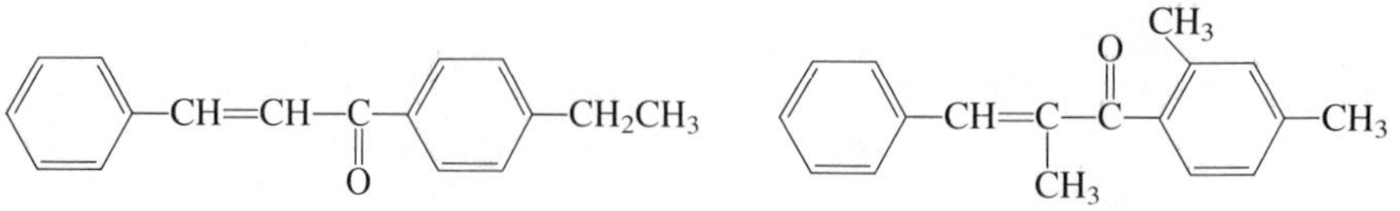

7. Interpret the infrared spectrum of the product that you obtained in this experiment.

Experiment 49
5,5-Dimethyl-1,3-cyclohexanedione (Dimedone)

Michael Reaction (conjugate addition)
Claisen Condensation (ketone + ester)
Decarboxylation Reaction
Handling Sodium Metal
Keto-Enol Tautomerism
Derivative Formation

In this experiment, you prepare 5,5-dimethyl-1,3-cyclohexanedione (dimedone) by a series of important organic reactions: Michael addition reaction, Claisen-type reaction of a ketone with an ester, and decarboxylation. In addition, you will learn how to handle sodium metal.

The first step involves the preparation of the base, sodium ethoxide. This strong base is used to deprotonate diethyl malonate to give the sodium salt that is used in the Michael addition reaction.

$$2\ CH_3CH_2\ddot{O}H + 2\ Na° \longrightarrow 2\ CH_3CH_2\ddot{O}:^-\ Na^+ + H_2$$

Sodium ethoxide

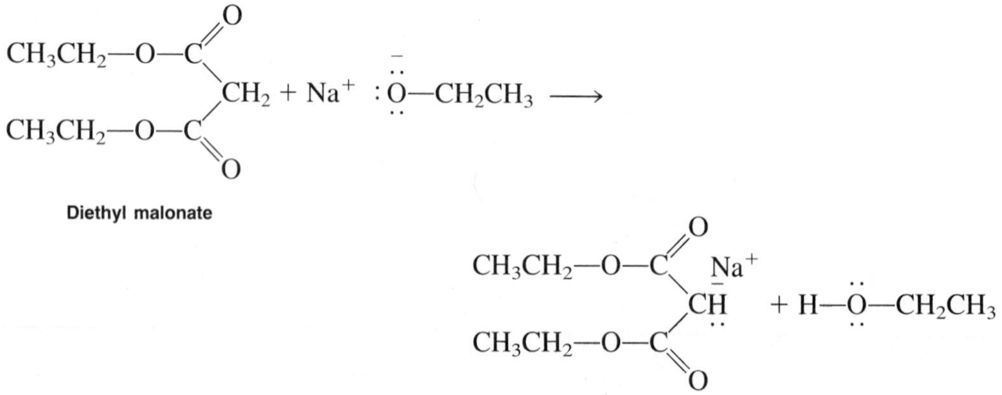

Diethyl malonate

Sodium salt

The sodium salt of diethylmalonate nucleophilically attacks the β position of the α,β unsaturated ketone, 4-methyl-3-penten-2-one (mesityl oxide). This reaction is an example of a Michael reaction or a conjugate addition reaction. The carbonyl group in the unsaturated ketone stabilizes the intermediate conjugate base by resonance. Nucleophilic addition to a C=C double bond cannot proceed without the carbonyl group.

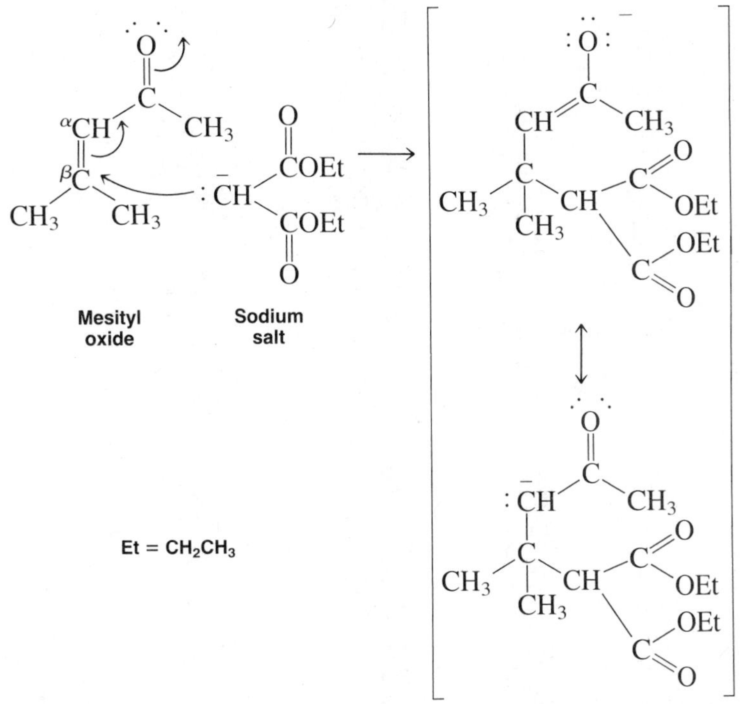

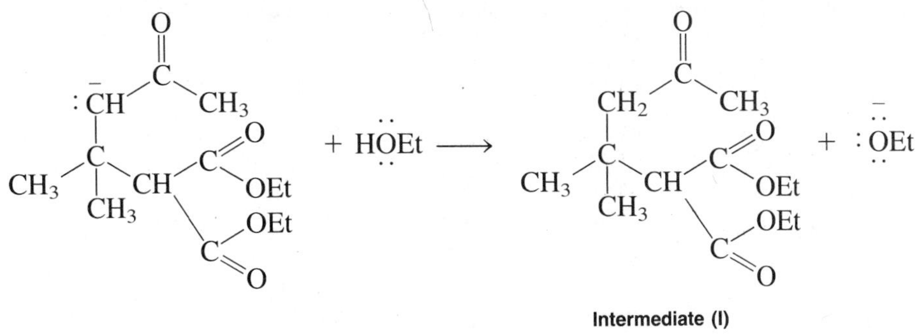

The intermediate (I) is now cyclized by a Claisen-type condensation reaction. Sodium ethoxide is again used as a base to generate a nucleophile. Intramolecular attack by this nucleophile on the ester carbonyl group gives a six-membered ring compound.

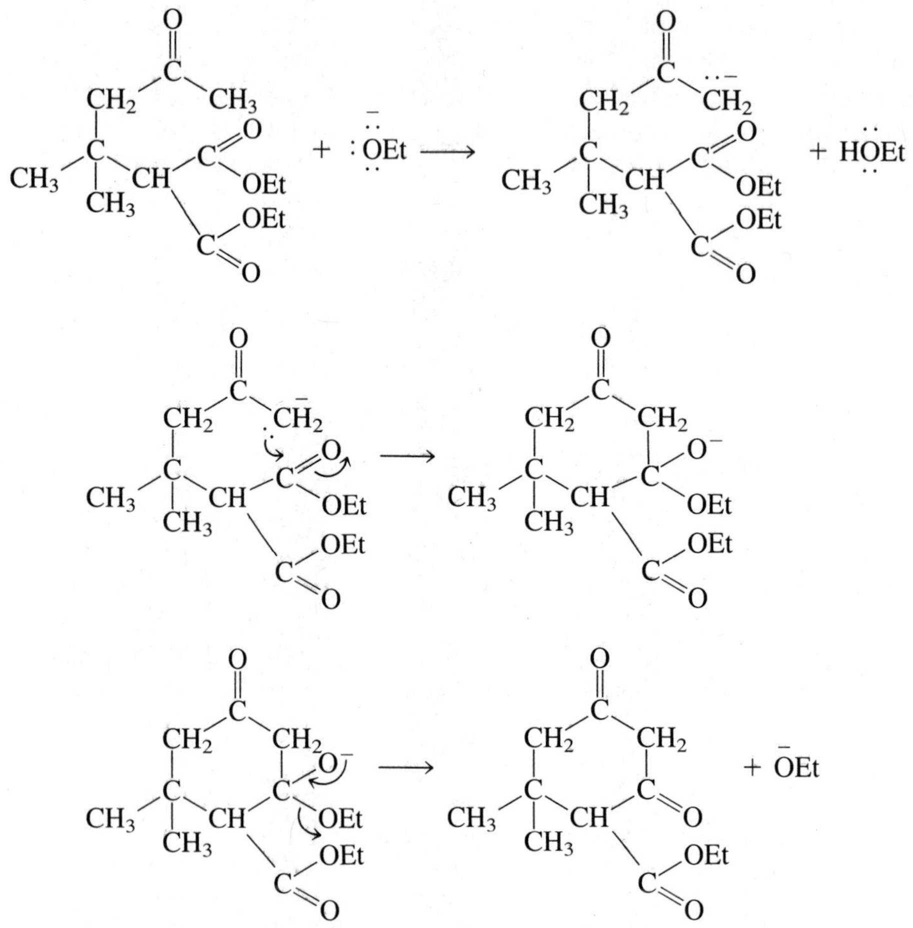

At this point in the synthetic scheme, the ester functional group is hydrolyzed with aqueous potassium hydroxide. Once the mixture is neutralized with acid, the resulting β-keto carboxylic acid (II) readily decarboxylates (loses CO_2) to give the desired product, 3,3-dimethyl-1,3-cyclohexanedione (dimedone).

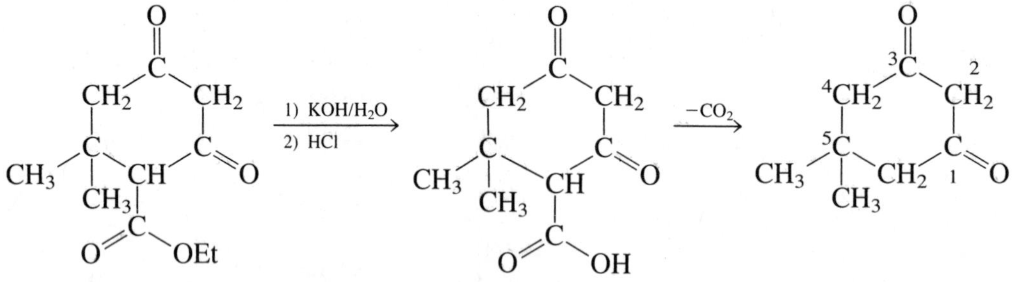

Intermediate (II)

3,3-Dimethyl-1,3-cyclo-hexanedione (dimedone)

Ordinarily, carboxylic acids do not decarboxylate easily. However, when the compound has a keto carbonyl group in the β position (a β-keto acid), it activates the loss of a carboxyl group by way of a cyclic mechanism. The enol tautomer is isomerized to the more stable keto tautomer.

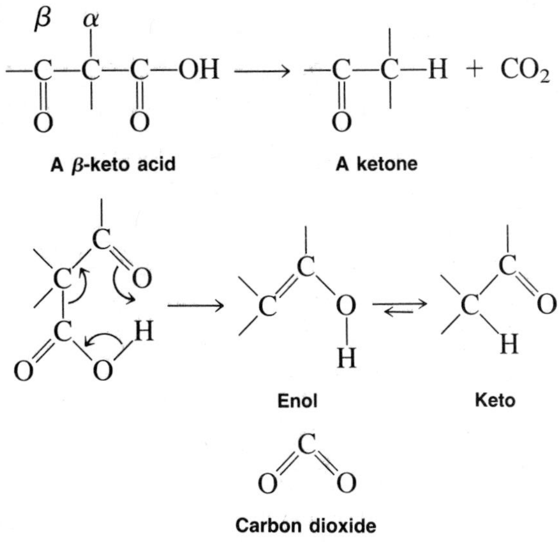

A β-keto acid   A ketone

Enol   Keto

Carbon dioxide

Dimedone, like most 1,3-diketones, exists partially in the enolic form, and both the enol and keto tautomers can be observed by means of proton nmr and infrared spectroscopy.

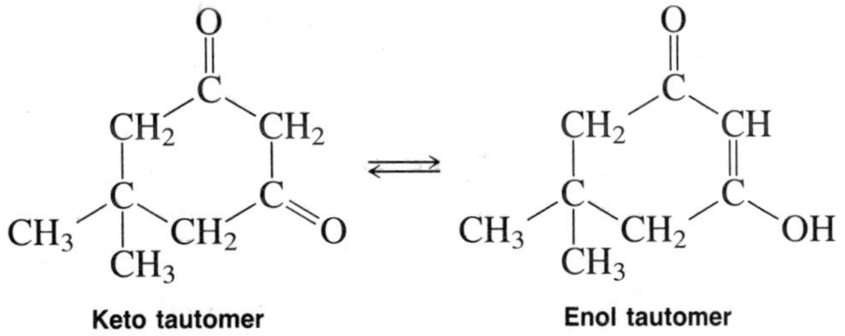

Keto tautomer   Enol tautomer

Dimedone can be used to form derivatives of aldehydes. For example, dimedone reacts readily with formaldehyde in aqueous ethanol to give a solid product. Ketones do not form this derivative.

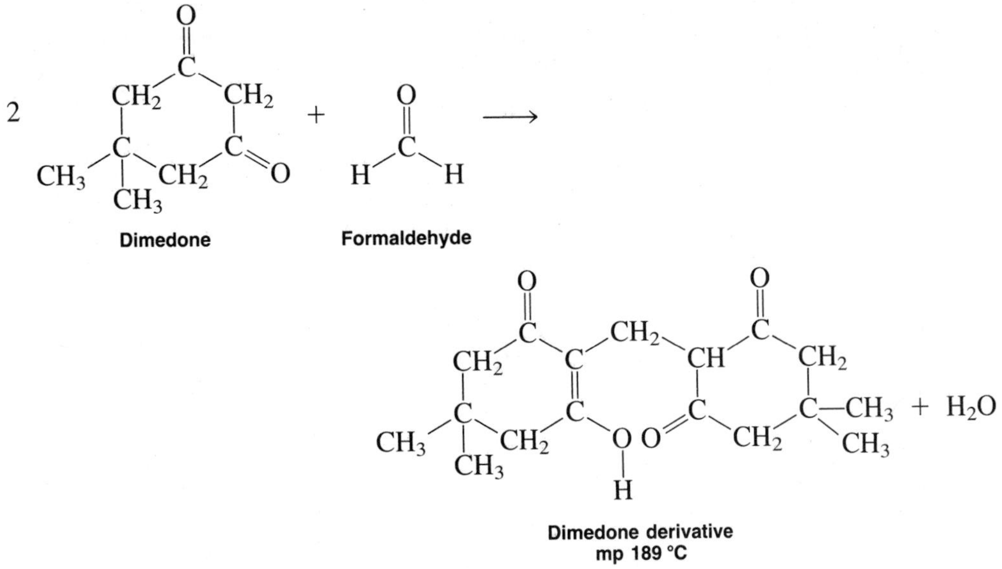

Dimedone   Formaldehyde

Dimedone derivative
mp 189 °C

SPECIAL INSTRUCTIONS

Before beginning this experiment, you should read Technique 14, which discusses techniques for handling sodium metal. When sodium reacts with ethanol, hydrogen gas is evolved. Be certain that all flames in the laboratory are extinguished before dissolving the sodium in ethanol. The reaction must be carried out under strict anhydrous conditions; the apparatus and alcohol must be dry. Safety glasses and gloves should be worn when handling sodium.

PROCEDURE

It is important to maintain anhydrous conditions while following the procedures below. Place 20 mL of absolute ethanol into a **dry** 100-mL round-bottomed flask and attach a **dry** condenser to the flask. Fill a drying tube with calcium chloride and attach it to the top of the condenser. ·

In the area provided in the laboratory and using the methods described in Technique 14, weigh out a 1.15-g (0.05-mole) "block" of sodium metal.[1] Use gloves and tweezers to handle the highly reactive metal, and avoid any contact with water. Safety glasses must be worn at all times. Cut the "block" of sodium metal into about 10 smaller pieces and keep them under xylene until they are used.

Add the sodium to the ethanol over a period of 0.5 hour. Each time that a piece of sodium is to be added, wipe off the xylene, remove the condenser, and drop the metal directly into the flask. Reattach the condenser between additions. The solution will warm as the sodium reacts, but do not cool the mixture. After the addition of sodium is completed, reflux the mixture with a heating mantle until the sodium has dissolved completely.

Remove the flask from the mantle so that it can cool slightly. Remove the condenser and add 8.5 g (0.053 mole) of diethyl malonate to the flask, with swirling. Remove the drying tube and add 5.0 g (0.051 mole) of 4-methyl-3-penten-2-one (mesityl oxide) through the top of the condenser, using a Pasteur pipet. Swirl the contents of the flask and reflux the mixture with a heating mantle for 0.5 hour. During that period, a solid forms and it may be necessary to adjust the temperature of the mantle to prevent bumping. Swirl the mixture occasionally.

Cool the mixture slightly by removing the flask from the mantle, remove the condenser, and add a solution of 6 g of potassium hydroxide in 30 mL of water to hydrolyze the ester. The solid dissolves to yield a homogeneous solution. Add a few boiling stones and reflux the mixture for 1 hour. Following the reflux, the mixture may be stored at this point or the work-up may be continued as indicated in the next paragraph.

Prepare about 30 mL of dilute hydrochloric acid (1 volume of concentrated hydrochloric acid to 2 volumes of water), and add portions of the acid to the mixture until a pH of 7 is reached. Save the remaining acid for later. Assemble the appropriate apparatus for simple distillation (Technique 6, Figure 6–6, p. 557), and distill 25 mL of solvent.

---

[1]Note to the instructor: You may want to cut "blocks" of sodium that are slightly larger than the correct weight in advance of the laboratory period. Students can then trim them to give the correct weight.

Transfer the mixture to a 250-mL Erlenmeyer flask, add 0.75 g of decolorizing carbon, and boil the contents of the flask on a hot plate for a few minutes. Vacuum filter the hot solution through a Büchner funnel, using a piece of Whatman #1 filter paper (Technique 2, Figure 2–4, p. 519). After cooling to room temperature, add dilute hydrochloric acid to the solution until the pH is about 4. As the acid is added, the product, dimedone, begins to precipitate. Vacuum filter the mixture and wash the nearly colorless solid with cold water. Dry the solid at room temperature in air. Weigh the product and determine the percentage yield. This material may be used without further purification for the preparation of the derivative.

Recrystallize a small smount of the product from hot acetone, and determine the melting point of the solid (literature, 148 °C). At the option of the instructor, obtain the infrared spectrum (KBr) and the nmr spectrum (CDCl$_3$). Assign the peaks in each to the enol and keto tautomers. Submit the crude and purified samples of dimedone in a labeled vial with your laboratory report.

DIMEDONE DERIVATIVE

Dissolve 0.2 g of dimedone in 2 mL of 95% ethanol. Place 5 drops of 38% aqueous formaldehyde solution into a test tube and add 3 mL of 95% ethanol and 3 mL of water. Combine the two solutions, shake or mix them, and allow the mixture to stand for 15 minutes. During that time, observe the formation of the fine needle crystals. Collect the crystalline solid by vacuum on a small Büchner funnel, wash them with about 20 mL of water, and allow them to dry until the next laboratory period. Determine the melting point (literature, 189 °C), and submit the sample in a labeled vial with your laboratory report.[2]

QUESTIONS

1. Give the products that you would expect from the following Michael addition reactions.

a.

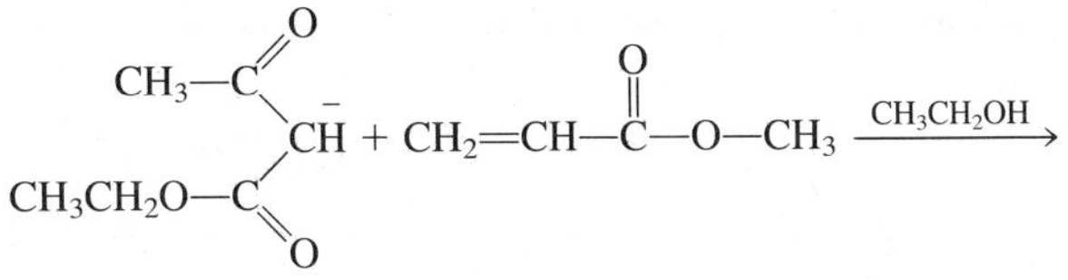

b.

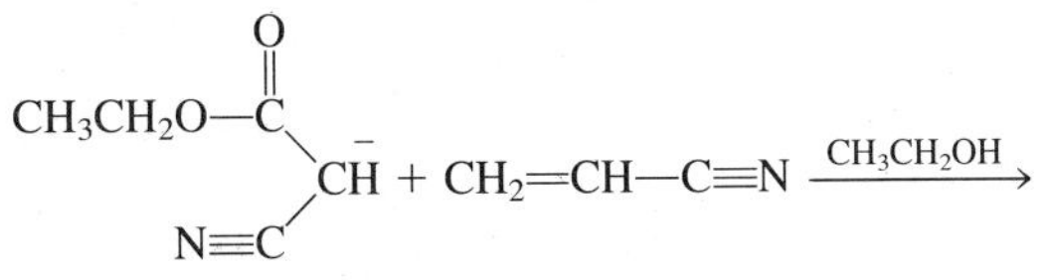

---

[2]Derivatives of other aldehydes may be prepared: Ethanal (141 °C), propanal (155 °C), butanal (142 °C), pentanal (105 °C), hexanal (109 °C), heptanal (103 °C), and octanal (90 °C). Use about 0.1 g of an aldehyde. If a solid does not appear within the 15-minute period, heat the solution for a few minutes. If this does not give a precipitate, add hot water until the solution becomes cloudy and cool it in an ice bath. Collect the solid and recrystallize it from aqueous ethanol.

c.

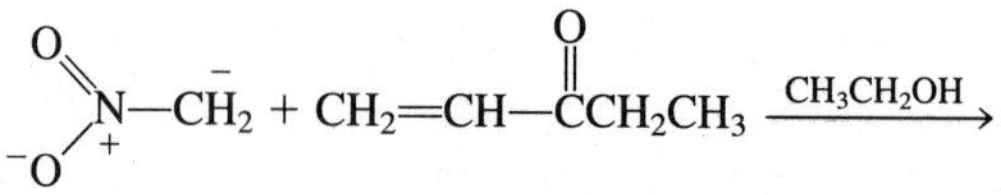

$$\underset{-O}{\overset{O}{\underset{+}{N}}}-\overline{CH}_2 + CH_2{=}CH{-}\overset{O}{\overset{\|}{C}}CH_2CH_3 \xrightarrow{CH_3CH_2OH}$$

2. Give the structure of the products expected from the following reactions.

a.

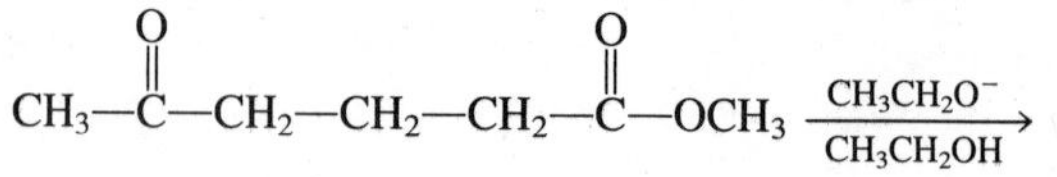

$$CH_3{-}\overset{O}{\overset{\|}{C}}{-}CH_2{-}CH_2{-}CH_2{-}\overset{O}{\overset{\|}{C}}{-}OCH_3 \xrightarrow[CH_3CH_2OH]{CH_3CH_2O^-}$$

b.

$$CH_3CH_2\overset{O}{\overset{\|}{C}}{-}CH_2{-}CH_2{-}CH_2{-}\overset{O}{\overset{\|}{C}}{-}OCH_3 \xrightarrow[CH_3CH_2OH]{CH_3CH_2O^-}$$

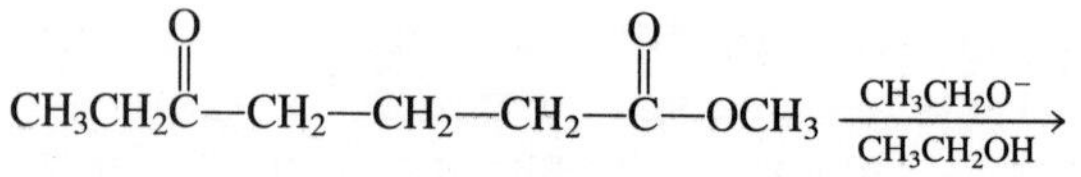

3. What would happen if water is present in the alcohol used in preparing sodium ethoxide?
4. Give a mechanism for the preparation of the dimedone derivative of formaldehyde.
5. Give a structure for the dimedone derivative of propanal.

Essay
PORPHYRINS

A very important class of compounds that finds important roles in biology is the **porphyrins.** The porphyrins perform very diverse functions. These include transporting oxygen from air to the cells, acting as intermediates in the transfer of electrons from oxidizing to reducing agents within a cell, converting light into chemical energy in photosynthesis, and playing a number of roles in biosynthetic pathways. A few of the most important members of this class of compounds will be introduced in this essay.

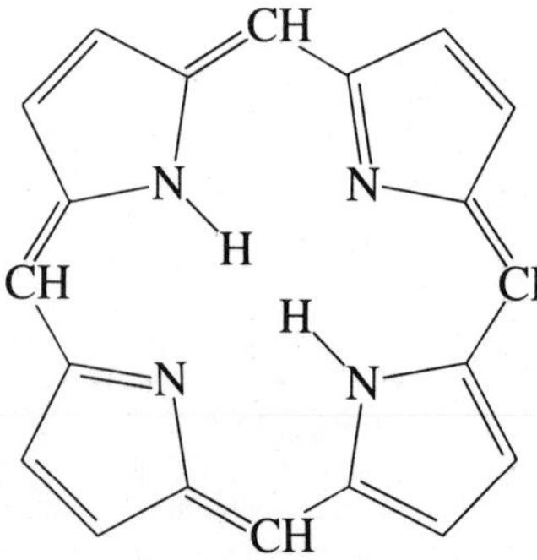

Porphyrin

The porphyrins are derivatives of the basic molecule, **porphyrin,** also known as **porphin.** Porphyrin consists of four pyrrole rings (five-membered heterocyclic rings containing a nitrogen atom) linked by one-carbon bridges. The location of the double bonds in the porphyrin structure results in a highly conjugated system of alternating double and single bonds that extends throughout the entire molecule. As would be expected for such an extended system of conjugation, the porphyrins are intensely colored.

CHLOROPHYLL

Chlorophyll is the green pigment found in plants. It acts as the principal photoreceptor molecule of plants. It is capable of absorbing visible light in order to convert its energy into chemical energy. Chlorophyll is a derivative of porphyrin, with side-chains attached to the pyrrole rings. A magnesium atom is located in the cavity in the center of the ring and is coordinated to the four nitrogen atoms. Thus, chlorophyll is a magnesium porphyrin. Chlorophyll is a **substituted** porphyrin with some interesting differences from its parent ring system: one of the pyrrole rings of chlorophyll is partially reduced; a cyclopentane ring is fused to one of the pyrrole rings; and one of the side-chains is an ester of **phytol,** a long-chain alcohol that contains 20 carbon atoms.

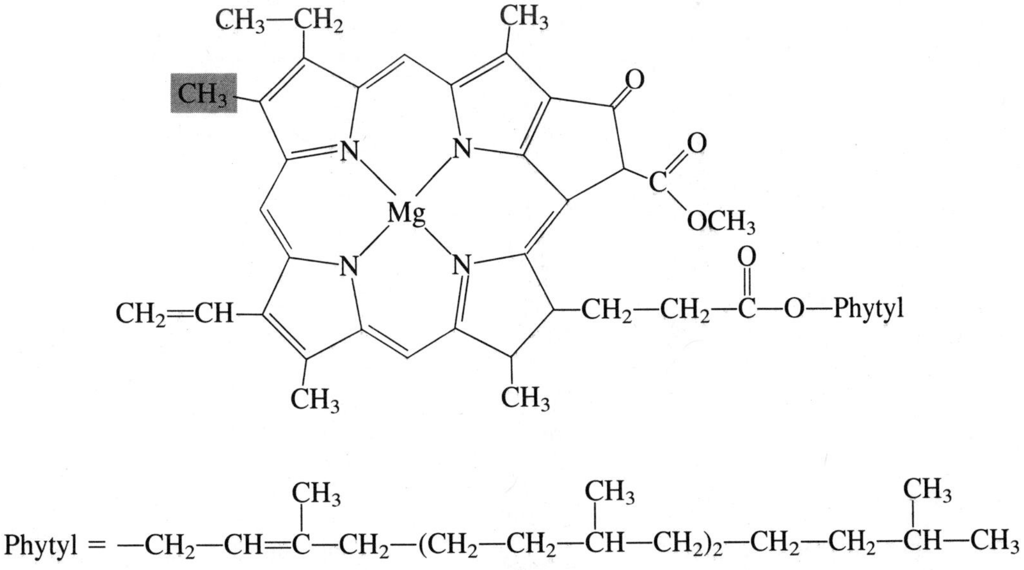

$Phytyl = -CH_2-CH=C(CH_3)-CH_2-(CH_2-CH_2-CH(CH_3)-CH_2)_2-CH_2-CH_2-CH(CH_3)-CH_3$

Chlorophyll a

There are two forms of chlorophyll: **chlorophyll a** and **chlorophyll b.** The two forms are identical, except that the methyl group that is shaded in the structural formula of chlorophyll a is replaced by a —CHO group in chlorophyll b.

The extended chains of conjugated double bonds that characterize the chlorophylls make them very effective in absorbing light. These compounds have very strong absorption bands in the visible region of the spectrum, which is the part of the spectrum in which solar radiation is most intense; in fact, the chlorophylls are among the most effective of all organic compounds at absorbing visible light (molar absorptivities are greater than 10^5 L/cm-mole). Interestingly, chlorophyll *a* and chlorophyll *b* do not have identical absorption spectra. Chlorophyll *a* absorbs light more effectively in the 600- to 700-nm region than chlorophyll *b*, while the *b* form has more intense absorption in the 400- to 500-nm region than the *a* form. Thus, the two forms complement each other, and together they are capable of absorbing sunlight very efficiently.

HEME

Throughout the animal kingdom, substances are required that are capable of binding oxygen from the air and transporting it to the cells, where it can be utilized in metabolism. The most common oxygen-transporting substances are the proteins, **hemoglobin** and **myoglobin.**

Many proteins require some form of nonpeptide structure to provide them with their biological activity. Such a structure, attached to the protein, is called a **prosthetic group.** Hemoglobin and myoglobin possess prosthetic groups that consist of a porphyrin containing an iron atom bound to the four nitrogens of the central cavity. This iron porphyrin is known as **heme.**

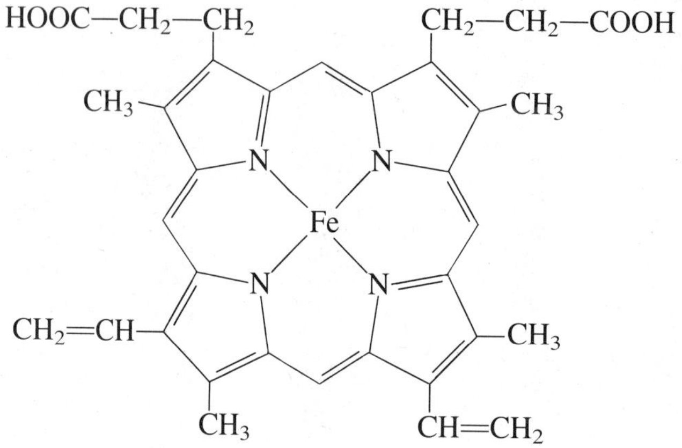

Heme

The iron atom of heme must be in the Fe(II) state for oxygen binding. Heme groups that contain Fe(III) are not capable of binding oxygen.

The protein myoglobin contains one heme group. It acts to store oxygen in muscle tissue and to promote the transfer of oxygen to the mitochondria of muscle cells. Hemoglobin, which binds oxygen from the air and transports it through the blood to the cells, contains four heme groups. Hemoglobin therefore can be considered a tetramer of myoglobin. The structures of these proteins were first reported in 1959 by J. C. Kendrew and M. F. Perutz. Their studies showed that the structures of these two proteins were very similar.

In the protein, four of the six coordination sites of the iron atom of heme are occupied by the four nitrogen atoms of the pyrrole rings. The fifth coordination site is occupied by a nitrogen atom of an imidazole group of the protein. In this way, the prosthetic group is attached to the peptide chain of the protein. The sixth coordination site may be empty or it may be occupied by an O_2 molecule, depending on whether or not the blood is oxygenated. A significant color change takes place when oxygen is bound to the iron, which is why oxygenated blood is redder than deoxygenated blood.

The normal human red blood cell has a lifetime of about 4 months before it is removed from circulation and degraded in the spleen. In the degradation process, heme is degraded first to biliverdin and then to bilirubin. Biliverdin has a green color; bilirubin is brown. The steps in the degradation of heme can be observed easily by noting the dramatic color changes that accompany a serious bruise.

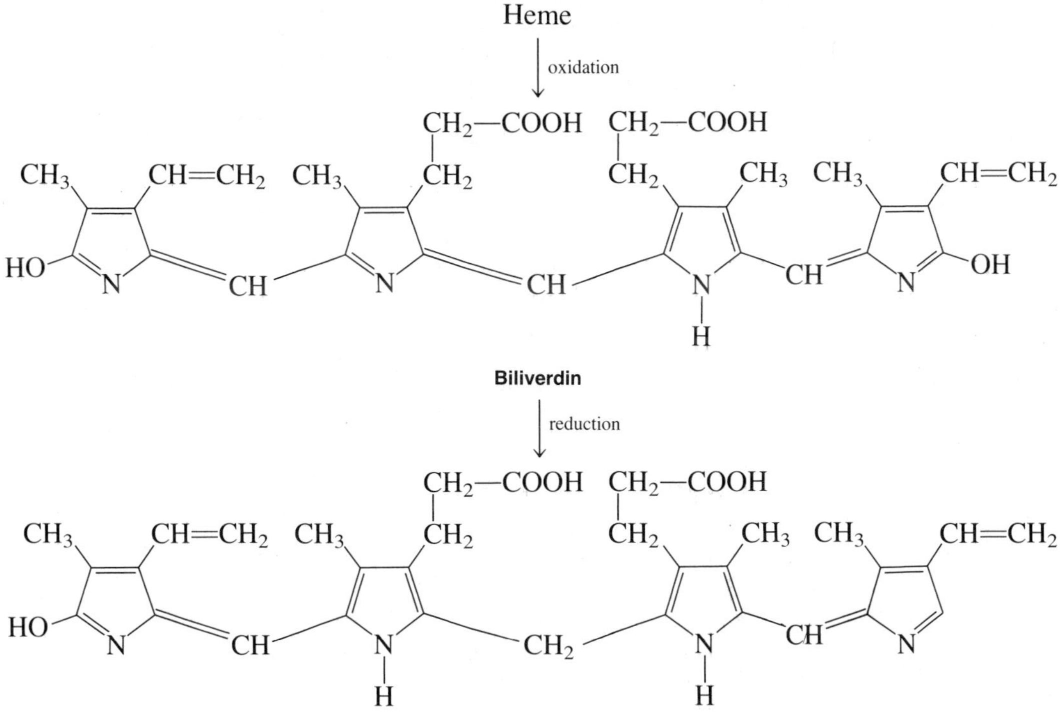

Biliverdin is responsible for the greenish coloration that is observed in the flesh of certain species of fish. Bilirubin is responsible for the characteristic color of feces. It is also responsible for the yellow color observed in newborn infants who suffer from jaundice. In these cases, the newborn child does not yet have sufficient liver function developed to metabolize bilirubin; therefore, bilirubin accumulates in the blood. The prescribed treatment is to expose the baby to ultraviolet light; this is usually accomplished by using a sunlamp that contains a component of ultraviolet light on the baby as it lies in the incubator. The light catalyzes the photooxidation of bilirubin, which degrades it. This treatment is continued until the liver function of the infant increases to the point where metabolism of bilirubin can take place efficiently.

THE CYTOCHROMES

The **cytochromes** are proteins that contain iron porphyrins as prosthetic groups. The cytochromes play important roles in the biological electron-transport chain. They act as intermediates in the transfer of electrons from reducing agents to oxidizing agents in the cell. There are several types of cytochromes, and they play different roles in electron transport. They all possess a heme unit, or a derivative of heme, as a prosthetic group.

VITAMIN B$_{12}$

The structure of **Vitamin B$_{12}$**, more correctly known as **cobalamin**, was first determined by Dorothy Hodgkin in 1952. It proved to be a very complex structure that was very similar to a porphyrin and that contained a cobalt atom. Cobalamin is not a true porphyrin; rather it is a **corrin**. It differs from a porphyrin in that two of the pyrrole rings are directly bonded to each other, rather than being linked by a one-carbon bridge. The structure of the corrin part of cobalamin is shown below. As can be seen, a cobalt atom occupies the cavity in the center of the molecule.

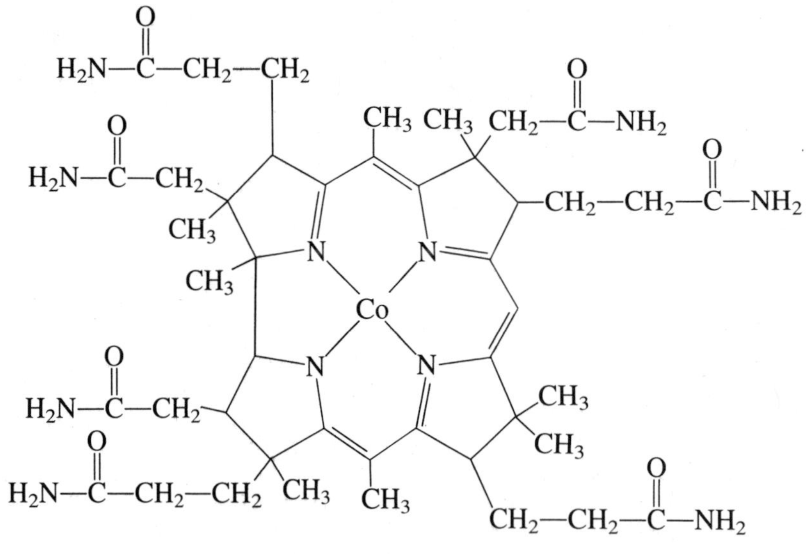

Corrin portion of cobalamin (Vitamin B$_{12}$)

Cobalamin is used as a cofactor for enzymes that catalyze important rearrangement and methylation reactions in the cell. Thus, cobalamin is an essential nutrient. Plants and animals are unable to synthesize cobalamin; hence, it is a **vitamin**. It appears to be synthesized only by microorganisms.

REFERENCES

Alkema, J., and Seager, S. L. "The Chemical Pigments of Plants." *Journal of Chemical Education, 59* (1982): 183.

Dolphin, D., editor. *The Porphyrins*. New York: Academic Press, 1978–1979. Volumes I to VII.

Senozan, N. M., and Hunt, R. L. "Hemoglobin: Its Occurrence, Structure, and Adaptation." *Journal of Chemical Education, 59* (1982): 173.

Stryer, L. *Biochemistry,* 2nd ed. San Francisco: W. H. Freeman and Co., 1981.

Youvan, D. C., and Marrs, B. L. "Molecular Mechanisms of Photosynthesis." *Scientific American, 256* (June 1987): 42.

Wilbraham, A. C. "Phototherapy and the Treatment of Hyperbilirubinemia." *Journal of Chemical Education, 61* (June 1984): 540.

Experiment 50
Preparation of *meso-*
Tetraphenylporphyrin and Some
Metalloporphyrins

Condensation Reaction
Metal Insertion
Crystallization
Ultraviolet-Visible Spectroscopy

The basic structure common to all porphyrins is that of **porphyrin**. Important porphyrins such as chlorophyll and heme are derivatives of the basic porphyrin molecule. In this experiment, a porphyrin will be prepared and various metal ions will be inserted into the ring as a study of compounds that are analogous to the biologically important porphyrins.

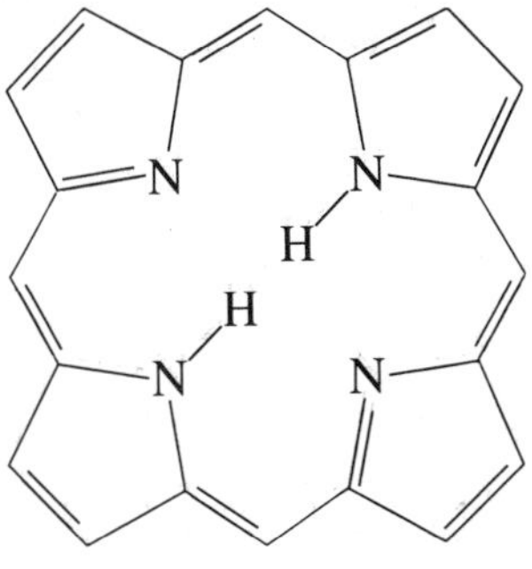

Porphyrin

Porphyrins and their metallo derivatives are interesting substances, since they are deeply colored. Their spectroscopic properties can be examined and compared. The condensation of pyrrole and benzaldehyde in an acidic solvent yields ***meso*-tetraphenylporphyrin**, a deep purple, crystalline solid.

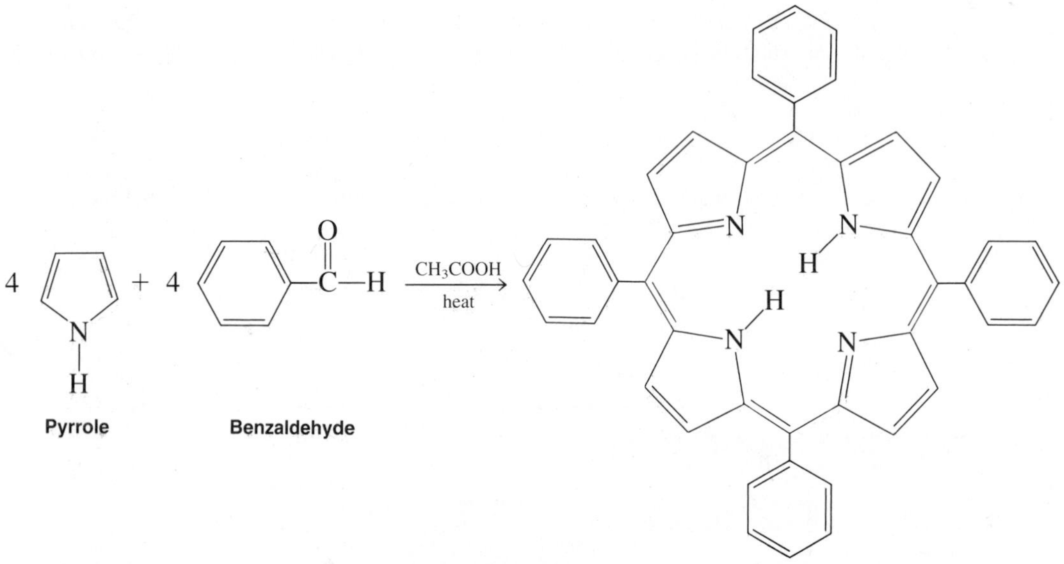

Pyrrole Benzaldehyde

meso-Tetraphenylporphyrin

Once the crystals are isolated and purified, various metals can be inserted into the central cavity of the ring. The inserted metal is coordinated to the four nitrogen atoms. An example of such a **metalloporphyrin** is shown in the accompanying figure.

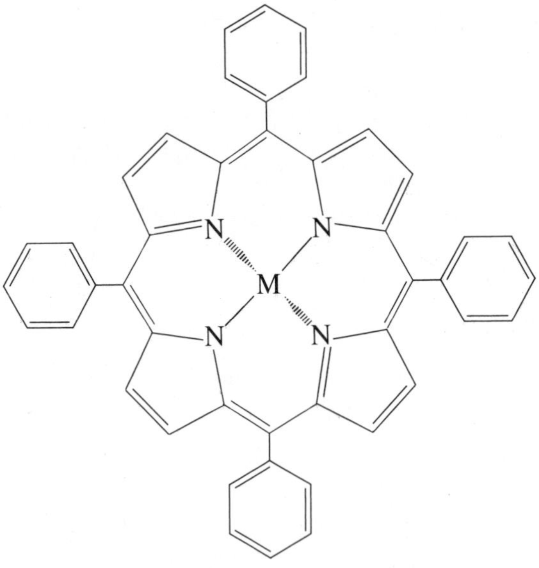

A metalloporphyrin
M = a metal (Cu, Fe, or Ni)

SPECIAL INSTRUCTIONS

Before beginning this experiment read the preceding essay, ''Porphyrins.'' The preparation of tetraphenylporphyrin requires the reaction to be conducted in refluxing glacial acetic acid. This hot acid is very corrosive; special care must be taken to avoid skin contact. Safety goggles must be worn. Toluene is used as an extraction solvent for the metalloporphyrins. Avoid inhaling this solvent. You may choose to insert either copper, iron, or nickel into the tetraphenylporphyrin to form the corresponding metalloporphyrin.

PROCEDURE

TETRAPHENYLPORPHYRIN

Place 5.6 mL of freshly distilled pyrrole (density 0.97 g/mL) in a 500-mL round-bottomed flask, along with 8 mL of reagent-grade benzaldehyde (density 1.04 g/mL) and 300 mL of glacial acetic acid. Attach a reflux condenser to the flask, and add several boiling stones to the contents of the flask. Heat the mixture with a heating mantle or an oil bath until the mixture begins to reflux gently. Heat the mixture under reflux for an hour (Technique 1, Section 1.7, pp. 509–511).

At the end of this reflux period, allow the solution to cool to room temperature, and then place the flask in an ice bath. Dark purple crystals should form. Remove the crystals by vacuum filtration, using a small Büchner funnel (Technique 2, Section 2.3, pp. 518–520). Wash the collected crystals thoroughly with ice-cold methanol and remove the boiling stones from the crystals. Allow the crystals to dry thoroughly at room temperature. When they are dry, weigh the product and calculate the percentage yield. Save approximately 0.10 g of product for later spectrophotometric analysis.

METALLOPORPHYRINS

Place 0.007 g of tetraphenylporphyrin in a 100-mL volumetric flask, and fill the flask to the mark with glacial acetic acid. The concentration of the solution should thus be approximately 10^{-4} M. Withdraw about 10 mL of this solution and combine it in a 25-mL round-bottomed flask with the weight of metal acetate shown below:

Copper(II) acetate	0.07 g
Iron(II) acetate	0.06 g
Nickel(II) acetate	0.06 g

Place a few boiling stones in the flask and attach a reflux condenser. Using a heating mantle or oil bath, heat the mixture gently under reflux for 5 minutes. Remove the heat from the flask and allow the mixture to cool to room temperature. Transfer the mixture to a small separatory funnel, using 10 mL of toluene to ensure complete transfer. Extract the reaction mixture with the toluene. Wash the toluene layer three or four times with distilled water and dry the toluene layer over anhydrous sodium sulfate.

Place the dried solution in a 25-mL volumetric flask and dilute the solution to the mark with toluene. Determine the visible spectrum of the metalloporphyrin solution in the range from 400 to 700 nm. Dilute the sample of tetraphenylporphyrin with toluene in a 25-mL volumetric flask and determine its visible spectrum over the same range. Compare the spectra of the pure tetraphenylporphyrin and the metalloporphyrin. Submit the spectra with your laboratory report. Compare the spectra of tetraphenylporphyrin and the copper, iron, and nickel metalloporphyrins.

REFERENCES

Adler, A. D., Longo, F. R., Finarelli, J. D., Goldmacher, J., Assour, J., and Korsakoff, L. "A Simplified Synthesis for *meso*-Tetraphenylporphyrin." *Journal of Organic Chemistry, 32* (1967): 476.

Adler, A. D., Longo, F. R., Kampas, F., and Kim, J. "On the Preparation of Metalloporphyrins." *Journal of Inorganic and Nuclear Chemistry, 32* (1970): 2443.

Dorough, G. D., Miller, J. R., and Huennekens, F. M. "Spectra of the Metallo-derivatives of $\alpha,\beta,\gamma,\delta$-Tetraphenylporphine." *Journal of the American Chemical Society, 73* (1951): 4315.

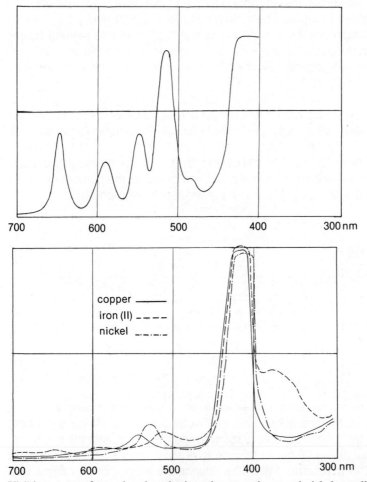

Visible spectra of tetraphenylporphyrin and copper, iron, and nickel metalloporphyrins

Essay
DIELS-ALDER REACTION AND INSECTICIDES

Since the 1930s, it has been known that the addition of an unsaturated molecule across a diene system forms a substituted cyclohexene. The original research dealing with this type of reaction was performed by Otto Diels and Kurt Alder in Germany, and the reaction has become known as the **Diels-Alder reaction.** The Diels-Alder reaction is the reaction of a **diene** with a species capable of reacting with the diene, the **dieno-phile.**

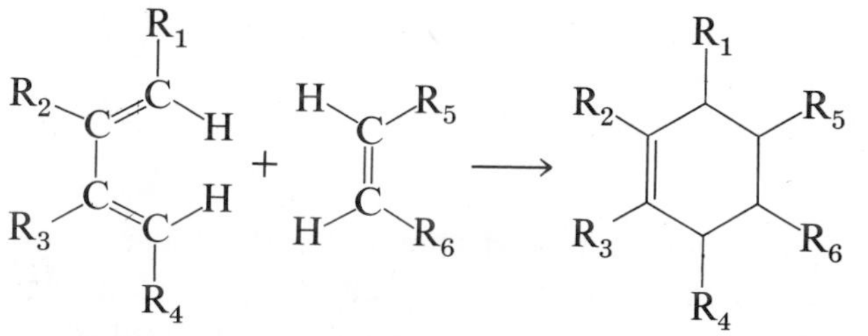

Diene   Dienophile

The product of the Diels-Alder reaction is usually a structure that contains a cyclohexene ring system. If the substituents as shown are simply alkyl groups or hydrogen atoms, the reaction proceeds only under extreme conditions of temperature and pressure. With more complex substituents, however, the Diels-Alder reaction may go on at low temperatures and under mild conditions. The reaction of tetraphenylcyclopentadienone with benzyne to form 1,2,3,4-tetraphenylnaphthalene, described in Experiment 51, is an example of a Diels-Alder reaction carried out under reasonably mild conditions.

A commercially important use of the Diels-Alder reaction involves using hexachlorocyclopentadiene as the diene. Depending on the dienophile, a variety of chlorine-containing addition products may be synthesized. Nearly all these products are powerful **insecticides.** Three insecticides synthesized by the Diels-Alder reaction are shown in the following figure.

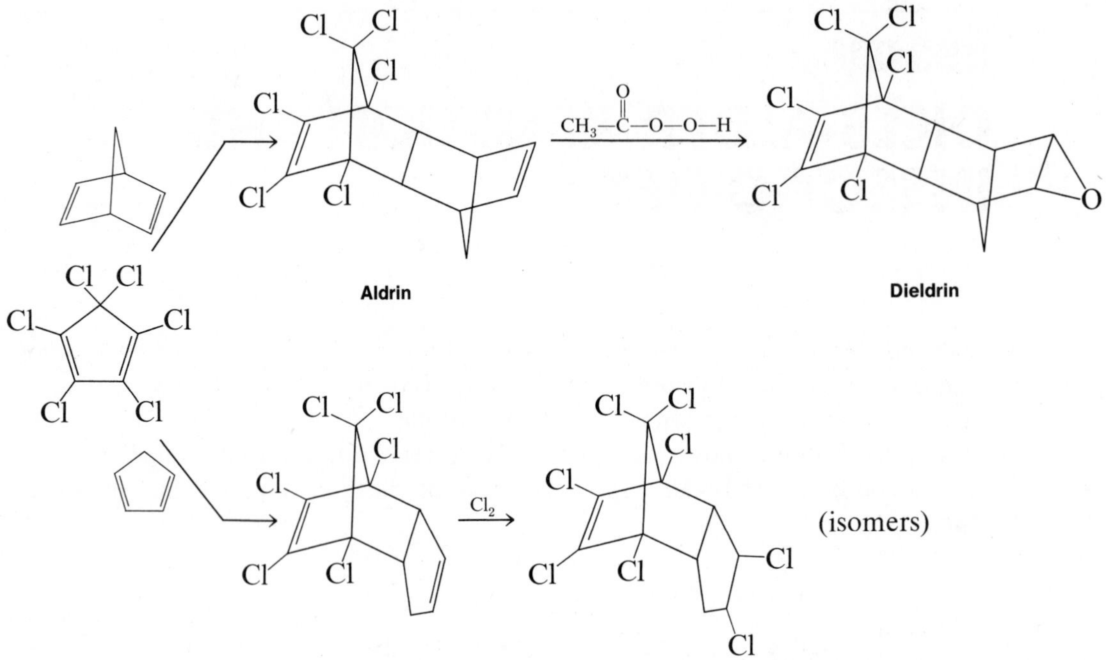

Aldrin

Dieldrin

Chlordane

(isomers)

Dieldrin and Aldrin are named after Diels and Alder. These insecticides are used against the insect pests of fruits, vegetables, and cotton; against soil insects, termites, and moths; and in treating seeds. Chlordane is used in veterinary medicine against insect pests of animals, including fleas, ticks, and lice.

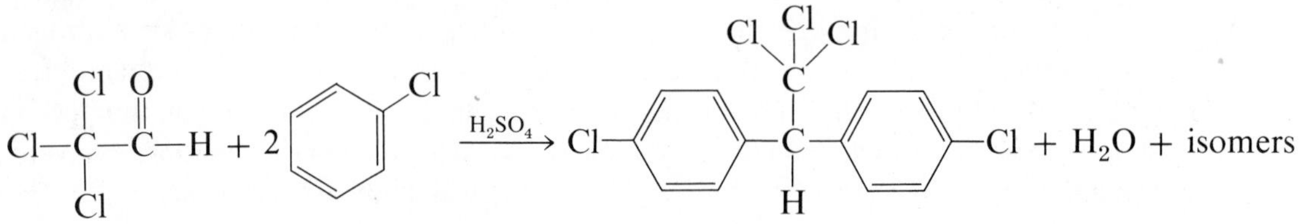

Chloral   **Chlorobenzene**   **DDT**

The best known insecticide, DDT, is not prepared by the Diels-Alder reaction but is nevertheless the best illustration of difficulties experienced when insecticides are used indiscriminately. DDT was first synthesized in 1874, and its insecticidal properties were first demonstrated in 1939. It is easily synthesized commercially, with inexpensive reagents.

At the time DDT was introduced, it was an important boon to mankind. It was effective in controlling lice, fleas, and malaria-carrying mosquitoes and thus helped to control human and animal disease. The use of DDT rapidly spread to the control of hundreds of insects that damage fruit, vegetable, and grain crops.

Pesticides that persist in the environment for a long time after application are called hard pesticides. Beginning in the 1960s, some of the harmful effects of such "hard" pesticides as DDT and the other chlorocarbon materials became known. DDT is a fat-soluble material and is therefore likely to collect in the fat, nerve, and brain tissues of animals. The concentration of DDT in tissues increases in animals high in the food chain. Thus, birds that eat poisoned insects accumulate large quantities of DDT. Animals that feed on the birds accumulate even more DDT. In birds, at least two undesirable effects of DDT have been recognized. First, birds whose tissues contain large amounts of DDT have been observed to lay eggs having shells too thin to survive until young birds are hatched. Second, large quantities of DDT in the tissues seem to interfere with normal reproductive cycles. The massive destruction of bird populations that sometimes occurs after heavy spraying with DDT has become an issue of great concern. The brown pelican and the bald eagle are in danger of extinction. The use of chlorocarbon insecticides has been identified as the principal reason for the decline in the numbers of these birds.

Because DDT is chemically inert, it persists in the environment without decomposing to harmless materials. It can decompose very slowly, but the decomposition products are every bit as harmful as DDT itself. Consequently, each application of DDT means that still more DDT will pass from species to species, from food source to predator, until it concentrates in the higher animals, possibly endangering their existence. Even humans may be threatened. As a result of evidence of the harmful effects of DDT, in the early 1970s the Environmental Protection Agency banned general use of DDT; it may still be used for certain purposes, although permission of the Environmental Protection Agency is required. In 1974, permission was granted for using DDT against the tussock moth in the forests of Washington and Oregon.

Because the life cycles of insects are short, they are able to evolve an immunity to insecticides within a short period of time. As early as 1948, several strains of DDT-resistant insects were identified. Today, the malaria-bearing mosquitoes are almost completely resistant to DDT, an ironic development. Other chlorocarbon insecticides have been used as alternatives to DDT against resistant insects. Examples of other chlorocarbon materials include Dieldrin, Aldrin, Chlordane, and the substances whose structures are shown below. Heptachlor and Mirex are also prepared using Diels-Alder reactions.

Lindane   Heptachlor   Mirex

In spite of structural similarity, Chlordane and Heptachlor show different behavior. Compared with Heptachlor, Chlordane is short-lived and less toxic to mammals. Nevertheless, all the chlorocarbon insecticides have been the objects of much suspicion. A ban on the use of Dieldrin and Aldrin has also been ordered by the Environmental Protection Agency. In addition, strains of insects resistant to Dieldrin, Aldrin, and other materials have been observed. Some insects become addicted to a chlorocarbon insecticide and thrive on it!

The problems associated with the chlorocarbon materials have led to the development of the "soft" insecticides. These usually are organophosphorus or carbamate derivatives, and they are characterized by a short lifetime before they are decomposed to harmless materials in the environment.

The organic structures of some organophosphorus insecticides are shown below.

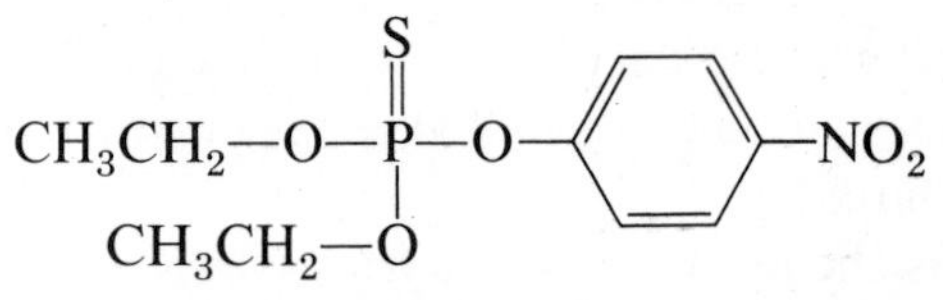

$$CH_3CH_2-O-\overset{\overset{\displaystyle S}{\|}}{P}-O-\!\!\!\!\bigcirc\!\!\!\!-NO_2$$
$$\underset{CH_3CH_2-O}{}$$

Parathion

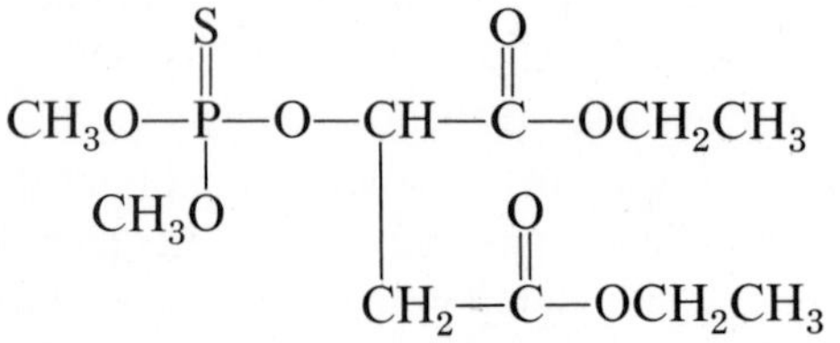

$$CH_3O-\overset{\overset{\displaystyle S}{\|}}{\underset{\underset{\displaystyle CH_3O}{|}}{P}}-O-\overset{\overset{\displaystyle }{\underset{\underset{\displaystyle CH_2-\overset{\overset{\displaystyle O}{\|}}{C}-OCH_2CH_3}{|}}{CH}}-\overset{\overset{\displaystyle O}{\|}}{C}-OCH_2CH_3 \qquad CH_3O-\overset{\overset{\displaystyle O}{\|}}{\underset{\underset{\displaystyle CH_3O}{|}}{P}}-O-CH\!=\!C\overset{\diagup Cl}{\diagdown Cl}$$

Malathion 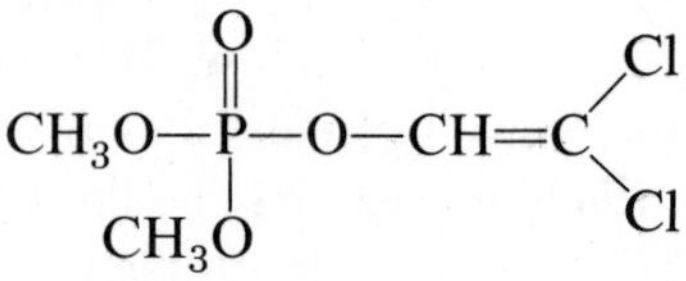   **DDVP or Dichlorvos**

Parathion and Malathion are used widely for agriculture. DDVP is used in "pest strips," which are used for combating household insect pests. The organophosphorus materials do not persist in the environment, and so they are not passed between species up the food chain, as the chlorocarbon compounds are. However, the organophosphorus compounds are highly toxic to humans. Some loss of life among migrant and other agricultural workers has been caused by accidents involving these materials. Stringent safety precautions must be applied when the organophosphorus insecticides are being used.

The carbamate derivatives, including Carbaryl, tend to be less toxic than the organophosphorus compounds. They are also readily degraded to harmless materials. Nevertheless, insects resistant to the soft insecticides have also been observed. Furthermore, the organophosphorus and carbamate derivatives destroy many more nontarget pests than the chlorocarbon compounds do. The danger to earthworms, mammals, and birds is very high.

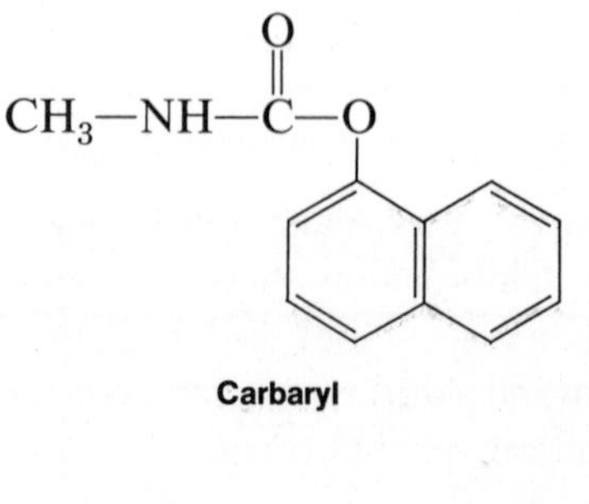

$$CH_3-NH-\overset{\overset{\displaystyle O}{\|}}{C}-O-\!\!\!\!\bigcirc\!\!\!\!\bigcirc$$

Carbaryl

ALTERNATIVES TO INSECTICIDES

Several alternatives to the massive application of insecticides have recently been explored. Insect attractants, including the pheromones (see the essay preceding Experiment 16), have been used in localized traps. Such methods have been effective against the gypsy moth. A "confusion technique," whereby a pheromone is sprayed into the air in such high concentrations that male insects are no longer able to locate females, has been studied. These methods are specific for the target pest and do not cause repercussions in the general environment.

Recent research has been focused on using an insect's own biochemical processes to control pests. Experiments with **juvenile hormone** have shown promise. Juvenile hormone is one of three internal secretions used by insects to regulate growth and metamorphosis from larva to pupa and thence to the adult. At certain stages in the metamorphosis from larva to pupa, juvenile hormone must be secreted; at other stages it must be absent, or the insect will either develop abnormally or fail to mature. Juvenile hormone is important in maintaining the juvenile, or larval, stage of the growing insect. The male cecropia moth, which is the mature form of the silkworm, has been used as a source of juvenile hormone. The structure of the cecropia juvenile hormone is shown below. This material has been found to prevent the maturation of yellow-fever mosquitoes and human body lice. Since insects are not expected to develop a resistance to their own hormones, it is hoped that insects will not be likely to develop a resistance to juvenile hormone.

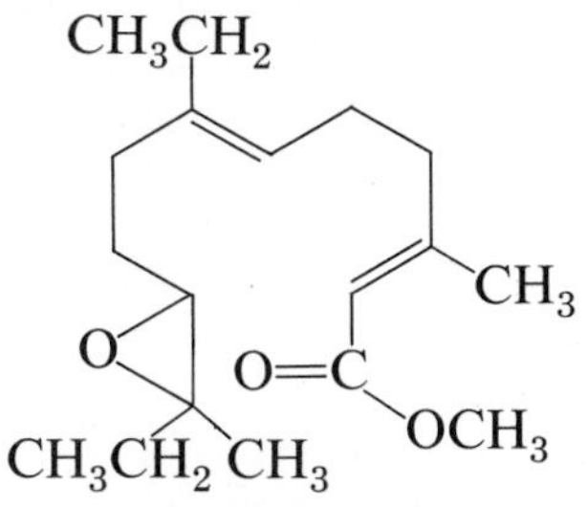

Cecropia juvenile hormone 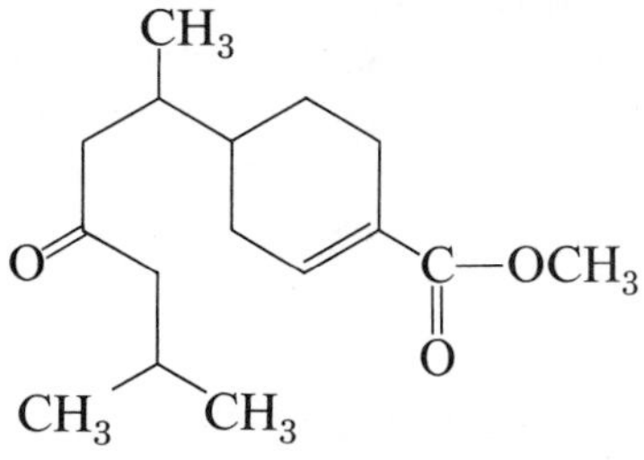   **Paper factor**

While it is very difficult to get enough of the natural substance for use in agriculture, synthetic analogs have been prepared, and they have been shown to be similar in properties and effectiveness to the natural substance. Williams, Sláma, and Bowers have identified and characterized a substance found in the American balsam fir (*Abies balsamea*), known as **paper factor,** which is active against the linden bug, *Pyrrhocoris apterus,* a European cotton pest. This substance is merely one of thousands of terpenoid materials synthesized by the fir tree. Other terpenoid substances are being investigated as potential juvenile hormone analogs.

Certain plants are capable of synthesizing substances that protect them against insects. Included among these natural insecticides are the **pyrethrins** and derivatives of **nicotine** (see the essay on nicotine preceding Experiment 5).

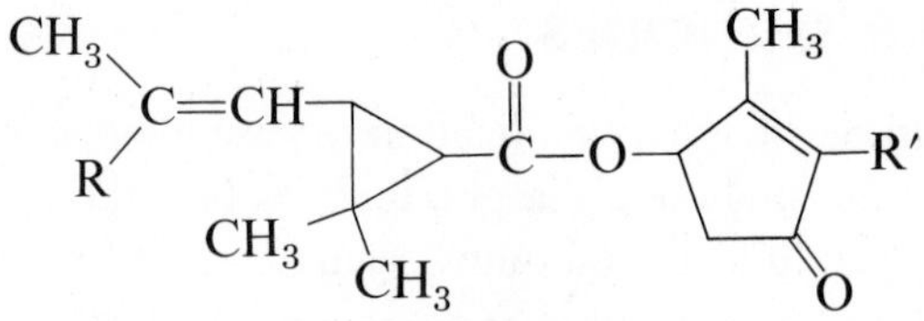

Pyrethrin

R = CH$_3$ or COOCH$_3$
R' = CH$_2$CH=CHCH=CH$_2$ or CH$_2$CH=CHCH$_3$
    or CH$_2$CH=CHCH$_2$CH$_3$

The search for environmentally suitable means of controlling agricultural pests continues with a great sense of urgency. Insects cause billions of dollars of damage to food crops each year. With food becoming increasingly scarce and with the world's population growing at an exponential rate, preventing such losses to food crops becomes absolutely essential.

REFERENCES

Berkoff, C. E. "Insect Hormones and Insect Control." *Journal of Chemical Education, 48* (1971): 577.

Bowers, W. S. and Nishida, R. "Juvocimenes: Potent Juvenile Hormone Mimics from Sweet Basil." *Science, 209* (1980): 1030.

Carson, R. *Silent Spring*. Boston: Houghton Mifflin, 1962.

Keller, E. "The DDT Story." *Chemistry, 43* (February 1970): 8.

O'Brien, R. D. *Insecticides: Action and Metabolism*. New York: Academic Press, 1967.

Peakall, D. B. "Pesticides and the Reproduction of Birds." *Scientific American, 222* (April 1970): 72.

Pryde, L. T. *Environmental Chemistry: An Introduction*. Menlo Park, Calif.: Cummings, 1973. Chap. 7.

Pyle, J. L. *Chemistry and the Technological Backlash*. Englewood Cliffs, N.J.: Prentice-Hall, 1974. Chap. 4.

Rudd, R. L. *Pesticides and the Living Landscape*. Madison: University of Wisconsin Press, 1964.

Saunders, H. J. "New Weapons Against Insects." *Chemical and Engineering News, 53* (July 28, 1975): 18.

Williams, C. M. "Third-Generation Pesticides." *Scientific American, 217* (July, 1967): 13.

Williams, W. G., Kennedy, G. G., Yamamoto, R. T., Thacker, J. D., and Bordner, J. "2-Tridecanone: A Naturally Occurring Insecticide from the Wild Tomato." *Science, 207* (1980): 888.

Experiment 51
Benzyne Formation and the Diels-Alder Reaction: Preparation of 1,2,3,4-Tetraphenylnaphthalene

Benzyne Formation
Diels-Alder Reaction

In this experiment, you prepare 1,2,3,4-tetraphenylnaphthalene. In the first step, benzyne is produced via the unstable diazonium salt. Benzyne is also unstable and cannot be isolated. In the second step, tetraphenylcyclopentadienone (Experiment 46) "traps" the reactive benzyne as it is formed by a Diels-Alder reaction, to give an unstable intermediate. This intermediate readily loses carbon monoxide and yields the fully aromatized naphthalene system. The reaction can be followed easily, since the reaction mixture changes from a purple to a yellow orange solution when the tetraphenylcyclopentadienone is consumed and 1,2,3,4-tetraphenylnaphthalene is produced.

STEP 1 BENZYNE FORMATION

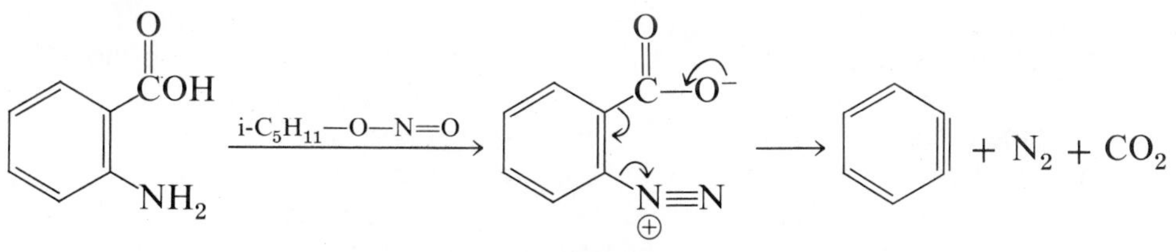

Anthranilic acid

Unstable diazonium salt

Benzyne

STEP 2 DIELS-ALDER REACTION

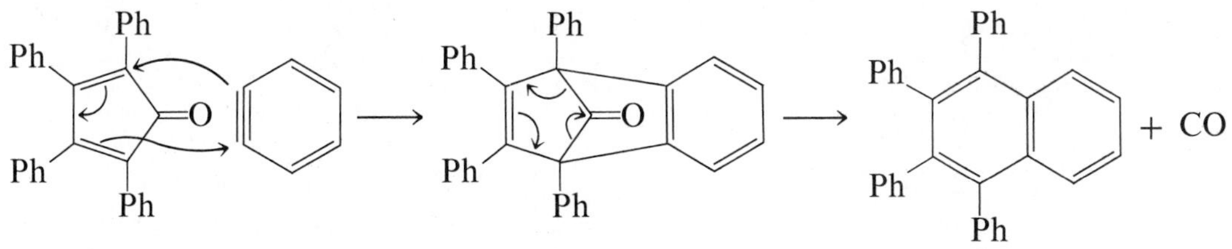

Diels-Alder reaction

Unstable intermediate

1,2,3,4-Tetraphenyl-naphthalene

SPECIAL INSTRUCTIONS

Read the essay that precedes this experiment. Special care should be taken to avoid breathing carbon monoxide gas and isopentyl nitrite (isoamyl nitrite). Some explosions have been observed with unstable diazonium salts. This has usually happened when a large concentration of diazonium salt was present. In this experiment, however, the salt reacts quickly to form benzyne, and it is trapped rapidly with tetraphenylcyclopentadienone. The isopentyl nitrite must be stored in a refrigerator when not in use. Restopper the bottle after the liquid has been removed to minimize contact with air.

PROCEDURE

Place 1.9 g of tetraphenylcyclopentadienone (Experiment 46) and 15 mL of 1,2-dimethoxyethane into a 100-mL round-bottomed flask. Add a Claisen adapter, addition funnel, and reflux condenser to the round-bottomed flask, and bring the solution to reflux on a steam bath (see illustration). Since carbon monoxide is evolved in this reaction, the reaction should be conducted in a hood, if possible.[1] Dissolve 0.75 g of anthranilic acid in 7 mL of 1,2-dimethoxyethane and place it in the addition funnel. Using the pipet provided with the reagent bottle, transfer 1 mL of isopentyl nitrite (isoamyl nitrite, density

> **CAUTION: Do not breathe the isopentyl nitrite vapor as it is a powerful heart stimulant.**

0.875 g/mL) to 7 mL of 1,2-dimethoxyethane in a small Erlenmeyer flask. Replace the cap on the reagent bottle as soon as possible to minimize the exposure to air.[2]

Add the isopentyl nitrite solution **dropwise** with an eye dropper or pipet through the top of the condenser. **Add simultaneously** the anthranilic acid solution from the addition funnel and adjust the rates of addition so that the materials are added at about the same rate. Add the two solutions over a 45- to 60-minute period (three to four drops per minute). Continue the reflux for an additional 15 minutes. The progress of the reaction can be monitored by the slow fading of the deep purple color of the tetraphenylcyclopentadienone. When the dienone is consumed, the color of the solution will be yellow orange. If the color of the solution has not changed following the 15-minute reflux, add 10 drops of pure isopentyl nitrite (no solvent) and a small amount (about 0.05 g) of anthranilic acid, and continue the reflux until the solution is yellow orange.[3]

---

[1]The amount of carbon monoxide formed in this reaction is very small. If the room is well ventilated, the reaction may be conducted at the laboratory bench (see Question 1). At the instructor's option, an alternative apparatus may be used in this experiment to avoid the evolution of carbon monoxide into the laboratory. Use a three-necked flask equipped with two addition funnels and a reflux condenser. The addition funnels contain the two solutions mentioned in the experiment. A trap for carbon monoxide is made by filling a large test tube half-full with a solution of cuprous chloride and ammonium chloride in aqueous ammonia (see below). Secure a short section of glass tubing with a clamp so that the end is immersed several centimeters below the surface of the solution in the test tube. Fasten the other end of the glass tubing to an aspirator trap with rubber tubing (Technique 2, Figure 2–4, p. 519). Connect this trap to the top of the condenser with a piece of rubber tubing. The aspirator trap is a safety device to prevent the copper solution from accidentally backing up into the reaction mixture. A large amount of trapping agent can be prepared as follows. Dissolve 200 g of cuprous chloride and 250 g of ammonium chloride in 700 mL of water. Add to the solution a third of its volume of concentrated (28%) ammonium hydroxide.

[2]Isopentyl nitrite (isoamyl nitrite) must be stored in a refrigerator. It decomposes in the presence of light and air. This reaction has the best results if the material has been bought recently. Material from Aldrich Chemical Co. (#15,049-5) works well.

[3]Sometimes the solution does not change color even after the extra isopentyl nitrite and anthranilic acid have been added. If the mixture has refluxed for an additional 30 minutes without change, add more isopentyl nitrite (up to 0.5 mL) and reflux the mixture for another 15 minutes. If no change has occurred after that time, continue the work-up as written in the procedure. The small amount of remaining purple tetraphenylcyclopentadienone will be removed during purification.

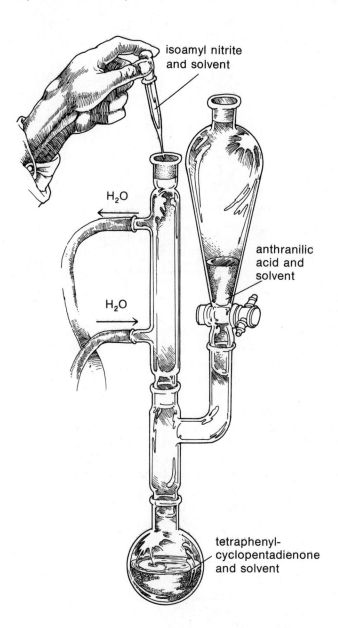

isoamyl nitrite and solvent

H₂O

H₂O

anthranilic acid and solvent

tetraphenyl-cyclopentadienone and solvent

Transfer the solution to a small filter flask, add a wooden applicator stick, and stopper the opening. Connect the sidearm to an aspirator by a trap assembly and evaporate the solvent by heating the flask on a steam bath under vacuum. The apparatus is shown in Technique 1, Figure 1–7, p. 513. Care should be taken to avoid foaming. When the solvent has been removed, an oily solid will remain.[4]

Add 10 mL of methanol to the residue in the filter flask and **triturate** the residue while heating it in a steam bath until it crystallizes fully. Trituration involves stirring and scraping the residue with a spatula. Cool the mixture in an ice bath for at least 15

[4]In some cases a viscous oil develops. This is usually found when the isopentyl nitrite is of poor quality (decomposed). Methanol dissolves the impurities and leaves the solid product. Yields of 1,2,3,4-tetraphenylnaphthalene are poor (about 15%) in this case.

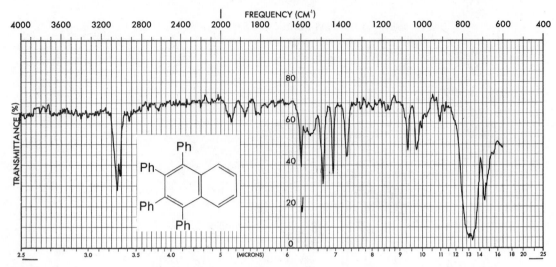

Ir spectrum of 1,2,3,4-tetraphenylnaphthalene, CCl$_4$

minutes before vacuum-filtering the solid on a small Büchner funnel. Wash the 1,2,3,4-tetraphenylnaphthalene with 5 to 10 mL of cold methanol to remove any colored impurities that may be present in the product. Allow the solid to dry fully, weigh the product, and determine the percentage yield. This material should be pure enough for spectroscopy.[5]

Determine the melting point of your 1,2,3,4-tetraphenylnaphthalene (literature value 196–199 °C). When the material has melted, remove the capillary tube from the melting-point apparatus and cool the tube until the material solidifies. Redetermine the melting point (literature value 203–205 °C).[6] The tetraphenylnaphthalene exists in two crystalline forms, each with a different melting point. At the instructor's option, determine the infrared spectrum in CCl$_4$ (Technique 17, Method A, p. 663) and the nmr spectrum in CCl$_4$ or CDCl$_3$. Submit the sample to the instructor in a labeled vial.

REFERENCES

Dougherty, C. M., Baumgarten, R. L., Sweeney, A., and Concepcion, E. "Phthalimide, Anthranilic Acid and Benzyne." *Journal of Chemical Education, 54* (1977): 643.

Fieser, L. F., and Haddadin, M. J. "Isobenzofurane, A Transient Intermediate." *Canadian Journal of Chemistry, 43* (1965): 1599.

[5]Further purification can be achieved by recrystallizing a 0.5-g sample from a minimum amount (about 18 mL) of hot 1:1 95% ethanol–1,2-dimethoxyethane in a 25-mL Erlenmeyer flask. Use a hot plate. Cool the solution in an ice bath for at least 30 minutes to complete the crystallization. Collect the solid on a small Büchner funnel under vacuum, allow the material to dry, and determine the melting points as detailed in the procedure.

[6]The compound exhibits a double melting point. The initial melting point varies with state of subdivision and is not a reliable index of purity. The remelt melting point is more reproducible and reliable.

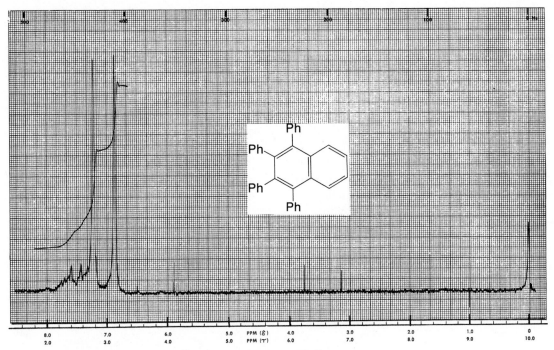

Nmr spectrum of 1,2,3,4-tetraphenylnaphthalene, CDCl₃

QUESTIONS

1. Calculate the number of moles and milliliters of carbon monoxide gas theoretically produced in this reaction.

2. Draw the structures of the products that would result from the following reactions.

(a) 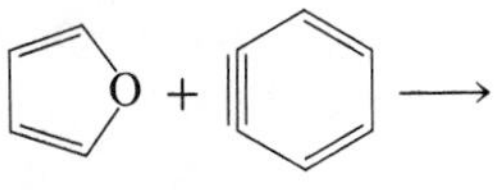 [furan structure] O + [benzene structure] ⟶

(c) 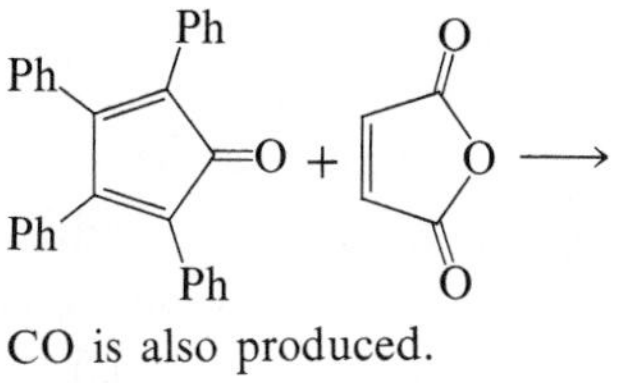 [Ph-substituted cyclopentadienone structure] Ph, Ph, Ph, Ph, =O + [maleic anhydride structure] O, O ⟶

CO is also produced.

(b) 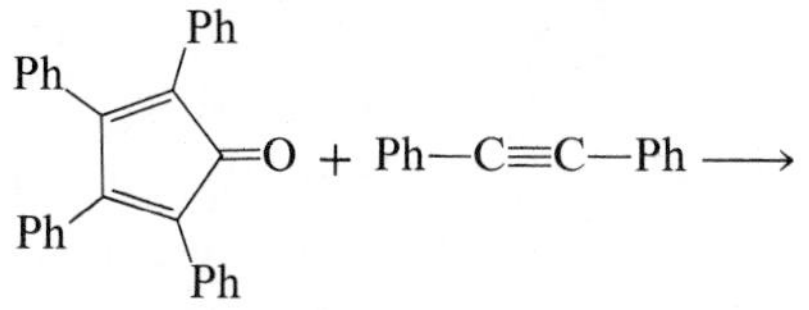 [Ph-substituted cyclopentadienone structure] Ph, Ph, Ph, Ph, =O + Ph—C≡C—Ph ⟶

CO is also produced.

(d) 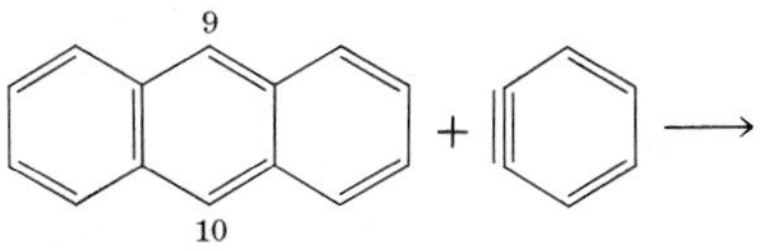 [anthracene structure, positions 9, 10] + [benzene structure] ⟶

Benzyne adds to the 9, 10 positions on anthracene.

3. Interpret the principal absorption bands in the infrared spectrum of 1,2,3,4-tetraphenylnaphthalene.

4. Interpret the nmr spectrum of 1,2,3,4-tetraphenylnaphthalene. In interpreting the nmr spectrum, notice that the molecule is symmetrical and that each of the singlets integrates for 10 hydrogens. The multiplet at 7.2 to 7.8 δ represents four hydrogens.

5. Draw a mechanism for the formation of the diazonium salt from anthranilic acid and isopentyl nitrite.

6. What is the ultimate fate in the reaction of the isopentyl group from the isopentyl nitrite? That is, what compound or compounds are formed?

Experiment 52
Photoreduction of Benzophenone

Photochemistry
Photoreduction
Energy Transfer

The photoreduction of benzophenone is one of the oldest and most thoroughly studied photochemical reactions. Early in the history of photochemistry, it was discovered that solutions of benzophenone are unstable to light when certain solvents are used. If benzophenone is dissolved in a ''hydrogen-donor'' solvent, such as 2-propanol, and exposed to ultraviolet light, an insoluble dimeric product, benzpinacol, will form.

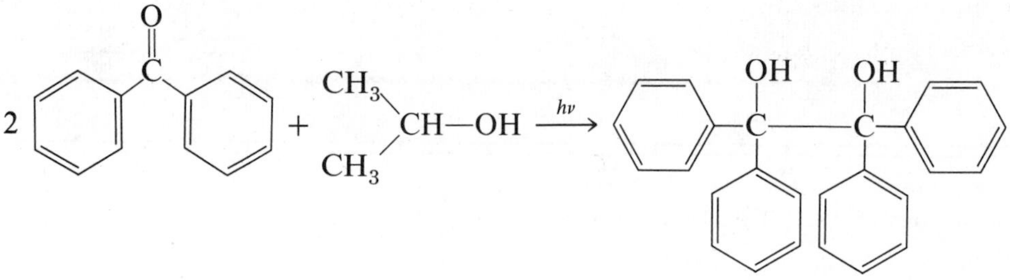

| Benzophenone | 2-Propanol | Benzpinacol |

To understand this reaction, one has to review some simple photochemistry as it relates to aromatic ketones. In the typical organic molecule, all the electrons are paired in the occupied orbitals. When such a molecule absorbs ultraviolet light of the appropriate wavelength, an electron from one of the occupied orbitals, usually the one of highest energy, is excited to an unoccupied molecular orbital, usually to the one of lowest energy. During this transition, the electron must retain its spin value, because a change of spin is a quantum-mechanically forbidden process during an electronic transition. Therefore, just as the two electrons in the highest occupied orbital of the molecule originally had their spins paired (opposite), so they will retain paired spins in the first electronically excited state of the molecule. This is true even though the two electrons will be in **different** orbitals after the transition. This first excited state of a molecule is called a **singlet state** (S_1), since its spin multiplicity ($2S + 1$) is one. The original unexcited state of the molecule is also a singlet state, since its electrons are paired, and it is called the **ground-state** singlet state (S_0) of the molecule.

The excited state singlet, S_1, may return to the ground state, S_0, by reemission of the absorbed photon of energy. This process is called **fluorescence.** Alternatively, the excited electron may undergo a change of spin to give a state of higher multiplicity, the excited **triplet state,** so called because its spin multiplicity ($2S + 1$) is three. The conversion from the first excited singlet state to the triplet state is called **intersystem crossing.** Because the triplet state has a higher multiplicity, it will inevitably be a lower energy state than the excited singlet state (Hund's rule). Normally this change of spin

384

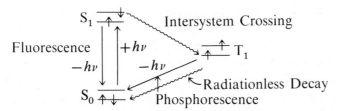

Electronic states of a typical molecule and the possible interconversions. In each state (S_0, S_1, T_1), the lower line represents the highest occupied orbital and the upper line represents the lowest unoccupied orbital of the unexcited molecule. Straight lines represent processes in which a photon is absorbed or emitted. Wavy lines represent radiationless processes—those that occur *without* emission or absorption of a photon.

(intersystem crossing) is a quantum-mechanically forbidden process, just as a direct excitation of the ground state (S_0) to the triplet state (T_1) is forbidden. However, in those molecules in which the singlet and triplet states lie close to one another in energy, the two states inevitably have several overlapping vibrational states, that is, states in common, a situation that allows the "forbidden" transition. In many molecules in which S_1 and T_1 have similar energy ($\Delta E < 10$ Kcal/mole), intersystem crossing occurs faster than fluorescence, and the molecule is rapidly converted from its excited singlet state to its triplet state. In benzophenone, S_1 undergoes intersystem crossing to T_1 with a rate of $k_{isc} = 10^{10}$ sec^{-1}, meaning that the lifetime of S_1 is only 10^{-10} second. The rate of fluorescence for benzophenone is $k_f = 10^6$ sec^{-1}, meaning that intersystem crossing occurs at a rate that is 10^4 times faster than fluorescence. Thus, the conversion of S_1 to T_1 in benzophenone is essentially a quantitative process. In molecules that have a wide energy gap between S_1 and T_1, this situation would be reversed. As you will see shortly, the naphthalene molecule presents a reversed situation.

Since the excited triplet state is lower in energy than the excited singlet state, the molecule cannot easily return to the excited singlet state. Nor can it easily return to the ground state by returning the excited electron to its original orbital. Once again, the transition $T_1 \longrightarrow S_0$ would require a change of spin for the electron, and this is a forbidden process. Hence, the triplet excited state usually has a long lifetime (relative to other excited states) since it generally has nowhere that it can easily go. Even though the process is forbidden, the triplet, T_1, may eventually return to the ground state, S_0, by a process called a **radiationless transition.** In this process, the excess energy of the triplet is lost to the surrounding solution as heat, thereby "relaxing" the triplet back to the ground state, S_0. This process is the study of much current research and is not well understood. In the second process, in which a triplet state may revert to the ground state, **phosphorescence,** the excited triplet emits a photon to dissipate the excess energy and returns directly to the ground state. Although this process is "forbidden," it nevertheless occurs when there is no other open pathway by which the molecule can dissipate its excess energy. In benzophenone, radiationless decay is the faster process, with rate $k_d = 10^5$ sec^{-1}, and phosphorescence, which is not observed, has a lower rate of $k_p = 10^2$ sec^{-1}.

Benzophenone is a ketone. Ketones have **two** possible excited singlet states and, consequently, two excited triplet states as well. This occurs since two relatively low energy transitions are possible in benzophenone. It is possible to excite one of the π electrons in the carbonyl π bond to the lowest-energy unoccupied orbital, a π^* orbital. It is also possible to excite one of the nonbonded or n electrons on oxygen to the same orbital. The first type of transition is called a $\pi-\pi^*$ transition, while the second is called an $n-\pi^*$ transition. These transitions and the states that result are illustrated pictorially.

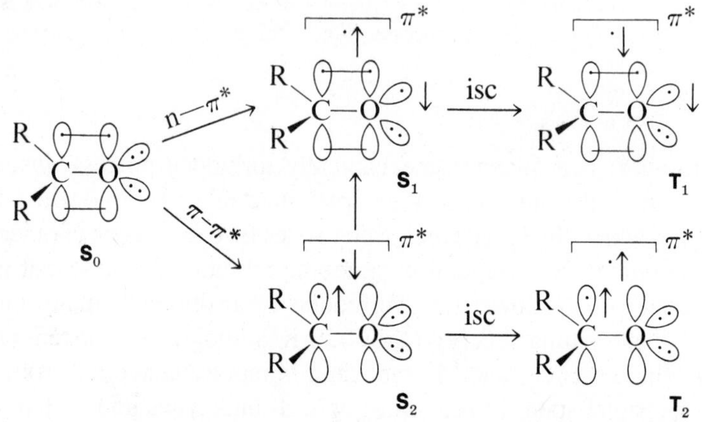

Spectroscopic studies show that for benzophenone and most other ketones, the $n-\pi^*$ excited states S_1 and T_1 are of lower energy than the $\pi-\pi^*$ excited states. An energy diagram depicting the excited states of benzophenone (along with one that depicts those of naphthalene) is shown.

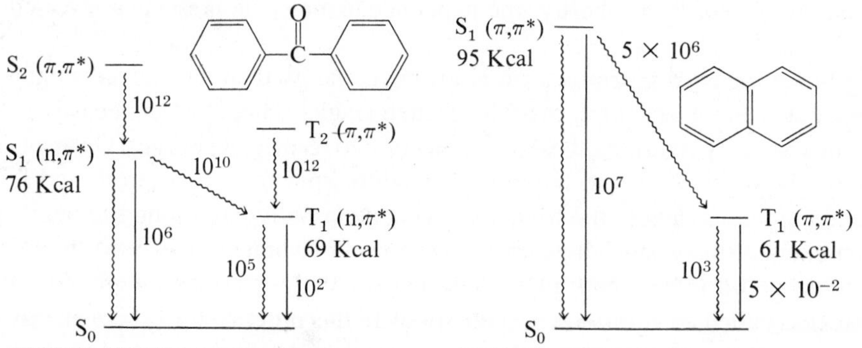

EXCITED STATES OF BENZOPHENONE   EXCITED STATES OF NAPHTHALENE

It is now known that the photoreduction of benzophenone is a reaction of the $n-\pi^*$ triplet state (T_1) of benzophenone. The $n-\pi^*$ excited states have radical character at the carbonyl oxygen atom because of the unpaired electron in the nonbonding orbital. Thus, the radical-like and energetic T_1 excited state species can abstract a hydrogen atom from a suitable donor molecule to form the diphenylhydroxymethyl radical. Two of these radicals, once formed, may coupled to form benzpinacol. The complete mechanism for photoreduction is outlined in the following diagram.

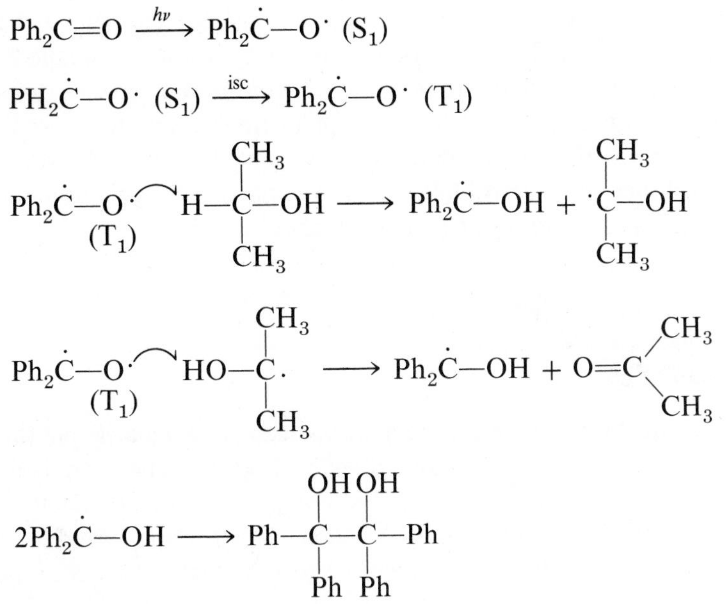

$$Ph_2C{=}O \xrightarrow{h\nu} Ph_2\dot{C}{-}O\cdot \ (S_1)$$

$$PH_2\dot{C}{-}O\cdot \ (S_1) \xrightarrow{isc} Ph_2\dot{C}{-}O\cdot \ (T_1)$$

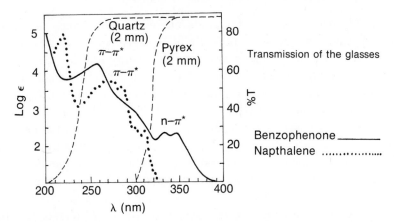

Many photochemical reactions must be carried out in a quartz apparatus because they require ultraviolet radiation of shorter wavelengths (higher energy) than the wavelengths that can pass through Pyrex. Benzophenone, however, requires radiation of approximately 350 nm (350 mμ, or 3500 Å) to become excited to its n–π^* singlet state, S_1, a wavelength that readily passes through Pyrex. In the accompanying figure, the ultraviolet absorption spectra of benzophenone and naphthalene are given. Superimposed on their spectra are two curves, which show the wavelengths that can be transmitted by Pyrex and quartz, respectively. Pyrex will not allow any radiation of wavelength shorter than approximately 300 nm to pass, whereas quartz will allow wavelengths as short as 200 nm to pass. Thus, when benzophenone is placed in a Pyrex flask, the only electronic transition possible is the n–π^* transition, which occurs at 350 nm.

However, even if it were possible to supply benzophenone with radiation of the appropriate wavelength to produce the second excited singlet state of the molecule, this singlet would rapidly convert to the lowest singlet state, S_1. The state S_2 has a lifetime of less than 10^{-12} second. The conversion process $S_2 \longrightarrow S_1$ is called an **internal conversion.** Internal conversions are processes of conversion between excited states of the same multiplicity (singlet–singlet, or triplet–triplet), and they usually are very rapid. Thus, when a S_2 or T_2 is formed, it readily converts to S_1 or T_1, respectively. As a consequence of their very short lifetimes, very little is known about the properties or the exact energies of S_2 and T_2 of benzophenone.

ENERGY TRANSFER

Using a simple **energy-transfer** experiment, one can show that the photoreduction of benzophenone proceeds via the T_1 excited state of benzophenone, rather than the S_1 excited state. If naphthalene is added to the reaction, the photoreduction is stopped because the excitation energy of the benzophenone triplet is transferred to naphthalene. The naphthalene is said to have **quenched** the reaction. This occurs in the following way.

When the excited states of molecules have long enough lifetimes, they often can transfer their excitation energy to another molecule. The mechanisms of these transfers are complex and cannot be explained here; however, the essential requirements can be outlined. First, for two molecules to exchange their respective states of excitation, the process must occur with an overall decrease in energy. Second, the spin multiplicity of the total system must not change. These two features can be illustrated by the two most common examples of energy transfer—singlet transfer and triplet transfer. In these two examples, the superscript 1 denotes an excited singlet state, the superscript 3 denotes a triplet state, and the subscript 0 denotes a ground state molecule. The designations A and B represent different molecules.

$$A^1 + B_0 \longrightarrow B^1 + A_0 \quad \text{SINGLET ENERGY TRANSFER}$$
$$A^3 + B_0 \longrightarrow B^3 + A_0 \quad \text{TRIPLET ENERGY TRANSFER}$$

In singlet energy transfer, excitation energy is transferred from the excited singlet state of A to a ground-state molecule of B, converting it to its excited singlet state and returning A to its ground state. In triplet energy transfer, there is a similar interconversion of excited state and ground state. Singlet energy is transferred through space by a dipole-dipole coupling mechanism, but triplet energy transfer requires the two molecules involved in the transfer to collide. In the usual organic medium, about 10^9 collisions occur per second. Thus, if a triplet state A^3 has a lifetime longer than 10^{-9} second, and if an acceptor molecule, B_0, which has a lower triplet energy than that of A^3 is available, energy transfer can be expected. If the triplet A^3 undergoes a reaction (like photoreduction) at a rate lower than the rate of collisions in the solution, and if an acceptor molecule is added to the solution, the reaction can be **quenched.** The acceptor molecule, which is called a **quencher,** deactivates, or "quenches," the triplet before it has a chance to react. Naphthalene has the ability to quench benzophenone triplets in this way and to stop the photoreduction.

Naphthalene cannot quench the excited-state singlet S_1 of benzophenone, since its own singlet has an energy (95 Kcal/mole) that is higher than the energy of benzophenone (76 Kcal/mole). In addition, the conversion $S_1 \longrightarrow T_1$ is very rapid (10^{-10} second) in benzophenone. Thus, naphthalene can intercept only the triplet state of benzophenone. The triplet excitation energy of benzophenone (69 Kcal/mole) is transferred to naphthalene ($T_1 = 61$ Kcal/mole) in an exothermic collision. Finally, the naphthalene molecule does not absorb light of the wavelengths transmitted by Pyrex (see spectra above), and therefore benzophenone is not inhibited from absorbing energy when naphthalene is present in solution. Thus, since naphthalene quenches the photoreduction reaction of benzophenone, we can infer that this reaction proceeds via the triplet state, T_1, of benzophenone. If naphthalene did not quench the reaction, the singlet state of benzophenone would be indicated as the reactive intermediate. In the experiment that follows, the photoreduction of benzophenone is attempted both in the presence and in the absence of added naphthalene.

SPECIAL INSTRUCTIONS

This experiment may be performed concurrently with some other experiment. It requires only 15 minutes during the first laboratory period, and only about 15 minutes in a subsequent laboratory period about 1 week later.

It is important that the reaction mixture be left where it will receive **direct** sunlight. If it does not, the reaction will be slow and may need more than a week for completion.

PROCEDURE

Label two 20 × 150-mm test tubes "No. 1" and "No. 2" near the top. Place 2.5 g of benzophenone in the first tube. Place 2.5 g of benzophenone and 0.25 g of naphthalene in the second tube. Add about 10 mL of 2-propanol (isopropyl alcohol) to each tube, and warm each on the steam bath to dissolve the solids. When the solids have dissolved, add 1 drop of glacial acetic acid to each tube, and then fill each tube nearly to the top with more 2-propanol. Stopper the tubes tightly with rubber stoppers, shake them well, and place them in a beaker on a window sill where they will receive direct sunlight. The reaction will need about 1 week for completion. If the reaction has occurred during this period, the product will have crystallized from the solution. Observe the result in each test tube. Collect the product by vacuum filtration, dry it, and determine its melting point and percentage yield. At your instructor's option, determine the infrared spectrum of the benzpinacol as a KBr mull. Submit the product to your instructor along with the report.

REFERENCE

Vogler, A., and Kunkely, H. "Photochemistry and Beer." *Journal of Chemical Education, 59* (January 1982): 25.

QUESTIONS

1. Can you think of a way to produce the benzophenone $n-\pi^*$ triplet T_1 **without** having benzophenone pass through its first singlet state? Explain.

2. A reaction similar to the one here described occurs when benzophenone is treated with the metal magnesium (pinacol reduction).

$$2Ph_2C{=}O \xrightarrow{\text{Mg}} Ph_2\overset{\overset{\displaystyle OH}{|}}{C}{-}\overset{\overset{\displaystyle OH}{|}}{C}Ph_2$$

Compare the mechanism of this reaction with the photoreduction mechanism. What are the differences?

3. Which of the following molecules do you expect would be useful in quenching benzophenone photoreduction? Explain.

Oxygen	$(S_1 = 22$ Kcal/mole$)$	Biphenyl	$(T_1 = 66$ Kcal/mole$)$
9,10-Diphenylanthracene	$(T_1 = 42$ Kcal/mole$)$	Toluene	$(T_1 = 83$ Kcal/mole$)$
trans-1,3-Pentadiene	$(T_1 = 59$ Kcal/mole$)$	Benzene	$(T_1 = 84$ Kcal/mole$)$
Naphthalene	$(T_1 = 61$ Kcal/mole$)$		

Essay
FIREFLIES AND PHOTOCHEMISTRY

The production of light as a result of a chemical reaction is called **chemiluminescence.** A chemiluminescent reaction generally produces one of the product molecules in an electronically excited state. The excited state emits a photon, and light is produced. If a reaction that produces light is biochemical, occurring in a living organism, the phenomenon is called **bioluminescence.**

The light produced by fireflies and other bioluminescent organisms has fascinated observers for many years. Many different organisms have developed the ability to emit light. They include bacteria, fungi, protozoans, hydras, marine worms, sponges, corals, jellyfishes, crustaceans, clams, snails, squids, fishes, and insects. Curiously, among the higher forms of life, only fish are included on the list. Amphibians, reptiles, birds, mammals, and the higher plants are excluded. Among the marine species, none is a freshwater organism. The excellent *Scientific American* article by McElroy and Seliger (see References) delineates the natural history, characteristics, and habits of many bioluminescent organisms.

The first significant studies of a bioluminescent organism were by the French physiologist Raphael Dubois in 1887. He studied the mollusk *Pholas dactylis,* a bioluminescent clam indigenous to the Mediterranean Sea. Dubois found that a cold-water extract of the clam was able to emit light for several minutes following the extraction. When the light emission ceased, it could be restored, he found, by a material extracted from the clam by hot water. A hot-water extract of the clam alone did not produce the luminescence. Reasoning carefully, Dubois concluded that there was an enzyme in the cold-water extract that was destroyed in hot water. The luminescent compound, however, could be extracted without destruction in either hot or cold water. He called the luminescent material **luciferin,** and the enzyme that induced it to emit light, **luciferase;** both names were derived from *Lucifer,* a Latin name meaning ''bearer of light.'' Today the luminescent materials from all organisms are called luciferins, and the associated enzymes are called luciferases.

The most extensively studied bioluminescent organism is the firefly. Fireflies are found in many parts of the world and probably thus represent the most familiar example of bioluminescence. In such areas, on a typical summer evening, fireflies, or ''lightning bugs,'' can frequently be seen to emit flashes of light as they cavort over the lawn or in the garden. It is now universally accepted that the luminescence of fireflies is a mating device. The male firefly flies about 2 ft above the ground and emits flashes of light at regular intervals. The female, who remains stationary on the ground, waits a characteristic interval and then flashes a response. In return, the male will reorient his direction of flight, toward her, and flash a signal once again. The entire cycle is rarely repeated more than 5 to 10 times before the male reaches the female. Fireflies of different species can recognize one another by their flash patterns, which vary in number, rate, and duration among species.

Although the total structure of the luciferase enzyme of the American firefly, *Photinus pyralis,* is unknown, the structure of the luciferin has been established. In spite of a large amount of experimental work, however, the complete nature of the chemical reactions that produce the light is still subject to some controversy. It is possible, nevertheless, to outline the most salient details of the reaction.

$$LUCIFERASE + ATP \longrightarrow LUCIFERASE\text{-}ATP$$

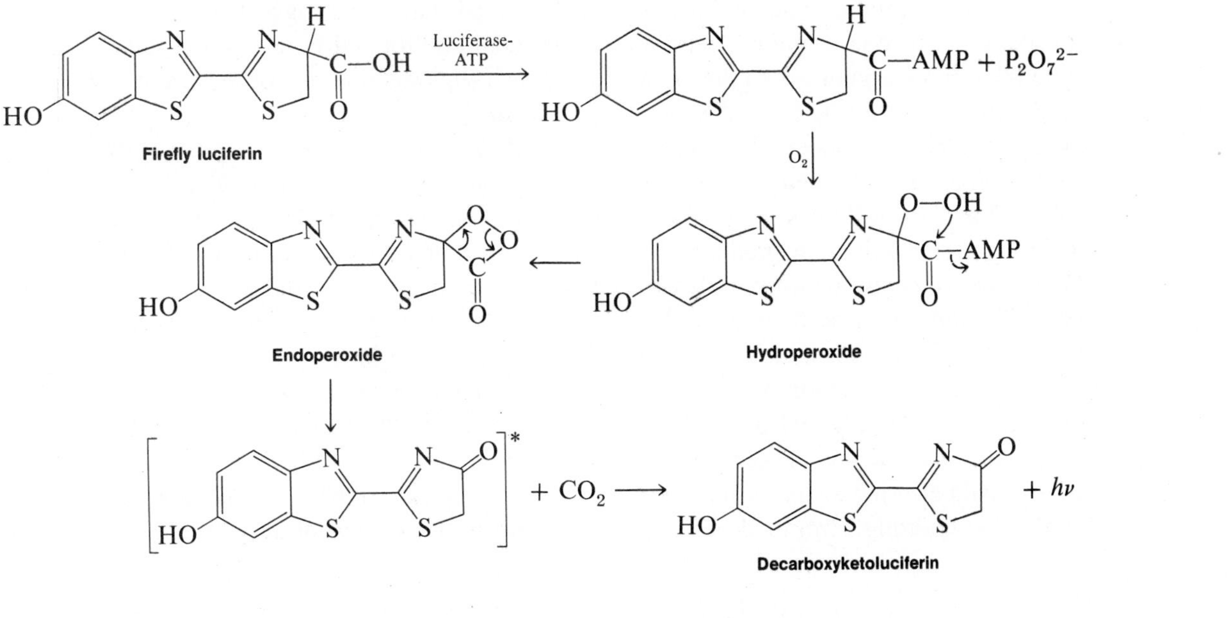

Firefly luciferin

Endoperoxide

Hydroperoxide

Decarboxyketoluciferin

Besides the luciferin and the luciferase, other substances—magnesium(II), ATP (adenosine triphosphate), and molecular oxygen—are needed to produce the luminescence. In the postulated first step of the reaction, the luciferase complexes with an ATP molecule. In the second step, the luciferin binds to the luciferase and reacts with the already bound ATP molecule to become "primed." In this reaction, pyrophosphate ion is expelled, and AMP (adenosine monophosphate) becomes attached to the carboxyl group of the luciferin. In the third step, the luciferin-AMP complex is oxidized by molecular oxygen to form a hydroperoxide; this cyclizes with the carboxyl group, expelling AMP and forming the cyclic endoperoxide. This reaction would be difficult if the carboxyl group of the luciferin had not been primed with ATP. The endoperoxide is unstable and readily decarboxylates, producing decarboxyketoluciferin in an **electronically excited state,** which is deactivated by the emission of a photon (fluorescence). Thus, it is the cleavage of the four-membered-ring endoperoxide that leads to the electronically excited molecule and hence the bioluminescence.

$$\text{endoperoxide} \longrightarrow \left[\begin{array}{c} {>}{=}O \\ {>}{=}O^* \end{array} \right] \longrightarrow 2 \; {>}{=}O + h\nu$$

That one of the two carbonyl groups, either that of the decarboxyketoluciferin or that of the carbon dioxide, should be formed in an excited state can be readily predicted from the orbital symmetry conservation principles of Woodward and Hoffmann. This reaction is formally like the decomposition of a cyclobutane ring to yield two ethylene molecules. In analyzing the forward course of that reaction, that is, two ethylene $\longrightarrow$ cyclobutane, one can easily show that the reaction, which involves four π electrons, is forbidden for two ground-state ethylenes but allowed for only one ethylene in the ground state and the other in an excited state. This suggests that in the reverse process, one of the ethylene molecules should be formed in an excited state. Extending these arguments to the endoperoxide also suggests that one of the two carbonyl groups should be formed in its excited state.

The emitting molecule, decarboxyketoluciferin, has been isolated and synthesized. When it is excited photochemically by photon absorption in basic solution (pH > 7.5–8.0), it fluoresces, giving a fluorescence emission spectrum that is identical to the emission spectrum produced by the interaction of firefly luciferin and firefly luciferase. The emitting form of decarboxyketoluciferin has thus been identified as the **enol dianion.** In neutral or acidic solution, the emission spectrum of decarboxyketoluciferin does not match the emission spectrum of the bioluminescent system.

The exact function of the enzyme firefly luciferase is not yet known, but it is clear that all these reactions occur while luciferin is bound to the enzyme as a substrate. Also, since the enzyme undoubtedly has several basic groups (—COO^-, —NH_2, and so on), the buffering action of those groups would easily explain why the enol dianion is also the emitting form of decarboxyketoluciferin in the biological system.

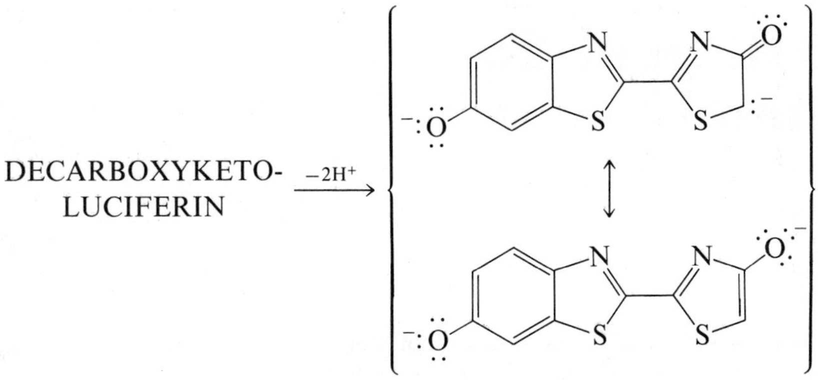

DECARBOXYKETO-
LUCIFERIN $\xrightarrow{-2H^+}$

Enol dianion

Most chemiluminescent and bioluminescent reactions require oxygen. Likewise, most produce an electronically excited emitting species through the decomposition of a **peroxide** of one sort or another. In the experiment that follows, a **chemiluminescent** reaction that involves the decomposition of a peroxide intermediate is described.

REFERENCES

Clayton, R. K. *Light and Living Matter*. Vol. 2: *The Biological Part*. New York: McGraw-Hill, 1971. Chap. 6, "The Luminescence of Fireflies and Other Living Things."

Fox, J. L. "Theory May Explain Firefly Luminescence." *Chemical and Engineering News* (March 6, 1978): 17.

Harvey, E. N. *Bioluminescence*. New York: Academic Press, 1952.

Hastings, J. W. "Bioluminescence." *Annual Review of Biochemistry, 37* (1968): 597.

McCapra, F. "Chemical Mechanisms in Bioluminescence." *Accounts of Chemical Research, 9* (1976): 201.

McElroy, W. D., and Seliger, H. H. "Biological Luminescence." *Scientific American, 207* (December 1962): 76.

McElroy, W. D., Seliger, H. H., and White, E. H. "Mechanism of Bioluminescence, Chemiluminescence and Enzyme Function in the Oxidation of Firefly Luciferin." *Photochemistry and Photobiology, 10* (1969): 153.

Seliger, H. H., and McElroy, W. D. *Light: Physical and Biological Action*. New York: Academic Press, 1965.

Experiment 53
Luminol

Chemiluminescence
Energy Transfer
Reduction of a Nitro Group
Amide Formation

In this experiment, the chemiluminescent compound luminol, or 5-aminophthal-hydrazide, will be synthesized from 3-nitrophthalic acid.

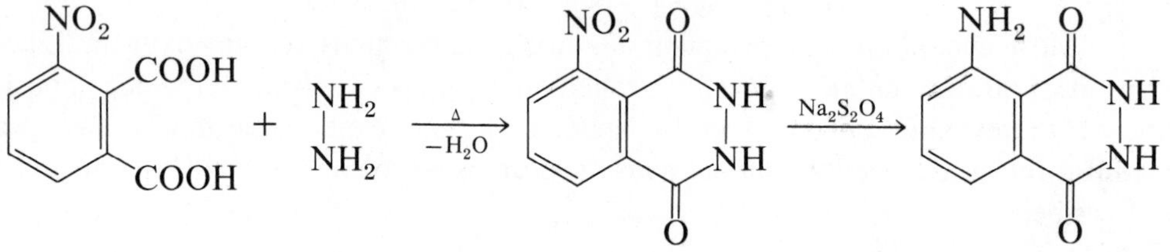

3-Nitrophthalic acid   Hydrazine   5-Nitrophthalhydrazide   Luminol

The first step of the synthesis is the simple formation of a cyclic diamide, 5-nitrophthalhydrazide, by reaction of 3-nitrophthalic acid with hydrazine. Reduction of the nitro group with sodium dithionite affords luminol.

In neutral solution, luminol exists largely as a dipolar ion (zwitterion). This dipolar ion itself exhibits a weak blue fluorescence after being exposed to light. However, in alkaline solution, luminol is converted to its dianion, which may be oxidized by molecular oxygen to give an intermediate that is chemiluminescent. The reaction is thought to have the following sequence:

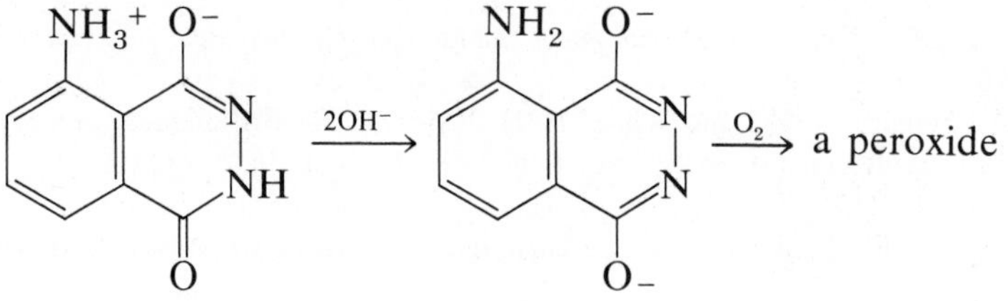

Luminol   Dianion

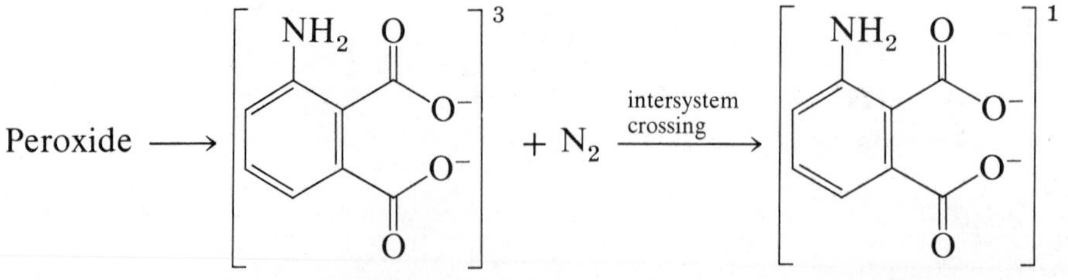

3-Aminophthalate
Triplet dianion (T_1)

3-Aminophthalate
Singlet dianion (S_1)

394

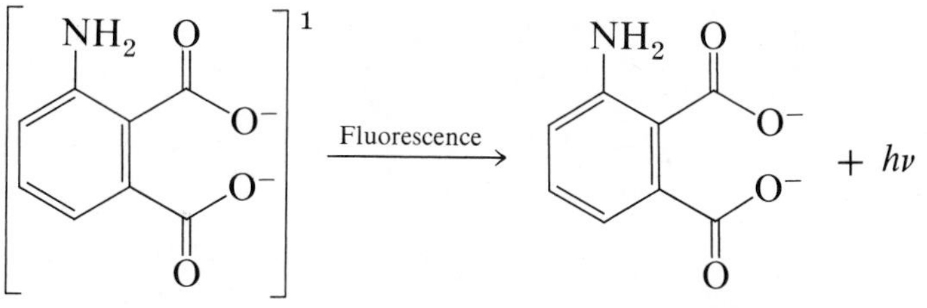

S₁

**3-Aminophthalate
Ground-state dianion, S₀**

The dianion of luminol undergoes a reaction with molecular oxygen to form a peroxide of unknown structure. This peroxide is unstable and decomposes with the evolution of nitrogen gas, producing the 3-aminophthalate dianion in an electronically excited state. The excited dianion emits a photon that is visible as light. One very attractive hypothesis for the structure of the peroxide postulates a cyclic endoperoxide that decomposes by the following mechanism:

A POSTULATE

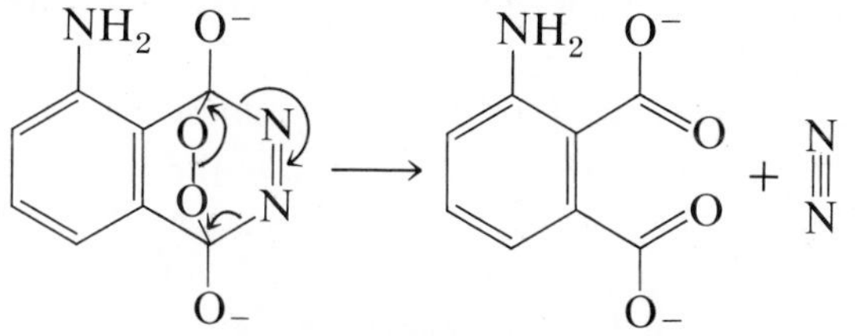

Certain experimental facts argue against this intermediate, however. For instance, certain acyclic hydrazides that cannot form a similar intermediate have also been found to be chemiluminescent.

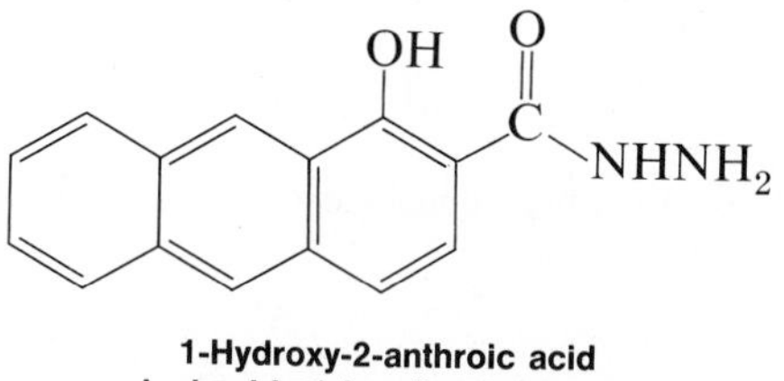

**1-Hydroxy-2-anthroic acid
hydrazide (chemiluminescent)**

Although the nature of the peroxide is still debatable, the remainder of the reaction is well understood. The chemical products of the reaction have been shown to be the 3-aminophthalate dianion and molecular nitrogen. The intermediate that emits light has been identified definitely as the **excited state singlet** of the 3-aminophthalate dianion.[1] Thus, the fluorescence emission spectrum of the 3-aminophthalate dianion (produced by photon absorption) is identical to the spectrum of the light emitted from the chemiluminescent reaction. However, for numerous complicated reasons, it is believed that the 3-aminophthalate dianion is formed first as a vibrationally excited triplet

---

[1]The terms *singlet, triplet, intersystem crossing, energy transfer,* and *quenching* are explained in the introduction to Experiment 52.

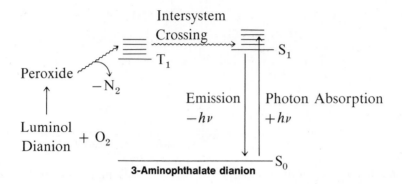

state molecule, which makes the intersystem crossing to the singlet state before emission of a photon.

The excited state of the 3-aminophthalate dianion may be quenched by suitable acceptor molecules, or the energy (about 50–80 Kcal/mole) may be transferred to give emission from the acceptor molecules. Several such experiments are described in the following procedure.

The system chosen for the chemiluminescence studies of luminol in this experiment uses dimethylsulfoxide, $(CH_3)_2SO$, as the solvent; potassium hydroxide as the base required for the formation of the dianion of luminol; and molecular oxygen. Several alternative systems have been used, substituting hydrogen peroxide and an oxidizing agent for molecular oxygen. An aqueous system using potassium ferricyanide and hydrogen peroxide is an alternative system used frequently.

REFERENCES

Rahaut, M. M. "Chemiluminescence from Concerted Peroxide Decomposition Reactions." *Accounts of Chemical Research, 2* (1969): 80.

White, E. H., and Roswell, D. F. "The Chemiluminescence of Organic Hydrazides." *Accounts of Chemical Research, 3* (1970): 54.

SPECIAL INSTRUCTIONS

Before doing this experiment, it will be helpful to read the introduction to Experiment 52, which defines many photochemical terms. The essay "Fireflies and Photochemistry," which precedes this experiment, also contains many relevant details.

This entire experiment can be completed in less than 1 hour. When you are working with hydrazine, you should remember that it is toxic and should not be spilled on the skin. It is also a suspect carcinogen. Dimethylsulfoxide may also be toxic; avoid breathing the vapors or spilling it on your skin.

A darkened room is required to observe adequately the chemiluminescence of luminol. A darkened hood that has had its window covered with butcher paper or aluminum foil also works well. Other fluorescent dyes besides those mentioned (for instance, 9,10-diphenylanthracene) can also be used for the energy-transfer experiments. The dyes selected may depend on what is immediately available. The instructor may have each student use one dye for the energy-transfer experiments, with one student making a comparison experiment without a dye.

PROCEDURE

5-NITROPHTHALHYDRAZIDE

Place 1.3 g of 3-nitrophthalic acid and 2 mL of a 10% aqueous solution of hydrazine (use gloves) in a large sidearm test tube.[2] At the same time, heat 20 mL of water in a beaker on a steam bath. Heat the test tube over a microburner until the solid dissolves. Add 4 mL of triethylene glycol and clamp the test tube in an upright position on a ring stand. Place a thermometer (do not seal the system) and a boiling stone in the test tube and attach a piece of pressure tubing to the sidearm. Connect this tubing to an aspirator (use a trap). Heat the solution with a microburner until the liquid boils vigorously and the refluxing water vapor is drawn away by the aspirator vacuum (the temperature will rise to about 120 °C). Continue heating and allow the temperature to increase rapidly until it rises above 200 °C. About 5 minutes will be required for this heating. Remove the burner briefly when this temperature has been achieved, and then resume gentle heating to maintain a fairly constant temperature of 210 to 220 °C for about 2 minutes. Allow the test tube to cool to about 100 °C, add the 20 mL of hot water that was prepared previously, and cool the test tube to room temperature by allowing tap water to flow over the outside of the test tube. Collect the light yellow crystals of 5-nitrophthalhydrazide by vacuum filtration, using a small Büchner funnel. It is not necessary to dry the product before you go on with the next reaction step.

LUMINOL (5-AMINOPHTHALHYDRAZIDE)

Transfer the moist 5-nitrophthalhydrazide to a 20 × 150-mm test tube. Add 6.5 mL of a 10% sodium hydroxide solution, and stir the mixture until the hydrazide dissolves. Add 4 g of sodium dithionite dihydrate (sodium hydrosulfite dihydrate, $Na_2S_2O_4 \cdot 2H_2O$). Using a capillary pipet, add enough water to wash the solid from the walls of the test tube. Heat the test tube until the solution boils, stir the solution, and maintain the boiling, with stirring, for 5 minutes. Add 2.6 mL of glacial acetic acid and cool the test tube to room temperature by allowing tap water to flow over the outside of it. Stir the mixture during this cooling step. Collect the light yellow crystals of luminol by vacuum filtration, using a small Büchner funnel. Save a small sample of this product, allow it to dry overnight, and determine its melting point (mp 319–320 °C). The remainder of the luminol may be used without drying for the chemiluminescence experiments.

---

[2]A 10% aqueous solution of hydrazine can be prepared by diluting 15.6 g of a commercial 64% hydrazine solution to a volume of 100 mL using water.

FLUORESCENT DYE	COLOR
No dye	Faint bluish white
2,6-Dichloroindophenol	Blue
9-Aminoacridine	Blue green
Eosin	Salmon pink
Fluorescein	Yellow green
Dichlorofluorescein	Yellow orange
Rhodamine B	Green
Phenolphthalein	Purple

CHEMILUMINESCENCE EXPERIMENTS

Cover the bottom of a 125-mL Erlenmeyer flask with a layer of potassium hydroxide pellets. Add enough dimethylsulfoxide to cover the pellets and 0.2 to 0.3 g of the moist luminol to the flask, stopper it, and shake it vigorously to mix air into the solution.[3] In a dark room, a faint glow of blue-white light will be visible. The intensity of the glow will increase with continued shaking of the flask and occasional removal of the stopper to admit more air.

To observe energy transfer to a fluorescent dye, dissolve a few crystals (1–5 mg) of the indicator dye in 2 to 3 mL of water. Add the dye solution to the dimethylsulfoxide solution of luminol, stopper the flask, and shake the mixture vigorously. Observe the intensity and the color of the light produced.

A table of some dyes and the colors produced when they are mixed with luminol is given above. Other dyes not included on this list may also be tested in this experiment.

Experiment 54
cis-,trans-, and trans,trans-1,4-Diphenyl-1,3-butadiene

Wittig Reaction
Phase-Transfer Catalysis
Thin-Layer Chromatography
Geometric Isomerism
Uv/nmr Spectroscopy

The Wittig reaction is often used to form alkenes from carbonyl compounds. In this experiment the isomeric dienes *cis-,trans-*, and *trans,trans*-1,4-diphenyl-1,3-butadiene

---

[3]An alternative method for demonstrating chemiluminescence, using potassium ferricyanide and hydrogen peroxide as oxidizing agents, is described in E. H. Huntress, L. N. Stanley, and A. S. Parker, *Journal of Chemical Education, 11* (1934): 142.

will be formed from cinnamaldehyde and benzyltriphenylphosphonium chloride Wittig reagent.

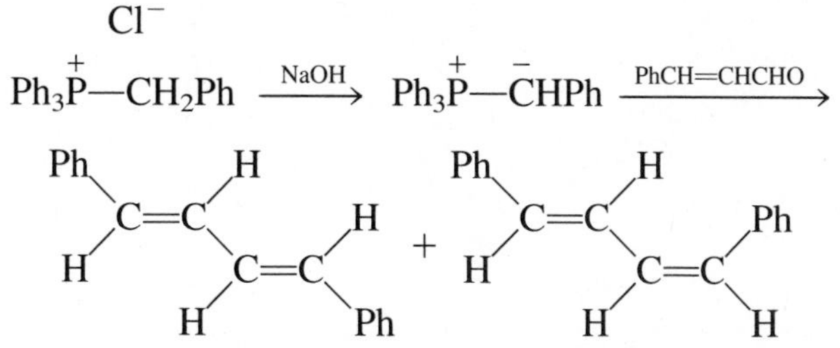

$$Cl^-$$

$$Ph_3\overset{+}{P}-CH_2Ph \xrightarrow{\text{NaOH}} Ph_3\overset{+}{P}-\overset{-}{C}HPh \xrightarrow{\text{PhCH=CHCHO}}$$

trans,trans   *cis,trans*

The reaction is carried out in two steps. First, the Wittig reagent (a salt) is formed by reaction of triphenylphosphine with benzyl chloride. The reaction is a simple nucleophilic displacement of chloride ion by triphenylphosphine. The salt that is formed is called the "Wittig reagent" or "Wittig salt."

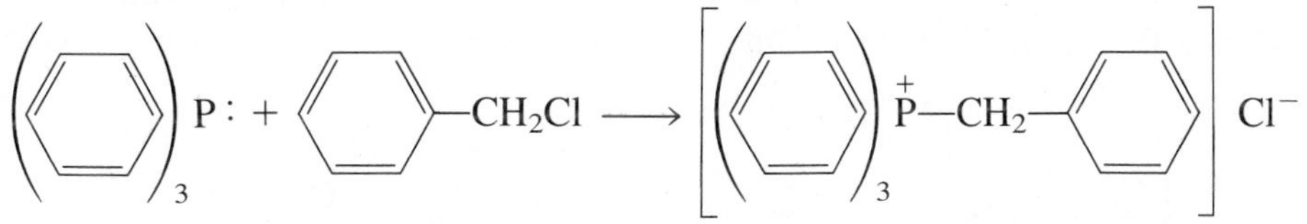

Benzyltriphenylphosphonium chloride

When treated with base, the Wittig salt forms an **ylide.** An ylide is a species having adjacent atoms oppositely charged. The ylide is stabilized due to the ability of

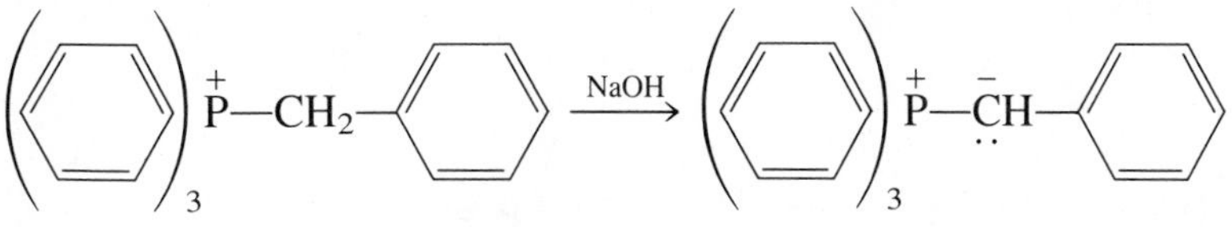

An ylide

phosphorus to accept more than eight electrons in its valence shell. Phosphorus uses its 3d orbitals to form the overlap with the 2p orbital of carbon, which is necessary for resonance stabilization. This stabilizes the carbanion.

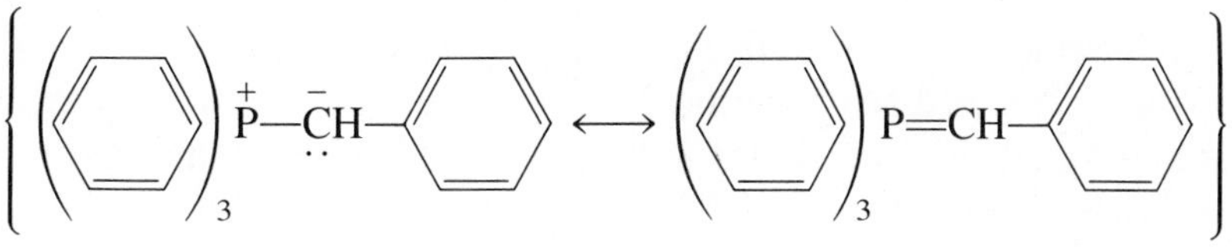

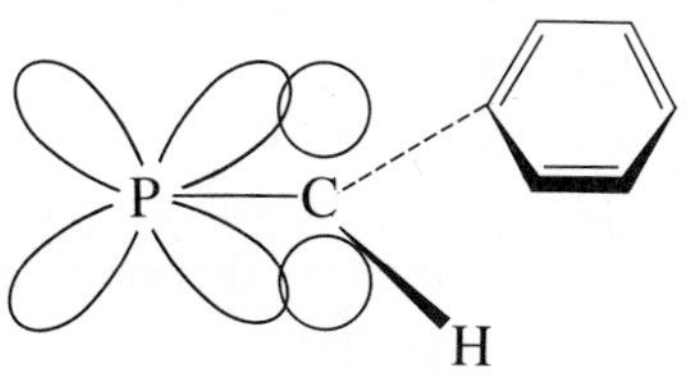

The ylide is a carbanion that acts as a nucleophile, and it adds to the carbonyl group.

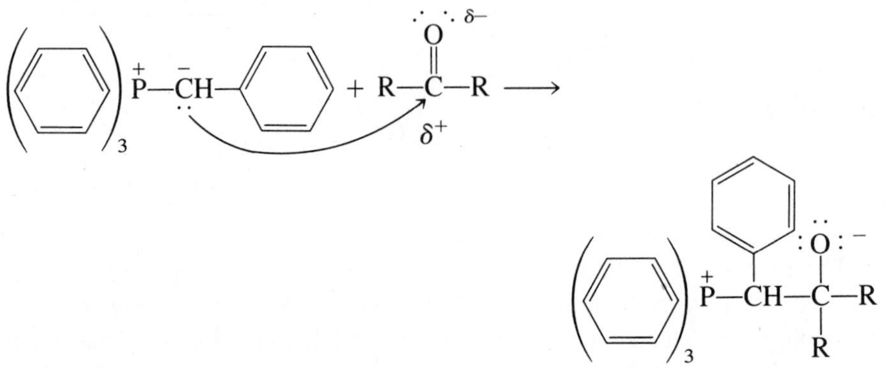

Following this initial nucleophilic addition, a remarkable sequence of events occurs, as outlined in the following mechanism:

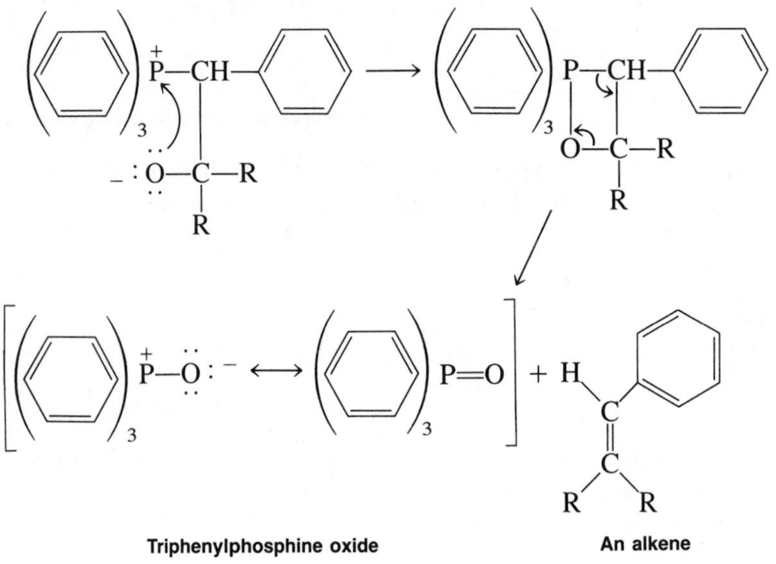

Triphenylphosphine oxide   **An alkene**

The addition intermediate, formed from the ylide and the carbonyl compound, cyclizes to form a four-membered-ring intermediate. This new intermediate is unstable and fragments into an alkene and triphenylphosphine oxide. Notice that the ring breaks open in a different way than it was formed. This is due primarily to the fact that triphenylphosphine oxide is very stable thermodynamically. A large decrease in potential energy is achieved upon its formation.

In this experiment, cinnamaldehyde is used as the carbonyl compound and gives mainly *trans,trans*-1,4-diphenyl-1,3-butadiene. The *cis,trans* isomer is obtained in smaller amounts. The *trans,trans* isomer is thermodynamically the most stable isomer of this compound and is formed preferentially.

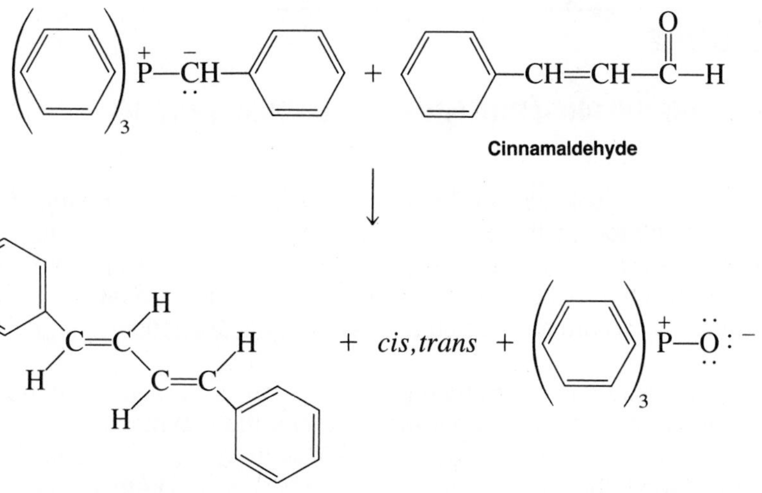

Cinnamaldehyde

trans,trans-1,4-Diphenyl-1,3-butadiene + *cis,trans* + Triphenylphosphine oxide

This reaction involves the use of phase-transfer catalysis, which is described in a general way in Experiment 28 on pages 206 to 208. A two-phase reaction mixture is used where the aqueous phase initially contains the Wittig salt and sodium hydroxide and the methylene chloride (organic) phase contains cinnamaldehyde. The Wittig salt, a phosphonium salt, acts as the phase-transfer catalyst, and "transfers" the ylide produced in the aqueous phase to the organic phase, where the isomeric dienes are produced.

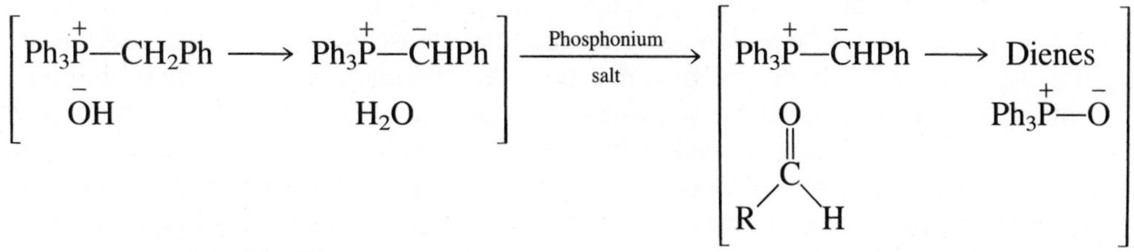

Aqueous phase   Organic phase

The reaction occurs in a "one-pot" reaction vessel, namely an Erlenmeyer flask.

SPECIAL INSTRUCTIONS

You should read Technique 11 (Thin-layer Chromatography), Sections 11.3 and 11.4, pp. 620 to 623. In addition, you should read about phase-transfer catalysis in Experiment 28, pages 206 to 208. Avoid contact with the 50% sodium hydroxide solution, because it is very caustic and can cause burns. Triphenylphosphine is rather toxic. Be careful not to inhale the dust. Benzyl chloride is a skin irritant and a lachrymator. It should be handled in the hood with care.

PROCEDURE

BENZYLTRIPHENYLPHOSPHONIUM CHLORIDE (WITTIG SALT)

Place 3.8 g of benzyl chloride, 11.0 g of triphenylphosphine, and 60 mL of p-cymene in a 500-mL round-bottomed, three-necked flask fitted with a reflux condenser. Stopper the unused openings. The large flask prevents loss of material resulting from bumping when solid forms during the reaction. Add a boiling stone. Using a heating mantle as a heat source, reflux the mixture for 1.5 hours. If reflux continues longer, more product will be obtained.

Cool the mixture in an ice bath for 15 minutes and collect the phosphonium salt by vacuum filtration using a Büchner funnel. Since the p-cymene is somewhat expensive, place the filtrate in the container provided for recycling.[1] Reassemble the filtration apparatus and wash the crystals with a 5-mL portion of cold petroleum ether to remove the residual p-cymene. Dry the crystals, weigh them, and calculate the percentage yield of phosphonium salt. At the option of the instructor, obtain the proton nmr spectrum of the salt in $CDCl_3$. The methylene group appears as a doublet (J = 14 Hz) at 5.5δ because of ^1H-^{31}P coupling.

1,4-DIPHENYL-1,3-BUTADIENE

Place 7.8 g (0.02 mole) of benzyltriphenylphosphonium chloride, 2.64 g (0.02 mole) of cinnamaldehyde, and 10 mL of methylene chloride in a 250-mL Erlenmeyer flask. Add 10 mL of 50% aqueous sodium hydroxide,[2] add a magnetic stirring bar, and stir the mixture vigorously for 10 minutes using a magnetic stirrer. After this period, add 40 mL of methylene chloride and 30 mL of water; separate the layers in a separatory funnel. If you have an emulsion, allow the separatory funnel to stand for about 5 minutes.[3] Save the lower layer. Dry the organic layer with magnesium sulfate, filter off the drying agent, and remove the methylene chloride by distillation on a steam bath. An oil is obtained, which solidifies as it cools. This mixture contains triphenylphosphine oxide, *trans,trans*-1,4-diphenyl-1,3-butadiene, and the *cis,trans* isomer. Analyze the mixture by thin-layer chromatography using silica gel plates, with a fluorescent indicator (Eastman Chromagram Sheet, No. 13181). Dissolve a small amount of the solid in methylene chloride and spot the plate in the usual manner (Technique 11, Section 11.3, pp. 620–621). Use petroleum ether (30–60 °C) as a solvent to develop (run) the plate, and visualize the spots with a uv lamp. A typical set of R_f values is presented below:

Ph_3P^+—O^-	0
trans,trans diene	0.36 (fluoresces brilliantly)
cis,trans diene	0.47

---

[1] The laboratory assistant should distill the recovered solvent for reuse.

[2] This solution should be prepared in large quantities by the laboratory assistant, using 50 g of sodium hydroxide for every 50 mL of water.

[3] Sometimes a more difficult emulsion has formed (three layers). If this is the case, pass the two lower layers through a cotton plug in a funnel. Add about 8 g of anhydrous magnesium sulfate, swirl, and allow the mixture to stand for about 10 minutes or until the layer clears. Gravity-filter the mixture through filter paper. The solution should be clear at this point. If it is not, add a little more magnesium sulfate and repeat the filtration procedure.

Triphenylphosphine oxide can be removed by adding 70 mL of 60% aqueous ethanol to the crude solid. The solution of this concentration may be prepared by mixing 60 mL of 95% ethanol with 40 mL of water. Break up the solid with a spatula or the bottom of a test tube, so that the solid is as finely divided as possible. Collect the crude solid *trans,trans* isomer by vacuum filtration on a Büchner funnel and wash it with small portions of ice-cold 60% aqueous ethanol. Transfer the solid to an Erlenmeyer flask. Most of the liquid *cis,trans* isomer passes through the filter and settles out in the filtrate. Be sure to save the filtrate so that the *cis,trans* isomer can be isolated by the procedure given below (see Isolation of the *cis,trans* Isomer).

ISOLATION OF THE *TRANS,TRANS* ISOMER

The pure *trans,trans* isomer can be isolated by adding 15 mL of 95% ethanol to the solid obtained above. Heat the mixture on a steam bath and break up all of the pieces. The solid will not dissolve completely. This procedure removes any remaining *cis,trans* isomer since it is soluble in 95% ethanol. Place the mixture in an ice/salt bath for 10 minutes, collect the iridescent crystals on a small Büchner funnel, and rinse the solid with 5 mL of ice-cold 95% ethanol to give nearly pure *trans,trans*-1,4-diphenyl-1,3-butadiene.[4] Dry the solid at room temperature and determine the melting point (literature, 152–153 °C). The solid can be recrystallized from hot 95% ethanol. Weigh the solid and determine the percentage yield.

ISOLATION OF THE *CIS,TRANS* ISOMER

The filtrate that was saved contains the *cis,trans* isomer. Decant the filtrate away from the insoluble oil remaining in the bottom of the flask. Dissolve the oil in methylene chloride, dry it over magnesium sulfate, filter the organic phase away from the drying agent, and remove the solvent by evaporation in a hood to give *cis,trans*-1,4-diphenyl-1,3-butadiene. It should be kept in the dark, because it is converted to the solid *trans,trans* isomer upon standing in light. Weigh the oil and determine the percentage yield.

Analyze the two isomers by thin-layer chromatography during the same laboratory to check for their purity. At the option of the instructor, obtain the proton nmr spectrum of each isomer. Some triphenylphosphine oxide may be observed at 7.5 to 7.8 δ. The isomers can be differentiated easily by nmr.

ISOMERIZATION OF THE *CIS,TRANS* ISOMER
TO *TRANS,TRANS* (OPTIONAL)

Obtain the uv spectra of the two isomers in hexane. Dissolve 10-mg samples in 100 mL of hexane in a volumetric flask. Remove 10 mL of this solution and dilute it to 100 mL in

---

[4]The filtrate contains some of the *cis,trans* isomer, which can be isolated by adding an equal volume of water and setting the mixture aside for about 10 minutes. The *cis,trans* isomer collects at the bottom as an oil. Decant the majority of the liquid away from the oil, and add methylene chloride to dissolve the oil. Dry the liquid with magnesium sulfate, and remove the solvent in a hood. This oil can be combined with the other *cis,trans* isomer isolated in the normal way.

another volumetric flask. This concentration should be adequate for analysis. The *trans,trans* isomer absorbs at 328 nm and possesses fine structure, while the *cis,trans* isomer absorbs at 313 nm and has a smooth curve.[5] Isomerize the *cis,trans* isomer to the *trans,trans* isomer by adding a small crystal of iodine and illuminating the mixture with a 150-watt floodlamp for about 10 minutes. Analyze the results by uv spectroscopy. Alternatively, the analysis may be done by proton nmr. Submit the spectral data with your laboratory report.

QUESTIONS

1. There is an additional isomer of 1,4-diphenyl-1,3-butadiene (mp 70 °C) that has not been shown in this experiment. Draw the structure and name it. Why is it not produced in this experiment? (*Hint:* the cinnamaldehyde has *trans* stereochemistry.)

2. Why should the *trans,trans* isomer be the thermodynamically most stable one?

3. A lower yield of phosphonium salt is obtained in refluxing benzene than in *p*-cymene. Look up the boiling points for these solvents and explain why the difference in boiling points might influence the yield.

4. Outline a synthesis for *cis* and *trans* stilbene (the 1,2-diphenylethenes) using the Wittig reaction.

5. The sex attractant of the female housefly (*Musca domestica*) is called **muscalure**, and its structure is shown below. Outline a synthesis of muscalure, using the Wittig reaction. Will your synthesis lead to the required *cis* isomer?

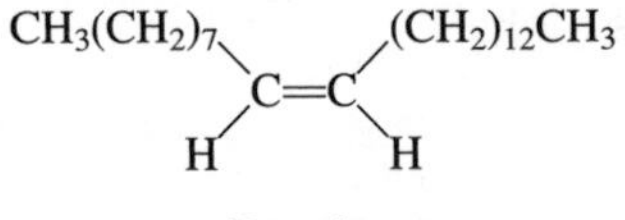

$$CH_3(CH_2)_7 \qquad (CH_2)_{12}CH_3$$
$$C=C$$
$$H \qquad\qquad H$$

Muscalure

Essay
POLYMERS AND PLASTICS

Chemically, plastics are composed of chainlike molecules of high molecular weight, called polymers. Polymers have been built up from simpler chemicals, called monomers. A different monomer or combination of monomers is used to manufacture each different type or family of polymers. There are many polymers around us that are familiar. Examples of synthetic polymers are Teflon, nylon, Dacron, polyethylene,

---

[5]The comparative study of the stereoisomeric 1,4-diphenyl-1,3-butadienes has been published: Pinckard, J. H., Wille, B., and Zechmeister, L. *Journal of the American Chemical Society, 70* (1948): 1938.

polyester, Orlon, epoxy, vinyl, polyurethane, silicones, Lucite, and boat resin. Examples of natural polymers are starch and cellulose (from glucose), rubber (from isoprene), and proteins (from amino acids). Certainly, polymers have had a great influence on our society. They are rapidly replacing many metals in manufacturing. In addition, synthetic polymeric textiles are replacing natural fibers for making cloth. As these materials have been created, a problem has arisen in disposing of them, since many are not biodegradable.

CHEMICAL STRUCTURES OF POLYMERS

Basically a polymer is made up of many repeating molecular units formed by sequential addition of monomer molecules to one another. Many monomer molecules of A, say 1000 to 1 million, can be linked to form a gigantic polymeric molecule:

$$\text{Many A} \longrightarrow \text{etc.—A-A-A-A-A—etc.} \qquad \text{or} \qquad -(A)_n$$

Monomer molecules   **Polymer molecule**

Monomers that are different can also be linked to form a polymer with an alternating structure. This type of polymer is called a copolymer.

$$\text{Many A + many B} \longrightarrow \text{etc.—A-B-A-B-A-B—etc.} \qquad \text{or} \qquad -(A\text{-}B)_n$$

Monomer molecules   **Polymer molecule**

TYPES OF POLYMERS

For convenience, chemists classify polymers in several main groups, depending on method of synthesis.

1. Addition polymers are formed by a reaction in which monomer units simply add to one another to form a long-chain (generally linear or branched) polymer. The monomers usually contain carbon-carbon double bonds. Familiar examples of addition polymers are polyethylene and Teflon. The process can be represented as follows:

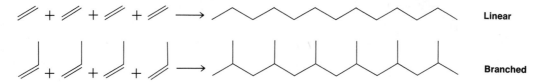

2. Condensation polymers are formed by reaction of bifunctional or polyfunctional molecules, with the elimination of some small molecule (such as water, ammonia, or hydrogen chloride) as a by-product. Familiar examples of condensation polymers are nylon, Dacron, and polyurethane. The process can be represented as follows:

$$\text{H}—\boxed{}—\text{X} + \text{H}—\boxed{}—\text{X} \longrightarrow \text{H}—\boxed{}—\boxed{}—\text{X} + \text{HX}$$

3. Cross-linked polymers are formed when long chains are linked in one gigantic, three-dimensional structure with tremendous rigidity. Addition and condensation polymers can exist with a cross-linked network, depending on the monomers used in the synthesis. Familiar examples of cross-linked polymers are Bakelite, rubber, and casting (boat) resin. The process can be represented as follows:

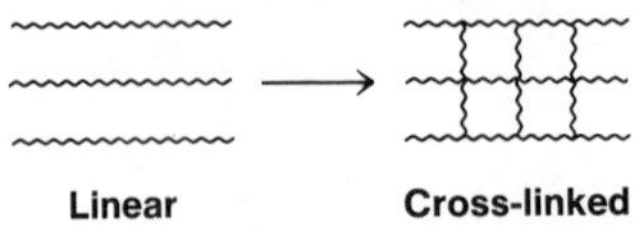

Linear   Cross-linked

Industrialists and technologists often classify polymers in other categories also:

1. Thermoplastics are materials that can be softened (melted) by heat and re-formed (molded) into another shape. Weaker, noncovalent bonds are broken during the heating. Technically, thermoplastics are the materials we call plastics. Both addition and condensation polymers can be so classified. Familiar examples include polyethylene (addition polymer) and nylon (condensation polymer).

2. Thermoset plastics are materials that melt initially but on further heating become permanently hardened. They cannot be softened and remolded without destruction of the polymer because covalent bonds are broken. Chemically, thermoset plastics are cross-linked polymers. Bakelite is an example of a thermoset plastic.

Polymers can also be classified in other ways; for example, many varieties of rubber are often referred to as elastomers, Dacron is a fiber, and polyvinyl acetate is an adhesive. The addition and condensation classification will be used in this essay.

ADDITION POLYMERS

Most of the polymers made are of the addition type. The monomers generally contain a carbon-carbon double bond. The most important example of an addition polymer is the well-known polyethylene, for which the monomer is ethylene. Countless numbers of ethylene molecules are linked in long-chain polymeric molecules by breaking the pi bond and creating two new single bonds between the monomer units. The number of recurring units may be large or small, depending on the polymerization conditions.

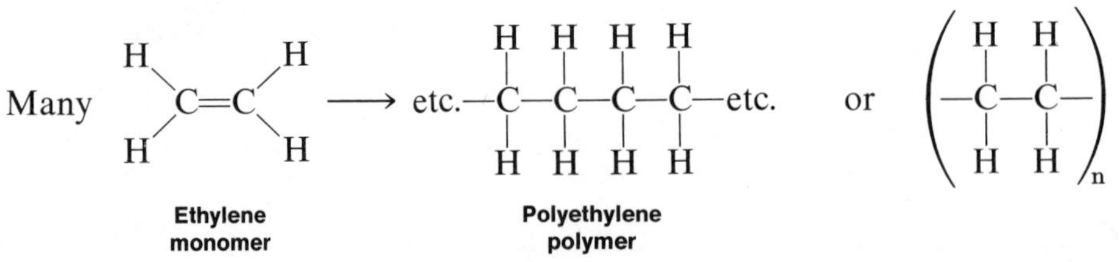

Ethylene
monomer

Polyethylene
polymer

This reaction can be promoted by heat, pressure, and a chemical catalyst. The molecules produced in a typical reaction vary in the number of carbon atoms in their chains. In other words, a mixture of polymers of varying length is produced, rather than a pure compound.

Polyethylenes, with linear structures, can pack together easily and are referred to as high-density polyethylenes. They are fairly rigid materials. Low-density polyethylenes consist of branched-chain molecules, with some cross-linking in the chains. They are more flexible than the high-density polyethylenes. The reaction conditions and the catalysts that produce polyethylenes of low and high density are quite different. The monomer, however, is the same in each case.

Another example of an addition polymer is polypropylene. In this case, the monomer is propylene. The polymer that results has a branched methyl on alternate carbon atoms of the chain.

Many
$$\underset{\substack{\text{Propylene}\\\text{monomer}}}{\text{H}\!\!\diagdown\!\!\underset{\text{H}}{\overset{\text{H}}{\text{C}}}\!\!=\!\!\underset{\text{CH}_3}{\overset{\text{H}}{\text{C}}}\!\!\diagup} \longrightarrow \text{etc.} \underset{\substack{\text{Polypropylene}\\\text{polymer}}}{-\overset{\text{H}}{\underset{\text{H}}{\text{C}}}\!-\!\overset{\text{H}}{\underset{\text{CH}_3}{\text{C}}}\!-\!-\!\overset{\text{H}}{\underset{\text{H}}{\text{C}}}\!-\!\overset{\text{H}}{\underset{\text{CH}_3}{\text{C}}}\!-\text{etc.}} \quad \text{or} \quad \left(\!-\overset{\text{H}}{\underset{\text{H}}{\text{C}}}\!-\!\overset{\text{H}}{\underset{\text{CH}_3}{\text{C}}}\!-\!\right)_{\!n}$$

Several common addition polymers are shown in Table 1. Some of their principal uses are also listed. The last three entries in the table all have a carbon-carbon double bond remaining after the polymer is formed. These bonds activate or participate in a further reaction to form cross-linked polymers called elastomers; this term is almost synonymous with *rubber*, since they designate materials with common characteristics.

CONDENSATION POLYMERS

Condensation polymers, for which the monomers contain more than one type of functional group, are more complex than addition polymers. In addition, most condensation polymers are copolymers made from more than one type of monomer. You will recall that addition polymers, in contrast, are all prepared from substituted ethylene molecules. The single functional group in each case is one or more double bonds, and a single type of monomer is generally used.

Dacron, a polyester, can be prepared by causing a dicarboxylic acid to react with a bifunctional alcohol (a diol):

$$\underset{\substack{\text{Terephthalic}\\\text{acid}}}{\text{HO}-\overset{\text{O}}{\overset{\|}{\text{C}}}\!\!-\!\!\bigcirc\!\!-\!\overset{\text{O}}{\overset{\|}{\text{C}}}\!-\!\boxed{\text{OH} \quad \text{H}}\!-} \underset{\substack{\text{Ethylene}\\\text{glycol}}}{\text{OCH}_2\text{CH}_2\text{OH}} \longrightarrow$$

$$\underset{\text{Dacron}}{-\overset{\text{O}}{\overset{\|}{\text{C}}}\!-\!\bigcirc\!-\!\overset{\text{O}}{\overset{\|}{\text{C}}}\!-\!\text{OCH}_2\text{CH}_2\!-\!\text{O}\!-} + \text{H}_2\text{O}$$

TABLE 1. Addition Polymers

EXAMPLE	MONOMER(S)	POLYMER	USE
Polyethylene	$CH_2{=}CH_2$	$-CH_2-CH_2-$	Most common and important polymer; bags, insulation for wires, squeeze bottles
Polypropylene	$CH_2{=}CH$ $\quad\quad CH_3$	$-CH_2-CH-$ $\quad\quad CH_3$	Fibers, indoor-outdoor carpets, bottles
Polystyrene	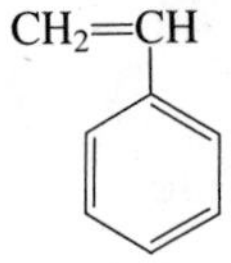 $CH_2{=}CH$ ⬡	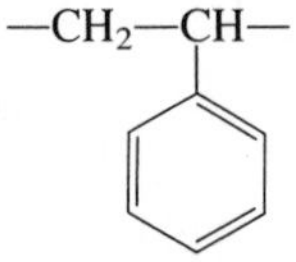 $-CH_2-CH-$ ⬡	Styrofoam, inexpensive household goods, inexpensive molded objects
Polyvinyl chloride (PVC)	$CH_2{=}CH$ $\quad\quad Cl$	$-CH_2-CH-$ $\quad\quad Cl$	Synthetic leather, clear bottles, floor covering. phonograph records, water pipe
Polytetrafluoroethylene (Teflon)	$CF_2{=}CF_2$	$-CF_2-CF_2-$	Nonstick surfaces, chemically resistant films
Polymethyl methacrylate (Lucite, Plexiglas)	$\quad\quad CO_2CH_3$ $CH_2{=}C$ $\quad\quad CH_3$	$\quad\quad CO_2CH_3$ $-CH_2-C-$ $\quad\quad CH_3$	Unbreakable "glass," latex paints
Polyacrylonitrile (Orlon, Acrilan, Creslan)	$CH_2{=}CH$ $\quad\quad CN$	$-CH_2-CH-$ $\quad\quad CN$	Fiber used in sweaters, blankets, carpets
Polyvinyl acetate (PVA)	$CH_2{=}CH$ $\quad\quad OCCH_3$ $\quad\quad\ \ |\ $ $\quad\quad\ \ O$	$-CH_2-CH-$ $\quad\quad OCCH_3$ $\quad\quad\ \ |\ $ $\quad\quad\ \ O$	Adhesives, latex paints, chewing gum, textile coatings
Natural rubber	$\quad\quad\ \ CH_3$ $CH_2{=}CCH{=}CH_2$	$\quad\quad\ \ CH_3$ $-CH_2-C{=}CH-CH_2-$	The polymer is cross-linked with sulfur (vulcanization).
Polychloroprene (neoprene rubber)	$\quad\quad\ \ Cl$ $CH_2{=}CCH{=}CH_2$	$\quad\quad\ \ Cl$ $-CH_2-C{=}CH-CH_2-$	Cross-linked with ZnO; resistant to oil and gasoline
Styrene butadiene rubber (SBR)	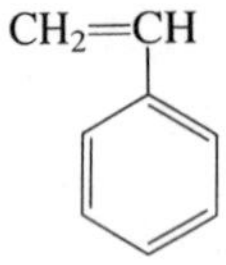 $CH_2{=}CH$ ⬡ $CH_2{=}CHCH{=}CH_2$	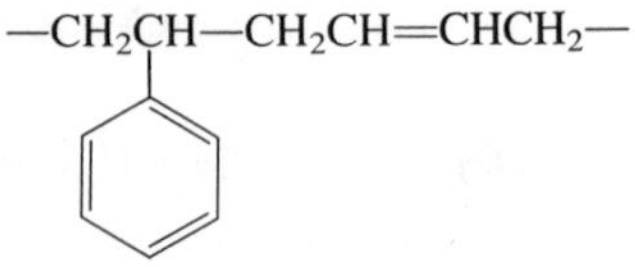 $-CH_2CH-CH_2CH{=}CHCH_2-$ ⬡	Cross-linked with peroxides; most common rubber; used for tires; 25% styrene 75% butadiene

Nylon 6-6, a polyamide, can be prepared by causing a dicarboxylic acid to react with a bifunctional amine:

$$
\underset{\substack{\text{Adipic}\\\text{acid}}}{\text{HO}-\overset{\displaystyle O}{\overset{\|}{\text{C}}}(\text{CH}_2)_4\overset{\displaystyle O}{\overset{\|}{\text{C}}}\boxed{-\text{OH} \quad \text{H}-}\underset{\substack{\text{Hexamethylene-}\\\text{diamine}}}{\overset{\displaystyle H}{\overset{|}{\text{N}}}(\text{CH}_2)_6\overset{\displaystyle H}{\overset{|}{\text{N}}}\text{H}} \longrightarrow \underset{\text{Nylon}}{-\overset{\displaystyle O}{\overset{\|}{\text{C}}}(\text{CH}_2)_4\overset{\displaystyle O}{\overset{\|}{\text{C}}}-\overset{}{\underset{\displaystyle H}{\underset{|}{\text{N}}}}(\text{CH}_2)_6\overset{}{\underset{\displaystyle H}{\underset{|}{\text{N}}}}-} + \text{H}_2\text{O}
$$

Notice, in each case, that a small molecule, water, is eliminated as a product of the reaction. Several other condensation polymers are listed in Table 2. Linear (or branched) chain polymers as well as cross-linked polymers are produced in condensation reactions.

The nylon structure contains the amide linkage at regular intervals,

$$
-\overset{\displaystyle O}{\overset{\|}{\text{C}}}-\overset{\displaystyle H}{\overset{|}{\text{N}}}-
$$

This type of linkage is extremely important in nature because of its presence in proteins and polypeptides. Proteins are gigantic polymeric substances made up of monomer units of amino acids. They are linked by the peptide (amide) bond.

Other important natural condensation polymers are starch and cellulose. They are polymeric materials made up of the sugar monomer glucose. Another important natural condensation polymer is the DNA molecule. A DNA molecule is made up of the sugar deoxyribose linked with phosphates to form the backbone of the molecule. A portion of a DNA molecule is shown in the essay that precedes Experiment 6.

PROBLEMS WITH PLASTICS

Plastics have certainly become very common in our society. However, they are not without problems. There are disposal problems, health hazards, littering problems, fire hazards, and energy shortages associated with their manufacture and use.

Plasticizers and Health Hazards

Certain types of plastics such as polyvinyl chloride (PVC) are mixed with plasticizers that soften the plastic so that it is more pliable. If plasticizers were not added, the plastic would be hard and brittle. Some of the plasticizers used in vinyl plastics are phthalate esters. The structure of a phthalate ester is shown on page 411. These esters are volatile compounds of low molecular weight. Part of the new car ''smell'' comes from the odor of these esters as they evaporate from the vinyl upholstery. The vapor often condenses on the windshield as an oily, insoluble film. After some time, the vinyl material may lose enough plasticizer to cause it to crack. Phthalate esters may constitute a health hazard. Sometimes vinyl containers incorporating phthalate plasticizers are used to store blood. The esters are leached from blood bags made of PVC and may be partly responsible for shock lung, a condition that sometimes leads to death during a blood transfusion. The long-term effects of these plasticizers are not known.

TABLE 2. Condensation Polymers

EXAMPLE	MONOMERS	POLYMER	USE
Polyamides (nylon)	$HOC(CH_2)_nCOH$ $H_2N(CH_2)_nNH_2$	$-C(CH_2)_nC-NH(CH_2)_nNH-$	Fibers, molded objects
Polyesters (Dacron, Mylar, Fortrel)	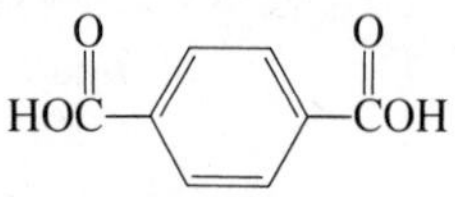 $HOC\text{—}\bigcirc\text{—}COH$ $HO(CH_2)_nOH$	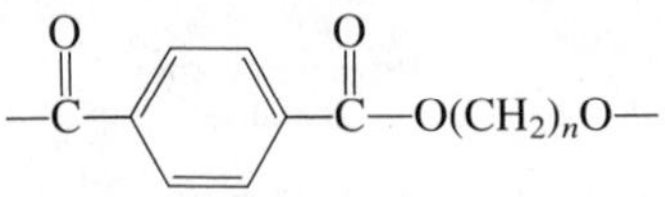 $-C\text{—}\bigcirc\text{—}C-O(CH_2)_nO-$	Linear polyesters; fibers, recording tape
Polyesters (Glyptal resin)	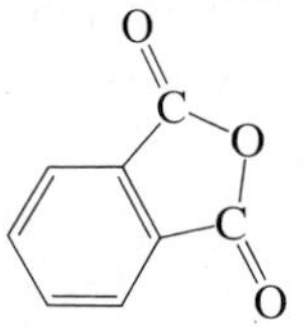 (phthalic anhydride) $HOCH_2CHCH_2OH$ $\;\;\;\;\;\;\;\;OH$	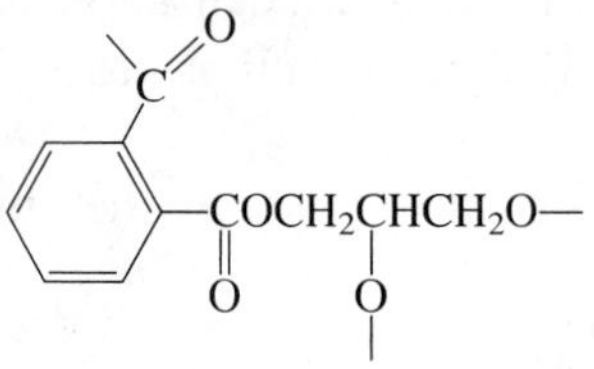 $-COCH_2CHCH_2O-$	Cross-linked polyester; paints
Polyesters (casting resin)	$HOCCH=CHCOH$ $HO(CH_2)_nOH$	$-CCH=CHC-O(CH_2)_nO-$	Cross-linked with styrene and peroxide; fiberglass boat resin
Phenol-formaldehyde resin (Bakelite)	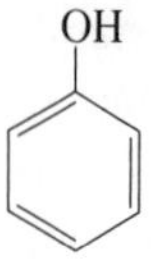 (phenol) $\;\;\;CH_2=O$	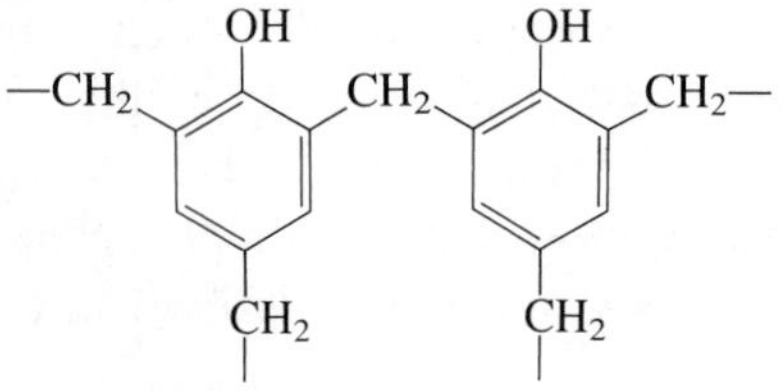 $-CH_2-$ (phenolic network) $-CH_2-$	Mixed with fillers; molded electrical goods, adhesives, laminates, varnishes
Cellulose acetate*	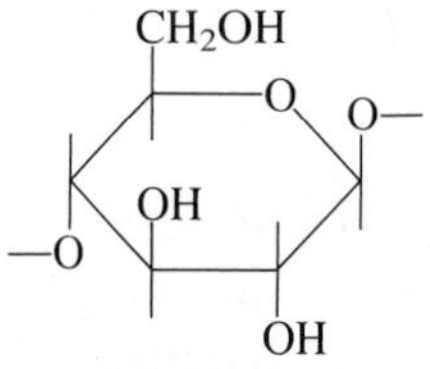 (glucose unit) CH_3COOH	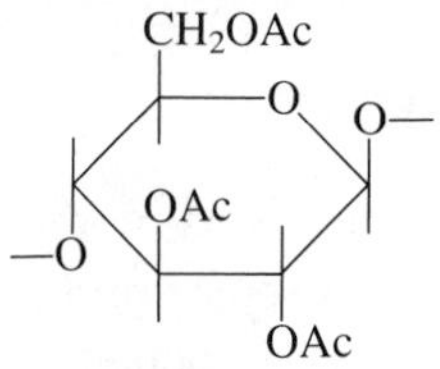 (acetylated glucose unit)	Photographic film
Silicones	CH_3 $Cl-Si-Cl\;\;\;H_2O$ CH_3	CH_3 $-O-Si-O-$ CH_3	Water-repellent coatings, temperature-resistant fluids and rubbers (CH_3SiCl_3 cross-links in water)
Polyurethanes	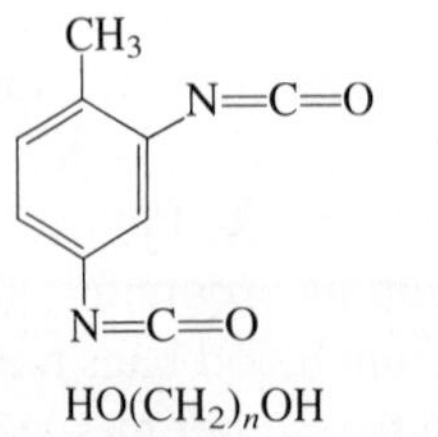 (toluene diisocyanate) $N=C=O$ $HO(CH_2)_nOH$	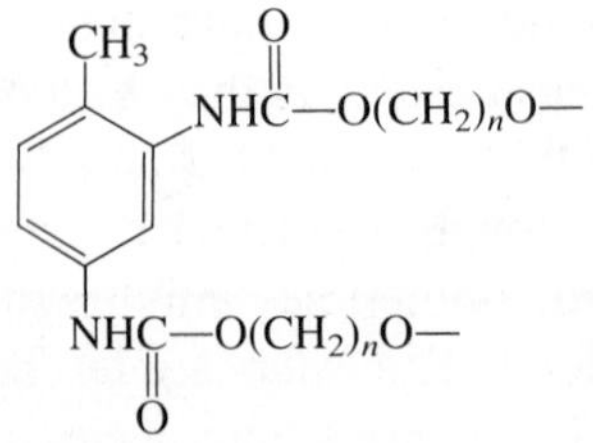 $NHC-O(CH_2)_nO-$ $NHC-O(CH_2)_nO-$	Rigid and flexible foams, fibers

*Cellulose, a polymer of glucose, is used as the monomer.

Recently, a rare and fatal form of liver cancer (angiosarcoma) was discovered among small numbers of workers in chemical companies making polyvinyl chloride. The monomer used in making PVC is vinyl chloride, a gas. The structure is shown in Table 1. Currently, industry is required to eliminate this health hazard by reducing or eliminating vinyl chloride from the atmosphere.

Other types of plasticizers once used were the polychlorinated biphenyls (PCB). These compounds and DDT have similar physiological effects, and they are even more persistent in the environment! The PCBs are actually a mixture of compounds that have had the hydrogens on the basic hydrocarbon structure, biphenyl, replaced with chlorines (from 1 to 10 hydrogens can be replaced). One typical PCB that may be present in a plasticizer mixture is shown. PCBs are no longer being sold except for use in closed systems, where they cannot leak into the environment.

Dibutyl phthalate   Vinyl chloride   A polychlorinated biphenyl (PCB)

Disposability Problems

What do we do with all our wastes? Currently, the most popular method is to bury our garbage in sanitary landfills. However, as we run out of good places to bury our garbage, incineration appears to be an attractive method for solving the solid waste problem. Plastics, which compose about 2% of our garbage, burn readily. The new high-temperature incinerators are extremely efficient and can be operated with very little air pollution. It should also be possible to burn our garbage and generate electrical power from it.

Ideally, we should either recycle all our wastes or not produce the waste in the first place. Plastic waste consists of about 55% polyethylene and polypropylene, 20% polystyrene, and 11% PVC. All these polymers are thermoplastics and can be recycled. They can be resoftened and remolded into new goods. Unfortunately, thermosetting plastics (cross-linked polymers) cannot be remelted. They decompose on high-temperature heating. Thus, thermosetting plastics should not be used for ''disposable'' purposes. To recycle plastics effectively, we must sort the materials according to the various types. This requires will power as well as knowledge about the plastics that we are discarding. Neither requirement is easily effected.

Littering Problems

Plastics, if they are well made, will not corrode or rust, and they last almost indefinitely. Unfortunately, these desirable properties also lead to a problem when plastics are buried in a landfill or thrown on the landscape—they do not decompose. Currently, research is being undertaken to discover plastics that are biodegradable or photodegradable, so that either microorganisms or light from the sun can decompose our litter and garbage. Some success has been achieved.

Fire Hazards

Numerous injuries are caused by clothing made of polymers, especially children's clothing. Many of these organic fibers burn readily. To combat this problem, chemists have developed flame-retardant fabrics, especially for children's sleepwear.

Toxic gases are sometimes liberated when plastics burn. For example, hydrogen chloride is generated when PVC is burned, and hydrogen cyanide when poly-acrylonitriles are burned. This presents a problem that compounds the fire danger.

Energy Shortage

The demand for energy has increased at an alarming rate, leading to the energy crisis. The production of polymers requires petroleum as a raw material and as a source of energy to conduct manufacturing. Unfortunately, fossil fuels are a nonrenewable resource, and as their availability decreases, we shall have an even greater problem. On the other hand, natural substances, such as cotton, are renewable resources; perhaps for some uses they would actually be better and less costly than the synthesized polymers. There are many plastics, however, that are superior to natural materials. The answer lies in using plastics wisely.

REFERENCES

Anderson, B. C., Bartron, L. R., and Collette, J. W. "Trends in Polymer Development." *Science, 208* (1980): 807.

"Biodegradability: Lofty Goal for Plastics." *Chemical and Engineering News* (September 11, 1972): 37–38.

Carraher, C. E., Jr. "What Are Polymers?" *Chemistry, 51* (June 1978): 6. Other articles in the same issue of *Chemistry* deal with polymers and the quality of life, sizes and shapes of giant molecules, biopolymers, fire and polymers, polymers in medicine and surgery, and polymers for the future.

"Degradable Plastic Lids Are Now Commercial." *Chemical and Engineering News* (June 19, 1972): 32.

Flory, P. J. "Understanding Unruly Molecules." *Chemistry* (May 1964): 6–13.

Heckert, W. W. "Synthetic Fibers." *Journal of Chemical Education, 30* (1953): 166.

Kaufman, M. *Giant Molecules*. New York: Doubleday, 1968.

Keller, E. "Nylon—from Test Tube to Counter." *Chemistry* (September 1964): 8–23.

McGrew, F. C. "Structure of Synthetic High Polymers." *Journal of Chemical Education, 35* (1958): 178.

Mark, H. F. "Giant Molecules." *Scientific American, 197* (1957): 80.

Mark, H. F. "The Nature of Polymeric Materials." *Scientific American, 217* (1967): 149.

"More Trouble Brewing for Vinyl Chloride." *Chemical and Engineering News* (September 2, 1974): 12–13.

Morton, M. "Big Molecules." *Chemistry* (January 1964): 13–17.

Morton, M. "Design and Formation of Long Chain Polymers." *Chemistry* (March 1964): 6–11.

Oster, B. "Polyethylene." *Scientific American, 197* (1957): 139.

Natta, G. "How Giant Molecules Are Made." *Scientific American, 197* (1957): 98.

"Plasticizers: Found in Heart Cells." *Chemical and Engineering News* (November 8, 1971): 7–8.

"Plastics Face Growing Pressure from Ecologists." *Chemical and Engineering News* (March 29, 1971): 12–13.

"PVC Makers Hit with Health Hazard Suit." *Chemical and Engineering News* (September 2, 1974): 9–10.

Sanders, H. J. "Flame Retardants." *Chemical and Engineering News, 56* (April 24, 1978): 22.

Several issues of *Journal of Chemical Education* have been devoted to aspects of polymer chemistry. Articles on this topic may be found in:

Journal of Chemical Education, 58 (November 1981)
Journal of Chemical Education, 58 (July 1981)
Journal of Chemical Education, 61 (March 1984)
Journal of Chemical Education, 63 (May 1986)
Journal of Chemical Education, 64 (January 1987)

Experiment 55
Preparation of Polymers: Polyester, Nylon, Polystyrene, and Polyurethane

Condensation Polymers
Addition Polymers
Cross-linked Polymers

In this experiment the syntheses of two polyesters (Procedure 55A), nylon (Procedure 55B), polystyrene (Procedure 55C), and polyurethane (Procedure 55D) are described. These polymers represent the most important commercial plastics. They also represent the main classes of polymers: condensation (linear polyester, nylon), addition (polystyrene), and cross-linked (Glyptal polyester, polyurethane).

SPECIAL INSTRUCTIONS

Before performing this experiment, you should read the essay on polymers and plastics, which precedes this experiment. Toluene diisocyanate (TDI) is toxic. It will irritate the skin and eyes. Avoid breathing the vapor.

Procedure 55A
Polyesters

Linear and cross-linked polyesters are prepared in this experiment. Both are examples of condensation polymers.

The linear polyester is prepared as follows:

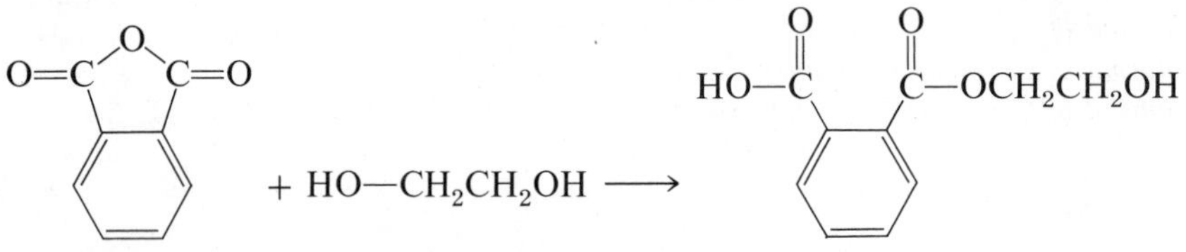

Phthalic anhydride **Ethylene glycol (A diol)**

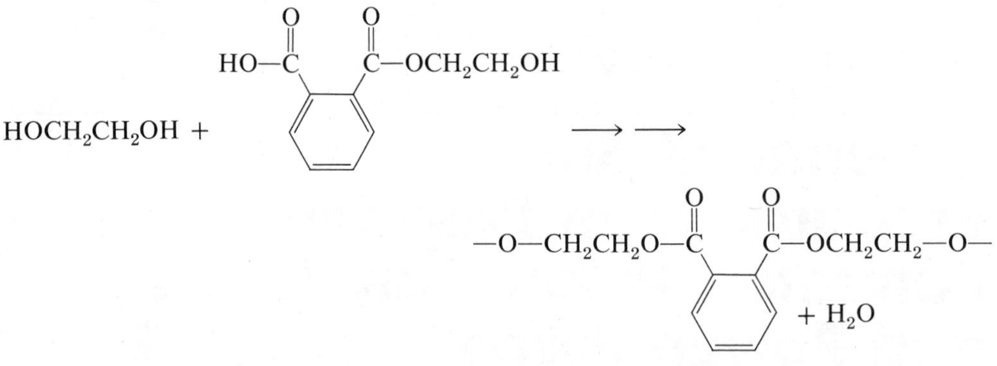

Linear polyester

This linear polyester is isomeric with Dacron, which is prepared from terephthalic acid and ethylene glycol (see the preceding essay). Dacron and the linear polyester made in this experiment are both thermoplastics.

If more than two functional groups are present in one of the monomers, the polymer chains can be linked to one another (cross-linked) to form a three-dimensional network. Such structures are usually more rigid than linear structures and are useful in making paints and coatings. They may be classified as thermosetting plastics. The polyester Glyptal is prepared as follows:

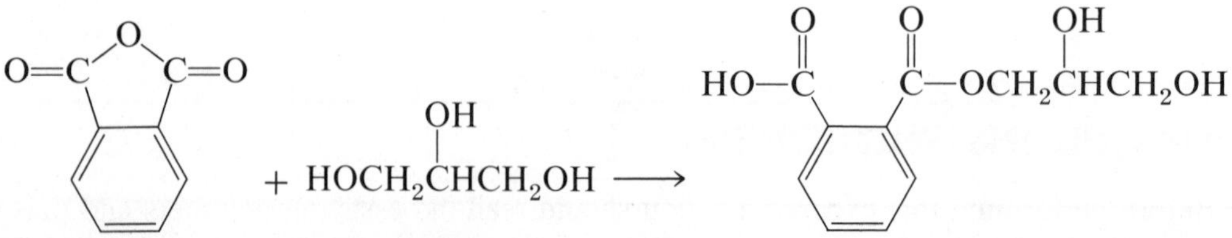

Phthalic anhydride **Glycerol (A triol)**

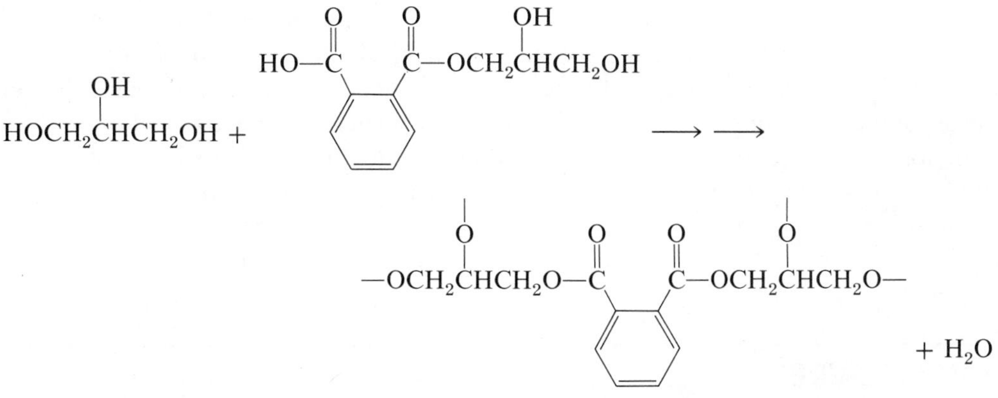

$$HOCH_2\overset{\overset{\displaystyle OH}{|}}{C}HCH_2OH \; + \; \underset{\text{(structure of phthalate diester)}}{\text{HO}-\overset{\overset{\displaystyle O}{\|}}{C}\;\;\overset{\overset{\displaystyle O}{\|}}{C}-OCH_2\overset{\overset{\displaystyle OH}{|}}{C}HCH_2OH} \longrightarrow \longrightarrow$$

$$-OCH_2\overset{\overset{\displaystyle |}{O}}{C}HCH_2O-\overset{\overset{\displaystyle O}{\|}}{C}\;\;\overset{\overset{\displaystyle O}{\|}}{C}-OCH_2\overset{\overset{\displaystyle |}{O}}{C}HCH_2O- \qquad + \; H_2O$$

Cross-linked polyester
(Glyptal resin)

The reaction of phthalic anhydride with a diol (ethylene glycol) is described in the procedure. This linear polyester is compared with the cross-linked polyester (Glyptal) prepared from phthalic anhydride and a triol (glycerol).

PROCEDURE

Place 2 g of phthalic anhydride and 0.1 g of sodium acetate in each of two test tubes. To one tube add 0.8 mL of ethylene glycol, and to the other add 0.8 mL of glycerol. Clamp both tubes so that they can be heated simultaneously with a flame. Heat the tubes gently until the solutions appear to boil (water is eliminated during the esterification), then continue the heating for 5 minutes. Allow the tubes to cool and compare the viscosity and brittleness of the two polymers.

Procedure 55B
Polyamide (Nylon)

Reaction of a dicarboxylic acid, or one of its derivatives, with a diamine leads to a linear polyamide through a condensation reaction. Commercially, nylon 6-6 (so called

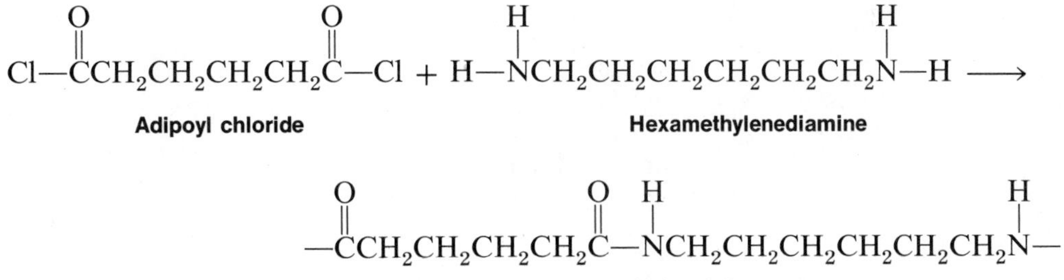

$$\underset{\textbf{Adipoyl chloride}}{Cl-\overset{\overset{\displaystyle O}{\|}}{C}CH_2CH_2CH_2CH_2\overset{\overset{\displaystyle O}{\|}}{C}-Cl} \; + \; \underset{\textbf{Hexamethylenediamine}}{H-\overset{\overset{\displaystyle H}{|}}{N}CH_2CH_2CH_2CH_2CH_2CH_2\overset{\overset{\displaystyle H}{|}}{N}-H} \longrightarrow$$

$$-\overset{\overset{\displaystyle O}{\|}}{C}CH_2CH_2CH_2CH_2\overset{\overset{\displaystyle O}{\|}}{C}-\overset{\overset{\displaystyle H}{|}}{N}CH_2CH_2CH_2CH_2CH_2CH_2\overset{\overset{\displaystyle H}{|}}{N}-$$

Nylon 6-6

because each monomer has six carbons) is made from adipic acid and hexamethylenediamine. In this experiment, you use the acid chloride instead of adipic acid. The acid chloride is dissolved in cyclohexane and this is added **carefully** to hexamethylenediamine dissolved in water. These liquids do not mix, and two layers will form. It can then be drawn out continuously to form a long strand of nylon. Imagine how many molecules have been linked in this long strand! It is a fantastic number.

PROCEDURE

Pour 10 mL of a 5% aqueous solution of hexamethylenediamine (1,6-hexanediamine) into a 50-mL beaker. Add 10 drops of 20% sodium hydroxide solution. Carefully add 10 mL of a 5% solution of adipoyl chloride in cyclohexane to the solution by pouring it down the wall of the slightly tilted beaker. Two layers will form (see figure), and there will be an immediate formation of a polymer film at the liquid-liquid interface. Using a copper-wire hook (a 6-in. piece of wire bent at one end), gently free the walls of the beaker from polymer strings. Then hook the mass at the center, and slowly raise the wire so that polyamide forms continuously, producing a rope that can be drawn out for many feet. The strand can be broken by pulling it faster. Rinse the rope several times with water and lay it on a paper towel to dry. With the piece of wire, vigorously stir the

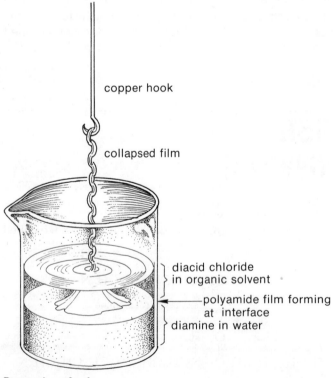

copper hook

collapsed film

diacid chloride
in organic solvent

polyamide film forming
at interface
diamine in water

Preparation of nylon

remainder of the two-phase system to form additional polymer. Decant the liquid and wash the polymer thoroughly with water. Allow the polymer to dry. Do not discard the nylon in the sink. Use a waste container.

Procedure 55C
Polystyrene

An addition polymer, polystyrene, is prepared in this experiment. Reaction can be brought about by free-radical, cationic, or anionic catalysts, the first of these being most common. In this experiment, polystyrene is prepared by free-radical-catalyzed polymerization.

The reaction is initiated by a free-radical source. The initiator will be benzoyl peroxide, a relatively unstable molecule, which at 80 to 90 °C decomposes with homolytic cleavage of the oxygen-oxygen bond:

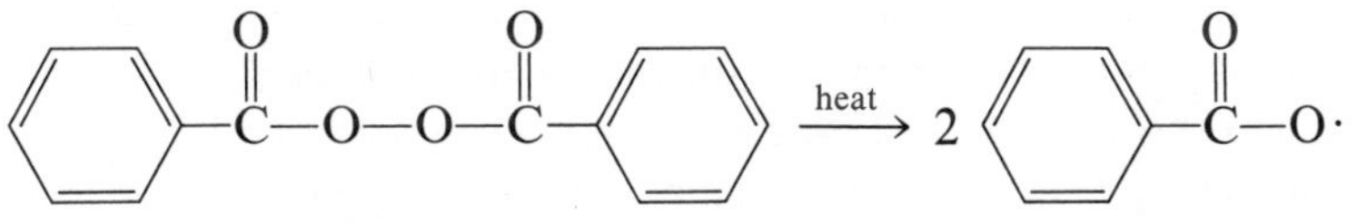

Benzoyl peroxide   **Benzoyl radical**

If an unsaturated monomer is present, the catalyst radical adds to it, initiating a chain reaction by producing a new free radical. If we let R stand for the catalyst radical, the reaction with styrene can be represented as

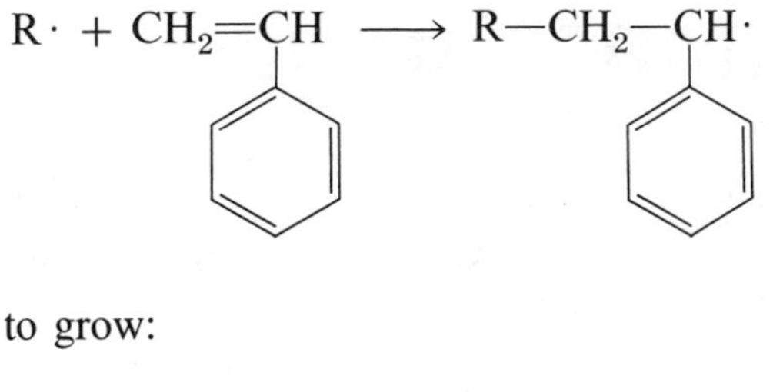

The chain continues to grow:

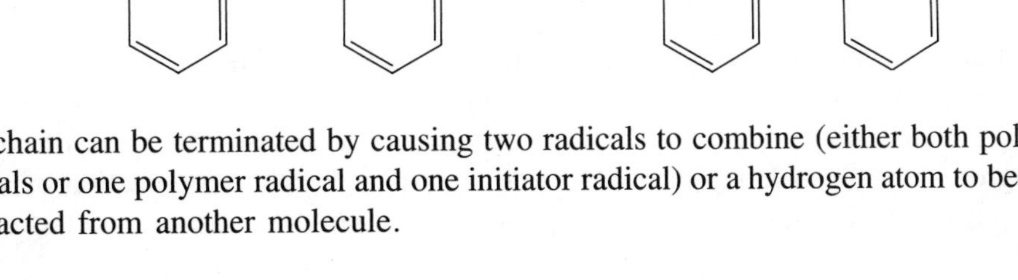

The chain can be terminated by causing two radicals to combine (either both polymer radicals or one polymer radical and one initiator radical) or a hydrogen atom to become abstracted from another molecule.

PROCEDURE

Since it is difficult to clean the glassware, this experiment is best performed by the laboratory instructor. One large batch should be made for the entire class (at least 10 times the amounts given). Perform the experiment in a hood. Place several thicknesses of newspaper in the hood.

> **CAUTION:** Benzoyl peroxide is flammable and may detonate on impact or on heating (or grinding). It should be weighed on glassine (glazed, not ordinary) paper. Clean up *all* spills with water. Wash the glassine paper with water before discarding it.

Place 25 to 30 mL of styrene monomer in a 150-mL beaker and add 0.7 g of benzoyl peroxide. Heat the mixture on a hot plate until the mixture turns yellow. When the color disappears and bubbles begin to appear, immediately take the beaker of styrene off the hot plate since the reaction is exothermic (use tongs or an insulated glove). After the reaction subsides, put the beaker of styrene back on the hot plate and continue heating it until the liquid becomes very syrupy. With a stirring rod, draw out a long filament of material from the beaker. If this filament can be cleanly snapped after a few seconds of cooling, the polystyrene is ready to be poured. If the filament does not break, continue heating the mixture and repeat the above process until the filament breaks easily. Pour the syrupy liquid on a watch glass. After being cooled, the polystyrene can be lifted from the glass surface by gentle prying with a spatula.

Procedure 55D
Polyurethane Foam

A cross-linked polymer, polyurethane foam, is prepared in this experiment from a diisocyanate and a triol. The main reaction is the addition of the alcohol across the $-N=C-$ bond of an isocyanate:

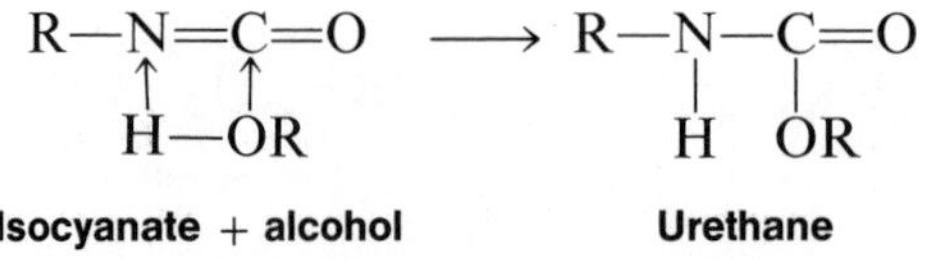

$$R-N=C=O \longrightarrow R-N-C=O$$
$$\quad\;\; H-OR \qquad\qquad\quad H \;\; OR$$

Isocyanate + alcohol   **Urethane**

With a diisocyanate and a triol, the reaction can proceed in **three** directions, leading to a large molecule that is rigidly held into a three-dimensional structure.

The foaming is caused by the evolution of carbon dioxide, much as in the baking of bread. In baking, carbon dioxide is evolved by the fermentation of sugars with yeast, which causes the bread to rise. In the present preparation, the carbon dioxide is produced by the small amount of water present, which decomposes a small amount of the isocyanate:

$$R-N=C=O + H-O-H \longrightarrow R-\overset{\displaystyle H}{\underset{\displaystyle \underset{\parallel}{O}}{\overset{|}{N}}}-C-OH \longrightarrow R-\overset{\displaystyle H}{\overset{|}{N}}-H + CO_2$$

The evolution of carbon dioxide bubbles creates pores in the viscous mixture as the foam sets into a rigid mass. Thus, the foam has excellent buoyant properties. The cell size and structure of the foams are controlled by adding silicone oil.

The structure of the polymer is as follows:

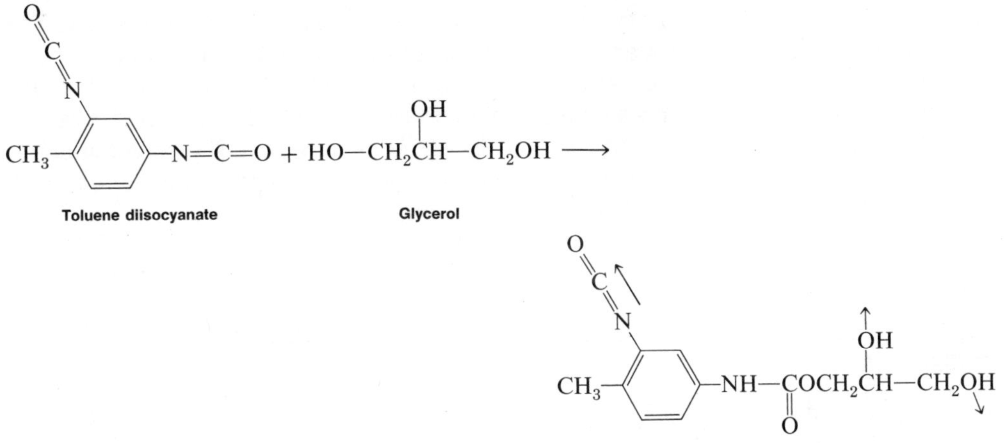

Toluene diisocyanate   Glycerol

The polymer can grow in all the indicated directions.

Castor oil, a triol, can also react in the same way. It is a triglyceride (fat) of ricinoleic acid (see the essay that precedes Experiment 13):

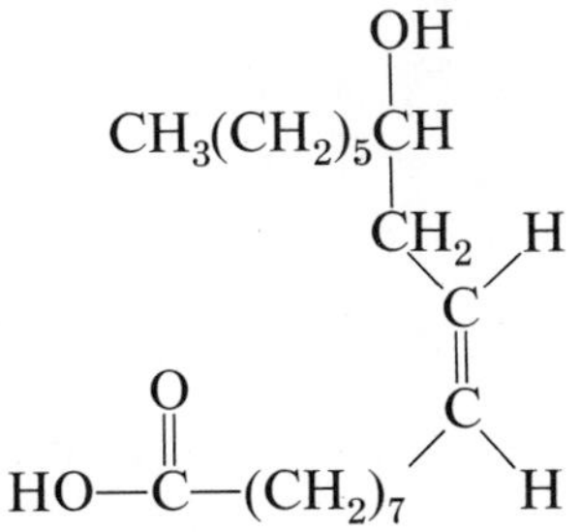

Ricinoleic acid

In the present experiment, glycerol, castor oil, small amounts of water, silicone oil (a foaming agent), and stannous octoate (a catalyst) are mixed together. The diisocyanate (TDI) is added to this mixture. The mixture is stirred and foaming begins. Commercial foams are not usually prepared from these simple materials. A polymeric diol or triol is generally used instead of glycerol and castor oil.

PROCEDURE

> **CAUTION:** Toluene diisocyanate (TDI) is toxic. It will irritate the skin and eyes. Avoid breathing the vapor. It may cause an allergic respiratory response. Work in a hood or in an area with adequate ventilation. Keep the container tightly closed when it is not in use (TDI reacts with moisture in the air). After handling TDI, wash your hands thoroughly.

Pour 17 mL of mixture A (shake well before using) into a waxed soft-drink cup.[1] Then add 10 mL of toluene diisocyanate (tolylene-2,4-diisocyanate). Stir the mixture rapidly and thoroughly with a stirring rod until the mixture is smooth and creamy. The mixture should become warm and should begin to evolve bubbles of carbon dioxide after about 1 minute. When the gas begins to evolve, immediately stop stirring (foaming will be spontaneous). Do not breathe the vapors. Place the mixture in the hood. After the foaming has ceased, allow the material to cool and set thoroughly. The polyurethane is initially sticky, but after several hours it will become firm. The paper container can then be removed. The material will shrink noticeably on standing.

QUESTIONS

1. Ethylene dichloride, $ClCH_2CH_2Cl$, and sodium polysulfide, Na_2S_4, react to form a chemically resistant rubber, Thiokol A. Give the structure of the rubber.

2. Vinylidene chloride, $CH_2{=}CCl_2$, is polymerized with vinyl chloride to make Saran. Give a structure, which includes at least two units, for the copolymer formed.

3. Isobutylene, $CH_2{=}C(CH_3)_2$, is used to prepare cold-flow rubber. Give a structure for the addition polymer formed from this alkene.

4. Kel-F is an addition polymer with the following partial structure. What is the monomer used to prepare it?

$$-\overset{\displaystyle F}{\underset{\displaystyle F}{C}}-\overset{\displaystyle F}{\underset{\displaystyle Cl}{C}}-\overset{\displaystyle F}{\underset{\displaystyle F}{C}}-\overset{\displaystyle F}{\underset{\displaystyle Cl}{C}}-\overset{\displaystyle F}{\underset{\displaystyle F}{C}}-\overset{\displaystyle F}{\underset{\displaystyle Cl}{C}}-$$

5. Maleic anhydride reacts with ethylene glycol to produce an alkyd resin. Give the structure of the condensation polymer produced.

Maleic anhydride

[1] Mixture A is prepared as follows: Place 350 g of castor oil, 100 g of glycerol, 50 drops of stannous octoate (stannous 2-ethylhexanoate), 50 drops of Dow-Corning 200 silicone oil (this is estimated since it is difficult to measure), and 150 drops of water in a bottle. Cap the bottle and shake it thoroughly. Allow this mixture to stand no more than 12 hours before use.

6. Kodel is a condensation polymer made from terephthalic acid and 1,4-cyclohexanedimethanol. Give the structure of the resulting polymer.

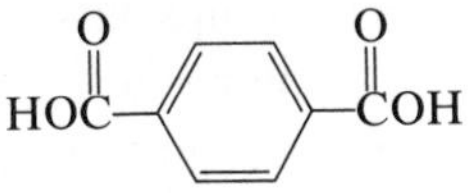

Terephthalic acid 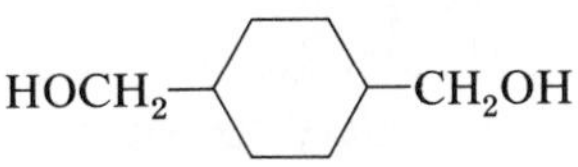   **1,4-Cyclohexanedimethanol**

Experiment 56
Identification of Unknowns

Qualitative organic analysis, the identification and characterization of unknown compounds, is an important part of organic chemistry. Every chemist must learn the appropriate methods for establishing the identity of a compound. In this experiment you will be issued an unknown compound and asked to identify it through chemical and spectroscopic methods. Your instructor may give you a general unknown or a specific unknown. With a **general unknown,** you must first determine the class of compound to which the unknown belongs, that is, identify its main functional group; and then you must determine the specific compound in that class that corresponds to the unknown. With a **specific unknown,** the class of compound (ketone, alcohol, amine, and so on) will be known in advance, and it will only be necessary to determine whatever specific member of that class was issued to you as an unknown. This experiment is designed so that the instructor can issue several general unknowns or as many as six successive specific unknowns, each having a different main functional group.

Although there are well over a million organic compounds that an organic chemist might be called upon to identify, the scope of this experiment is necessarily limited. In this textbook, just over 300 compounds are included in the tables of possible unknowns given for the experiment (see Appendix 1). Your instructor may wish to expand the list of possible unknowns, however. In such a case you will have to consult more extensive tables, such as those found in the work compiled by Rappoport (see References). In addition, the experiment is restricted to include only seven important functional groups:

Aldehydes	Amines
Ketones	Alcohols
Carboxylic acids	Esters
Phenols	

Even though this list of functional groups omits some of the important types of compounds (alkyl halides, alkenes, alkynes, aromatics, ethers, amides, mercaptans, nitriles, acid chlorides, acid anhydrides, nitro compounds, and so on), the methods introduced here can be applied equally well to these other classes of compounds. The list is sufficiently broad to illustrate all the principles involved in identifying an unknown compound.

In addition, although many of the functional groups listed as being excluded will not appear as the major functional group in a compound, several of them will frequently appear as secondary, or subsidiary, functional groups. Three examples of this are presented below.

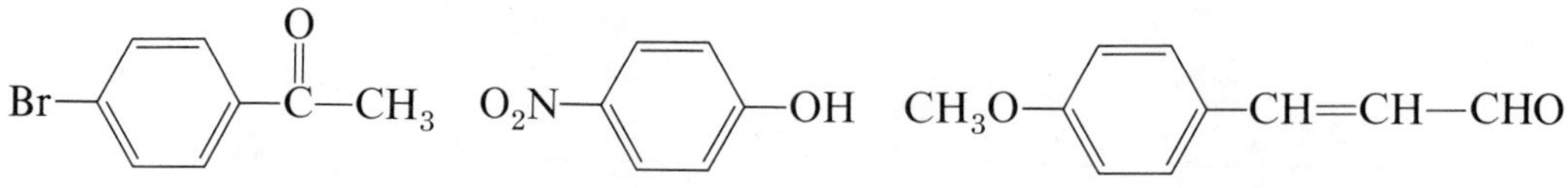

| MAJOR
SUBSIDIARY: | KETONE
Halide
Aromatic | PHENOL
Nitro
Aromatic | ALDEHYDE
Alkene Aromatic
Ether |

The groups included that have subsidiary status are

—Cl	Chloro	—NO$_2$	Nitro	C=C	Double Bond
—Br	Bromo	—C≡N	Cyano	C≡C	Triple Bond
—I	Iodo	—OR	Alkoxy	 ⬡	Aromatic

The experiment presents all of the chief chemical and spectroscopic methods of determining the main functional groups, and it includes methods for verifying the presence of the subsidiary functional groups as well. It will usually not be necessary to determine the presence of the subsidiary functional groups to identify the unknown compound correctly. **Every** piece of information helps the identification, however, and if these groups can be detected easily, one should not hesitate to determine them. Finally, complex bifunctional compounds are generally avoided in this experiment; only a few are included.

HOW TO PROCEED

Fortunately, one can detail a fairly straightforward procedure for determining all the necessary pieces of information. This procedure consists of the following steps:

PART ONE

1. Preliminary classification by physical state, color, and odor
2. Melting-point or boiling-point determination; other physical data
3. Purification, if necessary
4. Determination of solubility behavior in water and in acids and bases
5. Simple preliminary tests: Beilstein, ignition (combustion)
6. Application of relevant chemical classification tests

PART TWO

7. Determination of infrared and nmr spectra
8. Elemental analysis, if necessary
9. Preparation of derivatives
10. Confirmations of identity

Each of these steps is discussed briefly in the sections below.

PRELIMINARY CLASSIFICATION

One should note the physical characteristics of the unknown. These include its color, its odor, and its physical state (liquid, solid, crystalline form). Many compounds have characteristic colors or odors, or they crystallize with a specific crystal structure. This information can often be found in a handbook and can be checked later. Compounds with a high degree of conjugation are frequently yellow to red. Amines often have a fishlike odor. Esters have a pleasant fruity or floral odor. Acids have a sharp and pungent odor. A part of the training of every good chemist includes a cultivation of the ability to recognize familiar or typical odors. As a note of caution, many compounds have distinctly unpleasant or nauseating odors. Some have corrosive vapors. Any unknown substance should be sniffed with the greatest caution. As a first step, open the container, hold it away from you, and using your hand, carefully waft the vapors toward your nose. If you get past this stage, a closer inspection will be possible.

MELTING-POINT OR BOILING-POINT DETERMINATION

The single most useful piece of information to have for an unknown compound is its melting point or boiling point. Either piece of data will drastically limit the compounds that are possible. The electric melting-point apparatus gives a rapid and accurate measurement (see Technique 4, Sections 4.5 and 4.6). To save time, you can often determine two separate melting points. The first determination can be made rapidly to get an approximate value. Then, you can determine the second melting point more carefully.

The boiling point is easily obtained by a simple distillation of the unknown (Technique 6, Section 6.4) or by a micro boiling-point determination (Technique 6, Section 6.2). The simple distillation has the advantage that it also purifies the compound. The smallest distilling flask available should be used if a simple distillation is performed, and you should be sure that the thermometer bulb is fully immersed in the vapor of the distilling liquid. For an accurate boiling-point value, the liquid should be distilled rapidly.

If the solid is high-melting (>200 °C), or the liquid high-boiling (>200 °C), a thermometer correction may be needed (Technique 4, Section 4.8 and Technique 6, Section 6.8). In any event, allowance should be made for errors of as large as ±5 degrees in these values.

PURIFICATION

If the melting point of a solid has a wide range (>4–5 degrees), it should be recrystallized and the melting point redetermined.

If a liquid was highly colored before distillation, if it yielded a wide boiling-point range, or if the temperature did not hold constant during the distillation, it should be redistilled to determine a new temperature range. A reduced-pressure distillation is in order for high-boiling liquids or for those that show any sign of decomposition on heating.

Occasionally, column chromatography may be necessary to purify solids that have large amounts of impurities and do not yield satisfactory results on crystallization.

Acidic or basic impurities that contaminate a neutral compound may often be removed by dissolving the compound in a low-boiling solvent, such as CH_2Cl_2 or ether, and extracting with 5% $NaHCO_3$ or 5% HCl, respectively. Conversely, acidic or basic compounds can be purified by dissolving them in 5% $NaHCO_3$ or 5% HCl, respectively, and extracting them with a low-boiling organic solvent to remove impurities. After neutralization of the aqueous solution, the desired compound can be recovered by extraction.

SOLUBILITY BEHAVIOR

Tests on solubility are described fully in Procedure 56A. They are extremely important. The solubility of small amounts of the unknown in water, 5% HCl, 5% $NaHCO_3$, 5% NaOH, concentrated H_2SO_4, and organic solvents is determined. This information reveals whether a compound is an acid, a base, or a neutral substance. The sulfuric acid test reveals whether a neutral compound has a functional group that contains an oxygen, a nitrogen, or a sulfur atom that can be protonated. This information allows one to eliminate or to choose various functional-group possibilities. The solubility tests must be made on **all** unknowns.

PRELIMINARY TESTS

The two combustion tests, the Beilstein test (Procedure 56B) and the ignition test (Procedure 56C), can be performed easily and quickly, and they often give valuable information. It is recommended that they be performed on all unknowns.

CHEMICAL CLASSIFICATION TESTS

The solubility tests usually suggest or eliminate several possible functional groups. The chemical classification tests listed in Procedures 56D to 56I allow one to distinguish among the possible choices. Choose only those tests the solubility tests suggest might be meaningful. Time will be wasted doing unnecessary tests. There is no substitute for a firsthand thorough knowledge of these tests. Each of the sections should be studied carefully until the significance of each test is understood. Also, it will be helpful to actually try the tests on **known** substances. In this way, it will be easier to recognize a positive test. Appropriate test compounds are listed for many of the tests. When you are performing a test that is new to you, it is always good practice to run the test separately on both a known substance and the unknown **at the same time.** This lets you compare results directly.

Once the melting or boiling point, the solubilities, and the main chemical tests have been made, it will be possible to identify the class of compound. At this stage, with the melting point or boiling point as a guide, it will be possible to compile a list of possible compounds. Inspection of this list will suggest additional tests that must be performed to distinguish among the possibilities. For instance, one compound may be a methyl ketone and the other may not. The iodoform test is called for to distinguish the two possibilities. The tests for the subsidiary functional groups may also be required. These are described in Procedures 56B and 56C. These tests should also be studied carefully; there is no substitute for firsthand knowledge about these either.

One should not perform the chemical tests either willynilly or in a methodical, comprehensive sequence. One should use the tests selectively. Solubility tests automatically eliminate the need for some of the chemical tests. Each successive test will either eliminate the need for another test or dictate its use. One should also examine the tables of unknowns carefully. The boiling point or the melting point of the unknown may eliminate the need for many of the tests. For instance, the possible compounds may simply not include one with a double bond. Efficiency is the keyword here. You should not waste time in doing nonsensical or unnecessary tests. Many possibilities may be eliminated on the basis of logic alone.

How you proceed with the following steps may be limited by your instructor's wishes. Many instructors may restrict your access to infrared and nmr spectra until you have narrowed your choices to a few compounds **all within the same class.** Others may have you determine these data routinely. Some instructors may want students to perform elemental analysis on all unknowns; others may restrict it to only the most essential situations. Most unknowns can be identified without either spectroscopy or elemental analysis. Again, some instructors may require derivatives as a final confirmation of the compound's identity; others may not wish to use them at all.

SPECTROSCOPY

Spectroscopy is probably the most powerful and modern tool available to the chemist for determining the structure of an unknown compound. It is often possible to determine structure through spectroscopy alone. On the other hand, there are also situations for which spectroscopy is not of much help and the traditional methods must be relied on. For this reason, you should use spectroscopy not to the exclusion of the more traditional tests but rather as a confirmation of those results. Nevertheless, the main functional groups and their immediate environmental features can be determined quickly and accurately with spectroscopy.

ELEMENTAL ANALYSIS

Elemental analysis, which allows one to determine the presence of nitrogen, sulfur, or a specific halogen atom (Cl, Br, I) in a compound is often useful; however, other information often renders these tests unnecessary. A compound identified as an amine by solubility tests obviously contains nitrogen. Many nitrogen-containing groups (for instance, nitro groups) can be identified by infrared spectroscopy. Finally, it is not usually necessary to identify a specific halogen. The simple information that the compound contains a halogen (any halogen) may be enough information to distinguish between two compounds. A simple Beilstein test provides this information.

DERIVATIVES

One of the principal tests of the correct identification of an unknown compound comes in trying to convert the compound by a chemical reaction to another known compound. This second compound is called a **derivative.** The best derivatives are solid compounds, since the melting point of a solid provides an accurate and reliable identification of most compounds. Solids are also easily purified through crystallization. The derivative is a way of distinguishing two otherwise very similar compounds. Usually they will have derivatives (both prepared by the same reaction) that have different melting points. Tables of unknowns and derivatives are listed in Appendix 1, at the end of the book. Procedures for preparing derivatives are given in Appendix 2.

CONFIRMATION OF IDENTITY

A rigid and final test for identifying an unknown can be made if an "authentic" sample of the compound is available for comparison. One can compare infrared and nmr spectra of the unknown compound with the spectra of the known compound. If the spectra match, peak for peak, then the identity is probably certain. Other physical and chemical properties can also be compared. If the compound is a solid, a convenient test is the mixed melting point (Technique 4, Section 4.4). Thin-layer or gas chromatographic comparisons may also be useful. For thin-layer analysis, however, it may be necessary to experiment with several different development solvents to reach a satisfactory conclusion about the purity of the substance in question.

While we cannot be complete in this experiment in terms of the functional groups covered, or the tests described, the experiment should give a good introduction to the methods and the techniques chemists use to identify unknown compounds. Textbooks that cover the subject more thoroughly are listed in the references. You are encouraged to consult these for more information, including specific methods and classification tests.

REFERENCES

Comprehensive Textbooks

Cheronis, N. D., and Entrikin, J. B. *Identification of Organic Compounds*. New York: Wiley-Interscience, 1963.

Pasto, D. J., and Johnson, C. R. *Laboratory Text for Organic Chemistry*. Englewood Cliffs, N.J.: Prentice-Hall, 1979.

Shriner, R. L., Fuson, R. C., Curtin, D. Y., and Morrill, T. C. *The Systematic Identification of Organic Compounds*. 6th ed. New York: Wiley, 1980.

Spectroscopy

Bellamy, L. J. *The Infra-red Spectra of Complex Molecules*. 3rd ed. New York: Methuen, 1975.

Dyer, J. R. *Applications of Absorption Spectroscopy of Organic Compounds*. Englewood Cliffs, N.J.: Prentice-Hall, 1965.

Nakanishi, K. *Infrared Absorption Spectroscopy*. San Francisco: Holden-Day, 1962.

Pavia, D. L., Lampman, G. M., and Kriz, G. S., Jr. *Introduction to Spectroscopy: A Guide for Students of Organic Chemistry*. Philadelphia: W. B. Saunders, 1979.

Silverstein, R. M., Bassler, G. C., and Morrill, T. C. *Spectrometric Identification of Organic Compounds*. 4th ed. New York: Wiley, 1981.

Extensive Tables of Compounds and Derivatives

Rappoport, Z., ed. *Handbook of Tables for Organic Compound Identification*. Cleveland: Chemical Rubber Company, 1967.

Procedure 56A
Solubility Tests

Solubility tests should be performed on **every unknown.** They are extremely important in determining the nature of the main functional group of the unknown compound. The tests are very simple and require only small amounts of the unknown. In addition, solubility tests reveal whether the compound is a strong base (amine), a weak acid (phenol), a strong acid (carboxylic acid), or a neutral substance (aldehyde, ketone, alcohol, ester). The common solvents used to determine solubility types are

5% HCl	Concentrated H_2SO_4
5% $NaHCO_3$	Water
5% NaOH	Organic solvents

The solubility chart on p. 428 indicates solvents in which compounds containing the various functional groups are likely to dissolve. The summary charts in sections 56D through 56I repeat this information for each functional group included in this experiment. In this section, the correct procedure for determining whether a compound is soluble in a test solvent is given. Also given is a series of explanations detailing the reasons that compounds having specific functional groups are soluble in only specific solvents. This is accomplished by indicating the type of chemistry or the type of chemical interaction that is possible in each solvent.

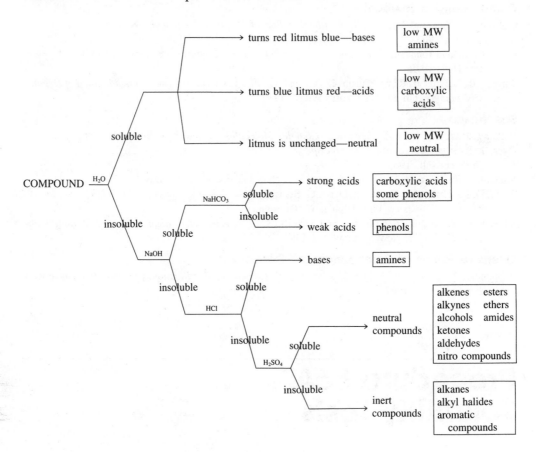

SOLUBILITY TESTS

Procedure. Place about 1 mL of the solvent in a small test tube. Add one drop of an unknown liquid from an eyedropper, or a few crystals of an unknown solid from the end of a spatula, directly into the solvent. Gently tap the test tube with your finger to ensure mixing, and then observe whether any mixing lines appear in the solution. The disappearance of the liquid or solid, or the appearance of the mixing lines, indicates that solution is taking place. Add several more drops of the liquid, or a few more crystals of the solid, to determine the extent of the compound's solubility. A common mistake in determining the solubility of a compound is testing with a quantity of the unknown too large to dissolve in the chosen solvent. Use small amounts. It may take several minutes to dissolve solids. Compounds in the form of large crystals will need more time to dissolve than powders or very small crystals. In some cases it is helpful to pulverize a compound with large crystals using a mortar and pestle. Sometimes gentle heating helps, but strong heating is discouraged, as it often leads to reaction. When colored compounds dissolve, the solution often assumes the color.

By the above procedure, the solubility of the unknown should be determined in each of the following solvents: water, 5% NaOH, 5% $NaHCO_3$, 5% HCl, and concentrated H_2SO_4. With sulfuric acid, a color change may be observed rather than solution. A color change should be regarded as a positive solubility test. Solid unknowns that do not dissolve in any of the test solvents may be inorganic substances. To eliminate this possibility, one must determine the solubility of the unknown in several organic solvents, like ether. If the compound is organic, a solvent that will dissolve it can usually be found.

If a compound is found to dissolve in water, the pH of the aqueous solution should be estimated with pH paper or litmus. Compounds soluble in water are usually soluble in **all** the aqueous solvents. If a compound is only slightly soluble in water, it may be **more** soluble in another aqueous solvent. For instance, a carboxylic acid may be only slightly soluble in water but very soluble in dilute base. It will often not be necessary to determine the solubility of the unknown in every solvent.

Test Compounds.   Five solubility unknowns will be found on the supply shelf. The five unknowns include a base, a weak acid, a strong acid, a neutral substance with an oxygen-containing functional group, and a neutral substance that is inert. Using solubility tests, distinguish these unknowns by type. Verify your answer with the instructor.

SOLUBILITY IN WATER

Compounds that contain four or fewer carbons and also contain oxygen, nitrogen, or sulfur are often soluble in water. Almost any functional group containing these elements will lead to water solubility for low-molecular-weight (C_4) compounds. Compounds having five or six carbons and any of those elements are often insoluble in water or have borderline solubility. Branching of the alkyl chain in a compound lowers the intermolecular forces between its molecules. This is usually reflected in a lowered boiling point or melting point and a greater solubility in water for the branched compound than for the corresponding straight-chain compound. This occurs simply because the molecules of the branched compound are more easily separated from one another. Thus, *t*-butyl alcohol would be expected to be more soluble in water than *n*-butyl alcohol.

When the ratio of the oxygen, nitrogen, or sulfur atoms in a compound to the carbon atoms is increased, the solubility of that compound in water often increases. This is due to the increased number of polar functional groups. Thus, 1,5-pentanediol would be expected to be more soluble in water than 1-pentanol.

As the size of the alkyl chain of a compound is increased beyond about four carbons, the influence of a polar functional group is diminished and the water solubility begins to decrease. A few examples of these generalizations are given below.

Soluble	Borderline	Insoluble

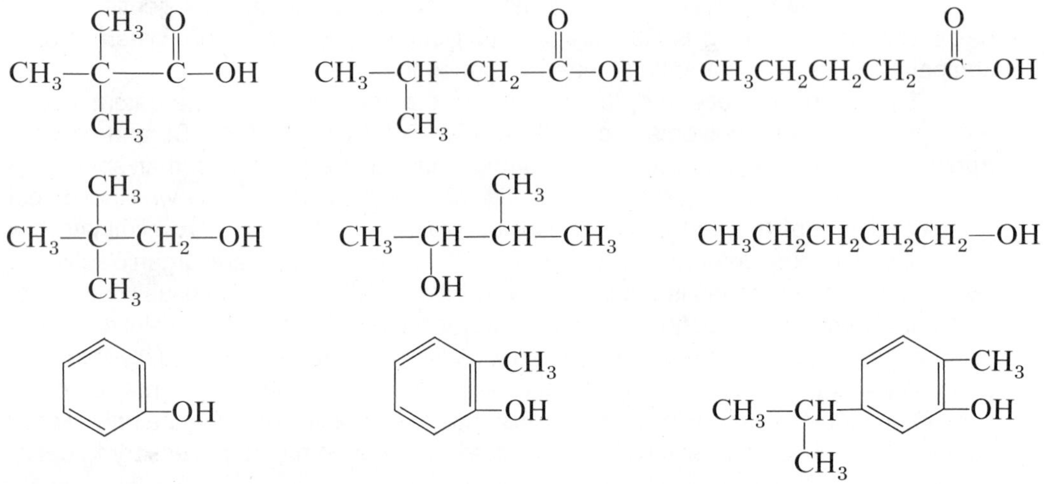

SOLUBILITY IN 5% HCL

The possibility of an amine should be considered immediately if a compound is soluble in dilute acid (5% HCl). Aliphatic amines (RNH_2, R_2NH, R_3N) are basic compounds that readily dissolve in acid because they form hydrochloride salts that are soluble in the aqueous medium:

$$R-NH_2 + HCl \longrightarrow R-NH_3^+ + Cl^-$$

The substitution of an aromatic ring, Ar, for an alkyl group, R, reduces the basicity of an amine somewhat, but the amine will still protonate, and it will still generally be soluble in dilute acid. The reduction in basicity in an aromatic amine is due to the resonance delocalization of the unshared electrons on the amino nitrogen of the free base. The delocalization is lost on protonation, a problem that does not exist for aliphatic amines. The substitution of two or three aromatic rings on an amine nitrogen reduces the basicity of the amine even further. Diaryl and triaryl amines do not dissolve in dilute HCl since they do not protonate easily. Thus, Ar_2NH and Ar_3N are insoluble in dilute acid. Some amines of very high molecular weight, like tribromoaniline (molecular weight [MW] 330), may also be insoluble in dilute acid.

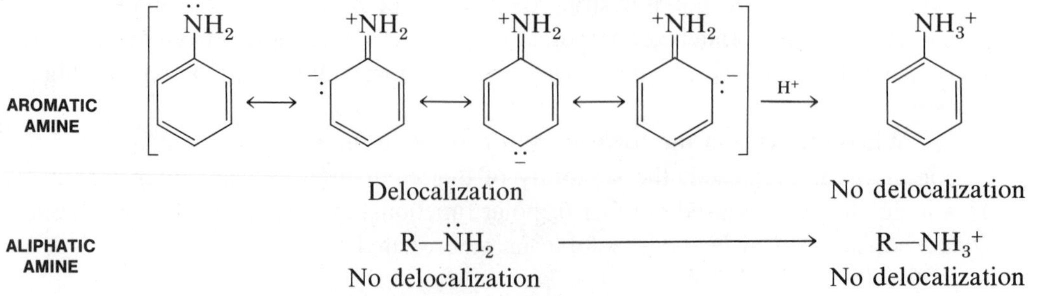

SOLUBILITY IN 5% NaHCO$_3$ AND 5% NaOH

Compounds that dissolve in sodium bicarbonate, a weak base, are strong acids. Compounds that dissolve in sodium hydroxide, a strong base, may be either strong or weak acids. Thus, one can distinguish weak and strong acids by determining their solubility in both strong (NaOH) and weak (NaHCO$_3$) base. The classification of some functional groups as either weak or strong acids is given in the accompanying table.

STRONG ACIDS (Soluble in both NaOH and NaHCO$_3$)	WEAK ACIDS (Soluble in NaOH but not NaHCO$_3$)

Sulfonic acids   RSO$_3$H
 Carboxylic acids   RCOOH
 Ortho- and *para*-substituted
 di- and trinitrophenols
 	
 (structures)
	Phenols ArOH
	 Nitroalkanes RCH$_2$NO$_2$
	   R$_2$CHNO$_2$
	
	 β-Diketones (structure) R—C(=O)—CH$_2$—C(=O)—R
	
	 β-Diesters (structure) RO—C(=O)—CH$_2$—C(=O)—OR
	
	 Imides (structure) R—C(=O)—NH—C(=O)—R
	
	 Sulfonamides ArSO$_2$NH$_2$
	   ArSO$_2$NHR

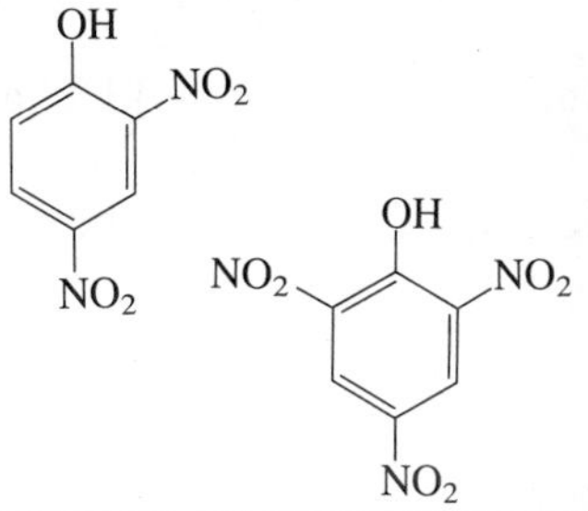

In this experiment, carboxylic acids (pK_a~5) are generally indicated when a compound is soluble in both bases, while phenols (pK_a~10) are indicated when it is soluble in NaOH only.

Compounds dissolve in base because they form sodium salts that are soluble in the aqueous medium. The salts of some high-molecular-weight compounds are not soluble, however, and precipitate. The salts of the long-chain carboxylic acids, such as myristic (C$_{14}$), palmitic (C$_{16}$), and stearic (C$_{18}$) acids, which form soaps, are in this category. Some phenols also produce insoluble sodium salts, and often these are colored due to resonance in the anion.

Both phenols and carboxylic acids produce resonance-stabilized conjugate bases. Thus, bases of the appropriate strength may easily remove their acidic protons to form the sodium salts.

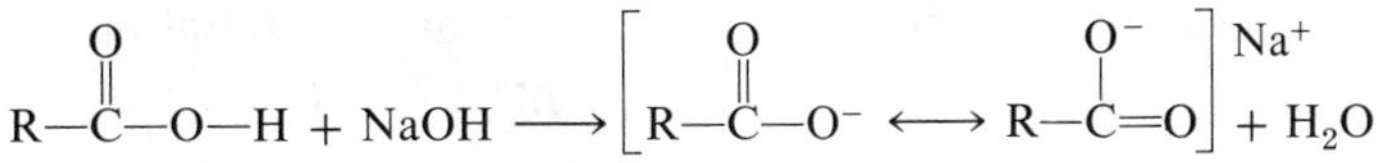

R—C(=O)—O—H + NaOH ⟶ [R—C(=O)—O$^-$ ⟷ R—C(—O$^-$)=O] Na$^+$ + H$_2$O

Delocalized anion

(phenol) →NaOH→ [delocalized anion structures] Na$^+$ + H$_2$O

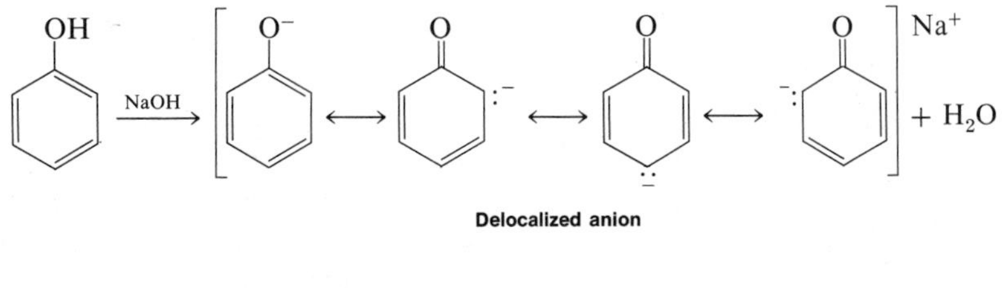

Delocalized anion

In phenols, substitution of nitro groups in the *ortho* and *para* positions of the ring increases the acidity. Nitro groups in these positions provide additional delocalization in the conjugate anion. Phenols that have two or three nitro groups in the *ortho* and *para* positions often dissolve in **both** sodium hydroxide and sodium bicarbonate solutions.

SOLUBILITY IN CONCENTRATED SULFURIC ACID

Many compounds are soluble in cold concentrated sulfuric acid. Of the compounds included in this experiment, alcohols, ketones, aldehydes, and esters are in this category. Other compounds that also dissolve include alkenes, alkynes, ethers, nitroaromatics, and amides. Since several different kinds of compounds are soluble in sulfuric acid, further chemical tests and spectroscopy will be needed to differentiate among them.

Compounds that are soluble in concentrated sulfuric acid but not in dilute acid are extremely weak bases. Almost any compound containing a nitrogen, an oxygen, or a sulfur atom can be protonated in concentrated sulfuric acid. The ions produced are soluble in the medium.

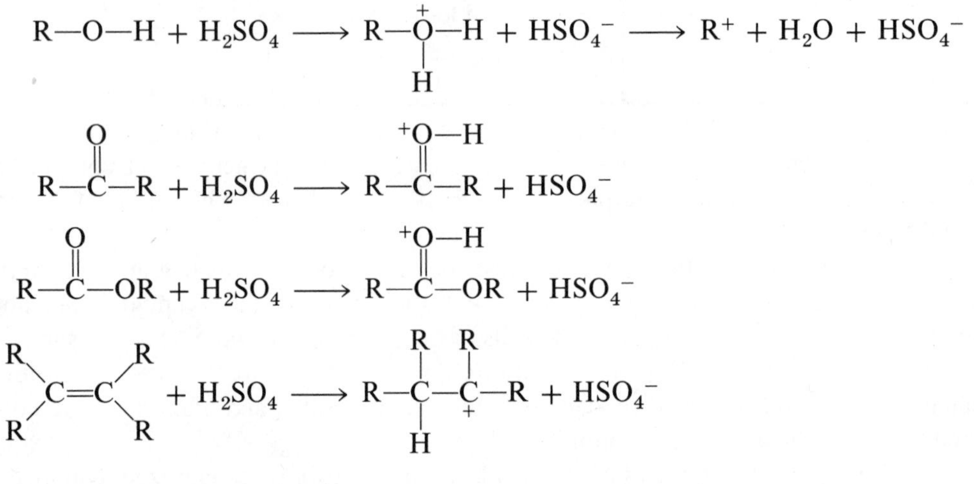

$$R\text{—}O\text{—}H + H_2SO_4 \longrightarrow R\overset{+}{\underset{H}{\text{—}O}}\text{—}H + HSO_4^- \longrightarrow R^+ + H_2O + HSO_4^-$$

$$R\overset{O}{\overset{\|}{\text{—}C}}\text{—}R + H_2SO_4 \longrightarrow R\overset{+O\text{—}H}{\overset{\|}{\text{—}C}}\text{—}R + HSO_4^-$$

$$R\overset{O}{\overset{\|}{\text{—}C}}\text{—}OR + H_2SO_4 \longrightarrow R\overset{+O\text{—}H}{\overset{\|}{\text{—}C}}\text{—}OR + HSO_4^-$$

$$\underset{R}{\overset{R}{{}}}C{=}C\underset{R}{\overset{R}{{}}} + H_2SO_4 \longrightarrow R\overset{R}{\underset{H}{\text{—}C}}\text{—}\overset{R}{\underset{+}{\text{C}}}\text{—}R + HSO_4^-$$

INERT COMPOUNDS

Compounds not soluble in concentrated sulfuric acid or any of the other solvents are said to be **inert**. Compounds not soluble in concentrated sulfuric acid include the alkanes, most simple aromatics, and the alkyl halides. Some examples of inert compounds are hexane, benzene, chlorobenzene, chlorohexane, and toluene.

Procedure 56B
Tests for the Elements (N, S, X)

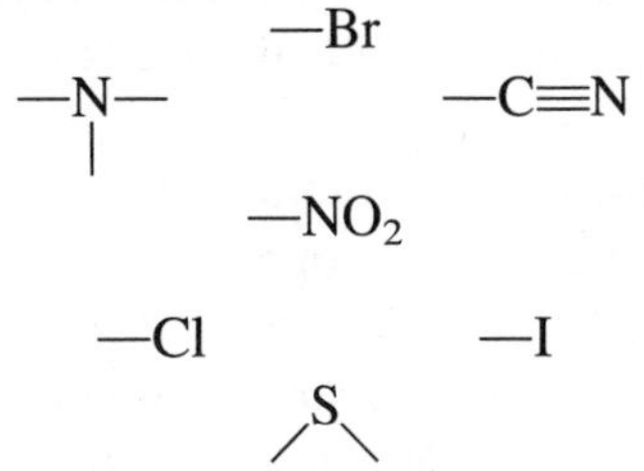

$$-\text{N}-\quad \overset{\displaystyle -\text{Br}}{\underset{\displaystyle -\text{NO}_2}{}}\quad -\text{C}\equiv\text{N}$$

$$-\text{Cl}\qquad -\text{I}$$

$$\diagup\text{S}\diagdown$$

Except for amines (Procedure 56G), which are easily detected by their solubility behavior, all compounds issued in this experiment will contain heteroelements (N, S, Cl, Br, I) only as **secondary** functional groups. These will be subsidiary to some other important functional group. Thus, no alkyl or aryl halides, nitro compounds, thiols, or thioethers will be issued. However, some of the unknowns may contain a halogen or a nitro group. Less frequently, they may contain a sulfur atom or a cyano group.

Consider as an example *p*-bromobenzaldehyde, an **aldehyde** that contains bromine as a ring substituent. The identification of this compound would hinge on whether the investigator could identify it as an aldehyde. It could probably be identified **without** proving the existence of bromine in the molecule. That information, however, could make the identification easier. In this experiment, methods are given for identifying the presence of a halogen or a nitro group in an unknown compound. Also given is a general method (sodium fusion) for detecting the principal heteroelements that may exist in organic molecules.

CLASSIFICATION TESTS

HALIDES	NITRO GROUPS	N, S, X(Cl, Br, I)
Beilstein test Silver nitrate Sodium iodide/acetone	Ferrous hydroxide	Sodium fusion

TESTS FOR A HALIDE

BEILSTEIN TEST

Procedure. Bend a small loop in the end of a short length of copper wire. Heat the loop end of the wire in a Bunsen burner flame. After cooling, dip the wire directly into a small sample of the unknown. Now, heat the wire in the Bunsen burner flame again. The compound will first burn. After the burning, a green flame will be produced if a halogen is present.

Test Compounds. Try this test on bromobenzene and benzoic acid.

Halogens can be detected easily and reliably by the Beilstein test. It is the simplest method for determining the presence of a halogen, but it does not differentiate among chlorine, bromine, and iodine, any one of which will give a positive test. However, when the identity of the unknown has been narrowed to two choices, of which one has a halogen and one does not, the Beilstein test will often be enough to distinguish between the two.

A positive Beilstein test results from the production of a volatile copper halide when an organic halide is heated with copper oxide. The copper halide imparts a blue-green color to the flame.

This test can be very sensitive to small amounts of halide impurities in some compounds. Therefore, one should use caution in interpreting the results of the test, especially when only a weak color has been obtained.

SILVER NITRATE TEST

Procedure. Add 1 drop of a liquid or 5 drops of a concentrated ethanolic solution of a solid unknown to 2 mL of a 2% ethanolic silver nitrate solution. If no reaction is observed after 5 minutes at room temperature, heat the solution on a steam bath and note whether a precipitate forms. If a precipitate forms, add 2 drops of 5% nitric acid and note whether the precipitate dissolves. Carboxylic acids give a false test by precipitating in silver nitrate, but they dissolve when nitric acid is added. Silver halides, on the other hand, do not dissolve in nitric acid.

Test Compounds. Apply this test to benzyl bromide (α-bromotoluene) and bromobenzene. Discard all waste reagents into a suitable waste container in the hood, since benzyl bromide is a lachrymator.

This test depends on the formation of a white or an off-white precipitate of silver halide when silver nitrate is allowed to react with a sufficiently reactive halide.

$$RX + Ag^+NO_3^- \longrightarrow \underset{\text{Precipitate}}{AgX} + R^+NO_3^- \xrightarrow{CH_3CH_2OH} R\!-\!O\!-\!CH_2CH_3$$

The test does not distinguish among chlorides, bromides, and iodides but does distinguish **labile** (reactive) halides from halides that are unreactive. Halides substituted on an aromatic ring will not usually give a positive silver nitrate test; however, alkyl halides of many types will give a positive test.

The most reactive compounds are those able to form stable carbonium ions in solution and those equipped with good leaving groups (X = I, Br, Cl). Benzyl, allyl, and tertiary halides give immediate reaction with silver nitrate. Secondary and primary halides do not react at room temperature but readily react when heated. Aryl and vinyl halides do not react at all, even at elevated temperatures. This pattern of reactivity fits the stability order for various carbonium ions quite well. Compounds that produce stable carbonium ions react at higher rates than those that do not.

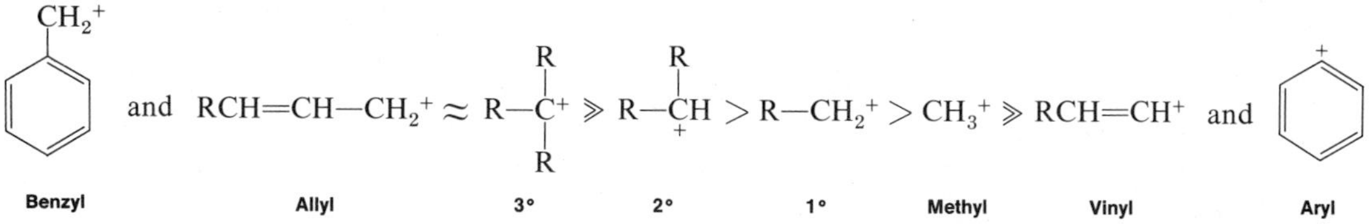

The fast reaction of benzylic and allylic halides is a result of the resonance stabilization that is available to the intermediate carbonium ions formed. Tertiary halides are more reactive than secondary halides, which are in turn more reactive than primary or methyl halides, since alkyl substituents are able to stabilize the intermediate carbonium ions by an electron-releasing effect. The methyl carbonium ion has no alkyl groups and is the least stable of all the carbonium ions mentioned thus far. Vinyl and aryl carbonium ions are extremely unstable since the charge is localized on an sp^2-hybridized carbon (double-bond carbon) rather than on one that is sp^3-hybridized.

SODIUM IODIDE IN ACETONE

Procedure. This test is described in Experiment 22.

DETECTION OF NITRO GROUPS

Although nitro compounds will not be issued as distinct unknowns, many of the unknowns may have a nitro group as a secondary functional group. The presence of a nitro group, and hence nitrogen, in an unknown compound is determined most easily by infrared spectroscopy. However, many nitro compounds give a positive result in the following test.

FERROUS HYDROXIDE

Procedure. Place 1.5 mL of freshly prepared 5% aqueous ferrous ammonium sulfate in a small test tube and add about 10 mg of the unknown compound. Mix the solution well and then add first one drop of $3N$ sulfuric acid, and then 1 mL of $2N$ potassium hydroxide in methanol. Stopper the test tube and shake it vigorously. A positive test is indicated by the formation of a red-brown precipitate, usually within 1 minute.

Most nitro compounds oxidize ferrous hydroxide to ferric hydroxide, which is a red-brown solid. A precipitate indicates a positive test.

$$R{-}NO_2 + 4H_2O + 6Fe(OH)_2 \longrightarrow R{-}NH_2 + 6Fe(OH)_3$$

SPECTROSCOPY

INFRARED

The nitro group gives two strong bands near 1560 and 1350 cm^{-1} (6.4 and 7.4 μ).

DETECTION OF A CYANO GROUP

Although nitriles will not be given as unknowns in this experiment, the cyano group may be a subsidiary functional group whose presence or absence is important to the final identification of an unknown compound. The cyano group can be hydrolyzed in strong base, with vigorous heating, to give a carboxylic acid and ammonia gas:

$$R-C{\equiv}N + 2H_2O \xrightarrow[\Delta]{NaOH} R-COOH + NH_3$$

The ammonia can be detected by its odor or by moist pH paper. However, this method is somewhat difficult, and the presence of a nitrile group is confirmed most easily by infrared spectroscopy. No other functional groups (except some C$\equiv$C) absorb in the same region of the spectrum as C$\equiv$N.

SPECTROSCOPY

INFRARED

C$\equiv$N stretch is a very sharp band of medium intensity near 2250 cm^{-1} (4.5 μ).

SODIUM FUSION TESTS (OPTIONAL) (DETECTION OF N, S, AND X)

When an organic compound containing nitrogen, sulfur, or halide atoms is fused with sodium metal, a reductive decomposition of the compound takes place, which converts these atoms to the sodium salts of the inorganic ions CN$^-$, S^{2-}, and X$^-$.

$$[\text{N, S, X}] \xrightarrow[\Delta]{\text{Na,}} \text{NaCN, Na}_2\text{S, NaX}$$

When the fusion mixture is dissolved in distilled water, the cyanide, sulfide, and halide ions can be detected by standard qualitative inorganic tests.

> **CAUTION: Read Technique 14 before proceeding. Always remember to manipulate the sodium metal with a knife or a forceps, and not to touch it with your fingers. Keep sodium away from water. Destroy all waste sodium with 1-butanol or ethanol. WEAR SAFETY GLASSES.**

SODIUM FUSION

Procedure. Using the procedures outlined in Technique 14, cut a small piece of sodium metal about the size of a small pea (3 mm on a side) and dry it on a paper towel. Place this small piece of sodium in a clean and dry small test tube (10 × 75 mm). Clamp the test tube to a ring stand and heat the bottom of the tube with a microburner until the sodium melts and its metallic vapor can be seen to rise about a third of the way up the tube. The bottom of the tube will probably have a dull red glow. Remove the burner and **immediately** drop the sample directly into the tube. About 10 mg of a solid placed on the end of a spatula or 2 to 3 drops of a liquid should be used. Be sure to drop the sample directly down the center of the tube so that it touches the hot sodium metal and does not adhere to the side of the test tube. There will usually be a flash or a small explosion if the fusion is successful. If the reaction is not successful, the tube should be heated to red heat for a few seconds to ensure complete reaction.

Allow the test tube to cool to room temperature and then carefully add 10 drops of methanol, a drop at a time, to the fusion mixture. Using a spatula or a long glass rod, reach into the test tube and stir the mixture to ensure complete reaction of any excess sodium metal. The fusion will have destroyed the test tube for other uses. Thus, the easiest way to recover the fusion mixture is to crush the test tube into a small beaker containing 5 to 10 mL of **distilled** water. The tube is easily crushed if it is placed in the angle of a clamp holder. Tighten the clamp until the tube is securely held near its bottom, and then—standing back from the beaker and holding the clamp at its opposite end—continue tightening the clamp until the test tube breaks and the pieces fall into the beaker. Stir the solution well, heat it to boiling, and then filter it by gravity through a fluted filter. Portions of this solution will be used for the tests to detect nitrogen, sulfur, and the halogens.

ALTERNATIVE METHOD

Procedure. With some volatile liquids, the method given above will not work. The compounds volatilize before they reach the sodium vapors. For such compounds, place 4 or 5 drops of the pure liquid in the clean, dry test tube, clamp it, and cautiously add the small piece of sodium metal. If there is any reaction, wait until it subsides. Then heat the test tube to red heat and continue according to the instructions in the second paragraph above.

NITROGEN TEST

Procedure. Using pH paper and a 10% sodium hydroxide solution, adjust the pH of about 1 mL of the stock solution to pH 13. Add 2 drops of saturated ferrous ammonium sulfate solution and 2 drops of 30% potassium fluoride solution. Boil the solution for about 30 seconds. Then acidify the hot solution by adding 30% sulfuric acid dropwise until the iron hydroxides dissolve. Avoid using excess acid. If nitrogen is present, a dark blue (not green) precipitate of Prussian blue, $NaFe_2(CN)_6$, will form or the solution will assume a dark blue color.

Reagents. Dissolve 5 g of ferrous ammonium sulfate in 100 mL of water and 30 g of potassium fluoride in 100 mL of water.

SULFUR TEST

Procedure. Acidify about 1 mL of the test solution with acetic acid and add a few drops of a 1% lead acetate solution. The presence of sulfur is indicated by a black precipitate of lead sulfide, PbS.

> **CAUTION: Many compounds of lead(II) are suspect carcinogens (see p. 12) and should be handled with care. Avoid contact.**

HALIDE TESTS

Procedure. Cyanide and sulfide ions interfere with the test for halides. If such ions are present, they must be removed. To accomplish this, acidify the solution with dilute nitric acid and boil it for about 2 minutes. This will drive off any HCN or H_2S that is formed. When the solution cools, add a few drops of a 5% silver nitrate solution. A **voluminous** precipitate indicates a halide. A faint turbidity **does not** mean a positive test. Silver chloride is white. Silver bromide is off-white. Silver iodide is yellow. Silver chloride will readily dissolve in concentrated ammonium hydroxide, whereas silver bromide is only slightly soluble.

DIFFERENTIATION OF CHLORIDE, BROMIDE, AND IODIDE

Procedure. Acidify 2 mL of the test solution with 10% sulfuric acid and boil it for about 2 minutes. Cool the solution and add about 0.5 mL of methylene chloride. Add a few drops of chlorine water or 2 to 4 mg of calcium hypochlorite.[1] Check to be sure that the solution is still acidic. Then stopper the tube, shake it vigorously, and set it aside to allow the layers to separate. An orange to brown color in the methylene chloride layer indicates bromine. Violet indicates iodine. No color or a **light** yellow indicates chlorine.

Procedure 56C
Tests for Unsaturation

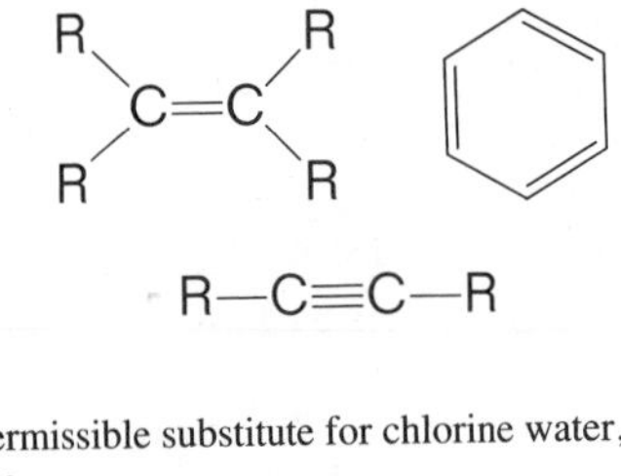

---

[1]Clorox, the commercial bleach, is a permissible substitute for chlorine water, as is any other brand of bleach, provided that it is based on sodium hypochlorite.

The unknowns to be issued for this experiment have neither a double bond nor a triple bond as their **only** functional group. Hence, simple alkenes and alkynes can be ruled out as possible compounds. Some of the unknowns may have a double or a triple bond, however, **in addition to** another more important functional group. The tests described allow one to determine the presence of a double bond or a triple bond (unsaturation) in such compounds.

CLASSIFICATION TESTS

UNSATURATION	AROMATICITY
Bromine–carbon tetrachloride Potassium permanganate	Ignition test

TESTS FOR SIMPLE MULTIPLE BONDS

BROMINE IN CARBON TETRACHLORIDE OR METHYLENE CHLORIDE

Procedure.   Dissolve 50 mg of a solid unknown or 2 drops of a liquid unknown in 1 mL of carbon tetrachloride (or 1,2-dimethoxyethane). Add a 2% (by volume) solution of bromine in carbon tetrachloride, dropwise with shaking, until the bromine color persists. The test is positive if more than 5 drops of the bromine solution are needed so that the color remains for 1 minute. Usually, many drops of the bromine solution will be needed if unsaturation is present. Hydrogen bromide should not be evolved. If hydrogen bromide gas is evolved, one will note a "fog" while blowing across the mouth of the test tube. The HBr can also be detected by a moistened piece of litmus or pH paper. If hydrogen bromide is evolved, the reaction is a **substitution reaction** and not an **addition reaction,** and a double or triple bond is probably not present.

Methylene Chloride.   Even though carbon tetrachloride is used in very small quantities in this test, it poses certain health hazards (see p. 7), and another solvent may be preferable. Methylene chloride (dichloromethane) can be substituted for carbon tetrachloride. Certain problems arise, however, because methylene chloride slowly reacts with bromine, presumably by a light-induced free-radical process, to produce HBr. After about a week, the color of a 2% solution of bromine in methylene chloride fades noticeably, and the odor of HBr can be detected in the reagent. Although the decolorization tests still work satisfactorily, the presence of HBr makes it difficult to distinguish between addition and substitution reactions. A freshly prepared solution of bromine in methylene chloride must be used to make this distinction. Deterioration of the reagent can be forestalled by storing it in a brown glass bottle. Most other substitute solvents also present problems. Ethers, for instance, react slowly in the same way as methylene chloride, and hydrocarbons, like hexane, are not general enough solvents to be able to dissolve all the possible test compounds.

Test Compounds.   Try this test with cyclohexene, cyclohexane, toluene, and acetone.

A successful test depends on the addition of bromine, a red liquid, to a double or a triple bond to give a colorless dibromide:

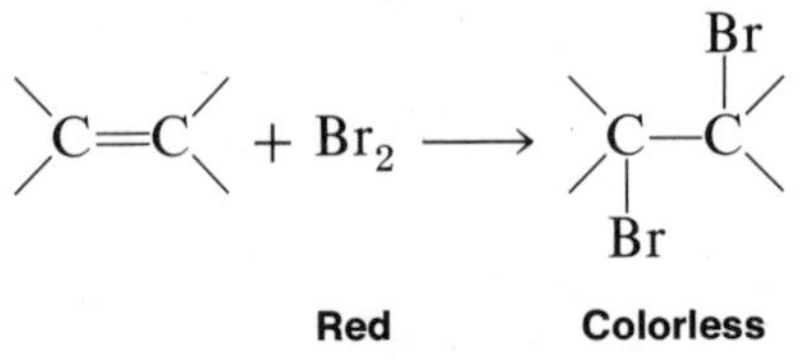

$$\text{C=C} + Br_2 \longrightarrow \text{C–C(Br)(Br)}$$

Red   **Colorless**

Not all double bonds react with bromine–carbon tetrachloride solution. Only those that are electron-rich are sufficiently reactive nucleophiles to initiate the reaction. A double bond that is substituted by electron-withdrawing groups often fails to react or reacts slowly. Fumaric acid is an example of a compound that fails to give the reaction.

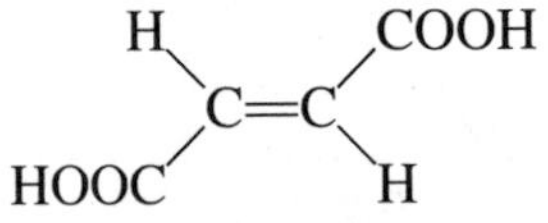

$$\text{HOOC}\diagdown\text{C=C}\diagup\text{COOH}$$

Fumaric acid

Aromatic compounds either do not react with bromine–carbon tetrachloride reagent or they react by **substitution.** Only the aromatic rings that have activating groups as substituents (—OH, —OR, or —NR$_2$) give the substitution reaction.

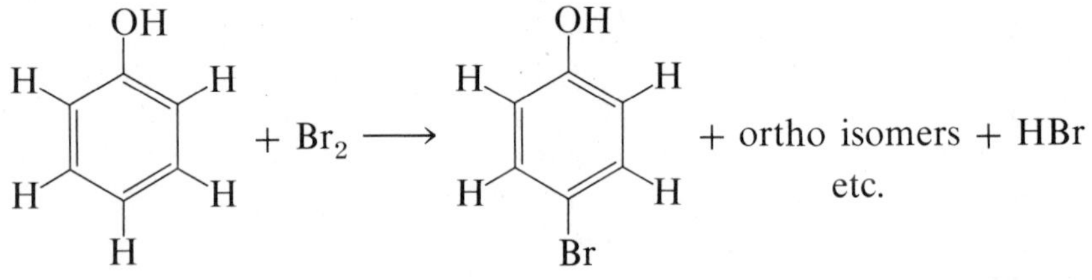

$$\text{phenol} + Br_2 \longrightarrow \text{bromophenol} + \text{ortho isomers} + HBr$$
$$\text{etc.}$$

Some ketones and aldehydes react with bromine to give a **substitution** product, but this reaction is slow except for acetone and some aldehydes and ketones that have a high enol content. When substitution occurs, not only is the bromine color discharged, but hydrogen bromide gas is also evolved.

POTASSIUM PERMANGANATE (BAEYER TEST)

Procedure. Dissolve 25 mg of a solid unknown or 2 drops of the liquid unknown in 2 mL of water or 95% ethanol (1,2-dimethoxyethane may also be used). Slowly add a 1% aqueous solution (weight/volume) of potassium permanganate, drop by drop with shaking, to the unknown. In a positive test, the purple color of the reagent is discharged, and a brown precipitate of manganese dioxide forms, usually within 1 minute. If alcohol was the solvent, the solution should not be allowed to stand for more than 5 minutes, since oxidation of the alcohol will begin slowly. Since permanganate solutions undergo some decomposition to manganese dioxide on standing, any small amount of precipitate should be interpreted with caution.

Test Compounds. Try this test on cyclohexene and toluene.

This test is positive for double and triple bonds but not for aromatic rings. It depends on the conversion of the purple ion MnO_4^- to a brown precipitate of MnO_2 following the oxidation of an unsaturated compound.

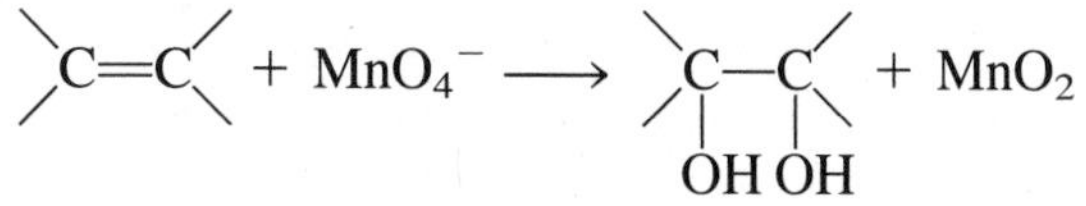

$$\overset{}{\underset{}{>}}C=C\overset{}{\underset{}{<}} + MnO_4^- \longrightarrow \underset{\underset{OH\ OH}{|\ \ \ |}}{>C-C<} + MnO_2$$

Purple   **Brown**

Other easily oxidized compounds also give a positive test with potassium permanganate solution. These substances include aldehydes, some alcohols, phenols, and aromatic amines. If you suspect that any of these functional groups are present, you should interpret the test with caution.

SPECTROSCOPY

INFRARED

DOUBLE BONDS (C=C)

C=C stretch usually occurs near 1680 to 1620 cm^{-1} (5.95–6.17 μ). Symmetrical alkenes may have no absorption.

C—H stretch of vinyl hydrogens occurs >3000 cm^{-1} (3.33 μ), but usually not higher than 3150 cm^{-1} (3.18 μ).

C—H out-of-plane bending occurs near 1000 to 700 cm^{-1} (10.0–14.3 μ).

TRIPLE BONDS (C≡C)

C≡C stretch usually occurs near 2250 to 2100 cm^{-1} (4.44–4.76 μ). The peak is usually sharp. Symmetrical alkynes show no absorption.

C—H stretch of terminal acetylenes occurs near 3310 to 3200 cm^{-1} (3.02–3.12 μ).

NUCLEAR MAGNETIC RESONANCE

Vinyl hydrogens have resonance near 5 to 7δ and have coupling values as follows: J_{trans} = 11–18 Hz, J_{cis} = 6–15 Hz, $J_{geminal}$ = 0–5 Hz. Allylic hydrogens have resonance near 2δ. Acetylenic hydrogens have resonance near 2.8 to 3.0δ.

TESTS FOR AROMATICITY

None of the unknowns to be issued for this experiment will be simple aromatic hydrocarbons. All aromatic compounds will have a principal functional group as a part of their structure. Nevertheless, in many cases it will be useful to be able to recognize the presence of an aromatic ring. Although spectroscopy provides the easiest method of determining aromatic systems, often they can be detected by a simple ignition test.

IGNITION TEST

Procedure. Place a small amount of the compound on a spatula and place it in the flame of a Bunsen burner. Observe whether a sooty flame is the result. Compounds giving a sooty yellow flame have a high degree of unsaturation and may be aromatic.

Test Compound. Try this test with naphthalene.

The presence of an aromatic ring or other centers of unsaturation will lead to the production of a sooty yellow flame in this test. Compounds that contain little oxygen, and have a high carbon-to-hydrogen ratio, burn at a low temperature with a yellow flame. Much carbon is produced when they are burned. Compounds that contain oxygen generally burn at a higher temperature with a clean blue flame.

SPECTROSCOPY

INFRARED

C=C aromatic ring double bonds appear in the 1650 to 1450 cm^{-1} (6–7 μ) region. There are often four sharp absorptions that occur in pairs near 1600 cm^{-1} (6.3 μ) and 1450 cm^{-1} (6.9 μ), which are characteristic of an aromatic ring.

Special ring absorptions: There are often weak ring absorptions around 2000 to 1600 cm^{-1} (5–6 μ). These are often obscured, but when they can be observed, the relative shapes and numbers of these peaks can often be used to ascertain the type of ring substitution (see Appendix 3, ''Infrared Spectroscopy'').

=C—H stretch, aromatic ring: The aromatic C—H stretch always occurs at a higher frequency than 3000 cm^{-1} (shorter wavelength than 3.33 μ).

=C—H out-of-plane bending peaks appear in the region 900 to 690 cm^{-1} (11–15 μ). The number and position of these peaks can be used to determine the substitution pattern of the ring (see Appendix 3, ''Infrared Spectroscopy'').

NUCLEAR MAGNETIC RESONANCE

Hydrogens attached to an aromatic ring usually have resonance near 7δ. Monosubstituted rings not substituted by anisotropic or electronegative groups usually give a single resonance for all the ring hydrogens. Monosubstituted rings with isotropic or electronegative groups usually have the aromatic resonances split into two groups integrating either 3:2 or 2:3. A nonsymmetric, *para*-disubstituted ring has a characteristic four-peak splitting pattern (see Appendix 4, ''Nuclear Magnetic Resonance'').

Procedure 56D
Aldehydes and Ketones

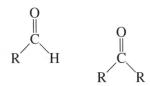

Compounds containing the carbonyl functional group, $\diagup$C=O, where it has only hydrogen atoms or alkyl groups as substituents are called aldehydes, RCHO, or ketones, RCOR′. The chemistry of these compounds is primarily due to the chemistry of the carbonyl functional groups. These compounds are identified by the distinctive reactions of the carbonyl function.

SOLUBILITY CHARACTERISTICS	CLASSIFICATION TESTS

HCl NaHCO$_3$ NaOH H$_2$SO$_4$ Ether
(−)   (−)   (−)   (+)   (+)
Water: <C$_5$ and some C$_6$ (+)
           >C$_5$ (−)
	ALDEHYDES AND KETONES
	  2,4-Dinitrophenylhydrazine
	ALDEHYDES ONLY METHYL KETONES
	  Chromic acid Iodoform test
	  Tollens reagent
	COMPOUNDS WITH HIGH ENOL CONTENT
	  Ferric chloride test

CLASSIFICATION TESTS

Most aldehydes and ketones give a solid, orange-to-red precipitate when mixed with 2,4-dinitrophenylhydrazine. However, only aldehydes will reduce chromium(VI) or silver(I). By this difference in behavior, you can differentiate between aldehydes and ketones.

2,4-DINITROPHENYLHYDRAZINE

Procedure. Place 1 drop of the liquid unknown in a small test tube and add 1 mL of the 2,4-dinitrophenylhydrazine reagent. If the unknown is a solid, dissolve about 10 mg (estimate) in a minimum amount of 95% ethanol or bis(2-ethoxyethyl) ether before adding the reagent. Shake the mixture vigorously. Most aldehydes and ketones will

443

give a yellow-to-red precipitate immediately. However, some compounds will require up to 15 minutes, or even **gentle** heating, to give a precipitate. A precipitate indicates a positive test.

Test Compounds. Try this test on cyclohexanone, benzaldehyde, and benzophenone.

> **CAUTION: Many derivatives of phenylhydrazine are suspect carcinogens (see p. 12) and should be handled with care. Avoid contact.**

Reagent. Dissolve 3.0 g of 2,4-dinitrophenylhydrazine in 15 mL of concentrated sulfuric acid. In a beaker mix 20 mL of water and 70 mL of 95% ethanol. With vigorous stirring, slowly add the 2,4-dinitrophenylhydrazine solution to the aqueous ethanol mixture. After thorough mixing, filter the solution by gravity through a fluted filter.

Most aldehydes and ketones give a precipitate, but esters generally do not give this result. Thus, an ester usually can be eliminated by this test. The color of the

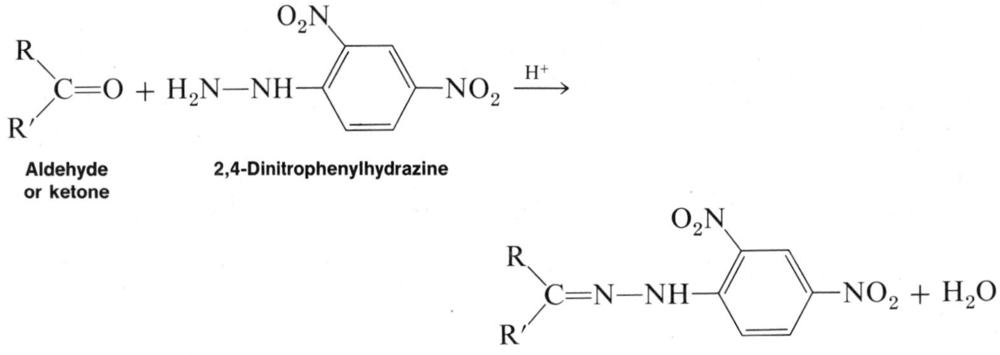

Aldehyde or ketone   2,4-Dinitrophenylhydrazine

2,4-Dinitrophenylhydrazone

2,4-dinitrophenylhydrazone (precipitate) formed is often a guide to the amount of conjugation in the original aldehyde or ketone. Unconjugated ketones, such as cyclohexanone, give yellow precipitates, whereas conjugated ketones, such as benzophenone, give orange-to-red precipitates. Compounds that are highly conjugated give red precipitates. However, the 2,4-dinitrophenylhydrazine reagent is itself orange-red, and the color of any precipitate must be judged cautiously. Occasionally, compounds that are either strongly basic or strongly acidic precipitate the unreacted reagent.

Some allylic and benzylic alcohols give this test because the reagent can oxidize them to aldehydes and ketones, which subsequently react. Some alcohols may be contaminated with carbonyl impurities, either because of their method of synthesis (reduction) or because they have become air-oxidized. A precipitate formed from small amounts of impurity in the solution will be formed in small amount. With some caution, a test that gives only a slight amount of precipitate can usually be ignored. The infrared spectrum of the compound should establish its identity and identify any impurities present.

CHROMIC ACID TEST

Procedure. Dissolve 1 drop of a liquid or 10 mg (approximate) of a solid aldehyde in 1 mL of **reagent-grade** acetone. Add several drops of the chromic acid reagent, a drop at a time, with shaking. A positive test is indicated by a green precipitate and a loss of the orange color in the reagent. With aliphatic aldehydes, RCHO, the solution turns cloudy within 5 seconds and a precipitate appears within 30 seconds. With aromatic aldehydes, ArCHO, it generally takes 30 to 120 seconds for a precipitate to form; but with some, it may take even longer.

In a negative test, usually there is no precipitate. In some cases, however, a precipitate forms, but the solution remains orange.

In performing this test, one should make quite sure that the acetone used for the solvent does not give a positive test with the reagent. Add several drops of the chromic acid reagent to a few drops of the reagent acetone contained in a small test tube. Allow this mixture to stand for 3 to 5 minutes. If no reaction has occurred by this time, the acetone is pure enough to use as a solvent for the test. If a positive test resulted, try another bottle of acetone, or distill some acetone from potassium permanganate to purify it.

Test Compounds. Try this test on benzaldehyde, butanal (butyraldehyde), and cyclohexanone.

> **CAUTION: Many compounds of chromium(VI) are suspect carcinogens (see p. 12) and should be handled with care. Avoid contact.**

Reagent. Dissolve 1.0 g of chromic oxide, CrO_3, in 1 mL of concentrated sulfuric acid. Then dilute this mixture carefully with 3 mL of water.

This test has as its basis the fact that aldehydes are easily oxidized to the corresponding carboxylic acid by chromic acid. The green precipitate is due to chromous sulfate.

$$2CrO_3 + 2H_2O \xrightleftharpoons{H^+} 2H_2CrO_4 \xrightleftharpoons{H^+} H_2Cr_2O_7 + H_2O$$

$$3RCHO + \underset{\text{Orange}}{H_2Cr_2O_7} + 3H_2SO_4 \longrightarrow 3RCOOH + \underset{\text{Green}}{Cr_2(SO_4)_3} + 4H_2O$$

Primary and secondary alcohols are also oxidized by this reagent (see Procedure 56H). Therefore, this test is not useful in identifying aldehydes **unless** a positive identification of the carbonyl group has already been made. Aldehydes give a 2,4-dinitrophenylhydrazine test, whereas alcohols do not.

There are numerous other tests used to detect the aldehyde functional group. Most are based on an easily detectible oxidation of the aldehyde to a carboxylic acid. The most common tests are the Tollens, Fehling, and Benedict tests. The Benedict test is described in Experiment 57. Only the Tollens test is described here.

TOLLENS TEST

Procedure. The reagent must be prepared immediately before use. To prepare the reagent, mix 1 mL of Tollens solution A with 1 mL of Tollens solution B. A precipitate of silver oxide will form. Add enough dilute (10%) ammonia solution (dropwise) to the mixture to **just** dissolve the silver oxide. The reagent so prepared can be used immediately for the test below.

Dissolve 1 drop of a liquid aldehyde or 10 mg (approximate) of a solid aldehyde in the minimum amount of bis(2-ethoxyethyl) ether. Add this solution, a little at a time, to the 2 or 3 mL of reagent contained in a small test tube. Shake the solution well. If a mirror of silver is deposited on the inner walls of the test tube, the test is positive. In some cases it may be necessary to warm the test tube in a bath of warm water.

> **CAUTION: The reagent should be prepared immediately before use and all residues disposed of immediately after use. Wash any residues down a sink with a large quantity of water. On standing, the reagent tends to form silver fulminate, a _very explosive_ substance. Solutions containing the mixed Tollens reagent should never be stored.**

Test Compounds. Try the test on acetone and benzaldehyde.

Reagents. Solution A: Dissolve 3.0 g of silver nitrate in 30 mL of water. Solution B: Prepare a 10% sodium hydroxide solution.

Most aldehydes reduce ammoniacal silver nitrate solution to give a precipitate of silver metal. The aldehyde is oxidized to a carboxylic acid:

$$RCHO + 2Ag(NH_3)_2OH \longrightarrow 2Ag + RCOO^-NH_4^+ + H_2O + NH_3$$

Ordinary ketones do not give a positive result in this test. The test should be used only if it has already been shown that the unknown compound is either an aldehyde or a ketone.

IODOFORM TEST

Procedure. Using a Pasteur pipet, add 4 drops of a liquid unknown to a large test tube (20 × 150 mm). Alternatively, 0.1 g of a solid unknown may be used. Add to the test tube 2 mL of bis(2-ethoxyethyl) ether, 5 mL of water, and 1 mL of 10% sodium hydroxide solution. Obtain 3 mL of the iodine-potassium iodide test solution, and add it with a dropper in about six portions to the solution in the test tube. Shake the tube after each addition. During the early additions, the intense iodine color is decolorized rapidly, but near the end the color may not be discharged as rapidly. In any case, **cork** the test tube and shake it vigorously until the intensely colored solution has decolorized completely to give a pale yellow solution. It may be necessary to heat the solution slightly on a steam bath or in a water bath at 60 °C to aid in the discharge of the color. Again, shake the stoppered test tube vigorously after each heating period.

After the solution has been decolorized or nearly so, fill the test tube with water, cork the tube, and shake it vigorously. Allow the tube to stand for 15 minutes. A yellow precipitate of iodoform will form if the unknown was a methyl ketone or a compound easily oxidized to a methyl ketone. Other ketones will also decolorize the iodine solution, but they will not give a precipitate of iodoform **unless** there is an impurity of a methyl ketone in the unknown. To prove the identity of the yellow precipitate as iodoform, one should collect and dry the solid and determine its melting point (iodoform, mp 119 to 121 °C).

Test Compound.   Try the test on acetone.

Reagent.   Dissolve 20 g of potassium iodide and 10 g of iodine in 100 mL of water.

The basis of this test is the ability of certain compounds to form a precipitate of iodoform when treated with a basic solution of iodine. Methyl ketones are the most common type of compounds that give a positive result in this test. However, acetaldehyde, CH_3CHO, and alcohols with the hydroxyl group at the 2-position of the chain also give a precipitate of iodoform. Alcohols of the type described are easily oxidized to methyl ketones under the conditions of the reaction. The other product of the reaction, besides iodoform, is the potassium salt or the sodium salt of a carboxylic acid.

$$R\text{—}\underset{\underset{OH}{|}}{C}H\text{—}CH_3 \xrightarrow[NaOH]{I_2} R\text{—}\underset{\underset{O}{\|}}{C}\text{—}CH_3 \xrightarrow[NaOH]{I_2} R\text{—}\underset{\underset{O}{\|}}{C}\text{—}CI_3 \xrightarrow{OH^-} R\text{—}\underset{\underset{O}{\|}}{C}\text{—}O^- + HCI_3$$

$\quad\quad$ **Alcohol** $\quad\quad\quad\quad\quad\quad$ **Methyl ketone** $\quad\quad\quad\quad\quad\quad\quad\quad\quad\quad\quad\quad\quad$ **Iodoform**

FERRIC CHLORIDE TEST

Procedure.   Some aldehydes and ketones, those that have a high **enol content,** give a positive ferric chloride test, as described for phenols in Procedure 56F.

SPECTROSCOPY

INFRARED

The carbonyl group is usually one of the strongest-absorbing groups in the infrared spectrum, with a very broad range: 1800 to 1650 cm^{-1} (5.85–6.20 μ). The aldehyde functional group has **very characteristic** CH stretch absorptions: two sharp peaks that lie **far outside** the usual region for —C—H, $=$C—H, or $\equiv$C—H.

ALDEHYDES

C=O stretch at approximately 1725 cm^{-1} (5.80 μ) is normal. 1725 to 1685 cm^{-1} (5.80–5.95 μ).[2]

C—H stretch (aldehyde —CHO). Two weak bands at about 2750 and 2850 cm^{-1} (3.65 and 3.50 μ).

KETONES

C=O stretch at approximately 1715 cm^{-1} (5.85 μ) is normal. 1780 to 1665 cm^{-1} (5.62–6.01 μ).[2]

NUCLEAR MAGNETIC RESONANCE

Hydrogens alpha to a carbonyl group have resonance in the region between 2 and 3δ. The hydrogen of an aldehyde group has a characteristic resonance between 9 and 10δ. In aldehydes, there is coupling between the aldehyde hydrogen and any alpha hydrogens (J = 1–3 Hz).

DERIVATIVES

The most common derivatives of aldehydes and ketones are the 2,4-diphenyl-hydrazones, oximes, and semicarbazones. Procedures for preparing these derivatives are given in Appendix 2 at the end of the book.

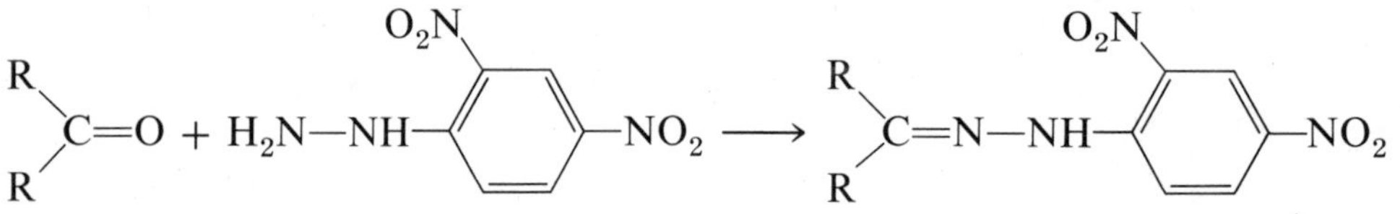

2,4-Dinitrophenylhydrazine   2,4-Dinitrophenylhydrazone

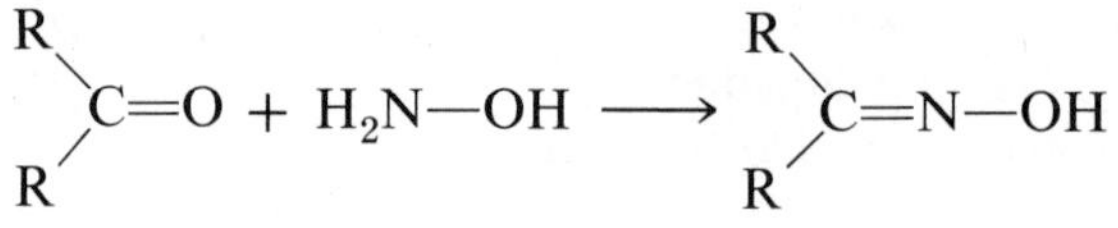

Hydroxylamine   Oxime

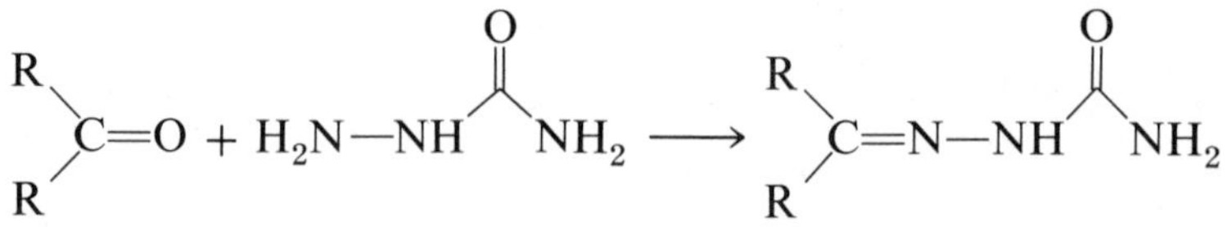

Semicarbazide   Semicarbazone

[2]**Conjugation** moves the absorption to lower frequencies (higher wavelength). **Ring strain** (cyclic ketones) moves the absorption to higher frequencies (lower wavelength).

Procedure 56E
Carboxylic Acids

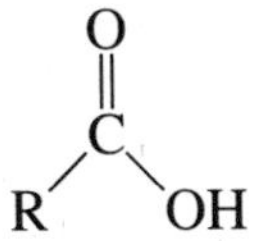

$$\underset{R}{\overset{\displaystyle\overset{\textstyle O}{\|}}{\underset{}{C}}}\text{—OH}$$

Carboxylic acids are detectable mainly by their solubility characteristics. They are soluble in **both** dilute sodium hydroxide and sodium bicarbonate solutions.

SOLUBILITY CHARACTERISTICS					CLASSIFICATION TESTS
HCl	NaHCO$_3$	NaOH	H$_2$SO$_4$	Ether	pH of an aqueous solution
(−)	(+)	(+)	(+)	(+)	Sodium bicarbonate
Water: <C$_6$ (+)					Silver nitrate
>C$_6$ (−)					Neutralization equivalent

CLASSIFICATION TESTS

pH OF AN AQUEOUS SOLUTION

Procedure. If the compound is soluble in water, simply prepare an aqueous solution and check the pH with pH paper. If the compound is an acid, the solution will have a low pH.

Compounds that are insoluble in water can be dissolved in ethanol (or methanol) and water. First dissolve the compound in the alcohol and then add water until the solution **just** becomes cloudy. Clarify the solution by adding a few drops of the alcohol, and then determine its pH, using pH paper.

SODIUM BICARBONATE

Procedure. Dissolve a small amount of the compound in a 5% aqueous sodium bicarbonate solution. Observe the solution carefully. If the compound is an acid, bubbles of carbon dioxide will be seen to form.

$$\text{RCOOH} + \text{NaHCO}_3 \longrightarrow \text{RCOO}^-\text{Na}^+ + \text{H}_2\text{CO}_3 \text{ (unstable)}$$

$$\text{H}_2\text{CO}_3 \longrightarrow \text{CO}_2 + \text{H}_2\text{O}$$

SILVER NITRATE

Procedure. Acids give a false silver nitrate test, as described in Procedure 56B.

NEUTRALIZATION EQUIVALENT

Procedure. Accurately weigh (three significant figures) approximately 0.2 g of the acid into a 125-mL Erlenmeyer flask. Dissolve the acid in about 50 mL of water or aqueous ethanol (the acid need not dissolve completely, since it will dissolve as it is titrated). Titrate the acid, using a solution of sodium hydroxide of known normality (about 0.1N) and a phenolphthalein indicator.

Calculate the neutralization equivalent (NE) from the equation

$$NE = \frac{mg \ acid}{normality \ of \ NaOH \times mL \ of \ NaOH \ added}$$

The NE is identical to the equivalent weight of the acid. If the acid has only one carboxyl group, the neutralization equivalent and the molecular weight of the acid are identical. If the acid has more than one carboxyl group, the neutralization equivalent equals the molecular weight of the acid divided by the number of carboxyl groups, that is, the equivalent weight. The NE can be used much like a derivative to identify a specific acid.

Many phenols are acidic enough to behave much like carboxylic acids. This is especially true of those substituted with electron-withdrawing groups at the *ortho* and *para* ring positions. These phenols, however, can usually be eliminated either by the ferric chloride test (Experiment 56F) or by spectroscopy (phenols have no carbonyl group).

SPECTROSCOPY

INFRARED

C=O stretch is very strong and often broad in the region between 1725 and 1690 cm^{-1} (5.8–5.9 μ).

O—H stretch is a very broad absorption in the region between 3300 and 2500 cm^{-1} (3.0–4.0 μ); it usually overlaps the CH stretch region.

NUCLEAR MAGNETIC RESONANCE

The acid proton of a —COOH group usually has resonance near 12.0δ.

DERIVATIVES

Derivatives of acids are usually amides. They are prepared via the corresponding acid chloride:

$$R-\overset{\overset{\displaystyle O}{\|}}{C}-OH + SOCl_2 \longrightarrow R-\overset{\overset{\displaystyle O}{\|}}{C}-Cl + SO_2 + HCl$$

The most common derivatives are the amides, the anilides, and the *p*-toluidides.

$$R-\overset{\overset{\displaystyle O}{\|}}{C}-Cl + \;2NH_4OH \longrightarrow R-\overset{\overset{\displaystyle O}{\|}}{C}-NH_2 + 2H_2O + NH_4Cl$$

Ammonia (aq.)   **Amide**

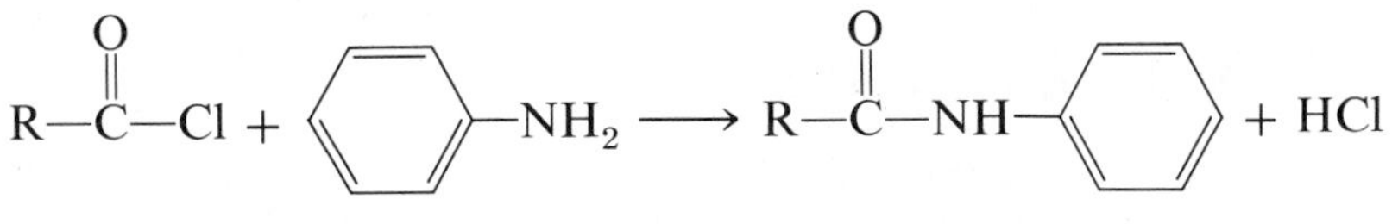

Aniline   **Anilide**

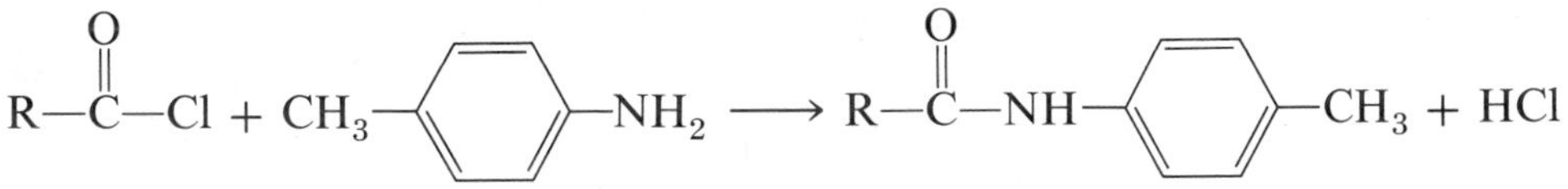

p-Toluidine   **p-Toluidide**

Procedures for the preparation of these derivatives are given in Appendix 2.

Procedure 56F
Phenols

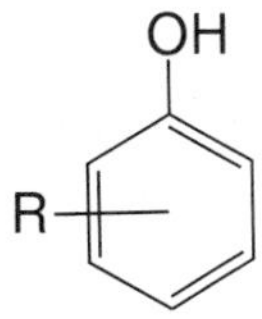

Like carboxylic acids, phenols are acidic compounds. However, except for the nitrosubstituted phenols (discussed in the section covering solubilities), they are not as acidic as the carboxylic acids. The pK_a of a typical phenol is 10, whereas the pK_a of a carboxylic acid is usually near 5. Hence, phenols are generally not soluble in the weakly basic sodium bicarbonate solution but do dissolve in sodium hydroxide solution, which is more strongly basic.

SOLUBILITY CHARACTERISTICS	CLASSIFICATION TESTS

HCl   NaHCO$_3$ NaOH   H$_2$SO$_4$ Ether
(−)   (−)   (+)   (+)   (+)
Water: Most are insoluble, although
   phenol itself and the nitrophenols
   are soluble
	Colored phenolate anion
	Ferric chloride
	Bromine/water

CLASSIFICATION TESTS

SODIUM HYDROXIDE SOLUTION

Procedure.   With phenols that have a high degree of conjugation possible in their conjugate base (phenolate ion), the anion is often colored. To observe the color, it is necessary only to dissolve a small amount of the phenol in 10% aqueous sodium hydroxide solution. Some phenols do not give a color. Others have an insoluble anion and give a precipitate. The more acidic phenols, like the nitrophenols, tend more toward colored anions.

FERRIC CHLORIDE

Procedure 1 (Water-Soluble Phenols).   Add several drops of a 2.5% aqueous solution of ferric chloride to 1 mL of a dilute aqueous solution (about 1–3% by weight of the phenol). Most phenols produce an intense red, blue, purple, or green color. Some colors are transient, and it may be necessary to observe the solution carefully just as the solutions are mixed. The formation of a color is usually immediate, but the color may not be permanent over any great period. Some phenols do not give a positive result in this test, so a negative test must not be taken as significant without other adequate evidence.

Test Compound.   Try this test on phenol.

Procedure 2 (Water-Insoluble Phenols).   Many phenols do not give a positive result when procedure 1 is used. Often the procedure now to be described will give a positive result with such phenols. Dissolve or suspend 20 mg of a solid phenol or 1 drop of a liquid phenol in 1 mL of methylene chloride. Add one drop of pyridine and 3 to 5 drops of a 1% (weight/volume) solution of ferric chloride in methylene chloride.

The colors observed in this test result from the formation of a complex of the phenols with Fe(III) ion. Carbonyl compounds that have a high enol content also give a positive result in this test.

BROMINE/WATER

Procedure. Prepare a 1% aqueous solution of the unknown, and then add a saturated solution of bromine in water to it, drop by drop with shaking, until the bromine color is no longer discharged. A positive test is indicated by the precipitation of a substitution product at the same time that the bromine color of the reagent is discharged.

Test Compound. Try this test on phenol.

Aromatic compounds with ring-activating substituents give a positive test with bromine in water. The reaction is an aromatic substitution reaction, which introduces bromine atoms into the aromatic ring at the positions *ortho* and *para* to the hydroxyl group. All available positions are usually substituted. The precipitate is the brominated phenol, which is generally insoluble because of its large molecular weight.

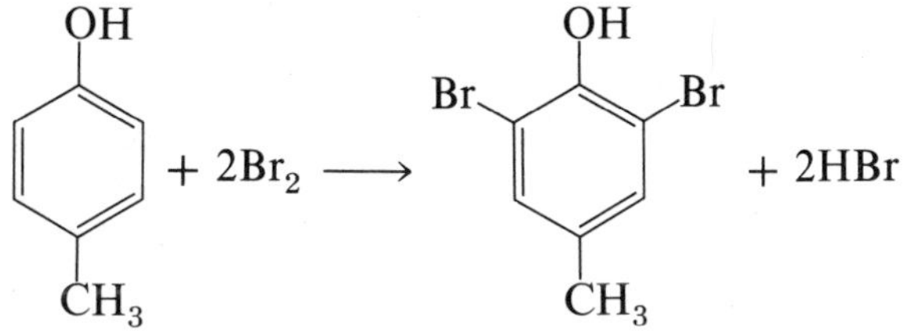

Other compounds that give a positive result with this test include aromatic compounds that have activating substituents other than hydroxyl. These compounds include anilines and alkoxyaromatics.

SPECTROSCOPY

INFRARED

O—H stretch is observed near 3600 cm^{-1} (2.8 μ).
C—O stretch is observed near 1200 cm^{-1} (8.3 μ).
The typical aromatic ring absorptions between 1650 and 1400 cm^{-1} (6–7 μ) are also found.
Aromatic CH is observed near 3100 cm^{-1} (3.2 μ).

NUCLEAR MAGNETIC RESONANCE

Aromatic protons are observed near 7δ. The hydroxyl proton has a resonance position that is concentration-dependent.

DERIVATIVES

Phenols form the same derivatives as alcohols do (Procedure 56H). They form urethanes by reaction with isocyanates, but whereas phenylurethanes are used for alcohols, the α-naphthylurethanes are more useful for phenols. Like alcohols, phenols also yield 3,5-dinitrobenzoates.

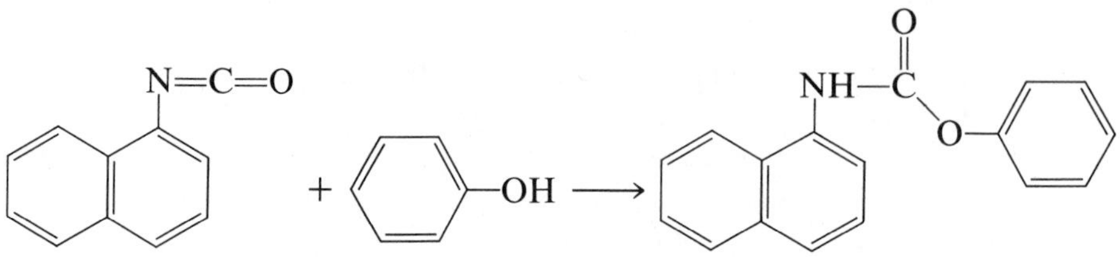

α-Naphthyl isocyanate   An α-naphthylurethane

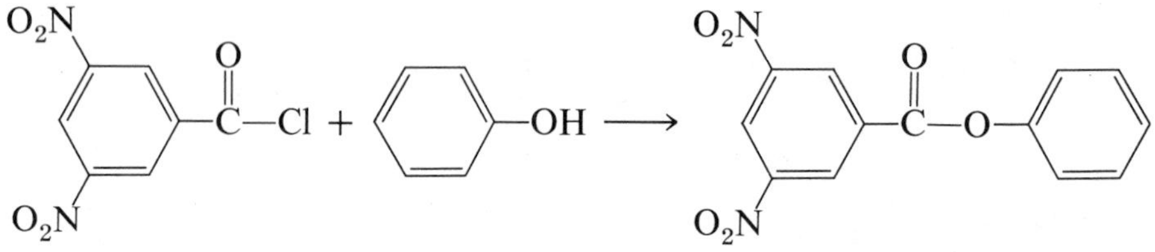

3,5-Dinitrobenzoyl
chloride   A 3,5-dinitrobenzoate

The bromine-water reagent yields solid bromo derivatives of phenols in several cases. These solid derivatives can be used to characterize an unknown phenol.

Procedures for preparing these derivatives are given in Appendix 2.

Procedure 56G
Amines

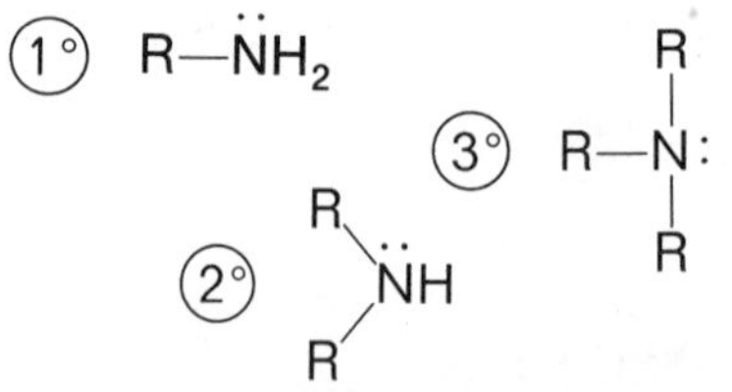

Amines are detected best by their solubility behavior and their basicity. They are the only basic compounds that will be issued for this experiment. Hence, once the compound has been identified as an amine, the main problem that remains is to decide whether it is primary (1°), secondary (2°), or tertiary (3°). This can usually be decided by the Hinsberg and nitrous acid tests.

SOLUBILITY CHARACTERISTICS	CLASSIFICATION TESTS
HCl NaHCO$_3$ NaOH H$_2$SO$_4$ Ether (+) (−) (−) (+) (+) Water: <C$_6$ (+) 　　　>C$_6$ (−)	pH of an aqueous solution Hinsberg test Nitrous acid test Acetyl chloride

CLASSIFICATION TESTS

HINSBERG TEST

Procedure. Place 0.1 mL of a liquid amine or 0.1 g of a solid amine, 0.2 g of *p*-toluenesulfonyl chloride, and 5 mL of 10% potassium hydroxide solution in a small test tube. Stopper the test tube tightly and shake it intermittently for 3 to 5 minutes. Remove the stopper and warm the test tube, with shaking, on a steam bath for 1 minute. Cool the solution and test a drop of it with pH paper to see whether it is still basic; if it is not, add more potassium hydroxide. If a precipitate has formed, dilute the basic mixture with 5 mL of water and shake it well. If the precipitate is insoluble, a disubstituted sulfonamide is probably present, which indicates that the unknown was a 2° amine.

> **NOTE: The precipitate may also be unreacted *p*-toluenesulfonyl chloride, leading to confusing results.**

If no precipitate remains after you dilute the mixture, or if none formed initially, carefully add 5% hydrochloric acid until the solution is just acidic to litmus (avoid excess acid). If a precipitate forms at this point, it should be the monosubstituted sulfonamide, indicating that the original compound was a 1° amine. If no reaction was apparent during the test, the original compound was probably a 3° amine.

If the above procedure gives confusing results(!), the procedure can be repeated, using 0.2 mL of benzenesulfonyl chloride instead of the *p*-toluenesulfonyl chloride. However, this reagent is likely to lead to the production of oils instead of solids.

Test Compounds. Try this test on aniline, *N*-methylaniline, and *N*,*N*-dimethylaniline. All three tests should be run simultaneously to allow easy comparison of the results.

This test is based on the production of monosubstituted and disubstituted sulfonamides from primary and secondary amines, respectively. Monosubstituted sulfonamides are soluble in base, while disubstituted sulfonamides are not soluble since they have no acidic hydrogens that can be removed to form a soluble salt.

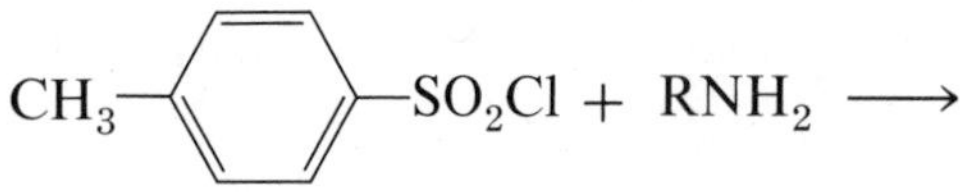

$$CH_3-\langle\hspace{-4pt}\bigcirc\hspace{-4pt}\rangle-SO_2Cl + RNH_2 \longrightarrow$$

p-Toluenesulfonyl
chloride (pTsCl)   **1° Amine**

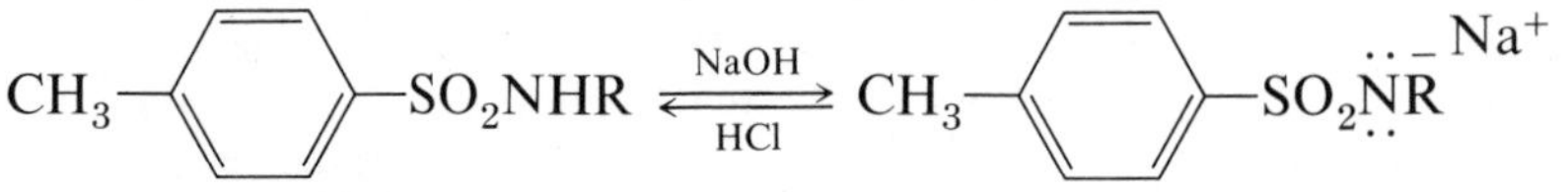

$$CH_3-\langle\hspace{-4pt}\bigcirc\hspace{-4pt}\rangle-SO_2NHR \underset{HCl}{\overset{NaOH}{\rightleftarrows}} CH_3-\langle\hspace{-4pt}\bigcirc\hspace{-4pt}\rangle-SO_2\overset{..}{\underset{..}{N}}R^{\;-}Na^+$$

Insoluble   **Soluble salt**
(but converted to
soluble salt in base)

$$CH_3-\langle\hspace{-4pt}\bigcirc\hspace{-4pt}\rangle-SO_2Cl + R-\underset{\underset{R}{|}}{N}H \longrightarrow$$

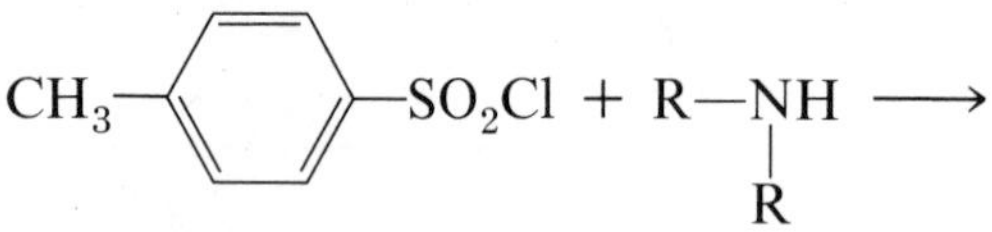

p-Toluenesulfonyl
chloride (pTsCl)   **2° Amine**

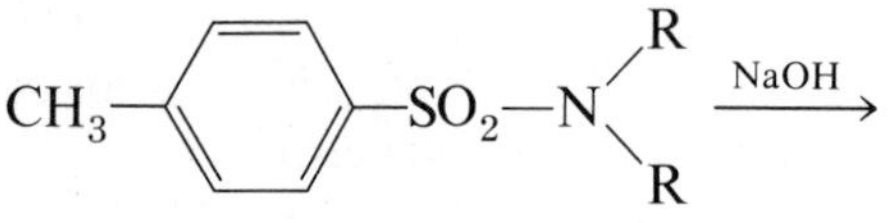

$$CH_3-\langle\hspace{-4pt}\bigcirc\hspace{-4pt}\rangle-SO_2-N\underset{R}{\overset{R}{\big\langle}} \overset{NaOH}{\longrightarrow}$$ **Will not dissolve**
since there is no
hydrogen on N to
be removed

Insoluble

Some sulfonamides of primary amines form insoluble sodium salts. This may lead to the mistaken assumption that the amine is secondary.

Tertiary amines seem to be unreactive under these conditions, presumably because a tertiary amine has **no** amino hydrogens to be replaced. Actually, tertiary amines do react with benzene- and p-toluenesulfonyl chlorides; in most cases, however, they yield an overall result that gives the impression that no reaction has occurred. Two types of behavior can be observed for tertiary amines, one characteristic of tertiary alkylamines, R_3N, and the other characteristic of tertiary arylamines (that is, anilines).

Most tertiary **alkylamines** react according to the following pattern:

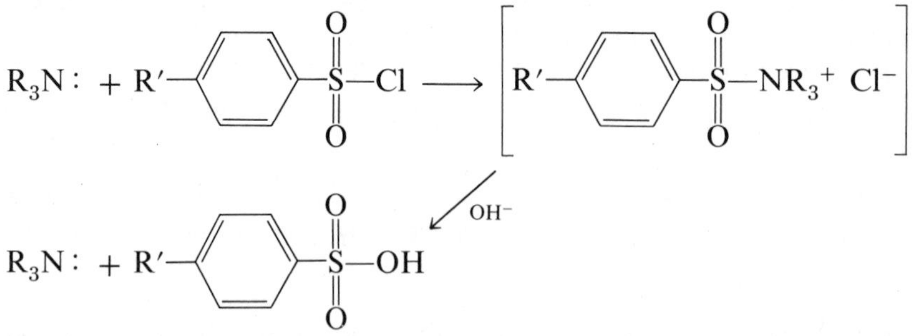

$$R_3N: + R'-\langle\hspace{-4pt}\bigcirc\hspace{-4pt}\rangle-\overset{O}{\underset{O}{\overset{\|}{\underset{\|}{S}}}}-Cl \longrightarrow \left[R'-\langle\hspace{-4pt}\bigcirc\hspace{-4pt}\rangle-\overset{O}{\underset{O}{\overset{\|}{\underset{\|}{S}}}}-NR_3^+\;Cl^- \right]$$

$$\overset{\displaystyle\swarrow OH^-}{}$$

$$R_3N: + R'-\langle\hspace{-4pt}\bigcirc\hspace{-4pt}\rangle-\overset{O}{\underset{O}{\overset{\|}{\underset{\|}{S}}}}-OH$$

This leads to an observation of "no reaction." However, reaction has definitely occurred. In addition, many tertiary alkylamines give complex precipitates when left standing in the reaction medium. Therefore, the reaction time allowed for the test must be relatively short, or reaction may really be observed.

Tertiary **arylamines** are usually not very soluble in the reaction medium and often form an "oil" in the bottom of the test tube. Under these conditions, hydroxide ions of the reaction medium react more rapidly with the sulfonyl chloride than the insoluble amines can, and this also leads to an observation of "no reaction."

$$(ArNR_2) + R' \underset{\text{Insoluble oil}}{\underbrace{}} \!\!\!\! \begin{array}{c} O \\ \| \\ S \\ \| \\ O \end{array} \!\!\!\! Cl \xrightarrow{OH^-} (ArNR_2) + R' \underset{\text{Insoluble oil}}{\underbrace{}} \!\!\!\! \begin{array}{c} O \\ \| \\ S \\ \| \\ O \end{array} \!\!\!\! OH$$

When tertiary arylamines do dissolve in the medium, complicated secondary reactions may occur, especially if there is either excess amine or sulfonyl chloride present or if the reaction is heated. Often these secondary products include intensely colored dyes.[3]

The accompanying table describes the results generally observed when the traditional Hinsberg test is used.

HINSBERG TEST SUMMARY

Primary amine + pTsCl $\xrightarrow{\text{NaOH}}$ solution $\xrightarrow{\text{HCl}}$ precipitate

Secondary amine + pTsCl $\xrightarrow{\text{NaOH}}$ precipitate

Tertiary amine + pTsCl $\xrightarrow{\text{NaOH}}$ no reaction $\xrightarrow{\text{HCl}}$ clear solution
(apparent)

Two cautions should be observed when you are performing the Hinsberg test. First, these tests work well with reagent-grade amines; however, practical grades often contain impurities. For instance, secondary amines are often made from primary amines that may be contaminants. Similarly, tertiary amines may contain traces of secondary amines. Therefore, trace precipitates should not be considered as definitive results. Second, reaction times should be short, and any heating should be gentle—many tertiary amines will react under more vigorous conditions.

NITROUS ACID TEST

Procedure. Dissolve 0.1 g of an amine in 2 mL of water to which 8 drops of concentrated sulfuric acid have been added. Use a large test tube. Cool the solution to 5 °C or less in an ice bath. Also cool 2 mL of 10% aqueous sodium nitrite in another test tube. In a third test tube, prepare a solution of 0.1 g β-naphthol in 2 mL of aqueous sodium hydroxide, and place it in an ice bath to cool. Add the cold sodium nitrite solution, drop by drop with shaking, to the cooled solution of the amine. Look for bubbles of nitrogen gas. Be careful not to confuse the evolution of the **colorless** nitrogen gas with

---

[3]A complete discussion of this chemistry is in C. R. Gambill, T. D. Roberts, and H. Shechter, "Benzenesulfonyl Chloride Does React with Tertiary Amines," *Journal of Chemical Education, 49*:4 (April 1972), p. 287. These authors also give an alternative version of the Hinsberg test, which may help students who have difficulty in interpreting their results using the tests described above.

an evolution of **brown** nitrogen oxide gas. Substantial evolution of gas at 5 °C or below indicates a primary aliphatic amine, RNH_2. The formation of a yellow oil or a yellow solid usually indicates a secondary amine, R_2NH. Either tertiary amines do not react, or they behave like secondary amines.

If little or no gas evolves at 5 °C, take **half** the solution and warm it gently to about room temperature. Nitrogen gas bubbles at this elevated temperature indicate that the original compound was a primary **aromatic** $ArNH_2$. Take the remaining solution and drop by drop add the solution of β-naphthol in base. If a red dye precipitates, the unknown has been conclusively shown to be a primary aromatic amine, $ArNH_2$.

Test Compounds.   Try this test with aniline, N-methylaniline, and butylamine.

> **CAUTION: The products of this reaction may include nitrosamines. Nitrosamines are suspect carcinogens. Avoid contact and dispose of all residues in designated waste containers.**

Before you make this test, it should definitely be proved by some other method that the unknown is an amine. Many other compounds react with nitrous acid (phenols, ketones, thiols, amides), and a positive result with one of these could lead to an incorrect interpretation.

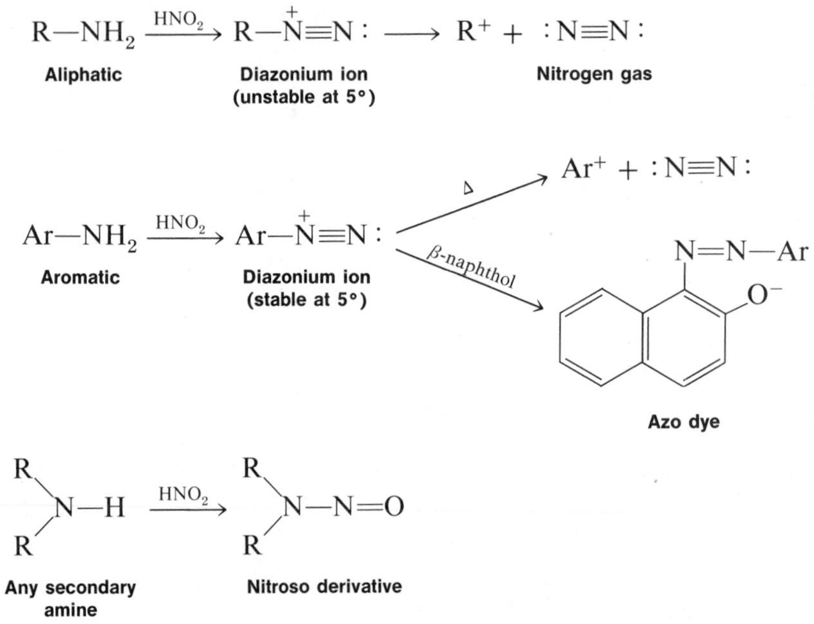

The test is best used to distinguish **primary** aromatic and **primary** aliphatic amines from secondary and tertiary amines. It also differentiates aromatic and aliphatic primary amines. It cannot distinguish between secondary and tertiary amines. Primary aliphatic amines lose nitrogen gas at low temperatures under the conditions of this test. Aromatic amines yield a more stable diazonium salt and do not lose nitrogen until the temperature is elevated. In addition, aromatic diazonium salts produce a red azo dye when β-naphthol is added. Secondary and tertiary amines produce yellow nitroso compounds, which may be soluble or may be oils or even solids. Many nitroso compounds have been shown to be carcinogenic. Avoid contact and immediately dispose of all such solutions in an appropriate waste container.

pH OF AN AQUEOUS SOLUTION

Procedure. If the compound is soluble in water, simply prepare an aqueous solution and check the pH with pH paper. If the compound is an amine, it will be basic, and the solution will have a high pH. Compounds that are insoluble in water can be dissolved in ethanol-water or 1,2-dimethoxyethane–water.

ACETYL CHLORIDE

Procedure. Amines give a positive acetyl chloride test (liberation of heat). This test is described for alcohols in Experiment 56H. When the test mixture is diluted with water, primary and secondary amines often give a solid acetamide derivative; tertiary amines do not.

SPECTROSCOPY

INFRARED

N—H stretch. Both aliphatic and aromatic **primary** amines show two absorptions (doublet due to symmetric and asymmetric stretches) in the region 3500 to 3300 cm^{-1} (2.86–3.03 μ).
  Secondary amines show a single absorption in this region.
  Tertiary amines have no N—H bonds.
N—H bend. **Primary** amines have a strong absorption at 1640 to 1560 cm^{-1} (6.10–6.41 μ).

  Secondary amines have an absorption at 1580 to 1490 cm^{-1} (6.33–6.71 μ). Aromatic amines show bands typical for the aromatic ring in the region 1650 to 1400 cm^{-1} (6–7 μ). Aromatic CH is observed near 3100 cm^{-1} (3.2 μ).

NUCLEAR MAGNETIC RESONANCE

The resonance position of amino hydrogens is extremely variable. The resonance may also be very broad (quadrupole broadening). Aromatic amines give resonances near 7δ due to the aromatic ring hydrogens.

DERIVATIVES

The derivatives of amines that are most easily prepared are the acetamides and the benzamides. These derivatives work well for both primary and secondary amines but not tertiary amines.

$$CH_3-\overset{\overset{\displaystyle O}{\|}}{C}-Cl + RNH_2 \longrightarrow CH_3-\overset{\overset{\displaystyle O}{\|}}{C}-NH-R + HCl$$

Acetyl chloride   **An acetamide**

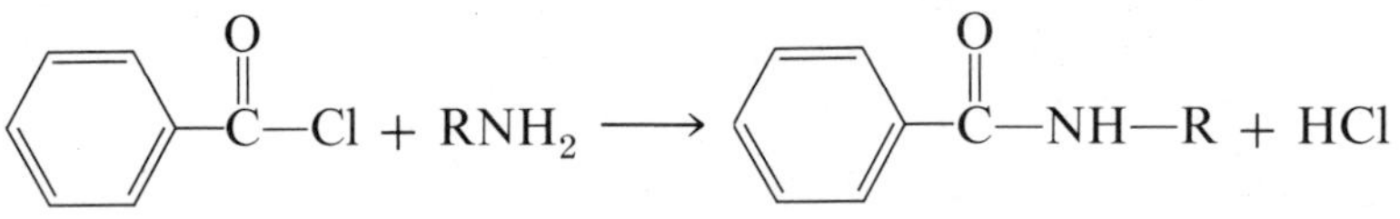

$$\langle \rangle -\overset{\overset{\displaystyle O}{\|}}{C}-Cl + RNH_2 \longrightarrow \langle \rangle -\overset{\overset{\displaystyle O}{\|}}{C}-NH-R + HCl$$

Benzoyl chloride   **A benzamide**

In some cases, the solid *p*-toluenesulfonamides and benzenesulfonamides prepared from primary and secondary amines in the Hinsberg test (see above) can be used as derivatives.

The most general derivative that can be prepared is the picric acid salt, or picrate, of an amine. This derivative can be used for primary, secondary, and tertiary amines.

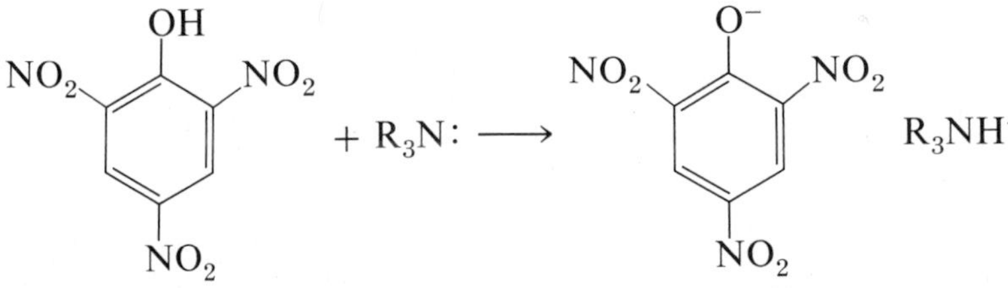

Picric acid   **A picrate**

For tertiary amines, the methiodide salt is often useful.

$$CH_3I + R_3N: \longrightarrow CH_3-NR_3{}^+I^-$$

A methiodide

Procedures for preparing derivatives from amines can be found in Appendix 2.

Procedure 56H
Alcohols

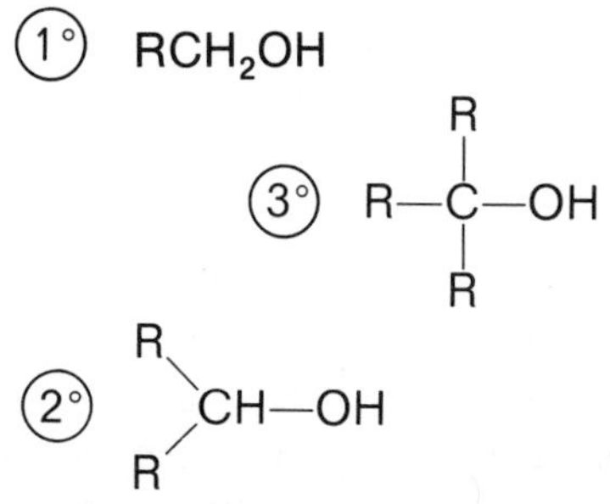

$$1° \quad RCH_2OH$$

$$3° \quad R-\underset{\underset{R}{|}}{\overset{\overset{R}{|}}{C}}-OH$$

$$2° \quad \underset{R}{\overset{R}{\diagdown}}CH-OH$$

Alcohols are neutral compounds. The only other classes of neutral compounds used in this experiment are the aldehydes and ketones and the esters. Alcohols and esters usually do not give a positive 2,4-dinitrophenylhydrazine test, whereas aldehydes and ketones do. Esters do not react with acetyl chloride or with Lucas reagent, as alcohols do, and they are easily distinguished from alcohols on this basis. Primary and secondary alcohols are easily oxidized, whereas esters and tertiary alcohols are not. A combination of the Lucas test and the chromic acid test will differentiate among primary, secondary, and tertiary alcohols.

SOLUBILITY CHARACTERISTICS					CLASSIFICATION TESTS
HCl	NaHCO$_3$	NaOH	H$_2$SO$_4$	Ether	Acetyl chloride
(−)	(−)	(−)	(+)	(+)	Lucas test
Water: <C$_6$ (+)					Chromic acid test
>C$_6$ (−)					Iodoform test

CLASSIFICATION TESTS

ACETYL CHLORIDE

Procedure. Cautiously add about 10 to 15 drops of acetyl chloride, drop by drop, to about 0.5 mL of the alcohol contained in a small test tube. Evolution of heat and hydrogen chloride gas indicates a positive reaction. Addition of water will sometimes precipitate the acetate.

Acid chlorides react with alcohols to form esters. Acetyl chloride forms acetate esters.

$$CH_3-\overset{\overset{O}{\|}}{C}-Cl + ROH \longrightarrow CH_3-\overset{\overset{O}{\|}}{C}-O-R + HCl$$

Usually the reaction is exothermic, and the heat evolved is easily detected. Phenols react with acid chlorides somewhat as alcohols do. Hence, phenols should be eliminated as possibilities before this test is attempted. Amines also react with acetyl chloride to evolve heat (see Procedure 56G).

LUCAS TEST

Procedure. Place 2 mL of Lucas reagent in a small test tube and add 3 to 4 drops of the alcohol. Stopper the test tube and shake it vigorously. Tertiary (3°), benzylic, and allylic alcohols give an immediate cloudiness in the solution as the insoluble alkyl halide separates from the aqueous solution. After a short time, the immiscible alkyl halide will form a separate layer. Secondary (2°) alcohols produce a cloudiness after 2 to 5 minutes. Primary (1°) alcohols dissolve in the reagent to give a clear solution. Some secondary alcohols may have to be heated slightly to encourage reaction with the reagent.

This test works only for alcohols that are soluble in the reagent. This often means that alcohols with more than six carbon atoms cannot be tested.

Test Compounds. Try this test with 1-butanol (*n*-butyl alcohol), 2-butanol (*sec*-butyl alcohol), and 2-methyl-2-propanol (*t*-butyl alcohol).

Reagent. Cool 10 mL of concentrated hydrochloric acid in a beaker, using an ice bath. While still cooling, and with stirring, dissolve 16 g of anhydrous zinc chloride in the acid.

This test depends on the appearance of an alkyl chloride as an insoluble second layer when an alcohol is treated with a mixture of hydrochloric acid and zinc chloride (Lucas reagent):

$$R\text{—}OH + HCl \xrightarrow{\text{ZnCl}_2} R\text{—}Cl + H_2O$$

Primary alcohols do not react at room temperature; therefore, the alcohol is seen simply to dissolve. Secondary alcohols react slowly, whereas tertiary, benzylic, and allylic alcohols react instantly. These relative reactivities are explained on the same basis as the silver nitrate reaction, which is discussed in Procedure 56B. Primary carbonium ions are unstable and do not form under the conditions of this test. Hence, no results are observed for primary alcohols.

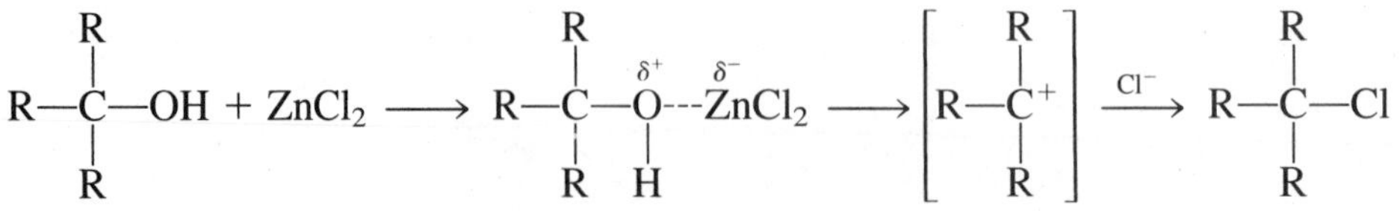

CHROMIC ACID TEST

Procedure. Dissolve 1 drop of a liquid or about 10 mg of a solid alcohol in 1 mL of **reagent-grade** acetone. Add 1 drop of the chromic acid reagent and note the result that occurs within 2 seconds. A positive test for a primary or a secondary alcohol is the appearance of a blue-green color. Tertiary alcohols do not give the test within 2 seconds, and the solution remains orange. To make sure that the acetone solvent is pure and does not give a positive test, add 1 drop of chromic acid to 1 mL of acetone that does not have an unknown dissolved in it. The orange color of the reagent should persist for **at least** 3 seconds. If it does not, a new bottle of acetone should be used.

Test Compounds. Try this test with 1-butanol (*n*-butyl alcohol), 2-butanol (*sec*-butyl alcohol), and 2-methyl-2-propanol (*t*-butyl alcohol).

> **CAUTION: Many compounds of chromium(VI) are suspect carcinogens (see p. 12) and should be handled with care. Avoid contact.**

Reagent. Dissolve 1 g of chromic oxide in 1 mL of concentrated sulfuric acid. Carefully add the mixture to 3 mL of water.

This test is based on the reduction of chromium(VI), which is orange, to chromium(III), which is green, when an alcohol is oxidized by the reagent. A change in color of the reagent from orange to green represents a positive test. Primary alcohols are oxidized by the reagent to carboxylic acids; secondary alcohols are oxidized to ketones.

$$2CrO_3 + 2H_2O \xrightarrow{H^+} 2H_2CrO_4 \xrightarrow{H^+} H_2Cr_2O_7 + H_2O$$

$$\underset{\text{Primary alcohol}}{R-\overset{\overset{\displaystyle H}{|}}{\underset{\underset{\displaystyle OH}{|}}{C}}-H} \xrightarrow{Cr_2O_7{}^{2-}} R-\overset{\overset{\displaystyle H}{|}}{\underset{\underset{\displaystyle O}{||}}{C}}-H \xrightarrow{Cr_2O_7{}^{2-}} R-\overset{}{\underset{\underset{\displaystyle O}{||}}{C}}-OH$$

Primary alcohol

$$R-\overset{\overset{\displaystyle H}{|}}{\underset{\underset{\displaystyle OH}{|}}{C}}-R \xrightarrow{Cr_2O_7{}^{2-}} R-\overset{}{\underset{\underset{\displaystyle O}{||}}{C}}-R$$

Secondary alcohol

Although primary alcohols are first oxidized to aldehydes, the aldehydes are further oxidized to carboxylic acids. The ability of chromic acid to oxidize aldehydes but not ketones is taken advantage of in a test that uses chromic acid to distinguish between aldehydes and ketones (Procedure 56D). Secondary alcohols are oxidized to ketones, but no further. Tertiary alcohols are not oxidized at all by the reagent. Hence, this test can be used to distinguish primary and secondary alcohols from tertiary alcohols. Unlike the Lucas test, this test can be used with all alcohols regardless of molecular weight and solubility.

IODOFORM TEST

Methyl carbinols give a positive iodoform test. See the discussion in Section 56D (Aldehydes and Ketones).

SPECTROSCOPY

INFRARED

O—H stretch. A medium to strong, and usually broad, absorption occurs in the region 3600 to 3200 cm^{-1} (2.8–3.1 μ). In dilute solutions or with little hydrogen bonding, it is a sharp absorption near 3600 cm^{-1} (2.8 μ). In more concentrated solutions, or with considerable hydrogen bonding, it is a broad absorption near 3400 cm^{-1} (2.9 μ). Sometimes both bands appear.

C—O stretch. There is strong absorption in the region 1200 to 1050 cm^{-1} (9.5–8.3 μ). Primary alcohols absorb nearer 1050 cm^{-1} (9.5 μ), tertiary alcohols and phenols nearer 1200 cm^{-1} (8.3 μ). Secondary alcohols absorb in the middle of this range.

NUCLEAR MAGNETIC RESONANCE

The hydroxyl resonance is extremely concentration-dependent, but it is usually found between 1 and 5δ. Under normal conditions, the hydroxyl proton does not couple with protons on adjacent carbon atoms.

DERIVATIVES

The most common derivatives for alcohols are the 3,5-dinitrobenzoate esters and the phenylurethanes. Occasionally, the α-naphthylurethanes (Procedure 56F) are also prepared, but these latter derivatives are more often used for phenols.

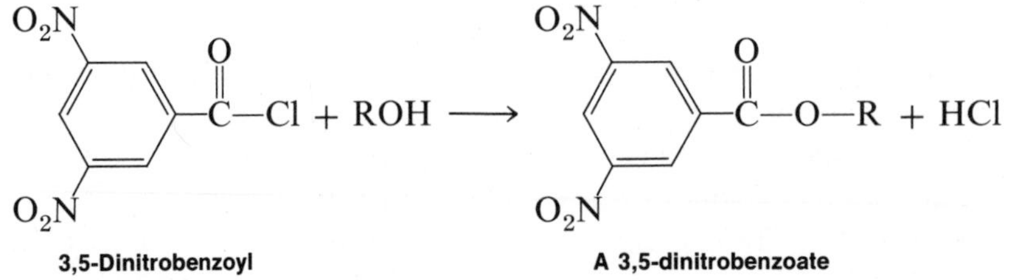

**3,5-Dinitrobenzoyl
chloride**   **A 3,5-dinitrobenzoate**

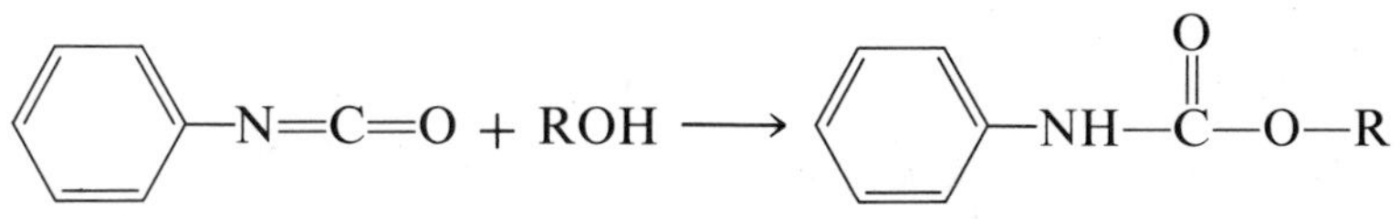

Phenyl isocyanate   **A phenylurethane**

Procedures for preparing these derivatives are given in Appendix 2.

Procedure 56I
Esters

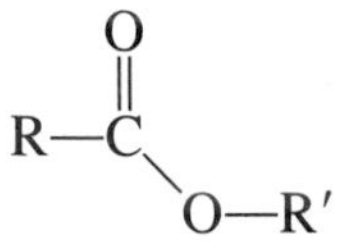

Esters are formally considered "derivatives" of the corresponding carboxylic acid. They are frequently synthesized from the carboxylic acid and the appropriate alcohol:

$$R—COOH + R'—OH \xrightarrow{H^+} R—COOR' + H_2O$$

Thus, esters are sometimes referred to as though they were composed of an acid part and an alcohol part.

Although esters, like aldehydes and ketones, are neutral compounds that have a carbonyl group, they do not usually give a 2,4-dinitrophenylhydrazine test. The two most common tests for identifying esters are the basic hydrolysis and ferric hydroxamate tests. The **saponification equivalent** is also used. However, it usually requires a difficult and time-consuming procedure and will not be considered here. Procedures for determining the saponification equivalent can be found in the references at the beginning of this experiment.

SOLUBILITY CHARACTERISTICS	CLASSIFICATION TESTS
HCl NaHCO$_3$ NaOH H$_2$SO$_4$ Ether (−) (−) (−) (+) (+) Water: <C$_4$ (+) Water: >C$_5$ (−)	Ferric hydroxamate test Basic hydrolysis

CLASSIFICATION TESTS

FERRIC HYDROXAMATE TEST

Procedure. Before starting, you must determine whether the compound to be tested already has enough enolic character in acid solution to give a positive ferric chloride test. Dissolve 1 drop of a liquid unknown or a few crystals of a solid unknown in 1 mL of 95% ethanol and add 1 mL of $1N$ hydrochloric acid. Add a drop or two of 5% ferric chloride solution. If a definite color, except yellow, appears, the ferric hydroxamate test (described below) cannot be used.

If the compound did not show enolic character, continue as follows. Dissolve 2 or 3 drops of a liquid ester, or about 40 mg of a solid ester, in a mixture of 1 mL of $0.5N$ hydroxylamine hydrochloride (dissolved in 95% ethanol) and 0.2 mL of $6N$ sodium hydroxide. Heat the mixture to boiling for a few minutes. Cool the solution and then add 2 mL of $1N$ hydrochloric acid. If the solution becomes cloudy, add 2 mL of 95% ethanol to clarify it. Add a drop of 5% ferric chloride solution and note whether a color is produced. If the color fades, continue to add ferric chloride until the color persists. A positive test should give a deep burgundy or magenta color.

On being heated with hydroxylamine, esters are converted to the corresponding hydroxamic acids:

$$\underset{}{R-\overset{\overset{\textstyle O}{\|}}{C}-O-R'} + H_2N-OH \longrightarrow \underset{\text{A hydroxamic acid}}{R-\overset{\overset{\textstyle O}{\|}}{C}-NH-OH} + R'-OH$$

$$\quad\quad\quad\quad\quad\quad\text{Hydroxylamine}$$

The hydroxamic acids form strong, colored complexes with ferric ion.

$$R-\overset{\overset{\textstyle O}{\|}}{C}-NH-OH + FeCl_3 \longrightarrow \left(R-C\overset{\diagup O}{\underset{NH-O}{\diagdown}}\right)_3 Fe + 3HCl$$

BASIC HYDROLYSIS

Procedure. Place 1 g of the ester in a small flask with 10 mL of 25% aqueous sodium hydroxide. Add a boiling stone and attach the reflux condenser. Reflux the mixture for about 30 minutes. Stop the heating and observe the solution to determine whether the oily ester layer has disappeared or whether the odor of the ester (usually pleasant) has disappeared. Low-boiling esters (below 110 °C) usually dissolve within 30 minutes if the alcohol part has a low molecular weight. If the ester has not dissolved, reheat the mixture to reflux for 1 to 2 hours. After that time, the oily ester layer should have disappeared along with the characteristic odor. Esters boiling up to 200 °C should hydrolyze during this time. Compounds remaining after this extended period of heating either are unreactive esters or are **not** esters.

For esters derived from solid acids, the acid part can, if desired, be recovered after hydrolysis. Extract the basic solution with ether to remove any unreacted ester (even if it appears to be gone), acidify the basic solution with hydrochloric acid, and extract the acidic phase with ether to remove the acid. Dry the ether layer over anhydrous sodium sulfate, decant, and evaporate the solvent to obtain the parent acid from the original ester. The melting point of the parent acid can provide valuable information in the identification process.

This procedure converts the ester to its separate acid and alcohol parts. The ester dissolves because the alcohol part (if small) is usually soluble in the aqueous medium, as is the sodium salt of the acid. Acidification produces the parent acid:

$$\underset{\text{Ester}}{R-\overset{\overset{\textstyle O}{\|}}{C}-O-R'} \xrightarrow{\text{NaOH}} \underset{\substack{\text{Salt of}\\\text{acid part}}}{R-\overset{\overset{\textstyle O}{\|}}{C}-O^-Na^+} + \underset{\substack{\text{Alcohol}\\\text{part}}}{R'OH} \xrightarrow{\text{HCl}} R-\overset{\overset{\textstyle O}{\|}}{C}-O-H + R'OH$$

All derivatives of carboxylic acids are converted to the parent acid on basic hydrolysis. Thus, amides, which are not covered in this experiment, would also dissolve in this test, liberating the free amine and the sodium salt of the carboxylic acid.

SPECTROSCOPY

INFRARED

The ester-carbonyl group (C=O) peak is usually a strong absorption, and so is the absorption of the carbonyl-oxygen link (C—O) to the alcohol part. C=O stretch at approximately 1735 cm^{-1} (5.75 μ) is normal.[4] C—O stretch usually gives two or more absorptions, one stronger than the others, in the region 1280 to 1050 cm^{-1} (7.8–9.5 μ).

NUCLEAR MAGNETIC RESONANCE

Hydrogens that are alpha to an ester carbonyl group have resonance in the region 2 to 3δ. Hydrogens alpha to the alcohol oxygen of an ester have resonance in the region 3 to 5δ.

DERIVATIVES

Esters present a double problem when one is trying to prepare derivatives. To characterize an ester completely, one needs to prepare derivatives of **both** the acid part and the alcohol part.

---

[4]Conjugation with the carbonyl group moves the carbonyl absorption to lower frequencies (longer wavelength). Conjugation with the alcohol oxygen raises the carbonyl absorption to higher frequencies (shorter wavelength). Ring strain (lactones) moves the carbonyl absorption to higher frequencies (shorter wavelength).

ACID PART

The most common derivative of the acid part is the *N*-benzylamide derivative.

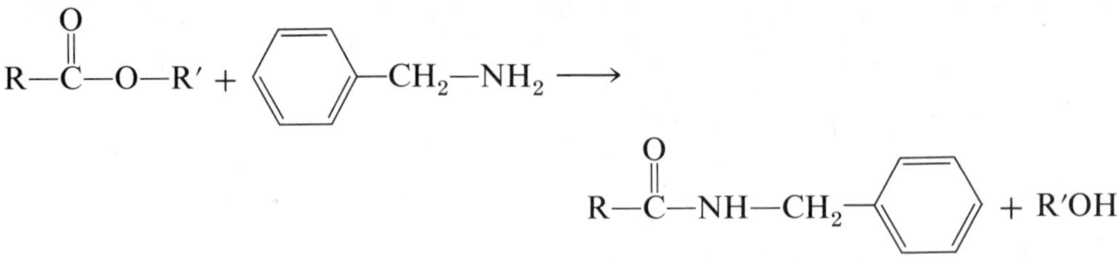

$$R-\overset{\overset{\textstyle O}{\|}}{C}-O-R' + \langle\!\langle \bigcirc \rangle\!\rangle-CH_2-NH_2 \longrightarrow$$

$$R-\overset{\overset{\textstyle O}{\|}}{C}-NH-CH_2-\langle\!\langle \bigcirc \rangle\!\rangle + R'OH$$

An *N*-benzylamide

The reaction does not proceed well unless R′ is methyl or ethyl. For alcohol portions that are larger, the ester must be transesterified to a methyl or an ethyl ester before preparing the derivative.

$$R-\overset{\overset{\textstyle O}{\|}}{C}-OR' + CH_3OH \xrightarrow{H^+} R-\overset{\overset{\textstyle O}{\|}}{C}-O-CH_3 + R'OH$$

Hydrazine also reacts well with methyl and ethyl esters to give acid hydrazides.

$$R-\overset{\overset{\textstyle O}{\|}}{C}-OR' + NH_2NH_2 \longrightarrow R-\overset{\overset{\textstyle O}{\|}}{C}-NHNH_2 + R'OH$$

An acid hydrazide

The saponification equivalent (mentioned above) is also sometimes used. This value gives the molecular weight of the ester divided by the number of its ester groups.

ALCOHOL PART

The best derivative of the alcohol part of an ester is the 3,5-dinitrobenzoate ester, which is prepared by an acyl interchange reaction:

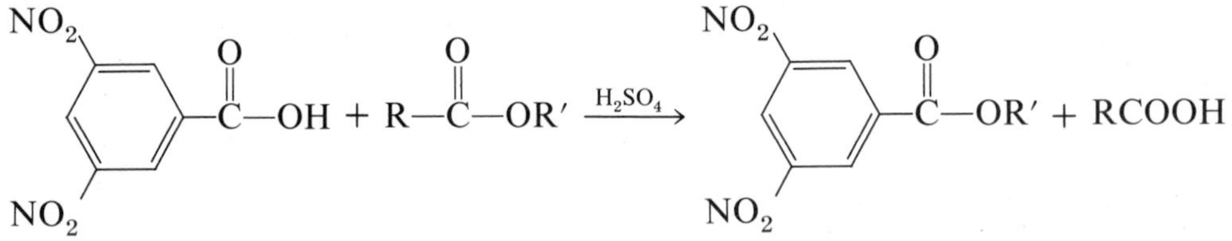

$$\underset{NO_2}{\overset{NO_2}{\bigcirc}}-\overset{\overset{\textstyle O}{\|}}{C}-OH + R-\overset{\overset{\textstyle O}{\|}}{C}-OR' \xrightarrow{H_2SO_4} \underset{NO_2}{\overset{NO_2}{\bigcirc}}-\overset{\overset{\textstyle O}{\|}}{C}-OR' + RCOOH$$

A 3,5-dinitrobenzoate ester

Most esters are composed of very simple acid and alkyl portions. For this reason, spectroscopy is usually a better method of identification than is the preparation of derivatives. Not only is it necessary to prepare two derivatives with an ester, but all esters with the same acid portion, or all those with the same alcohol portion, give identical derivatives of those portions.

Experiment 57
Carbohydrates

In this experiment, you perform tests that distinguish among various carbohydrates. The carbohydrates included and the classes they represent are as follows:

Aldopentoses: xylose and arabinose
Aldohexoses: glucose and galactose
Ketohexoses: fructose
Disaccharides: lactose and sucrose
Polysaccharides: starch and glycogen

The structures of these carbohydrates can be found in your lecture textbook.
The tests are classified in the following groups:

A. Tests based on the production of furfural or a furfural derivative: Molisch's test, Bial's test, and Seliwanoff's test
B. Tests based on the reducing property of a carbohydrate (sugar): Benedict's test and Barfoed's test
C. Osazone formation
D. Iodine test for starch
E. Hydrolysis of sucrose
F. Mucic acid test for galactose and lactose
G. Tests on unknowns

SPECIAL INSTRUCTIONS

All the procedures in this experiment involve simple test-tube reactions. Most of the tests are short; Seliwanoff's test, osazone formation, and the mucic acid test, however, take relatively longer to complete. You will need a minimum of 10 test tubes, numbered in order. Clean them carefully each time they are used. The 1% solutions of carbohydrates and the reagents needed for the tests have been prepared in advance by the laboratory instructor or assistant. Be sure to shake the starch solution before using it.

A. TESTS BASED ON PRODUCTION OF FURFURAL OR A FURFURAL DERIVATIVE

Under acidic conditions, aldopentoses and ketopentoses **rapidly** undergo dehydration to give furfural (equation 1). Ketohexoses **rapidly** yield 5-hydroxymethylfurfural (equation 2). Disaccharides and polysaccharides can first be hydrolyzed in an acid medium to produce monosaccharides, which then react to give furfural or 5-hydroxymethylfurfural.

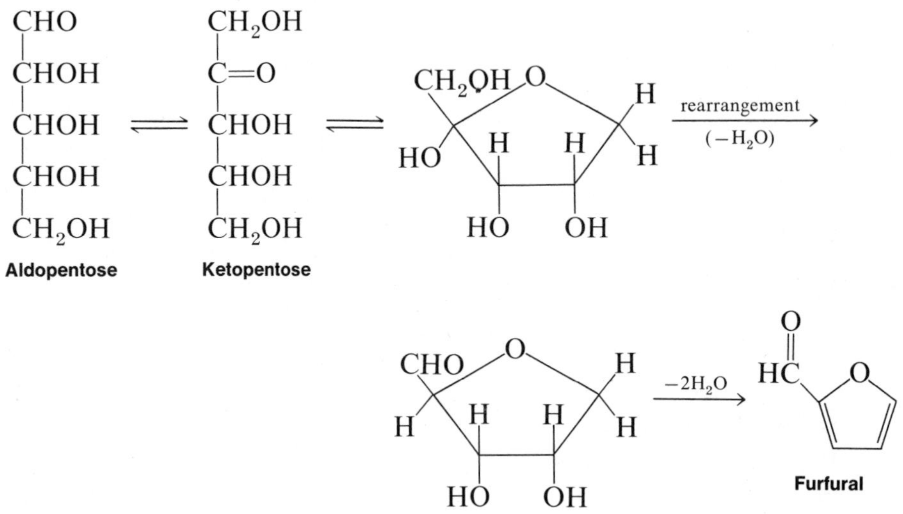

(1)

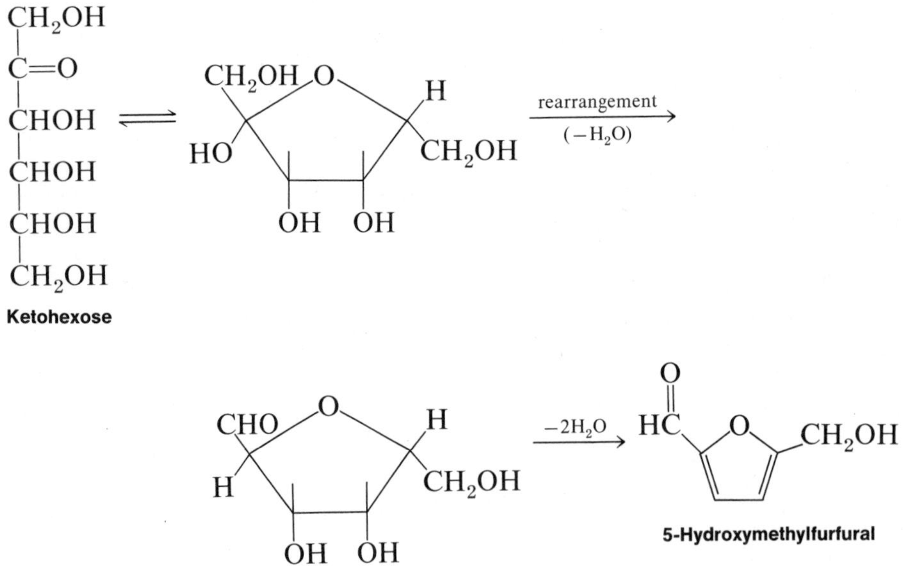

(2)

Aldohexoses are **slowly** dehydrated to 5-hydroxymethylfurfural. One possible mechanism is shown in equation 3. The mechanism is different from that given in equations 1 and 2 in that dehydration is at an early step and the rearrangement step is absent.

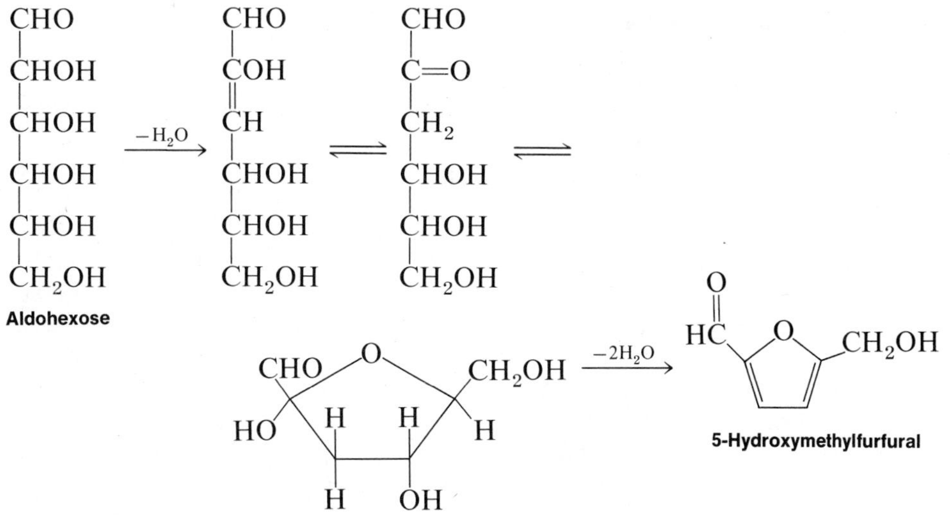

$$\begin{array}{c}
\text{CHO} \\
\text{CHOH} \\
\text{CHOH} \\
\text{CHOH} \\
\text{CHOH} \\
\text{CH}_2\text{OH}
\end{array}
\xrightarrow{-\text{H}_2\text{O}}
\begin{array}{c}
\text{CHO} \\
\text{COH} \\
\text{CH} \\
\text{CHOH} \\
\text{CHOH} \\
\text{CH}_2\text{OH}
\end{array}
\rightleftharpoons
\begin{array}{c}
\text{CHO} \\
\text{C}=\text{O} \\
\text{CH}_2 \\
\text{CHOH} \\
\text{CHOH} \\
\text{CH}_2\text{OH}
\end{array}
\rightleftharpoons \qquad (3)$$

Aldohexose

$$\xrightarrow{-2\text{H}_2\text{O}}$$

5-Hydroxymethylfurfural

Once furfural or 5-hydroxymethylfurfural is produced by equations 1, 2, or 3, either will then react with a phenol to produce a colored condensation product. The substance α-naphthol is used in the Molisch test, orcinol in Bial's test, and resorcinol in Seliwanoff's test.

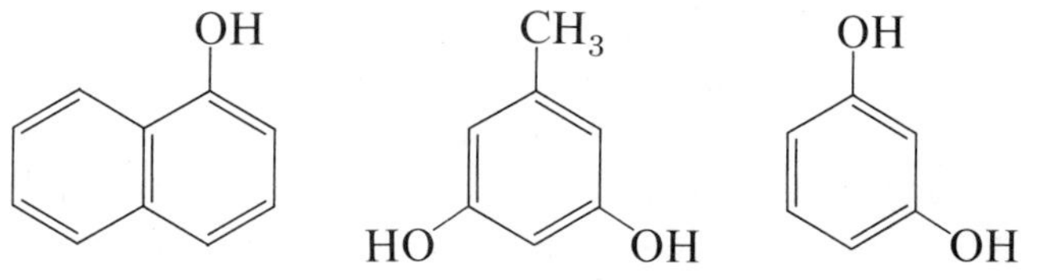

α-**Naphthol**	**Orcinol**	**Resorcinol**
(Molisch test)	**(Bial test)**	**(Seliwanoff test)**

The colors and the rates of formation of these colors are used to differentiate between the carbohydrates. The various color tests are discussed in sections 1, 2, and 3. A typical colored product formed from furfural and α-naphthol (Molisch's test) is the following (equation 4):

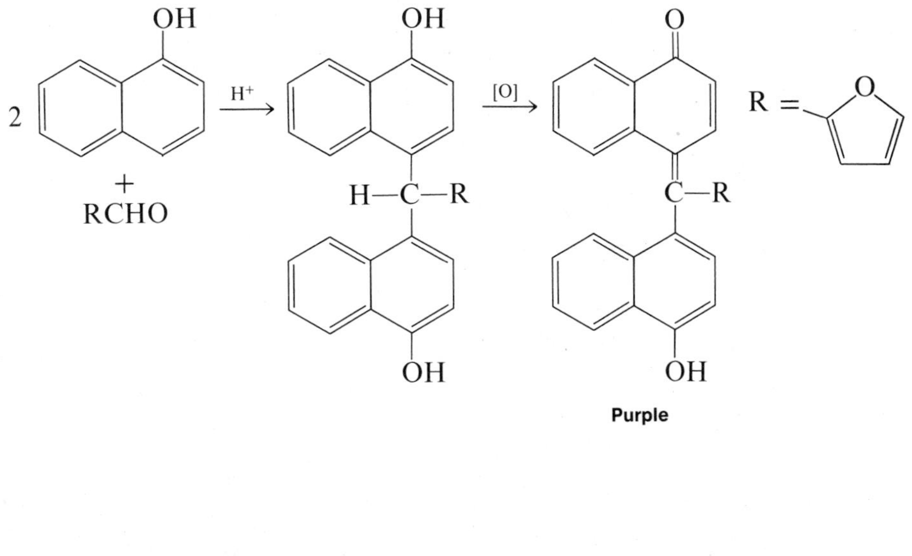

Purple

1. MOLISCH TEST FOR CARBOHYDRATES

The Molisch test is a **general** test for carbohydrates. Most carbohydrates are dehydrated with concentrated sulfuric acid to form furfural or 5-hydroxyfurfural. These furfurals react with the α-naphthol in the test reagent to give a purple product. Compounds other than carbohydrates may react with the reagent to give a positive test. A negative test usually indicates that there is no carbohydrate.

Procedure for the Molisch Test. Place 4 mL of each of the following 1% carbohydrate solutions in nine separate test tubes: xylose, arabinose, glucose, galactose, fructose, lactose, sucrose, starch (shake it), and glycogen. Also add 4 mL of distilled water to another tube to serve as a control.

Add 2 drops of the Molisch reagent[1] to each test tube and thoroughly mix the contents of the tube. Tilt each test tube slightly, and cautiously add 5 mL of concentrated sulfuric acid down the sides of the tubes. An acid layer forms at the bottom of the tubes. Note and record the color at the interface between the two layers in each tube. A purple color constitutes a positive test.

2. BIAL TEST FOR PENTOSES

The Bial test is used to differentiate pentose sugars from hexose sugars. Pentose sugars yield furfural on dehydration in acidic solution. Furfural reacts with orcinol and ferric chloride to give a blue-green condensation product. Hexose sugars give 5-hydroxy-methylfurfural, which reacts with the reagent to yield colors such as green, brown, and reddish brown.

Procedure for Bial's Test. Place 2 mL of each of the following 1% carbohydrate solutions in separate test tubes: xylose, arabinose, glucose, galactose, fructose, lactose, sucrose, starch (shake it), and glycogen. Also add 2 mL of distilled water to another tube to serve as a control.

Add 3 mL of Bial's reagent[2] to each test tube. Carefully heat each tube over a Bunsen burner flame until the mixture just begins to boil. Note and record the color produced in each test tube. If the color is not distinct, add 5 mL of water and 1 mL of 1-pentanol to the test tube. After shaking them, again observe and record the color. The colored condensation product will be concentrated in the 1-pentanol layer.

3. SELIWANOFF TEST FOR KETOHEXOSES

The Seliwanoff test depends on the relative rates of dehydration of carbohydrates. A ketohexose reacts rapidly by equation 2 to give 5-hydroxymethylfurfural, whereas an

---

[1]Dissolve 5 g of α-naphthol in 100 mL of 95% ethanol.

[2]Dissolve 3 g of orcinol in 1 L of concentrated hydrochloric acid and add 3 mL of 10% aqueous ferric chloride.

aldohexose reacts more slowly, by equation 3, to give the same product. Once 5-hydroxymethylfurfural is produced, it reacts with resorcinol to give a dark red condensation product. If the reaction is followed for some time, it will be found that sucrose hydrolyzes to give fructose, which eventually reacts to produce a dark red color.

Procedure for Seliwanoff's Test. Prepare a boiling-water bath for this experiment. Place 1 mL of each of the following 1% carbohydrate solutions in separate test tubes: xylose, arabinose, glucose, galactose, fructose, lactose, sucrose, starch (shake it), and glycogen. Add 1 mL of distilled water to another tube to act as a control.

Add 4 mL of Seliwanoff's reagent[3] to each test tube. Place all 10 tubes in a beaker of boiling water for **60 seconds.** Remove them and note the results in the notebook.

For the remainder of Seliwanoff's test, it is convenient to place a group of three or four tubes in the boiling water bath and to complete the observations before going on to the next group of tubes. Place three or four tubes in the boiling-water bath. Observe the color in each of the tubes at 1-minute intervals for 5 minutes beyond the original minute. Record the results at each 1-minute interval. Leave the tubes in the boiling-water bath during the entire 5-minute period. After the first group has been observed, remove that set of test tubes, and place the next group of three or four tubes in the bath. Follow the color changes as before. Finally, place the last group of tubes in the bath and follow the color changes over the 5-minute period.

B. TESTS BASED ON THE REDUCING PROPERTY OF A CARBOHYDRATE (SUGAR)

Monosaccharides, and those disaccharides that have a potential aldehyde group, will reduce reagents such as Benedict's solution to produce a red precipitate of cuprous oxide:

$$RCHO + 2Cu^{2+} + 4OH^- \longrightarrow RCOOH + Cu_2O + 2H_2O$$

(Red precipitate)

Glucose, for example, is a typical aldohexose, showing reducing properties. The two diastereomeric α- and β-D-glucoses are in equilibrium with each other in aqueous solution. The α-D-glucose opens at the anomeric carbon atom (hemiacetal) to produce the free aldehyde. This aldehyde rapidly closes to give β-D-glucose, and a new hemiacetal is produced. It is the presence of this free aldehyde that makes glucose a reducing carbohydrate (sugar). It reacts with Benedict's reagent to produce a red precipitate, the basis of the test. Carbohydrates that have the hemiacetal functional group show reducing properties.

---

[3]Dissolve 0.5 g of resorcinol in 1 L of dilute hydrochloric acid (1 volume of concentrated hydrochloric acid and 2 volumes of distilled water).

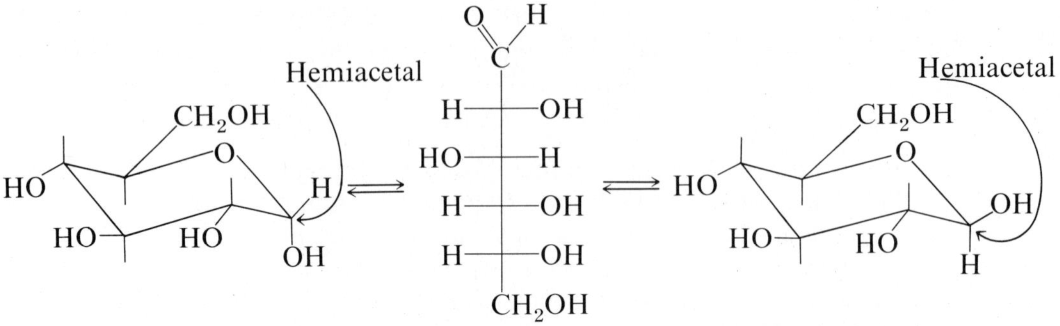

α-D-Glucose   D-Glucose   β-D-Glucose

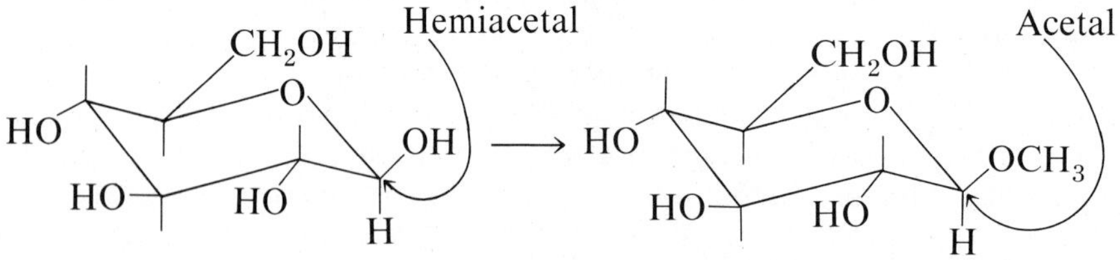

If the hemiacetal is converted to an acetal by methylation, the carbohydrate (sugar) will no longer reduce Benedict's reagent.

With disaccharides, two situations may arise. If the anomeric carbon atoms are bonded (head to head) to give an acetal, then the sugar will not reduce Benedict's reagent. If, however, the sugar molecules are joined head to tail, then one end will still be able to equilibrate through the free aldehyde form (hemiacetal). Examples of a reducing and a nonreducing disaccharide are shown in the accompanying figure.

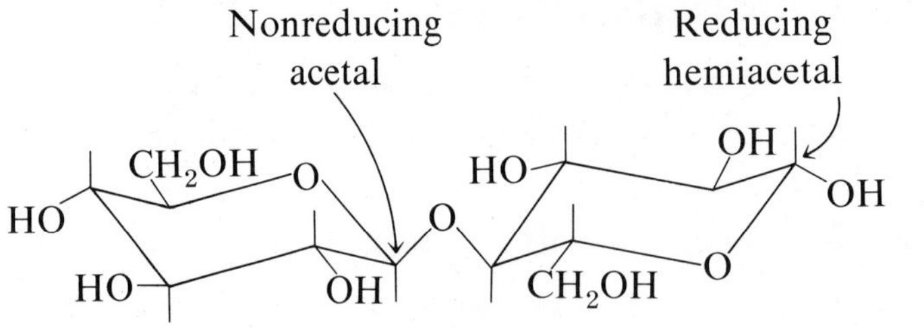

Cellobiose
(Reducing sugar)

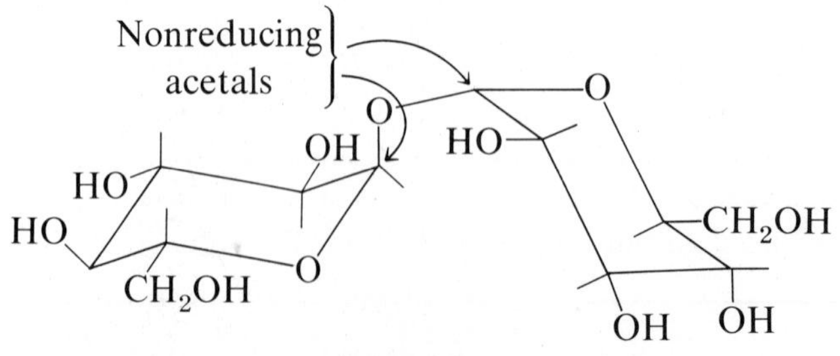

Trehalose
(Nonreducing sugar)

1. BENEDICT'S TEST FOR REDUCING SUGARS

Benedict's test is performed under mildly basic conditions. The reagent reacts with all reducing sugars to produce the red precipitate cuprous oxide, as shown on page 473. It also reacts with water-soluble aldehydes that are not sugars. Ketoses, such as fructose, also react with Benedict's reagent. Benedict's test is considered one of the classical tests for determining the presence of an aldehyde functional group.

Procedure for Benedict's test. Prepare a boiling-water bath for this experiment. Place 1 mL of each of the following 1% carbohydrate solutions in separate test tubes: xylose, arabinose, glucose, galactose, fructose, lactose, sucrose, starch (shake it), and glycogen. Add 1 mL of distilled water to another tube to serve as a control.

Add 5 mL of Benedict's reagent[4] to each test tube. Place the test tubes in a boiling-water bath for 2 to 3 minutes. Remove the tubes and note the results in a notebook. A red, brown, or yellow precipitate indicates a positive test for a reducing sugar. Ignore a change in color of the solution. A precipitate must form for the test to be positive.

2. BARFOED'S TEST FOR REDUCING MONOSACCHARIDES

Barfoed's test distinguishes reducing monosaccharides and reducing disaccharides by a difference in rate of reaction. The reagent consists of cupric ions, like Benedict's reagent. In this test, however, Barfoed's reagent reacts with reducing monosaccharides to produce cuprous oxide faster than with reducing disaccharides.

$$RCHO + 2Cu^{2+} + 2H_2O \longrightarrow RCOOH + Cu_2O + 4H^+$$

$$\text{(Red precipitate)}$$

Procedure for Barfoed's Test. Place 1 mL of each of the following 1% carbohydrate solutions in separate test tubes: xylose, arabinose, glucose, galactose, fructose, lactose, sucrose, starch (shake it), and glycogen. Add 1 mL of distilled water to another tube to function as a control.

Add 5 mL of Barfoed's reagent[5] to each test tube. Place the tubes in a boiling-water bath for 10 minutes. Remove the tubes and note the results in a notebook.

C. OSAZONE FORMATION

Carbohydrates react with phenylhydrazine to form crystalline derivatives, called **osazones.**

---

[4]Dissolve 173 g of hydrated sodium citrate and 100 g of anhydrous sodium carbonate in 800 mL of distilled water, with heating. Filter the solution. Add to it a solution of 17.3 g of cupric sulfate ($CuSO_4 \cdot 5H_2O$) dissolved in 100 mL of distilled water. Dilute the combined solutions to 1 L.

[5]Dissolve 66.6 g of cupric acetate in 1 L of distilled water. Filter the solution, if necessary, and add 9 mL of glacial acetic acid.

$$
\begin{array}{ccccc}
\text{CHO} & & \text{HC=NNHPh} & & \text{HC=NNHPh} \\
| & & | & & | \\
\text{CHOH} & & \text{CHOH} & & \text{C=NNHPh} \\
| & \xrightarrow{\text{PhNHNH}_2} & | & \xrightarrow{\text{2PhNHNH}_2} & | \\
\text{CHOH} & & \text{CHOH} & & \text{CHOH} \qquad + \text{NH}_3 + \text{PhNH}_2 \\
| & & | & & | \\
\text{CHOH} & & \text{CHOH} & & \text{CHOH} \\
| & & | & & | \\
\text{CHOH} & & \text{CHOH} & & \text{CHOH} \\
| & & | & & | \\
\text{CH}_2\text{OH} & & \text{CH}_2\text{OH} & & \text{CH}_2\text{OH}
\end{array}
$$

An osazone can be isolated as a derivative and its melting point determined. However, some of the monosaccharides give **identical** osazones (glucose, fructose, and mannose). Also, the melting points of different osazones are often in the same range. This limits the usefulness of an isolation of the osazone derivative.

A good experimental use for the osazone is to observe its rate of formation. The rates of reaction vary greatly even though the **same** osazone may be produced from different sugars. For example, fructose forms a precipitate in about 2 minutes, whereas glucose forms a precipitate about 5 minutes later. The osazone is the same in each case. The crystal structure of the osazone is often distinctive. Arabinose, for example, produces a fine precipitate, whereas glucose produces a coarse precipitate.

> **CAUTION: Phenylhydrazine is a suspected carcinogen. Handle with gloves.**

Procedure for Osazone Formation.   A boiling-water bath is needed for this experiment. Place 0.1 g of each of the following carbohydrates in separate test tubes: xylose, arabinose, glucose, galactose, fructose, lactose, sucrose, starch (shake it), and glycogen. Add 1 mL of distilled water to each tube. Then add 5 mL of phenylhydrazine reagent[6] to each tube. Place the tubes in a boiling-water bath simultaneously. Watch for a precipitate, or in some cases, cloudiness. Note the time at which the precipitate begins to form. After 30 minutes, cool the tubes and record the crystalline form of the precipitates. Reducing disaccharides will not precipitate until the tubes are cooled. Nonreducing disaccharides will hydrolyze first, and then the osazones will precipitate.

D. IODINE TEST FOR STARCH

Starch forms a typical blue color with iodine. This color is due to the absorption of iodine into the open spaces of the amylose molecules (helices) present in starch. Amylopectins, which are the other types of molecules present in starch, form a red to purple color with iodine.

---

[6]Dissolve 50 g of phenylhydrazine hydrochloride and 75 g of sodium acetate trihydrate in 500 mL of distilled water. The reagent deteriorates over time and should be prepared fresh.

Procedure for the Iodine Test. Place 2 mL of each of the following 1% carbohydrate solutions in three separate test tubes: glucose, starch (shake it), and glycogen. Add 2 mL of distilled water to another tube to act as a control.

Add 1 drop of iodine solution to each test tube and observe the results.[7] Add a few drops of sodium thiosulfate to the solutions and note the results.[8]

E. HYDROLYSIS OF SUCROSE

Sucrose can be hydrolyzed in acid solution to its component parts, fructose and glucose. The component parts can then be tested with Benedict's reagent.

Procedure for the Hydrolysis of Sucrose. Place 5 mL of a 1% solution of sucrose in a test tube. Add 2 drops of concentrated hydrochloric acid and heat the tube in a boiling water bath for 10 minutes. Cool the tube and neutralize the contents with 10% sodium hydroxide solution until the mixture is just basic to litmus (about 20 drops are needed). Test the mixture with Benedict's reagent (Part B). Note the results and compare them with the results obtained on sucrose that has not been hydrolyzed.

F. MUCIC ACID TEST FOR GALACTOSE AND LACTOSE

Procedures are given in Experiment 58 for the oxidation of galactose and lactose to mucic acid. This test confirms the presence of galactose or a galactose moiety in a carbohydrate (sugar).

G. TESTS ON UNKNOWNS

Obtain an unknown solid carbohydrate from the laboratory instructor or assistant. The unknown will be one of the following carbohydrates: xylose, arabinose, glucose, galactose, fructose, lactose, sucrose, starch, or glycogen. Carefully dissolve part of the unknown in distilled water to prepare a 1% solution (0.25 g carbohydrate in 25 mL water). Save the remainder for the tests requiring solid material. Apply whatever tests are necessary to identify the unknown.

At the instructor's option, the optical rotation can be determined as part of the experiment. Experimental details are given in Experiment 59 and Technique 15. Optical rotation data and decomposition points for carbohydrates and osazones are given in the standard reference works on identification of organic compounds (Experiment 56).

---

[7]The iodine solution is prepared as follows. Dissolve 2 g of potassium iodide in 50 mL of distilled water. Add 1 g of iodine and shake the solution until the iodine dissolves. Dilute the solution to 100 mL.

[8]The sodium thiosulfate solution is prepared as follows. Dissolve 2.5 g of sodium thiosulfate in 100 mL of water.

QUESTIONS

1. Find the structures for the following carbohydrates (sugars) in a reference work or a textbook, and decide whether they are reducing or nonreducing carbohydrates (sugars): sorbose, mannose, ribose, maltose, raffinose, and cellulose.

2. Mannose gives the same osazone as glucose. Explain.

3. Predict the results of the following tests with the carbohydrates listed in question 1: Molisch, Bial, Seliwanoff (after 1 minute and 6 minutes), Barfoed, and mucic acid tests.

4. Give a mechanism for the hydrolysis of the acetal linkage in sucrose.

5. The rearrangement in equations 1 and 2 can be considered a type of pinacol rearrangement. Give a mechanism for that step.

6. Give a mechanism for the acid-catalyzed condensation of furfural with 2 moles of α-naphthol, shown in equation 4.

ESSAY
Chemistry of Milk

Milk is a food of exceptional interest. Not only is milk an excellent food for the very young, but humans have also adapted milk, specifically cow's milk, as a food substance for persons of all ages. Many specialized milk products like cheese, yogurt, butter, and ice cream are staples of our diet.

Milk is probably the most nutritionally complete food that can be found in nature. This property is important for milk, since it is the only food young mammals consume in the nutritionally significant weeks following birth. Whole milk contains vitamins (principally thiamine, riboflavin, pantothenic acid, and vitamins A, D, and K), minerals (calcium, potassium, sodium, phosphorus, and trace metals), proteins (which include all the essential amino acids), carbohydrates (chiefly lactose), and lipids (fats). The only important elements in which milk is seriously deficient are iron and Vitamin C. Infants are usually born with a storage supply of iron large enough to meet their needs for several weeks. Vitamin C is easily secured through an orange juice supplement. The average composition of the milk of each of several mammals is summarized in the accompanying table.

**Average Percentage Composition of Milk
from Various Mammals**

	COW	HUMAN	GOAT	SHEEP	HORSE
Water	87.1	87.4	87.0	82.6	90.6
Protein	3.4	1.4	3.3	5.5	2.0
Fats	3.9	4.0	4.2	6.5	1.1
Carbohydrates	4.9	7.0	4.8	4.5	5.9
Minerals	0.7	0.2	0.7	0.9	0.4

FATS

Whole milk is an oil-water type of emulsion, containing about 4% fat dispersed as very small (5–10 microns in diameter) globules. The globules are so small that a drop of milk contains about a million of them. Because the fat in milk is so finely dispersed, it is digested more easily than fat from any other source. The fat emulsion is stabilized to some extent by complex phospholipids and proteins that are adsorbed on the surfaces of the globules. The fat globules, which are lighter than water, coalesce on standing and eventually rise to the surface of the milk, forming a layer of **cream.** Since vitamins A and D are fat-soluble vitamins, they are carried to the surface with the cream. Commercially, the cream is often removed by centrifugation and skimming and is either diluted to form coffee cream (''half and half''), sold as **whipping cream,** converted to **butter,** or converted to **ice cream.** The milk that remains is called **skimmed milk.** Skimmed milk, except for lacking the fats and vitamins A and D, has approximately the same composition as whole milk. If milk is **homogenized,** its fatty content will not separate. Milk is homogenized by forcing it through a small hole. This breaks up the fat globules and reduces their size to about 1 to 2 microns in diameter.

The structure of fats and oils is discussed in the essay that precedes Experiment 13. The fats in milk are primarily triglycerides. For the saturated fatty acids, the following percentages have been reported:

C_2 (3%) C_8 (2.7%) C_{14} (25.3%) $>C_{18}$ (~5%)
C_4 (1.4%) C_{10} (3.7%) C_{16} (9.2%)
C_6 (1.5%) C_{12} (12.1%) C_{18} (1.3%)

Thus, about two thirds of all the fatty acids in milk are saturated, and about one third are unsaturated. Milk is unusual in that about 12% of the fatty acids are **short**-chain fatty acids (C_2–C_{10}) like butyric, caproic, and caprylic acids.

Additional lipids (fats and oils) in milk include small amounts of cholesterol (see the essay preceding Experiment 8), phospholipids, and lecithins (phospholipids conjugated with choline). The structures of phospholipids and lecithins are shown. The phospholipids help to stabilize the whole milk emulsion; the phosphate groups help to achieve partial water solubility for the fat globules. All the fat can be removed from milk by extraction with petroleum ether or a similar organic solvent.

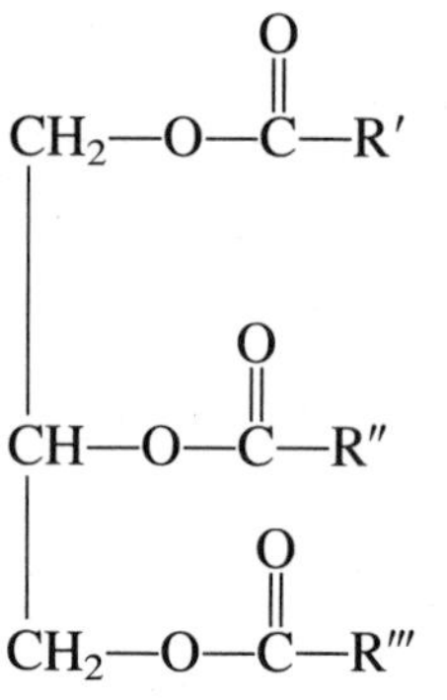

$$CH_2-O-\overset{\overset{\displaystyle O}{\|}}{C}-R' \qquad CH_2-O-\overset{\overset{\displaystyle O}{\|}}{P}-OR \qquad CH_2-O-\overset{\overset{\displaystyle O}{\|}}{P}-O-CH_2CH_2-N(CH_3)_3{}^+$$

A triglyceride 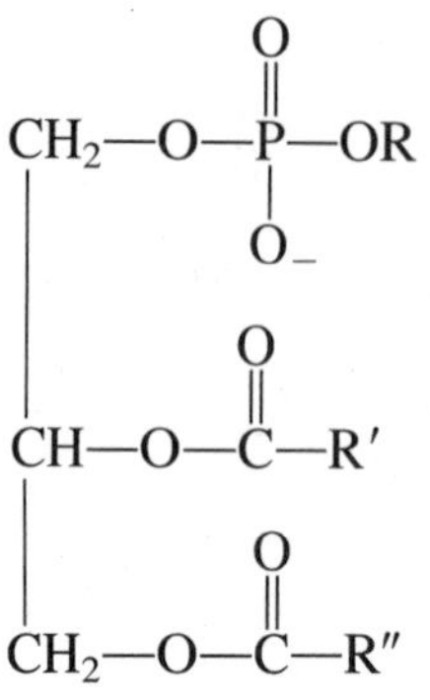   **A phospholipid** 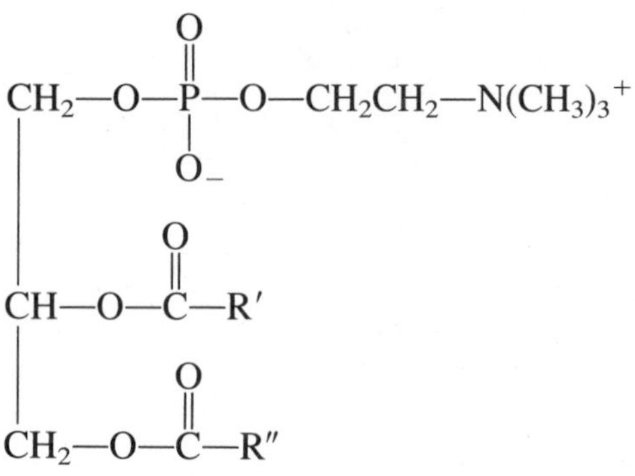   **A lecithin**

PROTEINS

Proteins may be classified broadly in two general categories: fibrous and globular. Globular proteins are those that tend to fold back on themselves into compact units that approach nearly spheroidal shapes. These types of proteins do not form intermolecular interactions between protein units (H bonds, and so on) as fibrous proteins do, and they are more easily solubilized as colloidal suspensions. There are three kinds of proteins in milk: **caseins, lactalbumins,** and **lactoglobulins.** All are globular.

Casein is a phosphoprotein, meaning that phosphate groups are attached to some of the amino acid side-chains. These are attached mainly to the hydroxyl groups of the serine and threonine moieties. Actually, casein is a mixture of at least three similar proteins, principally α, β, and κ caseins. These three proteins differ primarily in molecular weight and amount of phosphorus they contain (number of phosphate groups).

CASEIN	MW	PHOSPHATE GROUPS/MOLECULE
α	27,300	~9
β	24,100	~4–5
κ	~8,000	~1.5

Casein exists in milk as the calcium salt, **calcium caseinate.** This salt has a complex structure. It is composed of α, β, and κ caseins, which form a **micelle,** or a solubilized unit. Neither the α nor the β casein is soluble in milk, and neither is soluble either singly or in combination. If κ casein is added to either one, or to a combination of the two, however, the result is a casein complex that is soluble owing to the formation of the micelle.

A structure proposed for the casein micelle is shown on the following page. The κ casein is thought to stabilize the micelle. Since both α and β casein are phosphoproteins, they are precipitated by calcium ions. Recall that $Ca_3(PO_4)_2$ is fairly insoluble.

$$\boxed{\text{PROTEIN}}\text{—O—}\overset{\overset{\displaystyle O}{\|}}{\underset{\underset{\displaystyle O^-}{|}}{P}}\text{—O}^- + \text{Ca}^{2+} \longrightarrow \boxed{\text{PROTEIN}}\text{—O—}\overset{\overset{\displaystyle O}{\|}}{\underset{\underset{\displaystyle O^-}{|}}{P}}\text{—O}^-\text{Ca}^{2+}\downarrow$$

Insoluble

The κ casein protein, however, has fewer phosphate groups and a high content of carbohydrate bound to it. It is also thought to have all its serine and threonine residues (which have hydroxyl groups), as well as its bound carbohydrates, on only one side of its outer surfaces. This portion of its outer surface is easily solubilized in water since these polar groups are present. The other portion of its surface binds well to the water-insoluble α and β caseins and solubilizes them by forming a protective colloid or micelle around them. Since the entire outer surface of the micelle can be solubilized in water, the unit is solubilized **as a whole,** thus bringing the α and β caseins, as well as κ casein, into solution.

Calcium caseinate has its isoelectric (neutrality) point at pH 4.6. Therefore, it is insoluble in solutions of pH less than 4.6. The pH of milk is about 6.6; therefore casein has a negative charge at this pH and is solubilized as a salt. If acid is added to milk, the negative charges on the outer surface of the micelle are neutralized (the phosphate groups are protonated) and the neutral protein precipitates:

$$\text{Ca}^{2+} \text{ Caseinate} + 2\text{HCl} \longrightarrow \text{Casein} \downarrow + \text{CaCl}_2$$

The calcium ions remain in solution. When milk sours, lactic acid is produced by bacterial action (see below), and the consequent lowering of the pH causes the same **clotting** reaction. The isolation of casein from milk is described in Experiment 58.

The casein in milk can also be clotted by the action of an enzyme called **rennin.** Rennin is found in the fourth stomach of young calves. However, both the nature of the clot and the mechanism of clotting differ when rennin is used. The clot formed using rennin, **calcium paracaseinate,** contains calcium.

$$\text{Ca}^{2+} \text{ Caseinate} \xrightarrow{\text{rennin}} \text{Ca}^{2+} \text{ Paracaseinate} + \text{a small peptide}$$

Rennin is a hydrolytic enzyme (peptidase) and acts specifically to cleave peptide bonds between phenylalanine and methionine residues. It attacks the κ casein, breaking the peptide chain so as to release a small segment of it. This destroys the water-solubilizing

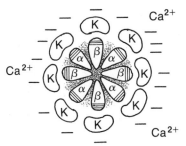

A casein micelle (average diameter, 1200 Å)

surface of the κ casein, which protects the inner α and β caseins, and causes the entire micelle to precipitate as calcium paracaseinate. Milk can be decalcified by treatment with oxalate ion, which forms an insoluble calcium salt. If the calcium ions are removed from milk, a clot will not be formed when the milk is treated with rennin.

The clot, or **curd,** formed by the action of rennin is sold commercially as **cottage cheese.** The liquid remaining is called the **whey.** The curd can also be used in producing various types of **cheese.** It is washed, pressed to remove any excess whey, and chopped. After this treatment, it is melted, hardened, and ground. The ground curd is then salted, pressed into molds, and set aside to age.

Albumins are globular proteins that are soluble in water and in dilute salt solutions. They are, however, denatured and coagulated by heat. The second most abundant protein types in milk are the **lactalbumins.** Once the caseins have been removed, and the solution has been made acidic, the lactalbumins can be isolated by heating the mixture to precipitate them. The typical albumin has a molecular weight of about 41,000.

A third type of protein in milk is the **lactoglobulins.** They are present in smaller amounts than the albumins and generally denature and precipitate under the same conditions as the albumins. The lactoglobulins carry the immunological properties of milk. They protect the young mammal until its own immune systems have developed.

CARBOHYDRATES

When the fats and the proteins have been removed from milk, the carbohydrates remain, as they are soluble in aqueous solution. The main carbohydrate in milk is lactose.

Lactose, a disaccharide, is the **only** carbohydrate that mammals synthesize. Hydrolyzed, it yields one molecule of D-glucose and one of D-galactose. It is synthesized in the mammary glands. In this process, one molecule of glucose is converted to galactose and joined to another of glucose. The galactose is apparently needed by the developing infant to build developing brain and nervous tissue. Brain cells contain **glycolipids** as a part of their structure. A glycolipid is a triglyceride in which one of the fatty acid groups has been replaced by a sugar, in this case galactose. Galactose is more stable (to metabolic oxidation) than glucose and affords a better material for forming structural units in cells.

Lactose
D-Galactose + D-Glucose

A glycolipid

Although almost all human infants can digest lactose, some adults lose this ability on reaching maturity, since milk is no longer an important part of their diet. An enzyme called **lactase** is necessary to digest lactose. Lactase is secreted by the cells of the small intestine, and it cleaves lactose into its two component sugars, which are easily digested. Persons lacking the enzyme lactase do not digest lactose properly. As it is poorly absorbed by the small intestine, it remains in the digestive tract, where its osmotic potential causes an influx of water. This results in cramps and diarrhea for the affected individual. Persons with a lactase deficiency cannot tolerate more than one glass of milk a day. The deficiency is most common among blacks, but it is also quite common among older whites.

Lactose can be removed from whey by adding ethanol. Lactose is insoluble in ethanol, and when the ethanol is mixed with the aqueous solution, the lactose is forced to crystallize. The isolation of lactose from milk is described in Experiment 58.

When milk is allowed to stand at room temperature, it sours. Many bacteria are present in milk, particularly **lactobacilli.** These bacteria act on the lactose in milk to produce the sour **lactic acid.** These microorganisms actually **hydrolyze** lactose and produce lactic acid only from the galactose portion of the lactose. Since the production of the lactic acid also lowers the pH of the milk, the milk clots when it sours:

$$C_{12}H_{22}O_{11} + H_2O \longrightarrow C_6H_{12}O_6 + C_6H_{12}O_6$$

Lactose   **Galactose**   **Glucose**

$$C_6H_{12}O_6 \xrightarrow{\text{lactobacilli}} CH_3\!-\!\underset{\underset{\textstyle OH}{|}}{CH}\!-\!COOH$$

Galactose   **Lactic acid**

Many "cultured" milk products are manufactured by allowing milk to sour before it is processed. For instance, milk or cream is usually allowed to sour somewhat by lactic acid bacteria before it is churned to make butter. The fluid left after the milk is churned is sour and is called **buttermilk.** Other cultured milk products include sour cream, yogurt, and certain types of cheese.

REFERENCES

Fox, B. A., and Cameron, A. G. *Food Science—A Chemical Approach.* New York: Crane, Russak, 1973. Chap. 6, ''Oils, Fats, and Colloids.''

Kleiner, I. S., and Orten, J. M. *Biochemistry.* 7th ed. St. Louis: C. V. Mosby, 1966. Chap. 7, ''Milk.''

McKenzie, H. A., ed. *Milk Proteins.* 2 volumes. New York: Academic Press, 1970.

Oberg, C. J., ''Curdling Chemistry—Coagulated Milk Products.'' *Journal of Chemical Education, 63* (September 1986): 770.

Experiment 58
Isolation of Casein and Lactose from Milk

Isolation of a Protein
Isolation of a Sugar

In this experiment you isolate several of the chemical substances found in milk. You first isolate a phosphorus-containing protein, casein (Procedure 58A). The remaining milk mixture will then be used as a source of a sugar, α-lactose (Procedure 58B). After you have isolated the milk sugar, several chemical tests can be made on this material. Fats, which are present in whole milk, are not isolated in this experiment, since nonfat (skimmed) milk is used.

Here is the procedure you will follow. First, the casein is precipitated by warming the nonfat milk and adding dilute acetic acid. It is important that the heating not be excessive or the acid too strong, since these conditions also hydrolyze lactose into its components, glucose and galactose. After the casein has been removed, the excess acetic acid is neutralized with calcium carbonate, and the solution is heated to its boiling point to precipitate the initially soluble protein, albumin. The albumin is removed by filtration, and the filtrate is concentrated. Alcohol is added to the concentrate, and the solution is decolorized. After vacuum filtration of the solution through Filter Aid, α-lactose crystallizes on cooling. The crystallization of sugars is sensitive to trace colloidal impurities, and the final filtrate must be clear to obtain crystalline lactose. The impurities are removed by treating the solution with decolorizing carbon and filtering it through Filter Aid.

Lactose is an example of a disaccharide. It is made up of two sugar molecules (moieties): galactose and glucose. In the structures shown in the diagram, the galactose moiety is on the left and glucose is on the right.

Galactose is bonded through an acetal linkage to glucose. One should notice that the glucose moiety can exist in one of two isomeric hemiacetal structures: α-lactose and β-lactose. The glucose moiety can also exist in a free aldehyde form. This aldehyde form (open form) is an intermediate in the equilibration (interconversion) of α- and β-lactose. Very little of this free aldehyde form exists in the equilibrium mixture. The isomeric α- and β-lactoses are diastereomers, since they differ in the configuration at one carbon atom, called the anomeric carbon atom.

The sugar α-lactose is easily obtainable by crystallization from a water-ethanol mixture at room temperature. On the other hand, β-lactose must be obtained by a more difficult process, which involves crystallization from a concentrated solution of lactose at temperatures above 93.5 °C. In the present experiment, α-lactose is isolated by the simpler experimental procedure indicated above.

It has been found that α-lactose undergoes numerous interesting reactions. First, α-lactose interconverts, via the free aldehyde form, to a large extent, to the β

484

Acetal (β linkage)

α-LACTOSE

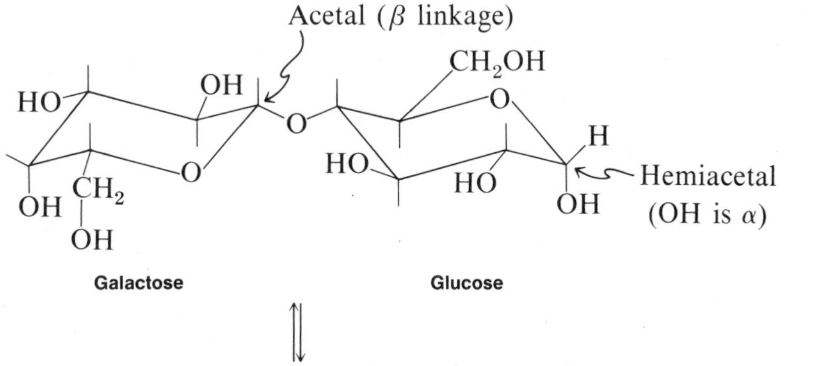

Galactose   Glucose

Hemiacetal
(OH is α)

LACTOSE
(ALDEHYDE FORM)

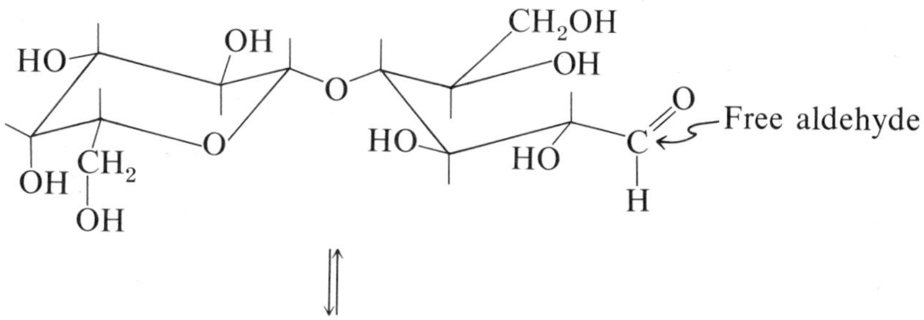

Free aldehyde

β-LACTOSE

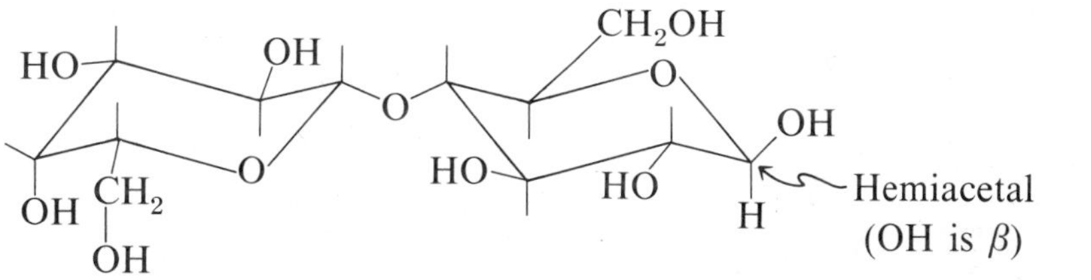

Hemiacetal
(OH is β)

isomer in aqueous solution. This causes a change in the rotation of polarized light from $+92.6°$ to $+52.3°$ with increasing time. The process that causes change in optical rotation with time is called **mutarotation.** Mutarotation of lactose can be studied in Experiment 59.

A second reaction of lactose is the oxidation of the free aldehyde form by Benedict's reagent. Lactose is referred to as a reducing sugar because it reduces Benedict's reagent (cupric ion to cuprous ion) and produces a red precipitate (Cu_2O). In the process, the aldehyde group is oxidized to a carboxyl group. The reaction that takes place in Benedict's test is

$$R\text{—}CHO + 2Cu^{2+} + 4OH^- \longrightarrow RCOOH + Cu_2O + 2H_2O$$

A third reaction of lactose is the oxidation of the galactose moiety by the mucic acid test. In this test, the acetal linkage between galactose and glucose moieties is cleaved by the acidic medium to give free galactose and glucose. Galactose is oxidized with nitric acid to the dicarboxylic acid, galactaric acid (mucic acid). Mucic acid is an insoluble, high-melting solid, which precipitates from the reaction mixture. On the other hand, glucose is oxidized to a diacid (glucaric acid), which is more soluble in the oxidizing medium and does not precipitate.

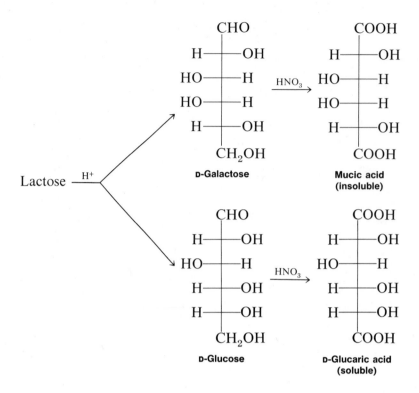

SPECIAL INSTRUCTIONS

It is necessary to read Techniques 1, 2, and 3 before starting this experiment. Read the essay that precedes this experiment. Procedures 58A and 58B should both be performed during one laboratory period. The lactose solution must be allowed to stand until the following laboratory period.

Procedure 58A
Isolation of Casein from Milk

Place 200 mL of nonfat (skimmed) milk in a 600-mL beaker.[1] Warm the milk to about 40 °C and add dropwise a dilute acetic acid solution (1 volume of glacial acetic acid to 10 volumes of water) with a pipet. Stir the mixture continuously with a glass rod during the addition. Continue to add dilute acetic acid until the casein no longer separates. Excess acid should be avoided because it may hydrolyze some of the lactose to glucose and

---

[1]The milk should not be allowed to stand too long before being used in this experiment. Lactose may be slowly converted to lactic acid even if the milk is stored in a refrigerator.

galactose. Stir the casein until it forms a large amorphous mass. With the stirring rod, remove the casein and place it in another beaker.[2] Immediately add 5 g of **powdered** calcium carbonate to the **first** beaker containing the liquid from which the casein was removed. Stir this mixture for a few minutes and save it for use in Procedure 58B. Use it as soon as possible during the same laboratory period. This mixture contains the lactose.

Vacuum-filter the casein mass for about 15 minutes to remove as much liquid as possible. Press the casein with a spatula during the filtering. Place the product between several layers of paper towels to help dry it (transfer the product at least three or four times to fresh paper towels). Allow the casein to dry completely in air for 1 or 2 days and weigh it. Submit the casein in a labeled vial to the instructor or save it for Experiment 60. The density of milk is 1.03 g/mL. Using this value, calculate the weight percentage yield of casein isolated from milk.

Procedure 58B
Isolation of Lactose from Milk

Heat the mixture saved from Procedure 58A to a gentle boil for about 10 minutes. This heating results in a nearly complete precipitation of the albumins. Filter the hot mixture by vacuum to remove the precipitated albumins and the remaining calcium carbonate. With a Bunsen burner, concentrate the filtrate in a 600-mL beaker to about 30 mL. Use several applicator sticks to help control the bumping caused by further precipitation of the albumins. The solution may also tend to foam out of the beaker if the mixture boils too vigorously. The foaming action can be controlled by blowing gently on the surface of the lactose solution.

Add 175 mL of 95% ethanol **(no flames!)** and 1 or 2 g of decolorizing carbon to the hot solution. After it has been mixed well, filter the warm solution by a gentle vacuum through a layer of wet Filter Aid (Technique 2, Section 2.4, p. 520). The filtrate should be clear.[3] Lactose is unlikely to crystallize unless the solution is clear. However, one may be deceived by cloudiness caused by rapid crystallization of lactose after the vacuum filtration. If, on standing, the solution shows relatively rapidly increasing cloudiness, further filtration should be avoided since it will remove the product.

Transfer the solution to an Erlenmeyer flask, stopper the flask, and allow the solution to stand overnight or until the next laboratory period. In some cases, it takes several days for complete crystallization. Lactose crystallizes on the wall and the bottom of the flask. Dislodge the crystals and vacuum-filter them. Wash the product with a few

---

[2]Even though the crude casein is pure enough for chromatographic analysis in Experiment 60, a small amount of fat may be isolated from casein at this stage. Extract the casein with two successive 15-mL portions of ether. Break up any lumps with a spatula. Stir the casein in the ether with a stirring rod for about 10 minutes. Following each extraction, decant the ether extracts into another container and remove the solvent by distillation in a hood. About 0.1 g of fat is obtained.

[3]Add 1 g of Filter Aid to the solution and refilter it by gravity if the solution is not clear or has a dark appearance caused by traces of decolorizing carbon. To keep the lactose in solution, the solution should be heated before it is filtered.

milliliters of cold 25% aqueous ethanol. Lactose crystallizes with one water molecule of hydration, $C_{12}H_{22}O_{11} \cdot H_2O$. Weigh the product after it is thoroughly dry. The density of milk is 1.03 g/mL. Using this value, calculate the weight percentage yield of lactose isolated.

At the instructor's option, make the following tests on the isolated lactose. Submit the remaining sample to the instructor or save it for the mutarotation studies (Experiment 59).

BENEDICT'S TEST (OPTIONAL)

Benedict's test is described in Experiment 57. Perform the test on a 1% solution of the isolated lactose. At the same time perform the test on 1% solutions of glucose and galactose.

MUCIC ACID TEST (OPTIONAL)

Place 0.2 g of the isolated lactose, 0.1 g of glucose (dextrose), and 0.1 g of galactose in three separate test tubes. Add 2 mL of water to each tube and dissolve the solids, with heating if necessary. Add 2 mL of concentrated nitric acid to each of the tubes. Heat the tubes in a boiling-water bath (hot plate) for 1 hour in a hood (nitrogen oxide gases are evolved). Remove the tubes and allow them to cool slowly after the heating period. Scratch the test tubes with clean stirring rods to induce crystallization. After the test tubes are cooled to room temperature, place them in an ice bath. A fine precipitate of mucic acid should begin to form in the galactose and lactose tubes about one-half hour after the tubes are removed from the water bath. Allow the test tubes to stand until the next laboratory period to complete the crystallization. Confirm the insolubility of the solid formed by adding about 2 mL of water, then shaking the resulting mixture. If the solid remains, it is mucic acid.

QUESTIONS

1. A student decided to determine the optical rotation of mucic acid. What should be expected as a value? Why?

2. Give a mechanism for the acid-catalyzed hydrolysis of the acetal bond in lactose.

3. β-Lactose is present to a larger extent in an aqueous solution when the solution is at equilibrium. Why is this to be expected?

4. Very little of the free aldehyde form is present in an equilibrium mixture of lactose. However, a positive test is obtained with Benedict's reagent. Explain.

5. Outline a separation scheme for isolating casein and lactose from milk. Use a flowchart like that shown in the **Advance Preparation and Laboratory Records** section at the beginning of the book.

Experiment 59
Mutarotation of Lactose

Polarimetry

In this experiment the mutarotation of lactose is studied by polarimetry. The disaccharide α-lactose, made up of galactose and glucose, can be isolated from milk (Experiment 58). As you can see in the structures drawn in Experiment 58, the glucose moiety can exist in one of two isomeric hemiacetal structures: α- and β-lactose. These isomers are diastereomers since they differ in configuration at one carbon atom. The glucose moiety can also exist in a free aldehyde form. This aldehyde form (open form in the equation below) is an intermediate in the equilibration of α- and β-lactose. Very little of this free aldehyde form exists in the equilibrium mixture.

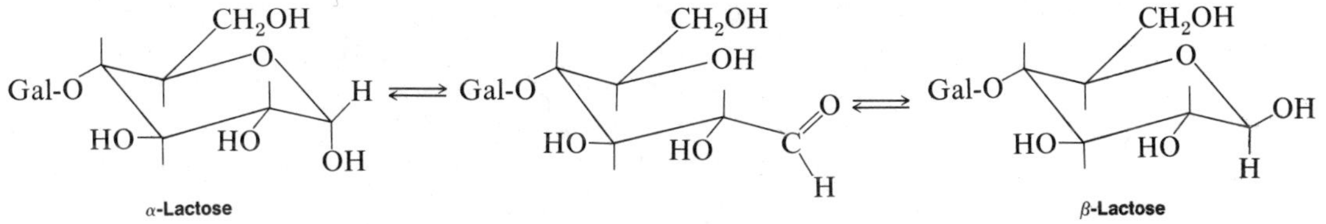

α-Lactose has a specific rotation at 20 °C of +92.6°. However, when it is placed in water, the optical rotation **decreases** until it reaches an equilibrium value of +52.3°. β-Lactose has a specific rotation of +34°. The optical rotation of β-lactose **increases** in water until it reaches the same equilibrium value obtained for α-lactose. At the equilibrium point, both the α and β isomers are present. However, since the equilibrium rotation is closer in value to the initial rotation of β-lactose, the mixture must contain more of this isomer. The process, which results in a change in optical rotation over time to approach an equilibrium value, is called **mutarotation.**

SPECIAL INSTRUCTIONS

Read Technique 15, especially Sections 15.2, 15.3, and 15.4, pp. 650–653, before starting this experiment. The procedure for preparing the cells and for operating the instrument are those appropriate for the Zeiss polarimeter. Your instructor will provide instructions for the use of another type of polarimeter if a Zeiss instrument is not available. A 2-dm cell is used for this experiment. If a cell of a different path length is used, adjust the concentrations appropriately. About 1 hour is needed to complete the mutarotation study.

PROCEDURE

Turn on the polarimeter to warm the sodium lamp. After about 10 minutes, adjust the instrument so that the scale reads about $+9°$. This scale reading provides an adjustment of the instrument to the approximate range of rotation that will be observed at the initial reading. Set the timer to zero. Clean and dry a 2-dm cell (Technique 15, Figure 15–6, p. 652). Weigh 1.25 g of α-lactose (Experiment 58) and transfer it completely to a **dry** 25-mL volumetric flask.

The operations described in the next paragraph should be studied carefully **before** starting this part of the experiment. It is essential to complete carefully the described operations in **2 minutes** or less. The reason for speedy operation is that the α-lactose immediately begins to mutarotate when it comes in contact with water. The initial rotations obtained are necessary to get a precise value of the rotation at zero time. You should practice with the necessary equipment in a place near the polarimeter before performing the actual operations. Study the scale on the polarimeter so that you can read it rapidly.

Add about half the volume of distilled water to the volumetric flask containing the α-lactose and swirl it to dissolve the solid. When about half the solid is dissolved (a rough estimate), start the timer. As soon as the solid is dissolved (about **20–25** seconds), carefully fill the flask to the mark with distilled water. Use an eye dropper to finish adding the water. Stopper the flask and shake it about five times to mix the contents. Using a funnel, fill the polarimeter cell with the lactose solution. Screw the end piece on the cell and tilt it to transfer any remaining bubbles to the enlarged ring. Place the cell in the polarimeter, close the cover, and adjust the analyzer until the split field is of uniform density (Technique 15, Figure 15–7, p. 653). Record the time and rotation in the notebook.

Obtain the optical rotation at 1-minute intervals for 8 additional minutes (10 minutes total from the time of initial mixing) and record these values, along with the times at which they were determined. After the 10-minute period, obtain readings at 2-minute intervals for the next 20 minutes. Record the optical rotations and times.

Remove the cell from the polarimeter and add 2 drops of concentrated ammonium hydroxide to the lactose solution. The ammonia rapidly catalyzes the mutarotation of lactose to its equilibrium value. If the ammonia is not added, the equilibrium value will not be obtained until after about 22 hours. Shake the tube and replace it in the polarimeter. Follow the decrease in rotation until there is no longer a change with time. This final value, which is the equilibrium optical rotation, should remain constant for about 5 minutes. Place a thermometer in the polarimeter and determine the temperature in the cell compartment.

Plot the data on a piece of graph paper ruled in millimeters, with the optical rotation plotted on the vertical axis and time plotted (up to 30 minutes) on the horizontal axis. Draw the best possible curved line through the points and extrapolate the line to $t = 0$. Remember that there may be some scattering of points about the line, especially at the values for the longer times. The extrapolated value at $t = 0$ corresponds to the optical rotation of α-lactose at the time of initial mixing.

Using the equation in Technique 15, Section 15.2, p. 650, calculate the specific rotation, $[\alpha]_D$, of α-lactose at $t = 0$. Likewise, calculate the specific rotation of the equilibrium mixture of α- and β-lactose.

Calculate the percentage of each of the diastereomers at equilibrium, using the experimentally determined specific rotation values for α-lactose and the equilibrium mixture and the literature value for the specific rotation of β-lactose ($+34°$). Assume a linear relation between the specific rotations and the concentrations of the species.

QUESTIONS

1. Explain why β-lactose predominates in the equilibrium mixture of α- and β-lactose.
2. The following rotation data have been obtained for D-glucose at 20 °C:

α-D-glucose	$+112.2°$
β-D-glucose	$+18.7°$
Equilibrium mixture	$+52.7°$

Using these values, calculate the percentage composition of the α and β isomers at equilibrium. Inspect the structures of α- and β-D-glucose in Experiment 57 and rationalize the values obtained in the calculations.

Experiment 60
Paper Chromatography of Amino Acids

Paper Chromatography

In this experiment, you determine the composition of proteins by paper chromatographic analysis of the hydrolysates.

Proteins can be hydrolyzed in acidic or basic solution or with enzymes. During the hydrolysis, the peptide bonds break to give shorter polymers (polypeptides), which in turn are further degraded to amino acids. Total hydrolysis of protein can be achieved in 20% hydrochloric acid solution at 100 °C for 12 to 48 hours. However, adequate hydrolysis for purposes of this experiment can be achieved in refluxing acid in less than 1 hour.

$$\begin{array}{c} R \quad O \quad\quad R \quad O \quad\quad R \quad O \quad\quad\quad\quad R \\ | \quad\ \| \quad\quad | \quad\ \| \quad\quad | \quad\ \| \quad\quad\quad\quad | \\ -\text{NH}-\text{CH}-\text{C}-\text{NH}-\text{CH}-\text{C}-\text{NH}-\text{CH}-\text{C}- \longrightarrow n\ \text{NH}_2\text{CHCOOH} \end{array}$$

Protein   **Amino acids**

These hydrolysates can be analyzed for their amino acid contents by chromatographic methods, such as paper chromatography. The most common amino acids are listed in the accompanying table. They are listed in the order of increasing R_f values. The amino acid contents for the proteins casein, gelatin, silk, and hair are also listed. The chief amino acid constituents in each of the proteins are italicized. Note particularly the large differences in the amino acid content of the various proteins. Since there are variations in amino acid content among samples, the values in the table are to be considered **approximate.**

Amino Acids, R_f Values, and Approximate Compositions of Casein, Gelatin, Silk, and Hair

AMINO ACID	FORMULA	R_f VALUE	APPROXIMATE PERCENTAGE COMPOSITION OF PROTEINS			
			Casein	*Gelatin*	*Silk*	*Hair*
Cystine	$S-CH_2CHCOOH$ (with NH_2), $S-CH_2CHCOOH$ (with NH_2)	0.16	0.4	0.1		*18.0*
Aspartic acid	$HOOCCH_2CHCOOH$ (with NH_2)	.32	*6.8*	*6.7*		3.9
Glutamic acid	$HOOCCH_2CH_2CHCOOH$ (with NH_2)	.40	*22.4*	*11.5*		*13.1*
Glycine	H_2NCH_2COOH	.42	2.6	*25.5*	*42.3*	4.1
Serine	$HOCH_2CHCOOH$ (with NH_2)	.43	*7.4*	0.4	*12.6*	10.6
Threonine	$CH_3CH-CHCOOH$ (with OH NH_2)	.51	4.7	1.9	1.5	8.5
Lysine (cation)	$H_3\overset{+}{N}CH_2CH_2CH_2CH_2CHCOOH$ (with NH_2)	.53	*7.9*	4.1	0.4	1.9
Alanine	$CH_3CHCOOH$ (with NH_2)	.59	2.9	8.7	*24.5*	2.8
Arginine (cation)	$H_2\overset{+}{N}=CNHCH_2CH_2CH_2CHCOOH$ (with NH_2 NH_2)	.60	3.9	*8.0*	1.1	*8.9*
Tyrosine	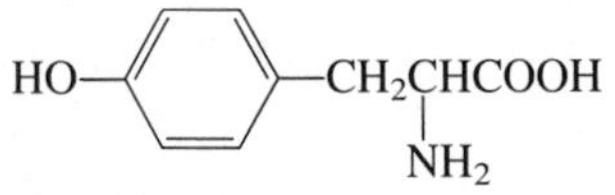 $HO-\text{(ring)}-CH_2CHCOOH$ (with NH_2)	.62	*6.1*	0.4	*10.6*	2.2
Valine	$CH_3CH-CHCOOH$ (with CH_3 NH_2)	.75	*6.9*	2.5	3.2	5.5
Methionine	$CH_3SCH_2CH_2CHCOOH$ (with NH_2)	.77	3.3	1.0		0.7
Leucine	$CH_3CHCH_2CHCOOH$ (with CH_3 NH_2)	.79	*8.8*	4.6	0.8	*11.2*
Phenylalanine	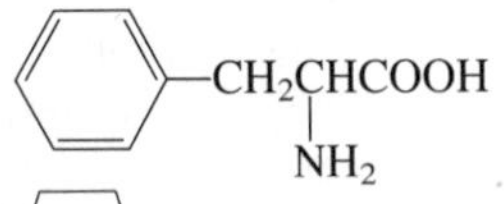 $\text{(ring)}-CH_2CHCOOH$ (with NH_2)	.82	4.8	2.2		2.4
Proline	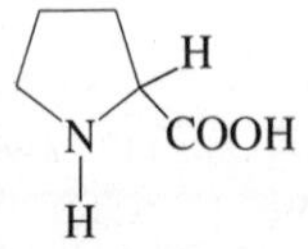 pyrrolidine ring with N, H, $COOH$	.85	*10.9*	*18.0*	1.5	4.3

NOTE: Principal constituents are italicized.

The individual amino acids on a chromatogram are made visible with ninhydrin. Ninhydrin reacts with amino acids to produce characteristic deep blue colors. A few amino acids produce a different color, however; proline, for example, produces a pale yellow color with ninhydrin. The reactions involved in the production of the color are as follows:

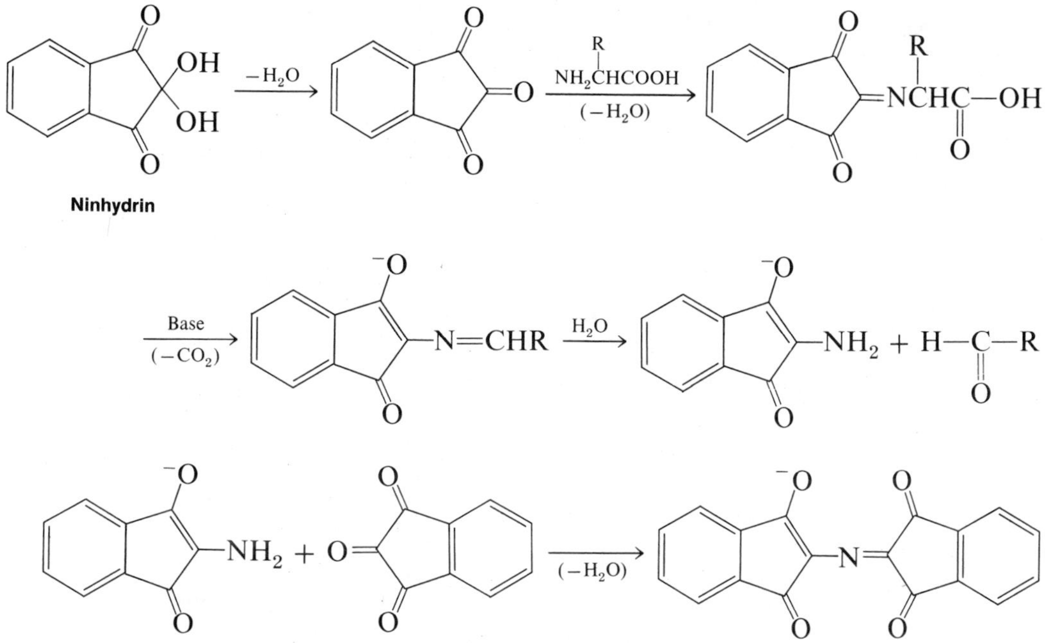

SPECIAL INSTRUCTIONS

It is necessary to read about amino acids and proteins in your lecture textbook. Also read Technique 11, "Thin-Layer Chromatography." Paper chromatography is similar to thin-layer chromatography. The casein isolated in Procedure 58A can be used in this experiment. The hydrolysates of casein, gelatin, silk, and hair are best provided by the laboratory instructor or assistant. The procedure for preparing them is given below. The chromatographic development takes about 4 hours to complete. The laboratory instructor or assistant may need to be responsible for removing the chromatogram at the end of the 4-hour development period.

The purpose of this experiment is to identify individual, unknown amino acids and to identify the constituent amino acids in the hydrolysates of some common proteins. This identification is accomplished by comparing R_f values of known amino acids with R_f values of the unknown. If the hydrolysates are provided by the laboratory instructor, go directly to the paper chromatography part of the experiment.

NOTE TO THE INSTRUCTOR: Tests on amino acids and proteins are given in the instructor's manual.

PROCEDURES

HYDROLYSIS OF A PROTEIN

Assemble a small-scale reflux apparatus, using a 100-mL round-bottomed flask. Place 20 mL of 19% hydrochloric acid (equal volumes of concentrated hydrochloric acid and distilled water), 0.5 g of the protein (casein, gelatin, hair, or silk), and a boiling stone in the flask. For casein or gelatin, carefully heat the mixture under reflux for 35 minutes using a heating mantle or a very small flame. For hair or silk, heat the mixture under reflux for 50 minutes. After the reflux period is completed, the hydrolysate must be decolorized. Add about 0.5 g of decolorizing carbon to the hot hydrolysate and swirl the mixture. Gravity-filter the hydrolysate into a 50-mL Erlenmeyer flask. The filtrate should be colorless or pale yellow. Check the hydrolysate with the biuret test to see whether the reaction is complete.[1] The test should be negative. If hydrolysis is not complete (a positive biuret test), add 3 to 5 mL of 19% hydrochloric acid and heat the mixture for an additional 15 minutes. Again check the hydrolysate with the biuret test to see whether it is negative. After the hydrolysis is complete, cool the hydrolysate, stopper it, and save the solution for the analysis by paper chromatography.

SEPARATION OF AMINO ACIDS BY PAPER CHROMATOGRAPHY

Obtain a 24 × 15.5-cm rectangle of Whatman No. 1 filter paper. Handle this paper by the top edge only. If the paper is handled carelessly, fingerprints may appear as colored spots on the chromatogram. Place the rectangle on a clean paper towel or notebook paper with the "x" in the upper right corner (see the figure). This "x" was placed on the rectangle when the paper was cut. Since the thickness of the paper is uneven, this marking procedure ensures that the direction of the solvent flow on each piece of chromatography paper is the same for all students. The R_f values are more reproducible when the chromatograms are developed in the same direction.

     Cut a 1-cm square from the two lower corners. About 1.5 cm from the bottom of the paper, pencil a line (do not use a pen) across the paper. Also pencil 13 dots, evenly spaced 1.5 cm from each other along this line, with the first dot 3 cm from the left-hand edge. Nine standard amino acids, two hydrolysates, and two unknown amino acids will be placed on the dots. We suggest you use the sixth and seventh dots for the unknowns, the eighth and ninth dots for the hydrolysates, and the rest of the dots for the individual standard amino acids. One possible sequence is shown in the figure. The standard amino acids that will be used are 0.1M solutions of aspartic acid (ASP), alanine (ALA), glycine (GLY), tyrosine (TYR), proline (PRO), leucine (LEU), glutamic acid (GLU), cystine (CYS), and arginine (ARG). Each of these 0.1M solutions has been acidified with 10 drops of 19% hydrochloric acid for each 10 mL of solution. Record the sequence in the laboratory notebook. Place your name or initials in the upper right-hand corner of the chromatogram.

---

[1]Place 5 drops of the hydrolysate and 10 drops of 10% sodium hydroxide in a test tube. Check the solution with litmus paper to make sure that it is definitely alkaline. Add more 10% sodium hydroxide if necessary. Add 4 or 5 drops of 2% copper sulfate solution. A blue color indicates that the hydrolysis is complete. A pink or violet color indicates that the hydrolysis is not complete.

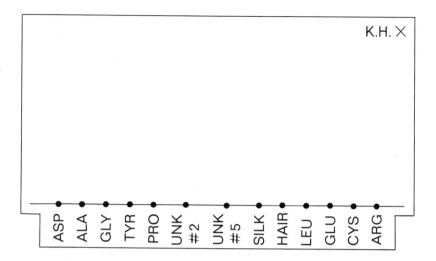

Prepare capillary micropipets for applying the solutions (Technique 11, Section 11.3, p. 620) or use the micropipets inserted in the standard amino acids, unknowns, or hydrolysate samples. It may be wise to practice the spotting technique on a small piece of Whatman No. 1 paper before trying to spot the actual chromatogram. The correct method of spotting is described in Technique 11, Section 11.3, p. 620. It is important that the spots should be made as small as possible and that the paper should not be over-loaded. If either of these conditions is not observed, the spots will tail and overlap after development. The applied spot should be 1 to 2 mm in diameter. Apply the appropriate samples to the dots with the micropipets, taking extreme care to avoid contaminating the samples. After applying the samples to each dot, allow the spots to dry completely and then make a second application at the same point. The hydrolysates should be spotted a third time. Allow the spots to dry completely before the chromatogram is developed.

> **CAUTION: Phenol will cause burns if it touches the skin. Immediately wash any affected area with copious quantities of soap and water. Clean up all spills.**

Use a 32-oz wide-mouthed screw-cap jar for a developing chamber. Insert a pipet **below** the protective layer of ligroin and remove 20 mL of an 80% aqueous phenol solution with a bulb.[2] Carefully transfer the phenol solution to the developing chamber so that it is not splattered on the sides of the jar. Check to see that the depth of the phenol solution does not exceed 1.5 cm. If it does, carefully pipet enough of the solution so that the depth is less than 1.5 cm. Curl the paper into a large cylinder so that the line of spots is on the bottom and on the inner surface. Overlap the top about 0.5 cm and make sure that the bottom edge is even. Hold the chromatogram together with a paper clip or a staple. Insert the cylinder into the jar so that the bottom end of the paper is immersed in the solvent. Tighten the lid securely and allow the development to proceed for 4 hours without disturbing the jar.

---

[2]The 80% aqueous phenol is prepared by mixing 80 g of phenol per 20 mL of distilled water. Heat the mixture until it dissolves completely. Add a protective layer of ligroin so that air is excluded from the aqueous phenol. If the ligroin is not added, the solution should be used as soon as possible.

Remove the chromatogram (no fingers), mark the solvent front with a pencil, and allow the solvent to evaporate in air.[3] In the next laboratory period, spray the paper uniformly with ninhydrin spray and place the paper in a 110 °C oven. Colored spots will begin to appear within 5 minutes. Remove the chromatogram and outline all the spots with a pencil. With a millimeter ruler, measure the distance each spot has traveled from the point of origin to the **front (top)** of the spot. Also measure the distance that the solvent traveled. Calculate the R_f value for each of the spots (Technique 11, Section 11.8, p. 625). In some cases, the separation may not be complete enough to calculate the R_f value accurately. Record the calculated values.

From the R_f values and colors of the standard amino acids, identify the two unknown amino acids and the principal constituents of the two protein hydrolysates. The table of protein composition may prove useful as you analyze the hydrolysates. Remember that amino acids with similar R_f values may not separate well enough to be seen on the chromatogram. You may have to place the amino acids in groups. Record your findings. Submit the chromatogram with your laboratory report.

Experiment 61
Resolution of (±)-α-Phenylethylamine

Resolution
Polarimetry

Although racemic α-phenylethylamine is readily obtainable commercially, it is much more difficult to obtain one of the enantiomers in optically pure form. We obtain the pure form through **resolution,** or separation of the enantiomers. In this experiment, you isolate only one of the enantiomers, the levorotatory one, since it can be more easily isolated than the dextrorotatory enantiomer. The resolving agent to be used is (+)-tartaric acid, which forms diastereomeric salts with racemic α-phenylethylamine. The important reactions for this experiment are indicated in the accompanying figure.

Optically pure (+)-tartaric acid is abundant in nature. It is readily obtained as a by-product of wine-making. The (−)-amine-(+)-tartrate diastereomeric salt has a lower solubility than the (+)-amine-(+)-tartrate salt, and it separates from the solution as crystals. The crystals are removed by filtration and purified. The salt is then treated with dilute base to regenerate the free (−)-amine. In principle, the mother liquor,

---

[3]This part of the procedure may have to be done by the laboratory instructor or assistant if there is not enough time.

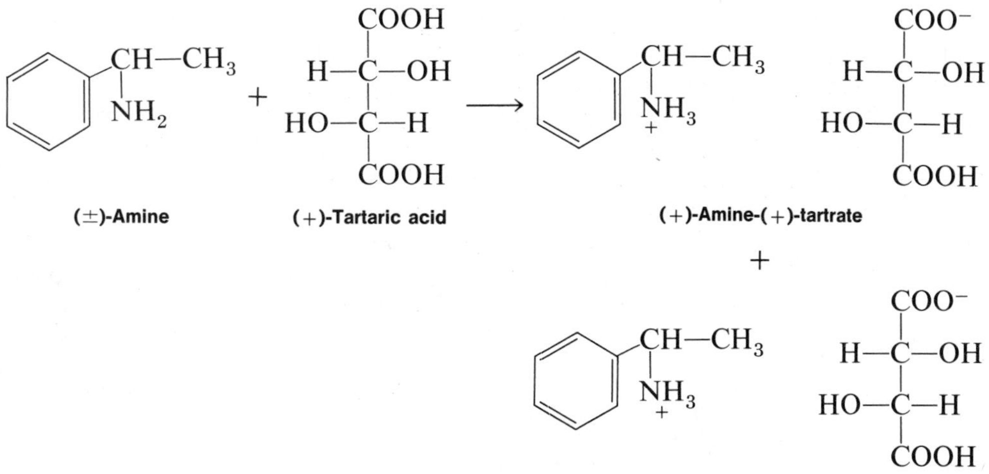

$$
\underset{(\pm)\text{-Amine}}{\text{C}_6\text{H}_5\text{–CH–CH}_3,\ \text{NH}_2}
\ +\
\underset{(+)\text{-Tartaric acid}}{\begin{array}{c}\text{COOH}\\ \text{H–C–OH}\\ \text{HO–C–H}\\ \text{COOH}\end{array}}
\longrightarrow
\underset{(+)\text{-Amine-}(+)\text{-tartrate}}{\text{C}_6\text{H}_5\text{–CH–CH}_3,\ \overset{+}{\text{NH}_3}
\ \begin{array}{c}\text{COO}^-\\ \text{H–C–OH}\\ \text{HO–C–H}\\ \text{COOH}\end{array}}
$$

+

$$
\underset{(-)\text{-Amine-}(+)\text{-tartrate}}{\text{C}_6\text{H}_5\text{–CH–CH}_3,\ \overset{+}{\text{NH}_3}
\ \begin{array}{c}\text{COO}^-\\ \text{H–C–OH}\\ \text{HO–C–H}\\ \text{COOH}\end{array}}
$$

which contains mostly the (+)-amine-(+)-tartrate salt, can also be purified to yield the other diastereomeric salt eventually. Hydrolysis of this salt would yield the (+)-amine.

SPECIAL INSTRUCTIONS

Before beginning this experiment, it is important to read Technique 15. Technique 6 should also be read. Once the amine-tartrate salt is precipitated, the instructor may have students work in pairs to complete the experiment.

PROCEDURE

Place 31.25 g of L-(+)-tartaric acid and 450 mL of methanol in a 1-L Erlenmeyer flask. Heat this mixture on a steam bath until the solution is nearly boiling. Slowly add 25 g of racemic α-phenylethylamine to this hot solution.

> **NOTE:** Caution should be exercised at this step, because the mixture is very likely to froth and boil over.

Next, stopper the flask and allow it to stand overnight. The crystals that form should be **prismatic.** If needles are obtained, they should be dissolved and recrystallized; you can do this by seeding the mixture with a prismatic crystal, if one is available. Alternatively, the mixture may be heated until **most** of the solid has dissolved. The needle crystals dissolve easily and usually a small amount of the prismatic crystals remain to seed the solution. Allow the solution to cool slowly to form prismatic crystals. The needles are not optically pure enough to give a complete resolution of enantiomers, and they must be dissolved and the material crystallized again. Filter the crystals, using a Büchner funnel, and wash them with a few portions of cold methanol.

At this point, two students should combine their yields of amine-tartrate adduct so that 25 g or more of crystals will be available for the next step. An alternative approach would be to reduce the volume of mother liquor to 250 mL by evaporation (in the hood; no flames) and to allow an additional amount (about 3.5 g) of adduct to crystallize. Partially dissolve the salt in 100 mL of water, add 15 mL of 50% sodium hydroxide solution (more if over 25 g of crystals were used), and extract this mixture with three 30-mL portions of diethyl ether. Dry the ether layer over anhydrous magnesium sulfate for about 10 minutes. Assemble a simple distillation apparatus, using a 100-mL distillation flask (Technique 6, Figure 6–6, p. 557). Remove the ether by simple distillation, using a heating mantle or an oil bath as a heat source (**no flames!**). Then, allow the temperature to rise, and distill the amine over a temperature range of 180 to 190 °C.

> **NOTE: Frothing can sometimes be a problem at this stage.**

Use a 25-mL Erlenmeyer flask that has previously been weighed, with its stopper, to within ± 0.003 g, to collect the amine. Stopper the flask tightly as soon as the product is collected. Weigh the flask to determine the weight of amine collected.

Since this reaction does not generally yield enough pure amine to fill a 2-dm polarimeter cell, the solution will have to be diluted with methanol.[1] Using a pipet, add 10 mL of absolute methanol to the stoppered flask and shake the solution to mix the contents thoroughly. At this point the total volume of the solution very nearly equals 10 mL plus the volume of the amine, or its weight divided by its density (0.9395 g/mL). The approximation of additivity of volumes introduces a negligible error in this procedure. This total volume of the solution is needed for calculating the new concentration (in g/mL) of the solution. Transfer the solution to a 2-dm polarimeter tube and determine its observed rotation (Technique 15, Sections 15.3 and 15.4, pp. 651–653). The published value is $[\alpha]_D^{22} = -40.3°$. Be sure to keep the pure amine tightly stoppered. Report the value of the specific rotation and the optical purity to the instructor. Calculate the percentage of **each** of the enantiomers in the sample.

REFERENCES

Ault, A. "Resolution of D,L-α-Phenylethylamine." *Journal of Chemical Education, 42* (1965): 269.
Jacobus, J., and Raban, M. "An NMR Determination of Optical Purity." *Journal of Chemical Education, 46* (1969): 351.

QUESTIONS

1. Using a reference text, find examples of reagents used in performing chemical resolutions of acidic, basic, and neutral racemic compounds.

---

[1]In some cases there may be enough amine to fill a smaller cell (i.e., 0.5 dm). In this event, the sample can be used neat (no methanol). This simplifies the procedure and calculations considerably.

2. Propose methods of resolving each of the following racemic compounds:

(a) 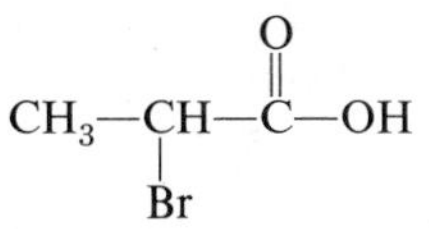 CH_3—CH—C—OH
        | ‖
        Br   O

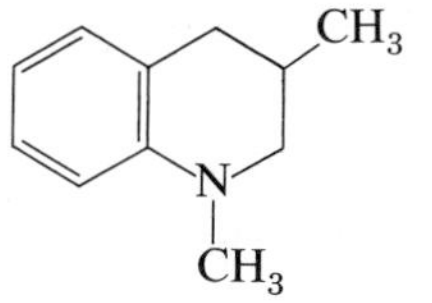

(b) [structure: 3-methyl-1-methyl-1,2,3,4-tetrahydroquinoline with CH₃ on carbon and N—CH₃]

(c) [structure: phenyl—CH—CH₃ with OH]

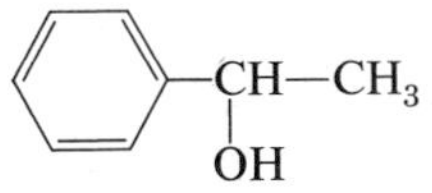

Experiment 62
NMR Determination of the Optical Purity of (−)-α-Phenylethylamine Using a Chiral Shift Reagent

Nuclear Magnetic Resonance
Chemical Shift Reagents
Optical Purity

In Experiment 61, a method for the resolution of the two enantiomers of α-phenylethylamine was given. In this experiment, we will use nmr to determine the relative success of the method. An nmr spectrum of racemic (±)-phenylethylamine is shown in the figure. In this spectrum there is no discernible difference between the two enantiomers. The methyl group appears as a doublet at about 1.4δ and the methine proton as a quartet at 4.0δ. The amino group hydrogens and the phenyl ring are the only other visible groups.

Although the normal spectrum shows no visible difference for the two enantiomers, there is a method that will allow the spectra of the two enantiomers to be distinguished. This method uses a chiral shift reagent. Chemical shift reagents are discussed in general in Appendix 4, Section NMR.13. These reagents "spread out" the resonances of the compound with which they are used, increasing the chemical shifts of the protons that are nearest the center of the metal complex by the largest amount. Since the spectra of both (+)- and (−)-α-phenylethylamine are identical, the usual chemical shift reagent would not help our analysis. However, if one uses a chemical shift reagent that is itself chiral, one can begin to distinguish the two separate enantiomers by their nmr spectra. The two enantiomers, which are chiral, will interact differently with the chiral shift reagent. The complexes formed from the R- and S-isomers and the (+)-camphor-containing shift reagent will be diastereomers. Diasteromers usually have different physical properties, and the nmr spectra are no exception. The two complexes will end up with slightly differing geometries. Although the effect is small, it is large enough to begin to see differences in the nmr spectra of the two enantiomers. In particular, the originally superimposed methyl doublets will begin to be resolved.

The chiral shift reagent used in this experiment is tris(heptafluorobutyryl-d-camphorato) europium(III), or Eu(hfbc)$_3$. In this complex, the europium is in a chiral environment because it is complexed to camphor, which is a chiral molecule.

Eu(hbfc)$_3$ has the following structure:

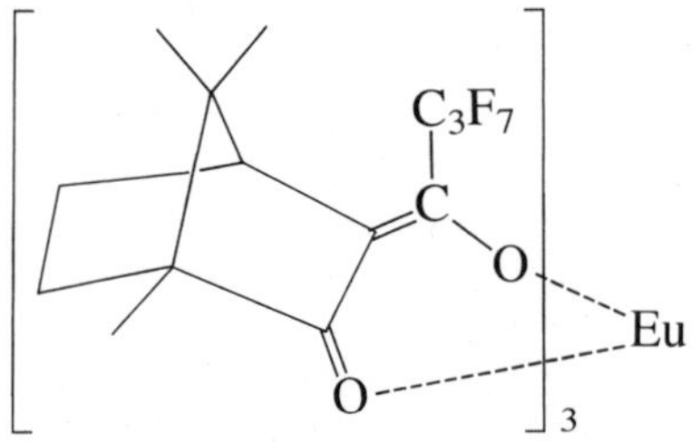

PROCEDURE

Place approximately 0.050 g of the amine in an nmr tube. Use a disposable pipet and an analytical balance to perform this operation. It is not important to weigh an exact quantity of the amine; any amount from 0.0500 to 0.0750 g will suffice, but you must know the exact weight.

Divide the quantity of amine that you weighed by 2.5 to determine the amount of Eu(hbfc)$_3$ that you will need. Using smooth weighing paper, use the analytical balance to weigh out this quantity of shift reagent. Again, it is not necessary to be perfectly exact, but you must record the amount. Carefully add this shift reagent to the nmr sample. Add a small quantity of CDCl$_3$ solvent containing tetramethylsilane (TMS), but do not more than double the initial volume of the sample of the amine. Allow the sample to stand for 20 minutes.

Determine the nmr spectrum of the sample. The peaks of interest are the methyl peaks, since they show the greatest enantiomeric distinction. Two sets of doublets should be seen. If you do not see this distinction, the amount of one of the enantiomers may be too small, or it may just appear as shoulders on the base of the peaks from the larger doublet. If you wish, a second portion of shift reagent, similar to the one above, may be added.

It is a good idea to test the ability of your instrumentation by preparing a reference sample containing equal quantities (about 1.00 g each) of racemic (±)-α-phenyl-ethylamine and your resolved (−)-α-phenylethylamine. This sample should contain about 75% (−)-isomer and 25% (+)-isomer. As before, use about 0.500 g of this sam-

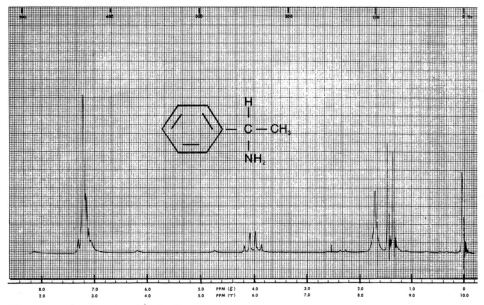

Nmr spectrum of α-phenylethylamine

ple. In this way you can tell how well your method is working and also assign the upfield and downfield peaks to the correct enantiomer.

    Determine the percentage of each isomer in both of your samples in the following manner. Measure the combined heights of the two peaks in the larger doublet and compare this total with the total height of the two peaks (or shoulders) from the smaller doublet. This represents the mole ratio of the two enantiomers. On most instruments, the peaks will be so sharp (narrow) that this method will be a good approximation to the areas underneath. If you have more capable instrumentation, determine the integrals in the usual fashion. On a Varian 60 MHz T-60 nmr spectrometer, the comparison-of-peak-heights method gives results that are off by no more than 2 to 3% when measuring accurately prepared reference samples.

Part Two

The Techniques

Technique 1
Solvents and Methods of Heating Reaction Mixtures

1.1 SAFETY PRECAUTIONS WITH SOLVENTS

Always remember that organic solvents are all at least somewhat toxic and that many are flammable. Great care must be exercised in using these materials. You should also be thoroughly familiar with the section entitled "Laboratory Safety," located in the introductory part of this book.

> **Read "Laboratory Safety," pages 4 to 13.**

The most common organic solvents are listed in Table 1-1, along with their boiling points. Solvents marked in boldface type will burn. Ether, pentane, and hexane are especially dangerous, since if they are combined with the correct amount of air, they will explode.

The terms **petroleum ether** and **ligroin** are often confusing. Petroleum ether is a mixture of hydrocarbons, with isomers of the formulas C_5H_{12} and C_6H_{14} predominating; it is not an ether at all, since there are no oxygen-bearing compounds in the mixture. Care should be exercised when instructions call for either **ether** or **petroleum**

TABLE 1–1. Common Organic Solvents

SOLVENT	BP (°C)	SOLVENT	BP (°C)
HYDROCARBONS		**ETHERS**	
Pentane	36	**Ether** (diethyl)	35
Hexane	69	**Dioxane***	101
Benzene*	80	**1,2-Dimethoxyethane**	83
Toluene	111	**OTHERS**	
HYDROCARBON MIXTURES		Acetic acid	118
		Acetic anhydride	140
Petroleum ether	30–60	**Pyridine**	115
Ligroin	60–90	**Acetone**	56
CHLOROCARBONS		**Ethyl acetate**	77
Methylene chloride	40	Dimethylformamide	153
Chloroform*	61	Dimethylsulfoxide	189
Carbon tetrachloride*	77		
ALCOHOLS			
Methanol	65		
Ethanol	78		

NOTE: Boldface type indicates flammability.
*Suspect carcinogen (see page 12).

ether; the two must not become accidentally confused. Confusion is particularly easy when one is selecting the container of solvent from the supply shelf.

Ligroin, or high-boiling petroleum ether, is like petroleum ether in composition, except that compared with petroleum ether, ligroin has alkanes that generally include higher-boiling alkane isomers. Depending on the supplier, ligroin may have different boiling ranges. While some brands of ligroin have boiling points ranging from about 60 °C to about 90 °C, other brands have boiling points ranging from about 60 °C to about 75 °C.

1.2 FLAMES

The simplest technique for heating mixtures is by the Bunsen burner. However, because of the high danger of fires, the use of the Bunsen burner should be strictly limited to those cases for which the danger of fire is low or for which no reasonable alternative source of heat is available. A flame should generally be used only to heat aqueous or very-high-boiling solutions. Even under these circumstances, great care should be taken to ensure that persons in the vicinity are not using flammable solvents.

In heating a flask with a Bunsen burner, you will find that using wire gauze can produce more even heating over a broader area. The wire gauze, when placed under the object being heated, spreads the flame to keep the flask from being heated in one small area only.

1.3 STEAM CONES

The steam cone or steam bath is a safe source of heat for most reaction mixtures. The steam cone is depicted in Figure 1–1. This device has the disadvantage that water vapor may be introduced, through condensation of the steam, into the mixture being heated. A slow flow of steam may minimize this difficulty.

Because water condenses in the steam line when it is not in use, it is necessary to purge the line of water before the steam will begin to flow. This purging should be accomplished before the flask to be heated is placed on the steam cone. The steam flow should be started with a high rate to purge the line; then the flow should be reduced to the desired rate. Once the steam cone is heated, a slow steam flow will maintain the temperature of the mixture being heated unless that mixture contains a high-boiling solvent. There is no advantage to having a Vesuvius on your desk! An excessive steam flow may cause problems with condensation, both in the flask being heated and elsewhere in the laboratory. This condensation problem can often be avoided by selecting the correct place to locate the flask on top of the steam cone.

The top of the steam cone, as shown in Figure 1–1, consists of several flat concentric rings. The amount of heat delivered to the flask being heated can be controlled by selecting the correct sizes of these rings. Heating is most efficient when the largest opening that will still support the flask is used. Heating large flasks on a steam bath while using the smallest opening provides slow heating and wastes laboratory time.

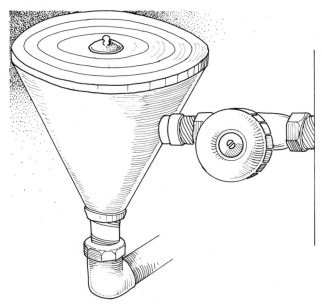

FIGURE 1–1. Steam cone

1.4 OIL BATHS

In some laboratories oil baths may be available. These can be used instead of a flame when one is carrying out a distillation or when one is heating a reaction mixture that needs a temperature above 100 °C. An oil bath with a variable transformer is shown in Figure 1–2. Since the oil is heated electrically by an immersion coil, the danger of fires is minimized. The heating coil must be plugged into a variable transformer, or Variac, which is a source of variable AC voltage. Controlling the immersion coil voltage controls the degree of heating. Because oil baths have a high heat capacity and heat slowly, it is advisable to heat the oil bath partially before the actual time at which it is to be used.

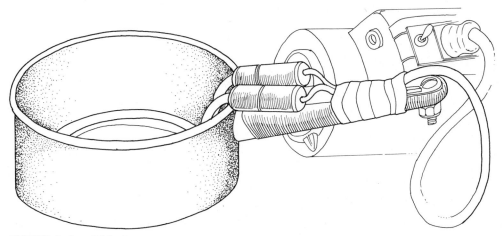

FIGURE 1–2. Oil bath

An oil bath with ordinary mineral oil cannot be used above 200 to 220 °C. Above this temperature the oil bath may "flash," or suddenly burst into flame. A hot oil fire is not extinguished easily. If the oil starts smoking, it may be near its flash temperature; discontinue heating. Old oil, which is dark, is more likely to flash than new oil is. Also, hot oil causes bad burns. Water should be kept away from a hot oil bath, since water in the oil will cause it to splatter. Never use an oil bath when it is obvious that there is water in the oil. If there should be water present, replace the oil before using the heating bath. An oil bath has only a finite lifetime. New oil is clear and colorless but, after extended use, becomes dark brown and gummy from oxidation.

Besides ordinary mineral oil, a variety of other types of oils can be used in an oil bath. Silicone oil does not begin to decompose at as low a temperature as mineral oil does. When silicone oil is heated high enough to decompose, however, its vapors are far more hazardous than mineral oil vapors. The polyethylene glycols may be used in oil baths. They are water-soluble, which makes cleaning up after using an oil bath much easier than with mineral oil. One may select any one of a variety of polymer sizes of polyethylene glycol, depending on the temperature range required. The polymers of large molecular weight are often solid at room temperature. Wax may also be used for higher temperatures, but this material also becomes solid at room temperature. Some workers prefer to use a material that solidifies when not in use since it minimizes both storage and spillage problems. Vegetable shortening is occasionally used in heating baths.

1.5 HEATING MANTLES

A useful source of heat for situations that require temperatures above 100 °C is the heating mantle, illustrated in Figure 1–3. The heating mantle consists of a blanket of spun fiberglass with electric heating coils embedded within the blanket. This blanket fits snugly around the flask, providing an even source of heat. The temperature of a heating mantle is controlled by a Variac. Some heating mantles are designed to fit only

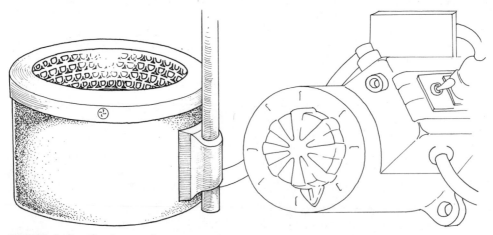

FIGURE 1–3. Heating mantle

specific sizes of flasks; other types fit a range of flask sizes. The heating mantle is safe, because it does not produce flames. It is rapid, since it does not have the high thermal inertia of the oil bath. It is convenient because it is not subject to contamination as the oil bath is, and it does not produce the problem of oil clinging to the outside of the flask. One must be careful, however, in using heating mantles, since there is the danger that the heating mantle may burn out if it is used to heat an empty flask. One should avoid spilling chemicals into the heating mantle. Finally, one additional disadvantage to the heating mantle is its great initial cost.

1.6 HOT PLATES

Occasionally, a hot plate is useful for heating small quantities of solvents when temperatures around 100 °C are needed. Care must be taken with flammable solvents to ensure against fires caused by ''flashing,'' when solvent vapors come into contact with the hot plate surface. One should never evaporate large quantities of a solvent by this method; the fire hazard is too large.

Some hot plates also have built-in magnetic stirring motors. Their use is described in Section 1.8.

1.7 HEATING UNDER REFLUX

Often it is desired to heat a mixture for a long time and to be able to leave it untended. The reflux apparatus (Figure 1–4) allows such heating. It also keeps solvent from evaporating. A condenser is attached to the boiling flask, and cooling water is circulated to condense escaping vapors. One should always use a boiling stone or a magnetic stirrer (see Section 1.8) to keep the boiling solution from ''bumping.'' The direction of the water flow should be such that the condenser will fill with cooling water; the water should enter the bottom of the condenser and leave from the top. The water should flow fast enough to withstand any changes in pressure in the water lines but should not flow any faster than absolutely necessary. An excessive flow rate greatly increases the chance of a flood, and high water pressure may force the hose from the condenser. When a flame is the source of heat, it is convenient to use a wire gauze beneath the flask to provide an even distribution of heat from the flame. In most cases, a heating mantle or a steam cone is preferable. It is essential that the cooling water be flowing before the heating has begun! If the water is to remain flowing overnight, it is advisable to fasten the rubber tubing securely with wire to the condenser.

If the heating rate has been correctly adjusted, the liquid being heated under reflux will travel only partly up the condenser tube before condensing. Below the condensation point, solvent will be seen running back into the flask; above it, the condenser will appear dry. The boundary between the two zones will be clearly demarcated, and a **reflux ring** or a ring of liquid will appear there. In heating under reflux, the rate of heating should be adjusted so that the reflux ring is no higher than a third to a half the distance to the top of the condenser.

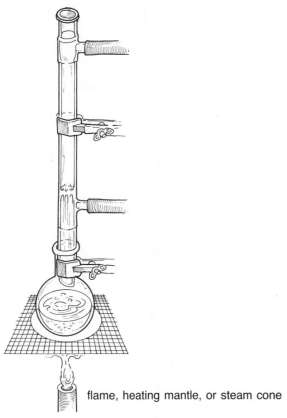

flame, heating mantle, or steam cone

FIGURE 1–4. Heating under reflux

It is possible to heat small amounts of a solvent under reflux in an Erlenmeyer flask. With gentle heating, the evaporated solvent will condense in the relatively cold neck of the flask and return to the solution. This technique, illustrated in Figure 1–5, requires constant attention. The flask must be swirled frequently and removed from the heating stage for a short period if the boiling becomes too vigorous. When heating is in progress, the reflux ring should not rise above the base of the flask's neck.

In Figure 1–5, still another technique for heating small amounts of solvent under reflux is illustrated. A **cold-finger condenser** is inserted into a test tube or a small flask. As the vapors rise, they touch the cold surface of the condenser. The vapors are condensed, and the resulting liquid drips back to the bottom of the container. Some commercial cold-finger condensers are designed to rest on top of the test tube or flask. If pressure builds up in the container, it is released when the condenser is lifted slightly. With cold-finger condensers that fit through cork stoppers, such as the one in the illustration, it is necessary to provide a slot in the stopper to prevent pressure from building within the container. Without the slotted stopper, one would be heating a closed system, creating a potential "bomb"!

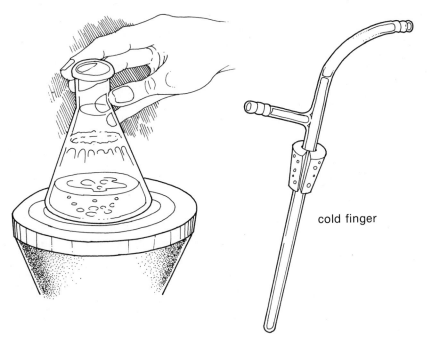

cold finger

FIGURE 1–5.   Tended reflux of small quantities

1.8 BOILING STONES

A boiling stone, also known as a boiling chip or Boileezer, is a small lump of porous material that produces a steady stream of fine air bubbles when it is heated in a solvent. This stream of bubbles, and the turbulence that accompanies it, breaks up large bubbles of gases in the liquid. It also reduces the tendency of the liquid to become superheated, and it promotes the smooth boiling of the liquid. If the liquid becomes superheated, very large bubbles may erupt violently from the solution; this is called **bumping.** The boiling stone, by its action, decreases the chances for bumping.

Boiling stones are generally made from pieces of pumice, carborundum, or marble. Wooden applicator sticks are also used.

Because boiling stones act to promote the smooth boiling of liquids, one should always make certain that the boiling stone has been placed in the liquid **before** the heating is begun. If one waits until the liquid is hot, it may have become superheated, at which time a boiling stone would cause all the liquid to try to boil at once. The liquid, as a result, would erupt entirely out of the flask, or at least froth violently. As soon as boiling ceases, the liquid is drawn into the pores of the boiling stone. When this happens, the boiling stone no longer can produce a fine stream of bubbles; it is spent. A new boiling stone must be added each time the boiling stops.

Magnetic stirrers provide much the same action as boiling stones, since they produce turbulence in the solution. The turbulence breaks up the large bubbles that form in hot solutions. A magnetic stirring system consists of a magnet that is rotated by an electric motor. The rate at which this magnet rotates can be adjusted by a potentiometer. One places a small bar magnet, which is coated with some nonreactive material such as Teflon or glass, into the flask. The magnet within the flask rotates in response to the rotating magnetic field caused by the motor-driven magnet. The result is that the inner bar magnet stirs the solution as it rotates. As previously mentioned, some available hot plates incorporate a magnetic stirring motor, so that heating and stirring can be done simultaneously.

1.9 EVAPORATION TO DRYNESS

Often one wants to evaporate a solution to dryness or to concentrate a solution by removing the solvent either completely or by a particular amount. This can be done by evaporating the solvent from an open Erlenmeyer flask. Such an evaporation must be conducted in a hood, since many solvent vapors are toxic or flammable. A boiling stone must be used. A gentle stream of air directed toward the surface of the liquid will remove vapors that are in equilibrium with the solution and accelerate the evaporation. An eyedropper tube or capillary pipet, connected by a short piece of rubber tubing to the compressed air line, will act as a convenient air nozzle. A tube or an inverted funnel connected to an aspirator may also be used. In this case, vapors are removed by suction. Both methods are illustrated in Figure 1–6. It is better to use an Erlenmeyer flask than a beaker for this procedure, since deposits of solid will usually build up on the sides of the beaker where the solvent evaporates. The refluxing action in an Erlenmeyer flask does not allow this build-up.

It is also possible to remove low-boiling solvents under reduced pressure. In this method, the solution is placed in a filter flask, along with a wooden applicator

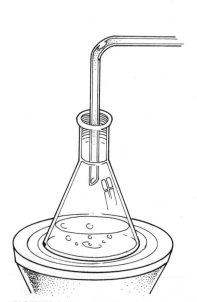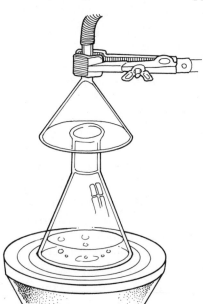

FIGURE 1–6. Rapid evaporation of a solvent

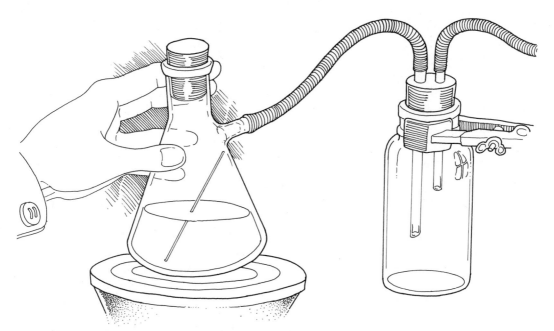

FIGURE 1–7.   Reduced-pressure solvent evaporation

stick. The flask is stoppered, and the sidearm is connected to an aspirator (by a trap), as described in Technique 2, Section 2.3, p. 518. Under reduced pressure, the solvent begins to boil. The wooden stick serves the same function as a boiling stone. By this method, solvent can be evaporated from a solution without heat. This technique, illustrated in Figure 1–7, is often used when heating the solution might decompose thermally sensitive substances. The method has the disadvantage that when low-boiling solvents are used, solvent evaporation cools the flask below the freezing point of water. When this happens, a layer of frost forms on the outside of the flask. Since frost is insulating, it must be removed to keep evaporation proceeding at a reasonable rate. The solution must take heat from the surrounding air to evaporate. The frost prevents this necessary heat transfer. Frost is best removed by one of two methods: either the flask is placed in a bath of warm water (with constant swirling) or it is heated on the steam bath (again with swirling). Either method promotes efficient heat transfer.

Large amounts of a solvent should be removed by distillation (see Technique 6). ONE SHOULD NEVER EVAPORATE ETHER SOLUTIONS TO DRYNESS, except on a steam bath or by the reduced-pressure method. The tendency of ether to form explosive peroxides is a serious potential hazard. If peroxides should be present, the large and rapid temperature increase in the flask once the ether evaporates could bring about the detonation of any residual peroxides. The temperature of a steam bath is not high enough to cause such a detonation.

1.10 COLD BATHS

At times, one may need a medium in which a flask or some other piece of apparatus is to be cooled rapidly to a temperature below room temperature. A **cold bath** is used for this purpose. The most common cold bath is an **ice bath.** An ice bath is a highly convenient source of 0 °C temperatures. The ice bath should actually be called an ice-water bath, since it requires water to work well. An ice bath made up of nothing but ice is not very effective, because the large pieces of ice do not make good contact with the outside walls of the vessel immersed in the bath. Some liquid water must be added to the ice to create an efficient cooling medium. There must be enough water to ensure that the flask being cooled is totally surrounded by water but not so much that the amount of ice present is no longer enough to keep a temperature of 0 °C. If too much water is added to the ice bath, the buoyancy of a flask resting in the ice bath may cause it to tip over. There should be enough ice in the bath to permit the flask to rest firmly.

When temperatures lower than 0 °C are desired, one may add some solid sodium chloride to the ice-water mixture. The ionic salt lowers the freezing point of the ice, so that temperatures in the range 0 to −10 °C can be reached. The lowest temperatures are reached with ice-water mixtures that contain relatively little water.

Very low temperatures can be secured with solid carbon dioxide, or Dry Ice, whose temperature is −78.5 °C. The large chunks of Dry Ice do not provide uniform contact with a flask being cooled. Liquid is required, along with the Dry Ice, to provide uniform contact with the flask and efficient cooling. The liquid used most often with Dry Ice is isopropyl alcohol, although acetone or ethanol can also be used. One should be cautious when handling Dry Ice or cooling baths made from it, since Dry Ice can inflict very severe frostbite.

Extremely low temperatures can be obtained with liquid nitrogen (−195.8 °C). It is not used in organic chemistry as often as Dry Ice.

Technique 2
Filtration

Filtration is a technique used for two main purposes. The first is to remove solid impurities from a liquid or a solution. The second is to collect a solid product from the solution from which it was precipitated or crystallized. Two different kinds of filtration are in general use: gravity filtration and vacuum (or suction) filtration.

2.1 GRAVITY FILTRATION

The most familiar filtration technique is probably filtration of a solution through a paper filter held in a funnel, allowing gravity to draw the liquid through the paper. In general, it is best to use a short-stemmed or a wide-stemmed funnel. In these types of funnels there is less likelihood that the stem of the funnel will become clogged by solid material accumulated in the stem. This clogging is a particular problem if a hot solution saturated with a dissolved solid is being filtered. If the hot saturated solution comes in contact with a relatively cold funnel (or a cold flask, for that matter), the solution is cooled. The rapidly cooled solution will be supersaturated, and crystallization will begin. Crystals will form in the filter, and either they will fail to pass through the filter paper or they will clog the stem of the funnel.

Four other measures are feasible for preventing clogging of the filter. The first is to keep the solution to be filtered at or near its boiling point at all times. The second measure is to preheat the funnel by pouring hot solvent through it before the actual filtration. This keeps the cold glass from causing instantaneous crystallization. The third way is to keep the **filtrate** (filtered solution) in the receiver hot enough to continue boiling **slightly** (by setting it on a steam bath, for example). The refluxing solvent heats the receiving flask and the funnel stem and washes them clean of solids. This boiling of the filtrate also keeps the liquid in the funnel warm. And fourth, it is useful to accelerate filtration by using **fluted filter paper,** as described below. A gravity filtration is shown in Figure 2–1.

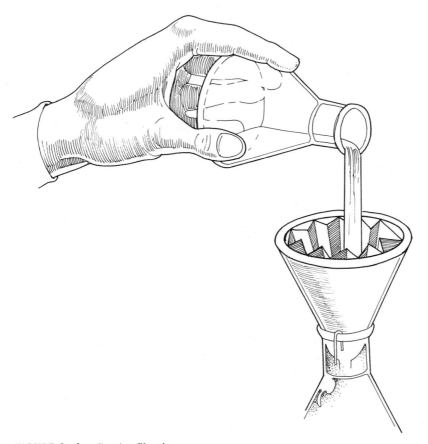

FIGURE 2–1.   Gravity filtration

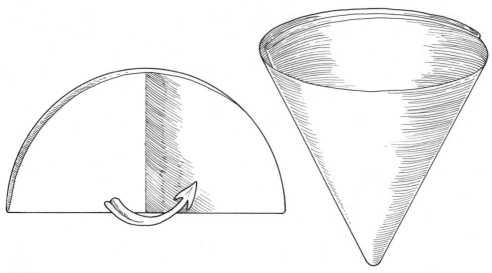

FIGURE 2–2.   Folding a filter cone

A. Filter Cones

The simplest way to prepare filter paper for gravity filtration is to prepare a filter cone, as indicated in Figure 2–2. The filter cone is particularly useful when the solid material being filtered from a mixture is to be collected and used later. The filter cone, because of its smooth sides, can easily be scraped free of collected solids. Because of the many folds, fluted filter paper, described in the next section, cannot be scraped easily.

        With filtrations using a simple filter cone, solvent may form seals between the filter and the funnel and between the funnel and the lip of the flask. When a seal forms, the filtration stops because the displaced air has no possibility of escaping. To avoid the solvent seal, you can insert a small piece of paper or a paper clip or other bent wire between the funnel and the lip of the flask to let the displaced air escape. Alternatively, you can support the funnel by a ring clamp fixed **above** the flask, rather than by placing it in the neck of the flask.

B. Fluted Filters

The technique for folding a fluted filter paper is shown in Figure 2–3. The fluted filter increases the speed of filtration in two ways. First, it increases the surface area of the filter paper through which the solvent seeps; second, it allows air to enter the flask along its sides to permit rapid pressure equalization. If pressure builds up in the flask from hot vapors, filtering slows down. This problem is especially pronounced with filter cones. The fluted filter tends to reduce this problem considerably, but it may be a good idea to use a piece of paper, paper clip, or wire between the funnel and the lip of the flask as an added precaution against solvent seals. Fluted filters are used when the desired material is expected to remain in solution. These filters are used to remove undesired solid materials, such as dirt particles, activated charcoal, and undissolved impure crystals.

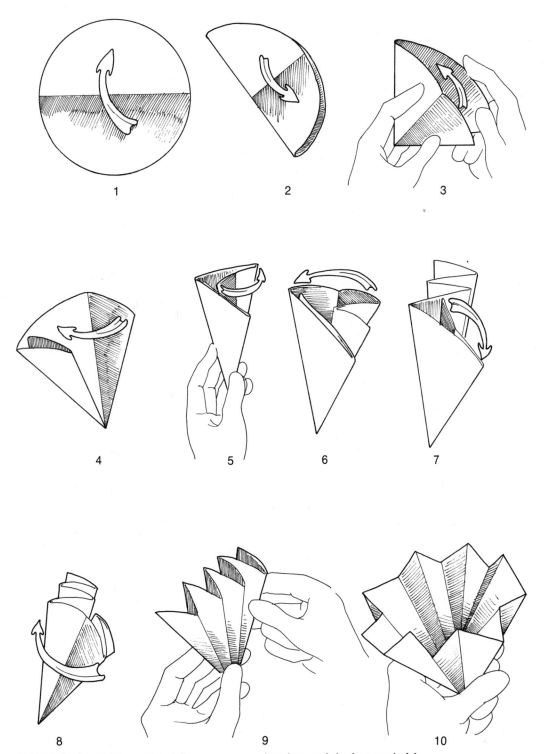

FIGURE 2–3. Folding a fluted filter paper, or origami at work in the organic lab

**TABLE 2–1. Some Common Qualitative Filter Paper Types
and Approximate Relative Speeds and Retentivities**

Fine   High   Slow

Porosity   Retentivity   Speed

Coarse   Low   Fast

SPEED	TYPE (by number)		
	E&D	*S&S*	*Whatman*
Very slow	610	576	5
Slow	613	602	3
Medium	615	597	2
Fast	617	595	1
Very fast	…	604	4

2.2 FILTER PAPER

Many kinds and grades of filter paper are available. Paper is generally available in fine, medium, and coarse porosities. Fine-porosity paper will catch very fine solid particles but generally gives very slow filtration. Coarse paper increases the rate of filtration but may not catch all the particles. The paper must be correct for a given application. In choosing, one should be aware of the various properties of filter paper. **Porosity** is a measure of the size of the particles the paper allows through. Highly porous paper does not remove small particles from solution; paper with low porosity removes very small particles. **Retentivity** is a property that is the opposite of porosity. Paper with low retentivity does not remove small particles from the filtrate. The **speed** of filter paper is a measure of the time it takes a liquid to drain through the filter. Fast paper allows the liquid to drain quickly; with slow paper, it takes much longer to complete the filtration. Since all these properties are related, fast filter paper usually has a low retentivity and high porosity, and slow filter paper usually has high retentivity and low porosity.

Table 2–1 compares some commonly available qualitative filter paper types and ranks them according to porosity, retentivity, and speed. Eaton-Dikeman (E&D), Schleicher and Schuell (S&S), and Whatman are the most common brands of filter paper. The numbers in the table refer to the grades of paper used by each company.

2.3 VACUUM FILTRATION

Vacuum, or suction, filtration is more rapid than gravity filtration, but without specially prepared filter media it does not catch fine particles without clogging the paper pores. In this technique, a receiver flask with a sidearm, a **filter flask,** is used. The sidearm is connected by **heavy-walled** rubber tubing to a source of vacuum. Thin-walled tubing will collapse under vacuum, due to atmospheric pressure on its outside walls, and will seal the vacuum source from the flask. A **Büchner funnel** (see Figure 2–4) is sealed to the filter flask by a rubber stopper or a rubber gasket (Neoprene adapter) cone. The flat bottom of the Büchner funnel is covered with an unfolded piece of circular filter paper, which is held in place by suction. To keep the unfiltered mixture from passing around the edges of the filter paper and contaminating the filtrate in the

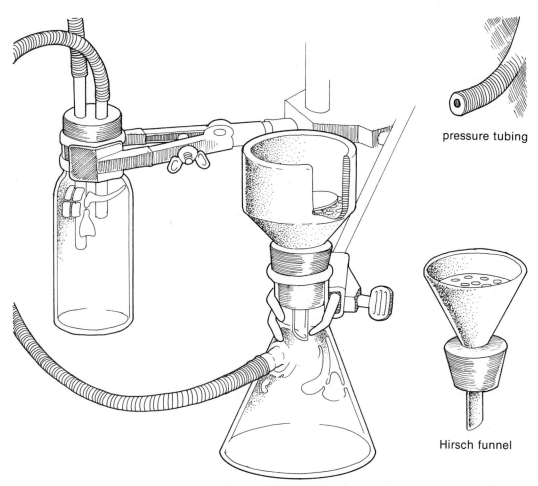

pressure tubing

Hirsch funnel

FIGURE 2–4. Vacuum filtration

flask below, it is advisable to moisten the paper with a small amount of solvent before beginning the filtration. The moistened filter paper adheres more strongly to the bottom of the Büchner funnel. Since the filter flask is attached to a source of vacuum, a solution poured into the Büchner funnel is literally ''sucked'' rapidly through the filter paper. To prevent the escape of solid materials from the Büchner funnel, one must be certain that the filter paper fits the Büchner funnel exactly. The paper must be neither too big nor too small. It must cover all the holes in the bottom of the funnel but not extend up the sides of the funnel.

Two types of funnel are useful for vacuum filtration. The Büchner funnel, which has already been considered, is used for filtering a large amount of crystals from solution. The Hirsch funnel, which is also shown in Figure 2–4, operates on the same principle as the Büchner funnel, except that it is smaller and its sides are sloped rather than vertical. The Hirsch funnel is used for isolating smaller quantities of solid materials from a solution (smaller sizes of the Büchner funnel are also available for this purpose). In the Hirsch funnel also, the filter paper must cover all the holes in the bottom but must not extend up the sides.

2.4 FILTER AID

It was mentioned that specially prepared filter beds are needed to separate fine particles when using vacuum filtration. Often, fine particles either pass right through a paper filter or they clog it so completely that the filtering stops. This is avoided by using a substance called Filter Aid, or Celite. This material is also called **diatomaceous earth,** because of its source. It is a finely divided inert material derived from the microscopic shells of dead diatoms (a type of phytoplankton that grows in the sea).

> **WARNING: LUNG IRRITANT**
>
> **When using Filter Aid, take care not to breathe the dust.**

Filter Aid will not clog the fiber pores of filter paper. It is slurried, or mixed with a solvent to form a rather thin paste, and filtered through a Büchner funnel (with filter paper in place) until a layer of diatoms about 3 mm thick is formed on top of the filter paper. The solvent in which the diatoms were slurried is poured from the filter flask, and if necessary, the filter flask is cleaned before the filtration is begun. Finely divided particles can now be suction-filtered through this layer and will be caught in the Filter Aid. This technique is used for removing impurities and not for collecting a product. The filtrate (filtered solution) is the desired material in this procedure. If the material caught in the filter was the desired material, one would have to try to separate the product from all those diatoms! Filtration with Filter Aid is not appropriate when the desired substance is likely to precipitate or crystallize from solution.

2.5 THE ASPIRATOR

The most common source of vacuum (approximately 10–20 mmHg) in the laboratory is the water aspirator, or "water pump," illustrated in Figure 2–5. This device passes water rapidly past a small hole to which a sidearm is attached. The Bernoulli effect causes a reduced pressure along the side of the rapidly moving water stream and creates a partial vacuum in the sidearm.

A water aspirator can never lower the pressure beyond the vapor pressure of the water used to create the vacuum. Hence, there is a lower limit to the pressure (on cold days) of 9 to 10 mmHg. A water aspirator does not provide as high a vacuum in the summer as in the winter, due to this water-temperature effect.

A trap must be used with an aspirator. A trap is illustrated in Figure 2–4. If the water pressure in the laboratory line drops suddenly, the pressure in the filter flask may suddenly become lower than the pressure in the water aspirator. This would cause water to be drawn from the aspirator stream into the filter flask and would contaminate the filtrate. The trap stops this reverse flow. A similar flow will occur if the water flow at the aspirator is stopped before the tubing connected to the aspirator sidearm is disconnected. ALWAYS DISCONNECT THE TUBING BEFORE STOPPING THE ASPIRATOR. If a "back-up" begins, disconnect the tubing as rapidly as possible

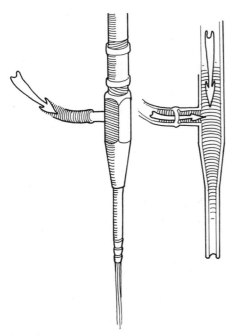

FIGURE 2–5. Aspirator

before the trap fills with water. Some workers like to fit a stopcock into the stopper on top of the trap. A three-hole stopper is required for this purpose. With a stopcock in the trap, the system can be vented before the aspirator is shut off. If the system is vented before the water is shut off, water cannot back up into the trap.

Aspirators do not work well if too many people use the water line at the same time, since the water pressure is lowered. Also, the sinks at the ends of the lab benches or the lines that carry away the water flow may have a limited capacity for draining the resultant water flow from many aspirators. Care must be taken to avoid floods.

2.6 CRUDE FILTRATIONS

Often one wants to make a very rapid filtration to remove dirt or impurities of large particle size from a solution. This is accomplished most easily by laying a loose mat of glass wool in the bottom of an ordinary funnel and pouring the solution through the mat. It may be helpful to decant, or pour off, the clear liquid gently before performing this crude filtration on the solid residue at the bottom of the flask.

Small amounts of solution can be filtered in a similar manner. A plug of glass wool is packed loosely into an eyedropper pipet, and the solution is dropped into the packed pipet from a second pipet.

Technique 3
Crystallization: Purification of Solids

Organic compounds that are solid at room temperature are usually purified by crystallization. The general technique involves dissolving the material to be crystallized in a **hot** solvent (or solvent mixture) and cooling the solution slowly. The dissolved material has a decreased solubility at lower temperatures and will precipitate from the solution as it is cooled. This phenomenon is called **crystallization** if the crystal growth is relatively slow and selective and **precipitation** if the process is rapid and nonselective. Crystallization is an equilibrium process and produces very pure material. A small seed crystal is formed initially, and it then grows layer by layer in a reversible manner. In a sense, the crystal "selects" the correct molecules from the solution. In precipitation, the crystal lattice is formed so rapidly that impurities are trapped within the lattice. Therefore, in any attempt at purification, too rapid a process should be avoided. Too slow a process should also be avoided. The time scale for crystal formation should cover tens of minutes or hours, rather than seconds or days. The two principal mistakes that can be made are (1) cooling the solution too rapidly and (2) suddenly adding an "incompatible" solvent to the solution. Both of these mistakes will be considered in this technique.

3.1 SOLUBILITY

The first problem in performing a crystallization is selecting a solvent in which the material to be crystallized shows the desired solubility behavior. Ideally, the material should be sparingly soluble at room temperature and yet quite soluble at the boiling point of the solvent selected. The solubility curve should be steep, as can be seen in line A of Figure 3–1. A curve with a low slope (line B, Figure 3–1) would not cause

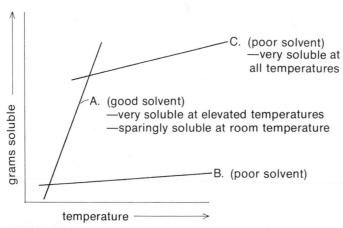

FIGURE 3–1. Graph of solubility versus temperature

TABLE 3–1. Solvents, in Decreasing Order of Polarity

	H_2O	Water
	RCOOH	Organic acids (acetic acid)
	$RCONH_2$	Amides (*N,N*-dimethylformamide)
	ROH	Alcohols (methanol, ethanol)
DECREASING POLARITY (APPROXIMATE)	RNH_2	Amines (triethylamine, pyridine)
	RCOR	Aldehydes, ketones (acetone)
	RCOOR	Esters (ethyl acetate)
	RX	Halides ($CH_2Cl_2 > CHCl_3 > CCl_4$)
	ROR	Ethers (diethyl ether)
	ArH	Aromatics (benzene, toluene)
↓	RH	Alkanes (hexane, petroleum ether)

significant crystallization when the temperature of the solution was lowered. A solvent in which the material was very soluble at all temperatures (line C, Figure 3–1) would not be a suitable crystallization solvent. The basic problem in performing a crystallization is to select a solvent (or mixed solvent) that will provide a steep solubility-versus-temperature curve for the material to be crystallized; that is, a solvent that allows the behavior shown in line A is an ideal crystallization solvent.

The solubility of organic compounds is a function of the polarities of both the solvent and the solute (dissolved material). A general rule states, "Like dissolves like." If the solute is very polar, a very polar solvent will be needed to dissolve it; if it is nonpolar, a nonpolar solvent will be needed. Usually compounds having functional groups that can form hydrogen bonds (for example, —OH, —NH, —COOH, —CONH) will be more soluble in hydroxylic solvents such as water or methanol than in hydrocarbon solvents such as benzene or hexane. However, if the functional group is not a major part of the molecules, this solubility behavior may be reversed. For instance, dodecyl alcohol, $CH_3(CH_2)_{10}CH_2OH$, is almost insoluble in water; its 12-carbon chain causes it to behave more like a hydrocarbon than an alcohol. The list found in Table 3–1 gives an approximate order for decreasing polarity of organic functional groups.

The stability of the crystal lattice also affects solubility. Often, other things being equal, the higher the melting point (the more stable the crystal), the less soluble the compound. For instance, *p*-nitrobenzoic acid (mp 242 °C) is, by a factor of 10, less soluble in a fixed amount of ethanol than the *ortho* (mp 147 °C) and *meta* (mp 141 °C) isomers.

3.2 THEORY OF CRYSTALLIZATION

A successful crystallization depends on a large difference in the solubility of a material in a hot solvent and its solubility in the same solvent cold. Naturally, when the impurities in a substance are equally soluble in both the hot and the cold solvent, an effective purification is not easily achieved through crystallization. A material can be purified by crystallization when both the desired substance and the impurity have similar solubilities, but only when the impurity represents a small fraction of the total solid. The

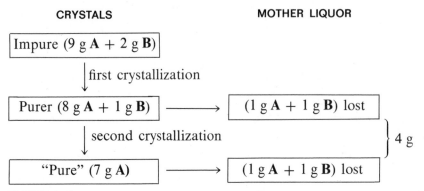

FIGURE 3–2.   Purification of a mixture by crystallization

desired substance will crystallize on cooling but the impurities will not. For example, consider a case in which the solubilities of substance A and its impurity B are both 1 g/100 mL of solvent at 20 °C and 10 g/100 mL of solvent at 100 °C. In an impure sample of A, the composition is given to be 9 g of A and 2 g of B for this particular example. At 20 °C this total amount of material will not dissolve. However, if the solvent is heated to 100 °C, all 11 g dissolve, since the solvent has the capacity to dissolve 10 g of A and 10 g of B at this temperature. If the solution is then cooled to 20 °C, only 1 g of each solute can remain dissolved, so 8 g of A and 1 g of B crystallize, leaving 2 g of material in the solution. This crystallization is shown in Figure 3–2. The solution that remains after a crystallization is called the **mother liquor.** If the process is now repeated by treating the crystals with 100 mL of fresh solvent, 7 g of A will crystallize again, leaving 1 g of A and 1 g of B in the mother liquor. As a result of these operations, 7 g of pure A are obtained, but with the loss of 4 g of material. Again, this second crystallization step is illustrated in Figure 3–2. This result illustrates an important aspect of crystallization—it is wasteful. Nothing can be done to prevent this waste; some A must be lost along with the impurity B for the method to be successful. Of course, if the impurity B were **more** soluble than A in the solvent, the losses would be reduced. Losses could also be reduced if the impurity were present in **much smaller** amounts than the desired material.

It should be noticed that for the above case the method operated successfully because A was present in substantially larger quantity than its impurity B. If there had been a 50/50 mixture of A and B initially, no separation would have been achieved. In general, a crystallization is successful only if there is a **small** amount of impurity. As the amount of impurity increases, the loss of material must also increase. Two substances with nearly equal solubility behavior, present in equal amounts, cannot be separated. If the solubility behavior of two components present in equal amounts is different, however, a separation or purification is frequently possible.

3.3   SELECTING A SOLVENT

A solvent that dissolves little of the material to be crystallized when it is cold but a great deal of the material when it is hot is a good solvent for the crystallization. But the realities of the matter are slightly more complicated than this statement would suggest.

Frequently, one selects a solvent for crystallization by experimenting with various solvents and a very small amount of the material to be crystallized. Such experiments are conducted on a small test-tube scale before the entire quantity of material is committed with a particular solvent. Such trial-and-error methods are common when one is trying to purify a solid material that has not been extensively studied.

With compounds that are well known, such as the compounds that are either isolated or prepared in this textbook, the correct crystallization solvent is already known through the experiments of earlier workers. In such cases, the chemical literature can be consulted to determine which solvent should be used. Such sources as handbooks or tables may also provide this information. Quite often, the correct crystallization solvent is indicated in the experimental procedures in this textbook.

One note of caution is pertinent to choosing a crystallization solvent. Care should be taken not to pick one whose boiling point is higher than the melting point of the substance to be crystallized. If the boiling point of the solvent is too high, the solid may melt in the solvent rather than dissolve. In such a case, the solid may "oil out." Oiling out occurs when the solid substance melts to form a liquid that is not soluble in the solvent. On cooling, the liquid refuses to crystallize; rather, it becomes a supercooled liquid, or oil. Such an oil may be solidified if the temperature is lowered sufficiently, but it will not crystallize. A solidified oil becomes an amorphous solid or a hardened mass—a condition that does not result in the purification of the substance. It is very difficult to deal with oils in the laboratory. One must try to redissolve them and hope that they will precipitate as crystals with careful cooling.

One additional criterion for selecting the correct crystallization solvent is the **volatility** of that solvent. A solvent with a low boiling point may be removed from the crystals through evaporation without much difficulty. It will be difficult to remove a solvent with a high boiling point from the crystals without heating them under vacuum.

Table 3–2 lists common crystallization solvents. The solvents used most commonly are listed first in the table.

TABLE 3–2. Common Solvents for Crystallization

	BOILS (°C)	FREEZES (°C)	SOLUBLE IN H_2O	FLAMMABILITY
Water	100	0	+	−
Methanol	65	*	+	+
95% Ethanol	78	*	+	+
Ligroin	60–90	*	−	+
Benzene†	80	5	−	+
Chloroform†	61	*	−	−
Acetic acid	118	17	+	+
Dioxane†	101	11	+	+
Acetone	56	*	+	+
Diethyl ether	35	*	Slightly	++
Petroleum ether	30–60	*	−	++
Methylene chloride	41	*	−	−
Carbon tetrachloride†	77	*	−	−

*Lower than 0 °C (ice temperature).
†Suspect carcinogen.

3.4 TECHNIQUE AND METHOD

Dissolving the Solid. To minimize losses of material to the mother liquor, it is desirable to **saturate** the boiling solvent with solute. This solution, when cooled, will return the maximum possible amount of solute as crystals. To achieve this high return, the solvent is brought to its boiling point, and the solute is dissolved in the MINIMUM AMOUNT(!) OF BOILING SOLVENT. For this procedure it is advisable to maintain a container of boiling solvent. From this container, a small portion (a few milliliters) of the solvent is added to the flask containing the solid to be crystallized, and this mixture is heated until it resumes boiling. If the solid has totally dissolved in this small portion of solvent, too much has been used and the solvent must be partially evaporated. If the solid has not dissolved in this portion of boiling solvent, then another small portion of boiling solvent is added to the flask. The mixture is heated again until it resumes boiling. If the solid has dissolved, no more solvent is added. But if the solid has not dissolved, another portion of boiling solvent is added, as before, and the process is repeated until the solid dissolves. It is important to stress that the portions of solvent added each time are small, so that only the **minimum** amount of solvent necessary for dissolving the solid is added. It is also important to stress that the procedure requires the addition of solvent to solid. One must never add portions of solid to a fixed quantity of boiling solvent. By this latter method it may be impossible to tell when saturation has been achieved.

Occasionally, one encounters an impure solid that contains small particles of insoluble impurities, pieces of dust, or paper fibers that will not dissolve in the hot crystallizing solvent. A common error is to add too much of the hot solvent in an attempt to dissolve these small particles, without realizing that they are not soluble. In such cases, one must be careful not to add too much solvent. It is probably better to add too little solvent and not dissolve all the desired solid than to add too much solvent and lower the yield of solid returned as crystals. Once the solid has dissolved, decolorizing charcoal is added **if it is required** (see Section 3.5).

Filtration. The hot solution is filtered if any insoluble matter remains or if charcoal has been used. A gravity filtration through a fluted filter paper is preferred for this step (see Technique 2, Section 2.1, p. 514). In some cases, especially if the solvent is low-boiling (boiling point lower than 50 °C), it may be advisable to add a small amount of extra solvent to the hot solution. This procedure ensures that crystals will not form in the filter paper during the gravity filtration. It will be necessary to remove this extra solvent by evaporation once the filtration is complete (the amount of extra solvent added should have been noted). For the gravity filtration, a stemless funnel is heated with boiling solvent before use (Figure 3–3). The funnel may be preheated by pouring hot solvent through it as it rests within a beaker. After this step, the funnel is fitted with a fluted filter as rapidly as possible and installed at the top of the Erlenmeyer flask to be used for the actual filtration. In this method, the solvent used to preheat the funnel is discarded. Alternatively, or in addition, the funnel and filter paper can be preheated by placing the assembly in the neck of the Erlenmeyer flask and resting the entire apparatus on top of a steam bath or hot plate. The material to be filtered is brought to its boiling point and poured into the filter. The solvent that first passes through the filter into the flask will begin to boil. The hot vapors will rise around the funnel and heat it.

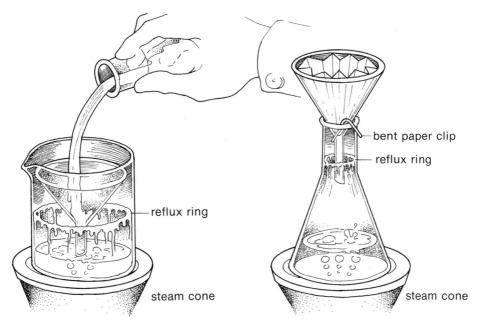

FIGURE 3–3. Methods of preheating a funnel

These preheating operations prevent the formation of crystals in the filter and also prevent the clogging of the filter that such crystallization would cause.

During the filtration, the **hot** solution is poured through the filter in portions. It is necessary to keep the solutions in both flasks at their boiling temperatures to prevent premature crystallization. The refluxing action of the filtrate keeps the funnel warm and reduces the chance that the filter will clog with crystals that may have formed during the filtration. With low-boiling solvents, one must be aware that some solvent may be lost through evaporation. Consequently, extra solvent must be added to make up for this loss. It is advisable to place a small piece of wire between the funnel and the mouth of the flask to relieve any buildup of pressure caused by the boiling solvent. This filtration procedure is illustrated in Figure 3–4. It should be stressed that if the solution is clear and colorless, if all the solid material has dissolved, or if decolorizing charcoal was not used, this gravity filtration is not necessary. Finally, if any extra solvent was added to prevent the formation of crystals in the filter, this extra solvent should be removed by evaporation before the crystals are allowed to form.

If the crystals begin to crystallize in the filter during gravity filtration, a minimum amount of boiling solvent is added to redissolve the crystals and to allow the solution to pass through the funnel. After the filtration, the filtrate is boiled until the extra amount of solvent required to redissolve crystals caught in the filter has evaporated. Once this extra amount of solvent has evaporated, the solution is set aside to cool until crystals are formed.

Crystallization. It is always a good idea to use an Erlenmeyer flask, not a beaker, for crystallization. The large open top of a beaker makes it an excellent dust catcher. The narrow opening of the Erlenmeyer flask reduces contamination by dust and allows the flask to be stoppered if it is to be set aside for a long period. Mixtures set

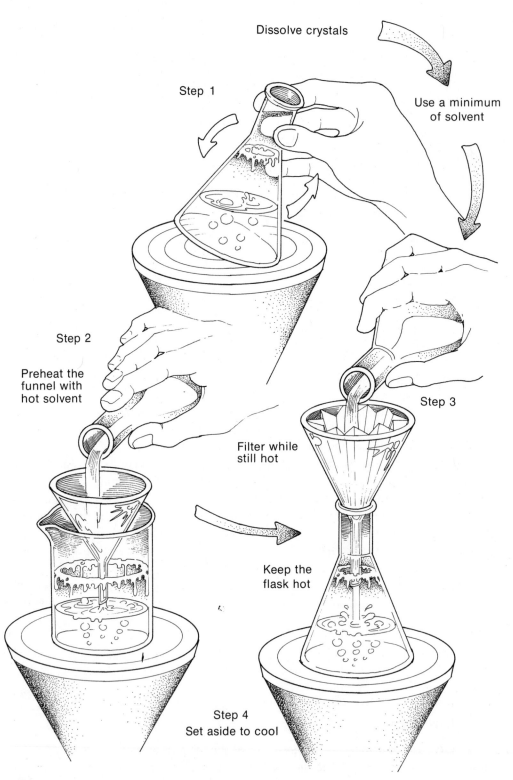

Dissolve crystals

Step 1

Use a minimum
of solvent

Step 2

Preheat the
funnel with
hot solvent

Filter while
still hot

Step 3

Keep the
flask hot

Step 4
Set aside to cool

FIGURE 3–4. Filtration and crystallization

aside for long periods must be stoppered to prevent evaporation of solvent. If all the solvent evaporates, no purification is achieved, and the crystals originally formed become coated with the dried contents of the mother liquor.

If the cooled solution does not crystallize, several techniques may be used to induce crystallization. First, one should try vigorous scratching of the inside surface of the flask with a glass rod that **has not been** fire-polished. The motion of the rod should be vertical (in and out of the solution) and should be vigorous enough to produce an audible scratching. Such scratching often induces crystallization. The effect is not well understood, although two explanations have been proposed. The high-frequency vibrations may have something to do with initiating crystallization; or perhaps—a more likely possibility—small amounts of solution dry by evaporation on the side of the flask, and the solute is pushed into the solution. These small amounts of material provide ''seed crystals,'' or nuclei, on which crystallization may begin.

A second technique that can be used to induce crystallization is to cool the solution in an ice bath. This decreases the solubility of the solute. A third technique is useful when small amounts of the original material to be crystallized are saved. The saved material can be used to ''seed'' the cooled solution. A small crystal dropped into the cooled flask often will start the crystallization—this is called **seeding.**

Isolation of Crystals.   The crystals are collected by vacuum filtration through a Büchner or a Hirsch funnel (see Technique 2, Section 2.3, p. 518, and Figure 2–4, p. 519). The crystals should be washed with a small amount of **cold** solvent to remove any mother liquor adhering to their surface. Hot or warm solvent will dissolve some of the crystals. The crystals should then be left for a short time in the funnel, where air as it passes will dry them free of solvent. It is often wise to cover the Büchner (or Hirsch) funnel with an oversize filter paper or towel during this air-drying. This prevents accumulation of dust in the crystals.

3.5 DECOLORIZATION

Small amounts of highly colored impurities may make the original crystallization solution appear colored; this color can often be removed by **decolorization,** using activated charcoal (called Norit). As soon as the solute is dissolved in the minimum amount of boiling solvent, a small amount of Norit is added to the boiling mixture. The Norit adsorbs the impurities. A reasonable amount of Norit would be what could be held on the end of a small nickel spatula, or about 0.25 g (⅛ tsp). If too much Norit is used, it will adsorb product as well as impurities. A small amount of Norit should be used, and its use should be repeated if necessary. Caution should be exercised that the solution does not froth or erupt when the finely divided charcoal is added. The boiling mixture is filtered by gravity, using a fluted filter (see Technique 2, Section 2.1, p. 514, and Technique 3, Section 3.4, p. 526), and the crystallization is carried forward as described in Section 3.4. The Norit is very finely divided and is removed most effectively by gravity filtration through fluted filter paper. If a Büchner funnel is used to suction-filter the mixture, a layer of Celite should be used to trap the fine particles of charcoal (see Technique 2, Section 2.4, p. 520). Decolorizing charcoal usually passes right through filter paper when suction is used. Filtration by suction through Celite is used only when the solute cannot precipitate or if excess solvent has been used. If excess solvent has been used, it is removed by evaporation after the filtration has been completed. The Norit preferentially adsorbs the colored impurities and removes them from the solution. The technique seems to be most effective with hydroxylic solvents.

In using Norit, one should be careful not to breathe the dust. Normally, such small quantities are used that little risk of lung irritation exists.

3.6 DRYING CRYSTALS

The most common method of drying crystals involves placing them in a watch glass or an evaporating dish and allowing them to dry in air. While the advantage of this method is that heat is not required, thus reducing the danger of decomposition or melting, exposure to atmospheric moisture may cause the hydration of strongly **hygroscopic** materials. A hygroscopic substance is one that absorbs moisture from the air.

Another method of drying crystals is to place the crystals on a watch glass or on a piece of absorbent paper in an oven. Although this method is simple, some possible difficulties deserve mention. Crystals that sublime readily should not be dried in an oven, since they might pass into the vapor state and disappear. Care should be taken that the temperature of the oven does not exceed the melting point of the crystals. One must remember that the melting point of crystals is lowered by the presence of solvent, and one must allow for this melting-point depression when selecting a suitable oven temperature. Some materials decompose on exposure to heat, and they should not be dried in an oven. Finally, when many different samples are being dried in the same oven, crystals might be lost due to confusion or reaction with another person's sample. It is important to label the crystals when they are placed in the oven.

A third method, which requires neither heat nor exposure to atmospheric moisture, is the use of a vacuum desiccator. In a desiccator, the sample is placed under vacuum in the presence of a drying agent. Two potential problems must be noted. The first deals with samples that sublime readily. Under vacuum, the likelihood of sublimation is increased. The second problem deals with the vacuum desiccator itself. Since the surface area of glass that is under vacuum is large, there is some danger that the desiccator could implode. A vacuum desiccator should never be used unless it has been placed within a protective metal container (cage).

3.7 MIXED SOLVENTS

Often the desired solubility characteristics for a particular compound are not found in a single solvent. In these cases a mixed solvent may be used. One simply selects a first solvent in which the solute is soluble and a second solvent, miscible with the first, in

TABLE 3–3. Common Solvent Pairs for Crystallization

Methanol–Water	Ether–Acetone
Ethanol–Water	Ether–Petroleum ether
Acetic acid–Water	Benzene*–Ligroin
Acetone–Water	Methylene chloride–Methanol
Ether–Methanol	Dioxane*–Water

*Suspect carcinogen.

which the solute is relatively insoluble. The compound is dissolved in a minimum amount of the boiling solvent in which it is soluble. Following this, the second hot solvent is added to the boiling mixture, dropwise, until the mixture barely becomes cloudy. The cloudiness indicates precipitation. At this point, more of the first solvent should be added. Just enough is added to clear the cloudy solution. At that point the solution is saturated, and as it is cooled, crystals should separate. Common solvent mixtures are listed in Table 3–3. Benzene–ligroin and methanol–water are the most common solvent mixtures.

It is important not to add an excess of the second solvent or to cool the solution too rapidly. Either of these actions may cause the solute to oil out, or separate as a viscous liquid. If this happens, one should reheat the solution and add more of the first solvent.

Steps in a Crystallization

A. DISSOLVING
1. Find a solvent with a steep solubility-vs-temperature characteristic. (Done by trial and error, using small amounts of material, or by consulting a handbook.)
2. Heat the desired solvent to its boiling point.
3. Dissolve the solid in a **minimum** of boiling solvent.
4. Add decolorizing charcoal if necessary.
5. Filter the hot solution through a preheated funnel to remove insoluble impurities or charcoal. If no decolorizing charcoal has been added or if there are no undissolved particles, this filtration may be omitted.
6. Allow the solution to cool.
7. If crystals do not appear, go to part B; if crystals do appear, go to part C.

B. INDUCING CRYSTALLIZATION
1. Scratch the flask with a glass rod or
2. Seed the solution or
3. Cool the solution in an ice-water bath.

C. COLLECTING
1. Collect crystals by vacuum filtration, using a Büchner (or a Hirsch) funnel.
2. Rinse crystals with a small portion of **cold** solvent.
3. Continue suction until crystals are dry.

D. DRYING
1. Air-dry the crystals or
2. Place the crystals in a drying oven or
3. Dry the crystals in a vacuum desiccator.

PROBLEMS

1. Listed below are solubility-vs-temperature data for an organic substance A dissolved in water.

TEMPERATURE (°C)	SOLUBILITY OF A IN 100 mL OF WATER
0	1.5 g
20	3.0
40	6.5
60	11.0
80	17.0

   (a) Graph the solubility of A vs temperature. Use the data given above. Connect the data points with a smooth curve.
   (b) Suppose 0.5 g of A and 5 mL of water were mixed and heated to 80 °C. Would all the substance A dissolve?
   (c) The solution prepared in (b) is cooled. At what temperature will crystals of A appear?
   (d) Suppose the cooling described in (c) were continued to 0 °C. How many grams of A would come out of solution? Explain how you obtained your answer.

2. What would be likely to happen if a hot saturated solution were filtered by vacuum filtration using a Büchner funnel?

3. (a) Draw a graph of a cooling curve (temperature vs time) for a solution of a solid substance that shows no supercooling effects. Assume that the solvent does not freeze.
   (b) Repeat the above instructions for a solution of a solid substance that shows some supercooling behavior but eventually yields crystals if the solution is cooled sufficiently.

4. A solid substance A is soluble in water to the extent of 1 g/100 mL of water at 25 °C and 10 g/100 mL of water at 100 °C. You have a sample that contains 10 g of A and an impurity B.
   (a) Assuming that 0.2 g of the impurity B is present along with 10 g of A, describe how you could purify A if B is completely insoluble in water.
   (b) Assuming that 0.2 g of the impurity B is present along with 10 g of A, describe how you could purify A if B had the same solubility behavior as A. Would one crystallization produce absolutely pure A?
   (c) Assume that 3 g of the impurity B is present along with 10 g of A. Describe how you could purify A if B had the same solubility behavior as A. Each time, use the correct amount of water to just dissolve the solid. Would one crystallization produce absolutely pure A? How many crystallizations would be needed to produce pure A? How much A would have been recovered when the crystallizations had been completed?

Technique 4
Melting Point: An Index of Purity

4.1 INTRODUCTION

The primary index of purity used by an organic chemist for a crystalline compound is its melting point. A small amount of material is heated **slowly** in a special apparatus equipped with a thermometer, a heating coil or a heating bath, and usually a magnifying eyepiece for observing the sample. Two temperatures are noted. The first is the point at which the first drop of liquid forms among the crystals; the second is the point at which the whole mass of crystals turns to a **clear** liquid. The melting point is then recorded, giving this range of melting. One might say, for example, that the melting point of a substance is 51 to 54 °C. That is, the substance melted over a 3° range.

The melting point of a **pure** crystalline substance is a physical property of that substance. Since the vapor pressure of a solid, compared with a liquid, is low, the melting point is usually insensitive to changes of pressure (within reasonable limits). This melting point can be used to identify a given substance.

4.2 MELTING-POINT BEHAVIOR

The melting point (or range) indicates purity in two ways. First, the purer the material, the higher its melting point. Second, the purer the material, the narrower the melting-point range. Adding successive amounts of an impurity to a pure substance generally causes its melting point to decrease in proportion to the amount of impurity. The reason is that the freezing point of a substance is lowered when a foreign substance is added. The freezing point is simply the melting point (solid → liquid) being approached from the opposite direction (liquid → solid). In Figure 4–1, the usual melting-point behavior of various mixtures of two substances, A and B, is depicted. The two extreme tempera-

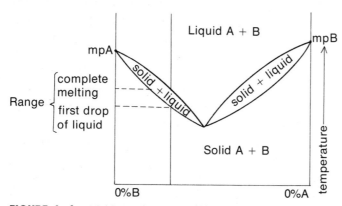

FIGURE 4–1. Melting point–composition curve

tures of the melting range for various mixtures of A and B are shown. The upper curves indicate the temperature at which all the sample has melted. The lower curve indicates the temperature at which melting is observed to begin. If one begins with pure A, the melting point decreases as impurity B is added. At some particular proportion of the two substances, a minimum temperature is reached, and the melting point begins to increase to that of a pure substance, B. In general, the melting-point depression curves for A + impurity B and B + impurity A always intersect at some particular composition, and a minimum melting point is reached. In the curve in Figure 4–1, the vertical distance between the convex and concave curves represents the **melting range.** One may consider a mixture containing substance A together with a relatively small amount of B. In this mixture, the melting point would be lowered, and the range of the melting point would be increased. This example illustrates that for mixtures containing small amounts of impurity (<15%), the **melting-point range often indicates purity.** A substance that melts within a narrow range should be pure. However, at the minimum point of the melting point–composition curve, the mixture often forms a **eutectic,** which also melts sharply. Not all binary mixtures form eutectics, and some caution must be exercised in assuming that every binary mixture follows the behavior described above. Some compounds form more than one eutectic. In spite of these variations, both the melting point and its range are useful indications of purity and are easily determined experimentally.

4.3 THEORY

Figure 4–2 is a phase diagram describing the behavior of the usual two-component mixture (A + B) on melting. The behavior on melting depends on the relative amounts of A and B in the mixture. If A is a pure substance (no B), then A melts sharply at its melting point, t_A. This is represented by point A on the left side of the diagram. When B is a pure substance, it melts at t_B; its melting point is represented by point B on the right side of the diagram. At either point A or point B, the pure solid passes cleanly, with a narrow range, from solid to liquid.

In mixtures of A and B, the behavior is different. Using Figure 4–2, consider a mixture of 80% A and 20% B on a mole per mole basis (that is, mole percentage). The melting point of this mixture is given by t_M at point M on the diagram. That is, adding B to A has lowered the melting point of A from t_A to t_M. It has also expanded the melting range. The temperature of t_M corresponds to the **upper limit** of the melting range.

Lowering the melting point of A by adding impurity B comes about in the following way. Substance A has the lower melting point in the phase diagram shown, and with heating, it begins to melt first. As A begins to melt, solid B begins to dissolve in the liquid A that is formed. When solid B dissolves in liquid A, the melting point is depressed. To understand this, consider the melting point from the opposite direction. When a liquid starts at a high temperature, as it cools it reaches a point at which it solidifies, or "freezes." The temperature at which the liquid freezes is identical to its melting point. It will be recalled that the freezing point of a liquid can be depressed by

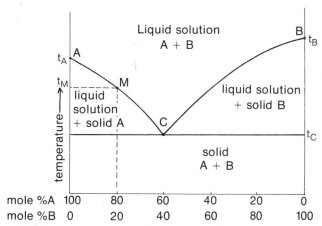

FIGURE 4–2. Phase diagram for melting in a two-component system

adding an impurity. Since the freezing point and the melting point are identical, lowering the freezing point corresponds to lowering the melting point. In parallel with lowering the freezing point, as more impurity is added to a solid, the lower its melting point becomes. There is, however, a limit to how far the melting point can be depressed. One cannot dissolve an infinite amount of impure substance in a liquid. At some point the liquid will become saturated with the impure substance. The solubility of B in A has an upper limit. In Figure 4–2, the solubility limit of B in liquid A is reached at point C, the **eutectic point.** The melting point of the mixture cannot be depressed below t_C, the melting temperature of the eutectic.

Now consider what happens when the melting point of a mixture of 80% A and 20% B is approached. As the temperature is increased, A begins to ''melt.'' This is not really a visible phenomenon; it happens before liquid is visible. It is a softening of the compound to a point at which it can begin to mix with the impurity. As A begins to soften, it dissolves B. As it dissolves B, the melting point is lowered. The lowering continues until all B is dissolved or until the eutectic composition is reached. When the maximum possible amount of B has been dissolved, actual melting begins, and one can observe the first appearance of liquid. The initial temperature of melting will be below t_A. The amount below t_A at which melting begins is determined by the amount of B dissolved in A but will never be below t_C. Once all B has been dissolved, the melting point of the mixture begins to rise as more A begins to melt. As more A melts, the semisolid solution is diluted by A, and its melting point rises. While all this is happening, one can observe **both** solid and liquid in the melting-point capillary. Once all A has begun to melt, the composition of the mixture M becomes uniform and will reach 80% A and 20% B. At this point, the mixture finally melts sharply, giving a clear solution. The maximum melting point will be t_M, since t_A is depressed by the impurity B that is present. The lower end of the melting range will always be t_C; however, melting will not always be observed at this temperature. An observable melting at t_C only comes about when a large amount of B is present. Otherwise, the amount of liquid formed at t_C will be too small to observe. Therefore, the melting behavior that is **actually** observed will have a smaller range, as shown in Figure 4–1.

4.4 MIXED MELTING POINTS

The melting point can be used as supporting evidence in identifying a given compound if an authentic sample of that compound is available for comparison. If the object of the experiment is to confirm that A and B are identical substances since each individually exhibits similar or identical melting points, they are pulverized finely and mixed in equal quantities. Then the melting point of the mixture is determined. If there is a melting-point depression or if the range of melting is expanded by a large amount, one can conclude that one substance has acted as an impurity toward the other and they are not the same compound. If there is no depression of the melting point for the mixture (the melting point is identical with that of either pure A or pure B), then A and B are almost certainly the same compound.

4.5 PACKING THE MELTING-POINT TUBE

Melting points usually are determined by heating the sample in a piece of thin-walled capillary tubing (1 mm × 100 mm) that has been sealed at one end. To pack the tube, one presses the open end gently into a pulverized sample of the crystalline material. Crystals will stick in the open end of the tube. The amount of solid pressed into the tube should correspond to a column no more than 1 to 2 mm high. To transfer the crystals to the closed end of the tube, one drops the capillary, closed end first, down a ⅔-m length of glass tubing, which is held upright on the desk top. When the capillary tube hits the desk top, the crystals will pack down into the bottom of the tube. This procedure is repeated if necessary. Tapping the capillary on the desk top with the fingers is not recommended, since it is easy to drive the small tubing into a finger if the tubing should break.

4.6 DETERMINING THE MELTING POINT

There are two principal types of melting-point apparatus available: the Thiele tube and a commercially available, electrically heated apparatus. The Thiele tube, shown in Figure 4–3, is the simpler device, and it is widely available. It is a glass tube designed to contain heating oil and a thermometer to which a capillary tube containing the sample is attached. The shape of the Thiele tube allows convection currents to form in the oil when it is heated. These currents maintain a uniform temperature distribution throughout the oil in the tube. The sidearm of the tube is designed to generate these convection currents and thus transfer the heat from the flame evenly and rapidly throughout the heating oil. The sample, which is in a capillary tube attached to the thermometer, is held by a rubber band or a small slice of rubber tubing. It is important that this rubber band be above the level of the oil (allowing for expansion of the oil on heating), so that the oil does not soften the rubber and allow the capillary tubing to fall into the oil.

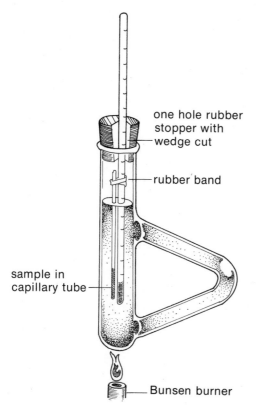

one hole rubber
stopper with
wedge cut

rubber band

sample in
capillary tube

Bunsen burner

FIGURE 4–3.   Thiele tube

The Thiele tube is usually heated by a microburner. During the heating, the rate of temperature increase should be regulated. Usually one holds the burner by its base and, using a gentle flame, moves the burner slowly back and forth along the bottom of the arm of the Thiele tube. If the heating is too fast, the burner is removed for a few seconds, and then the heating resumed. The rate of heating should be **low** near the melting point (about 1 °C per minute) to ensure that the temperature increase is not faster than the rate at which heat can be transferred to the sample being observed. At the melting point, it is necessary that the mercury in the thermometer and the sample in the capillary tube be at temperature equilibrium.

The need for careful heating near the melting point also applies when using the electrical apparatus. Such an apparatus is operated by moving the switch to the ON position, adjusting the potentiometric control for the desired rate of heating, and observing the sample through the magnifying eyepiece. The temperature is read from a thermometer. Two examples of such an apparatus are shown in Figure 4–4. There are many types and styles of melting-point apparatus. Your instructor will demonstrate and explain the type used in your laboratory.

Most electrical apparatuses do not heat or increase the temperature of the substance linearly. Although in the early heating stages the rate of heating is linear, it usually decreases and leads to a constant temperature at some upper limit. The upper-

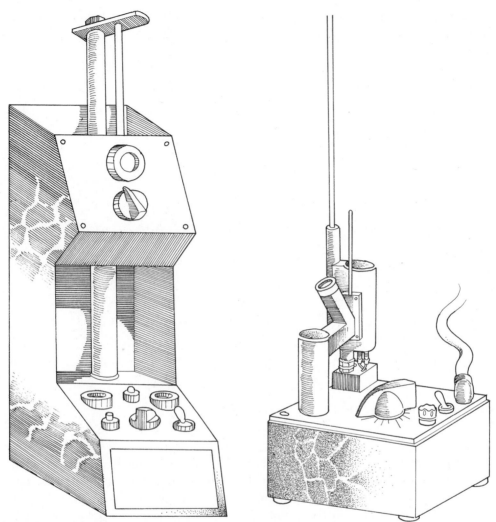

FIGURE 4–4.   Melting-point apparatus

limit temperature is determined by the setting of the heating control. Thus, a family of heating curves is usually obtained for various control settings, as shown in Figure 4–5. The four hypothetical curves (1 through 4) shown might correspond to different control settings. For a compound melting at temperature t_1, the setting corresponding to curve 3 would be ideal. In the beginning of the curve, the temperature is increasing too rapidly to allow determination of an accurate melting point, but after the change in slope, the temperature increase will have slowed to a more usable rate.

       If the melting point of the substance is unknown, one can often save time by preparing two samples for melting-point determination. With one sample, one rapidly determines a crude melting-point value. Then the experiment is repeated more carefully, using the second sample. For the second determination, one already has some approximate idea of what the melting temperature should be.

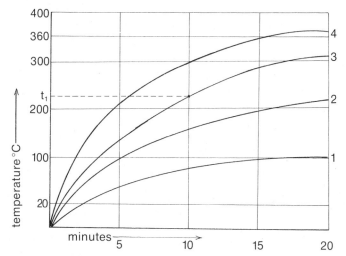

FIGURE 4–5. Heating-rate curves

4.7 DECOMPOSITION, DISCOLORATION, SOFTENING, AND SHRINKAGE

Many solid substances undergo some degree of unusual behavior before melting. At times it may be difficult to distinguish these other types of behavior from actual melting. One should learn, through experience, how to recognize melting and how to distinguish it from decomposition, discoloration, and particularly softening and shrinkage.

Some compounds decompose on melting. This decomposition is usually evidenced by discoloration of the sample. Frequently this decomposition point is a reliable physical property to be used in lieu of an actual melting point. Such decomposition points are listed in tables of physical properties. An example of a decomposition point is given for thiamine hydrochloride, whose melting point would be listed as 248° d, indicating that this substance melts with decomposition at 248 °C.

Some substances may begin to decompose at a temperature **below** their melting point. Thermally unstable substances may undergo elimination reactions or anhydride formation reactions during heating. The decomposition products thus formed represent impurities in the original sample, so the melting point of the substance is artificially lowered due to such decomposition.

It is normal for compounds to soften or shrink immediately before melting. Such behavior represents not decomposition but a change in the crystal structure. Some substances tend to "sweat," or release solvent of crystallization, before melting. Such changes do not indicate the beginning of melting. Actual melting begins when the first drop of liquid becomes visible, and the melting range continues until the temperature at which all the solid has been converted to the liquid state. With experience, one soon learns to distinguish between softening, or "sweating," and actual melting.

Some substances have such a high vapor pressure that they sublime at or before the melting point. In such cases, the melting-point determination must be conducted in sealed capillary tubes.

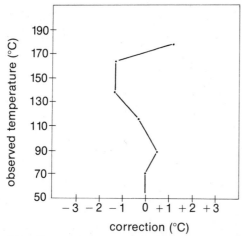

FIGURE 4–6.   Thermometer calibration curve

4.8   THERMOMETER DEVIATIONS

When the melting-point determination has been completed, one expects to obtain a result that exactly duplicates the result recorded in a handbook or in the original chemical literature. It is not infrequent, however, that there will be a discrepancy of one or two degrees from the literature value. Such a discrepancy does not necessarily indicate that the experiment was incorrectly performed; rather it may indicate that the thermometer used for the determination is a source of systematic error. Most thermometers used in the laboratory do not measure the temperature with perfect accuracy; their readings are more likely to be somewhat inaccurate.

To determine accurate melting-point values, one must calibrate the thermometer that is going to be used for the melting-point determination. This calibration is done by determining the melting point of a variety of standard substances with the thermometer. A plot is drawn of the observed temperature vs the correction needed to duplicate the published melting point of each standard substance, as in the example shown in Figure 4–6. The correction obtained in this way is applied to each melting point determined with that particular thermometer. A list of suitable standard substances for calibrating thermometers is given in Table 4–1.

TABLE 4–1.   Melting-Point Standards

COMPOUND	MELTING POINT (°C)
Ice (solid-liquid water)	0
Acetanilide	115
Benzamide	128
Urea	132
Succinic acid	189
3,5-Dinitrobenzoic acid	205

Technique 5
Extraction, The Separatory
Funnel, Drying Agents

5.1 DISTRIBUTION COEFFICIENT

When a solution (solute A in solvent 1) is shaken with a second solvent (solvent 2) with which it is immiscible, the solute distributes itself between the two liquid phases. When the two phases have separated again into two distinct solvent layers, an equilibrium will have been achieved such that the ratio of the concentrations of the solute in each layer defines a constant. This constant, called the **distribution coefficient** (or partition coefficient) K is then defined by

$$K = \frac{C_2}{C_1}$$

where C_1 and C_2 are the concentrations at equilibrium, in grams per liter, of the solute A in solvent 1 and in solvent 2, respectively. This relation is independent of the total concentration and the actual amounts of the two solvents mixed. The distribution coefficient has a constant value for each solute considered and depends on the nature of the solvents used in each case. The solute distributes itself between the two solvents so that its chemical activity (effective concentration) is the same in each phase.

5.2 EXTRACTION

Transferring a solute from one solvent to another is called **extraction.** The solute is extracted from one solvent into the other by the distribution described in Section 5.1 (see Figure 5–1). Extraction is used for many purposes in organic chemistry. Many **natural products** (chemicals that exist in nature) are present in animal and plant tissues having high water content. Extracting these tissues with a water-immiscible solvent is useful for isolating the natural products. Often ether is used for this purpose. Sometimes, alternative water-immiscible solvents such as hexane, petroleum ether, ligroin, benzene, chloroform, methylene chloride, and carbon tetrachloride are used. For instance, caffeine, a natural product, can be extracted from an aqueous tea solution by shaking it successively with several portions of methylene chloride. On the other hand, water can be used to extract impurities from an organic reaction mixture.

According to the distribution coefficient as described in Section 5.1, it is apparent that not all the solute will be transferred to solvent 2 in a single extraction unless K is very large. Usually several extractions are needed to remove all the solute from solvent 1.

FIGURE 5–1.   The extraction process. *A*. Solvent 1 contains a mixture of molecules (black and white). It is desired to separate the white molecules by extraction. A second solvent (shaded), which is immiscible with the first solvent, is added, and the two solvents are shaken together. *B*. After the separation of layers, most white molecules, but not all, have been extracted into the new solvent. *C*. With separation of the two layers, the black and white molecules have been partially separated.

In extracting a solute from a solution, it is always better to use several small portions of the second solvent than to make a single extraction with a large portion. Suppose, as an illustration, a particular extraction proceeds with a distribution coefficient of 10. The system consists of 5.0 g of organic compound dissolved in 100 mL of water (solvent 1). In this illustration, the effectiveness of three 50-mL extractions with ether (solvent 2) is compared with one 150-mL extraction with ether. In the first 50-mL extraction, the amount extracted into the ether layer is given by the following calcula-tion. The amount of compound remaining in the aqueous phase is given by x.

$$K = 10 = \frac{C_2}{C_1} = \frac{\left(\dfrac{5.0 - x}{50} \dfrac{\text{g}}{\text{mL ether}}\right)}{\left(\dfrac{x}{100} \dfrac{\text{g}}{\text{mL H}_2\text{O}}\right)}; \ 10 = \frac{(5.0 - x)(100)}{50x}$$

$$500x = 500 - 100x$$
$$600x = 500$$
$$x = 0.83 \text{ g remaining in the aqueous phase}$$
$$5.0 - x = 4.17 \text{ g in the ether layer}$$

As a check on the calculation, it is possible to substitute the value 0.83 g for x in the original equation and demonstrate that the concentration in the ether layer divided by the concentration in the water layer equals the distribution coefficient.

$$\frac{\left(\dfrac{5.0 - x}{50} \dfrac{\text{g}}{\text{mL ether}}\right)}{\left(\dfrac{x}{100} \dfrac{\text{g}}{\text{mL H}_2\text{O}}\right)} = \frac{\dfrac{4.17}{50}}{\dfrac{0.83}{100}} = \frac{0.083 \text{ g/mL}}{0.0083 \text{ g/mL}} = 10 = K$$

A second extraction with another 50-mL portion of fresh ether performed on the aqueous phase, which now contains 0.83 g of the solute, will extract an amount of solute given by the calculation

$$K = 10 = \frac{\left(\dfrac{0.83 - x}{50} \dfrac{\text{g}}{\text{mL ether}}\right)}{\left(\dfrac{x}{100} \dfrac{\text{g}}{\text{mL H}_2\text{O}}\right)}; \ 10 = \frac{(0.83 - x)(100)}{50x}$$

$$500x = 83 - 100x$$
$$600x = 83$$
$$x = 0.14 \text{ g remaining in the water layer}$$
$$0.83 - x = 0.69 \text{ g in the ether layer}$$

By a similar calculation, it can be shown that a third extraction with another fresh 50-mL portion of ether will remove 0.12 g of solute into the ether layer, leaving 0.02 g of solute remaining in the water layer. The total amount extracted into the combined ether layers, $4.17 + 0.69 + 0.12$, equals 4.98 g of solute.

If an extraction were performed using an equivalent amount of ether (150 mL) in **one** extraction, the amount extracted would be given by

$$K = 10 = \frac{\left(\dfrac{5.0 - x}{150} \dfrac{\text{g}}{\text{mL ether}}\right)}{\left(\dfrac{x}{100} \dfrac{\text{g}}{\text{mL H}_2\text{O}}\right)}; \ 10 = \frac{(5.0 - x)(100)}{150x}$$

$$1500x = 500 - 100x$$
$$1600x = 500$$
$$x = 0.31 \text{ g remaining in the water layer}$$
$$5.0 - x = 4.69 \text{ g in the ether layer}$$

One can see that the three extractions that used smaller amounts of ether succeeded in extracting 0.29 g **more** solute from the aqueous phase than one large extraction could remove. This differential represents 5.8% of the total material. If the material of interest is present in only small quantities or is very expensive, such a difference in efficiency of extraction becomes important.

5.3   PURIFICATION AND SEPARATION METHODS

In nearly all the synthetic experiments undertaken in the first part of this textbook, a series of operations involving extractions are used after the actual reaction has been concluded. These extractions form an important part of the purification. Using them, the desired product is separated from unreacted starting materials or from undesired side products in the reaction mixture. These extractions may be grouped into three categories, depending on the nature of the impurity they are designed to remove.

The first category involves extracting an organic mixture with **water.** Water extractions are designed to remove such highly polar materials as inorganic salts, strong acids or bases, and such **low-molecular-weight,** polar substances as alcohols, carboxylic acids, and amines. Many organic compounds containing fewer than five carbons are water-soluble. Water extractions are also used immediately following extractions of the mixture with either acid or base to ensure that all traces of acid or base have been removed.

The second category concerns extraction of an organic mixture with a dilute **acid,** usually 5 or 10% hydrochloric acid. Such acid extractions are intended to remove basic impurities, especially such basic impurities as organic amines. The bases are converted to their corresponding cationic salts by the acid used in the extraction. If an amine is one of the reactants, or if pyridine or another amine is a solvent, such an extraction might be used to remove any excess amine present at the end of a reaction.

$$RNH_2 + HCl \longrightarrow RNH_3^+ \ Cl^-$$
(Water-soluble salt)

The cationic salts are usually soluble in the aqueous solution, and they are thus extracted from the organic material. A water extraction may be used immediately following the acid extraction to ensure that all traces of the acid have been removed from the organic material.

The third category is extraction of an organic mixture with a dilute base, usually 5% sodium bicarbonate, although extractions with dilute sodium hydroxide can also be used. Such basic extractions are intended to convert acidic impurities, such as organic acids, to their corresponding anionic salts. In the preparation of an ester, a sodium bicarbonate extraction might be used to remove any excess carboxylic acid that is present.

$$RCOOH + NaHCO_3 \longrightarrow RCOO^-Na^+ + H_2O + CO_2$$

(pK_a ~ 5)   **(Water-soluble salt)**

The anionic salts, being highly polar, would be expected to be soluble in the aqueous phase. As a result, these acidic impurities are extracted from the organic material into the basic solution. A water extraction may be used after the basic extraction to ensure that all the base has been removed from the organic material.

Occasionally, phenols may be present in a reaction mixture as impurities, and removing them by extraction may be desired. Because phenols, although they are acidic, are about 10^5 times less acidic than carboxylic acids, basic extractions may be used to separate phenols from carboxylic acids by a careful selection of the base. If sodium bicarbonate is used as a base, carboxylic acids are extracted into the aqueous base, but phenols are not. Phenols are not sufficiently acidic to be deprotonated to any substantial degree by the weak base, bicarbonate. Extraction with sodium hydroxide, on the other hand, extracts **both** carboxylic acids and phenols into the aqueous basic solution, since hydroxide ion is a sufficiently strong base to deprotonate phenols.

(pK_a ~ 10)   **(Water-soluble salt)**

It would be a useful exercise for you to examine the experimental instructions for some of the preparative experiments in Part One of this textbook. While you are examining these procedures, you should try to identify which impurities are being removed at each extraction step.

Mixtures of acidic, basic, and neutral compounds are easily separated by extraction techniques. One such example is shown in Figure 5–2.

Materials that have been extracted can be regenerated by neutralizing the extraction reagent. If an acidic material has been extracted with aqueous base, the material can be regenerated by acidifying the extract until the solution becomes acidic to blue litmus. The material will separate from the acidified solution. A basic material can be recovered from an acidic extract by adding base to the extract. These substances can then be removed from the neutralized aqueous solutions by extraction with an organic solvent such as ether. Evaporating the ether extracts yields the isolated compounds.

5.4 THE SEPARATORY FUNNEL

The separatory funnel is the piece of apparatus used in extraction. This apparatus is illustrated in Figure 5–3. There is an art to using a separatory funnel correctly, and it is best learned by observing a person, such as the instructor, who is thoroughly familiar with its use. To fill the separatory funnel, one usually supports it in an iron ring attached to a ring stand. Since it is easy to break a separatory funnel by "clanking" it against the metal ring, it is recommended that three short lengths (about 3 cm each) of rubber tubing be cut and split open along their length. When slipped over the inside of

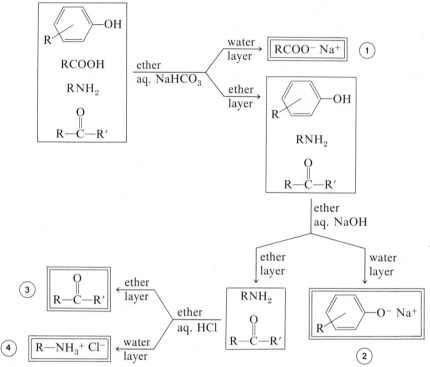

FIGURE 5–2. Separating a four-component mixture by extraction

the ring, these pieces of tubing will stay in place and cushion the funnel. Next the stopcock is closed. This simple maneuver can easily be forgotten—with disastrous consequences for the experiment! An ordinary funnel is placed in the top opening, and both the solution and the extraction solvent are poured into the funnel. The separatory funnel is swirled gently by holding it by its upper neck, and then it is stoppered. The separatory funnel is picked up with **two** hands and held as shown in Figure 5–4. It is essential to hold the stopper in place firmly because the two immiscible solvents build up pressure when they mix, and this pressure may force the stopper out of the separatory funnel. The increased pressure represents the value of the two partial vapor pressures added together during the mixing; the vapors of **both** solvents are now in equilibrium with the solution. The pressure problem becomes especially great with sodium bicarbonate extractions, in which the acidic impurities react with the sodium bicarbonate to produce carbon dioxide gas. As this gas is liberated, it causes increased pressure within the separatory funnel. To release this buildup of pressure, the funnel is vented by holding it upside-down (hold the stopper securely) and **slowly** opening the stopcock. Usually the rush of vapors out of the opening can be heard. Shaking and frequent venting should be continued until the "whoosh" is no longer audible. At this point the mixture is equilibrated, and further shaking will lead to no improvement. The funnel is then placed in the ring, and the top stopper is removed immediately. The two immiscible solvents separate into two layers after a short time, and they can be separated from one another by draining the lower layer through the stopcock. If the stopper is left in the separatory funnel, the resulting partial vacuum prevents drainage from the

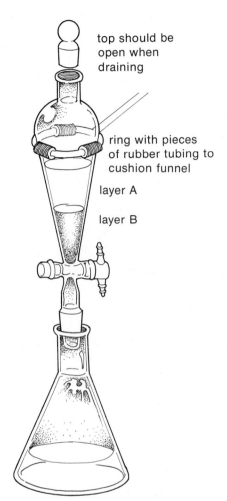

top should be
open when
draining

ring with pieces
of rubber tubing to
cushion funnel

layer A

layer B

FIGURE 5–3. The separatory funnel

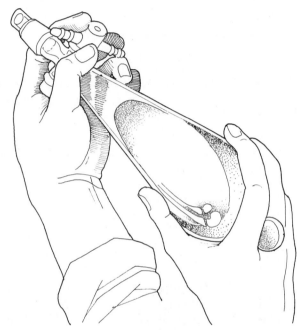

FIGURE 5–4. Correct way of shaking and venting the separatory funnel

funnel. Approximately three fourths of the lower layer is allowed to drain before the stopcock is closed. A few minutes are allowed to pass so that any of the lower phase adhering to the inner glass surfaces of the separatory funnel can drain down. The stopcock is again opened carefully, and the remainder of the lower layer is allowed to drain until the interface between the upper and lower phases just begins to enter the bore of the stopcock. At this moment, the stopcock is closed. The remaining upper layer is removed by pouring it from the top opening of the separatory funnel. To minimize contamination of the two layers, the lower layer should always be removed from the **bottom** of the separatory funnel and the upper layer from the **top** of the funnel.

At this point some caution should be exercised in identifying the aqueous and organic layers. Common sense usually will suffice if relative densities of the two solvents are considered. For example, in the extraction of an aqueous solution, the extraction solvent may be either ''lighter'' than water, as with benzene, or ''heavier,''

**TABLE 5–1. Densities of
Common Extraction Solvents**

SOLVENT	DENSITY (g/mL)
Ligroin	0.67–0.69
Diethyl ether	0.713
Benzene*	0.879
Water	1.000
Saturated NaCl	1.198
Methylene chloride	1.335
Chloroform*	1.498
Sulfuric acid (conc.)	1.84

*Suspect carcinogen.

as with chloroform. Ether or benzene layers float on top of water, whereas a chloroform layer sinks below the water. If a solvent dissolves a large amount of material, however, its density will be increased, and it may not have the relative density originally predicted. If there is any doubt, a few drops of each layer should be tested by adding them to a small test tube containing a little water. The water-miscible layer can be identified easily in this way. Table 5–1 lists the densities of common extraction solvents.

5.5 EMULSIONS

An **emulsion** is a colloidal suspension of one liquid in another. Minute droplets of an organic solvent often are held in suspension in an aqueous solution when the two are mixed vigorously; these droplets form an emulsion. This is especially true if any gummy or viscous material was present in the solution. Emulsions are encountered often in performing extractions. Emulsions may require a long time to separate into two layers and are a nuisance to the organic chemist. Fortunately, several "tricks" can be used to break up emulsions. If one of the solvents is water, adding a saturated aqueous sodium chloride solution will help destroy the emulsion. This makes the aqueous and organic layers, which are usually somewhat mutually soluble since they form the emulsion in the first place, less compatible, thereby forcing separation. Adding a very small amount of a water-soluble detergent may also help. This method has been described in the past for combating oil spills. The detergent helps to solubilize the tightly bound oil droplets. Often gravity filtration (see Technique 2, Section 2.1, p. 514) helps to destroy an emulsion by removing gum particles. In many cases, once the gum is removed, the emulsion breaks up rapidly. Getting the liquid to swirl in the separatory funnel sometimes helps. **Gentle** stirring can also be tried. If all else fails, this is definitely a case where patience, the correct vituperations, or both are beneficial in one way or another.

 If a solution is known, through prior experience, to have a tendency to form an emulsion, the mixing should be gentle and the shaking nonvigorous. Extractions should be performed with gentle swirling instead of shaking or with several gentle inversions of the separatory funnel. The separatory funnel must not be shaken vigorously in these cases. It is important to stress that observing these precautions involves more time to carry out an extraction than if the danger of forming an emulsion were lower.

5.6 DRYING AGENTS

After an organic solvent has been shaken with an aqueous solution, it will be "wet," that is, it will have dissolved some water even though its miscibility with water is not great. The amount of water dissolved varies from solvent to solvent; ether represents a solvent in which a fairly large amount of water dissolves. Ether will hold 1.5% of its weight of water. To remove water from the organic layer, a **drying agent** is used. A drying agent is an anhydrous inorganic salt that acquires waters of hydration when exposed to moist air or a wet solution. Anhydrous sodium sulfate crystals are added to the wet solution, which is usually allowed to stand for at least 15 minutes. Enough sodium sulfate is added to make a 2- to 3-mm layer in the bottom of the flask, depending on the volume of the solution. After a period of standing, the crystals are removed by filtration or decantation (see Technique 2, Sections 2.1, p. 514, and 2.6, p. 521), and the solution then is relatively free of water. At times the drying operation will have to be repeated more than once to obtain a relatively dry solution. In this case, the liquid is decanted into another **dry** flask, and more fresh drying agent is added.

Several simple observations allow one to determine whether a solution is "dry." If a solution is wet, the drying agent usually forms clumps and sticks to the flask. In extreme cases the drying agent may even be seen to dissolve in the aqueous phase that has formed at the bottom of the flask. If the solution is dry, the drying agent shifts or moves freely on the bottom of the flask. A wet solution usually appears cloudy; a dry one is clear.

Other drying agents frequently used are magnesium sulfate, calcium chloride, calcium sulfate (Drierite), and potassium carbonate. The **anhydrous** salts must be used. These have varying properties and applications. For instance, not all will absorb the same amount of water for a given weight, nor will they dry the solution to the same extent. **Capacity** is a term used to refer to the amount of water a drying agent absorbs per unit weight. Sodium and magnesium sulfates absorb a large amount of water (high capacity), but the magnesium compound dries a solution more completely. **Completeness** refers to a compound's effectiveness in removing all the water from a solution by the time equilibrium has been reached. Magnesium ion has the disadvantage that it sometimes causes rearrangements of compounds such as epoxides; it is a strong Lewis acid. Calcium chloride is a good drying agent but cannot be used with most compounds containing oxygen or nitrogen since it forms complexes. Calcium chloride does absorb methanol and ethanol in addition to water, so it is useful for removing these materials as well as being a drying agent. Potassium carbonate is a base and is used for drying basic solutions. Calcium sulfate dries completely but has a low total capacity.

Sodium sulfate is the best all-round drying agent. It is mild and effective but will not free a solution completely of water. It must also be used at room temperature to

TABLE 5–2. Solid Drying Agents Removable by Filtration

	ACIDITY	HYDRATED	CAPACITY*	COMPLETE-NESS†	RATE‡	USE
Magnesium sulfate	Neutral	$MgSO_4 \cdot 7H_2O$	High	Medium	Rapid	General
Sodium sulfate	Neutral	$Na_2SO_4 \cdot 7H_2O$ $Na_2SO_4 \cdot 10H_2O$	High	Low	Medium	General
Calcium chloride	Neutral	$CaCl_2 \cdot 2H_2O$ $CaCl_2 \cdot 6H_2O$	Low	High	Rapid	Hydrocarbons Halides
Calcium sulfate (Drierite)	Neutral	$CaSO_4 \cdot \frac{1}{2}H_2O$ $CaSO_4 \cdot 2H_2O$	Low	High	Rapid	General
Potassium carbonate	Basic	$K_2CO_3 \cdot 1\frac{1}{2}H_2O$ $K_2CO_3 \cdot 2H_2O$	Medium	Medium	Medium	Amines, esters Bases, ketones
Potassium hydroxide	Basic	...	...	...	Rapid	Amines only
Molecular sieves (3 Å or 4 Å)	Neutral	...	High	Extremely high	...	General

*Amount of water removed per given weight of drying agent.
†Refers to amount of H_2O still in solution at equilibrium with drying agent.
‡Refers to rate of action (drying).

be effective; it cannot be used with boiling solvents. Table 5–2 compares the various drying agents.

At room temperature, ether dissolves 1.5% by weight of water, and water dissolves 7.5% of ether. Ether, however, dissolves a much smaller amount of water from a saturated aqueous sodium chloride solution. Hence, the bulk of water in ether, or ether in water, can be removed by shaking it with a saturated aqueous sodium chloride solution. Any salt acts similarly, but not many are as cheap as sodium chloride or nearly as soluble in water. A solution of high ionic strength is usually not compatible with an organic solvent and forces separation of it from the aqueous layer.

PROBLEMS

1. Suppose solute A has a distribution coefficient of 1.0 between water and diethyl ether. Demonstrate that if 100 mL of a solution of 5 g of A in water were extracted with two 25-mL portions of ether, a smaller amount of A would remain in the water than if the solution were extracted with one 50-mL portion of ether.

2. For a preparative experiment from Part One of the textbook, outline the purification method described and explain the purpose of each extraction step.

Technique 6
Boiling Points, Simple Distillation, and Vacuum Distillation

Distillation is the process of vaporizing a substance, condensing the vapor, and collecting the condensate in another container. This technique is useful for **separating** a mixture when the components have different boiling points. It is the principal method of **purifying** a liquid. Four basic distillation methods are available to the chemist: simple distillation, vacuum distillation (distillation at reduced pressure), fractional distillation, and steam distillation. Technique 6 discusses simple distillation and vacuum distillation. Fractional distillation is discussed in Technique 7, and steam distillation is discussed in Technique 8.

6.1   BOILING POINTS

As a liquid is heated, the vapor pressure of the liquid increases to the point where it just equals the applied pressure (usually atmospheric pressure). At this point the liquid will be observed to boil. The normal boiling point is measured at 760 mmHg (1 atmosphere). At a lower applied pressure, the vapor pressure needed for boiling is also lowered, and the liquid boils at a lower temperature. The relation between the applied pressure and the temperature of boiling for a liquid is determined by its vapor pressure–temperature behavior. Figure 6–1 is an idealization of the typical vapor pressure–temperature behavior of a liquid.

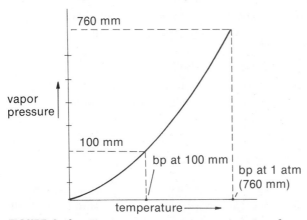

FIGURE 6–1.   The vapor pressure–temperature curve for a typical liquid

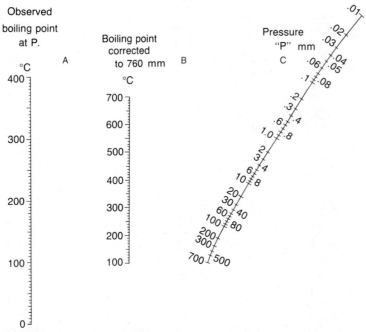

FIGURE 6–2. Pressure–temperature alignment nomograph. **How to use the nomograph:** Assume a reported boiling point of 100 °C at 1 mm. To determine the boiling point at 18 mm, connect 100 °C (column A) to 1 mm (column C) with a transparent plastic rule and observe where this line intersects column B (about 280 °C). This value would correspond to the normal boiling point. Next, connect 280 °C (column B) with 18 mm (column C) and observe where this intersects column A (151 °C). The approximate boiling point will be 151 °C at 18 mm. (Reprinted by courtesy of MC/B Manufacturing Chemists, Inc.)

Since the boiling point is sensitive to pressure, it is important to record the barometric pressure if the distillation is being conducted at an elevation significantly above or below sea level. If it is being conducted at a reduced pressure, such as that obtained with a vacuum pump or an aspirator, the pressure should be recorded.

As a rule of thumb, the boiling point of many liquids drops about 0.5 ° for a 10-mm decrease in pressure in the vicinity of 760 mmHg. At lower pressures, a 10 ° drop in boiling point is observed for each halving of the pressure. For example, if the observed boiling point for a liquid is 150 °C at 10-mm pressure, then the boiling point would be about 140 °C at 5 mmHg.

A more accurate estimate of the change in boiling point with a change of pressure can be made by using a **nomograph.** In Figure 6–2, a nomograph is given and a method is described for using it to obtain boiling points at various pressures when the boiling point is known at some other pressure.

6.2 DETERMINATION OF A BOILING POINT

Two experimental methods of determining boiling points are easily available. When you have large quantities of material, you can simply record the boiling point (or boiling range) on a thermometer as the substance distills during a simple distillation (see Section 6.4). With smaller amounts of material, you can carry out a micro determination of the boiling point by the apparatus illustrated in Figure 6–3.

To carry out the micro determination, a piece of 5-mm glass tubing sealed at one end is attached to a thermometer with a rubber band or a thin slice of rubber tubing. The liquid whose boiling point is to be determined is introduced by pipet or eyedropper into this piece of tubing, and a short piece of melting-point capillary tubing (sealed at one end) is dropped in with the open end down. The whole unit is then placed in a Thiele tube (see Technique 4, Figure 4–3, p. 537). The rubber band should be placed well above the level of the oil in the Thiele tube. If it is not, the band may soften in the hot oil. When positioning the band, one should bear in mind that the oil will expand when heated. Next, the Thiele tube is heated as if one were determining a melting point until a **rapid** and continuous stream of bubbles emerges from the inverted capillary tube. At this point the heating is stopped. Soon the stream of bubbles slows down and stops. When the bubbles stop, the liquid enters the capillary tube. The moment at which the liquid enters the capillary corresponds to the boiling point of the liquid, and the temperature is recorded.

The explanation of this method is simple. During the initial heating, the air trapped in the capillary tube expands and leaves the tube, giving rise to a stream of bubbles. Once the heating is stopped, the only vapor left in the capillary comes from the heated liquid that seals its open end. There is always vapor in equilibrium with a heated liquid. If the temperature of the liquid is above its boiling point, the pressure of the trapped vapor will either exceed or equal the atmospheric pressure. As the liquid cools, its vapor pressure decreases. When the vapor pressure drops just below atmospheric pressure, the liquid is forced into the capillary tube.

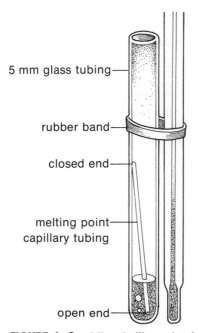

5 mm glass tubing

rubber band

closed end

melting point
capillary tubing

open end

FIGURE 6–3. Micro boiling-point determination

Two problems are common to this method. The first arises when the liquid is heated so strongly that it evaporates or is boiled away. The second arises when the liquid is not heated above its boiling point. If the heating is stopped at any point below the boiling point of the liquid, the liquid will enter the tube **immediately.** It will enter the tube because the trapped vapor will have a pressure less than that of the atmosphere.

The temperature observed on the thermometer during distillation or micro boiling-point determination may often be **lower** than the actual boiling point. This is especially true in measuring the temperatures higher than 150 °C. Manufacturers design most thermometers to read correctly only when they are immersed in the medium to be measured so as to at least cover all the mercury thread. Since this situation is rare in using a thermometer, a **stem correction** should be added to the observed temperature. This correction may be fairly large when high temperatures are being measured. (Stem correction is described in Section 6.8.)

6.3   SIMPLE DISTILLATION

When a pure liquid is distilled, vapor rises from the distilling flask and comes in contact with a thermometer. The vapor then passes through a condenser, which reliquefies the vapor and passes it into the receiving flask. The temperature observed during the distillation of a **pure substance** remains constant throughout the distillation so long as both vapor **and** liquid are present in the system (see Figure 6–5, Part A). When a **liquid mixture** is distilled, often the temperature does not remain constant but increases throughout the distillation. The reason is that the composition of the vapor that is distilling varies continuously during the distillation (see Figure 6–5, Part B).

For a liquid mixture, the composition of the vapor in equilibrium with the heated solution is different from the composition of the solution itself. This is shown in Figure 6–4, which is a phase diagram of the typical vapor–liquid relation for a two-component system (A + B).

On this diagram, horizontal lines represent constant temperatures. The upper curve represents vapor composition; the lower curve, liquid composition. For any horizontal line (constant temperature), like that shown at t, the intersections of the line with the curves give the compositions of the liquid and the vapor that are in equilibrium with one another at that temperature. In the diagram, at temperature t, the intersection of the curve at x indicates that liquid of composition w will be in equilibrium with vapor of composition z, which corresponds to the intersection at y. Composition is given as mole percentage of A and B in the mixture. Pure A, which boils at temperature t_A, is represented at the left. Pure B, which boils at temperature t_B, is represented on the right. For either pure A or pure B, the vapor and liquid curves meet at the boiling point. Thus, either pure A or pure B will distill at a constant temperature (t_A or t_B). Both the vapor and the liquid must have the same composition in either of these cases. This is not the case for mixtures of A and B.

A mixture of A and B of composition w will have the following behavior when heated. The temperature of the liquid mixture will increase until the boiling point of the mixture is reached. This corresponds to following line wx from w to x, the boiling point

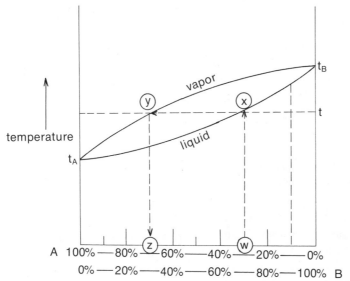

FIGURE 6–4. Phase diagram for a typical liquid mixture of two components

of the mixture (t). At temperature t the liquid begins to vaporize, which corresponds to line xy. The vapor will have the composition corresponding to z. In other words, the first vapor obtained in distilling a mixture of A and B does not consist of pure A. It is richer in A than the original mixture, but still contains a significant amount of the higher-boiling component B, **even from the very beginning of the distillation.** The result of this is that it is never possible to separate a mixture completely by a simple distillation. However, in two cases, it is possible to get an acceptable separation into relatively pure components. In the first case, if the boiling points of A and B differ by a large amount (>100 °), and if the distillation is carried out carefully, it will be possible to get a fair separation of A and B. In the second case, if A contains a fairly small amount of B ($<10\%$), a reasonable separation of A from B can be achieved. When boiling-point differences are not large, and when highly pure components are desired, it is necessary to do a **fractional distillation.** Fractional distillation is de-

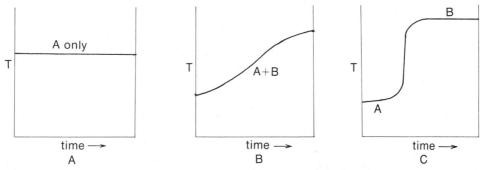

FIGURE 6–5. Three types of temperature behavior during a simple distillation: *A*, a relatively pure component is being distilled; *B*, a mixture of two components of similar boiling point is being distilled; *C*, a mixture of two components with widely differing boiling points is being distilled. Good separations are achieved in A and C.

scribed in Technique 7, where the behavior during a simple distillation is also considered in detail. Note only that as vapor distills from the mixture of composition w (Figure 6–4), it is richer in A than is the solution. Thus, the composition of the material left behind in the distillation becomes richer in B (moves to the right from w toward pure B in the graph). A mixture that is 90% B (dotted line to right of Figure 6–4) has a higher boiling point than at w. Hence, the temperature of the liquid in the distilling flask will increase during the distillation, and the composition of the distillate will change (as is shown in Figure 6–5, part B).

When two components that have a large boiling-point difference are distilled, the temperature remains constant while the first component distills. If the temperature remains constant, a relatively pure substance is being distilled. After the first substance distills, the temperature of the vapors rises, and then the second component distills, again at a constant temperature. This is shown in Figure 6–5, part C. A typical application of this type of distillation might be an instance of a reaction mixture containing the desired component A (bp 140 °C) contaminated with a small amount of undesired component B (bp 250 °C) and mixed with a solvent such as diethyl ether (bp 36 °C). The ether is easily removed at low temperature. Pure A is removed at a higher temperature and collected in a separate receiver. Component B could then be distilled, but it usually is left as a residue and not distilled. This separation would not be difficult and would represent a case where simple distillation might be used to advantage.

6.4 SIMPLE DISTILLATION: METHODS

For a simple distillation, the apparatus shown in Figure 6–6 is used. The distilling flask, condenser, and vacuum adapter should be clamped. It is best if the receiving flask is supported by wooden blocks or a wire gauze on an iron ring attached to a ring stand. Either method will facilitate removal or change of the receiving flask during the distillation. If an oil bath or heating mantle is used for heating, it should likewise be supported by wooden blocks adjusted to such a height that the heat source can be removed from the distillation flask by removing the blocks to stop the distillation.

Each standard-taper ground-glass joint should be lightly greased with stopcock grease to ensure that the joints do not "freeze" together. Excess grease should not be used. When properly seated, greased, and clamped, the joint should appear nearly transparent, with no air bubbles visible. The tightness of the joints should be checked frequently during the distillation. If the joints are not sealed tightly, material will be lost.

The thermometer is inserted through the rubber adapter so that the bulb extends to a depth just **below** the sidearm of the distilling head (see inset in Figure 6–6). The thermometer bulb must be **in the vapor stream,** not above it, to give a correct reading.

The water lines should be attached to the condenser so that the water passes into the lower end of the condenser and leaves from the upper end. The condenser does not remain full if the water flows from the top down.

Before the liquid to be distilled is heated, one or two boiling stones should be added. These boiling stones prevent superheating of the liquid and reduce the tendency of the liquid to "bump." (See Technique 1, Section 1.8, p. 511.)

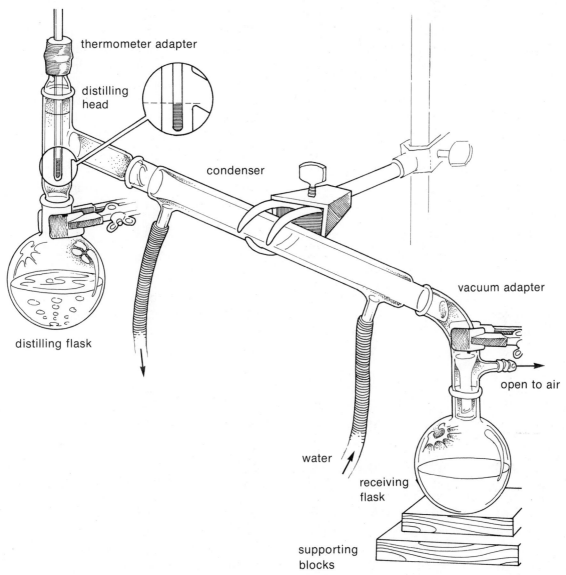

FIGURE 6–6. Apparatus for a simple distillation

The distilling flask must not be filled to more than two thirds of its capacity, since the surface area of boiling liquid should be kept as large as possible. On the other hand, too large a distilling flask should also be avoided. With too large a flask, the **holdup** is excessive; the holdup is the amount of material that cannot distill since some vapor must fill the empty flask.

From information provided in Technique 1, the appropriate heating source for a distillation can be chosen. In general, an oil bath or a heating mantle will be used. When the boiling point of the liquid to be distilled is reached, a ring of condensate (or **reflux ring**) moves up through the apparatus and comes in contact with the thermometer bulb. At this point, the thermometer records a rapid rise in temperature, and soon condensate begins to pass from the condenser into the receiver. The heating is then adjusted to provide the proper rate of **takeoff,** the rate at which distillate leaves the condenser. One drop per second is considered a proper rate of takeoff for most applications. At a greater rate, equilibrium is not established within the distillation apparatus, and the separation may be poor. A slower takeoff is also unsatisfactory since the temperature recorded on the thermometer is not maintained by a constant vapor stream, thus leading to an inaccurately low observation of the boiling temperature. Generally, the material is distilled, and **fractions** (portions collected in separate flasks) are collected over narrow ranges of temperature. If the separation has been good, the temperature of the vapors will drop between fractions and then rise dramatically when the next component begins to distill.

One should never distill to dryness. A dry residue may explode on overheating, or the flask may melt or crack. With no liquid in the flask the heat applied is not carried away, and the flask becomes **very** hot.

6.5 VACUUM DISTILLATION

Vacuum distillation (distillation at reduced pressures) is used when compounds have high boiling points (above 200 °C) and/or decompose at the high temperatures required for atmospheric-pressure distillation. As seen in Figure 6–2, the boiling point is reduced substantially by lowering the pressure. For example, a liquid with a boiling point of 200 °C at 760 mmHg would boil at about 90 °C at 20 mmHg. Thus, it may be advantageous to use a vacuum when distilling such a material. Counterbalancing this advantage is the fact that often a separation will not be as good under reduced pressure as it was at atmospheric pressure.

6.6 VACUUM DISTILLATION: METHODS

An apparatus such as that shown in Figure 6–7 (or Figure 9–6) is assembled for reduced-pressure distillation. It is important that all the rubber tubing (except for water connections to the condenser) be heavy-walled, so that the tubing does not collapse under vacuum. If the rubber tubing shows cracks when it is bent or extended by pulling from both ends, it should be discarded and replaced with new tubing. Similarly, rubber stoppers should be inspected to make sure that they fit well and are crack-free and pliable. The connections between glass and rubber tubing must be secure and airtight. It may be necessary to tighten the thermometer adapter with wire. It is advisable to use the shortest possible lengths of rubber tubing in the system.

All ground-glass equipment must be free of cracks. Careful examination of the distilling flask to make sure that it does not have small cracks is recommended. If there is doubt about the condition of the apparatus, the laboratory supervisor should be consulted. Any cracked equipment or, for that matter, thin-walled glassware of any type (large Erlenmeyer flasks, for example) may implode because of the large pressure differences between the outside and the inside of the glassware. SAFETY GLASSES MUST BE WORN AT ALL TIMES. The ground-glass joints should be lubricated as for the simple distillation, and one should determine that all joints fit well.

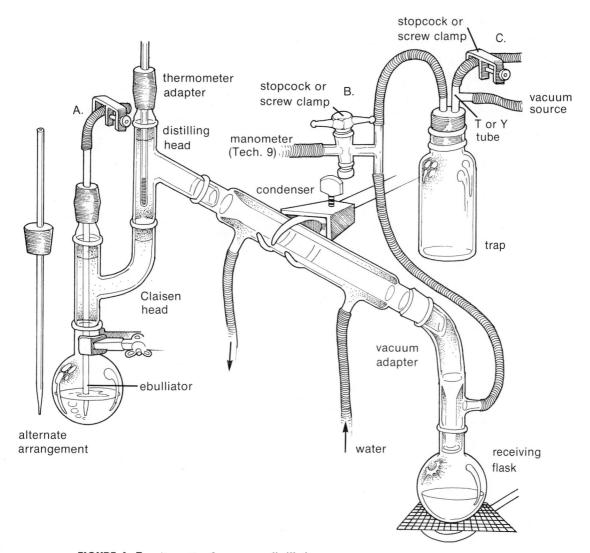

FIGURE 6–7. Apparatus for vacuum distillation

In distilling, a method is needed to form small bubbles to prevent superheating of the liquid or ''bumping.'' Conventional boiling stones do not work under a vacuum. Although wooden applicator sticks and microporous boiling stones have been used to prevent bumping, **ebulliators** (bubblers) are generally the choice in most vacuum systems. Most standard ground-glass kits contain an ebulliator. If one is not available, an ebulliator can be prepared easily by heating a section of glass tubing and drawing it out about 3 cm. The glass is then scored in the center of this drawn-out section and broken. The ebulliator can be inserted into a thermometer adapter, if available, or into a rubber stopper, as shown in Figure 6–7. The ebulliator tube is used in conjunction with a **Claisen head.** The Claisen head prevents carryover if there is bumping. The ebulliator is adjusted so that it comes close to the bottom of the flask. A short section of heavy-walled rubber tubing, with an open screw clamp, is attached to the ebulliator at *A*, as shown in Figure 6–7. This clamp is used to regulate the amount of air admitted to the system and the rate of production of bubbles.

Normally, water should be used as a coolant in the condenser. Some very high-boiling materials may crack the condenser because of the large differences in the temperature of the vapor and cold water. In those cases, an "air condenser" may be used. An air condenser is a normal condenser, but no water is allowed to pass through it. The condenser is cooled solely by the air in the room. Some high-boiling substances may solidify in the condenser. In these cases, the condenser should not be water-cooled. The use of an air condenser or the use of steam, rather than water, circulating in the condenser should be considered.

When more than one fraction is expected from a vacuum distillation, it is considered good practice to have several preweighed receiver flasks, including the original, available before the distillation is begun. Such preparation permits the rapid changing of receiving flasks during the distillation. The preweighing permits easy calculation of the weight of distillate in each fraction without the need for transferring the distillate to yet another flask.

With the apparatus shown in Figure 6–7, the vacuum must be stopped so receiving flasks can be changed when a new substance (fraction) begins to distill. Although this situation can be easily tolerated in distillations that involve only two substances in the initial mixture, it is inconvenient when more substances are present. In Figure 6–8, two pieces of apparatus are shown in which collecting fractions under vacuum is greatly simplified. All one has to do is rotate the device to collect the various fractions.

If the vacuum is interrupted while fractions are to be changed, the heat must be removed from the apparatus and the clamp controlling the air flow into the ebulliator

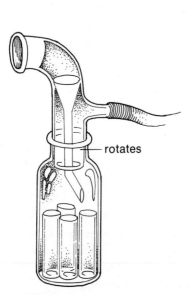

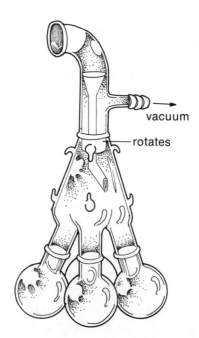

FIGURE 6–8.   Rotating receiving flasks

must be opened. If heating is continued while air has been admitted to the distilling flask, the material being distilled may undergo oxidation. If air is admitted into the apparatus at some other point while the screw clamp controlling the ebulliator remains closed, liquid may back up into the ebulliator tube.

The trap assembly, shown in Figure 6–7, or an alternative assembly, as shown in Technique 9 (Figure 9–6, p. 593), is necessary to keep water from backing up into the receiving flask and manometer when there is a change in pressure in the distillation system. A suitable manometer (pressure-measuring device) should be included in the system at least part of the time during the distillation to measure the pressure at which the distillation is being conducted. A boiling point is of little value unless the pressure is known! In Figure 6–7, the stopcock at point *B* may be closed when the manometer is not in use or not attached to the system. Several types of manometers are discussed in Technique 9. The screw clamp at point *C* can be opened slowly to allow air to pass into the apparatus to bring the system back to atmospheric pressure. The manometer may break if the screw clamp is opened too rapidly.

The most convenient source of vacuum for a reduced-pressure distillation is the aspirator. The aspirator, or other vacuum source, is attached to the trap. The aspirator theoretically can pull a vacuum equal to the vapor pressure of the water flowing through it. The vapor pressure of the flowing water depends on its temperature (24 mmHg at 25 °C; 18 mmHg at 20 °C; 9 mmHg at 10 °C). However, in the typical laboratory, the pressures attained are higher than expected due to reduced water pressure when many students are using their aspirators simultaneously. Good laboratory practice requires that only a few students on a given bench use the aspirator at any given time. In addition, the pressure should be checked early in the assembly procedure to make sure that the aspirator is working properly. Movement to another location in the laboratory may be necessary. Generally, a pressure of 10 to 30 mmHg is adequate. Lower pressures can be obtained with a mechanical vacuum pump.

6.7 VACUUM DISTILLATION: STEPWISE DIRECTIONS

The procedures in applying vacuum distillation are now described.

Evacuating the Apparatus

1. Assemble the apparatus as shown in Figure 6–7 and as described in Section 6.6. REMEMBER TO USE SAFETY GLASSES! Weigh each empty receiving flask to be used in collecting the various fractions during the distillation.
2. Concentrate the material to be distilled in an Erlenmeyer flask or beaker by removing all volatile solvents, such as ether, on a steam bath in the hood. Use boiling stones. Transfer the concentrate to a distilling flask, and complete the transfer with a small amount of solvent. Again concentrate the material until no additional volatile solvent can be removed (boiling will cease). The flask should be no more than half-full after concentration. Attach the flask to the apparatus and secure it with a clamp. Make sure all joints are tight.

3. Open the stopcock at B (Figure 6–7).

4. Turn on the aspirator to the maximum extent.

5. Tighten the screw clamp at A until the tubing is **nearly** closed.

6. Slowly tighten the screw clamp at C until the tubing is closed completely. Watch the bubbling action of the ebulliator to see that it is not too vigorous or too slow. Adjust A until a fine steady stream of bubbles is formed with C closed.

7. Record the pressure obtained, after waiting a few minutes to allow any residual solvent to be removed. Readjust A if necessary. If the pressure is not satisfactory, check all connections to see whether they are tight. **Do not proceed until you have a good vacuum.**

Beginning Distillation

8. Raise the heat source (see Technique 1) into position with wooden blocks, or other means, and begin to heat.

9. Increase the temperature of the heat source. Eventually a reflux ring will contact the thermometer bulb, and distillation will begin. Record the temperature range and the pressure range during the distillation. The distillate should be collected at the rate of about 1 drop per second. The Claisen head and distilling head may have to be wrapped with glass wool or aluminum foil (shiny side in) for insulation during the distillation if it is slow. The boiling point should be relatively constant so long as the pressure is constant. A rapid increase in pressure may be due to increased use of the aspirators in the laboratory or to rapid decomposition of the material being distilled. Decomposition will produce a dense white fog in the distilling flask. If this happens, reduce the temperature of the heat source, or remove it, and **stand back** until the system cools. Investigate the cause.

Changing Receiving Flasks

10. To change receiving flasks during distillation when a new component begins to distill (higher boiling point at the same pressure), open the clamp at C slowly, and immediately lower the heat source. (Watch the ebulliator for **excessive** backup. It may be necessary to open clamp A.) The wooden blocks under the receiver are removed, or the clamp is released, and the flask is replaced with a clean, preweighed receiver.

11. Reclose the clamp at C and allow several minutes for the system to reestablish the reduced pressure. Bubbling will commence after the liquid is drawn back out of the ebulliator. This liquid may have been forced into the ebulliator when the vacuum was interrupted.

12. Raise the heating source back into position under the distilling flask and continue with the distillation. When the temperature falls at the thermometer, this usually indicates that distillation is complete. If a significant amount of liquid remains, however, the bubbling may have stopped, the pressure may have risen, the heating source may not be hot enough, or perhaps insulation of the distillation head is required. Adjust accordingly.

Shutdown

13. At the end of the distillation, remove the heat source, and **slowly** open the clamps at *A* and *C*. **Then** turn off the water at the aspirator. Remove the receiving flask and clean all glassware as soon as possible after disassembly to keep the ground-glass joints from sticking.

6.8   THERMOMETER STEM CORRECTIONS

The temperature as you see it on the thermometer during distillation or micro boiling-point determination may read **lower** than the actual boiling point because of stem error in the thermometer. When you want to make a stem correction for a thermometer, the formula given below may be used. It is based on the fact that the portion of the mercury thread in the stem is cooler than the portion immersed in the vapor. The mercury will not have expanded in the cool stem to the same extent as in the warmed section of the thermometer; therefore, the reading must be corrected. The equation used is

$$(0.000154)(T - t_1)(T - t_2) = \text{correction to be added to observed temperature } T$$

1. The factor 0.000154 is a constant that corresponds to a coefficient of expansion for mercury in the thermometer.
2. The term $(T - t_1)$ corresponds to the length of the mercury thread not immersed in vapor. It is convenient to use the temperature scale for this measurement, where T is the observed temperature and t_1 is the **approximate** place where the heated part of the stem ends and the cooler part begins. In simple distillation cases, t_1 would correspond approximately to the lower part of the thermometer adapter, which is inserted into the distilling head. See Figure 6–6.
3. The term $(T - t_2)$ corresponds to the difference between the temperature of the mercury in the vapor, T, and the temperature of the mercury in the air outside the distillation apparatus, t_2. The term T represents the observed temperature, and t_2 is measured by hanging another thermometer so that the bulb is close to the stem of the main thermometer. Figure 6–9 illustrates the method of correcting a thermometer and indicates where the various temperatures are determined.

By the formula given above, it can be shown that high-boiling substances are the most likely to require a stem correction and that low-boiling substances need not be corrected. The calculations given below illustrate this point:

EXAMPLE 1
$T = 200\ °C$
$t_1 = 0\ °C$
$t_2 = 35\ °C$
$(0.000154)(200)(165) = 5.1\ °$ stem correction
$200\ °C + 5\ ° = 205\ °C$ corrected boiling point

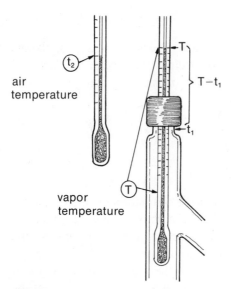

FIGURE 6–9. Measurement for a thermometer stem correction

EXAMPLE 2

$T = 100\ °C$
$t_1 = 0\ °C$
$t_2 = 35\ °C$
$(0.000154)(100)(65) = 1.0\ °$ stem correction (negligible)

PROBLEMS

1. Using the pressure-temperature alignment chart in Figure 6–2, answer the following questions.
   (a) What is the normal boiling point (at 760 mm) for a compound that boils at 150 °C at 10 mmHg pressure?
   (b) Where would the compound in (a) boil if the pressure were 40 mm?
   (c) A compound was distilled at atmospheric pressure and had a boiling range 290 to 300 °C. Since decomposition was extensive, it was decided that vacuum distillation was required. What would be the approximate boiling range observed at 15 mmHg?
2. Calculate the corrected boiling point for nitrobenzene by using the method given in Section 6.8. The apparatus used is shown in Figure 6–6. The observed boiling point was 205 °C. The thermometer was inserted in the thermometer adapter so that the 0 °C graduation was visible at the bottom of the adapter. The air temperature just above the distilling head was measured on a second thermometer and was 35 °C.
3. In Section 6.4 it was stated that the distilling flask should not be too large because of extensive holdup (material that cannot distill, since some vapor must fill the empty flask). Using the ideal gas equation, prove that the statement is correct by making the following calculations:
   (a) A 15-g sample of acetone (MW 58) was placed in a 1-L distilling flask. The acetone was distilled at 1 atm at the boiling point of 56 °C until the flask appeared empty. Assuming that the flask contained only acetone vapor at that point, calculate how many grams of acetone remained in the "empty" flask.
   (b) Repeat the above calculation in (a), using a distilling flask with a volume of 50 mL.
   (c) In each of the above cases, calculate the percentage loss due to holdup with the 15-g sample of acetone. Should you always use the smallest possible distilling flask?

Technique 7
Fractional Distillation, Azeotropes

Simple distillation, described in Technique 6, works well for most routine separation and purification of organic compounds. When the boiling-point differences of components to be separated are not large, however, **fractional distillation** must be used to achieve a good separation.

7.1 DIFFERENCES BETWEEN SIMPLE AND FRACTIONAL DISTILLATION

When an ideal solution of two liquids, such as benzene (bp 80 °C) and toluene (bp 110 °C), is distilled by **simple distillation** with the apparatus shown in Technique 6, Figure 6–6, p. 557, the first vapor produced will be enriched in the lower-boiling component (benzene). When the vapor is condensed and analyzed, however, it is unlikely that the distillate will be pure benzene, since the boiling-point difference of benzene and toluene is only 30 °C. Likewise, the liquid remaining in the distilling flask will contain a larger amount of the higher-boiling component but will be far from being pure toluene.

In principle, one could distill a solution of 50% benzene and 50% toluene by simple distillation and collect the distillate in **fractions** (portions collected in separate flasks). The first fraction would contain the largest amount of benzene and the least amount of toluene and would have the lowest boiling-point range. The second fraction would contain less benzene and more toluene than the previous one and would have a higher boiling-point range. The trend of decreasing benzene and increasing toluene would continue until the last fraction was removed. This last fraction would have the smallest amount of benzene, the largest amount of toluene, and the highest boiling-point range. The results of such a hypothetical distillation are given in Table 7–1.

TABLE 7–1. Simple Distillation of a Mixture of Benzene and Toluene

| FRACTION | BOILING RANGE (°C) | PERCENTAGE COMPOSITION | |
		Benzene	*Toluene*
1	80–85	90	10
2	85–90	72	28
3	90–95	55	45
4	95–100	45	55
5	100–105	27	73
6	105–110	10	90

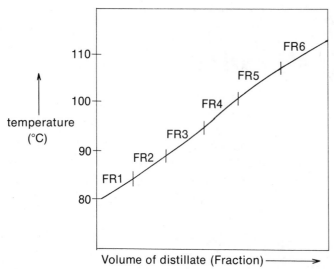

FIGURE 7–1. Temperature–distillate plot for simple distillation of a benzene–toluene mixture

As such a distillation progressed, a plot of boiling point vs volume of condensate (distillate) collected might appear as in Figure 7–1. Clearly, separation by this method would be poor. The continuously increasing temperature observed in Figure 7–1 indicates that the composition of the vapor itself was also continuously changing. At no time did a pure substance distill.

However, each of the fractions indicated in Table 7–1 could be redistilled. In redistillation, each of these fractions would yield vapor and a resulting condensate that would contain **more** benzene than what was initially present. The residue in the distilling flask would contain **more** toluene than what was initially present. The distillates and residues of similar composition (similar boiling ranges) could be combined and redistilled. Eventually, one should obtain distillate that would be essentially pure benzene and a residue that would be essentially pure toluene.

Obviously the above procedure would be very tedious; fortunately, it need not be done in usual laboratory practice. **Fractional distillation** accomplishes the same result. One simply has to use a column inserted between the distilling flask and the distilling head, as shown in Figure 7–2. The column is filled with a suitable packing, such as a stainless steel sponge. This packing allows a mixture of benzene–toluene to be subjected continuously to many vaporization–condensation cycles as the material moves up the column. **With each cycle within the column, the composition of the vapor is progressively enriched in the lower-boiling component (benzene).** Finally, nearly pure benzene (bp 80 °C) emerges from the top of the column, condenses, and passes into the receiving flask at the first fraction. The distillation must be carried out slowly to ensure numerous vaporization–condensation cycles. When nearly all the benzene is removed, the temperature begins to rise and a small amount of a second fraction, which contains some benzene and toluene, is collected. When the temperature reaches 110 °C, the boiling point of pure toluene, the vapor is condensed and collected in another receiving flask as the third fraction.

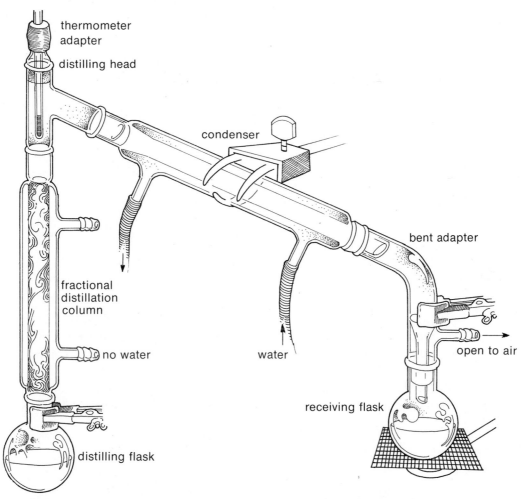

FIGURE 7–2. Fractional distillation apparatus

A plot of boiling point vs volume of condensate (distillate) would resemble Figure 7–3. Clearly the separation would be much better than that achieved by a simple distillation (Figure 7–1).

7.2 VAPOR–LIQUID COMPOSITION DIAGRAMS

A vapor–liquid composition phase diagram like the one in Figure 7–4 can be used to explain the operation of a fractionating column with an **ideal solution** of two liquids, A and B. An ideal solution is one in which the two liquids are chemically similar and totally miscible in all proportions but do not interact. Such solutions are said to obey Raoult's law. (This law will be explained in detail in Section 7.3.) In the example given, at atmospheric pressure, pure A boils at 50 °C while pure B boils at 90 °C.

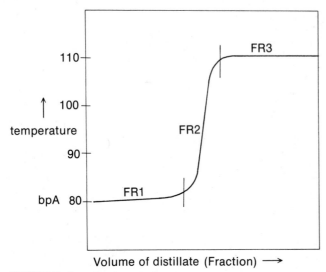

FIGURE 7–3. Temperature–distillate plot for fractional distillation of a benzene–toluene mixture

The phase diagram relates the compositions of the boiling liquid (lower curve) and its vapor (upper curve) as a function of temperature. Any horizontal line drawn across the diagram (a constant-temperature line) intersects the diagram in two places. These intersections relate the vapor composition to the composition of the boiling liquid that produces that vapor. By convention, composition is expressed either in **mole fraction** or in **mole percentage.** The mole fraction and mole percentage are defined as follows:

$$\text{Mole } \textbf{fraction } A = N_A = \frac{\text{moles A}}{\text{moles A} + \text{moles B}}$$

$$\text{Mole } \textbf{fraction } B = N_B = \frac{\text{moles B}}{\text{moles A} + \text{moles B}}$$

$$N_A + N_B = 1$$

$$\text{Mole } \textbf{percentage } A = N_A \times 100$$
$$\text{Mole } \textbf{percentage } B = N_B \times 100$$

The horizontal and vertical lines shown in Figure 7–4 represent the processes going on during a fractional distillation. Each of the **horizontal lines** (L_1V_1, L_2V_2, etc.) represents the **vaporization** step of a given vaporization–condensation cycle and indicates the composition of the vapor in **equilibrium** with liquid at a given temperature. For example, at 63 °C a liquid with a composition of 50% A (L_3 on diagram) would yield vapor of composition 80% A (V_3 on diagram) at equilibrium. The vapor is richer in the lower-boiling component A than the original liquid was.

Each of the **vertical lines** (V_1L_2, V_2L_3, etc.) represents the **condensation** step of a given vaporization–condensation cycle. The composition does not change as the temperature drops on condensation. The vapor at V_3, for example, condenses to give a liquid (L_4) of composition 80% A with a drop in temperature from 63 to 53 °C.

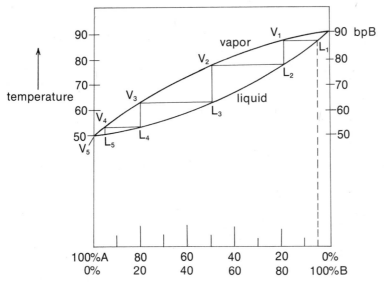

FIGURE 7—4. Phase diagram for a fractional distillation of an ideal two-component system

Now consider the phase diagram in Figure 7–4 for a solution initially 5% in component A. The solution of composition 5% A (95% B) is heated (following the dotted line) until it is observed to boil at 87 °C (L_1). The vapor has a composition (V_1) of 20% A (80% B). The vapor is richer in A than the original liquid was but is by no means pure A. In a simple distillation apparatus, this liquid would have been condensed and passed into the receiving flask in a highly unpurified state. However, with a fractionating column in place, the vapor is condensed in the column to give liquid (L_2) of composition 20% A (80% B). This liquid (L_2) is revaporized (bp 78 °C) to give vapor of composition V_2 (50% A), which is condensed to give liquid L_3. Liquid L_3 (50% A) is revaporized (bp 63 °C) to give vapor of composition V_3 (80% A), which is condensed to give liquid L_4. Liquid L_4 (80% A) is revaporized (bp 53 °C) to give vapor of composition V_4 (95% A), which is condensed to give liquid L_5. Finally, liquid L_5 (95% A) is once again vaporized (bp 51 °C) to give vapor (V_5) that is essentially pure A. The nearly pure liquid emerges from the column, is condensed, and passes into the receiving flask as a pure liquid A. In the meantime component B has been concentrated in the distilling flask and is now removed as a pure liquid by distillation.

The above vaporization–condensation in a fractional distillation column is shown in Figure 7–5. The composition of the liquids, their boiling points, and the composition of the vapor are shown.

7.3 RAOULT'S LAW

Two liquids (A and B) that are mutually soluble (miscible) and that do not interact form an **ideal solution** and follow Raoult's law. The law states that the partial vapor pressure of component A in the solution (P_A) equals the vapor pressure of pure A (P_A^0) times its mole fraction (N_A) in the solution (equation 1). Likewise, an expression can be

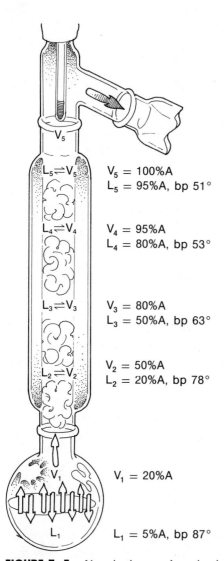

$V_5 = 100\%A$
$L_5 = 95\%A,$ bp $51°$

$V_4 = 95\%A$
$L_4 = 80\%A,$ bp $53°$

$V_3 = 80\%A$
$L_3 = 50\%A,$ bp $63°$

$V_2 = 50\%A$
$L_2 = 20\%A,$ bp $78°$

$V_1 = 20\%A$

$L_1 = 5\%A,$ bp $87°$

FIGURE 7–5. Vaporization–condensation in a fractionation column

written for component B (equation 2). The mole fractions, N_A and N_B, were defined in section 7.2.

$$\text{Partial vapor pressure of A in solution} = P_A = (P_A{}^0)(N_A) \tag{1}$$

$$\text{Partial vapor pressure of B in solution} = P_B = (P_B{}^0)(N_B) \tag{2}$$

$P_A{}^0$ = vapor pressure of pure A, independent of B

$P_B{}^0$ = vapor pressure of pure B, independent of A

The partial vapor pressures are **added** to give a total pressure above the solution (equation 3). When the applied pressure equals the sum of the partial vapor pressures (P_{total}), the solution will be observed to boil.

$$P_{\text{total}} = P_A + P_B = P_A{}^0 N_A + P_B{}^0 N_B \tag{3}$$

The composition of A and B in the vapor is given by:

$$N_A(\text{vapor}) = \frac{P_A}{P_{\text{total}}}; \; N_B(\text{vapor}) = \frac{P_B}{P_{\text{total}}}$$

For application of Raoult's law, consider the following problems.

1. What is the partial vapor pressure of A in the solution of $N_A = 0.5$, where the vapor pressure of pure A at 100 °C is 1020 mmHg? Answer: $P_A = P_A^0 N_A = (1020)(0.5) = 510$ mmHg.

2. What is the partial vapor pressure of B in a solution of $N_B = 0.5$, where the vapor pressure of pure B at 100 °C is 500 mmHg? Answer: $P_B = P_B^0 N_B = (500)(0.5) = 250$ mmHg.

3. Would the solution boil at 100 °C if the applied pressure were 760 mm? Answer: Yes, $P_{\text{total}} = P_A + P_B = 510 + 250 = 760$ mm.

4. What is the composition of the vapor at the boiling point (100 °C)? Answer: The fraction of each component is proportional to the partial vapor pressure of that component compared with the total pressure. Therefore, $N_A(\text{vapor}) = {}^{510}\!/_{760} = 0.67$; $N_B(\text{vapor}) = {}^{250}\!/_{760} = 0.33$.

One should note in Problem 4 that the **vapor is richer ($N_A = 0.67$) in the lower-boiling (higher-vapor-pressure) component (A) than the liquid was before vaporization ($N_A = 0.5$).** This proves mathematically what was described in Section 7.2.

The consequences of Raoult's law are shown schematically in Figure 7–6. In Figure 7–6A the boiling points are identical (vapor pressures are the same), and no separation is attained regardless of how the distillation is conducted. In Figure 7–6B a

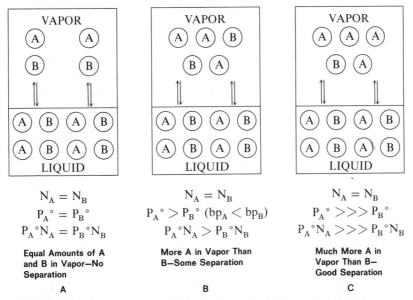

FIGURE 7–6. Consequences of Raoult's Law. *A,* Boiling points (vapor pressures) are identical—**no separation**; *B,* Boiling point somewhat less for A than for B—**requires fractional distillation**; *C,* Boiling point much less for A than for B—**simple distillation will suffice.**

fractional distillation is required, while in Figure 7–6C a simple distillation provides an adequate separation.

In the extreme case of a negligible vapor pressure for B, the vapor will be pure A. For example, the distillation of an aqueous salt solution will behave like this:

$$P_{\text{total}} = P^0_{\text{H}_2\text{O}} N_{\text{H}_2\text{O}} + P^0_{\text{salt}} N_{\text{salt}}$$

$$P^0_{\text{salt}} = 0$$

$$P_{\text{total}} = P^0_{\text{H}_2\text{O}} N_{\text{H}_2\text{O}}$$

A solution whose mole fraction of water is 0.7 will not boil at 100 °C, since $P_{\text{total}} = (760)(0.7) = 532$ mmHg and is less than atmospheric pressure. Eventually, the solution can be heated to the boiling point. At 110 °C, the solution boils because $P_{\text{total}} = (1085)(0.7) = 760$ mmHg. The vapor is pure water, and its observed boiling point is 100 °C. (The vapor pressure of pure water at 110 °C is 1085 mmHg.)

7.4 COLUMN EFFICIENCY

A measure of column efficiency is given by **theoretical plates.** A column would have 1 theoretical plate if the first distillate (condensed vapor) had the composition located at L_2 (20% A) when starting with a liquid with composition L_1 (5% A) as shown in Figure 7–4. This would correspond to a simple distillation, or one vaporization–condensation cycle. A column would have 2 theoretical plates if the distillate (vapor) had the composition L_3 (50% A) starting with a liquid with composition L_1 (5% A). The 2-theoretical-plate column essentially carries out ''two simple distillations,'' corresponding to lines $L_1V_1L_2$ and $L_2V_2L_3$ in Figure 7–4. Thus, **five theoretical plates** are necessary to separate nearly pure A from B (lines $L_1V_1L_2$, $L_2V_2L_3$, $L_3V_3L_4$, $L_4V_4L_5$, and L_5V_5 in the example shown in Figure 7–4). In effect, a 5-plate column corresponds to ''five simple distillations.'' In practice, as shown in Figure 7–5, the first theoretical plate corresponds to the initial vaporization from the distilling flask, and the remaining 4 plates are obtained from the condensation–vaporization cycles in the column.

Most columns do not allow distillation in **discrete steps,** as indicated in Figure 7–4. Instead, the process is **continuous,** allowing vapors to be continuously in contact with liquid of changing composition as it passes through the column. In principle, almost any material can be used to pack the column so long as it can be wetted by the liquid.

The relation between number of **theoretical plates** needed to separate an ideal two-component mixture and the difference between the boiling points of the components is given in Table 7–2. The values have been calculated assuming an average boiling point of 150 °C for each mixture and for a column operating at equilibrium. Since columns are seldom operated at equilibrium, more theoretical plates than listed may be necessary for a complete separation. For example, 3 or 4 plates rather than the listed 2 plates may be necessary to separate a mixture with a boiling-point difference of

TABLE 7–2. Theoretical Plates Required to Separate Mixtures, Based on Boiling-Point Differences of Components

BOILING-POINT DIFFERENCE	NUMBER OF THEORETICAL PLATES
108	1
72	2
54	3
43	4
36	5
20	10
10	20
7	30
4	50
2	100

72 °. The table still is useful in giving an approximate number of plates needed to separate two-component mixtures. For example, one can separate compound A (bp 100 °C) from compound B (bp 200 °C) by simple distillation (approximately 1 plate is required). A fractional distillation with a 5-plate column would be needed to separate a mixture of A (bp 130 °C) from B (bp 166 °C). For the separation to be adequate, the **efficiency** (number of theoretical plates) of the column must increase as the boiling-point differences between the components decrease.

7.5 TYPES OF FRACTIONATING COLUMNS AND PACKINGS

Numerous types of fractionating columns have been developed over the years to carry out separations. Only a few of the most common are considered here. In Table 7–3, three combinations of columns and packing are compared. A Vigreux column, shown in Figure 7–7A, has indentations that incline downward at angles of 45 ° and are in pairs on opposite sides of the column. A column of this type can be prepared by heating a small spot on a piece of glass tubing until it melts and then pushing the glass inward with the tip of a pencil with a long point. The projections into the column give increased possibilities for condensation and for the vapor to equilibrate with the liquid.

TABLE 7–3. Fractionating Columns and Packing

TYPE (20 cm long × 1 cm diameter)	THEORETICAL PLATES	BOILING-POINT DIFFERENCES (°C)	HOLDUP (mL)
Vigreux	2.5	60	1
Packed (glass helices)	5	36	4
Packed (Heli-Pak)	13	17	7

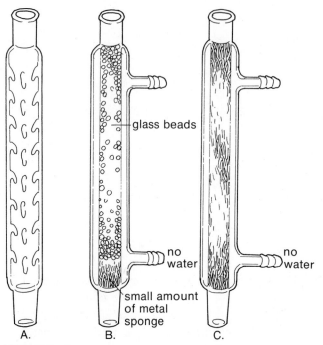

FIGURE 7–7. Types of fractionating columns: *A*, Vigreux; *B*, glass beads; *C*, metal sponge

With this type of column one can distill material rapidly, but the efficiency is not very high. For example, a 20-cm column would have only 2.5 plates and would separate only materials with a 60 ° boiling-point difference. A longer column, such as a 40-cm column, would give 5 plates and would allow a separation of compounds with a 36 ° boiling-point difference. Vigreux columns are often used in small-scale distillations because of their small **holdup** (the amount of liquid retained by the column). The 20-cm column mentioned above would only retain 1 mL of liquid at the end of the distillation.

The most common type of column is made by packing a tube with some material such as glass (beads, short sections of tubing) or metal (stainless steel sponge, copper scouring pad, or Heli-Pak). The glass has the advantage that it does not react with organic compounds, whereas the metal may react with such compounds as alkyl halides.

Two commercial packings (helices and Heli-Pak) listed in Table 7–3 are relatively expensive. Glass beads or short sections of glass tubing (Raschig rings) are often packing alternatives. These packings are held in place with tight-fitting sections of glass tubing or a small amount of stainless steel sponge (Figure 7–7B). An inexpensive metal packed column is prepared by inserting part of a stainless steel sponge or copper scouring pad in one of the condensers available in a typical organic kit and tamping it gently with a pencil (Figure 7–7C).

Many fractionating columns must be insulated so that temperature equilibrium is established at all times. When this is necessary, a simple way to prepare insulation is to sandwich fine glass wool between aluminum foil with the shiny side out. The edges of the foil are crimped so that the glass wool is not visible. The column is then wrapped with this insulation. A piece of cloth may also be used as insulation.

It is also important to distill the mixture as slowly as possible. Much of the liquid should be allowed to return through the column to establish good vapor–liquid equilibrium. The values given in Table 7–3 depend on establishing good equilibrium. The distillation must be conducted fast enough to maintain a constant takeoff (rate at which material collects in the receiver) so that the temperature at the thermometer bulb remains constant. In other words, a distillation conducted too slowly will be unsatisfactory.

7.6 FRACTIONAL DISTILLATION: METHODS

For a fractional distillation, the apparatus shown in Figure 7–2 and a column similar to one of those shown in Figure 7–7 are used. The distilling flask, condenser, and vacuum adapter should be clamped so that the column is perpendicular to the desk top. No cooling water is used in the fractionating column. The receiving flask will have to be supported by wire gauze on an iron ring attached to a ring stand. If an oil bath or a heating mantle is used, wooden blocks should be placed under the heat source so that it can be removed easily.

The steps given in Technique 6, Section 6.4, p. 556 for simple distillation are also used in fractional distillation. Besides those steps, care must often be taken to **insulate** the column with glass wool and aluminum foil, or a cloth towel if the material to be distilled has a high boiling point, and to distill as **slowly** as possible (Section 7.5).

For the best possible separation, the temperature of the material in the distilling flask should be raised slowly so that the liquid–vapor can move up the column and be equilibrated. If the contents of the flask are heated too quickly, the column fills with liquid (flooding). Flooding decreases the efficiency of the separation. If there is flooding, the heat source must be lowered so the liquid can return to the distilling flask.

The temperature at the thermometer bulb should remain constant as the pure low-boiling component is removed (Figure 7–3). When most of this component is distilled, the distillation rate decreases. At this point, the distilling flask is heated to a higher temperature, and an intermediate fraction is collected until the temperature of the vapor at the thermometer bulb stabilizes at the higher value. The higher-boiling component is collected in another container. Ideally, results should be as shown in Figure 7–3.

7.7 NONIDEAL SOLUTIONS: AZEOTROPES

Many mixtures of compounds, because of intermolecular attractions or repulsions, do not show ideal behavior. Because of this nonideal behavior, Raoult's law is not followed. There are two types of vapor–liquid composition diagrams that result from this nonideal behavior: the minimum-boiling-point and the maximum-boiling-point diagrams. The minimum or maximum points in such diagrams correspond to a constant-boiling mixture called an **azeotrope.** An azeotrope has a fixed composition, which cannot be altered by normal distillation (simple or fractional), and a fixed boiling point. Hence, an azeotrope acts as if it were a **pure** compound.

A. Minimum-Boiling-Point Diagrams

The most common two-component mixture that gives a minimum-boiling-point azeotrope is the ethanol–water system shown in Figure 7–8. A minimum-boiling-point azeotrope results from a slight incompatibility of the substances, which leads to higher-than-expected combined vapor pressures from the solution. The higher combined vapor pressures bring about a lower boiling point for the mixture than that of either of the two components. One notes that the azeotrope V_3 in Figure 7–8 has a composition of about 96% ethanol–4% water and a boiling point of 78.1 °C. This boiling point is not much lower than the boiling point of pure ethanol (78.3 °C). However, this small difference means that one can obtain only 96% ethanol–4% water in a simple or fractional distillation of an aqueous solution of ethanol. Even with the **best** fractionating column, one cannot obtain 100% ethanol! The remaining 4% of water can be removed by adding benzene and removing a different azeotrope, the benzene–water–ethanol azeotrope (bp 65 °C). Once the water is removed, the excess benzene is removed as an ethanol-benzene azeotrope (bp 68 °C). The resulting material will be free of water and is called "absolute" ethanol.

The behavior on fractional distillation of an ethanol–water mixture of composition X (Figure 7–8) can be described as follows. The mixture is heated (line XL_1) until it is observed to boil (L_1). The vapor (V_1) will be richer in the lower-boiling component, ethanol, than the original mixture. The condensate (L_2) is vaporized to give V_2. The process continues, following the lines to the right, until the azeotrope (V_3) is

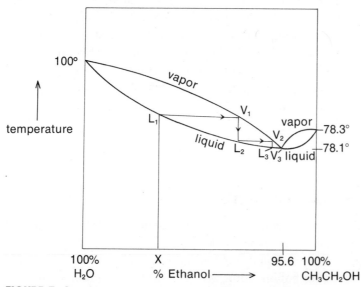

FIGURE 7–8. Ethanol–water minimum-boiling-point phase diagram

TABLE 7–4. Common Minimum-Boiling Azeotropes

AZEOTROPE	COMPOSITION (Weight percentage)	BOILING POINT (°C)
Ethanol–water	95.6% C_2H_5OH, 4.4% H_2O	78.17
Benzene–water	91.1% C_6H_6, 8.9% H_2O	69.4
Benzene–water–ethanol	74.1% C_6H_6, 7.4% H_2O, 18.5% C_2H_5OH	64.9
Methanol–carbon tetrachloride	20.6% CH_3OH, 79.4% CCl_4	55.7
Ethanol–benzene	32.4% C_2H_5OH, 67.6% C_6H_6	67.8
Methanol–toluene	72.4% CH_3OH, 27.6% $C_6H_5CH_3$	63.7
Methanol–benzene	39.5% CH_3OH, 60.5% C_6H_6	58.3
Cyclohexane–ethanol	69.5% C_6H_{12}, 30.5% C_2H_5OH	64.9
2-Propanol–water	87.8% $(CH_3)_2CHOH$, 12.2% H_2O	80.4
Butyl acetate–water	72.9% $CH_3COOC_4H_9$, 27.1% H_2O	90.7
Phenol–water	9.2% C_6H_5OH, 90.8% H_2O	99.5

obtained. The distillate is not pure ethanol but contains 96% ethanol and 4% water. The contents of the distilling flask become progressively richer in the higher-boiling component, water, as the distillation proceeds. When all the alcohol is removed as the azeotrope, pure water remains, distilling at 100 °C. A 96% ethanol solution acts as if it were a pure liquid and boils at 78.1 °C with no change in composition of the vapor.

Some common minimum-boiling azeotropes are given in Table 7–4. Numerous other azeotropes are formed in two- and three-component systems; such azeotropes are common. Water forms azeotropes with many substances; thus, water must be carefully removed with **drying agents** before compounds are distilled. Extensive azeotropic data are available in references such as the *Handbook of Chemistry and Physics*.[1]

B. Maximum-Boiling-Point Diagrams

A two-component maximum-boiling-point phase diagram is shown in Figure 7–9. A maximum-boiling-point azeotrope results from a slight attraction between the component molecules, which leads to lower combined vapor pressures in the solution. The lower combined vapor pressures cause a higher boiling point than what would be characteristic of either of the two components. Since the azeotrope has a higher boiling point than any of the components, it will be concentrated in the distilling flask, as the distillate (pure B) is removed. The distillation of a solution of composition X would follow to the right along the lines in Figure 7–9. Once the composition of the material remaining in the distilling flask has reached that of the azeotrope, the temperature will rise and the azeotrope will begin to distill. The azeotrope will continue to distill until the material in the distillation flask has been exhausted.

Some maximum-boiling azeotropes are listed in Table 7–5. They are not nearly as common as minimum-boiling azeotropes.[1]

---

[1]More examples of azeotropes, with their compositions and boiling points, can be found in the *CRC Handbook of Chemistry and Physics;* also in L. H. Horsley, ed., *Advances in Chemistry Series*, no. 116. *Azeotropic Data, III*, (Washington: American Chemical Society, 1973).

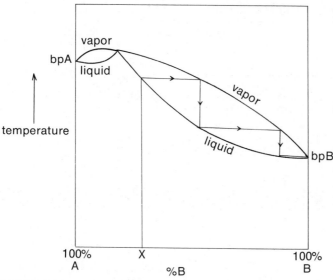

FIGURE 7–9. A maximum-boiling-point phase diagram

7.8 AZEOTROPIC DISTILLATION: APPLICATIONS

Examples are numerous of chemical reactions in which the amount of product is low because of an unfavorable equilibrium. An example is the direct acid-catalyzed esterification of an acid with an alcohol:

$$R-\overset{\overset{\textstyle O}{\|}}{C}-OH + ROH \underset{}{\overset{H^+}{\rightleftharpoons}} R-\overset{\overset{\textstyle O}{\|}}{C}-OR + H_2O$$

Since the equilibrium does not favor formation of the ester, it must be shifted to the right, in favor of the product, by using an excess of one of the starting materials. In most cases, the alcohol is the least expensive reagent and is the material used in excess. Isopentyl acetate (Experiment 10), methyl salicylate (Experiment 11), and benzocaine (Experiment 44) are examples of esters prepared by using one of the starting materials in excess.

Another way of shifting the equilibrium to the right is to remove one of the products from the reaction mixture as it is formed. In the above example, water can be

TABLE 7–5. Maximum-Boiling Azeotropes

AZEOTROPE	COMPOSITION (Weight percentage)	BOILING POINT (°C)
Acetone–chloroform	20.0% CH_3COCH_3, 80.0% $CHCl_3$	64.7
Chloroform–methyl ethyl ketone	17.0% $CHCl_3$, 83.0% $CH_3COCH_2CH_3$	79.9
Hydrochloric acid	20.2% HCl, 79.8% H_2O	108.6
Acetic acid–dioxane	77.0% CH_3COOH, 23.0% $C_4H_8O_2$	119.5
Benzaldehyde–phenol	49.0% C_6H_5CHO, 51.0% C_6H_5OH	185.6

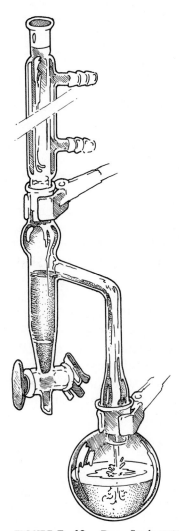

FIGURE 7–10.   Dean-Stark water separator

removed as it is formed by **azeotropic distillation,** using the Dean-Stark water separator shown in Figure 7–10. In this technique, an inert solvent, commonly benzene or toluene, is added to the reaction mixture contained in a round-bottomed flask. The sidearm of the water separator is also filled with this solvent. If benzene is used, as the mixture is refluxed the benzene–water azeotrope (bp 69.4 °C, Table 7–4) distills out of the flask.[2] When the vapor condenses, it enters the sidearm directly below the condenser, and water separates from the benzene-water condensate. Once the water (lower phase) separates from the benzene (upper phase), liquid benzene overflows from the sidearm back into the flask. The cycle is repeated continuously until no more water

[2]Actually, with ethanol a lower-boiling three-component azeotrope distills at 64.9 °C (see Table 7–4). It consists of benzene–water–ethanol. Because some ethanol is lost in the azeotropic distillation, a large excess of ethanol is used in esterification reactions. The excess also helps to shift the equilibrium to the right.

forms in the sidearm. One may calculate the weight of water that should theoretically be produced and compare this value with the amount of water collected in the sidearm. Since the density of water equals unity, the volume of water collected can be compared directly with the calculated amount of water obtained, assuming 100% yield.

The most important theoretical consideration in using azeotropic distillation is that the azeotrope containing water must have a **lower boiling point** than the alcohol used in preparing the ester. With ethanol, the benzene–water azeotrope boils at a much lower temperature (69.4 °C) than ethanol (78.3 °C), and the technique described above works well. With higher-boiling alcohols, azeotropic distillation also works well because of the large boiling-point difference between the azeotrope and the alcohol.

However, with methanol (bp 65 °C), the boiling point of the benzene–water azeotrope is actually **higher** by about 5 °, and methanol distills first. Thus, in esterifications involving methanol, a totally different approach must be taken. For example, one can mix carboxylic acid, methanol, acid catalyst, and **1,2-dichloroethane** in a conventional reflux apparatus (Technique 1, Figure 1–4, p. 510) without a Dean-Stark trap. During the reaction, water separates from the 1,2-dichloroethane, because it is not soluble. However, the remainder of the components are soluble in that solvent, so the reaction can continue. The equilibrium is shifted to the right by the "removal" of the water from the reaction mixture.

Azeotropic distillation is also used in other types of reactions, such as ketal-acetal formation and enamine formation. The use of azeotropic distillation is illustrated in the formation of 2-acetylcyclohexanone (Experiment 47) via the enamine intermediate. Toluene is used in the azeotropic distillation of water. The apparatus shown in Experiment 47 is an adaptation of the Dean-Stark water separator.

$$\text{ACETAL FORMATION} \qquad R-\overset{\overset{\textstyle O}{\|}}{C}-H + 2ROH \underset{H^+}{\rightleftharpoons} R-\overset{\overset{\textstyle OR}{|}}{\underset{\underset{\textstyle OR}{|}}{C}}-H + H_2O$$

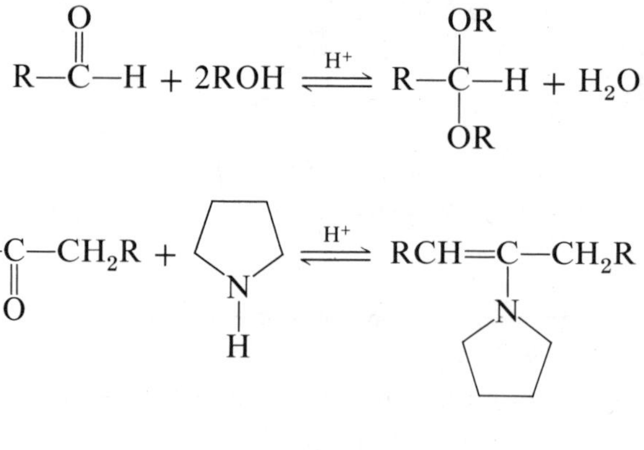

$$\text{ENAMINE FORMATION} \qquad RCH_2-\overset{\|}{\underset{O}{C}}-CH_2R + \Big\langle \overset{}{\underset{\overset{N}{\underset{H}{|}}}{}} \Big\rangle \underset{H^+}{\rightleftharpoons} RCH=\underset{N}{\overset{|}{C}}-CH_2R + H_2O$$

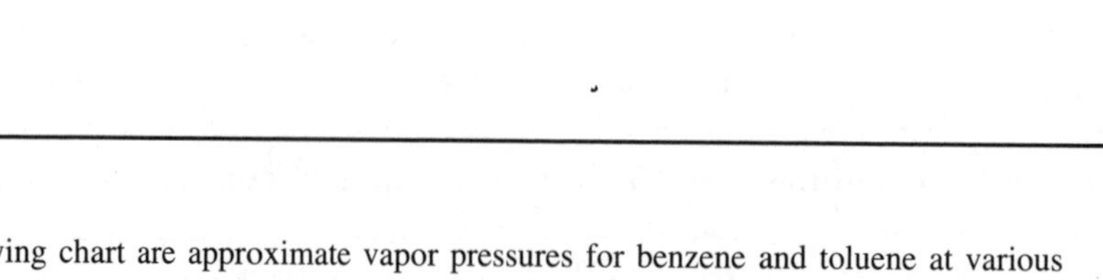

PROBLEMS

1. In the accompanying chart are approximate vapor pressures for benzene and toluene at various temperatures.

TEMP. (°C)	mmHg	TEMP. (°C)	mmHg
Benzene 30	120	Toluene 30	37
40	180	40	60
50	270	50	95
60	390	60	140
70	550	70	200
80	760	80	290
90	1010	90	405
100	1340	100	560
		111	760

(a) What is the mole fraction of each component if 39 g of benzene, C_6H_6, is dissolved in 46 g of toluene, C_7H_8?

(b) Assuming that this mixture is ideal, that is, it follows Raoult's law, what is the partial vapor pressure of benzene in this mixture at 50 °C?

(c) Estimate to the nearest degree the temperature at which the vapor pressure of the solution equals 1 atm (bp of the solution).

(d) Calculate the composition of the vapor (mole fraction of each component) that is in equilibrium with the solution at the boiling point of this solution.

(e) Calculate the composition in weight percentage of the vapor that is in equilibrium with the solution.

2. Estimate how many theoretical plates are needed to separate a mixture that has a mole fraction of B equal to 0.70 (70% B) in Figure 7–4.

3. Two moles of sucrose are dissolved in eight moles of water. Assume that the solution follows Raoult's law and that the vapor pressure of sucrose is negligible. The boiling point of water is 100 °C. The distillation is carried out at 1 atm (760 mmHg).

(a) Calculate the vapor pressure of the solution when the temperature reaches 100 °C.

(b) What temperature would be observed during the entire distillation?

(c) What would be the composition of the distillate?

(d) If a thermometer were immersed below the surface of the liquid of the boiling flask, what would be observed?

4. Explain why the boiling point of a two-component mixture rises slowly throughout a simple distillation when the boiling-point differences are not large.

5. Given the boiling points of several known mixtures of A and B (mole fractions are known) and the vapor pressures of A and B in the pure state (P_A^0 and P_B^0) at these same temperatures, how would you construct a boiling point–composition phase diagram for A and B?

6. Describe the behavior on distillation of a 98% ethanol solution through an efficient column. Refer to Figure 7–8.

7. Construct an approximate boiling point–composition diagram for a benzene–methanol system. The mixture shows azeotropic behavior. Include on the graph the boiling points of pure benzene and pure methanol and the boiling point of the azeotrope. Qualitatively describe the behavior for a mixture that is initially rich in benzene and a mixture initially rich in methanol.

8. Construct an approximate boiling point–composition diagram for an acetone–chloroform system (maximum-boiling mixture). Include on the graph the boiling points of each of the pure components and the boiling point of the azeotrope. Qualitatively describe the behavior on distillation of a mixture that is rich in acetone (90%) and of a mixture rich in chloroform (90%).

9. Two compounds have a difference in boiling point of 20 °C. How many theoretical plates may be needed to separate these substances? How long must a Vigreux column be to separate this mixture? How many milliliters of liquid will be held up in the column?

Technique 8
Steam Distillation

The simple and fractional distillations described in Techniques 6 and 7 are applicable to completely soluble (miscible) mixtures only. When liquids are not mutually soluble (that is, are immiscible), they can also be distilled. A mixture of immiscible liquids will boil at a lower temperature than the boiling points of any of the separate components as pure compounds. When steam is used to provide one of the immiscible phases, the process is called **steam distillation.** The advantage of this technique is that the desired material distills at a temperature below 100 °C. Thus, if unstable or very high-boiling substances are to be removed from a mixture, decomposition is avoided. Since all gases mix, the two substances can mix in the vapor and codistill. Once the distillate is cooled, the desired component, which is not miscible, separates from the water. Steam distillation is used widely in isolating liquids and solids from natural sources. It is also used in removing a reaction product from a tarry reaction mixture.

8.1 DIFFERENCES BETWEEN DISTILLATION OF MISCIBLE AND IMMISCIBLE MIXTURES

$$\text{MISCIBLE LIQUIDS} \qquad P_{\text{total}} = P_A{}^0 N_A + P_B{}^0 N_B \qquad (1)$$

Two liquids A and B that are mutually soluble (miscible) and that do not interact form an ideal solution and follow Raoult's law, as shown in equation (1). One notes that the vapor pressures of pure liquids $P_A{}^0$ and $P_B{}^0$ are not added directly to give the total pressure P_{total} but are reduced by the respective mole fractions N_A and N_B. The total pressure above a miscible or a homogeneous solution will depend on $P_A{}^0$ and $P_B{}^0$ and also N_A and N_B. Thus, the composition of the vapor will also depend on **both** the vapor pressures and the mole fractions of each component.

$$\text{IMMISCIBLE LIQUIDS} \qquad P_{\text{total}} = P_A{}^0 + P_B{}^0 \qquad (2)$$

On the other hand, when two mutually insoluble (immiscible) liquids are "mixed" to give a heterogeneous mixture, each exerts its own vapor pressure, independently of the other, as shown in equation (2). The mole fraction term does not appear in this equation, since the compounds are not miscible. One simply adds the vapor pressures of the pure liquids $P_A{}^0$ and $P_B{}^0$ at a given temperature to obtain the total pressure above the mixture. When the total pressure equals 760 mm, the mixture boils. The composition of the vapor from an immiscible mixture, in contrast to the miscible mixture, is determined only by the vapor pressures of the two substances codistilling. Equation (3) defines the composition of the vapor from an immiscible mixture. Calculations involving this equation are given in Section 8.2.

$$\frac{\text{Moles A}}{\text{Moles B}} = \frac{P_A{}^0}{P_B{}^0} \qquad (3)$$

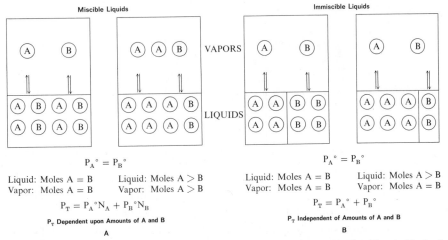

FIGURE 8–1.   **Total pressure behavior for miscible and immiscible liquids.** *A*, Ideal miscible liquids follow Raoult's law: P_T depends on the mole fractions and vapor pressures of A and B; *B*, Immiscible liquids do not follow Raoult's law: P_T depends only on the vapor pressures of A and B.

A mixture of two immiscible liquids boils at a lower temperature than the boiling points of either component. The explanation for this behavior is like that given for minimum-boiling azeotropes (Technique 7, Section 7.7). One may hold that immiscible liquids behave as they do because an extreme incompatibility between the two liquids leads to higher combined vapor pressures than Raoult's law would predict. The higher combined vapor pressures cause a lower boiling point for the mixture than for either single component. Thus, one may think of steam distillation as a special type of azeotropic distillation, in which the substance is completely insoluble in water.

The differences in the behavior of miscible and immiscible liquids, where P_A^0 equals P_B^0, are shown in Figure 8–1. One notes that with miscible liquids, the composition of the vapor depends on the relative amounts of A and B present (Figure 8–1A). Thus, the composition of the vapor must change during a distillation. In contrast, the composition of the vapor with immiscible liquids is independent of the amount of A and B present (Figure 8–1B). Hence, the vapor composition must remain **constant** during the distillation of such liquids, as predicted by equation (3). Immiscible liquids act as if they were being distilled simultaneously from separate compartments, as shown in Figure 8–1B, even though in practice they are "mixed" during a steam distillation. Since all gases mix, they do give rise to a homogeneous vapor and codistill.

8.2   IMMISCIBLE MIXTURES: CALCULATIONS

We have stated that the composition of a distillate is constant during a steam distillation. This would also require that the boiling point of the mixture be constant. The boiling point of the mixtures will be below the boiling points of water (100 °C) and the pure substance. Some representative boiling points and compositions of steam distill-

TABLE 8-1. Boiling Points and Compositions of Steam Distillates

MIXTURE	BOILING POINT OF PURE SUBSTANCE (°C)	BOILING POINT OF MIXTURE (°C)	COMPOSITION (% WATER)
Benzene–water	80.1	69.4	8.9%
Toluene–water	110.6	85.0	20.2%
Hexane–water	69.0	61.6	5.6%
Heptane–water	98.4	79.2	12.9%
Octane–water	125.7	89.6	25.5%
Nonane–water	150.8	95.0	39.8%
1-Octanol–water	195.0	99.4	90.0%

ates are given in Table 8–1. Note that the higher the boiling point of a pure substance, the more closely the temperature of the steam distillate approaches, but does not exceed, 100 °C. A substance may be codistilled with water at a temperature below 100 °C. This avoids the decomposition that might otherwise result at high temperatures with a simple distillation.

For immiscible liquids, the molar proportions of two components in a distillate equal the ratio of their vapor pressures in the boiling mixture, as given in equation (3). When the general equation (3) is rewritten for an immiscible mixture involving water, equation (4) results. Equation (4) can be modified by substituting the relation moles = (weight/molecular weight) to give equation (5).

$$\frac{\text{Moles substance}}{\text{Moles water}} = \frac{P^0_{\text{substance}}}{P^0_{\text{water}}} \tag{4}$$

$$\frac{\text{Wt substance}}{\text{Wt water}} = \frac{(P^0_{\text{substance}})(\text{Molecular weight}_{\text{substance}})}{(P^0_{\text{water}})(\text{Molecular weight}_{\text{water}})} \tag{5}$$

Consider, as a sample calculation, the steam distillation of 1-octanol. From Table 8–1, you see that the mixture boils at 99.4 °C. From a handbook, one obtains the vapor pressure of pure water at 99.4 °C (744 mm). Since P_{total} must equal 760 mm, the vapor pressure of 1-octanol at 99.4 °C must equal 16 mm. Equation (5) is now solved as follows:

$$\frac{\text{Wt 1-octanol}}{\text{Wt water}} = \frac{(16)(130)}{(744)(18)} = 0.155 \text{ g/g water}$$

Thus, 0.155 g of 1-octanol codistills with each gram of water. It will require 100 g of water to remove 15.5 g of 1-octanol from a distilling flask. The distillate will have the calculated composition of 87% water and 13% 1-octanol, which is very close to the experimental value given in Table 8–1.

8.3 STEAM DISTILLATION: METHODS

Two methods for steam distillation are generally used in the laboratory. The first method uses live steam, from a steam line, passed into the flask containing the compound. In the second method, steam is generated in situ by heating a flask containing the compound and water.

A. Live Steam Method

The method with live steam is the most widely used, especially with high-molecular-weight (low-vapor-pressure) substances. It can even be used with volatile solids. The apparatus is assembled as shown in Figure 8–2. A piece of 6-mm glass tubing is bent to fit the flask and serves as a steam-inlet tube. The ground-glass joints must be lubricated, and the apparatus must be fastened securely to a ring stand. During the distillation, the joint between the condenser and the distilling head sometimes separates due to vibration. If this happens, vapor will be lost unless the joint is resealed immediately. The distilling flask should never be filled more than half full. The flask should be placed about 15 cm above the desk top. This distillation method does not usually require an external heating source, since the mixture is maintained at the boiling point with steam. Sometimes the distilling flask may begin to fill with water during the distillation. If this happens, it may be necessary to use a heating source such as a Bunsen burner to heat the liquid to its boiling point. The heat from the burner should be dispersed by means of a wire gauze. The entering steam helps to prevent bumping of the mixture. The Claisen head helps to reduce the possibility that material will be transferred to the receiving flask through excessive bumping.

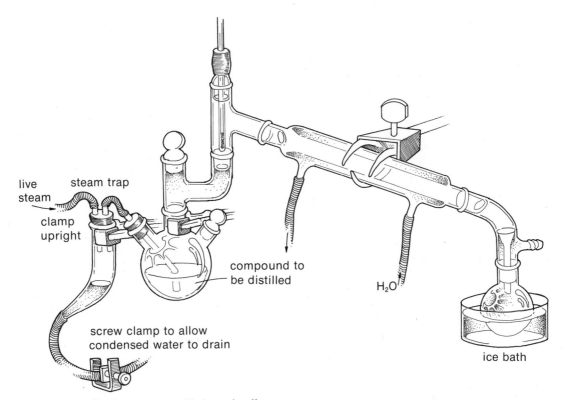

FIGURE 8–2. Steam distillation using live steam

As shown in Figure 8–2, a steam-water trap must be placed in the steam line. When the steam valve is first turned on, a large amount of water (condensate) comes out of the valve. To prevent this condensate from entering the distilling flask and filling it with water, one must open the screw clamp on the trap **fully** so water can drain before reaching the flask. Once the water is drained, steam will begin to enter the distilling flask. The screw clamp on the trap may then be closed. Occasionally, the clamp should be opened to allow condensate to be removed during the distillation. An ice bath increases the efficiency of the condensation of the distillate in the receiving flask. One should adjust the rate of flow of the steam so that vapor passes into the condenser as quickly as possible but continues to be condensed. The flow of cooling water through the condenser should be faster than in other types of distillations to help cool the vapor. The vacuum adapter should be cool to the touch during the distillation. The condensate will be cloudy while the steam volatile material is distilling. The substances that codistill will separate on cooling, to give this cloudiness. Once the distillate is clear, the distillation is nearly complete. It is considered good practice to remove at least 10 mL more distillate after this point. When the distillation is to be stopped, the screw clamp on the steam trap must be opened completely and the steam inlet tube must be removed from the distilling flask. If this is not done, the liquid will back up into the tube and steam trap.

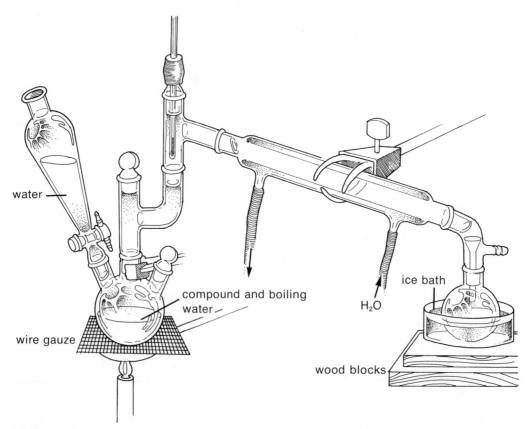

FIGURE 8–3.   Direct steam distillation

B. Direct Method

The direct method is experimentally more convenient. Steam is produced in the distilling flask (in situ) by heating water to the boiling point in the presence of the compound to be distilled (Figure 8–3). As the steam is removed with the compound, water is added dropwise from the separatory funnel. This method works well for volatile liquids or for small amounts of material. It is mainly used for mixtures that do not have solids present. Solids may cause excessive bumping. In addition, foaming may be more troublesome. The live-steam method often eliminates these two problems.

PROBLEMS

1. Calculate the weight of benzene codistilled with each gram of water and the percentage composition of the vapor produced during a steam distillation. The boiling point of the mixture is 69.4 °C. The vapor pressure of water at 69.4 °C is 227.7 mmHg. Compare the result with the data in Table 8–1.
2. Calculate the approximate boiling point of a mixture of bromobenzene and water at atmospheric pressure. A table of vapor pressures of water and bromobenzene at various temperatures is given.

TEMPERATURE (°C)	VAPOR PRESSURES (mmHg)	
	Water	Bromobenzene
93	588	110
94	611	114
95	634	118
96	657	122
97	682	127
98	707	131
99	733	136

3. Calculate the weight of nitrobenzene that codistills (bp 99 °C) with each gram of water during a steam distillation. You may need the data given in the previous problem.
4. A mixture of *p*-nitrophenol and *o*-nitrophenol can be separated by steam distillation. The *o*-nitrophenol is steam-volatile while the *para* isomer is not volatile. Explain. Base your answer on the ability of the isomers to form hydrogen bonds internally.

Technique 9
Manometers: Measuring Pressure

The earth's atmosphere is a mixture of gases, primarily nitrogen (78.08%), oxygen (20.95%), argon (0.9%), and carbon dioxide (0.03%). There is also water vapor, the amount depending on weather conditions. In various localities, pollutants are present also. This layer of gases that make up the atmosphere is more dense near the surface of the earth than at higher elevations, mainly from the influence of gravitational attraction. These gaseous constituents of air exert pressure on all objects within the earth's atmosphere. The pressure arises from the bombardment of objects by the atoms and molecules in the air.

9.1 THE BAROMETER

Atmospheric pressure is most often measured with a device known as a **barometer.** The barometer was invented by Torricelli in the seventeenth century. It is constructed by taking a glass tube that is sealed at one end, filling it with mercury, and then inverting the tube in another container of mercury so that the open end is below the surface of the mercury in the second container. When this is done, the level of the

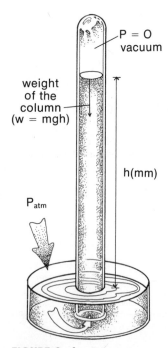

FIGURE 9–1. Barometer

mercury in the sealed tube falls because gravity attracts the column of mercury. Recall that

$$\text{weight} = mgh = (\text{mass})(\text{gravitational force constant})(\text{height})$$

The tube of mercury does not empty completely. At some point, the weight of the descending column of mercury (its downward force) is equaled by the force that the atmosphere exerts on the surface of the mercury in the container. This force is transmitted through the liquid and pushes upward on the descending column. Since the voided space at the top of the column contains no air, it is a vacuum, and the pressure is zero. The only downward force comes from the weight of the column of mercury. Hence, as shown in Figure 9–1, the atmospheric pressure supports the column of mercury, and the height of the column supported is directly proportional to the pressure the atmosphere exerts.

The height of the column of mercury in a barometer varies both with elevation (it goes down at high elevation, where air is less dense) and prevailing weather conditions. Under normal conditions and at sea level, the pressure atmospheric gases exert supports a column of mercury 760 mm high. Hence, 1 atmosphere is defined to correspond to 760 mm of mercury, or 760 Torr (1 mmHg = 1 Torr), a unit named in honor of Torricelli.

9.2   THE OPEN-END MANOMETER

To measure the pressure of a gas sample, one uses a **manometer.** There are two basic types of manometers, the **open-end** and the **closed-end** manometer. The closed-end manometer is described in Section 9.3.

An open-end manometer is basically a U tube filled with mercury. One end is opened to the atmosphere (hence, open end), and the other end is connected to a flask or bulb filled with a gas at a pressure higher or lower than that exerted by the atmosphere. In Figure 9–2, atmospheric pressure supports the column on the right side of the U tube, and the gas pressure in the bulb supports the column on the left side. The column on the left side is higher than the right column by exactly the pressure difference between the sample of gas and the atmosphere: Δh (mm) = $(P_{gas} - P_{atm})$. The atmospheric pressure is obtained by reading a barometer. The gas pressure P_{gas} is then calculated as follows: $P_{gas} = P_{atm} + \Delta h$. Thus, the gas pressure in the bulb is **greater** than the pressure of the atmosphere by the amount Δh. On the other hand, if the pressure in the bulb were lower than the pressure of the atmosphere, then the right column would be higher than the left. For this circumstance, the gas pressure in the bulb is calculated as follows: $P_{gas} = P_{atm} - \Delta h$.

A simple type of open-end manometer is shown in Figure 9–3. It is simply a glass U tube mounted on a board to which a meter stick is affixed for measuring Δh. Using such a device, one can measure the pressure of a gas sample that has pressure either greater than atmospheric pressure or less by adding or subtracting, respectively, Δh from the known barometric pressure.

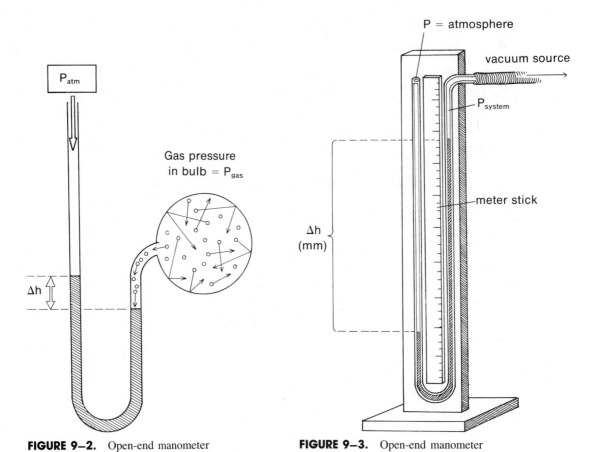

FIGURE 9–2. Open-end manometer

FIGURE 9–3. Open-end manometer

9.3 THE CLOSED-END MANOMETER

The other principal type of manometer is the **closed-end manometer.** It can only be used to measure pressures that are lower than atmospheric pressure. Two basic types are shown in Figure 9–4. The manometer shown in Figure 9–4A consists of a U tube that is closed at one end and mounted on a wooden support. It is more difficult to construct this type of manometer than an open-end manometer, because it must be filled under vacuum. One can construct the manometer from 9-mm capillary tubing and fill it as shown in Figure 9–5. The U tube is evacuated with a good vacuum pump, and then the mercury is introduced by tilting the mercury reservoir. When the vacuum is interrupted by admitting air, the mercury is forced by atmospheric pressure to the end of the evacuated U tube. The manometer is then ready for use. The constriction shown in Figure 9–5 helps to protect the manometer against breakage when the pressure is released. When an aspirator or any other vacuum source is used, a manometer can be connected into the system. As the pressure is lowered, the mercury rises in the right tube and drops in the left tube until Δh corresponds to the pressure of the system (see Figure 9–4A). A short piece of a metric ruler or a piece of graph paper ruled in millimeter squares is mounted on the support board to allow Δh to be read. No addition

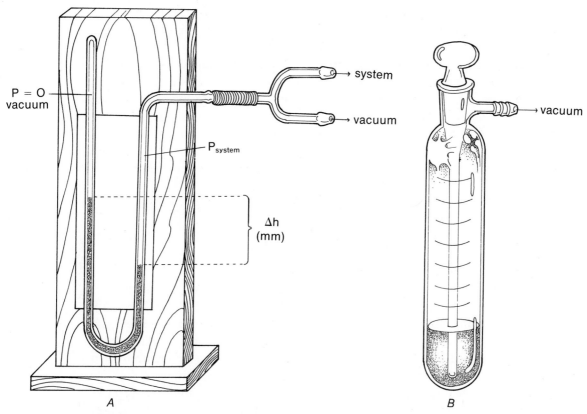

FIGURE 9–4. Closed-end manometers: *A*, U-tube type; *B*, Commercial type

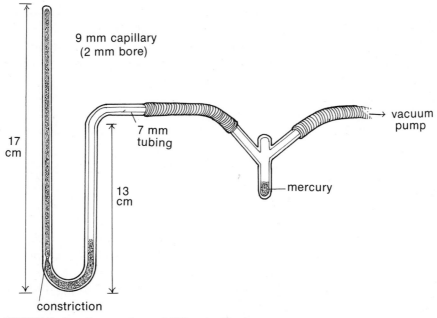

FIGURE 9–5. Constructing and filling the U-tube manometer

or subtraction is necessary, since the reference pressure (created by the initial evacuation when filling) is zero. Hence, the height difference Δh gives the pressure in the system directly.

A barometer must be at least 760 mm tall; an open-end manometer is often 1 m tall. A closed-end manometer is often less than 20 cm long and is thus very convenient for laboratory use. These manometers are conveniently used with an aspirator that only rarely creates pressures lower than 10 to 20 mmHg. Since the open-end and the closed-end manometers cannot be read more finely than ±1 mmHg, they are not suitable for very low pressures. They should not be used for high-vacuum applications. Other types of manometers must be used with high vacuum. The manometer common for high-vacuum systems is the McLeod gauge, which will not be discussed here.

9.4 CONNECTING AND USING A MANOMETER

The most common use of a closed-end manometer is to monitor pressure during a reduced-pressure distillation, as discussed in Technique 6, Sections 6.5, 6.6, and 6.7, pp. 558–563. The manometer is placed in a vacuum-distillation system as shown in Figure 9–6. Generally an aspirator is the source of vacuum. Both the manometer and the distillation apparatus should be protected by a trap from possible backups in the water line. An alternative trap arrangement is shown in Technique 6, Figure 6–7, p. 559. The trap assembly shown in Figure 9–6 is more easily constructed. You may notice in the figure that the trap has a device for opening the system to the atmosphere. This is especially important in using a manometer, since one must always make pressure changes slowly. If this is not done, there is danger of spraying mercury throughout the system, breaking the manometer, or spurting mercury into the room. In the closed-end manometer, if the system is opened suddenly, the mercury will rush to the closed end of the U tube with such speed and force that the end will often be broken out of the manometer. Air should be admitted **slowly** by opening the valve (Figure 9–6) on the trap cautiously. Similarly, when the vacuum is being started, the valve should be closed slowly so that mercury will not suddenly be sucked out of the gauge. Rapid opening of an open-end manometer system inevitably results in spraying a fountain of mercury through the open end.

When one is conducting a reduced-pressure distillation, if the pressure is lower than desired, it is possible to adjust it by means of a **bleed** valve. One removes the screw clamp on the valve shown in Figure 9–6 and connects the base of a Tirrill-style Bunsen burner to the water trap, as shown in the alternative arrangement. The needle valve in the burner can be used to adjust precisely the amount of air that is bleeding into the system and hence the pressure. When the valve is opened fully, atmospheric pressure is admitted; when it is fully closed, maximum vacuum is achieved.

The boiling point of a liquid is a function of the applied pressure. The boiling point is often reported at atmospheric pressure or some other pressure that cannot be exactly reproduced by using the aspirator. By the nomograph chart in Technique 6 (Figure 6–2, p. 552), it is possible to determine the boiling point of a liquid at any given pressure if the value is known for at least one other pressure.

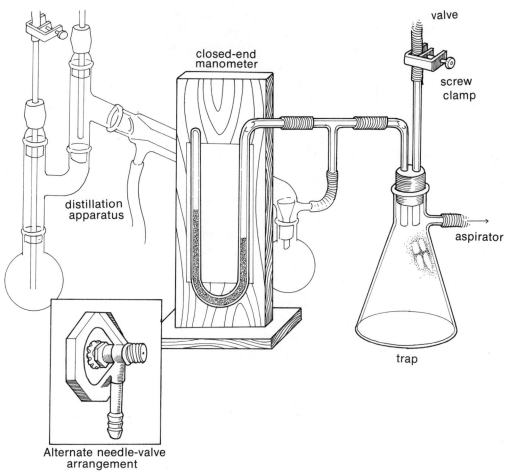

valve

screw
clamp

closed-end
manometer

distillation
apparatus

aspirator

trap

Alternate needle-valve
arrangement

FIGURE 9–6.   Method of connecting the manometer to a vacuum-distillation system

Technique 10
Column Chromatography

The most modern and sophisticated methods of separating mixtures that the organic chemist has available all involve **chromatography.** Chromatography is defined as the separation of a mixture of two or more different compounds (in some cases, ions) by distribution between two phases, one of which is stationary and one of which is moving. Various types of chromatography are possible, depending on the nature of the two phases involved: **solid-liquid** (column, thin-layer, and paper), **liquid-liquid,** and **gas-liquid** (vapor-phase) chromatographic methods are common.

All chromatography works on much the same principle as solvent extraction (Technique 5). Basically, the methods depend on differential solubilities (or adsorptivities) of the substances to be separated relative to the two phases between which they are to be partitioned. In this section, column chromatography, a solid-liquid method, is considered. Thin-layer chromatography is examined in Technique 11; gas chromatography, a gas-liquid method, is discussed in Technique 12.

10.1 ADSORBENTS

Column chromatography is a technique based on both adsorptivity and solubility. It is a solid-liquid phase-partitioning technique. The solid may be almost any material that does not dissolve in the associated liquid phase; those solids most commonly used are silica gel, $SiO_2 \cdot xH_2O$, also called silicic acid, and alumina, $Al_2O_3 \cdot xH_2O$. These compounds are used in their powdered or finely ground (usually 200- to 400-mesh) forms.

Most alumina used for chromatography is prepared from the impure ore bauxite, $Al_2O_3 \cdot xH_2O + Fe_2O_3$. The bauxite is dissolved in hot sodium hydroxide and filtered to remove the insoluble iron oxides; the alumina in the ore forms the soluble amphoteric hydroxide $Al(OH)_4^-$. The hydroxide is precipitated by CO_2 (which reduces the pH) as $Al(OH)_3$. When heated, the $Al(OH)_3$ loses water to form pure alumina, Al_2O_3.

$$\text{Bauxite (crude)} \xrightarrow{\text{hot NaOH}} Al(OH)_4^- \text{ (aq)} + Fe_2O_3 \text{ (insoluble)}$$

$$Al(OH)_4^- \text{ (aq)} + CO_2 \longrightarrow Al(OH)_3 + HCO_3^-$$

$$2Al(OH)_3 \xrightarrow{\Delta} Al_2O_3(s) + 3H_2O$$

Alumina prepared in this way is called **basic alumina,** because it still contains some hydroxides. Basic alumina cannot be used for chromatography of compounds that are base-sensitive. Therefore, it is washed with acid to neutralize the base, giving **acid-washed alumina.** This material is unsatisfactory unless it has been washed with enough water to remove **all** the acid; on being so washed, it becomes the best chromatographic material, called **neutral alumina.** If a compound is acid-sensitive, either basic or neutral alumina must be used. One should be careful to ascertain what type of alumina is being used for chromatography. Silica gel is not available in any form other than that suitable for chromatography.

10.2 INTERACTIONS

If powdered or finely ground alumina (or silica gel) is added to a solution containing an organic compound, some of the organic compound will **adsorb** onto or stick to the fine particles of alumina. Many kinds of intermolecular forces cause organic molecules to bind to alumina. These forces vary in strength according to their type. Nonpolar compounds bind to the alumina using only Van der Waals forces. These are weak forces,

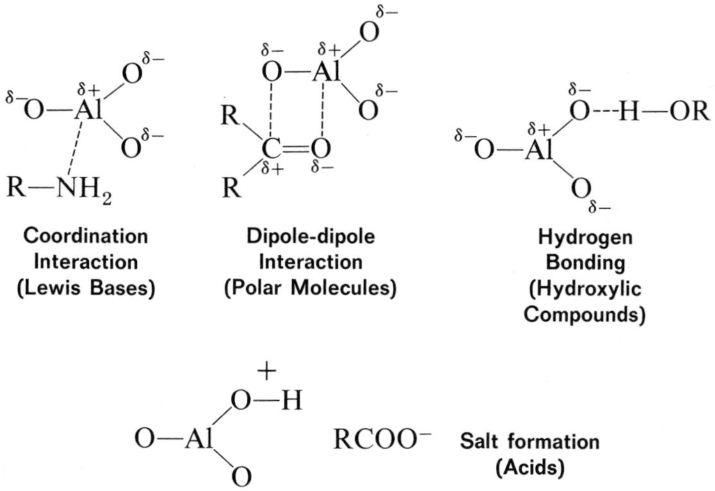

FIGURE 10–1. Possible interactions of organic compounds with alumina

and nonpolar molecules do not bind strongly unless they have extremely high molecular weights. The most important interactions are those typical of polar organic compounds. Either these forces are of the dipole-dipole type or they involve some direct interaction (coordination, hydrogen bonding, or salt formation). These types of interactions are illustrated in Figure 10–1, in which for convenience only a portion of the alumina structure is shown. Similar interactions occur with silica gel. The strengths of such interactions vary in the approximate order

Salt formation > coordination > hydrogen bonding > dipole-dipole > Van der Waals

Strength of interaction varies among compounds. For instance, a strongly basic amine would bind more strongly than a weakly basic one (by coordination). In fact, strong bases and strong acids often interact so strongly that they **dissolve** alumina to some extent. One can use the following rule of thumb: THE MORE POLAR THE FUNCTIONAL GROUP, THE STRONGER THE BOND TO ALUMINA (OR SILICA GEL).

A similar rule holds for solubility. Polar solvents dissolve polar compounds more effectively than nonpolar solvents; nonpolar compounds are dissolved best by nonpolar solvents. Thus, the extent to which any given solvent can wash an adsorbed compound from alumina depends almost directly on the solvent's relative polarity. For example, while a ketone adsorbed on alumina might not be removed by hexane, it might be removed completely by chloroform. For any adsorbed material, a kind of **distribution** equilibrium can be envisioned between the adsorbent material and the solvent. This is illustrated in Figure 10–2.

The distribution equilibrium is **dynamic,** with molecules constantly **adsorbing** from the solution and **desorbing** into it. The average number of molecules remaining adsorbed on the particle at equilibrium depends both on the particular molecule (RX) involved and the dissolving power of the solvent with which the adsorbent must compete.

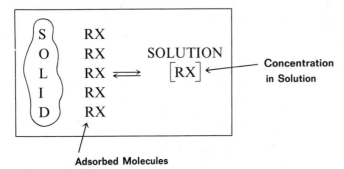

FIGURE 10–2. Dynamic adsorption equilibrium

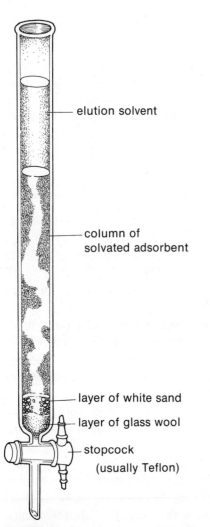

FIGURE 10–3. Chromatographic column

10.3 PRINCIPLE OF COLUMN CHROMATOGRAPHIC SEPARATION

The dynamic equilibrium mentioned above, and the variations in the extent to which different compounds adsorb on alumina (or silica gel), underlie a versatile and ingenious method for **separating** mixtures of organic compounds. In this method, the mixture of compounds to be separated is introduced onto the top of a cylindrical glass column (Figure 10–3) **packed** (filled) with fine alumina particles (stationary solid phase). The adsorbent is then continuously washed by a flow of solvent (moving phase) passing through the column.

Initially the components of the mixture adsorb onto the alumina particles at the top of the column. The continuous flow of solvent through the column **elutes,** or washes, the solutes off the alumina and sweeps them down the column. The solutes (or materials to be separated) are called **eluates** or **elutants;** and the solvents, **eluents.** As the solutes pass down the column to fresh alumina, new equilibria are established between the adsorbent, the solutes, and the solvent. The constant equilibration means that different compounds will move down the column at differing rates depending on their relative affinity for the adsorbent on one hand and for the solvent on the other. Since the number of alumina particles is large, since they are closely packed, and since fresh solvent is being added continuously, the number of equilibrations between adsorbent and solvent that the solutes experience is enormous.

As the components of the mixture are separated, they begin to form moving bands (or zones), each band containing a single component. If the column is long enough and the various other parameters (column diameter, adsorbent, solvent, and rate of flow) are correctly chosen, the bands separate from one another, leaving gaps of pure solvent in between. As each band (solvent and solute) passes out the bottom of the column, it can be collected completely before the next band arrives. If the parameters mentioned are poorly chosen, the various bands either overlap or coincide, in which case either a poor separation or no separation at all is the result. The chromatographic separation is illustrated in Figure 10–4.

10.4 PARAMETERS AFFECTING SEPARATION

Chromatography is truly a sophisticated method of separating mixtures. Its versatility results from the many factors that can be adjusted. These include the

1. Adsorbent chosen
2. Polarity of the solvent or solvents chosen
3. Size of the column (both length and diameter) relative to the amount of material to be chromatographed
4. Rate of elution (or flow)

By careful choice of conditions, almost any mixture can be separated. Recently the technique has even been used to separate optical isomers. An optically active solid-phase adsorbent was used to separate the enantiomers.

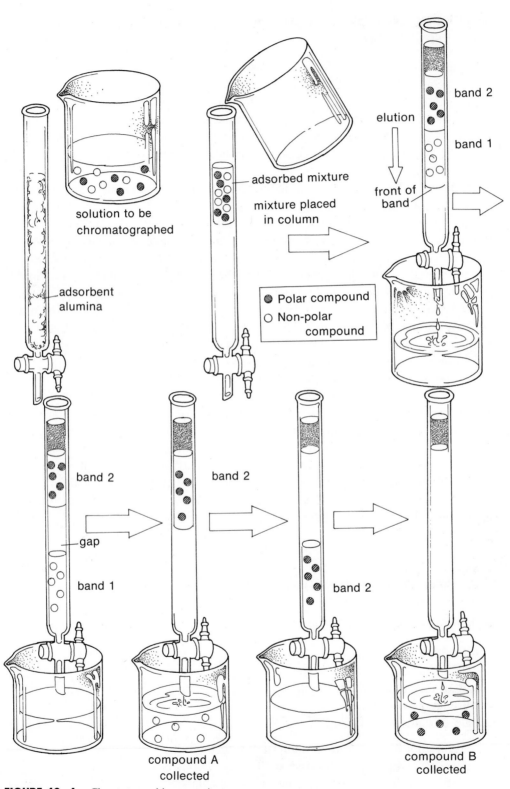

FIGURE 10—4. Chromatographic separation

TABLE 10–1. Solid Adsorbents for Column Chromatography

Paper	
Cellulose	
Starch	
Sugars	
Magnesium silicate	
Calcium sulfate	**Increasing strength of**
Silicic acid	**binding interactions**
Silica gel	**toward polar compounds**
Florisil	
Magnesium oxide	
Aluminum oxide (Alumina)*	
Activated charcoal (Norit)	

*Basic, acid-washed, and neutral.

Two fundamental choices for anyone attempting a chromatographic separation are the kind of adsorbent and the solvent system. In general, nonpolar compounds pass through the column faster than polar compounds, since they have a smaller affinity for the adsorbent. If the adsorbent chosen binds all the solute molecules (both polar and nonpolar) strongly, they will not move down the column. On the other hand, if too polar a solvent is chosen, all the solutes (polar and nonpolar) may simply be washed through the column, with no separation taking place. The adsorbent and the solvent should be chosen so that neither is excessively favored in the equilibrium competition for solute molecules.[1]

A. Adsorbents

In Table 10–1 various kinds of adsorbents (solid phases) used in column chromatography are listed. The choice of adsorbent often depends on the types of compounds to be separated. Cellulose, starch, and sugars are used for polyfunctional plant and animal materials (natural products) very sensitive to acid-base interactions. Magnesium silicate is often used for separating acetylated sugars, steroids, and essential oils. Silica gel and Florisil are relatively mild toward most compounds and are widely used for a variety of functional groups—hydrocarbons, alcohols, ketones, esters, acids, azo compounds, amines. Alumina is the most widely used adsorbent and is obtained in the three forms mentioned in Section 10.1: acidic, basic, and neutral. The pH of acidic or acid-washed alumina is approximately 4. This adsorbent is particularly useful for separating acidic materials such as carboxylic acids and amino acids. Basic alumina has a pH of 10 and is useful in separating amines. Neutral alumina can be used to separate a variety of nonacidic and nonbasic materials.

The approximate strength of the various adsorbents listed in Table 10–1 is also given. The order is only approximate, and therefore it may vary. For instance, the

---

[1]Often the chemist uses thin-layer chromatography (tlc), which is described in Technique 11, to arrive at the best choices of solvents and adsorbents for the best separation. The tlc experimentation can be done quickly and with extremely small amounts (microgram quantities) of the mixture to be separated. This brings great savings of time and materials. Technique 11 describes this use of tlc.

strength, or separating abilities, of alumina and silica gel largely depend on the amount of water present. Water binds very tightly to either adsorbent, taking up sites on the particles that could otherwise be used for equilibration. If one adds water to the adsorbent it is said to have been **deactivated.** Anhydrous alumina or silica gel are said to be highly **activated.** High activity is usually avoided in these adsorbents since use of very active forms may lead to certain types of destruction or decomposition of the compounds to be separated. Use of the highly active forms of either alumina or silica gel, or of the acidic or basic forms of alumina, can often lead to molecular rearrangement in certain types of solute compounds.

B. Solvents

In Table 10–2 some common chromatographic solvents are listed along with their relative ability to dissolve polar compounds. Sometimes a single solvent can be found that will separate all the components of a mixture. Sometimes a mixture of solvents can be found that will achieve separation. More often, one must start elution with a nonpolar solvent to remove relatively nonpolar compounds from the column and then gradually increase the solvent polarity to force compounds of greater polarity to come down the column, or elute. The approximate order in which various classes of compounds elute by this procedure is given in Table 10–3. In general, nonpolar compounds travel through the column faster (elute first), and polar compounds travel more slowly (elute last). However, molecular weight is also a factor in determining order of elution. A nonpolar compound of high molecular weight travels more slowly than a nonpolar compound of low molecular weight and may even be passed by some polar compounds.

When the polarity of the solvent has to be changed during a chromatographic separation, some precautions must be taken. Rapid changes from one solvent to another are to be avoided (especially when silica gel or alumina is involved). Usually, small percentages of a new solvent are mixed slowly into the one in use until the percentage

TABLE 10–2. Solvents (Eluents) for Chromatography

Petroleum ether	
Cyclohexane	
Carbon tetrachloride*	
Benzene*	
Chloroform*	**Increasing polarity**
Methylene chloride	**and "solvent power"**
Diethyl ether	**toward polar functional**
Ethyl acetate	**groups**
Acetone	
Pyridine	
Ethanol	
Methanol	
Water	
Acetic acid	↓

*Suspect carcinogens.

TABLE 10–3. Elution Sequence for Compounds

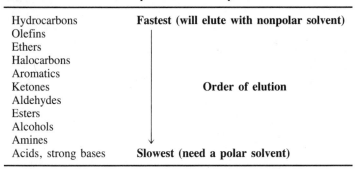

Hydrocarbons	**Fastest (will elute with nonpolar solvent)**
Olefins	
Ethers	
Halocarbons	
Aromatics	
Ketones	**Order of elution**
Aldehydes	
Esters	
Alcohols	
Amines	
Acids, strong bases	**Slowest (need a polar solvent)**

reaches the desired level. If this is not done, the column packing often ''cracks'' as a result of the heat liberated when alumina or silica gel is mixed with a solvent. The solvent solvates the adsorbent and the formation of a weak bond generates heat.

$$\text{Solvent} + \text{alumina} \longrightarrow (\text{alumina} \cdot \text{solvent}) + \text{heat}$$

Often enough heat is generated locally to evaporate the solvent. The formation of vapor creates bubbles, which force a separation of the column packing; this is called cracking. A cracked column does not give a good separation since it has discontinuities in the **packing** (column of adsorbent). The way in which a column is packed or filled with adsorbent is very important.

That the solvent itself has a tendency to adsorb on the alumina is an important factor in how compounds move down the column. The solvent can displace the adsorbed compound if it is more polar than the compound and hence can move it down the column. Thus, a more polar solvent not only dissolves more compound but also is effective in removing the compound from the alumina since it displaces the compound from its site of adsorption.

Certain solvents should be avoided with alumina or silica gel, especially the acidic, basic, and highly active forms. For instance, with any of these adsorbents, acetone dimerizes via an aldol condensation to give diacetone alcohol. Mixtures of esters transesterify (exchange their alcoholic portions) when ethyl acetate or an alcohol is the eluant. Finally, the most active solvents (pyridine, methanol, water, and acetic acid) dissolve and elute some of the adsorbent itself. Generally, try to avoid going to solvents more polar than diethyl ether or methylene chloride in the eluent series (Table 10–2).

C. Column Size and Adsorbent Quantity

The column size and the amount of adsorbent must also be selected correctly to separate a given amount of sample well. As a rule of thumb, the amount of adsorbent should be 25 to 30 times, by weight, the amount of material to be separated by chromatography. Additionally, the column should have a height-to-diameter ratio about 8:1. Some typical relations of this sort are given in Table 10–4.

TABLE 10–4. Size of Column and Amount of Adsorbent for Typical Sample Sizes

AMOUNT OF SAMPLE (g)	AMOUNT OF ADSORBENT (g)	COLUMN DIAMETER (mm)	COLUMN HEIGHT (mm)
0.01	0.3	3.5	30
0.10	3.0	7.5	60
1.00	30.0	16.0	130
10.00	300.0	35.0	280

Note, as a caution, that the difficulty of the separation is also a factor in determining the size and length of the column to be used and in the amount of adsorbent needed. Compounds that do not separate easily may require larger columns and more adsorbent than specified in Table 10–4. For easily separated compounds, a smaller column and less adsorbent may suffice.

D. Flow Rate

The rate at which solvent flows through the column also is significant in the effectiveness of a separation. In general, the time the mixture to be separated remains on the column is directly proportional to the extent of equilibration between stationary and moving phases. Thus, similar compounds eventually separate if they remain on the column long enough. The time a material remains on the column depends on the flow rate of the solvent. If the flow is too slow, however, the substances in the mixture, when they are in the solvent, may diffuse faster than the rate at which they move down the column. Then the bands grow wider and more diffuse, and the separation becomes poorer.

10.5 PACKING THE COLUMN: TYPICAL PROBLEMS

The most critical operation in column chromatography is packing (filling with adsorbent) the column. The column of alumina (or other solid adsorbent), the **column packing,** must be evenly packed and free of irregularities, air bubbles, and gaps. As a compound travels down the column, it moves in an advancing zone, or **band.** It is important that the leading edge, or **front,** of this band be horizontal, or perpendicular to the long axis of the column. If two bands are close together and do not have horizontal band fronts, it is impossible to collect each band exclusive of the other. The leading edge of the second band begins to elute before the first band has finished eluting. This condition can be seen in Figure 10–5. There are two main reasons for this problem: First, if the top surface edge of the adsorbent packing is not level, nonhorizontal bands result. Second, bands may also be nonhorizontal if the column is not held in an exactly vertical position in both planes (front-to-back and side-to-side). When you are preparing a column for use, both of these factors must be watched carefully.

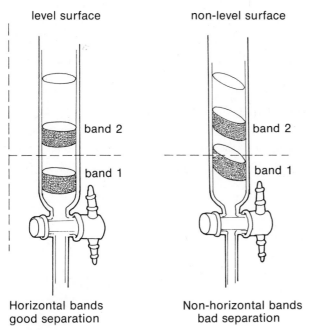

level surface   non-level surface

band 2

band 1

band 2

band 1

Horizontal bands
good separation

Non-horizontal bands
bad separation

FIGURE 10—5. Comparison of horizontal and nonhorizontal band fronts

Another phenomenon, called **streaming,** or **channeling,** occurs when part of the band front advances ahead of the major part of the band. Channeling occurs if there are any irregularities in the adsorbent surface or any irregularities of air bubbles in the packing. Often, large channels or air gaps form in the column packing, and a part of the advancing front gets ahead of the rest through the channel. Two examples of channeling are shown in Figure 10–6.

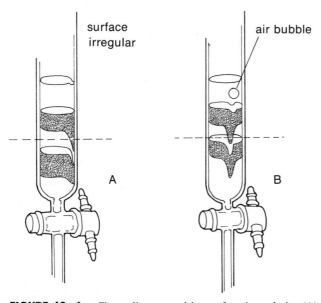

surface
irregular

air bubble

A

B

FIGURE 10—6. Channeling caused by surface irregularity (A) or an air bubble (B)

10.6 PACKING THE COLUMN: METHODS

The following methods are used to avoid the problems resulting from uneven packing and column irregularities. These procedures should be followed carefully in preparing a chromatographic column. Failure to pay close attention to the preparation of the column may well affect the quality of the separation.

      Preparation of a column involves two distinct stages. In the first stage, a support base on which the packing will rest is prepared. This must be done so that the packing, a finely divided material, does not wash out of the bottom of the column. In the second stage, the column of adsorbent is deposited on top of the supporting base. Several alternatives for the second process are detailed below.

A. Preparing the Support Base

First, the chromatographic column is clamped upright in a vertical position. The column (Figure 10–3) is a piece of cylindrical glass tubing with a stopcock attached at one end. The stopcock usually has a Teflon plug since stopcock grease (used on glass plugs) dissolves in many of the organic solvents used as eluents. Such admixture of stopcock grease in the eluent contaminates the eluates.

      Instead of a stopcock, a piece of flexible tubing is often attached to the bottom of the column, and a screw clamp is used to stop or regulate the flow (Figure 10–7). When this alternative arrangement is used, care must be taken that the tubing used is not dissolved by the chromatographic solvents that are to effect the separation. Rubber, for instance, dissolves in chloroform, benzene, methylene chloride, toluene, or tetrahy-

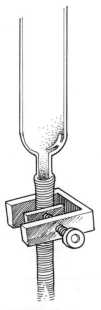

FIGURE 10–7. Tubing with screw clamp to regulate solvent flow on a chromatography column

drofuran (THF). Tygon tubing dissolves (actually, the plasticizer is removed) in many different solvents, including benzene, methylene chloride, chloroform, ether, ethyl acetate, toluene, and THF. Polyethylene tubing is the best choice, since it is inert with most solvents.

Next, the column is partially filled with a quantity of solvent, usually a non-polar solvent like hexane, and a support for the finely divided adsorbent is prepared in the following way. A loose plug of glass wool is tamped down into the bottom of the column with a long glass rod until all entrapped air is forced out as bubbles. Care should be taken not to plug the column totally by tamping the glass wool too hard. A small layer of clean, white sand is formed on top of the glass wool by pouring sand into the column. The column is tapped to level the surface of the sand. Any sand adhering to the side of the column is washed down with a small quantity of solvent. The sand forms a base that supports the column of adsorbent and prevents it from washing through the stopcock. Then the column is best packed in one of two ways: the ''slurry method'' or the ''dry pack method.''

B. Depositing the Adsorbent

Slurry Method

In the slurry method, the adsorbent is packed into the column as a **slurry.** A slurry is a mixture of a solvent and an undissolved solid. The slurry is prepared in a separate container by adding the solid adsorbent, a little at a time, to a quantity of the solvent. This order of addition (adsorbent to solvent) should be followed strictly, since the adsorbent solvates and liberates heat. If the solvent is added to the adsorbent, it may boil away almost as fast as it is added due to heat evolved, especially if ether or another low-boiling solvent is used. When this happens, the final mixture is uneven and lumpy. Enough adsorbent is added to the solvent, with swirling, to form a thick but flowing slurry. The slurry should be swirled until it is homogeneous and relatively free of entrapped air bubbles.

When the slurry has been prepared, the column is filled about half full with solvent, and the stopcock is opened to allow solvent to drain slowly into a large beaker. Alternately, the slurry is mixed by swirling and is then poured in portions into the top of the draining column (a wide-necked funnel may be useful here). The column is tapped constantly and **gently** on the side, during the pouring operation, with a pencil fitted with a rubber stopper. A short piece of large-diameter pressure tubing may also be used for tapping. The tapping promotes even settling and mixing and gives an evenly packed column free of air bubbles. Tapping is continued until all the material has settled, showing a well-defined level at the top of the column. Solvent from the collecting beaker may be re-added to the slurry if it becomes too thick to be poured into the column at one time. In fact, the collected solvent should be cycled through the column several times to ensure that settling is complete and that the column is firmly packed. The downward flow of solvent tends to compact the adsorbent. One should take care never to let the column run dry during packing.

Dry Pack Method

In the dry pack method, the column is filled with solvent and allowed to drain **slowly.** The dry adsorbent is added, a little at a time, from a beaker, while the column is tapped constantly, as described above. When the column has the desired length, no more adsorbent is added. This method also gives an evenly packed column. For the same reasons as those already described, solvent should be cycled through this column several times before each use.

In another dry pack method, the entire column is packed dry, that is, without any solvent. Then the solvent is allowed to percolate down the column slowly, until the column is entirely moistened. **This method is not recommended for use with silica gel or alumina,** since the combination leads to uneven packing, air bubbles, and cracking, especially if a solvent that has a highly exothermic heat of solvation is used.

10.7 APPLYING THE SAMPLE TO THE COLUMN

The solvent (or solvent mixture) used to pack the column is normally the least polar elution solvent one intends to use during the chromatography. The compounds to be chromatographed will not be infinitely soluble in this solvent. If they were, they would probably have a greater affinity for the solvent than for the adsorbent and would pass right through the column without equilibrating. Since it would take a large amount of the initial chromatographic solvent to dissolve the compound in this case, it would be difficult to get the mixture to form a narrow band on top of the column. A narrow band is ideal for an optimum separation of components. For the best separation, the compound is applied to the top of the column, undiluted if it is a liquid or in a **very small** amount of a highly polar solvent if it is a solid.

In adding the sample to the column, the following procedure should be used. The solvent level is lowered to the top of the adsorbent column by draining. The liquid (diluted or neat) or dissolved solid is added (usually with a small pipet) to form a small layer on top of the adsorbent. Care is taken not to disturb the surface. This is done best by touching the pipet to the inside of the glass column and slowly draining it so as to allow the sample to spread into a thin film, which slowly descends to cover the entire adsorbent surface. The pipet is drained close to the surface of the adsorbent. When all the sample has been added, this small layer of liquid is drained into the column until the top surface of the column **just begins** to dry. A small layer of the chromatographic solvent is then added carefully with a pipet, again with care not to disturb the surface. This small layer of solvent is drained into the column until the column just dries. Another small layer of fresh solvent is added, if necessary, and the process is repeated until it is clear that the sample is strongly adsorbed on the top of the column. If the sample is colored and the fresh layer of solvent acquires this color, the sample has not been properly adsorbed. Once the sample has been properly applied, the level surface of the adsorbent may be protected by carefully filling the top of the column with solvent and sprinkling clean, white sand into the column so as to form a small protective layer on top of the adsorbent.

Separations are often better if the sample is allowed to stand a short time on the column before elution. This allows a true equilibrium to be established. In columns that stand for too long, however, the adsorbent often compacts or even swells, and the flow can become annoyingly slow. Diffusion of the sample to widen the bands also becomes a problem if a column is allowed to stand over an extended period.

10.8 ELUTION TECHNIQUES

Solvents for analytical and preparative chromatography should be pure reagents. Commercial-grade solvents often contain small amounts of residue, which remain when the solvent is evaporated. For normal work, and for relatively easy separations that take only small amounts of solvent, the residue usually presents few problems. Commercial-grade solvents may have to be redistilled before use. This is especially true for hydrocarbon solvents, which tend to have more residue than other solvent types. Most of the experiments in this laboratory manual have been designed to avoid this particular problem.

One usually begins elution of the products with a nonpolar solvent, like hexane or petroleum ether. The polarity of the elution solvent can be increased gradually by adding successively greater percentages of either ether or toluene (for instance, 1%, 2%, 5%, 10%, 15%, 25%, 50%, 100%) or some other solvent of greater solvent power (polarity) than hexane. The transition from one solvent to another should not be too rapid in most solvent changes. If the two solvents to be changed differ greatly in their heats of solvation in binding to the adsorbent, enough heat can be generated to crack the column. Ether is especially bad in this respect, as it has both a low boiling point and a relatively high heat of solvation. Most organic compounds can be separated on silica gel or alumina using hexane–ether or hexane–toluene combinations for elution and following these by pure methylene chloride. Solvents of greater polarity are usually avoided for the various reasons mentioned above.

The flow of solvent through the column should not be too rapid, or the solutes will not have time to equilibrate with the adsorbent as they pass down the column. If the rate of flow is too low or stopped for a period, diffusion can become a problem— the solute band will diffuse, or spread out, in all directions. In either of these cases, separation will be poor. As a general rule (and only an approximate one), most columns are run with flow rates ranging from 5 to 50 drops of effluent per minute; a steady flow of solvent is usually avoided. To avoid diffusion of the bands, do not stop the column and do not set it aside overnight.

10.9 RESERVOIRS

When large quantities of solvent are used in a chromatographic separation, it is often convenient to use a solvent reservoir to forestall the need for continually adding small portions of fresh solvent. The simplest type of reservoir, a feature of many columns, is

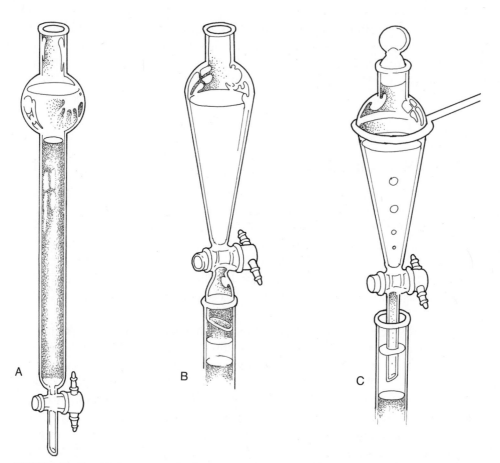

FIGURE 10–8. Various types of solvent-reservoir arrangements for chromatographic columns

created by fusing the top of the column to a round-bottomed flask (Figure 10–8A). If the column has a standard taper joint at its top, a reservoir can be created by joining a standard-taper separatory funnel to the column (Figure 10–8B). In this arrangement the stopcock is left open, and no stopper is placed in the top of the separatory funnel. A third common arrangement is shown in Figure 10–8C. A separatory funnel is filled with solvent; its stopper is wetted with solvent and put **firmly** in place. The funnel is inserted into the empty filling space at the top of the chromatographic column, and the stopcock is opened. Solvent flows out of the funnel, filling this space until the solvent level is well above the outlet at the end of the funnel's stem. As solvent leaves the capped separatory funnel, a partial vacuum is drawn in the funnel, and solvent cannot drain unless air is admitted. Air cannot be admitted until the solvent level drops below the opening in the funnel's stem. At that point air can enter the funnel, and a quantity of solvent will be discharged so that the solvent level in the column rises to cover the outlet once again. Air cannot enter the funnel again until the solvent level in the column drops. The column is thus self-filling. For this arrangement, the funnel should have a **long** stem, with a correspondingly long portion of the column unfilled to accommodate the stem and solvent discharges. Generally, the funnel discharges a constant, but fairly large, solvent volume each time the column is filled. The available volume at the top of the column must be large enough that the column will not overflow each time it is filled.

10.10 MONITORING THE COLUMN

It is a happy instance when the compounds to be separated are colored. This separation can then be followed visually and the various bands collected separately as they elute from the column. For the majority of organic compounds, however, this lucky circumstance does not exist, and other methods must be used to determine the positions of the various bands. The most common method of following a separation of uncolored compounds is to collect **fractions** of constant volume in preweighed flasks, to evaporate the solvent from each fraction, and to reweigh the flask plus any residue. A plot of fraction number versus the weight of the residues after evaporation of solvent gives a plot like that in Figure 10–9. Clearly, fractions 2 through 7 (Peak 1) may be combined as a single compound, and so can fractions 8 through 11 (Peak 2) and 12 through 15 (Peak 3). The size of the fractions collected (10 mL, 100 mL, or 500 mL) depends on the size of the column and the ease of separation.

Another common method of monitoring the column is to mix an inorganic phosphor into the adsorbent used to pack the column. When the column is illuminated with an ultraviolet light, the adsorbent treated in this way fluoresces. However, many solutes have the ability to **quench** the fluorescence of the indicator phosphor. In areas in which solutes are present, the adsorbent does not fluoresce, and a dark band is visible. Thus, in this type of column, the separation can also be followed visually.

Thin-layer chromatography is often used to monitor a column. This method is described in Technique 11 (Section 11.9, p. 626). Several sophisticated instrumental and spectroscopic methods, which we shall not detail, can also monitor a chromatographic separation.

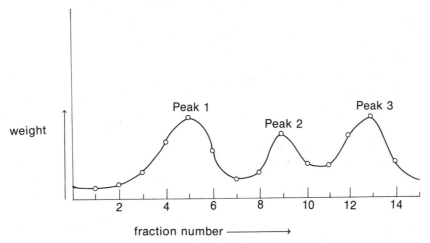

FIGURE 10–9. Typical elution graph

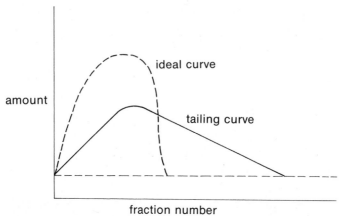

FIGURE 10–10.   Elution curves: one ideal and one that "tails"

10.11   TAILING

Often when a single solvent is used for elution, an elution curve (weight-vs-fraction) like that shown as a solid line in Figure 10–10 is observed. An ideal elution curve is shown by dashed lines. In the nonideal curve, the compound is said to be **tailing.** Tailing can interfere with the beginning of a curve or a peak of a second component and lead to a poor separation. One way to avoid this is to increase the polarity of solvent constantly while eluting. In this way, at the tail of the peak, where the solvent polarity is increasing, the compound will move slightly faster than at the front and allow the tail to squeeze forward, forming a more nearly ideal band.

10.12   RECOVERING THE SEPARATED COMPOUNDS

In recovering each of the separated compounds of a chromatographic separation when they are solids, the various correct fractions are combined, evaporated, and recrystallized. If the compounds are liquids, the correct fractions are combined, evaporated, and distilled. The combination of chromatography-crystallization or chromatography-distillation usually yields very pure compounds.

10.13   GEL CHROMATOGRAPHY

The stationary phase in gel chromatography consists of a cross-linked polymeric material. Molecules are separated according to their **size** by their ability to penetrate a sievelike structure. Molecules permeate the porous stationary phase as they move down the column. Small molecules penetrate the porous structure more easily than larger ones. Thus, the large molecules move through the column faster than the smaller ones and elute first. With adsorption chromatography using materials such as alumina or silica (Section 10.2), the order is usually the reverse: small molecules (of low molecular weight) pass through the column **faster** than large molecules (of high molecular weight) because large molecules are more strongly attracted to the polar stationary phase.

Equivalent terms used by chemists for the gel-chromatography technique are **gel filtration** (biochemistry term), **gel-permeation chromatography** (polymer chemistry term), and **molecular sieve chromatography. Size-exclusion chromatography** is a general term for the technique, and it is perhaps the most descriptive term for what occurs on a molecular level.

Sephadex is one of the most popular materials for gel chromatography. It is widely used by biochemists for separating proteins, nucleic acids, enzymes, and carbohydrates. Most often water or aqueous solutions of buffers are used as the moving phase. Chemically, Sephadex is a polymeric carbohydrate that has been cross-linked. The degree of cross-linking determines the size of the ''holes'' in the polymer matrix. In addition, the hydroxyl groups on the polymer can adsorb water, which causes the material to swell. As it expands, ''holes'' are created in the matrix. Several different gels are available from manufacturers, each with its own set of characteristics. For example, a typical Sephadex gel such as G-75 can separate molecules in the molecular-weight (MW) range 3000 to 70,000. Assume for the moment that one has a four-component mixture containing compounds with molecular weights of 10,000, 20,000, 50,000, and 100,000. The 100,000-MW compound would pass through first because it cannot penetrate the polymer matrix. The 50,000-, 20,000-, and 10,000-MW compounds penetrate the matrix to varying degrees and would be separated. The molecules would elute in the order given (decreasing order of molecular weights). The gel separates on the basis of molecular size and configuration rather than molecular weight. The molecular-weight relations usually parallel the relative ''size'' of the molecules, however.

Sephadex LH-20 has been developed for other solvents than water. Some of the hydroxyl groups have been alkylated, and thus the material can swell under both aqueous and nonaqueous conditions (it now has ''organic'' character). This material can be used with several organic solvents, such as alcohols, acetone, methylene chloride, and aromatic hydrocarbons.

Another type of gel is based on a polyacrylamide structure (Bio-Gel P and Poly-Sep AA). These gels can also be used in water and some polar organic solvents. They tend to be more stable than Sephadex, especially under acidic conditions. Polyacrylamides can be used for many biochemical applications involving macromolecules. For separating synthetic polymers, cross-linked polystyrene beads (copolymer of styrene and divinylbenzene) are common. Again the beads are swollen before use. Common organic solvents can be used to elute the polymers. As with other gels, the higher-molecular-weight compounds elute before the lower-molecular-weight compounds.

10.14   DRY-COLUMN CHROMATOGRAPHY

In dry-column chromatography, the entire column is packed dry without any solvent. The most common column for this method consists of a piece of flexible nylon tubing that has been sealed at one end (stapled). After some glass wool has been inserted into the bottom of the tubing, the bottom is perforated in several places so that air can be expelled. This column is then filled in portions with the adsorbent, usually alumina. After each addition, the adsorbent is compacted by dropping the tubing repeatedly on a hard surface.

Instead of applying the sample as described in Section 10.7, the material to be separated is dissolved in a volatile solvent and mixed with the adsorbent, and the solvent is removed by evaporation. The resulting dry powder is introduced into the packed column. Solvent is allowed to percolate down the column slowly until the solvent front is about 1 cm from the end of the column. In dry-column chromatography, in contrast to elution chromatography (Section 10.8), neither the solvent nor the separated materials actually leave the column. Only one solvent is used in dry-column chromatography.

The plastic tubing can now be cut in segments to recover the separated compounds. Each portion of adsorbent together with the pure compound can be swirled with a volatile solvent to remove the compound from the adsorbent. Once the adsorbent is removed by filtration, the solvent can be evaporated to give the purified compound. With colored compounds, one can easily decide where to cut the tubing. With colorless materials, an ultraviolet light (Technique 11, Section 11.6, p. 623) may be used to detect compounds before slicing the tubing. Alternatively, one may use a new technique, in which the tubing is punctured at 1-cm intervals with glass pipets. By this technique, a small amount of adsorbent is removed with the pipet and analyzed by thin-layer chromatography, as in the method described in Technique 11, Section 11.9, p. 626. Using these analyses, one can decide where the column should be cut for the best separation of components.

You may recognize the similarity of dry-column chromatography to thin-layer chromatography (tlc), described in Technique 11. The separation is virtually identical to what is obtained with tlc. One can expect the same results from both methods so long as the same adsorbent and the same solvent are used in each case. Thus, the R_f value (Technique 11, Section 11.8, p. 625) obtained by tlc correlates well with the value from the dry-column chromatographic method. The main advantage of this method over conventional tlc is that much larger samples can be separated.

10.15   HIGH PERFORMANCE LIQUID CHROMATOGRAPHY

If the column packing used in column chromatography is made more dense by using an adsorbent that has a smaller particle size, the separation that can be achieved is greater. The solute molecules encounter a much larger surface area on which they can be absorbed as they pass through the column packing. At the same time, the solvent spaces between the particles are reduced in size. As a result of this tight packing, equilibrium between the liquid and solid phases can be established very rapidly with a fairly short column, and the degree of separation is markedly improved. The disadvantage of making the column packing more dense is that the solvent flow rate becomes very slow or stops. Gravity is not strong enough to pull the solvent through a tightly packed column.

A recently developed technique can be applied to obtain much better separations with tightly packed columns. A pump is used to force the solvent through the column packing. As a result, solvent flow rate is increased, while the advantage of better separation is retained. This technique, called **high performance liquid chromatography (hplc)** is becoming widely applied to problems where separations by ordinary column chromatography are unsatisfactory. Because the pump often provides pressures in excess of 1000 pounds per square inch (psi), this method is also known as **high pressure liquid chromatography.** High pressures are not required, however, and satisfactory separations can be achieved with pressures as low as 100 psi.

The basic design of an hplc instrument is shown in Figure 10–11. The instrument contains the following essential components:

1. Solvent reservoir
2. Solvent filter and degasser
3. Pump
4. Pressure gauge
5. Sample injection system
6. Column
7. Detector
8. Amplifier and electronic controls
9. Chart recorder

There may be other variations on this simple design. Some instruments have heated ovens in order to maintain the column at a specified temperature, fraction collectors, and microprocessor-controlled data-handling systems. Additional filters for the solvent and sample may also be included.

The chromatography column is generally packed with silica or alumina adsorbents. Unlike column chromatography, however, the adsorbents used for hplc have a much smaller particle size. Typically, particle size ranges from 5 to 20 micrometers in diameter for hplc, while it is on the order of 100 micrometers for column chromatography. The actual column may be constructed of heavy glass or even stainless steel. A strong column is required to withstand the high pressures that may be used.

A flow-through **detector** must be provided to determine when a substance has passed through the column. In most applications, the detector used either detects the change in index of refraction of the liquid as its composition changes or detects the presence of solute by its absorption of ultraviolet light. The signal generated by the detector is amplified and treated electronically in a manner similar to that found in gas chromatographs (Technique 12, Section 12.6, page 635).

A variation unique to hplc is to use a nonpolar column packing with a polar solvent. This is called **reversed-phase chromatography.** In this method, the silica

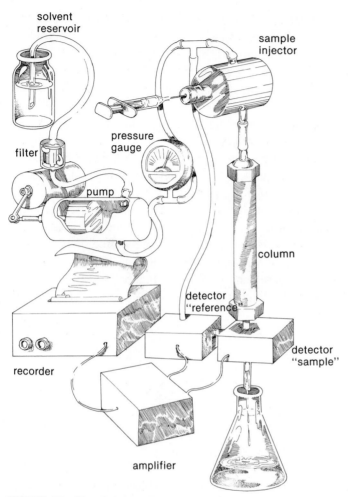

FIGURE 10–11. Schematic diagram of a high performance liquid chromatograph

column packing is treated with alkylating agents. As a result, nonpolar alkyl groups are bonded to the silica surface, making the adsorbent nonpolar. When a polar solvent is used, polar solutes are more strongly attracted to the mobile (solvent) phase than they are to the stationary phase (adsorbent). The order of elution, therefore, is reversed, with polar solutes passing through the column faster than nonpolar solutes. In many applications, better separations can be achieved with reversed-phase chromatography than with polar adsorbents.

If a single solvent (or solvent mixture) is used for the entire separation, the chromatogram is said to be **isochratic.** Special electronic units are available, which will allow one to program changes in the solvent composition from the beginning to the end of the chromatography. These are called **gradient elution systems.** With gradient elution, the time required for a separation may be shortened considerably.

High performance liquid chromatography is an excellent analytical technique, but the separated compounds may also be isolated. The technique can be used for preparative experiments. Just as in column chromatography, the fractions can be collected into individual receiving containers as they pass through the column. The solvents can be evaporated from these fractions, allowing one to isolate separated components of the original mixture.

REFERENCES

Deyl, Z., Macek, K., and Janák, J. *Liquid Column Chromatography*. Amsterdam: Elsevier, 1975.
Heftmann, E. *Chromatography*. 3rd ed. New York: Van Nostrand Reinhold, 1975.

Technique 11
Thin-Layer Chromatography

Thin-layer chromatography (tlc) is a very important technique for the rapid separation and qualitative analysis of small amounts of material. The technique is closely related to column chromatography. In fact, tlc can be considered simply column chromatography **in reverse,** with the solvent ascending the adsorbent rather than descending. Because of this close relation to column chromatography, and because the principles governing the two techniques are similar, Technique 10, on column chromatography, should be read first.

11.1 PRINCIPLES OF THIN-LAYER CHROMATOGRAPHY

Like column chromatography, tlc is a solid–liquid partitioning technique. However, the moving, liquid phase is not allowed to percolate down the adsorbent; it is caused to **ascend** a thin layer of adsorbent coated onto a backing support. The most typical backing is a glass plate, but other materials are also used. A thin layer of the adsorbent is spread onto the plate and allowed to dry. A coated and dried plate of glass is called a **thin-layer plate** or a **thin-layer slide.** (The reference to *slide* comes about because microscope slides are often used to prepare small thin-layer plates.) When a thin-layer plate is placed upright in a vessel that contains a shallow layer of solvent, the solvent ascends the layer of adsorbent on the plate by capillary action.

In tlc, the sample is applied to the plate before the solvent is allowed to ascend the adsorbent layer. The sample is usually applied as a small spot near the base of the plate, and this technique is often referred to as **spotting.** The plate is spotted by repeated applications of a sample solution from a small capillary pipet. When the filled pipet touches the plate, capillary action delivers its contents to the plate, and a small spot is formed.

As the solvent ascends the plate, the sample is partitioned between the moving, liquid phase and the stationary, solid phase. During this process, one is said to be **developing,** or **running,** the thin-layer plate. In development, the various components in the applied mixture are separated. The separation is based on the many equilibrations the solutes experience between the moving and the stationary phases. (The nature of these equilibrations was thoroughly discussed in Technique 10, Sections 10.2 and 10.3, pp. 594–597.) As in column chromatography, the least polar substances advance faster than the most polar substances. A separation results from the differences in the rates at which the individual components of the mixture advance upward on the plate. When many substances are present in a mixture, each has its own characteristic solubility and adsorptivity properties, depending on the functional groups in its structure. In general, the stationary phase is strongly polar and strongly binds polar substances. The moving liquid phase is usually less polar than the adsorbent and most easily dissolves substances that are less polar or even nonpolar. Thus, while substances that are the most polar travel slowly upward, or not at all, nonpolar substances travel more rapidly if the solvent is sufficiently nonpolar.

When the thin-layer plate has been developed, it is removed from the developing tank and allowed to dry until it is free of solvent. If the mixture that was originally spotted on the plate was separated, there will be a vertical series of spots on the plate. Each spot corresponds to a separate component or compound from the original mixture. If the components of the mixture are colored substances, the various spots will be clearly visible after development. More often, however, the "spots" will not be visible because they correspond to colorless substances. If spots are not apparent, they can be made visible only if a **visualization method** is used. Often spots can be seen when the thin-layer plate is held under ultraviolet light; the ultraviolet lamp is a common visualization method. Also common is the use of iodine vapor. The plates are placed in a chamber containing iodine crystals and left to stand for a short time. The iodine reacts with the various compounds adsorbed on the plate to give colored complexes that are clearly visible. Because iodine often changes the compounds by reaction, the mixture components cannot be recovered from the plate when this method is used. (Other methods of visualization are discussed in Section 11.6.)

11.2 PREPARATION OF THIN-LAYER SLIDES AND PLATES

The two adsorbent materials most often used for tlc are alumina G (aluminum oxide) and silica gel G (silicic acid). The G designation stands for gypsum (calcium sulfate). Calcined gypsum, $CaSO_4 \cdot \frac{1}{2}H_2O$, is better known as plaster of Paris. When exposed to water or moisture, gypsum sets in a rigid mass, $CaSO_4 \cdot 2H_2O$, which binds the adsorbent together and to the glass plates used as a backing support. In the adsorbents used for tlc, about 10 to 13% by weight of gypsum is added as a binder. The adsorbent materials are otherwise like those used in column chromatography; the adsorbents used in column chromatography have a larger particle size, however. The material for thin-layer work is a fine powder. The small particle size, along with the added gypsum, makes it impossible to use silica gel G or alumina G for column work. In a column, these adsorbents generally set so rigidly that solvent virtually stops flowing through the column.

A. Microscope Slide TLC Plates

For qualitative work such as identifying the number of components in a mixture or trying to establish that two compounds are identical, small tlc plates made from microscope slides are especially convenient. Coated microscope slides are easily made by dipping the slides into a container holding a slurry of the adsorbent material. Although numerous solvents can be used to prepare a slurry, methylene chloride is probably the most convenient solvent. It has the two advantages of low boiling point (40 °C) and inability to cause the adsorbent to set or form lumps. The low boiling point means that it is not necessary to dry the coated slides in an oven. Its inability to cause the gypsum binder to set means that slurries made with it are stable for several days. It has the disadvantage that the layer of adsorbent formed is fragile and must be treated carefully. For this reason, some persons prefer to add a small amount of methanol to the methylene chloride to enable the gypsum to set more firmly. The methanol solvates the calcium sulfate much as water does. More durable plates can be made by dipping plates into a slurry prepared from water. These plates must be oven-dried before use. Also, a slurry prepared from water must be used soon after its preparation. If it is not, it will begin to set and to form lumps. Thus, an aqueous slurry must be prepared immediately before use; it cannot be used after it has stood for any length of time. For microscope slides, a slurry of silica gel G in methylene chloride is not only convenient but also adequate for most purposes.

Preparing the Slurry

The slurry is most conveniently prepared in a 4-oz wide-mouthed screw-cap jar. About 3 mL of methylene chloride are required for each gram of silica gel G. For a smooth slurry without lumps, the silica gel should be added to the solvent while the mixture is being either stirred or swirled. Adding solvent to the adsorbent usually causes lumps to form in the mixture. When the addition is complete, the cap should be placed on the jar tightly and the jar shaken vigorously to ensure thorough mixing. The slurry may be stored, in the tightly capped jar, until it is to be used. More methylene chloride may have to be added to replace evaporation losses.

Preparing the Slides

If new microscope slides are available, they can be used without any special treatment. However, it is more economical to reuse or recycle used microscope slides. The slides should be washed with soap and water, rinsed with water, and then rinsed with 50%

aqueous methanol. The plates should be allowed to dry thoroughly on paper towels. They should be handled by the edge because fingerprints on the plate surface will make it difficult for the adsorbent to bind to the glass.

Coating the Slides

The slides are coated with adsorbent by dipping them into the container of slurry. Two slides can be coated simultaneously by sandwiching them together before dipping them in the slurry. The slurry should be shaken vigorously just before dipping the slides. Since the slurry settles on standing, it should be mixed in this way before each set of slides is dipped. The depth of the slurry in the jar should be about 3 in., and the plates should be dipped into the slurry until only about 0.25 in. at the top remains uncoated. The dipping operation should be done smoothly. The plates may be held at the top (see Figure 11–1), where they will not be coated. They are dipped into the slurry and withdrawn with a slow and steady motion. The dipping operation takes about 2 seconds. Some practice may be required to get the correct timing. After dipping, the cap should be replaced on the jar, and the plates should be held for a minute until most of the solvent has evaporated. The plates may then be separated and placed on paper towels to complete the drying.

The plates should have an even coating; there should be no streaks and no thin spots where glass shows through the adsorbent. The plates should not have a thick and lumpy coating. Two conditions cause thin and streaked plates: First, the slurry may not have been mixed thoroughly before the dipping operation; the adsorbent might then

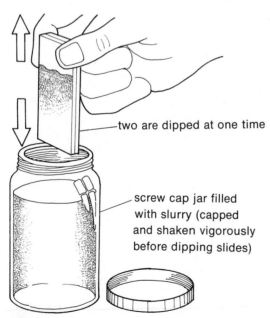

—two are dipped at one time

screw cap jar filled with slurry (capped and shaken vigorously before dipping slides)

FIGURE 11–1.   Dipping slides to coat them

have settled to the bottom of the jar, and the thin slurry at the top would not have coated the slides properly. Second, the slurry simply may not have been thick enough; more silica gel G must then be added to the slurry until the consistency is proper. If the slurry is too thick, the coating on the plates will be thick, uneven, and lumpy. To correct this, the slurry should be diluted with enough solvent to achieve the proper consistency.

Plates with an unsatisfactory coating may be wiped clean with a paper towel and redipped. Care must be taken to handle the plates only from the top or by the sides.

B. Larger Thin-Layer Plates

For separations involving large amounts of material, or for difficult separations, it may be necessary to use larger thin-layer plates. Plates with dimensions up to 20 to 25 sq cm are common. With larger plates, it is desirable to have a more durable coating, and a water slurry of the adsorbent should be used to prepare them. If silica gel is used, the slurry should be made up in the ratio about 1 g of silica gel G to each 2 mL of water. The glass plate used for the thin-layer plate should be washed, dried, and placed on a sheet of newspaper. Along two edges of the plate are placed two strips of masking tape. More than one layer of masking tape is used if a thicker coating is desired on the plate. A slurry is prepared, shaken well, and poured along one of the untaped edges of the plate. A heavy piece of glass rod, long enough to span the taped edges, is used to level and spread the slurry over the plate. While the rod is resting on the tape, it is pushed along the plate from the end at which the slurry was poured toward the opposite end of the plate. This is illustrated in Figure 11–2. After the slurry is spread, the masking-tape strips are removed, and the plates are dried in a 110 °C oven for about 1 hour. Plates of 20 to 25 sq cm are easily prepared by this method. Larger plates present more difficulties. Many laboratories have a commercially manufactured spreading machine that makes the entire operation simpler.

The dry-column method (Technique 10, Section 10.14, p. 612) offers an alternative for separating large quantities of material.

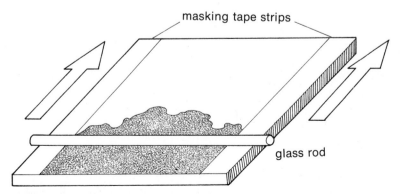

FIGURE 11–2. Preparing a large plate

C. Commercially Prepared Plates

Many manufacturers supply glass plates precoated with a durable layer of silica gel or alumina. More conveniently, plates are also available that have either a flexible plastic backing or an aluminum backing. The plates with flexible plastic backing are becoming increasingly common. They are expensive but are made quite uniformly and, being flexible, have the advantage that they do not flake easily. They can also be cut with a pair of scissors to whatever size is required.

11.3 SAMPLE APPLICATION: SPOTTING THE PLATES

Preparing a Micropipet

To apply the sample that is to be separated to the thin-layer plate, one uses a micropipet. A micropipet is easily made from a short length of thin-walled capillary tubing like that used for melting-point determinations. The capillary tubing is heated at its midpoint with a microburner and rotated until it is soft. When the tubing is soft, the heated portion of the tubing is drawn out until a constricted portion of tubing 4 to 5 cm long is formed. After cooling, the constricted portion of tubing is scored at its center with a file or scorer and broken. The two halves yield two capillary micropipets. Figure 11–3 shows how to make such pipets.

Spotting the Plate

To apply a sample to the plate, begin by placing about 1 mg of a solid test substance, or one drop of a liquid test substance, in a small container like a watch glass or a test tube. Dissolve the sample in a few drops of a volatile solvent. Acetone or methylene chloride

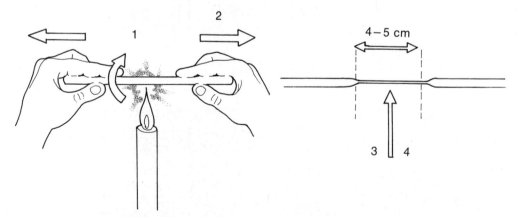

① Rotate in flame until soft.   ③ Score lightly in center of pulled section.

② Remove from flame and pull.   ④ Break in half to give two pipets

FIGURE 11–3. Construction of two capillary micropipets

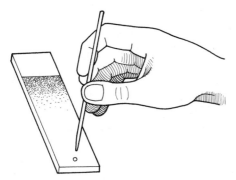

FIGURE 11–4. Spotting the plate with a drawn capillary pipet

is usually a suitable solvent. If a solution is to be tested, it can often be used directly. The small capillary pipet, prepared as described, is filled by dipping the pulled end into the solution to be examined. Capillary action fills the pipet. One empties the pipet by touching it **lightly** to the thin-layer plate at a point about 1 cm from the bottom (Figure 11–4). The spot must be high enough that it does not dissolve in the developing solvent. It is important to touch the plate very lightly and not to gouge a hole in the adsorbent. When the pipet touches the plate, solution is transferred to the plate as a small spot. The pipet should be touched to the plate **very briefly** and then removed. If the pipet is held to the plate, its entire contents will be delivered to the plate. Only a small amount of material is needed. It is often helpful to blow gently on the plate as the sample is applied. This helps to keep the spot small by evaporating the solvent before it can spread out on the plate. The smaller the spot formed, the better the separation obtainable. If needed, additional material can be applied to the plate by repeating the spotting procedure. It is best to repeat the procedure with several small amounts, rather than to apply one large amount. The solvent should be allowed to evaporate between applications. If the spot is not small (about 2 mm in diameter), a new plate should be prepared. The capillary pipet may be used several times if it is rinsed between uses. It is repeatedly dipped into a small portion of solvent to rinse it and touched to a paper towel to empty it.

  As many as three different spots may be applied to a microscope-slide tlc plate. Each spot should be about 1 cm from the bottom of the plate, and all spots should be evenly spaced, with one spot in the center of the plate. Due to diffusion, spots will often increase in diameter as the plate is developed. To keep spots containing different materials from merging, and to avoid confusing the samples, do not place more than three spots on a single plate. Larger plates can accommodate many more samples.

11.4 DEVELOPING (RUNNING) TLC PLATES

Preparing a Development Chamber

A convenient developing chamber for microscope-slide tlc plates can be made from a 4-oz wide-mouthed screw-cap jar. The inside of the jar should be lined with a piece of

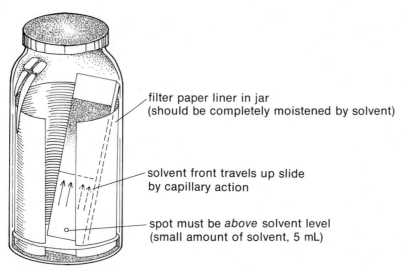

filter paper liner in jar
(should be completely moistened by solvent)

solvent front travels up slide
by capillary action

spot must be *above* solvent level
(small amount of solvent, 5 mL)

FIGURE 11–5. Development chamber with thin-layer plate undergoing development

filter paper, cut so that it does not quite extend around the inside of the jar. A small vertical opening (2–3 cm) should be left for observing the development. Before development, the filter paper liner inside the jar should be thoroughly moistened with the development solvent. The solvent-saturated liner helps to keep the chamber saturated with solvent vapors, thereby speeding development. Once the liner is saturated, the level of solvent in the bottom of the jar is adjusted to a depth of about 5 mm, and the jar is capped and set aside until it is to be used. A correctly prepared development chamber (with slide in place) is shown in Figure 11–5.

Developing the TLC Slide

Once the spot has been applied to the thin-layer plate and the solvent has been selected (see Section 11.5), the plate is placed in the chamber for development. The plate must be placed in the chamber carefully so that none of the coated portion touches the filter paper liner. In addition, the solvent level in the bottom of the chamber must not be above the spot that was applied to the plate, or the spotted material will dissolve in the pool of solvent instead of undergoing chromatography. Once the plate has been placed correctly, one replaces the cap on the developing chamber and waits for the solvent to advance up the plate by capillary action. This generally occurs rapidly, and one should watch carefully. As the solvent rises, the plate becomes visibly moist. When the solvent has advanced to within 5 mm of the end of the coated surface, the plate should be removed, and the position of the solvent front should be marked **immediately** by scoring the plate along the solvent line with a **pencil.** The solvent front must not be allowed to travel beyond the end of the coated surface. The plate should be removed before this happens. The solvent will not actually advance beyond the end of the plate, but spots allowed to stand on a completely moistened plate on which the solvent is not in motion expand by diffusion. Once the plate has dried, any visible spots should be outlined on the plate with a pencil. If no spots are apparent, a visualization method (Section 11.6) may be needed.

11.5 CHOOSING A SOLVENT FOR DEVELOPMENT

The development solvent used depends on the materials to be separated. One may have to try several solvents before a satisfactory separation is achieved. Since microscope slides can be prepared and developed rapidly, an empirical choice is usually not hard to make. A solvent that causes all the spotted material to move with the solvent front is too polar. One that does not cause any of the material in the spot to move is not polar enough. As a guide to the relative polarity of solvents, Table 10–2 in Technique 10 (p. 600) should be consulted.

Methylene chloride and toluene are solvents of intermediate polarity and good choices for a wide variety of functional groups to be separated. For hydrocarbon materials, good first choices are hexane, petroleum ether (ligroin), or toluene. Hexane or petroleum ether with varying proportions of toluene or ether gives solvent mixtures of moderate polarity that are useful for many common functional groups. Polar materials may require ethyl acetate, acetone, or methanol.

A rapid way to determine a good solvent is to apply several sample spots to a single plate. The spots should be placed a minimum of about 1 cm apart. A capillary pipet is filled with a solvent and gently touched to one of the spots. The solvent will expand outward in a circle. The solvent front should be marked with a pencil. A different solvent is applied to each spot. As the solvents expand outward, the spots expand as concentric rings. From the appearance of the rings, one can judge approximately the suitability of the solvent. Several types of behavior experienced with this method of testing are shown in Figure 11–6.

11.6 VISUALIZATION METHODS

If the compounds separated by tlc are colored, it is a fortunate result, because the separation can be followed visually. More often than not, however, the compounds are colorless. Then the separated materials must be made visible by some reagent or some

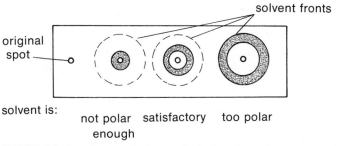

FIGURE 11–6. Concentric-ring method of testing solvents

method that makes the separated compounds visible. Reagents that give rise to colored spots are called **visualization reagents.** Methods of viewing that make the spots apparent are **visualization methods.**

The visualization reagent used most often is iodine. Iodine reacts with many organic materials to form complexes that are either brown or yellow. In this visualization method, the developed and dried tlc plate is placed in a 4-oz wide-mouthed screw-cap jar along with a few crystals of iodine. The jar is capped and gently warmed on a steam bath at low heat. The jar fills with iodine vapors, and the spots begin to appear. When the spots are sufficiently intense, the plate is removed from the jar, and the spots are outlined with a pencil. The spots are not permanent. Their appearance results from the formation of complexes the iodine makes with the organic substances. As the iodine sublimes off the plate, the spots fade. Hence, they should be marked immediately. Nearly all compounds except saturated hydrocarbons and alkyl halides form complexes with iodine. The intensities of the spots do not accurately indicate the amount of material present, except in the crudest way.

The second most common method of visualization is by an ultraviolet lamp. Under uv light, compounds often look like bright spots on the plate. This often suggests the structure of the compounds, because certain types of compounds shine very brightly under uv light since they fluoresce.

Another method with good results involves adding a fluorescent indicator to the adsorbent used to coat the plates. A mixture of zinc and cadmium sulfides is often used. When treated in this way and held under uv light, the entire plate fluoresces. However, dark spots appear on the plate where the separated compounds are seen to quench this fluorescence.

In addition to the above methods, several chemical methods are available that either destroy or permanently alter the separated compounds through reaction. Many of these methods are specific only for particular functional groups.

Alkyl halides can be visualized if a dilute solution of silver nitrate is sprayed on the plates. Silver halides are formed. These halides decompose if exposed to light, giving rise to dark spots (free silver) on the tlc plate.

Most organic functional groups can be made visible if they are charred with sulfuric acid. Concentrated sulfuric acid is sprayed on the plate, which is then heated in an oven at 110 °C to complete the charring. Permanent spots are thus created.

Colored compounds can be prepared from colorless compounds by making derivatives before spotting them on the plate. An example of this is the preparation of 2,4-dinitrophenylhydrazones from aldehydes and ketones to produce yellow and orange compounds. One may also spray the 2,4-dinitrophenylhydrazine reagent on the plate after the ketones or aldehydes have separated. Red and yellow spots form where the compounds are located. Other examples of this method are using ferric chloride for visualizing phenols and using bromocresol green for detecting carboxylic acids. Chromium trioxide, potassium dichromate, and potassium permanganate can be used for visualizing compounds that are easily oxidized. *p*-Dimethylaminobenzaldehyde easily detects amines. Ninhydrin reacts with amino acids to make them visible. Numerous other methods and reagents available from various supply outlets are specific for certain types of functional groups. These visualize only the class of compounds of interest.

11.7 PREPARATIVE PLATES

If large plates (Section 11.2B) are used, materials can be separated and the separated components can be recovered individually from the plates. Plates used in this way are said to be **preparative plates.** For preparative plates, a thick layer of adsorbent is generally used. Instead of being applied as a spot or a series of spots, the mixture to be separated is applied as a line of material about 1 cm from the bottom of the plate. As the plate is developed, the separated materials form bands. After development, the separated bands are observed, usually by uv light, and the zones are outlined in pencil. If the method of visualization is destructive, most of the plate is covered with paper to protect it, and the reagent is applied only at the extreme edge of the plate.

Once the zones have been identified, the adsorbent in those bands is scraped from the plate and extracted with solvent to remove the adsorbed material. Filtration removes the adsorbent, and evaporation of the solvent gives the recovered component from the mixture.

11.8 THE R_f VALUE

Thin-layer chromatography conditions include

1. Solvent system
2. Adsorbent
3. Thickness of the adsorbent layer
4. Relative amount of material spotted

Under an established set of such conditions, a given compound always travels a fixed distance relative to the distance the solvent front travels. This ratio of the distance the compound travels to the distance the solvent front travels is called the R_f value. The symbol R_f stands for "ratio to front," and it is expressed as a decimal fraction:

$$R_f = \frac{\text{distance traveled by substance}}{\text{distance traveled by solvent front}}$$

When the conditions of measurement are completely specified, the R_f value is constant for any given compound, and it corresponds to a physical property of that compound.

The R_f value can be used to identify an unknown compound; but like any other identification based on a single piece of data, the R_f value is best confirmed with some additional data. Many compounds can have the same R_f value, just as many different compounds have the same melting point.

It is not always possible, in measuring an R_f value, to duplicate exactly the conditions of measurement another worker has used. Therefore, R_f values tend to be of more use to a single worker in one laboratory than they are to workers in different laboratories. The only exception to this is when two workers use tlc plates from the same source, as in commercial plates, or know the **exact** details of how the plates were prepared. Nevertheless, the R_f value can be a useful guide. If exact values cannot be relied on, the relative values can provide another worker with useful information about

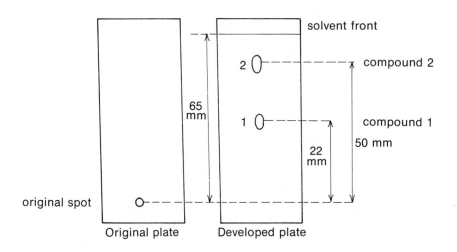

FIGURE 11–7. Sample calculation of R_f values

$$R_f \text{ (compound 1)} = \frac{22}{65} = 0.34 \qquad R_f \text{ (compound 2)} = \frac{50}{65} = 0.77$$

what to expect. Anyone using published R_f values will find it a good idea to check them by comparing them with standard substances whose identity and R_f values are known.

To calculate the R_f value for a given compound, one measures the distance that the compound has traveled from the point at which it was originally spotted. For spots that are not too large, one measures to the center of the migrated spot. For large spots, the measurement should be repeated on a new plate, using less material. For spots that show tailing, the measurement is made to the "center of gravity" of the spot; this first distance measurement is then divided by the distance the solvent front has traveled from the same original spot. A sample calculation of the R_f values of two compounds is illustrated in Figure 11–7.

11.9 THIN-LAYER CHROMATOGRAPHY APPLIED IN ORGANIC CHEMISTRY

Thin-layer chromatography has several important uses in organic chemistry. It can be used in the following applications:

1. To establish that two compounds are identical
2. To determine the number of components in a mixture
3. To determine the appropriate solvent for a column chromatographic separation
4. To monitor a column chromatographic separation
5. To check the effectiveness of a separation achieved on a column, by crystallization, or by extraction
6. To monitor the progress of a reaction

In all these applications, tlc has the advantage that only small amounts of material are necessary. Material is not wasted. With many of the visualization methods, less than a tenth of a microgram (10^{-7} g) of material can be detected. On the other hand, samples as large as a milligram may also be used. With preparative plates that are large (about 9 in. on a side) and have a relatively thick (>500 μm) coating of adsorbent, it is often possible to separate from 0.2 to 0.5 g of material at one time. The main disadvantage of tlc is that volatile materials cannot be used, since they evaporate from the plates.

Thin-layer chromatography can establish that two compounds suspected to be identical are in fact identical. One simply spots both compounds side by side on a single plate and develops the plate. If both compounds travel the same distance on the plate (have the same R_f value), they are probably identical. If the spot positions are not the same, the compounds are definitely not identical. It is important to spot both compounds **on the same plate.** This is especially important with hand-dipped microscope slides, since they vary widely from plate to plate, no two plates having exactly the same thickness of adsorbent. If commercial plates are used, this precaution is not necessary, although it is nevertheless a good idea.

Thin-layer chromatography can establish whether a compound is a single substance or a mixture. A single substance gives a single spot no matter what solvent is used to develop the plate. On the other hand, the number of components in a mixture can be established by trying various solvents on a mixture. A word of caution should be given. It may be difficult, in dealing with compounds of very similar properties, isomers for example, to find a solvent that will separate the mixture. Inability to achieve a separation is not absolute proof that a compound is a single pure substance. Many compounds can be separated only by **multiple developments** of the tlc slide with a fairly nonpolar solvent. In this method, the plate is removed after the first development and allowed to dry. After being dried, it is placed in the chamber again and developed once more. This effectively doubles the length of the slide. At times, several developments may be necessary.

When a mixture is to be separated, tlc can be used to choose the best solvent to separate it if column chromatography is contemplated. Various solvents can be tried on a plate coated with the same adsorbent as will be used in the column. The solvent that resolves the components best will probably work well on the column. These small-scale experiments are quick, use very little material, and save time that would be wasted by attempting to separate the entire mixture on the column. Similarly, tlc plates can **monitor** a column. A hypothetical situation is shown in Figure 11–8. A solvent was found that would separate the mixture into four components (A–D). A column was run using this solvent and 11 fractions of 15 mL each were collected. Thin-layer analysis of the various fractions showed that fractions 1 through 3 contained component A; fractions 4 through 7, component B; fractions 8 and 9, component C; and fractions 10 and 11, component D. A small amount of cross-contamination was observed in fractions 3, 4, 7, and 9.

In another tlc example, a worker found a product from a reaction to be a mixture. It gave two spots, A and B, on a tlc slide. After the product was crystallized, the crystals were found by tlc to be pure A, whereas the mother liquor was found to have a mixture of A and B. The crystallization was judged to have purified A satisfactorily.

Finally, it is often possible to monitor the progress of a reaction by tlc. At various points during a reaction, samples of the reaction mixture are taken and sub-

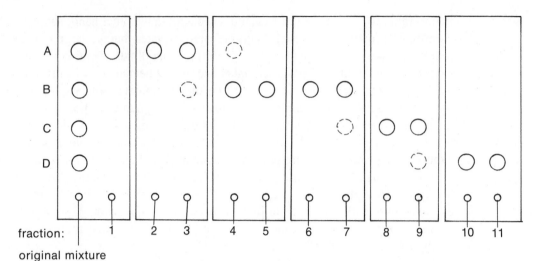

fraction:

original mixture

FIGURE 11–8. Monitoring a column

jected to tlc analysis. An example is given in Figure 11–9. In this case, the desired
reaction was the conversion of A to B. At the beginning of the reaction (0 hr), a tlc
slide was prepared that was spotted with pure A, pure B, and the reaction mixture.
Similar slides were prepared at 0.5, 1, 2, and 3 hours after the start of the reaction. The
slides showed that the reaction was complete in 2 hours. When the reaction was run
longer than 2 hours, a new compound, side-product C, began to appear. Thus, the
optimum reaction time was judged to be 2 hours.

11.10 PAPER CHROMATOGRAPHY

Paper chromatography is often considered related to thin-layer chromatography. The
experimental techniques are somewhat like those of tlc, but the principles are more

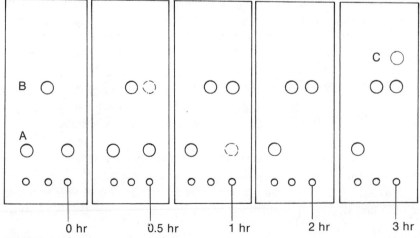

FIGURE 11–9. Monitoring a reaction

closely related to those of extraction. Paper chromatography is actually a liquid–liquid partitioning technique, rather than a solid–liquid technique. For paper chromatography, a spot is placed near the bottom of a piece of high-grade filter paper (Whatman No. 1 is often used). Then the paper is placed in a developing chamber. The development solvent ascends the paper by capillary action and moves the components of the spotted mixture upward at differing rates. Although paper consists mainly of pure cellulose, the cellulose itself does not function as the stationary phase. Rather, the cellulose absorbs water from the atmosphere, especially from an atmosphere saturated with water vapor. Cellulose can absorb up to about 22% of water. It is this water adsorbed on the cellulose that functions as the stationary phase. To ensure that the cellulose is kept saturated with water, many development solvents used in paper chromatography contain water as a component. As the solvent ascends the paper, the compounds are partitioned between the stationary water phase and the moving solvent. Since the water phase is stationary, the components in a mixture that are most highly water-soluble, or those that have the greatest hydrogen-bonding capacity, are the ones that are held back and move most slowly. Paper chromatography applies mostly to highly polar compounds or to those that are polyfunctional. The most common use of paper chromatography is for sugars, amino acids, and natural pigments. Since filter paper is manufactured with good uniformity, R_f values can often be relied on in paper chromatographic work. However, R_f values are customarily measured from the leading edge (top) of the spot—not from its center, as is customary in tlc.

Technique 12
Gas Chromatography

Gas chromatography resembles column chromatography in principle but differs in three respects. First, the partitioning processes for the compounds to be separated are carried out between a **moving gas phase** and a **stationary liquid phase.** (Recall that in column chromatography the moving phase is a liquid and the stationary phase is a solid adsorbent.) Second, the solubility of any given compound in the gas phase is a function of its vapor pressure only. And third, the temperature of the gas system can be controlled, since the column is contained in an insulated oven.

Gas chromatography (gc) is also known as vapor-phase chromatography (vpc) and as gas–liquid partition chromatography (glpc). All three names, as well as their indicated abbreviations, are often found in the literature of organic chemistry. In reference to the technique, the last term, glpc, is the most strictly correct and is preferred by most authors.

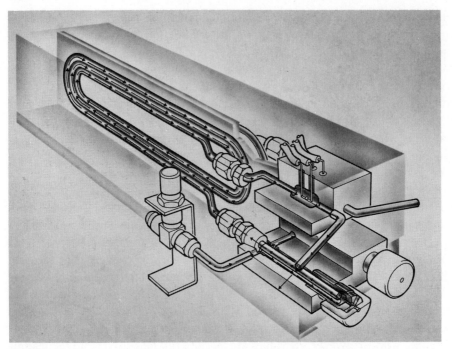

FIGURE 12–1.   Gas chromatograph (Reprinted by courtesy of Carle Instruments, Inc.)

12.1   THE GAS CHROMATOGRAPH

The apparatus used to carry out a gas–liquid chromatographic separation is generally called a **gas chromatograph.** A typical student-model gas chromatograph, the Carle model 8000, is illustrated (with a partial cutaway view) in Figure 12–1. A schematic block diagram of a basic gas chromatograph is shown in Figure 12–2. The basic elements of the apparatus are easily seen. In short, the sample is injected into the chromatograph, and it is immediately vaporized and introduced into a moving stream of gas, called the **carrier gas.** The vaporized sample is then swept into a column filled with particles coated with a liquid adsorbent. The column is contained in a temperature-controlled oven. As the sample passes through the column, it is subjected to many gas–liquid partitioning processes and is separated. As each component leaves the column, its presence is detected by an electrical detector that generates a signal, which is recorded on a strip chart recorder.

12.2   THE COLUMN

The heart of the gas chromatograph is the packed column. This column is usually made of copper or stainless steel tubing, but sometimes glass is used. Various diameters of tubing are used, but ⅛ in. (3 mm) and ¼ in. (6 mm) are probably most common. To construct a column, one selects the length of tubing desired (1 m, 2 m, 3 m, 10 m, and

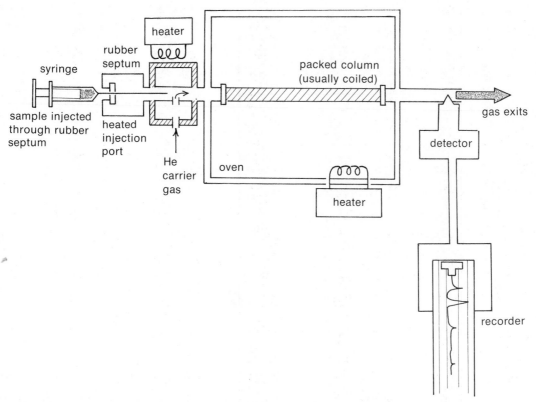

FIGURE 12–2.   Schematic diagram of gas chromatograph

so on), cuts it to the desired length, and attaches the proper fittings on each of the two ends to connect it to the apparatus.

Next, the tubing (column) is packed with the **stationary phase.** The material chosen for the stationary phase is usually a liquid, a wax, or a low-melting solid. This material should be relatively nonvolatile, that is, it should have a low vapor pressure and a high boiling point. Liquids commonly used are high-boiling hydrocarbons, silicone oils, waxes, and polymeric esters, ethers, and amides. Some typical substances are listed in Table 12–1.

The liquid phase is usually coated onto a **support material.** A common support material is crushed firebrick. Many methods exist for coating the high-boiling liquid phase onto the support particles. The easiest is to dissolve the liquid (or low-melting wax or solid) in a volatile solvent like methylene chloride (bp 40 °C). The firebrick (or other support) is added to this solution, which is then slowly evaporated (rotary evaporator) so as to leave each particle of support material evenly coated. Common support materials are listed in Table 12–2.

Finally, the liquid-phase-coated support material is packed into the tubing as evenly as possible. Then, the tubing is bent or coiled so that it fits into the oven of the gas chromatograph with its two ends connected to the gas entrance and exit ports.

Selection of a liquid phase usually revolves about two factors. First, most of them have an upper temperature limit above which they cannot be used. Above the

TABLE 12-1. Typical Liquid Phases

	TYPE	COMPOSITION	MAXIMUM TEMPERATURE (°C)	POLAR	TYPICAL USE
Apiezons (L,M,N,etc.)	Hydrocarbon greases (varying MW)	Hydrocarbon mixtures	250–300	No	Hydrocarbons
Carbowaxes (400–6000)	Polyethylene glycols (varying chain length)	Polyether $HO—(CH_2CH_2—O)_n—CH_2CH_2OH$	Up to 250	Yes	Alcohols Ethers
DC-200	Silicone oil (R=CH$_3$)	$R_3Si—O—\left[\begin{array}{c} R \\ \mid \\ Si—O \\ \mid \\ R \end{array}\right]_n—SiR_3$	225	Med.	Aldehydes Ketones Halocarbons
DEGS	Diethylene glycol succinate	Polyester $\left(CH_2CH_2—O—\overset{O}{\underset{\parallel}{C}}—(CH_2)_2—\overset{O}{\underset{\parallel}{C}}—O\right)_n$	225	Yes	Esters Acids
SE-30	Methyl silicone rubber	Like oil, but cross-linked	350	No	Steroids Pesticides
Nujol	Mineral oil	Hydrocarbon mixture	200	No	—

TABLE 12-2. Typical Solid Supports

Crushed firebrick	Chromosorb T
Nylon beads	(Teflon beads)
Glass beads	Chromosorb P
Silica	(Pink diatomaceous earth,
Alumina	highly absorptive, pH 6–7)
Charcoal	Chromosorb W
Molecular sieves	(White diatomaceous earth,
	medium absorptivity, pH 8–10)
	Chromosorb G
	(like the above,
	low absorptivity, pH 8.5)

specified limit of temperature, the liquid phase itself will begin to ''bleed'' off the column. Second, the materials to be separated must be considered. For polar samples, it is best to use a polar liquid phase; for nonpolar samples, a nonpolar liquid phase is indicated.

12.3 PRINCIPLES OF SEPARATION

After a column is selected, packed, and installed, the **carrier gas** (usually helium, argon, or nitrogen) is allowed to flow through the column supporting the liquid phase. The mixture of compounds to be separated is introduced into the carrier gas stream, where its components are equilibrated between the moving gas phase and the stationary liquid phase (Figure 12–3). The latter is held stationary because it is adsorbed onto the surfaces of the support material.

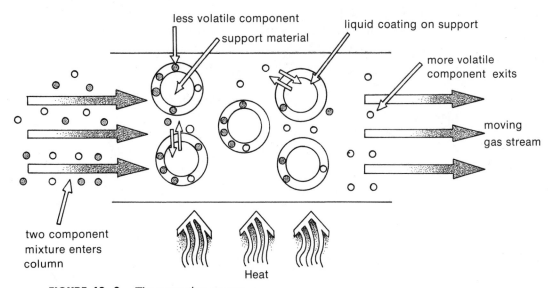

FIGURE 12-3. The separation process

At room temperature, most organic compounds are not very soluble in a gas like helium or nitrogen, and they would never make it through the column unless they were caused to develop a significant vapor pressure, usually by heating. The solubility of a compound in the gas phase is determined almost solely by its vapor pressure. You may remember from general chemistry that all gases mix with one another, and therefore, if the mixture to be separated is vaporized as it is introduced into the carrier gas stream, it is swept along with the carrier gas and introduced into the column. To accomplish this, the sample is introduced into the gas chromatograph by a hypodermic-like syringe. The sample is injected as a liquid or a solution through a rubber septum into a heated chamber, called the **injection port,** where it is vaporized and mixed with the carrier gas. As this mixture reaches the column, it begins to equilibrate between the liquid and gas phases. This process, however, cannot last very long unless the column is maintained at a temperature high enough to allow the components of the mixture to revaporize eventually back into the moving gas stream from the liquid phase in which they have become dissolved. For this reason, the column is placed in an insulated oven, whose temperature is adjustable. At the correct temperature, and if the correct liquid phase has been selected, the compounds in the injected mixture travel through the column at different rates and are separated.

12.4 FACTORS AFFECTING SEPARATION

Several factors determine the rate at which a given compound travels through a gas chromatograph. First of all, compounds of low boiling point will generally travel through the gas chromatograph faster than compounds of higher boiling point. Obviously, this is because the column is heated, and low-boiling compounds always have higher vapor pressures than compounds of higher boiling point. If the column is heated to a temperature that is too high, the entire mixture to be separated is flushed through the column at the same rate as the carrier gas, and no equilibration takes place with the liquid phase. At too low a temperature, the mixture dissolves in the liquid phase and never revaporizes. Thus, it is retained on the column.

Second, the rate of flow of the carrier gas can be important. The carrier gas must not pass over the liquid phase so rapidly that molecules ''dissolved'' in the gas phase cannot equilibrate with the liquid-coated particles. On the other hand, the rate of flow must be great enough to prevent holdup.

Third, the particular liquid phase chosen to prepare a column is important. The molecular weights, functional groups, and polarities of the component molecules in the mixture to be separated must be considered when a liquid phase is being chosen. One generally uses a different type of material for hydrocarbons, for instance, than for esters. The useful temperature limit of the liquid phase selected must also be considered.

Fourth, the length of the column is important. Compounds that resemble each other closely in general require longer columns than very dissimilar compounds. Many kinds of isomeric mixtures fit in the ''difficult'' category. The components of isomeric mixtures are so much alike that they travel through the column at very similar rates. A longer column, therefore, is needed to take advantage of any differences that may exist.

12.5 ADVANTAGES OF GAS CHROMATOGRAPHY

All factors that have been mentioned must be adjusted by the chemist for any mixture to be separated. Considerable preliminary investigation is required before a mixture can successfully be separated into its components by gas chromatography (or at least a goodly amount of experience). Nevertheless, the advantages of the technique are many.

First, many mixtures can be separated by this technique when no other method is adequate. Second, as little as 10 to 20 μL (1 μL = 10^{-6} L) of a mixture can be separated by this technique. It can be used on a very small scale indeed! Third, when gas chromatography is coupled with an electronic recording device (see below), the amount of each component present in the separated mixture can be estimated quantitatively.

The range of compounds that can be usefully separated by gas chromatography extends from gases, such as oxygen (bp -183 °C) and nitrogen (bp -196 °C), to organic compounds with boiling points over 400 °C. The only requirement for the compounds to be separated is that they have an appreciable vapor pressure at a temperature at which they can be separated and that they be thermally stable at this temperature.

12.6 MONITORING THE COLUMN (THE DETECTOR)

To follow the separation of the mixture injected into the gas chromatograph, it is necessary to use an electrical device called a **detector.** The most common detector is the **thermal conductivity detector.** This is simply a hot wire placed in the gas stream at the column exit. The wire is heated by constant electrical voltage. When a steady stream of carrier gas passes over this wire, the rate at which it loses heat and its electrical resistance have constant values. When the composition of the vapor stream changes, the rate of heat flow from the wire, and hence its resistance, changes. Helium, which has a higher thermal conductivity than most common organic substances, is a common carrier gas. Thus, when a substance elutes in the vapor stream, the wire generally heats up, and its resistance decreases.

A typical thermal conductivity detector operates by difference. Two detectors are used: one exposed to the actual effluent gas and the other exposed to a reference flow of carrier gas only. To achieve this situation, a portion of the carrier gas stream is diverted **before** it enters the injection port. The diverted gas is routed through a reference column into which no sample has been admitted. The detectors mounted in the sample and reference columns are arranged so as to form the arms of a Wheatstone bridge circuit, as shown in Figure 12–4. As long as the carrier gas alone flows over both detectors, the circuit is in balance. However, when a sample elutes from the sample column, the bridge circuit becomes unbalanced, creating an electrical signal. This signal can be amplified and used to activate a strip chart recorder. The recorder is an instrument that plots, by means of a moving pen, the unbalanced bridge current versus time on a continuously moving roll of chart paper. This record of detector

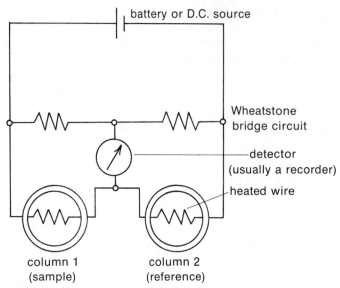

FIGURE 12–4. Typical thermal conductivity detector

response (current) versus time is called a **chromatogram.** A typical gas chromatogram is illustrated in Figure 12–5. Deflections of the pen are called peaks.

At the time the sample is injected, a peak, called the **air peak,** usually appears. This is because some air (CO_2, H_2O, N_2, and O_2) is introduced along with the sample. The air travels through the column almost as rapidly as the carrier gas; as it passes the detector, it causes a small pen response, thereby giving a peak. At later times (t_1, t_2, t_3), the components of the mixture also give rise to peaks on the chromatogram as they pass out of the column and past the detector.

12.7 RETENTION TIME

The period following injection that is required for a compound to pass through the column is called the **retention time** of that compound. For a given set of constant conditions (flow rate of carrier gas, column temperature, column length, liquid phase, injection port temperature, carrier gas), the retention time of any compound is always constant (much like the R_f value in thin-layer chromatography). The retention time is measured from the time of injection to the time of maximum pen deflection (detector current) for the component being observed. This value, when obtained under controlled conditions, can identify a compound by a direct comparison of it with values for known compounds determined under the same conditions. For easier measurement of retention times, most strip chart recorders are adjusted to move the paper at a rate that corresponds to time divisions calibrated on the chart paper. The retention times (t_1, t_2, t_3) are indicated in Figure 12–5 for the three peaks illustrated.

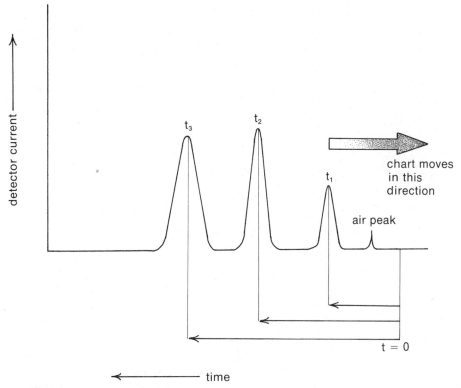

FIGURE 12–5. Typical chromatogram

12.8 QUALITATIVE ANALYSIS

Unfortunately, the gas chromatograph gives **no** information whatever about the identities of the substances it has separated. The little information it does provide is given by the retention time. It is hard to reproduce this quantity from day to day, however, and complete duplications of separations performed last month may be difficult to make this month. It is usually necessary to **calibrate** the column each time it is used. That is, one must run pure samples of all known and suspected components of a mixture individually, just before chromatographing the mixture, to obtain the retention time of each known compound. Alternatively, each suspected component can be added, one by one, to the unknown mixture while the operator looks to see which peak has its intensity increased relative to the unmodified mixture. Finally, the components can be individually **collected** as they emerge from the gas chromatograph. Each component can then be identified by other means, such as by infrared or nuclear magnetic resonance spectroscopy.

12.9 COLLECTING THE SAMPLE

Samples passed through the gas chromatograph can be collected or recovered by a cooled trap connected to the outlet of the column. The simplest form of trap is just a

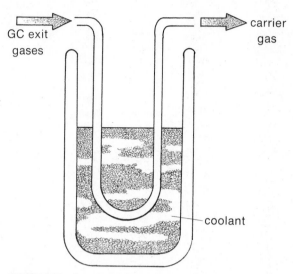

FIGURE 12–6. Collection trap

U-shaped tube dipped into a bath of coolant (ice water, liquid nitrogen, or Dry Ice–acetone), as shown in Figure 12–6.

For instance, if the coolant is liquid nitrogen (bp -196 °C) and the carrier gas is helium (bp -269 °C), compounds boiling above the temperature of liquid nitrogen generally are condensed or trapped, while the carrier gas continues to flow. However, if the compound leaving is frozen in the form of a fine mist, it may be carried through the trap even though frozen. When this happens, special trapping arrangements are required. To collect each component of the mixture, one must change the trap each time a new peak begins to appear on the chromatogram. Since even a gas chromatograph with large-diameter columns can rarely handle more than about 0.5 mL of a mixture at a time, a chemist or student having 50 mL of a mixture would have to make 50 to 100 separate injections, changing the trap for each component as it comes off the column. Fortunately, there are machines that can be attached to a gas chromatograph to do this task automatically.

12.10 QUANTITATIVE ANALYSIS

The area under a gas chromatograph peak is proportional to the amount (moles) of compound eluted. Hence, the molar percentage composition of a mixture can be approximated by comparing relative peak areas. This method of analysis does assume that the detector is equally sensitive to all compounds eluted and that it gives a linear response with regard to amount. Nevertheless, it gives reasonably accurate results.

The simplest method of measuring the area of a peak is by geometrical approximation, or triangulation. In this method, one multiplies the height h of the peak above the base line of the chromatogram by the width of the peak at half of its height, $w_{1/2}$. This is illustrated in Figure 12–7. The base line is approximated by drawing a line between the two "sidearms" of the peak. This method works well only if the peak is

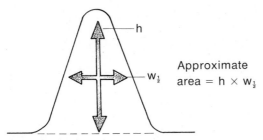

FIGURE 12–7. Triangulation of a peak

symmetrical. If the peak has tailed or is unsymmetrical, it is best to cut out the peaks with scissors and weigh them (the paper pieces) on an **analytical** balance. Since the weight per area of a piece of good chart paper is reasonably constant from place to place, the ratio of the areas is the same as the ratio of the weights. To obtain a percentage composition for the mixture, one first adds all the peak areas (weights). Then, to calculate the percentage of any component in the mixture, one divides its individual area by the total area and multiplies the result by 100. A sample calculation is illustrated in Figure 12–8. If peaks overlap (Figure 12–9A) or are not well resolved, either the gas chromatographic conditions must be readjusted to achieve better resolution of the peaks (Figure 12–9B) or the peak shape must be estimated.

Poor resolution is often caused by using too much sample, too high a column temperature, too short a column, a liquid phase that does not discriminate well between the two components, a column with too large a diameter, or, in short, almost any

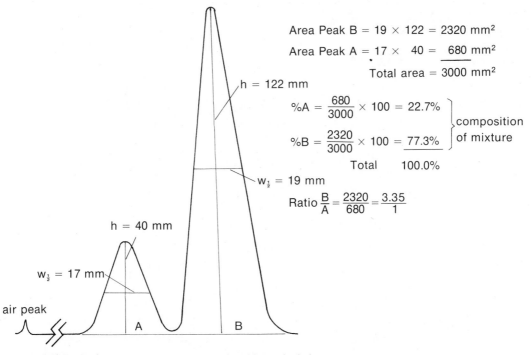

FIGURE 12–8. Sample percentage composition calculation

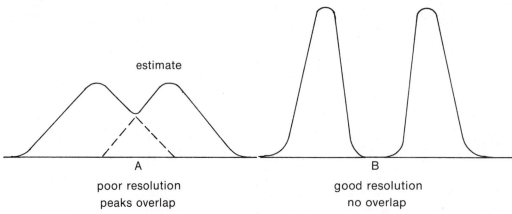

FIGURE 12–9.

wrongly adjusted parameter. The best resolution is always obtained with smaller-diameter columns; however, it is always necessary to use less sample with these. For analytical work, it is not uncommon to use Golay or capillary columns of diameter 0.1 to 0.2 mm. With these, no solid support is used; the liquid is coated directly on the inner walls of the tubing. Such columns are often long: lengths of 200 to 300 ft are common.

Technique 13
Sublimation

In Technique 6, the influence of temperature on the change in vapor pressure of a liquid was considered (see Figure 6–1, p. 551). It was shown that the vapor pressure of a liquid increases with temperature. At atmospheric pressure, the vapor pressure of a liquid equals 760 mmHg at its boiling point. The vapor pressure of a solid also varies with temperature. Because of this behavior, some solids can readily pass directly into the vapor phase without going through a liquid phase. This process is called **sublimation.** Since the vapor can be resolidified, the overall vaporization–solidification cycle can be used as a purification method. The purification can be successful only if the impurities have significantly lower vapor pressures than the material being sublimed.

13.1 VAPOR PRESSURE BEHAVIOR OF SOLIDS AND LIQUIDS

In Figure 13–1, vapor pressure curves for solid and liquid phases for two different substances are shown. Along lines **AB** and **DF,** the sublimation curves, the solid and

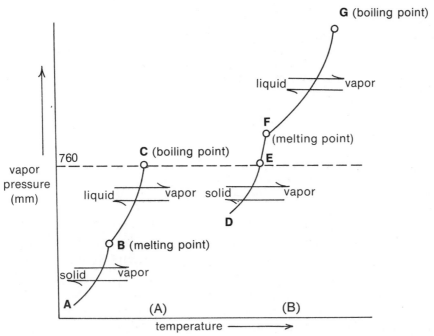

FIGURE 13–1. Vapor pressure curves for solids and liquids. *A,* Substance shows normal solid to liquid to gas transitions at 760 mmHg pressure; *B,* substance shows a solid to gas transition at 760 mmHg pressure.

vapor are at equilibrium. To the left of these lines, the solid phase exists, and to the right of these lines the vapor phase is present. Along lines **BC** and **FG,** the liquid and vapor are at equilibrium. To the left of these lines, the liquid phase exists, and to the right, the vapor is present. The two substances vary greatly in their physical properties, as seen in Figure 13–1.

In the first case (Figure 13–1A), the substance shows normal change-of-state behavior on being heated, going from solid to liquid to gas. The dashed line, which represents an atmospheric pressure of 760 mmHg, is drawn **above** the melting point, **B** in Figure 13–1A. Thus, the applied pressure is **greater** than the vapor pressure of the solid–liquid phase at the melting point. Starting at **A,** as the temperature of the solid is raised, the vapor pressure increases along **AB** until the solid is observed to melt at **B.** At **B,** the vapor pressures of **both** the solid and liquid are identical. As the temperature continues to rise, the vapor pressure will increase along **BC** until the liquid is observed to boil at **C.** The description given is for the "normal" behavior expected for a solid substance. All three states (solid, liquid, and gas) are observed sequentially during the change in temperature.

In the second case (Figure 13–1B), the substance develops enough vapor pressure to vaporize completely at a temperature below its melting point. The substance shows a solid-to-gas transition only. The dashed line is now drawn **below** the melting point, **F,** of this substance. Thus, the applied pressure is **less** than the vapor pressure of the solid–liquid phase at the melting point. Starting at **D,** the vapor pressure of the solid rises as the temperature increases along line **DF.** However, the vapor pressure of the solid reaches atmospheric pressure (point **E**) **before** the melting point (point **F**) is

TABLE 13–1. Vapor Pressures of Solids at their Melting Points

COMPOUND	VAPOR PRESSURE OF SOLID AT MP (mmHg)	MELTING POINT (°C)
Carbon dioxide	3876 (5.1 atm)	−57
Perfluorocyclohexane	950	59
Hexachloroethane	780	186
Camphor	370	179
Iodine	90	114
Naphthalene	7	80
Benzoic acid	6	122
p-Nitrobenzaldehyde	0.009	106

attained. Therefore, sublimation occurs at **E.** No melting behavior will be observed at atmospheric pressure for this substance. For a melting point to be reached and the behavior to go along line **FG,** a sealed pressure apparatus would have to be used.

The sublimation behavior just described is relatively rare for substances at atmospheric pressure. Several compounds exhibiting this behavior—carbon dioxide, perfluorocyclohexane, and hexachloroethane—are listed in Table 13–1. Notice that these compounds have vapor pressures **above** 760 mmHg at their melting point. In other words, their vapor pressures reach 760 mmHg below their melting points and they sublime rather than melt. Anyone trying to determine the melting point of hexachloroethane at atmospheric pressure will see vapor pouring from the end of the melting point tube! With a sealed capillary tube, the melting point of 186 °C is observed.

13.2 SUBLIMATION BEHAVIOR OF SOLIDS

Sublimation is usually a property of relatively nonpolar substances that also have a highly symmetrical structure. Symmetrical compounds have relatively high melting points and high vapor pressures. The ease with which a substance can escape from the solid is determined by the strength of intermolecular forces. Symmetrical molecular structures have a relatively uniform distribution of electron density and a small dipole moment. A smaller dipole moment means a higher vapor pressure because of lower electrostatic forces in the crystal.

Solids sublime if their vapor pressures are sizable at their melting points. Some compounds are listed in Table 13–1 with the vapor pressures at their melting points. The first three entries in the table were discussed in Section 13.1. At atmospheric pressure they would sublime rather than melt, as shown in Figure 13–1B.

The next four entries in Table 13–1—camphor, iodine, naphthalene, and benzoic acid—exhibit typical change-of-state behavior (solid, liquid, and gas) at atmospheric pressure, as shown in Figure 13–1A. These compounds sublime readily under reduced pressure, however. Vacuum sublimation is discussed in Section 13.3.

Compared with many other organic compounds, camphor, iodine, and naphthalene all have relatively high vapor pressures at relatively low temperatures. For example, they have a vapor pressure of 1 mmHg at 42°, 39°, and 53°, respectively. While this vapor pressure does not seem very large, it is high enough to lead, after a time, to **evaporation** of the solid from an open container. Mothballs (naphthalene and 1,4-dichlorobenzene) show this behavior. When iodine is allowed to stand in a closed container over a period, one observes movement of crystals from one part of the container to another.

Although chemists often refer to any solid–vapor transition as sublimation, the process described for camphor, iodine, and naphthalene is really an **evaporation** of a solid. Strictly speaking, a sublimation point is like a melting point or a boiling point. It is defined as the point at which the vapor pressure of the solid **equals** the applied pressure. Many liquids readily evaporate at temperatures far below their boiling points. It is, however, much less common for solids to evaporate. Solids that readily sublime (evaporate) must be stored in tightly stoppered containers. When the melting point of such a solid is being determined, some of the solid may sublime and collect toward the open end of the melting-point tube while the rest of the sample melts. To solve the sublimation problem, one seals the capillary tube or rapidly determines the melting point. It is possible to use the sublimation behavior to purify camphor. For example, at atmospheric pressure, camphor can be readily sublimed, just below its melting point, at 175 °C. At 175 °C, the vapor pressure of camphor is 320 mm. The vapor solidifies on a cool surface.

13.3   VACUUM SUBLIMATION

Many organic compounds sublime readily under reduced pressure. When the vapor pressure of the solid equals the applied pressure, sublimation occurs, and the behavior is identical to that shown in Figure 13–1B. The solid phase passes directly into the vapor phase. From the data given in Table 13–1, one expects camphor, naphthalene, and benzoic acid to sublime at or below the respective applied pressures of 370, 7, and 6 mmHg. In principle, one could sublime *p*-nitrobenzaldehyde (last entry in Table 13–1), but it would not be practical because of the low applied pressure required.

13.4   SUBLIMATION METHODS

Sublimation can be used to purify solids. The solid is warmed until its vapor pressure becomes high enough for it to vaporize and condense as a solid on a cooled surface placed close above. Several types of apparatus are illustrated in Figure 13–2. The upper surface provided for collection may be cooled by a continuous flow of water using a ''cold-finger'' condenser (Figure 13–2A, B), by an ice-water (or Dry Ice–acetone) mixture (Figure 13–2D), or by a flow of air directed from a nozzle (Figure 13–2C). The water hoses must be securely attached to the inlet and outlet of the cold-finger condenser. Otherwise, the connections may leak and allow water to pass into the sublimation apparatus. Many solids do not develop enough vapor pressure at 760 mmHg but can be sublimed at reduced pressure. Thus, most sublimation equipment has provision for connection to an aspirator or a vacuum pump. Reduction of the

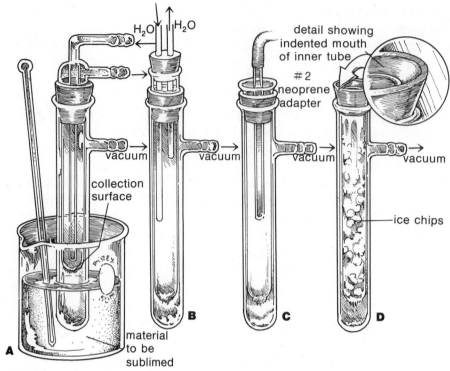

FIGURE 13–2.   Sublimation apparatus

pressure is also advantageous in preventing thermal decomposition of substances that require high temperatures to sublime at ordinary pressures.

An inexpensive apparatus, shown in Figure 13–2D, can be prepared from easily available equipment. A complete description of the assembly of this apparatus is given in Experiment 6, p. 65. Detailed procedures for using this apparatus are also given in that experiment.

In practice, the term *sublimation* is loosely applied and is often used to describe the process in which the solid actually melts before vaporization. This is not really a sublimation. It should be remembered that while performing a sublimation it is important to keep the temperature below the melting point of the solid.

After sublimation, the material that has collected on the cooled surface is recovered by removing the central tube (cold finger) from the apparatus. Care must be used in removing this tube to avoid dislodging the crystals that have collected. The deposit of crystals is scraped from the cold finger with a spatula. If reduced pressure has been used, the pressure must be released carefully to keep a blast of air from dislodging the crystals.

13.5   ADVANTAGES OF SUBLIMATION

One advantage of sublimation is that no solvent is used and therefore none needs to be removed later. Sublimation also removes occluded material, like molecules of solvation, from the sublimed substance. For instance, caffeine (sublimes at 178 °C, melts at 236 °C) absorbs water gradually from the atmosphere to form a hydrate. During sublimation this water is lost and anhydrous caffeine is obtained. If too much solvent is present in a sample to be sublimed, however, instead of becoming lost, it condenses on the cooled surface and thus interferes with the sublimation.

Sublimation is a faster method of purification than crystallization but not so selective. Similar vapor pressures are often a factor in dealing with solids that sublime; consequently, little separation can be achieved. For this reason, solids are far more often purified by crystallization. Sublimation is most effective in removing a volatile substance from a nonvolatile compound, particularly a salt or other inorganic material. Sublimation is also effective in removing highly volatile bicyclic or other symmetrical molecules from less volatile reaction products. Examples of volatile bicyclic compounds are borneol, isoborneol, and camphor.

PROBLEMS

1. Why is solid carbon dioxide called Dry Ice? How does it differ from solid water in behavior?
2. Under what conditions can one have **liquid** carbon dioxide?
3. A solid substance has a vapor pressure of 800 mmHg at its melting point (80 °C). Describe how this solid behaves as the temperature is raised at atmospheric pressure (760 mmHg).
4. A solid substance has a vapor pressure of 100 mmHg at the melting point (100 °C). Describe the behavior of this solid as the temperature is raised at atmospheric pressure (760 mmHg).
5. A substance has a vapor pressure of 50 mmHg at the melting point (100 °C). Describe how one would experimentally sublime this substance.

Technique 14
Techniques for Handling Sodium Metal

> **EXTREME CARE AND CAUTION SHOULD ALWAYS BE EXERCISED WHEN WORKING WITH SODIUM METAL**

Elemental sodium is a metal that is low-melting (98 °C) and has a low density (0.96 g/cc). It has a metallic silver color when pure. It is extremely soft, and it can be easily cut by a spatula or a dull knife. It is very malleable, that is, small pieces can be "squashed" easily by pressing against them with the flat side of a spatula and by thumb pressure only.

14.1   REACTIVITY OF SODIUM

Sodium is very reactive. It is one of the most electropositive elements and reacts explosively with elements such as fluorine, chlorine, and pure oxygen. Sodium easily gives up its lone valence electron to more electronegative elements or other suitable acceptors to form the ion Na^+. In particular, sodium reacts vigorously, often violently, with organic compounds having acidic hydrogens. Hydrogen is formed in these reactions.

$$Na \cdot + H^+ \longrightarrow [H \cdot] + Na^+$$

$$2[H \cdot] \longrightarrow H_2$$

With compounds that are moderately to very acidic, this reaction is often sufficiently exothermic to ignite the hydrogen as it forms. Even water is sufficiently acidic to fall in this category:

$$2Na + 2H_2O \longrightarrow 2NaOH + \underbrace{H_2 + heat}_{Ignites}$$

14.2   CAUTION WHEN HANDLING SODIUM

SODIUM REACTS VIOLENTLY WITH WATER, and care should always be taken when working with this metal to avoid contact with water or any wet apparatus. Note especially that SODIUM SHOULD NEVER BE DISPOSED OF IN THE SINK. Organic solvents may have been poured accidentally into the sink in the organic laboratory. If these solvents are flammable, throwing sodium into the sink, where it reacts with water to produce and ignite hydrogen, might cause a colossal fire in the sink.

$$2Na + 2ROH \longrightarrow 2NaOR + H_2$$

Sodium reacts less violently with alcohols than with water—the alcohols are less acidic. As the hydrocarbon chain length of the alcohols increases, the rate of reaction decreases. Sodium can be added to ethanol, for instance, if some care is taken. The ethanol becomes hot enough to boil but does not ignite if sodium is added slowly and if the solution is cooled. The reaction with 1-butanol is quite slow; therefore, excess sodium is disposed of most prudently by placing it in a beaker with enough 1-butanol to cover it. The sodium should be left until **all** the metal has dissolved and no bubbling action is apparent. Then either water or methanol is added to the 1-butanol solution, and the material is discarded. Other functional groups that react with sodium include acids, amines, terminal acetylenes, enols, phenols, and mercaptans.

Anyone working with sodium metal should avoid all contact of the sodium with the skin. Sodium reacts with moisture on the surface of the skin to produce sodium hydroxide, which is highly caustic.

14.3   STORING SODIUM

Freshly cut sodium reacts with the moisture and oxygen in air to form an oxide layer consisting of Na_2O and NaOH on the surface of the metal. A sample of freshly cut, silvery sodium quickly tarnishes to gray when exposed to moist air. Eventually a thick white crust forms on all the surfaces. To prevent this crust from forming, sodium metal usually is stored in a shallow, wide-mouthed jar under an inert, high-boiling solvent. Xylene or ligroin, or sometimes mineral oil, is used most often for this purpose. Even so, some oxidation takes place, and sodium usually must be "pared" of its oxide coating before use.

14.4   CUTTING AND WEIGHING SODIUM

Since sodium tarnishes and oxidizes when exposed to air, it is usually manipulated under xylene or ligroin. For sodium to be weighed, a beaker containing xylene or ligroin is placed on a balance and tared (its weight recorded). Then, pieces of sodium are added directly to this beaker, resting on the balance, until the proper weight of sodium has been obtained. The following procedure is used. The sodium is transferred from the storage container to a crystallizing dish or a paper towel, and it is cut as quickly as possible with a spatula or a knife so as to remove the oxide coating. The freshly cut pieces of sodium are transferred quickly to the tared beaker of xylene or ligroin to forestall oxidation. When enough sodium has been trimmed, as judged by the increase in weight of the beaker, the excess is returned to the storage container. Small pieces of sodium may be destroyed by treating them with 1-butanol, as described in Section 14.2. The crystallizing dish may be filled with 1-butanol to decompose the small amounts of metal that remain. If a paper towel was used, it should be placed in a beaker of 1-butanol. Additional 1-butanol should be poured over the paper towel; after this treatment, the towel may be removed and discarded as usual.

The sodium weighed in the beaker is usually cut into small pieces in the beaker under the covering liquid. A knife or a spatula is used. If mineral oil is used, the pieces of sodium must be rinsed before use; a good solvent for this rinsing is petroleum ether. The small pieces should be pulled from the liquid with forceps or tweezers (never the fingers), plunged into another beaker containing petroleum ether, and quickly transferred to the reaction mixture.

For use in organic reactions, the sodium should be cut into pieces about the size of a pea or a BB shot. Pieces that correspond to a cube of anywhere from 2 to 4 mm per side are satisfactory.

14.5   OTHER ALKALI METALS

Lithium (mp 186 °C, density 0.53 g/cc) should be handled much like sodium. It is **slightly** less reactive than sodium. Potassium (mp 62.3 °C, density 0.86 g/cc) is extremely reactive and must be handled with extra caution.

Lithium is sold often as a wire of known weight per length. It is stored in mineral oil. Weighing a given amount of lithium is easy when it comes in this form; one only needs to cut off a measured length of the wire.

14.6 THE SODIUM PRESS

Many labs have a **sodium press,** which extrudes the metal into a wire about ⅛ in. in diameter. This sodium wire is then handled easily, like lithium wire. The weight per centimeter is usually constant from one sodium wire to the next if they come from the same press, and measuring a length of wire is a good substitute for a weight determination.

Technique 15
Polarimetry

15.1 NATURE OF POLARIZED LIGHT

Light has a dual nature, since it shows properties of waves and also of particles. The wave nature of light can be demonstrated by two experiments: polarization and interference. Of the two, polarization is the more interesting to organic chemists, since they can take advantage of polarization experiments to learn something about the structure of an unknown molecule.

Ordinary white light consists of wave motion in which the waves have a variety of wavelengths and vibrate in all possible planes perpendicular to the direction of propagation. Light can be made to be **monochromatic** (of one wavelength or color) by the use of filters or special light sources. Frequently, a sodium lamp (sodium D line = 5893 Å) is used. Although the light from this lamp consists of waves of only one wavelength, the individual light waves still vibrate in all possible planes perpendicular to the beam. If we imagine that the beam of light is aimed directly at the viewer, ordinary light can be represented by showing the edges of the planes oriented randomly around the path of the beam, as in the left part of Figure 15–1.

A Nicol prism, which consists of a specially prepared crystal of Iceland spar (or calcite), has the property of serving as a screen that can restrict the passage of light waves. Waves that are vibrating in one plane are transmitted, while those in a perpendicular plane are rejected (either refracted in another direction or absorbed). The light

FIGURE 15–1. Ordinary versus plane-polarized light

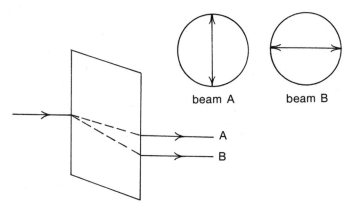

FIGURE 15–2. Double refraction

that passes through the prism is called **plane-polarized light,** and it consists of waves
that vibrate only in one plane. A beam of plane-polarized light aimed directly at the
viewer can be represented by showing the edges of the plane oriented in one particular
direction, as in the right portion of Figure 15–1.

Iceland spar has the property of **double refraction,** that is, it can split, or
doubly refract, an entering beam of ordinary light into two separate emerging beams of
light. Each of the two emerging beams (labeled A and B in Figure 15–2) has only a
single plane of vibration, and the plane of vibration in beam A is perpendicular to the
plane of beam B. In other words, the crystal has separated the incident beam of ordinary light into two beams of plane-polarized light, with the plane of polarization of
beam A perpendicular to the plane of beam B.

To generate a single beam of plane-polarized light, one can take advantage of
the double-refracting property of Iceland spar. A Nicol prism, invented by the Scottish
physicist William Nicol, consists of two crystals of Iceland spar cut to specified angles
and cemented by Canada balsam. This prism transmits one of the two beams of plane-
polarized light while reflecting the other at a sharp angle so that it does not interfere
with the transmitted beam. Plane-polarized light can also be generated by a Polaroid
filter, a device invented by E. H. Land, an American physicist. Polaroid filters consist
of certain types of crystals, embedded in transparent plastic and capable of producing
plane-polarized light.

After passing through a first Nicol prism, plane-polarized light can also pass
through a second Nicol prism, but only if the second prism has its axis oriented so that
it is **parallel** to the incident light's plane of polarization. Plane-polarized light is **absorbed** by a Nicol prism that is oriented so that its axis is **perpendicular** to the incident
light's plane of polarization. These situations can be illustrated by the picket-fence
analogy, as shown in Figure 15–3. Plane-polarized light can pass through a fence
whose slats are oriented in the proper direction but is blocked out by a fence whose slats
are oriented perpendicularly.

An optically active substance is one that interacts with polarized light to rotate
the plane of polarization through some angle α. Figure 15–4 illustrates this phenomenon.

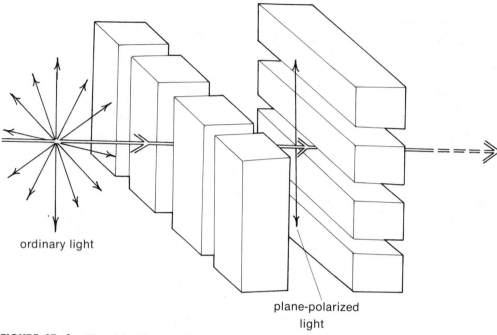

ordinary light

plane-polarized
light

FIGURE 15-3. The picket fence analogy

15.2 THE POLARIMETER

An instrument called a **polarimeter** is used to measure the extent to which a substance interacts with polarized light. A schematic diagram of a polarimeter is shown in Figure 15–5. The light from the source lamp is polarized by being passed through a fixed Nicol prism, called a polarizer. This light passes through the sample, with which it may or may not interact to have its plane of polarization rotated in one direction or the other. A second, rotatable Nicol prism, called the analyzer, is adjusted to allow the maximum amount of light to pass through. The number of degrees and the direction of rotation required for this adjustment are measured to give the **observed rotation** α.

So that data determined by several persons under different conditions can be compared, a standardized means of presenting optical rotation data is necessary. The

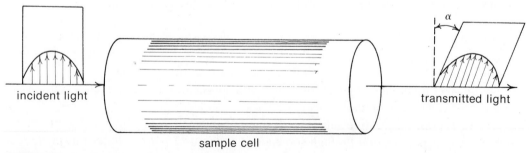

incident light   transmitted light

sample cell

FIGURE 15-4. Optical activity

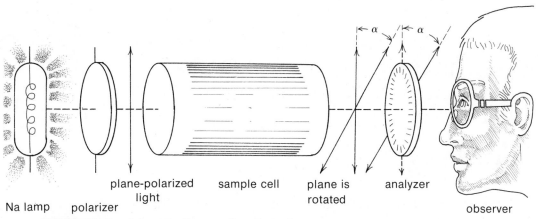

FIGURE 15—5. Schematic diagram of a polarimeter

most common way of presenting such data is by recording the **specific rotation** $[\alpha]_\lambda^t$, which has been corrected for differences in concentration, cell path length, temperature, solvent, and wavelength of the light source. The equation defining the specific rotation of a compound in solution is

$$[\alpha]_\lambda^t = \frac{\alpha}{cl}$$

where

α = observed rotation in degrees
c = concentration in grams per milliliter of solution
l = length of sample tube in decimeters
λ = wavelength of light (usually indicated as ''D,'' for the sodium D line)
t = temperature in degrees Celsius

For pure liquids, the density d of the liquid in grams per milliliter replaces c in the above formula. Occasionally one wants to compare compounds of different molecular weights, so a **molecular rotation,** based on moles instead of grams, is more convenient than a specific rotation. The molecular rotation M_λ^t is derived from the specific rotation $[\alpha]_\lambda^t$ by

$$M_\lambda^t = \frac{[\alpha]_\lambda^t \times \text{molecular weight}}{100}$$

Usually measurements are made at 25 °C with the sodium D line as a light source, so specific rotations are reported as $[\alpha]_D^{25°}$.

15.3   THE SAMPLE CELLS

It is important for the solution whose optical rotation is to be determined to contain no suspended particles of dust or dirt that might disperse the incident polarized light.

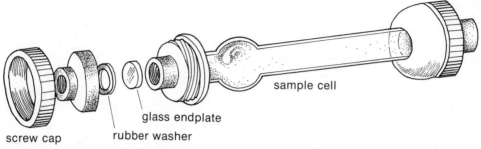

FIGURE 15–6. Polarimeter cell assembly

Therefore it is necessary to clean the sample cell carefully and to make certain that there are no air bubbles trapped in the path of the light. The sample cells contain an enlarged ring near one end, in which the air bubbles may be trapped. The sample cell, shown in Figure 15–6, is tilted upward and tapped until the air bubbles move into the enlarged ring. It is important not to get fingerprints on the glass endplate in reassembling the cell.

The sample is generally prepared by dissolving 0.1 to 0.5 g of the substance to be studied in 25 mL of solvent, usually water, ethanol, or methylene chloride; chloroform was used in the past. If the specific rotation of the substance is very high or very low, it may be necessary to make the concentration of the solution respectively lower or higher, but usually this is determined after first trying a concentration range such as that suggested above.

15.4   OPERATION OF THE POLARIMETER

The procedures given here for preparing the cells and for operating the instrument are appropriate for the Zeiss polarimeter with the circular scale; other models of polarimeter are operated similarly. It is necessary, before beginning the measurements, to turn the power switch to the ON position and wait 5 to 10 minutes until the sodium lamp is properly warmed.

The instrument should be checked initially by making a zero reading with a sample cell filled only with solvent. If the zero reading does not correspond with the zero-degree calibration mark, then the difference in readings must be used to correct all subsequent readings. The reading is determined by laying the sample tube in the cradle, enlarged end up (making sure that there are no air bubbles in the light path), closing the cover, and turning the knob until the proper angle of the analyzer is reached. Most instruments, including the Zeiss polarimeter, are of the double-field type, in which the eye sees a split field whose sections must be matched in light intensity. The value of the angle through which the plane of polarized light has been rotated (if any) is read directly from the scale that can be seen through the eyepiece directly below the split-field image. Figure 15–7 shows how this split field might appear.

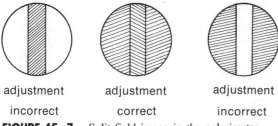

adjustment   adjustment   adjustment

incorrect   correct   incorrect

FIGURE 15–7.   Split-field image in the polarimeter

The cell containing the solution of the sample is then placed in the polarimeter, and the observed angle of rotation is measured in the same way. Be sure to record not only the numerical value of the angle of rotation in degrees but also the direction of rotation. Rotations clockwise are due to **dextrorotatory** substances and are indicated by the sign "+." Rotations counterclockwise are due to **levorotatory** substances and are indicated by the sign "−." It is best, in making a determination, to take several readings, including readings for which the actual value was approached from both sides. In other words, where the actual reading might be +75°, first approach this reading upward from a reading near zero, then on the next measurement approach this reading downward from an angle greater than +75°. Duplicating readings and approaching the observed value from both sides reduce errors. The readings are then averaged to get the observed rotation α. This rotation is then corrected by the appropriate factors, according to the formulas in Section 15.2, to provide the specific rotation. The specific rotation is always reported as a function of temperature, indicating the wavelength by "D" if a sodium lamp is used, and reporting the concentration and solvent used. For example, $[\alpha]_D^{20} = +43.8°$ ($c = 7.5$ g/100 mL in absolute ethanol).

15.5 OPTICAL PURITY

When one prepares a sample of an enantiomer by a resolution method, the sample is not always 100.0% pure enantiomer. It frequently is contaminated by residual amounts of the opposite stereoisomer. To determine the amount of the desired enantiomer in the sample, one calculates the **optical purity,** or the **excess** of one enantiomer in a mixture expressed as a percentage of the total. In a racemic (±) mixture, there is no excess enantiomer and the optical purity is zero; in a completely resolved material, the excess enantiomer equals the total material in weight, and the optical purity is 100%. Although the following is not the most precise equation for determining the optical purity, it should prove useful in most simple applications:

$$\text{Optical purity} = \frac{\text{observed specific rotation}}{\text{specific rotation of pure substance}} \times 100$$

A compound that is x% optically pure contains x% of one enantiomer and $(100 - x)$% of a **racemic mixture.**

If the optical purity is given, the relative percentages of the enantiomers can be calculated easily. If the predominant form in the impure, optically active mixture is assumed to be the (+) enantiomer, the percentage of the (+) enantiomer is

$$\left[x + \left(\frac{100 - x}{2} \right) \right] \%$$

and the percentage of the (−) enantiomer is $[(100 - x)/2]\%$. The relative percentages of (+) and (−) forms in a partially resolved mixture of enantiomers can be calculated as shown below. Consider a partially resolved mixture of camphor enantiomers. The specific rotation for pure (+)-camphor is +43.8° in absolute ethanol, but the mixture shows a specific rotation of +26.3°.

$$\text{Optical purity} = \frac{+26.3°}{+43.8°} \times 100 = 60\% \text{ optically pure}$$

$$\% \text{ (+) enantiomer} = 60 + \left(\frac{100 - 60}{2} \right) = 80\%$$

$$\% \text{ (−) enantiomer} = \left(\frac{100 - 60}{2} \right) = 20\%$$

PROBLEMS

1. Calculate the specific rotation of a substance that is dissolved in a solvent (0.4 g/mL) and that has an observed rotation of −10° as determined with a 0.5-dm cell.
2. Calculate the observed rotation for a solution of a substance (2.0 g/mL) that is 80% optically pure. A 2-dm cell is used. The specific rotation for the optically pure substance is +20°.
3. What is the optical purity of a partially racemized product if the calculated specific rotation is −8° and the pure enantiomer has a specific rotation of −10°? Calculate the percentage of each of the enantiomers in the partially racemized product.

Technique 16
Refractometry

The **refractive index** is a useful physical property of liquids. Often a liquid can be identified from a measurement of its refractive index. Alternatively, comparing the experimentally measured refractive index with the value reported in the literature for an ultrapure sample provides a measure of the purity of the sample being examined. The closer the measured sample's value to the literature value, the purer the sample.

16.1 THE REFRACTIVE INDEX

The refractive index has as its basis the fact that light travels at a different velocity in condensed phases (liquids, solids) than in air. The refractive index (n) is defined as the ratio of the velocity of light in air to the velocity of light in the medium being measured:

$$n = \frac{V_{\text{air}}}{V_{\text{liquid}}} = \frac{\sin \theta}{\sin \phi}$$

It is not difficult to measure experimentally the ratio of the velocities. It corresponds to $\sin \theta / \sin \phi$, where θ is the angle of incidence for a beam of light striking the surface of the medium and ϕ is the angle of refraction of the beam of light **within** the medium. This is illustrated in Figure 16–1.

The refractive index for a given medium depends on two variable factors. First, it is **temperature**-dependent. The density of the medium changes with temperature; hence, the speed of light in the medium also changes. Second, the refractive index is **wavelength**-dependent. Beams of light with different wavelengths are refracted to different extents in the same medium and give different refractive indices for that medium. It is usual to report refractive indices measured at 20 °C, with a sodium discharge lamp as the source of illumination. The sodium lamp gives off yellow light of 589-mm wavelength, the so-called sodium D line. Under these conditions, the refractive index is reported in the following form.

$$n_{\text{D}}^{20} = 1.4892$$

The superscript indicates the temperature, and the subscript indicates that the sodium D line was used for the measurement. If another wavelength is used for the determination, the D is replaced by the appropriate value, usually in nanometers (1 nm = 10^{-9} m).

Notice that the hypothetical value reported above has four decimal places. It is easy to determine the refractive index to within several parts in 10,000. Therefore, n_{D}

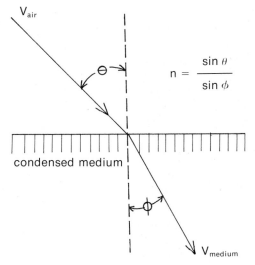

FIGURE 16–1. The refractive index

is a very accurate physical constant for a given substance and can be used for identification. However, it is sensitive to even small amounts of impurity in the substance measured. Unless the substance is purified **extensively,** it will not usually be possible to reproduce the last two decimal places given in a handbook or other literature source.

16.2 THE ABBÉ REFRACTOMETER

The instrument used to measure the refractive index is called a **refractometer.** Although many styles of refractometer are available, by far the most common instrument is the Abbé refractometer. This style of refractometer has the following advantages:

1. White light may be used for illumination; but the instrument is compensated, so that the index of refraction obtained is actually that for the sodium D line.
2. The prisms can be temperature-controlled.
3. Only a small sample is required (a few drops of a liquid).

A common type of Abbé refractometer is shown in Figure 16–2.

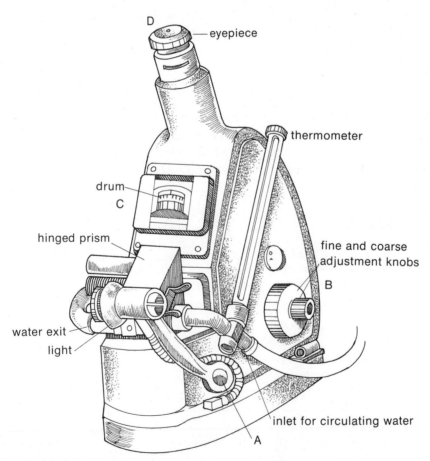

FIGURE 16–2.   Abbé refractometer (Bausch and Lomb Abbé 3L)

The optical arrangement of the refractometer is very complex; a simplified diagram of the internal workings is given in Figure 16–3. The letters A, B, C, and D label corresponding parts in both Figures 16–2 and 16–3. A complete description of refractometer optics is too difficult to attempt here, but Figure 16–3 gives a simplified diagram of the essential operating principles.

The sample to be measured is introduced between the two prisms. If it is a ''free-flowing'' liquid, it may be introduced into a channel along the side of the prisms, injected from a blunt pipet or an eyedropper. If it is a viscous sample, the prisms must be opened (they are hinged) by lifting the upper one; a few drops of liquid are applied to the lower prism with a wooden applicator or an eyedropper. If an eyedropper is used, care must be taken not to touch the prisms, since they become scratched easily. When the prisms are closed, the liquid should spread evenly to make a thin film.

Next, one turns on the light and looks into the eyepiece D. The hinged lamp and the coarse adjustment knob at B are adjusted to give the most uniform illumination to the visible field in the eyepiece (no dark areas). The light rotates at pivot A.

Once a uniform field is found, one rotates the coarse and fine adjustment knobs at B until the dividing line between the light and dark halves of the visual field coincides with the center of the cross hairs (Figure 16–4). If the cross hairs are not in sharp focus, it will be necessary to adjust the eyepiece to focus them. If the horizontal line

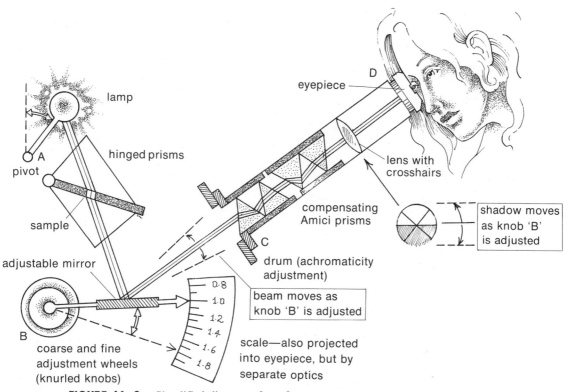

FIGURE 16–3. Simplified diagram of a refractometer

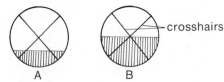

FIGURE 16–4. *A,* Refractometer incorrectly adjusted; *B,* correct adjustment

dividing the light and dark areas appears as a colored band, as in Figure 16–5, the refractometer is showing **chromatic aberration** (color dispersion). This can be adjusted with the knob labeled C. This knurled knob rotates a series of prisms, called Amici prisms, that color-compensate the refractometer and cancel out dispersion. The knob should be adjusted to give a sharp, uncolored division between the light and dark segments. When one has adjusted everything correctly (as in Figure 16–4B), the refractive index is read. In the instrument described here, a small button on the left side of the housing is pressed, and the scale becomes visible in the eyepiece. In other refractometers the scale is visible at all times, frequently through a separate eyepiece.

Occasionally the refractometer will be so far out of adjustment that it may be difficult to measure the refractive index of an unknown. When this happens, it is wise to place a pure sample of known refractive index in the instrument, set the scale to the correct value of refractive index, and adjust the controls for the sharpest line possible. Once this has been done, it will be easier to measure an unknown sample. Typical organic liquids have refractive index values between 1.3400 and 1.5600.

There are many styles of refractometer, but most have adjustments similar to those described here.

16.3   CLEANING THE REFRACTOMETER

You should always remember, in using the refractometer, that if the prisms are scratched the instrument will be ruined. DO NOT TOUCH THE PRISMS WITH ANY HARD OBJECT. This admonition includes eyedroppers and glass rods.

When measurements are completed, the prisms should be cleaned with ethanol or petroleum ether. **Soft** tissues are moistened with the solvent, and the prisms are wiped **gently.** When the solvent has evaporated from the prism surfaces, they should be locked together. The refractometer should be left with the prisms closed to avoid collection of dust in the space between them. The instrument should also be turned off when it is no longer in use.

FIGURE 16–5.   Refractometer showing chromatic aberration (color dispersion). The dispersion is incorrectly adjusted.

16.4 TEMPERATURE CORRECTIONS

If the refractive index is not determined in a room in which the temperature is 20 °C, or if 20 °C cooling water is not used to circulate through the instrument, a temperature correction must be made. Although the magnitude of the temperature correction may vary from one class of compound to another, a value of 0.00045 per degree Celsius is a useful approximation for most substances. The index of refraction of a substance **decreases** with **increasing** temperature. Therefore, one adds the correction to the observed n_D value for temperatures higher than 20 °C and subtracts it for temperatures lower than 20 °C. As an example, the reported n_D^{20} value for nitrobenzene is 1.5529. One would observe a value at 25 °C of 1.5506. The temperature correction would be made as follows:

$$n_D^{20} = 1.5506 + 5(0.00045) = 1.5529$$

Technique 17
Preparation of Samples for Spectroscopy

Modern organic chemistry requires sophisticated scientific instruments. Most important among these instruments are the two spectroscopic instruments: the infrared and nuclear magnetic resonance spectrometers. The instruments are indispensable to the modern organic chemist in proving the structures of unknown substances, in verifying that reaction products are indeed what had been predicted, and in characterizing organic compounds. The theory underlying these instruments can be found in most standard lecture textbooks in organic chemistry. Additional information, including correlation charts, to help in interpreting spectra are found in this textbook in Appendix 3 (Infrared Spectroscopy) and in Appendix 4 (Nuclear Magnetic Resonance). This technique chapter concentrates on the preparation of samples for these spectroscopic methods.

PART A. INFRARED

17.1 SAMPLE PREPARATION

To determine the infrared spectrum of a compound, one must place it in a sample holder or cell of some kind. In infrared spectroscopy this immediately poses a problem.

Glass, quartz, and plastics absorb strongly throughout the infrared region of the spectrum (any compound with covalent bonds usually absorbs) and cannot be used to construct sample cells. Ionic substances must be used in cell construction. Metal halides (sodium chloride, potassium bromide, silver chloride) are commonly used for this purpose. Single crystals of sodium chloride are cut and polished to give plates that are transparent throughout the infrared region. These plates are then used to fabricate sample cells. The only disadvantages are that the plates cleave easily (break) when too much pressure is applied and that sodium chloride is water-soluble. This means that samples must be **dry** before a spectrum can be obtained. Since water absorbs in the infrared region, water should be removed from samples if necessary. Silver chloride is water-insoluble, but expensive, and can be used for aqueous solutions.

17.2   LIQUID SAMPLES

Salt Plates

The simplest method of preparing the sample, if it is a liquid, is to place a thin layer of the liquid between two sodium chloride plates that have been flat-ground and polished. A drop of the liquid is placed on the surface of one plate, and the second plate then placed on top. The pressure of this second plate causes the liquid to spread out and form a thin, capillary film between the two plates. The plates are then mounted in a holder designed to allow them to be placed in the sample beam of the spectrophotometer. This holder is illustrated in Figure 17–1. All parts of the holder (plates and mounting holder)

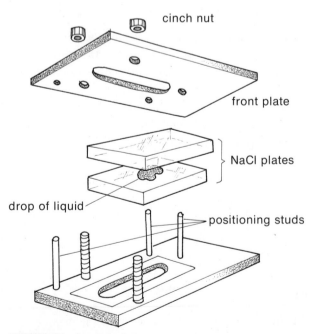

FIGURE 17–1.   Salt plates

are usually stored in a desiccator or an oven near the infrared spectrophotometer. The large polished NaCl plates are expensive because they are cut from a large, single crystal of NaCl, and break easily (specifically, they can be cleaved). They are also water-soluble. Handle these plates carefully; **they should be touched only on their edges.** Moisture from fingers will mar and occlude the polished surfaces. A spectrum of water or any solution containing water can obviously not be taken, as water will damage the plates. **It must be certain that the sample is dry or free from water.**

When you are assembling the holder, **do not tighten the cinch nuts excessively** because too great a pressure will cleave, or split, the NaCl plates. Just tighten the nuts firmly, but do not use any force at all to turn them. Spin them with the fingers until they stop; then turn them just another fraction of a full turn, and they will be tight enough.

Once the spectrum is determined, the NaCl plates should be washed with a volatile dry solvent such as methylene chloride. A soft tissue, moistened with solvent, may be used to wipe them off; this will be especially useful if one has run a Nujol mull (Section 17.4).

To some extent the thickness of the film obtained between the two plates is a function of two factors: (1) the amount of liquid placed on the first plate (one drop, two drops, etc.) and (2) the pressure used to hold the plates together. If more than 1 or 2 drops of liquid have been used, it will probably be too much, and the resulting spectrum will show such strong absorptions everywhere that they are off the scale of the chart paper. Only enough liquid is needed to wet both surfaces. The capillary layer of liquid can be made thinner by increasing the pressure on the two plates, but one must be cautious of exerting too much pressure, since then the plates will break.

Occasionally a sample will have too low a viscosity or be too volatile to give a practicable capillary film; the liquid will either form too thin a film or evaporate too quickly. Should this happen, the spectrum will have to be determined using solution cells. (These are described in Section 17.3.)

17.3 SOLID SAMPLES

KBr Pellets

The easiest method of preparing a solid sample is to make a KBr pellet. Effectively, KBr does not absorb in the infrared region, and it can be used to make a solid "solution" of an unknown sample whose spectrum is desired. A good grade of KBr must be used, and it must be dry. Potassium bromide acquires waters of hydration on standing, and it is therefore stored in an oven. One weighs out 1 to 2 mg of the solid sample and mixes it with about 100 mg of KBr. The first few times this mixture is prepared it should be weighed out on an analytical balance. After some experience, these quantities can be quite closely estimated by eye. The mixture is then ground with a pestle to a fine powder in an agate or a glass mortar. Be sure the sample is uniformly mixed into the KBr. The powder must be finely ground or it will scatter the infrared radiation excessively. The total grinding time will be 3 to 5 minutes.

A **portion** of the finely ground powder (usually not more than half) is placed into a die that compresses it into a translucent pellet. The simplest die consists of two

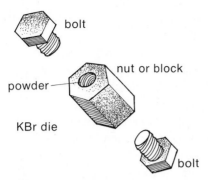

FIGURE 17–2. KBr pellet minipress

stainless steel bolts and a large nut, as illustrated in Figure 17–2. The bolts have their ends ground flat.[1] To use this die, screw one of the bolts into the nut, but not all the way; leave one or two turns. Carefully add the powder into the open end of the partly assembled die and tap it lightly on the desk top to give an even layer of filling. Then, carefully screw the second bolt into the open nut until it is just firm. Then transfer the die to a holder that is bolted to the table and keeps the head of one bolt from turning. The final tightening of the die to compress the KBr mixture is done with a torque wrench. Continue to turn the torque wrench until a loud click is heard (the ratchet mechanism makes softer clicks) or until the appropriate torque value (20 ft-lb) is reached. If the bolt is tightened beyond this point, the head may be twisted off one of the bolts or the block (nut) may be split. Leave the die under pressure for 30 to 60 seconds and then reverse the ratchet on the torque wrench or pull the torque wrench in the opposite direction to open the assembly. Remove the two bolts but leave the compressed KBr disc in the center of the block. The block rests in a holder, which allows the sample to sit in the sample beam of the spectrophotometer.

The KBr pellet should be either clear or translucent. If it is not, the sample will scatter the beam of infrared radiation to such an extent that a spectrum cannot be obtained. If the pellet is found to be cloudy, one of several things may have been wrong:

1. The KBr mixture may not have been ground finely enough, and the particle size may be too big.
2. The sample may not have been thoroughly dried.
3. The ratio of sample to KBr may have been too high, that is, too much sample was used.
4. The pellet may be too thick, that is, too much of the mixture was put into the die.
5. The KBr may have been "wet" or have acquired moisture from the air.
6. The die may not have been tightened enough (this is improbable when a torque wrench is used).
7. The sample may have a low melting point. Low-melting solids not only are difficult to dry but also melt under pressure.

[1]A minipress of this type is available from Wilks Scientific Company, a division of Analabs, 80 Republic Dr., North Haven, CT 06473.

Often a cloudy pellet cannot be avoided. When this happens, it is often possible to compensate (at least partially) for the scattering by placing a wire screen in the reference beam, thereby balancing the lowered transmittance of the pellet.

When the spectrum has been determined, punch the pellet out of the block and wash both the block and the bolts thoroughly with water. Then rinse them with acetone to dry them and return them to the oven for storage.

Solution Spectra

Method A

For substances that are soluble in carbon tetrachloride, a quick and easy method for determining the spectra of solids is available. The solid is dissolved in carbon tetrachloride (weight ratio $1:10$). One or two drops of the solution are placed between sodium chloride plates in precisely the same manner as that used for pure liquids (Section 17.2). The spectrum is determined as described for pure liquids using salt plates (Section 17.2). Since the spectrum contains the absorptions of the solute superimposed on the absorptions of carbon tetrachloride, it is important to remember that any absorption that appears to the right of about 900 cm^{-1} (11.1 μ) may be due to the stretching of the C–Cl bond of the solvent. Information contained to the right of 900 cm^{-1} is not usable in this method. Chloroform solutions cannot be studied by this method, since the solvent has too many interfering absorptions.

> **CAUTION: Persons using this method must be warned that carbon tetrachloride is a hazardous solvent.**

Carbon tetrachloride, besides being toxic, is suspected of being a carcinogen. In spite of the health problems associated with its use, there is no suitable alternative solvent for infrared spectroscopy. Other solvents have too many interfering infrared absorption bands. Therefore, it is necessary for the operator to handle carbon tetrachloride very carefully to minimize the adverse health effects. The spectroscopic-grade carbon tetrachloride should be stored in a 1-pt stoppered bottle in a hood. A capillary pipet should be attached to the bottle, possibly by storing it in a test tube taped to the side of the bottle. All sample preparation should be conducted in the hood. Rubber or plastic gloves should be worn. The cells should also be cleaned in the hood. All carbon tetrachloride used in preparing samples or in cleaning cells should be disposed of in an appropriately marked waste container.

Method B

The spectra of solids or of liquids that are too volatile or of two low a viscosity to be run between salt plates can be determined in a type of permanent sample cell called a **solution cell.** The solution cell, shown in Figure 17–3, is made from two salt plates, mounted with a Teflon spacer between them to control the thickness of the sample. The top sodium chloride plate has two holes drilled in it so that the sample can be intro-

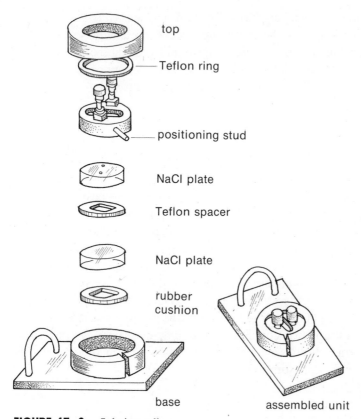

top

Teflon ring

positioning stud

NaCl plate

Teflon spacer

NaCl plate

rubber
cushion

base   assembled unit

FIGURE 17-3. Solution cell

duced into the cavity between the two plates. These holes are extended through the face
plate by two tubular extensions designed to hold Teflon plugs, which seal the internal
chamber and prevent evaporation. The tubular extensions are tapered so that a syringe
body (Luer lock without a needle) will fit snugly into them from the outside. The cells
are thus filled from a syringe; usually they are held upright and filled from the bottom
entrance port.

These cells are expensive (over $125 a pair), and hence your instructor may
limit their availability for use by large classes. It is strongly suggested that you obtain
the instructor's permission to use the solution cells before you begin to handle them.
The cells are bought in matched pairs. The dissolved sample is placed in one cell (the
sample cell), and the pure solvent is placed in the other cell (the reference cell). The
spectrum of the solvent is thus subtracted from the spectrum of the solution (not always
completely), and a spectrum of the solute is thus provided. For this solvent compensa-
tion to be as exact as possible, it is essential that the same cell be used as a reference
and that the other cell be used as a sample cell without ever being interchanged. When
the spectrum is determined, it is important to clean the cells by flushing them with
clean solvent.

Solvents most often used in determining infrared spectra are carbon tetrachloride, chloroform, and carbon disulfide. The spectra of these substances are shown in
Figures 17-4 through 17-6. For solution work, a 5 to 10% solution usually gives a

INFRARED SPECTRA OF SOLVENTS COMMONLY USED FOR SAMPLE PREPARATION

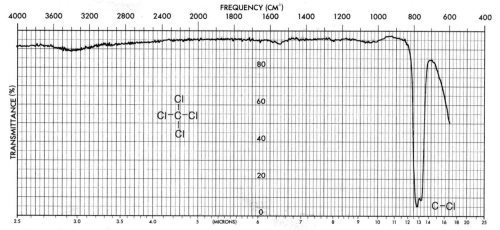

FIGURE 17–4. Carbon tetrachloride

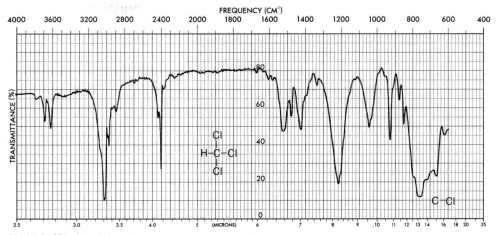

FIGURE 17–5. Chloroform

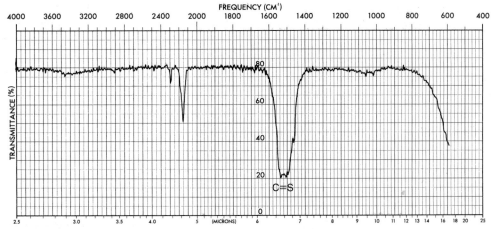

FIGURE 17–6. Carbon disulfide

good spectrum. Carbon tetrachloride and chloroform are suspect carcinogens. However, since there are no suitable alternative solvents, these compounds must be used in infrared spectroscopy. The procedure outlined on p. 663 for carbon tetrachloride should be followed. This procedure serves equally well for chloroform.

> **Before you use the solution cells, you must obtain the instructor's permission and instruction on how to fill and clean the cells.**

17.4 NUJOL MULLS

If an adequate KBr pellet cannot be obtained, the spectrum of a solid may be determined as a Nujol mull. In this method, about 5 mg of the solid sample are ground finely in a glass or an agate mortar with a pestle. One or two drops of Nujol mineral oil (white) are then added, and the mixture is ground to a very fine dispersion about the consistency of milk of magnesia. This mull is then placed between two salt plates, as is done with liquid samples (see Section 17.2).

Nujol is a mixture of high-molecular-weight hydrocarbons. Hence, it has absorptions in the C–H stretch and CH_2 and CH_3 bending areas of the spectrum (Figure 17–7). Clearly, if Nujol is used, no information can be obtained in these portions of the spectrum. In interpreting the spectrum, you must ignore these Nujol peaks. It is important to label the spectrum immediately after it was determined, noting that it was determined as a Nujol mull. Otherwise it might be forgotten that the C–H peaks belong to Nujol and not to the solute.

17.5 RECORDING THE SPECTRUM

The instructor will describe how to operate the infrared spectrophotometer, since the controls vary, depending on the manufacturer and model of the instrument. In all cases,

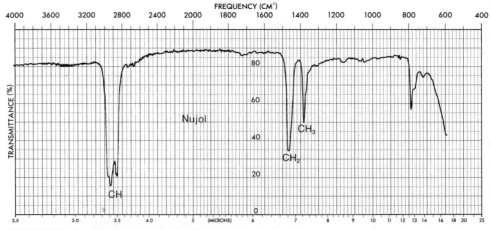

FIGURE 17–7. Nujol (mineral oil)

it is important that the sample, the solvent, the type of cell or method used, and any other pertinent information be written on the spectrum immediately after determination. This information may be important, and it is easily forgotten if not recorded.

17.6 CALIBRATION

To know the frequency, or wavelength, of each absorption peak precisely, the frequency scale of the spectrum must be calibrated. This calibration is accomplished by recording a partial spectrum of a standard substance, which is always polystyrene, over the spectrum of the compound being studied. The complete spectrum of polystyrene is shown in Figure 17–8. A thin film of polystyrene is mounted in a card designed to fit into the sample cell holder on the spectrophotometer. The card is inserted into the sample cell holder, and the tips (not the entire spectrum) of the important peaks are recorded over the sample spectrum. The most important of these peaks is at 1603 cm^{-1} (6.238 μ), while other useful peaks are at 2850 cm^{-1} (3.509 μ) and 906 cm^{-1} (11.035 μ).

Although some modern instruments record spectra so precisely that calibration is usually not necessary, it is always good practice to calibrate a spectrum whenever possible. The time spent in calibration is insignificant, and it minimizes frequency assignment errors caused by a poor fit of paper to the chart recorder bed.

PART B. NUCLEAR MAGNETIC RESONANCE
17.7 PREPARING AN NMR SAMPLE

The nmr sample tubes used in most 60-MHz instruments are approximately ⅜ × 6 in. in overall dimension and are fabricated of uniformly thin glass tubing. The sample tube is spun around its cylindrical axis while suspended in the magnet gap by a holder. The

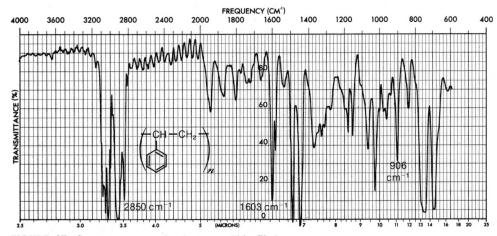

FIGURE 17–8. Ir spectrum of polystyrene (thin film)

lower tip of the tube is positioned between the magnet pole pieces and the oscillator and detector coils by a depth gauge. To be sure that the sample will be aligned correctly, fill the tube to a minimum depth of 1 to 1.5 in. from the bottom. This usually requires a 0.5- to 1.0-mL quantity of the sample solution (sample dissolved in a suitable solvent). Figure 17–9 depicts an nmr sample tube, showing the correct sample level.

Liquid compounds that are not viscous can be determined "neat," that is, without solvent. Viscous compounds must be determined in solution. It is generally best to determine all samples in solution. Solutions that are 10 to 30% sample (weight/weight) generally give satisfactory spectra. Typically, a 20% concentration is used. Usually the sample can be prepared "by eye" if a sample of a liquid compound is being prepared. Visually divide the length of the tube required for a 1.5-inch height of sample into fifths. Fill the first fifth with sample and the remaining four fifths with solvent.

A solid sample, however, is best weighed out when preparing the solution. A deceivingly large amount (by volume), or so it seems, of solid is required to make a 20% w/w solution of a solid. One should weigh out about 150 mg of solid per 0.5 mL of solvent used.

To prepare the solution, one must of course choose the appropriate solvent *first!* There are two requirements for the solvent: (1) the sample should dissolve in it and (2) the solvent should have no nmr absorption peaks of its own, that is, no protons. Carbon tetrachloride, CCl_4, is the most widely used solvent for this purpose. Most compounds dissolve in CCl_4. It has the ability to dissolve a wide variety of functional groups and has **no protons.** If the sample does not dissolve in CCl_4, the more polar solvent deuterochloroform, or chloroform-d, $CDCl_3$, can often be used to advantage. Deuterium, $_1^2H$, does not absorb in the proton region and is thus "invisible," or not seen, in the proton nmr spectrum. An even more polar solvent is deuterium oxide,

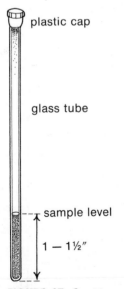

plastic cap

glass tube

sample level

$1 - 1\frac{1}{2}''$

FIGURE 17–9.   Nmr sample tube

D_2O. This solvent is often used for very polar compounds. Since the deuterated solvents are **expensive,** the spectrum is usually determined in carbon tetrachloride if possible.

> **CAUTION: Persons using this method must be warned that carbon tetrachloride is a hazardous solvent. See page 7.**

The chemical shift values commonly reported for various types of protons refer to the chemical shift as determined in either CCl_4 or $CDCl_3$. If another solvent is used, the chemical shift values can, and often do, change markedly, owing to the different solvent environment. When $CDCl_3$ is used, there is often a low-intensity peak in the nmr spectrum at 7.27δ due to an unavoidably small amount of $CHCl_3$ impurity. Spectra determined in D_2O often show a small peak because of OH impurity. If the sample compound has acidic hydrogens, these may **exchange** with D_2O, leading to the appearance of an OH peak in the spectrum and the **loss** of the original absorption from the acid proton, owing to the exchanged hydrogen. In many cases this will also alter the splitting patterns of a compound, since J_{HD} vicinal is often negligible ($J_{HD} \approx 0$ to 1.5 Hz).

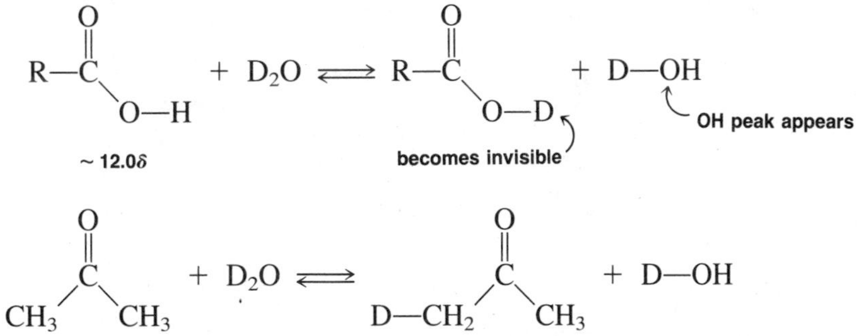

Most solid carboxylic acids do not dissolve in CCl_4, $CDCl_3$, or even D_2O. In such cases one adds a small piece of sodium metal to about 1 mL of D_2O. The acid is then dissolved in this solution. The resulting basic solution enhances the solubility of the carboxylic acid. In such a case, the hydroxyl proton of the carboxylic acid cannot be observed in the nmr spectrum, since it exchanges with the solvent. A large DOH peak is observed, however, due to the exchange and the H_2O impurity in the D_2O solvent.

When the above solvents fail, other special solvents can be used. Acetone, acetonitrile, dimethylsulfoxide, pyridine, benzene, and dimethylformamide can be used if one is not interested in the region or regions of the nmr spectrum in which they give rise to absorption. The deuterated (but expensive) analogs of these compounds are also used in special instances (for example, acetone-d_6, dimethylsulfoxide-d_6, dimethylformamide-d_7, and benzene-d_6). If the sample is not sensitive to acid, trifluoroacetic acid (which has no protons with $\delta < 12$) can be used. Once again, one has to be aware that these solvents often lead to different chemical shift values from those determined in CCl_4 or $CDCl_3$. Variations of as much as 0.5 to 1.0 ppm have been observed. In fact, it is sometimes possible, by switching to pyridine, benzene, acetone, or dimethyl-sulfoxide as solvents, to separate peaks that overlap when CCl_4 or $CDCl_3$ solutions are used.

Carbon tetrachloride, chloroform (and chloroform-d), and benzene (and ben-zene-d_6) are hazardous solvents. Besides being highly toxic, they are also suspect carcinogens. In spite of these health problems, these solvents are commonly used in nmr spectroscopy because there are no suitable alternatives. These solvents are used because they contain no protons and because they are excellent solvents for most organic compounds. Therefore it is necessary that one learn to handle these solvents with great care to minimize the hazard. These solvents should be stored either in a hood or in septum-capped bottles. If the bottles have screw caps, a pipet should be attached to each bottle. A recommended way of attaching the pipet is to store it in a test tube taped to the side of the bottle. Septum-capped bottles can be used only by withdrawing the solvent with a hypodermic syringe that has been designated solely for this use. All samples should be prepared in a hood, and solutions should be disposed of in an appropriately designated waste container that is stored in the hood. Rubber or plastic gloves should be worn when samples are being prepared or discarded.

> **Before using any deuterated solvent, check the solubility of the compound in its undeuterated analog. Deuterated compounds are expensive and should not be wasted.**

17.8 RECORDING THE SPECTRUM

In most instances the instructor or some qualified laboratory assistant will actually record the nmr spectrum of a student sample. If students are permitted to operate the nmr spectrometer, the instructor will provide instructions. Since the controls of nmr spectrometers vary, depending on the make or model of the instrument, we shall not try to describe these controls. **Do not operate the nmr spectrometer unless you have been properly instructed.**

17.9 REFERENCE SUBSTANCES

To provide the internal reference standard, tetramethylsilane (TMS) must be added to the sample solution. This substance has the formula $(CH_3)_4Si$. The concentration of TMS should be 1 to 3%. Some people prefer to add 1 to 2 drops of TMS to the sample just before determining the spectrum. (Not much TMS is needed since it has 12 equivalent protons!) For the addition, a small pipet or a syringe is used. It is far easier to make (in bulk) standard solvents already containing 1 to 3% of TMS dissolved in them. Since

TMS is highly volatile (bp 26.5 °C), such solutions must be stored, tightly stoppered, in a refrigerator. Tetramethysilane itself is best stored in a refrigerator as well.

Tetramethylsilane does not dissolve in D_2O. For spectra determined in D_2O, a different internal standard, sodium 2,2-dimethyl-2-silapentane-5-sulfonate, must be used. This standard is water-soluble and gives a resonance peak at essentially the same δ value as TMS.

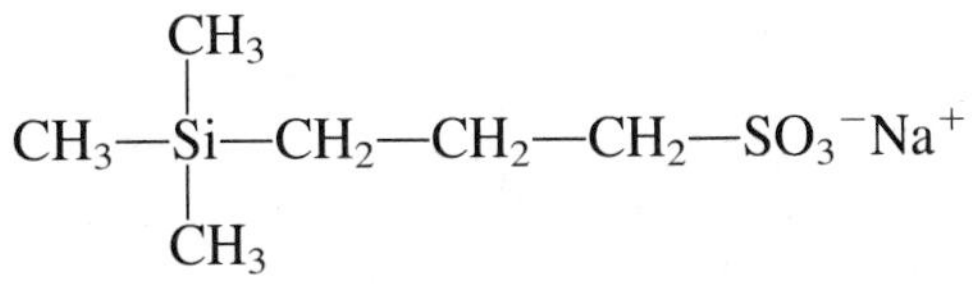

$$CH_3-\underset{\underset{CH_3}{|}}{\overset{\overset{CH_3}{|}}{Si}}-CH_2-CH_2-CH_2-SO_3^-Na^+$$

Sodium 2,2-dimethyl-2-silapentane-5-sulfonate (DSS)

Technique 18
Guide to the Chemical Literature

Often, it may be necessary to go beyond the information contained in the typical organic chemistry textbook and to use reference material in the library. At first glance, using library materials may seem formidable because of the numerous sources the library contains. If, however, one adopts a systematic approach, the task can prove rather useful. This description of various popular sources and an outline of logical steps to follow in the typical literature search should be helpful.

18.1 LOCATING PHYSICAL CONSTANTS: HANDBOOKS

To find information on routine physical constants, such as melting points, boiling points, indices of refraction, and densities, you should first consider a handbook. Examples of suitable handbooks are

R. C. Weast, ed. *CRC Handbook of Chemistry and Physics*. 66th ed. Boca Raton, FL: CRC Press, 1985. Revised annually.

J. A. Dean, ed. *Lange's Handbook of Chemistry*. 13th ed. New York: McGraw-Hill, 1985.

M. Windholz, ed. *The Merck Index*. 10th ed. Rahway, NJ: Merck & Co., 1983.

The *Handbook of Chemistry and Physics* is the handbook most often consulted. For organic chemistry, however, *The Merck Index* is probably better suited. This handbook also contains literature references on the isolation, structure determination, and synthesis of a substance, along with its molecular formula, elemental analysis, and certain properties of medicinal interest (e.g., toxicity and medicinal and veterinary uses).

A more complete handbook is

J. Buckingham, editor, *Dictionary of Organic Compounds,* Chapman & Hall/Methuen, New York: 1982.

This is a revised version of an earlier four-volume handbook edited by I. M. Heilbron and H. M. Bunbury. In its present form, it consists of five volumes with 15 supplements.

18.2 SIDE REACTIONS AND GENERAL SYNTHETIC METHODS

The easiest way of determining possible side reactions for a particular reaction is to examine all the possible reactions of the starting materials. Many standard introductory textbooks in organic chemistry provide tables that summarize most of the common reactions for a given class of compounds. These tables can be used to determine which possible side reactions the starting materials might undergo. Examples of textbooks that include tables of this type are

N. L. Allinger, M. P. Cava, D. C. DeJongh, C. R. Johnson, N. A. Lebel, and C. L. Stevens. *Organic Chemistry.* 2nd ed. New York: Worth Publishers, 1976.

R. F. Francis. *Student Guide and Solutions Manual to Accompany Ternay's Contemporary Organic Chemistry.* 2nd ed. Philadelphia: W. B. Saunders, 1979.

C. D. Gutsche and D. J. Pasto. *Fundamentals of Organic Chemistry.* Englewood Cliffs, NJ: Prentice-Hall, 1975.

D. S. Kemp and F. Vellaccio. *Organic Chemistry.* New York: Worth Publishers, 1980.

J. McMurry. *Organic Chemistry.* Monterey, CA: Brooks/Cole, 1984.

R. T. Morrison and R. N. Boyd. *Organic Chemistry.* 4th ed. Boston: Allyn and Bacon, 1983.

D. C. Neckers and M. P. Doyle. *Organic Chemistry.* New York: Wiley, 1977.

S. H. Pine, J. B. Hendrickson, D. J. Cram, and G. S. Hammond. *Organic Chemistry.* 4th ed. New York: McGraw-Hill, 1980.

T. W. G. Solomons. *Organic Chemistry.* 2nd ed. New York: Wiley, 1980.

R. J. Fessenden and J. S. Fessenden. *Introduction to Organic Chemistry.* 3rd ed. Monterey, CA: Brooks/Cole, 1986.

Similar tables may be used to identify alternative methods of preparing a particular compound. Tables of synthetic methods for each important class of compounds can be found in many of the textbooks from the above list, as well as in

J. March. *Advanced Organic Chemistry: Reactions, Mechanisms, and Structure.* 3rd ed. New York: Wiley, 1985.

W. H. Reusch. *An Introduction to Organic Chemistry.* San Francisco: Holden-Day, 1977.

J. D. Roberts and M. C. Caserio. *Basic Principles of Organic Chemistry.* 2nd ed. Menlo Park, CA: W. H. Benjamin, 1977.

A. Streitwieser, Jr., and C. H. Heathcock. *Introduction to Organic Chemistry.* 3rd ed. New York: Macmillan, 1985.

The textbooks cited above represent only a partial list of books in which information about side reactions or other methods of preparation may be found. This information is to be found in virtually any introductory textbook on organic chemistry, although it may not be presented in a convenient series of tables.

18.3 SEARCHING THE CHEMICAL LITERATURE

If the information one is seeking is not available in any of the handbooks mentioned above, or if one is searching for more detailed information than they can provide, then a proper literature search is in order. While an examination of standard textbooks can provide some help, often one must use all the resources of the library, including journals, reference collections, and abstracts. The following sections of this chapter outline how the various types of sources should be used and what sort of information can be obtained from them.

The methods for searching the literature discussed in this chapter use printed materials. Modern methods of literature searching also make use of computerized databases. These are vast collections of data and bibliographic materials that can be scanned very rapidly from remote computer terminals. While computerized searching is becoming more widely available, its use is not usually accessible to undergraduate students.

18.4 COLLECTIONS OF SPECTRA

Collections of infrared, nuclear magnetic resonance, and mass spectra can be found in the following catalogs of spectra:

A. Cornu and R. Massot. *Compilation of Mass Spectral Data*. 2nd ed. London: Heyden and Sons, Ltd., 1975.
High-Resolution NMR Spectra Catalog. Palo Alto, CA: Varian Associates. Volume 1, 1962; volume 2, 1963.
C. J. Pouchert. *Aldrich Library of Infrared Spectra*. 3rd ed. Milwaukee: Aldrich Chemical Co., 1981.
C. J. Pouchert. *Aldrich Library of NMR Spectra*. 2nd ed. Milwaukee: Aldrich Chemical Co., 1983.
Sadtler Standard Spectra. Philadelphia: Sadtler Research Laboratories. Continuing collection.

The American Petroleum Institute has also published collections of infrared, nuclear magnetic resonance, and mass spectra.

18.5 ADVANCED TEXTBOOKS

Much information about synthetic methods, reaction mechanisms, and reactions of organic compounds is available in any of the many current advanced textbooks in organic chemistry. Examples of such books are

F. A. Carey and R. J. Sundberg. *Advanced Organic Chemistry. Part A. Structure and Mechanisms; Part B. Reactions and Synthesis*. 2nd ed. New York: Plenum Press, 1983.
L. F. Fieser and M. Fieser. *Advanced Organic Chemistry*. New York: Reinhold, 1961.
I. L. Finar. *Organic Chemistry*. 6th ed. London: Longman Group, Ltd., 1986.

H. O. House. *Modern Synthetic Reactions*. 2nd ed. Menlo Park, CA: W. H. Benjamin, 1972.
J. March. *Advanced Organic Chemistry: Reactions, Mechanisms, and Structure*. 3rd ed. New York: Wiley, 1985.
C. R. Noller. *Chemistry of Organic Compounds*. 3rd ed. Philadelphia: W. B. Saunders, 1965.

These books often contain references to original papers in the literature for students wanting to follow the subject further. Consequently, the student obtains not only a review of the subject from such a textbook but also a key reference that is helpful toward a more extensive literature search. The textbook by March is particularly useful for this purpose.

18.6 SPECIFIC SYNTHETIC METHODS

Anyone interested in locating information about a particular method of synthesizing a compound should first consult one of the many general textbooks on the subject. Useful ones are

C. A. Buehler and D. E. Pearson. *Survey of Organic Syntheses*. New York: Wiley-Interscience, 1970.
F. A. Carey and R. J. Sundberg. *Advanced Organic Chemistry. Part B. Reactions and Synthesis*. 2nd ed. New York: Plenum Press, 1983.
L. F. Fieser and M. Fieser. *Reagents for Organic Synthesis*. New York: Wiley-Interscience, 1967–1986. This is a continuing series, now in twelve volumes.
I. T. Harrison et al. *Compendium of Organic Synthetic Methods*. New York: Wiley-Interscience, 1971–1977. Three-volume set.
H. O. House. *Modern Synthetic Reactions*. 2nd ed. Menlo Park, CA: W. H. Benjamin, 1972.
J. March. *Advanced Organic Chemistry: Reactions, Mechanisms, and Structure*. 3rd ed. New York: Wiley, 1985.
S. Patai, ed. *The Chemistry of the Functional Groups*. London: Interscience, 1964–present. This series consists of many volumes, each one specializing in a particular functional group.
A. I. Vogel. Revised by members of the School of Chemistry, Thames Polytechnic, *Vogel's Textbook of Practical Organic Chemistry, Including Qualitative Organic Analysis*. 4th ed. London: Longman Group, Ltd., 1978.
R. B. Wagner and H. D. Zook. *Synthetic Organic Chemistry*. New York: Wiley, 1956.

More specific information, including actual reaction conditions, exists in collections specializing in organic synthetic methods. The most important of these is

Organic Syntheses. New York: Wiley, 1921–present. Published annually.

One of the features of the advanced organic textbook by March is that it includes references to specific preparative methods contained in *Organic Syntheses*. While *Organic Syntheses* is published with annual volumes, older volumes are combined in groups of 10 into a series of collective volumes. Tables found at the end of each of the collective volumes classify methods according to the type of reaction, type of compound prepared, formula of compound prepared, preparation or purification of solvents and reagents, and use of various types of specialized apparatus.

More advanced material on organic chemical reactions and synthetic methods may be found in any one of a number of annual publications that review the original literature and summarize it. Examples include

Advances in Organic Chemistry: Methods and Results. New York: Wiley, 1960–present.
Annual Reports of the Chemical Society, Section B. London: Chemical Society, 1905–present. Specifically, the section on *Synthetic Methods.*
Progress in Organic Chemistry. New York: Wiley, 1952–1973.
Organic Reactions. New York: Wiley, 1942–present.

Each of these publications contains a great many citations referring the reader to the appropriate articles in the original literature.

18.7   ADVANCED LABORATORY TECHNIQUES

The student who is interested in reading about more advanced techniques than those described in this textbook, or in more complete descriptions of techniques, should consult one of the advanced textbooks specializing in organic laboratory techniques. Besides focusing on apparatus construction and the performance of complex reactions, these books also provide advice on purifying reagents and solvents. Useful sources of information on organic laboratory techniques include

R. B. Bates and J. P. Schaefer. *Research Techniques in Organic Chemistry.* Englewood Cliffs, NJ: Prentice-Hall, 1971.
A. J. Krubsack. *Experimental Organic Chemistry.* Boston: Allyn and Bacon, 1973.
T. S. Ma and V. Horak. *Microscale Manipulations in Chemistry.* New York: Wiley-Interscience, 1976.
R. S. Monson. *Advanced Organic Synthesis: Methods and Techniques.* New York: Academic Press, 1971.
A. Weissberger et al., eds. *Technique of Organic Chemistry.* 3rd ed. New York: Wiley-Interscience, 1959–1969. This work is in 14 volumes.
K. B. Wiberg. *Laboratory Technique in Organic Chemistry.* New York: McGraw-Hill, 1960.

Numerous works specialize in particular techniques, and so do other more general textbooks. The above list is only representative of the most common books in this category.

18.8   REACTION MECHANISMS

As with the case of locating information on synthetic methods, a great deal of information can be obtained about reaction mechanisms by consulting one of the common textbooks on physical organic chemistry. The textbooks listed here provide a general description of mechanisms, but they do not contain specific literature citations. Very general textbooks include

R. Breslow. *Organic Reaction Mechanism.* New York: Benjamin, 1966.
P. Sykes. *A Guidebook to Mechanism in Organic Chemistry.* 6th ed. London: Longman Group, Ltd., 1986.

More advanced textbooks include

F. A. Carey and R. J. Sundberg. *Advanced Organic Chemistry. Part A. Structure and Mechanisms.* 2nd ed. New York: Plenum Press, 1983.
R. D. Gilliom. *Introduction to Physical Organic Chemistry.* Reading, MA: Addison-Wesley, 1970.
L. P. Hammett. *Physical Organic Chemistry: Reaction Rates, Equilibria, and Mechanisms.* 2nd ed. New York: McGraw-Hill, 1970.

J. Hine. *Physical Organic Chemistry*. 2nd ed. New York: McGraw-Hill, 1962.

J. A. Hirsch. *Concepts in Theoretical Organic Chemistry*. Boston: Allyn and Bacon, 1974.

C. K. Ingold. *Structure and Mechanism in Organic Chemistry*. 2nd ed. Ithaca, NY: Cornell University Press, 1969.

R. A. Y. Jones. *Physical and Mechanistic Organic Chemistry*. 2nd ed. Cambridge: Cambridge University Press, 1984.

J. B. Leffler and E. Grunwald. *Rates and Equilibria of Organic Reactions*. New York: Wiley, 1963.

T. H. Lowry and K. S. Richardson. *Mechanism and Theory in Organic Chemistry*. 2nd ed. New York: Harper & Row, 1981.

J. W. Moore and R. G. Pearson. *Kinetics and Mechanism*. 3rd ed. New York: Wiley, 1981.

These books include extensive bibliographies that permit the reader to delve more deeply into the subject.

Most libraries also subscribe to annual series of publications that specialize in articles dealing with reaction mechanisms. Among these are

Advances in Physical Organic Chemistry. London: Academic Press, 1963–present.

Annual Reports of the Chemical Society. Section B. London: Chemical Society, 1905–present. Specifically, the section on *Reaction Mechanisms*.

Organic Reaction Mechanisms. Chichester, England: Wiley, 1965–present.

Progress in Physical Organic Chemistry. New York: Interscience, 1963–present.

These publications provide the reader with citations from the original literature that can be very useful in an extensive literature search.

18.9 ORGANIC QUALITATIVE ANALYSIS

Experiment 56 contains a procedure for identifying organic compounds through a series of chemical tests and reactions. Occasionally, one might require a more complete description of analytical methods or a more complete set of tables of derivatives. Textbooks specializing in organic qualitative analysis should fill this need. Examples of sources for such information include

N. D. Cheronis and J. B. Entriken. *Identification of Organic Compounds: A Student's Text Using Semimicro Techniques*. New York: Interscience, 1963.

D. J. Pasto and C. R. Johnson. *Laboratory Text for Organic Chemistry: A Source Book of Chemical and Physical Techniques*. Englewood Cliffs, N.J.: Prentice-Hall, 1979.

Z. Rappoport, ed. *Handbook of Tables for Organic Compound Identification,* 3rd ed. Cleveland: Chemical Rubber Co., 1967.

R. L. Shriner, R. C. Fuson, D. Y. Curtin, and T. C. Morrill. *The Systematic Identification of Organic Compounds: A Laboratory Manual*. 6th ed. New York: Wiley, 1980.

A. I. Vogel. *Elementary Practical Organic Chemistry*. Part 2. *Qualitative Organic Analysis*. 2nd ed. New York: Wiley, 1966.

A. I. Vogel. Revised by members of the School of Chemistry, Thames Polytechnic. *Vogel's Textbook of Practical Organic Chemistry, Including Qualitative Organic Analysis*. 4th ed. London: Longman Group, Ltd., 1978.

18.10 BEILSTEIN AND CHEMICAL ABSTRACTS

One of the most useful sources of information about the physical properties, synthesis, and reactions of organic compounds is *Beilsteins Handbuch der Organischen Chemie*.

This is a monumental work, initially edited by Friedrich Konrad Beilstein, and updated through several revisions by the Beilstein Institute in Frankfurt-am-Main, Germany. The original edition (the *Hauptwerk,* abbreviated H) was published in 1918 and covers completely the literature to 1909. Four supplementary series (*Ergänzungswerken*) have been published since that time. The first supplement (*Erstes Ergänzungswerk,* abbreviated E I) covers the literature from 1910 to 1919; the second supplement (*Zweites Ergänzungswerk,* E II) covers 1920 to 1929; the third supplement (*Drittes Ergänzungswerk,* E III) covers 1930 to 1949; and the fourth supplement (*Viertes Ergänzungswerk,* E IV) covers 1950 to 1959. Volumes 17 to 27 of supplementary series III and IV, covering heterocyclic compounds, are combined in a joint issue, E III/IV. Supplementary series III and IV are not complete, so the coverage of *Handbuch der Organischen Chemie* can be considered complete to 1929, with partial coverage to 1959.

Beilsteins Handbuch der Organischen Chemie, usually referred to simply as *Beilstein,* also contains two types of cumulative indices. The first of these is a name index (*Sachregister*) and the second is a formula index (*Formelregister*). These indices are particularly useful for a person wishing to locate a compound in *Beilstein.*

The principal difficulty in using *Beilstein* is that it is written in German. While some reading knowledge of German is useful, the beginner can obtain information from the work by learning a few key phrases. For example, *Bildung* is "formation" or "structure," *Darst.* or *Darstellung* is "preparation," K_P or *Siedepunkt* is "boiling point," and *F* or *Schmelzpunkt* is "melting point." Furthermore, the names of some compounds in German are not cognates of the English names. Some examples are *Apfelsäure* for "malic acid" (*Säure* means "acid"), *Harnstoff* for "urea," *Jod* for "iodine," and *Zimtsäure* for "cinnamic acid." If the student has access to a German-English dictionary for chemists, many of these difficulties can be overcome. The best such dictionary is

A. M. Patterson. *German-English Dictionary for Chemists.* 3rd ed. New York: Wiley, 1959.

Beilstein is organized according to a very sophisticated and complicated system. To locate a compound in *Beilstein,* one could learn all the intricacies of this system. However, most students do not wish to become experts on *Beilstein* to this extent. A simpler, though slightly less reliable, method is to look for the compound in the formula index that accompanies the second supplement. Under the molecular formula, one will find the names of compounds that have that formula. After that name will be a series of numbers that indicate the pages and volume in which that compound is listed. Suppose, as an example, that one is searching for information on *p*-nitroaniline. This compound has the molecular formula $C_6H_6N_2O_2$. Searching for this formula in the formula index to the second supplement, one finds

4-Nitro-anilin **12** 711, **I** 349, **II** 383.

This information tells us that *p*-nitroaniline is listed in the main edition (*Hauptwerk*) in volume 12, page 711. Locating this particular volume, which is devoted to isocyclic monoamines, we turn to page 711 and find the beginning of the section on *p*-nitroaniline. At the left side of the top of this page, we find "Syst. Nr. 1671." This is the system number given to compounds in this part of volume 12. The system number is useful, as it can help us find entries for this compound in subsequent supplements. The organization of *Beilstein* is such that all entries on *p*-nitroaniline in each of the supplements will always be found in volume 12. The entry in the formula index also indicated that material on this compound could be found in the first supplement on page 349 and in the second supplement on page 383. On page 349 of volume 12 of the first supplement, there is a heading, **"XII, 710–712,"** and on the left is "Syst. Nr. 1671." Material on *p*-nitroaniline will be found in each supplement on a page that is headed with the volume and page of the *Hauptwerk* in which the same compound is found. On page 383 of volume 12 of the second supplement, the heading in the center of the top of the page is **"H12, 710–712."** On the left we find "Syst. Nr. 1671." Again, because *p*-nitroaniline appeared in volume 12, page 711, of the main edition, it can be located by searching through volume 12 of any supplement until one finds a page with the heading corresponding to volume 12, page 711. Because the third and fourth supplements are not complete, there is no comprehensive formula index for these supplements. However, one can still find material on *p*-nitroaniline by using the system number and the volume and page in the main work. In the third supplement, because the amount of information available has grown so much since the early days of Beilstein's work, volume 12 has now expanded so that it is found in several bound parts. However, one selects the part that includes system number 1671. In this part of volume 12, one looks through the pages until one finds a page headed "Syst. Nr. 1671/H 711." The information on *p*-nitroaniline is found on this page (page 1580). If volume 12 of the fourth supplement were available, one would go on in the same way to locate more recent data on *p*-nitroaniline. This example is meant to illustrate how one can locate information on particular compounds without having to learn the Beilstein system of classification. You might do well to test your ability at finding compounds in *Beilstein* as we have described here.

Guidebooks to using *Beilstein,* which include a description of the Beilstein system, are recommended for anyone who wants to work extensively with *Beilstein.* Among such sources are

How to Use Beilstein. Beilstein Institute, Frankfurt-am-Main. Berlin: Springer-Verlag.

E. H. Huntress. *A Brief Introduction to the Use of Beilsteins Handbuch der Organischen Chemie.* 2nd ed. New York: Wiley, 1938.

O. Weissbach. *The Beilstein Guide: A Manual for the Use of Beilsteins Handbuch der Organischen Chemie.* New York: Springer-Verlag, 1976.

Beilstein reference numbers are listed in such handbooks as *CRC Handbook of Chemistry and Physics* and *Lange's Handbook of Chemistry.* Additionally, Beilstein numbers are included in the *Aldrich Catalog Handbook of Fine Chemicals,* issued by the Aldrich Chemical Company. If the compound one is seeking is listed in one of these handbooks, one will find that using *Beilstein* is simplified.

Another very useful publication that can help you in finding references to research on a particular topic is *Chemical Abstracts,* published by the Chemical Abstracts Service of the American Chemical Society. *Chemical Abstracts* contains abstracts of articles appearing in more than 10,000 journals from virtually every country conducting scientific research. These abstracts list the authors, the journal in which the article appeared, the title of the paper, and a short summary of the contents of the article.

Abstracts of articles that appeared originally in a foreign language are provided in English, with a notation describing the original language.

To use *Chemical Abstracts,* one must know how to use the various indices that accompany it. At the end of each volume there appears a set of indices, including a formula index, a general subject index, a chemical substances index, an author index, and a patent index. The listings in each index refer the reader to the appropriate abstract, according to the number assigned to it. There are also collective indices that combine all the indexed material appearing in a 5-year period (10-year period before 1956). In the collective indices, the listings include the volume number as well as the abstract number.

For material after 1929, *Chemical Abstracts* provides the most complete coverage of the literature. For material before 1929, use *Beilstein* before consulting *Chemical Abstracts*. *Chemical Abstracts* has the advantage that it is written entirely in English. Nevertheless, the most common requirement for a literature search by the student involves a search for a relatively simple compound. Finding the desired entry for a simple compound is much easier in *Beilstein* than in *Chemical Abstracts*. For simple compounds, the indices in *Chemical Abstracts* are likely to contain very many entries. To locate the desired information, the student will have to comb through this multitude of listings—potentially a very time-consuming task.

The opening pages of each index in *Chemical Abstracts* contain a brief set of instructions on using that index. Students who want a more complete guide to *Chemical Abstracts* may consult a textbook that contains problems designed to familiarize the reader with these abstracts and indices. An example of such a textbook is

CAS Printed Access Tools: A Workbook. Washington: Chemical Abstracts Service, American Chemical Society, 1977.

Chemical Abstracts Service maintains a computerized database that permits users to search through *Chemical Abstracts* very rapidly and thoroughly. This service is called *CAS On-Line,* and it is available in most major university and industrial libraries. The database can be searched from a remote terminal using telephone lines to connect to the main computer in Columbus, Ohio. The database extends from 1965 to the present, and it is being expanded currently. A modest charge is made for each search.

18.11 SCIENTIFIC JOURNALS

Ultimately, someone wanting information about a particular area of research will be required to read articles from the scientific journals. These journals are of two basic types: review journals and primary scientific journals. Journals that specialize in review articles summarize all the work that bears on the particular topic. These articles may focus on the contributions of one particular researcher but often consider the contributions of many different workers to the subject. These articles also contain extensive bibliographies, which refer the reader to the original research articles. Among the important journals devoted, at least partly, to review articles are

Accounts of Chemical Research
Angewandte Chemie (International Edition, in English)

Chemical Reviews
Chemical Society Reviews (formerly known as *Quarterly Reviews*)
Nature
Science

The details of the research of interest appear in the primary scientific journals. While there are thousands of different journals published in the world, a few important journals specializing in articles dealing with organic chemistry might be mentioned here. These are

Annalen der Chemie
Canadian Journal of Chemistry
Chemische Berichte
Journal of the American Chemical Society
Journal of the Chemical Society, Perkin Transactions (Parts I and II)
Journal of Organic Chemistry
Tetrahedron
Tetrahedron Letters

Articles devoted to topics of educational interest appear in *Journal of Chemical Education*.

Some journals specialize in news articles and articles focusing on current events in chemistry or in science in general. Articles in these journals can be useful in keeping the reader abreast of developments in science that are not part of his or her normal specialized scientific reading. Among the important journals of this type are

Chemical and Engineering News
Chemistry in Britain
Nature
Science

18.12 HOW TO CONDUCT A LITERATURE SEARCH

The easiest method to follow in searching the literature is to begin with secondary sources and then go on to the primary sources. In other words, one would try to locate material in a textbook, *Beilstein,* or *Chemical Abstracts.* From the results of that search, one would then consult one of the primary scientific journals.

A literature search that ultimately requires the reader to read one or more papers in the scientific journals is best conducted if one can identify a particular paper central to the study. Often this reference can be obtained from a textbook or a review article on the subject. If this is not available, a search through *Beilstein* is required. A search through one of the handbooks that provide *Beilstein* reference numbers (see Section 18.10) may be helpful. Searching through *Chemical Abstracts* would be considered the next logical step after *Beilstein.* From these sources, it should be possible to identify citations from the original literature on the subject.

Additional citations may be found in the references cited in the journal article. In this way, the background leading to the research can be examined. It is also possible to conduct a search forward in time from the date of the journal article through the

Science Citation Index. This publication provides the service of listing articles and the papers in which these articles were cited. While *Science Citation Index* consists of several types of indices, the *Citation Index* is most useful for the purposes described here. A person who knows of a particular key reference on a subject can examine *Science Citation Index* to obtain a list of papers that have used that seminal reference in support of the work described. The *Citation Index* lists papers by their senior author, and the journal, volume, page, and date, followed by citations of papers that have referred to that article, and the author, journal, volume, page, and date of each. The *Citation Index* is published in annual volumes, with quarterly supplements issued during the current year. Each volume contains a complete list of the citations of the key article made during that year. A disadvantage is that *Science Citation Index* has been available only since 1961. An additional disadvantage is that one may miss journal articles on the subject of interest if they failed to cite that particular key reference in their bibliographies—a reasonably likely possibility.

One can, of course, conduct a literature search by a ''brute force'' method, by beginning the search with *Beilstein* or even with the indices in *Chemical Abstracts*. However, the task can be made much easier by starting with a book or an article of general and broad coverage, which can provide a few citations for starting points in the search.

The following guides to using the chemical literature are provided for the reader who is interested in going further into this subject.

R. T. Bottle, ed. *The Use of Chemical Literature*. 3rd ed. London: Butterworths, 1979.

C. R. Burman. *How to Find Out in Chemistry*. 2nd ed. New York: Oxford University Press, 1966.

R. F. Gould, ed. *Advances in Chemistry Series*. No. 30. *Searching the Chemical Literature*. Washington: American Chemical Society, 1961.

R. E. Maizell. *How to Find Chemical Information: A Guide for Practicing Chemists, Teachers, and Students*. 2nd ed. New York: Wiley-Interscience, 1987.

M. G. Mellon. *Chemical Publications*. 4th ed. New York: McGraw-Hill, 1965.

H. M. Woodburn. *Using the Chemical Literature: A Practical Guide*. New York: Marcel Dekker, 1974.

Appendix 1
Tables of Unknowns and Derivatives

More extensive tables of unknowns may be found in Z. Rappoport, ed. *Handbook of Tables for Organic Compound Identification*, 3rd ed. Cleveland: Chemical Rubber Co., 1967

Aldehydes

COMPOUND	BP	MP	SEMI-CARBAZONE*	2,4-DINITRO-PHENYL-HYDRAZONE*
Ethanal (Acetaldehyde)	21	—	162	168
Propanal (Propionaldehyde)	48	—	89	148
Propenal (Acrolein)	52	—	171	165
2-Methylpropanal (Isobutyraldehyde)	64	—	125	187
Butanal (Butyraldehyde)	75	—	95	123
3-Methylbutanal (Isovaleraldehyde)	92	—	107	123
Pentanal (Valeraldehyde)	102	—	—	106
2-Butenal (Crotonaldehyde)	104	—	199	190
2-Ethylbutanal (Diethylacetaldehyde)	117	—	99	95
Hexanal (Caproaldehyde)	130	—	106	104
Heptanal (Heptaldehyde)	153	—	109	108
2-Furaldehyde (Furfural)	162	—	202	212
2-Ethylhexanal	163	—	254	114
Octanal (Caprylaldehyde)	171	—	101	106
Benzaldehyde	179	—	222	237
Phenylethanal (Phenylacetaldehyde)	195	33	153	121
2-Hydroxybenzaldehyde (Salicylaldehyde)	197	—	231	248
4-Methylbenzaldehyde (*p*-Tolualdehyde)	204	—	234	234
3,7-Dimethyl-6-octenal (Citronellal)	207	—	82	77
2-Chlorobenzaldehyde	213	11	229	213
4-Methoxybenzaldehyde (*p*-Anisaldehyde)	248	2.5	210	253
trans-Cinnamaldehyde	250 d.	—	215	255
3,4-Methylenedioxybenzaldehyde (Piperonal)	263	37	230	266 d.
2-Methoxybenzaldehyde (*o*-Anisaldehyde)	245	38	215 d.	254
4-Chlorobenzaldehyde	214	48	230	254
3-Nitrobenzaldehyde	—	58	246	293
4-Dimethylaminobenzaldehyde	—	74	222	325
Vanillin	285 d.	82	230	271
4-Nitrobenzaldehyde	—	106	221	320 d.
4-Hydroxybenzaldehyde	—	116	224	280 d.
(±)-Glyceraldehyde	—	142	160 d.	167

NOTE: "d" indicates decomposition.
*See "Procedures for Preparing Derivatives," Appendix 2.

Ketones

COMPOUND	BP	MP	SEMI-CARBAZONE*	2,4-DINITRO-PHENYL-HYDRAZONE*
2-Propanone (Acetone)	56	—	187	126
2-Butanone (Methyl ethyl ketone)	80	—	146	117
3-Methyl-2-butanone (Isopropyl methyl ketone)	94	—	112	120
2-Pentanone (Methyl propyl ketone)	101	—	112	143
3-Pentanone (Diethyl ketone)	102	—	138	156
Pinacolone	106	—	157	125
4-Methyl-2-pentanone (Isobutyl methyl ketone)	117	—	132	95
2,4-Dimethyl-3-pentanone (Diisopropyl ketone)	124	—	160	95
2-Hexanone (Methyl butyl ketone)	128	—	125	106
4-Methyl-3-penten-2-one (Mesityl oxide)	130	—	164	205
Cyclopentanone	131	—	210	146
2,3-Pentanedione	134	—	122 (mono) 209 (di)	209
2,4-Pentanedione (Acetylacetone)	139	—	—	122 (mono) 209 (di)
4-Heptanone (Dipropyl ketone)	144	—	132	75
2-Heptanone (Methyl amyl ketone)	151	—	123	89
Cyclohexanone	156	—	166	162
2,6-Dimethyl-4-heptanone (Diisobutyl ketone)	168	—	122	92
2-Octanone	173	—	122	58
Cycloheptanone	181	—	163	148
2,5-Hexanedione (Acetonylacetone)	191	−9	185 (mono) 224 (di)	257 (di)
Acetophenone (Methyl phenyl ketone)	202	20	198	238
Phenyl-2-propanone (Phenylacetone)	216	27	198	156
Propiophenone (Ethyl phenyl ketone)	218	21	182	191
4-Methylacetophenone	226	—	205	258
2-Undecanone	231	12	122	63
4-Chloroacetophenone	232	12	204	236
4-Phenyl-2-butanone (Benzylacetone)	235	—	142	127
4-Chloropropiophenone	—	36	176	223
4-Phenyl-3-buten-2-one	—	37	187	227
4-Methoxyacetophenone	258	38	198	228
Benzophenone	305	48	167	238
4-Bromoacetophenone	225	51	208	230
2-Acetonaphthone	—	54	235	262
Desoxybenzoin	320	60	148	204
3-Nitroacetophenone	202	80	257	228
9-Fluorenone	345	83	234	283
Benzoin	344	136	206	245
4-Hydroxypropiophenone	—	148	—	229
(±)-Camphor	205	179	237	177

*See ''Procedures for Preparing Derivatives,'' Appendix 2.

Carboxylic Acids

COMPOUND	BP	MP	*p*-TOLUIDIDE*	ANILIDE*	AMIDE*
Formic acid	101	8	53	47	43
Acetic acid	118	17	148	114	82
Propenoic acid (Acrylic acid)	139	13	141	104	85
Propanoic acid (Propionic acid)	141	—	124	103	81
2-Methylpropanoic acid (Isobutyric acid)	154	—	104	105	128
Butanoic acid (Butyric acid)	162	—	72	95	115
2-Methylpropenoic acid (Methacrylic acid)	163	16	—	87	102
Trimethylacetic acid (Pivalic acid)	164	35	—	127	178
Pyruvic acid	165 d.	14	109	104	124
3-Methylbutanoic acid (Isovaleric acid)	176	—	109	109	135
Pentanoic acid (Valeric acid)	186	—	70	63	106
2-Methylpentanoic acid	186	—	80	95	79
2-Chloropropanoic acid	186	—	124	92	80
Dichloroacetic acid	194	6	153	118	98
Hexanoic acid (Caproic acid)	205	—	75	95	101
2-Bromopropanoic acid	205	24	125	99	123
Octanoic acid (Caprylic acid)	237	16	70	57	107
Nonanoic acid	254	12	84	57	99
Decanoic acid (Capric acid)	268	32	78	70	108
4-Oxopentanoic acid (Levulinic acid)	246	33	108	102	108 d.
Dodecanoic acid (Lauric acid)	299	43	87	78	100
3-Phenylpropanoic acid (Hydrocinnamic acid)	279	48	135	98	105
Bromoacetic acid	208	50	—	131	91
Tetradecanoic acid (Myristic acid)	—	54	93	84	103
Trichloroacetic acid	198	57	113	97	141
Hexadecanoic acid (Palmitic acid)	—	62	98	90	106
Chloroacetic acid	189	63	162	137	121
Octadecanoic acid (Stearic acid)	—	69	102	95	109
trans-2-Butenoic acid (Crotonic acid)	—	72	132	118	158
Phenylacetic acid	—	77	136	118	156
2-Methoxybenzoic acid (*o*-Anisic acid)	200	101	—	131	129
2-Methylbenzoic acid (*o*-Toluic acid)	—	104	144	125	142
Nonanedioic acid (Azelaic acid)	—	106	201 (di)	107 (mono) 186 (di)	93 (mono) 175 (di)
3-Methylbenzoic acid (*m*-Toluic acid)	263 s.	110	118	126	94
(±)-Phenylhydroxyacetic acid (Mandelic acid)	—	118	172	151	133
Benzoic acid	249	122	158	163	130
2-Benzoylbenzoic acid	—	127	—	195	165
Maleic acid	—	130	142 (di)	198 (mono) 187 (di)	172 (mono) 260 (di)
Decanedioic acid (Sebacic acid)	—	133	201 (di)	122 (mono) 200 (di)	170 (mono) 210 (di)
Cinnamic acid	300	133	168	153	147

NOTE: "s" indicates "sublimation"; "d" indicates decomposition.
*See "Procedures for Preparing Derivatives," Appendix 2.

Carboxylic Acids (Continued)

COMPOUND	BP	MP	*p*-TOLUIDIDE*	ANILIDE*	AMIDE*
2-Chlorobenzoic acid	—	140	131	118	139
3-Nitrobenzoic acid	—	140	162	155	143
2-Aminobenzoic acid (Anthranilic acid)	—	146	151	131	109
Diphenylacetic acid	—	148	172	180	167
2-Bromobenzoic acid	—	150	—	141	155
Benzilic acid	—	150	190	175	154
Hexanedioic acid (Adipic acid)	—	152	239	151 (mono) 241 (di)	125 (mono) 220 (di)
Citric acid	—	153	189 (tri)	199 (tri)	210 (tri)
4-Chlorophenoxyacetic acid	—	158	—	125	133
2-Hydroxybenzoic acid (Salicyclic acid)	—	158	156	136	142
5-Bromo-2-hydroxybenzoic acid (5-Bromosalicyclic acid)	—	165	—	222	232
Methylenesuccinic acid (Itaconic acid)	—	166 d.	—	152 (mono)	191 (di)
(+)-Tartaric acid	—	169	—	180 (mono) 264 (di)	171 (mono) 196 (di)
4-Chloro-3-nitrobenzoic acid	—	180	—	131	156
4-Methylbenzoic acid (*p*-Toluic acid)	—	180	160	145	160
4-Methoxybenzoic acid (*p*-Anisic acid)	280	184	186	169	167
Butanedioic acid (Succinic acid)	235 d.	188	180 (mono) 255 (di)	143 (mono) 230 (di)	157 (mono) 260 (di)
3-Hydroxybenzoic acid	—	201	163	157	170
3,5-Dinitrobenzoic acid	—	202	—	234	183
Phthalic acid	—	210 d.	150 (mono) 201 (di)	169 (mono) 253 (di)	144 (mono) 220 (di)
4-Hydroxybenzoic acid	—	214	204	197	162
Pyridine-3-carboxylic acid (Nicotinic acid)	—	236	150	132	128
4-Nitrobenzoic acid	—	240	204	211	201
4-Chlorobenzoic acid	—	242	—	194	179
Fumaric acid	—	300	—	233 (mono) 314 (di)	270 (mono) 266 (di)

NOTE: ''d'' indicates decomposition.
*See ''Procedures for Preparing Derivatives,'' Appendix 2.

Phenols†

COMPOUND	BP	MP	α-NAPHTHYL-URETHANE*	BROMO DERIVATIVE*			
				Mono	Di	Tri	Tetra
2-Chlorophenol	176	7	120	48	76	—	—
3-Methylphenol (m-Cresol)	203	12	128	—	—	84	—
2-Methylphenol (o-Cresol)	191	32	142	—	56	—	—
2-Methoxyphenol (Guaiacol)	204	32	118	—	—	116	—
4-Methylphenol (p-Cresol)	202	34	146	—	49	—	198
Phenol	181	42	133	—	—	95	—
4-Chlorophenol	217	43	166	33	90	—	—
2,4-Dichlorophenol	210	45	—	68	—	—	—
4-Ethylphenol	219	45	128	—	—	—	—
2-Nitrophenol	216	45	113	—	117	—	—
2-Isopropyl-5-methylphenol (Thymol)	234	51	160	55	—	—	—
3,4-Dimethylphenol	225	64	141	—	—	171	—
4-Bromophenol	238	64	169	—	—	95	—
3,5-Dimethylphenol	220	68	109	—	—	166	—
2,5-Dimethylphenol	212	75	173	—	—	178	—
1-Naphthol (α-Naphthol)	278	96	152	—	105	—	—
2-Hydroxyphenol (Catechol)	245	104	175	—	—	—	192
3-Hydroxyphenol (Resorcinol)	281	109	275	—	—	112	—
4-Nitrophenol	—	112	150	—	142	—	—
2-Naphthol (β-Naphthol)	286	121	157	84	—	—	—
1,2,3-Trihydroxybenzene (Pyrogallol)	309	133	—	—	158	—	—
4-Phenylphenol	305	164	—	—	—	—	—

*See ''Procedures for Preparing Derivatives,'' Appendix 2.
†Also check:
   Salicylic acid (2-Hydroxybenzoic acid)
   Esters of salicylic acid (salicylates)
   Salicylaldehyde (2-Hydroxybenzaldehyde)
   4-Hydroxybenzaldehyde
   4-Hydroxypropiophenone
   3-Hydroxybenzoic acid
   4-Hydroxybenzoic acid
   4-Hydroxybenzophenone

Primary Amines

COMPOUND	BP	MP	BENZAMIDE*	PICRATE*	ACETAMIDE*
t-Butylamine	46	—	134	198	101
Propylamine	48	—	84	135	—
Allylamine	56	—	—	140	—
sec-Butylamine	63	—	76	139	—
Isobutylamine	69	—	57	150	—
Butylamine	78	—	42	151	—
Cyclohexylamine	135	—	149	—	104
Furfurylamine	145	—	—	150	—
Benzylamine	184	—	105	194	60
Aniline	184	—	163	198	114
2-Methylaniline (*o*-Toluidine)	200	—	144	213	110
3-Methylaniline (*m*-Toluidine)	203	—	125	200	65
2-Chloroaniline	208	—	99	134	87
2,6-Dimethylaniline	216	11	168	180	177
2-Methoxyaniline (*o*-Anisidine)	225	6	60	200	85
3-Chloroaniline	230	—	120	177	74
2-Ethoxyaniline (*o*-Phenetidine)	231	—	104	—	79
4-Chloro-2-methylaniline	241	29	142	—	140
4-Ethoxyaniline (*p*-Phenetidine)	250	2	173	69	137
4-Methylaniline (*p*-Toluidine)	200	43	158	182	147
2-Ethylaniline	210	47	147	194	111
2,5-Dichloroaniline	251	50	120	86	132
4-Methoxyaniline (*p*-Anisidine)	—	58	154	170	130
4-Bromoaniline	245	64	204	180	168
2,4,5-Trimethylaniline	—	64	167	—	162
4-Chloroaniline	—	70	192	178	179
2-Nitroaniline	—	72	110	73	92
Ethyl *p*-aminobenzoate	—	89	148	—	110
o-Phenylenediamine	258	102	301 (di)	208	185 (di)
2-Methyl-5-nitroaniline	—	106	186	—	151
2-Chloro-4-nitroaniline	—	108	161	—	139
3-Nitroaniline	—	114	157	143	155
4-Chloro-2-nitroaniline	—	118	—	—	104
2,4,6-Tribromoaniline	300	120	200	—	232 (mono) 127 (di)
2-Methyl-4-nitroaniline	—	130	—	—	202
2-Methoxy-4-nitroaniline	—	138	149	—	153
p-Phenylenediamine	267	140	128 (mono) 300 (di)	—	162 (mono) 304 (di)
4-Nitroaniline	—	148	199	100	215
4-Aminoacetanilide	—	162	—	—	304
2,4-Dinitroaniline	—	180	202	—	120

*See ''Procedures for Preparing Derivatives,'' Appendix 2.

Secondary Amines

COMPOUND	BP	MP	BENZAMIDE*	PICRATE*	ACETAMIDE*
Diethylamine	56	—	42	155	—
Diisopropylamine	84	—	—	140	—
Pyrrolidine	88	—	Oil	112	—
Piperidine	106	—	48	152	—
Dipropylamine	110	—	—	75	—
Morpholine	129	—	75	146	—
Diisobutylamine	139	—	—	121	86
N-Methylcyclohexylamine	148	—	85	170	—
Dibutylamine	159	—	—	59	—
Benzylmethylamine	184	—	—	117	—
N-Methylaniline	196	—	63	145	102
N-Ethylaniline	205	—	60	132	54
N-Ethyl-m-toluidine	221	—	72	—	—
Dicyclohexylamine	256	—	153	173	103
N-Benzylaniline	298	37	107	48	58
Indole	254	52	68	—	157
Diphenylamine	302	52	180	182	101
N-Phenyl-1-naphthylamine	335	62	152	—	115

*See ''Procedures for Preparing Derivatives,'' Appendix 2.

Tertiary Amines

COMPOUND	BP	MP	PICRATE*	METHIODIDE*
Triethylamine	89	—	173	280
Pyridine	115	—	167	117
2-Methylpyridine (α-Picoline)	129	—	169	230
3-Methylpyridine (β-Picoline)	144	—	150	92
Tripropylamine	157	—	116	207
N,N-Dimethylbenzylamine	183	—	93	179
N,N-Dimethylaniline	193	—	163	228 d.
Tributylamine	216	—	105	186
N,N-Diethylaniline	217	—	142	102
Quinoline	237	—	203	133

NOTE: ''d'' indicates decomposition.
*See ''Procedures for Preparing Derivatives,'' Appendix 2.

Alcohols

COMPOUND	BP	MP	3,5-DINITRO-BENZOATE*	PHENYL-URETHANE*
Methanol	65	—	108	47
Ethanol	78	—	93	52
2-Propanol (Isopropyl alcohol)	82	—	123	88
2-Methyl-2-propanol (*t*-Butyl alcohol)	83	26	142	136
2-Propen-1-ol (Allyl alcohol)	97	—	49	70
1-Propanol	97	—	74	57
2-Butanol (*sec*-Butyl alcohol)	99	—	76	65
2-Methyl-2-butanol (*t*-Pentyl alcohol)	102	−8.5	116	42
2-Methyl-3-butyn-2-ol	104	—	112	—
2-Methyl-1-propanol (Isobutyl alcohol)	108	—	87	86
2-Propyn-1-ol (Propargyl alcohol)	114		—	—
3-Pentanol	115	—	101	48
1-Butanol	118	—	64	61
2-Pentanol	119	—	62	—
3-Methyl-3-pentanol	123	—	96	43
2-Methoxyethanol	124	—	—	(113)‡
2-Chloroethanol	129	—	95	51
3-Methyl-1-butanol (Isoamyl alcohol)	130	—	70	31
4-Methyl-2-pentanol	132	—	65	143
1-Pentanol	138	—	46	46
Cyclopentanol	140	—	115	132
2-Ethyl-1-butanol	146	—	51	—
2,2,2-Trichloroethanol	151	—	142	87
1-Hexanol	157	—	58	42
Cyclohexanol	160	—	113	82
(2-Furyl)-methanol (Furfuryl alcohol)	170	—	80	45
1-Heptanol	176	—	47	60
2-Octanol	179	—	32	114
1-Octanol	195	—	61	74
3,7-Dimethyl-1,6-octadien-3-ol (Linalool)	196	—	—	66
Benzyl alcohol	204	—	113	77
1-Phenylethanol	204	20	92	95
2-Phenylethanol	219	—	108	78
1-Decanol	231	7	57	59
3-Phenylpropanol	236	—	45	92
1-Dodecanol (Lauryl alcohol)	—	24	60	74
3-Phenyl-2-propen-1-ol (Cinnamyl alcohol)	250	34	121	90
1-Tetradecanol (Myristyl alcohol)	—	39	67	74
(−)-Menthol	212	41	158	111
1-Hexadecanol (Cetyl alcohol)	—	49	66	73
1-Octadecanol (Stearyl alcohol)	—	59	77	79
Diphenylmethanol (Benzhydrol)	288	68	141	139
Benzoin	—	133	—	165
Cholesterol	—	147	—	168
(+)-Borneol	—	208	154	138

*See ''Procedures for Preparing Derivatives,'' Appendix 2.
‡α-Naphthylurethane

Esters

COMPOUND	BP	MP	COMPOUND	BP	MP
Methyl formate	34	—	Pentyl acetate	142	—
Ethyl formate	54	—	(n-Amyl acetate)		
Vinyl acetate	72	—	3-Methylbutyl acetate	142	—
Ethyl acetate	77	—	(Isoamyl acetate)		
Methyl propanoate	77	—	Ethyl chloroacetate	143	—
(Methyl propionate)			Ethyl lactate	154	—
Methyl acrylate	80	—	Ethyl hexanoate	168	—
2-Propyl acetate	85	—	(Ethyl caproate)		
(Isopropyl acetate)			Methyl acetoacetate	170	—
Ethyl chloroformate	93	—	Dimethyl malonate	180	—
Methyl 2-methylpropanoate	93	—	Ethyl acetoacetate	181	—
(Methyl isobutyrate)			Diethyl oxalate	185	—
2-Propenyl acetate	94	—	Methyl benzoate	199	—
(Isopropenyl acetate)			Ethyl octanoate	207	—
2-(2-Methylpropyl) acetate	98	—	(Ethyl caprylate)		
(t-Butyl acetate)			Ethyl cyanoacetate	210	—
Ethyl acrylate	99	—	Ethyl benzoate	212	—
Ethyl propanoate	99	—	Diethyl succinate	217	—
(Ethyl propionate)			Methyl phenylacetate	218	—
Methyl methacrylate	100	—	Diethyl fumarate	219	—
Methyl trimethylacetate	101	—	Methyl salicylate	222	—
(Methyl pivalate)			Diethyl maleate	225	—
Propyl acetate	102	—	Ethyl phenylacetate	229	—
Methyl butanoate	102	—	Ethyl salicylate	234	—
(Methyl butyrate)			Dimethyl suberate	268	—
2-Butyl acetate	111	—	Ethyl cinnamate	271	—
(sec-Butyl acetate)			Diethyl phthalate	298	—
Methyl 3-methylbutanoate	117	—	Dibutyl phthalate	340	—
(Methyl isovalerate)			Methyl cinnamate	—	36
Ethyl butanoate	120	—	Phenyl salicylate	—	42
(Ethyl butyrate)			Methyl p-chlorobenzoate	—	44
Butyl acetate	127	—	Ethyl p-nitrobenzoate	—	56
Methyl pentanoate	128	—	Phenyl benzoate	314	69
(Methyl valerate)			Methyl m-nitrobenzoate	—	78
Methyl chloroacetate	130	—	Methyl p-bromobenzoate	—	81
Ethyl 3-methylbutanoate	132	—	Ethyl p-aminobenzoate	—	90
(Ethyl isovalerate)			Methyl p-nitrobenzoate	—	94

Appendix 2
Procedures for Preparing Derivatives

> **CAUTION:** Some of the chemicals used in preparing derivatives are suspect carcinogens. The list of suspect carcinogens on p. 12 should be consulted before beginning any of these procedures. Care should be exercised in handling these substances.

ALDEHYDES AND KETONES

Semicarbazones. Place 0.5 mL of a 2M stock solution of semicarbazide hydrochloride (or 0.5 mL of a solution prepared by dissolving 1.11 g of semicarbazide hydrochloride [MW 111.5] in 5 mL of water) in a small test tube. Add an estimated 1 millimole (mmol) of the unknown compound to the test tube. If the unknown does not dissolve in the solution, or if the solution becomes cloudy, add enough methanol to dissolve the solid and produce a clear solution. Using a disposable capillary pipet, add 10 drops of pyridine and heat the mixture gently on a steam bath for about 5 minutes; by that time, the product should have begun to crystallize. Collect the product by vacuum filtration. The product can be recrystallized from ethanol if necessary.

Semicarbazones (Alternative Method). Dissolve 1 g of semicarbazide hydrochloride and 1.5 g of sodium acetate in 5 mL of water. Then dissolve 1.0 g of the unknown in 10 mL of ethanol. Mix the two solutions together in a 50-mL Erlenmeyer flask and heat the mixture to boiling for about 5 minutes. After heating, place the reaction flask in a beaker of ice and scratch the sides of the flask with a glass rod to induce crystallization of the derivative. Collect the derivative by vacuum filtration and recrystallize it from ethanol.

2,4-Dinitrophenylhydrazones. Place 10 mL of a solution of 2,4-dinitrophenylhydrazine (prepared as described for the classification test in Procedure 56D) in a test tube, and add an estimated 1 mmol of the unknown compound. If the unknown is a solid, it should be dissolved in the minimum amount of 95% ethanol or 1,2-dimethoxyethane before it is added. If crystallization is not immediate, gently warm the solution for a minute on a steam bath and then set it aside to crystallize. Collect the product by vacuum filtration.

CARBOXYLIC ACIDS

Using a reflux condenser and the smallest flask in the organic kit, heat a mixture of 0.5 g of the acid and 2 mL of thionyl chloride on a steam bath for about 30 minutes. Allow the mixture to cool and use it for one of the following three procedures:

691

Amides. Working in a hood, pour the reaction mixture into a beaker containing 10 mL of ice-cold concentrated ammonium hydroxide and stir it vigorously. When the reaction is complete, collect the product by vacuum filtration and recrystallize it from water or from water–ethanol, using the mixed-solvents method (Technique 3, Section 3.7).

Anilides. Dissolve 1 g of aniline in 25 mL of toluene and carefully add it to the reaction mixture. Warm the mixture for an additional 5 minutes on a hot plate. Then transfer the toluene solution to a separatory funnel and wash it sequentially with 5 mL of water, 5 mL of 5% hydrochloric acid, 5 mL of 5% sodium hydroxide, and a second 5-mL portion of water. Dry the toluene layer over a small amount of anhydrous sodium sulfate. Decant the toluene layer away from the drying agent into a small beaker and evaporate the toluene on a hot plate in the hood. Recrystallize the product from water or from ethanol–water, using the mixed-solvents method (Technique 3, Section 3.7).

***p*-Toluidides.** Use the same procedure as that described for the anilide, but substitute *p*-toluidine for aniline.

PHENOLS

α-Naphthylurethanes. Follow the procedure given below for preparing phenylurethanes from alcohols but substitute α-naphthylisocyanate for phenylisocyanate.

Bromo Derivatives. First, if a stock brominating solution is not available, prepare one by dissolving 0.75 g of potassium bromide in 5 mL of water and adding 0.5 g of bromine. Dissolve 0.1 g of the phenol in 1 mL of methanol or 1,2-dimethoxyethane and then add 1 mL of water. Add 1 mL of the brominating mixture to the phenol solution and swirl the mixture vigorously. Then, continue adding the brominating solution, dropwise with swirling, until the color of the bromine reagent persists. Finally, add 3 to 5 mL of water and shake the mixture vigorously. Collect the precipitated product by vacuum filtration and wash it well with water. Recrystallize the derivative from methanol–water, using the mixed-solvents method (Technique 3, Section 3.7).

AMINES

Acetamides. Place an estimated 1 mmol of the amine and 0.5 mL of acetic anhydride in a small Erlenmeyer flask. Heat the mixture for about 5 minutes; then add 5 mL of water and stir the solution vigorously to precipitate the product and hydrolyze the excess acetic anhydride. If the product does not crystallize, it may be necessary to scratch the walls of the flask with a glass rod. Collect the crystals by vacuum filtration and wash them with several portions of cold 5% hydrochloric acid. Recrystallize the derivative from methanol–water, using the mixed-solvents method (Technique 3, Section 3.7).

Aromatic amines, or those amines that are not very basic, may require pyridine (2 mL) as a solvent and a catalyst for the reaction. If pyridine is used, a longer period of heating will be required (up to 1 hour), and the reaction should be carried out in an apparatus equipped with a reflux condenser. After reflux, the reaction mixture must be extracted with 5 to 10 mL of 5% sulfuric acid to remove the pyridine.

Benzamides. Using a test tube, suspend an estimated 1 mmol of the amine in 1 mL of 10% sodium hydroxide solution and add 0.5 g of benzoyl chloride. Stopper the test tube with a cork and shake the mixture vigorously for about 10 minutes. After shaking, add enough dilute hydrochloric acid to bring the pH of the solution to pH 7 or 8. Collect the precipitate by vacuum filtration, wash it thoroughly with cold water, and recrystallize it from ethanol–water, using the mixed-solvents method (Technique 3, Section 3.7).

Benzamides (Alternative Method). Dissolve 0.5 g of the amine in a solution of 2.5 mL of pyridine and 5 mL of toluene. Add 0.5 mL of benzoyl chloride to the solution and heat the mixture under reflux about 30 minutes. Pour the cooled reaction mixture into 50 mL of water and stir the mixture vigorously to hydrolyze the excess benzoyl chloride. Separate the toluene layer and wash it, first with 3 mL of water and then with 3 mL of 5% sodium carbonate. Dry the toluene over anhydrous sodium sulfate, decant the toluene into a small beaker, and remove the toluene by evaporation on a hot plate in the hood. Recrystallize the benzamide from ethanol or ethanol–water, using the mixed-solvents method (Technique 3, Section 3.7).

Picrates. Dissolve 0.2 g of the unknown in about 5 mL of ethanol and add 5 mL of a saturated solution of picric acid in ethanol. Heat the solution to boiling and then allow it to cool slowly. Collect the product by vacuum filtration and rinse it with a small amount of cold ethanol.

Methiodides. Mix equal-volume quantities of the amine and methyl iodide in a large test tube (about 0.5 mL is convenient) and allow the mixture to stand for several minutes. Then heat the mixture under reflux on a steam bath for about 5 minutes. The methiodide should crystallize on cooling. If it does not, you can induce crystallization by scratching the walls of the tube with a glass rod. Collect the product by vacuum filtration and recrystallize it from ethanol or ethyl acetate.

ALCOHOLS

3,5-Dinitrobenzoates. *Liquid Alcohols.* Dissolve 0.5 g of 3,5-dinitrobenzoyl chloride[1] in 0.5 mL of the alcohol and heat the mixture for about 5 minutes. Allow the mixture to cool and add 3 mL of a 5% sodium carbonate solution and 2 mL of water. Stir the mixture vigorously and crush any solid that forms. Collect the product by vacuum filtration and wash it with cold water. Recrystallize the derivative from ethanol–water, using the mixed-solvents method (Technique 3, Section 3.7).

---

[1]This is an acid chloride and undergoes hydrolysis readily. The purity of this reagent should be checked before its use by determining its melting point. When the carboxylic acid is present, the melting point will be high.

Solid Alcohols.   Dissolve 0.5 g of the alcohol in 3 mL of dry pyridine and add 0.5 g of 3,5-dinitrobenzoyl chloride.[1] Heat the mixture under reflux for 15 minutes. Pour the cooled reaction mixture into a cold mixture of 5 mL of 5% sodium carbonate and 5 mL of water. Keep the solution cooled in an ice bath until the product crystallizes, and stir it vigorously during the entire period. Collect the product by vacuum filtration, wash it with cold water, and recrystallize it from ethanol–water, using the mixed-solvents method (Technique 3, Section 3.7).

Phenylurethanes.   Place 0.5 g of the **anhydrous** alcohol in a dry test tube and add 0.5 mL of phenylisocyanate (α-naphthylisocyanate for a phenol). If the compound is a phenol, add several drops of pyridine to catalyze the reaction. If the reaction is not spontaneous, heat the mixture on a steam bath for 5 to 10 minutes. Cool the test tube in a beaker of ice and scratch the tube with a glass rod to induce crystallization. Decant the liquid from the solid product or, if necessary, collect the product by vacuum filtration. Dissolve the product in 5 to 6 mL of hot ligroin or hexane and filter the mixture by gravity (preheat funnel) to remove any unwanted and insoluble diphenylurea present. Cool the filtrate to induce crystallization of the urethane. Collect the product by vacuum filtration.

ESTERS

Preparing the derivatives of esters is usually complicated. We recommend that esters be characterized by spectroscopic methods whenever possible. If a derivative must be prepared, consult a comprehensive textbook. Several are listed in the first section of Experiment 56.

Appendix 3
Infrared Spectroscopy

Almost any compound having covalent bonds, whether organic or inorganic, will be found to absorb various frequencies of electromagnetic radiation in the infrared region of the spectrum. The infrared region of the electromagnetic spectrum lies at wavelengths longer than those associated with visible light, which includes wavelengths from approximately 400 nm to 800 nm (1 nm $= 10^{-9}$ m), but at wavelengths shorter than those associated with radio waves, which have wavelengths longer than 1 cm. For chemical purposes, we are interested in the **vibrational** portion of the infrared region. This portion is defined as that including radiations with wavelengths (λ) between 2.5 μ and 15 μ (1 μ = 1 micron = 1 μm $= 10^{-6}$ m). Although the more technically correct unit for wavelength in the infrared region of the spectrum is micrometer, we shall follow common practice and use micron as the unit. The relation of the infrared region to others included in the electromagnetic spectrum is illustrated in Figure IR–1.

As with other types of energy absorption, molecules are excited to a higher energy state when they absorb infrared radiation. The absorption of infrared radiation is, like other absorption processes, a quantized process. Only selected frequencies (energies) of infrared radiation are absorbed by a molecule. The absorption of infrared radiation corresponds to energy changes on the order of 2 to 10 kcal/mol. Radiation in this energy range corresponds to the range encompassing the stretching and bending vibrational frequencies of the bonds in most covalent molecules. In the absorption process, those frequencies of infrared radiation that match the natural vibrational frequencies of the molecule in question are absorbed, and the energy absorbed increases the **amplitude** of the vibrational motions of the bonds in the molecule.

Many chemists refer to the radiation in the vibrational infrared region of the electromagnetic spectrum by units called **wavenumbers** ($\bar{\nu}$). Wavenumbers are expressed in reciprocal centimeters (cm^{-1}) and are easily computed by taking the reciprocal of the wavelength (λ) expressed in centimeters. This unit has the advantage, for

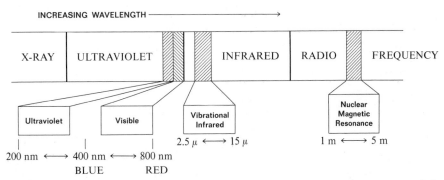

FIGURE IR–1. Portion of electromagnetic spectrum, showing relation of vibrational infrared to other types of radiation

those performing calculations, that it is directly proportional to energy. Thus, the vibrational infrared extends from about 4000 to 650 cm^{-1} (or wavenumbers).

Wavelengths (μ) and wavenumbers (cm^{-1}) can be interconverted by the following relationships:

$$\text{cm}^{-1} = \frac{1}{(\mu)} \times 10{,}000 \quad \text{and} \quad \mu = \frac{1}{(\text{cm}^{-1})} \times 10{,}000$$

IR.1 USES OF THE INFRARED SPECTRUM

Since every different type of bond has a different natural frequency of vibration, and since the same type of bond in two different compounds is in a slightly different environment, no two molecules of different structure have exactly the same infrared absorption pattern, or **infrared spectrum.** Although some of the frequencies absorbed in the two cases might be the same, in no case of two different molecules will their infrared spectra (the patterns of absorption) be identical. Thus, the infrared spectrum can be used for molecules much as a fingerprint can be used for humans. By comparing the infrared spectra of two substances thought to be identical, one can establish whether or not they are in fact identical. If the infrared spectra of two substances coincide peak for peak (absorption for absorption), in most cases, the substances are identical.

A second and more important use of the infrared spectrum is that it gives structural information about a molecule. The absorptions of each type of bond (N—H, C—H, O—H, C—X, C=O, C—O, C—C, C=C, C≡C, C≡N, and so on) are regularly found only in certain small portions of the vibrational infrared region. A small range of absorption can be defined for each type of bond. Outside this range, absorptions will normally be due to some other type of bond. Thus, for instance, any absorption in the range 3000 ± 150 cm^{-1} (around 3.33 μ) will almost always be due to the presence of a CH bond in the molecule; an absorption in the range 1700 ± 100 cm^{-1} (around 6.1 μ) will normally be due to the presence of a C=O bond (carbonyl group) in the molecule. The same type of range applies to each type of bond. The way these are spread out over the vibrational infrared is illustrated schematically in Figure IR–2. It is a good idea to try to remember this general scheme for future convenience.

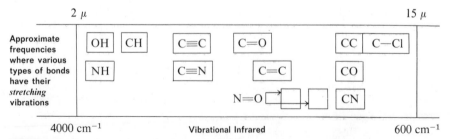

FIGURE IR–2. Approximate regions in which various common types of bonds absorb. (Bending and twisting and other types of bond vibration have been omitted for clarity.)

IR.2 MODES OF VIBRATION AND BENDING

The simplest types, or **modes,** of vibrational motion in a molecule that are **infrared-active,** that is, give rise to absorptions, are the stretching and bending modes.

$$C—H \qquad C \overset{O}{\underset{}{\diagup \quad \diagdown}} H$$

Stretching   Bending

Other, more complex types of stretching and bending are also active, however. To introduce several words of terminology, the normal modes of vibration for a methylene group are shown on the next page.

In any group of three or more atoms—at least two of which are identical—there are **two** modes of stretching or bending: the symmetric mode and the asymmetric mode. Examples of such groupings are $—CH_3$, $—CH_2—$, $—NO_2$, $—NH_2$, and anhydrides, $(CO)_2O$. For the anhydride, owing to asymmetric and symmetric modes of stretch, this functional group gives **two** absorptions in the $C=O$ region. A similar phenomenon is seen for amino groups, where primary amines usually have **two** absorptions in the NH stretch region while secondary amines (R_2NH) have only one absorption peak. Amides show similar bands. There are two strong $N=O$ stretch peaks for a nitro group, which are caused by asymmetric and symmetric stretching modes.

IR.3 WHAT TO LOOK FOR IN EXAMINING INFRARED SPECTRA

The instrument that determines the absorption spectrum for a compound is called an **infrared spectrophotometer.** The spectrophotometer determines the relative strengths and positions of all the absorptions in the infrared region and plots this information on a piece of calibrated chart paper. This plot of absorption intensity versus wavenumber or wavelength is referred to as the **infrared spectrum** of the compound. A typical infrared spectrum, that of methyl isopropyl ketone, is shown in Figure IR–3.

The strong absorption in the middle of the spectrum corresponds to $C=O$, the carbonyl group. Note that the $C=O$ peak is quite intense. In addition to the characteristic position of absorption, the **shape** and **intensity** of this peak are also unique to the $C=O$ bond. This is true for almost every type of absorption peak; both shape and intensity characteristics can be described, and these characteristics often allow one to distinguish the peak in a confusing situation. For instance, to some extent both $C=O$ and $C=C$ bonds absorb in the same region of the infrared spectrum:

$$C=O \quad 1850–1630 \text{ cm}^{-1} \ (5.5–6.1 \ \mu)$$
$$C=C \quad 1680–1620 \text{ cm}^{-1} \ (5.95–6.2 \ \mu)$$

However, the $C=O$ bond is a strong absorber, whereas the $C=C$ bond generally absorbs only weakly. Hence, a trained observer would not normally interpret a strong peak at 1670 cm^{-1} to be a carbon–carbon double bond nor a weak absorption at this frequency to be due to a carbonyl group.

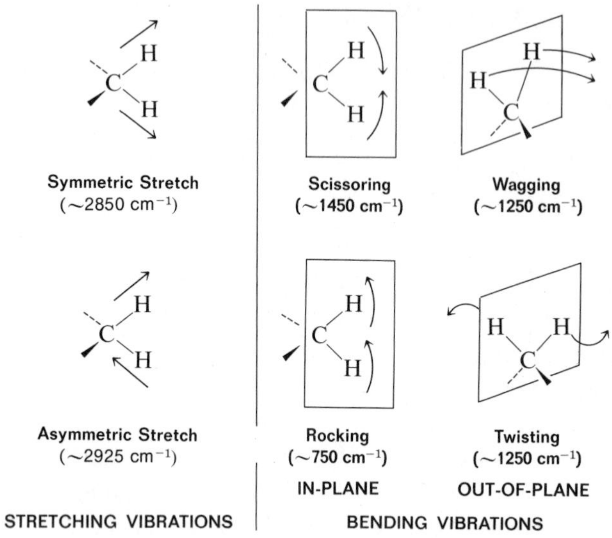

Symmetric Stretch
(~2850 cm⁻¹)
Asymmetric Stretch
(~2925 cm⁻¹)
STRETCHING VIBRATIONS
Scissoring
(~1450 cm⁻¹)
Wagging
(~1250 cm⁻¹)
Rocking
(~750 cm⁻¹)
Twisting
(~1250 cm⁻¹)
IN-PLANE
OUT-OF-PLANE
BENDING VIBRATIONS

The shape of a peak often gives a clue to its identity as well. Thus, while the NH and OH regions of the infrared overlap,

$$\text{OH } 3650\text{–}3200 \text{ cm}^{-1} \ (2.75\text{–}3.12\ \mu)$$
$$\text{NH } 3500\text{–}3300 \text{ cm}^{-1} \ (2.85\text{–}3.00\ \mu)$$

NH usually gives a **sharp** absorption peak (absorbs a very narrow range of frequencies), and OH, when it is in the NH region, usually gives a **broad** absorption peak. Primary amines give **two** absorptions in this region, whereas alcohols give only one.

Therefore, while you are studying the sample spectra in the pages that follow, you should also notice shapes and intensities. They are as important as the frequency at which an absorption occurs, and the eye must be trained to recognize these features. Often, in the literature of organic chemistry, one will find absorptions referred to as strong (s), medium (m), weak (w), broad, or sharp. The author is trying to convey some idea of what the peak looks like without actually drawing the spectrum.

FREQUENCY (CM⁻¹)
4000 3600 3200 2800 2400 2000 1800 1600 1400 1200 1000 800 600 400
TRANSMITTANCE (%)
80
60
40
20
0
CH Aliphatic
C=O
2.5 3.0 3.5 4.0 5 (MICRONS) 6 7 8 9 10 11 12 13 14 16 18 20 25

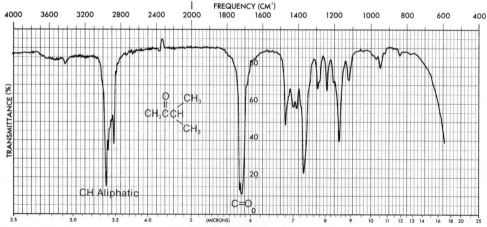

FIGURE IR–3. Ir spectrum of methyl isopropyl ketone (neat liquid, salt plates)

IR.4 CORRELATION CHARTS AND TABLES

To extract structural information from infrared spectra, one must know the frequencies or wavelengths at which various functional groups absorb. Infrared **correlation tables** present as much information as is known about where the various functional groups absorb. The books listed at the end of this Appendix present extensive lists of correlation tables. Sometimes the absorption information is given in a chart, called a **correlation chart.** A simplified correlation table is given in Table IR–1.

TABLE IR–1. A Simplified Correlation Table

TYPE OF VIBRATION			FREQUENCY (cm^{-1})	WAVELENGTH (μ)	INTENSITY
C—H	Alkanes	(stretch)	3000–2850	3.33–3.51	s
	—CH$_3$	(bend)	1450 and 1375	6.90 and 7.27	m
	—CH$_2$—	(bend)	1465	6.83	m
	Alkenes	(stretch)	3100–3000	3.23–3.33	m
		(bend)	1700–1000	5.88–10.0	s
	Aromatics	(stretch)	3150–3050	3.17–3.28	s
		(out-of-plane bend)	1000–700	10.0–14.3	s
	Alkyne	(stretch)	ca 3300	ca 3.03	s
	Aldehyde		2900–2800	3.45–3.57	w
			2800–2700	3.57–3.70	w
C—C	Alkane	Not interpretatively useful			
C=C	Alkene		1680–1600	5.95–6.25	m–w
	Aromatic		1600–1400	6.25–7.14	m–w
C≡C	Alkyne		2250–2100	4.44–4.76	m–w
C=O	Aldehyde		1740–1720	5.75–5.81	s
	Ketone (acyclic)		1725–1705	5.80–5.87	s
	Carboxylic acid		1725–1700	5.80–5.88	s
	Ester		1750–1730	5.71–5.78	s
	Amide		1700–1640	5.88–6.10	s
	Anhydride		ca 1810	ca 5.52	s
			ca 1760	ca 5.68	s
C—O	Alcohols, ethers, esters, carboxylic acids		1300–1000	7.69–10.0	s
O—H	Alcohols, phenols				
	Free		3650–3600	2.74–2.78	m
	H-Bonded		3400–3200	2.94–3.12	m
	Carboxylic acids		3300–2500	3.03–4.00	m
N—H	Primary and secondary amines		ca 3500	ca 2.86	m
C≡N	Nitriles		2260–2240	4.42–4.46	m
N=O	Nitro (R—NO$_2$)		1600–1500	6.25–6.67	s
			1400–1300	7.14–7.69	s
C—X	Fluoride		1400–1000	7.14–10.0	s
	Chloride		800–600	12.5–16.7	s
	Bromide, iodide		<600	>16.7	s

NOTE: s, strong; m, medium; w, weak.

TABLE IR–2.   Base Values for Absorptions of Bonds

OH	3600 cm^{-1}	2.78 μ	C≡C	2150 cm^{-1}	4.65 μ
NH	3500	2.86	C=O	1715	5.83
CH	3000	3.33	C=C	1650	6.06
C≡N	2250	4.44	C—O	1100	9.09

Although you may think assimilating the mass of data in Table IR–1 will be difficult, it is not if you make a modest start and then gradually increase your familiarity with the data. An ability to interpret the fine details of an infrared spectrum will follow. This is most easily done by first establishing the broad visual patterns of Figure IR–2 firmly in mind. Then, as a second step, a "typical absorption value" can be memorized for each of the functional groups in this pattern. This value will be a single number that can be used as a pivot value for the memory. For instance, start with a simple aliphatic ketone as a model for all typical carbonyl compounds. The typical aliphatic ketone has a carbonyl absorption of 1715 ± 10 cm^{-1}. Without worrying about the variation, memorize 1715 cm^{-1} as the base value for carbonyl absorption. Then, more slowly, learn the extent of the carbonyl range and the visual pattern of how the different kinds of carbonyl groups are arranged throughout this region. See, for instance, Figure IR–14, which gives typical values for carbonyl compounds. Also learn how factors like ring size (when the functional group is contained in a ring) and conjugation affect the base values (that is, in which direction the values are shifted). Learn the trends—always remembering the base value (1715 cm^{-1}). It might prove useful as a beginning to memorize the base values in Table IR–2 for this approach. Notice that there are only eight.

IR.5   ANALYZING A SPECTRUM (OR WHAT YOU CAN TELL AT A GLANCE)

In trying to analyze the spectrum of an unknown, you should concentrate first on trying to establish the presence (or absence) of a few major functional groups. The most conspicuous peaks are C=O, O—H, N—H, C—O, C=C, C≡C, C≡N, and NO$_2$. If they are present, they give immediate structural information. Do not try to analyze in detail the CH absorptions near 3000 cm^{-1} (3.33 μ); almost all compounds have these absorptions. Do not worry about subtleties of the exact type of environment in which the functional group is found. A checklist of the important gross features follows:

1. Is a carbonyl group present?
   The C=O group gives rise to a strong absorption in the region 1820 to 1660 cm^{-1} (5.5–6.1 μ).
   The peak is often the strongest in the spectrum and of medium width. You can't miss it.

2. If C=O is present, check the following types. (If it is absent, go to 3.)
   ACIDS   Is OH also present?
                      Broad absorption near 3300 to 2500 cm^{-1} (3.0–4.0 μ) (usually overlaps C—H).

AMIDES	Is NH also present?
	Medium absorption near 3500 cm^{-1} (2.85 μ), sometimes a double peak, equivalent halves.
ESTERS	Is C—O also present?
	Medium intensity absorptions near 1300 to 1000 cm^{-1} (7.7–10 μ).
ANHYDRIDES	Have **two** C=O absorptions near 1810 and 1760 cm^{-1} (5.5 and 5.7 μ).
ALDEHYDES	Is aldehyde CH present?
	Two weak absorptions near 2850 and 2750 cm^{-1} (3.50 and 3.65 μ) on the right side of CH absorptions.
KETONES	The above five choices have been eliminated.

3. If C=O is absent

ALCOHOLS	Check for OH.
or PHENOLS	**Broad** absorption near 3600 to 3300 cm^{-1} (2.8–3.0 μ).
	Confirm this by finding C—O near 1300 to 1000 cm^{-1} (7.7–10 μ).
AMINES	Check for NH.
	Medium absorption(s) near 3500 cm^{-1} (2.85 μ).
ETHERS	Check for C—O (and absence of OH) near 1300 to 1000 cm^{-1} (7.7–10 μ).

4. Double Bonds or Aromatic Rings or Both

C=C is a **weak** absorption near 1650 cm^{-1} (6.1 μ).

Medium to strong absorptions in the region 1650 to 1450 cm^{-1} (6–7 μ) often imply an aromatic ring.

Confirm the above by consulting the CH region.

Aromatic and vinyl CH occur to the left of 3000 cm^{-1} (3.33 μ) (aliphatic CH occurs to the right of this value).

5. Triple Bonds

C≡N is a medium, sharp absorption near 2250 cm^{-1} (4.5 μ).

C≡C is a weak but sharp absorption near 2150 cm^{-1} (4.65 μ).

Check also for acetylenic CH near 3300 cm^{-1} (3.0 μ).

6. Nitro Groups

Two strong absorptions 1600 to 1500 cm^{-1} (6.25–6.67 μ) and 1390 to 1300 cm^{-1} (7.2–7.7 μ).

7. Hydrocarbons

None of the above is found.

Main absorptions are in CH region near 3000 cm^{-1} (3.33 μ).

Very simple spectrum, only other absorptions near 1450 cm^{-1} (6.90 μ) and 1375 cm^{-1} (7.27 μ).

The beginning student should resist the idea of trying to assign or interpret **every** peak in the spectrum. You simply will not be able to do this. Concentrate first on learning the principal peaks and recognizing their presence or absence. This is best done by carefully studying the illustrative spectra in the section that follows.

> **NOTE:** In describing the shifts of absorption peaks or their relative positions, we have used the phrases "to the left" and "to the right." This was done to facilitate economy of use when using *both* microns and reciprocal centimeters. The meaning is clear, since all spectra are conventionally presented left to right from 4000 cm^{-1} to 600 cm^{-1} or from 2.5 μ to 16 μ. "To the right" avoids saying each time "to lower frequency (cm^{-1})" or "to longer wavelength (μ)," which is confusing, since cm^{-1} and μ have an inverse relation; as one goes up, the other goes down.

IR.6 SURVEY OF THE IMPORTANT FUNCTIONAL GROUPS

Alkanes

Spectrum is usually simple with few peaks.

C—H   Stretch occurs around 3000 cm^{-1} (3.33 μ).

    (a) In alkanes (except strained ring compounds) absorption always occurs to the right of 3000 cm^{-1} (3.33 μ).

    (b) If a compound has vinylic, aromatic, acetylenic, or cyclopropyl hydrogens, the CH absorption is to the left of 3000 cm^{-1} (3.33 μ).

CH$_2$ Methylene groups have a characteristic absorption at approximately 1450 cm^{-1} (6.90 μ).

CH$_3$ Methyl groups have a characteristic absorption at approximately 1375 cm^{-1} (7.27 μ).

C—C   Stretch—not interpretatively useful—has many peaks.

The spectrum of decane is shown in Figure IR–4.

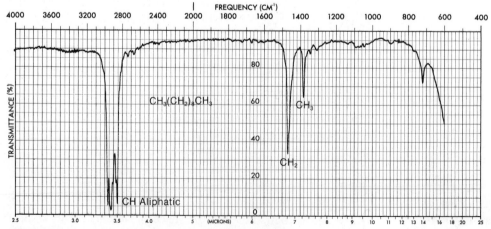

FIGURE IR–4. Ir spectrum of decane (neat liquid, salt plates)

Alkenes

=C—H   Stretch occurs to the left of 3000 cm^{-1} (3.33 μ).
=C—H   Out-of-plane (oop) bending at 1000 to 650 cm^{-1} (10–15 μ)
C=C   Stretch 1675 to 1600 cm^{-1} (5.95–6.25 μ), often weak
Conjugation moves C=C stretch to the right.
Symmetrically substituted bonds, for example, 2,3-dimethyl-2-butene, do not absorb in the infrared region (no dipole change). Highly substituted double bonds are often vanishingly weak in absorption.

The spectrum of styrene is shown in Figure IR–5. The spectrum of cyclohexene is shown in Experiment 27.

Aromatic Rings

=C—H   Stretch is always to the left of 3000 cm^{-1} (3.33 μ).
CH   Out-of-plane (oop) bending at 900 to 690 cm^{-1} (11.0–14.5 μ)
The CH out-of-plane absorptions often allow one to determine the type of ring substitution by their numbers, intensities, and positions. The correlation chart in Figure IR–6A indicates the positions of these bands.
The patterns are generally reliable—most particularly reliable for rings with alkyl substituents, least for polar substituents.

Ring Absorptions (C=C).   There are often four sharp absorptions that occur in pairs at 1600 cm^{-1} (6.25 μ) and 1450 cm^{-1} (6.90 μ) and are characteristic of an aromatic ring. See, for example, the spectra of anisole (Figure IR–10), benzonitrile (Figure IR–13) and methyl benzoate (Figure IR–17).

There are many weak combination and overtone absorptions that appear between 2000 and 1667 cm^{-1} (5 to 6 μ). The relative shapes and numbers of these peaks can be used to tell whether an aromatic ring is monosubstituted or di-, tri-, tetra-,

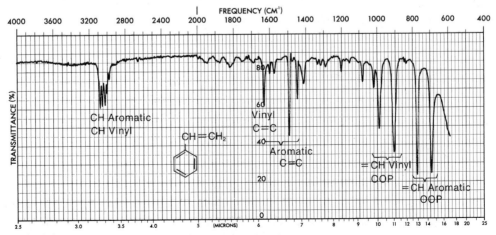

FIGURE IR–5.   Ir spectrum of styrene (neat liquid, salt plates)

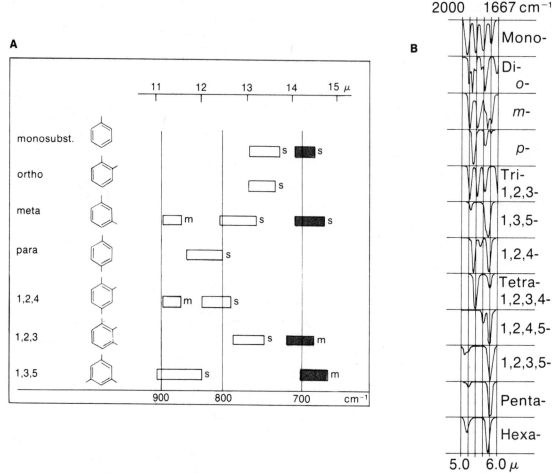

FIGURE IR–6. *A.* The C—H out-of-plane bending vibrations for substituted benzenoid compounds. *B.* The 2000 to 1667 cm^{-1} (5–6 μ) region for substituted benzenoid compounds. (From John R. Dyer, *Applications of Absorption Spectroscopy of Organic Compounds.* Englewood Cliffs, NJ: Prentice-Hall, 1965.)

penta-, or hexasubstituted. Positional isomers can also be distinguished. Since the absorptions are weak, these bands are best observed by using neat liquids or concentrated solutions. If the compound has a high-frequency carbonyl group, this absorption overlaps the weak overtone bands, so that no useful information can be obtained from analyzing this region. The various patterns that are obtained in this region are shown in Figure IR–6B.

The spectra of styrene and *o*-dichlorobenzene are shown in Figures IR–5 and IR–7.

Alkynes

$\equiv$C—H   Stretch is usually near 3300 cm^{-1} (3.0 μ).
C$\equiv$C   Stretch is near 2150 cm^{-1} (4.65 μ).
          Conjugation moves C$\equiv$C stretch to the right.

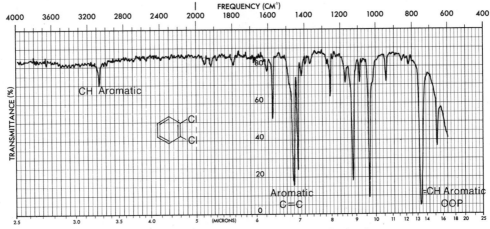

FIGURE IR–7.   Ir spectrum of *o*-dichlorobenzene (neat liquid, salt plates)

Disubstituted or symmetrically substituted triple bonds give either no absorption or weak absorption.

The spectrum of propargyl alcohol is shown in Figure IR–8.

Alcohols and Phenols

O—H   Stretch is a sharp peak at 3650 to 3600 cm^{-1} (2.74–2.78 μ) if no hydrogen bonding takes place. (This is usually only observed in dilute solutions.)

If there is hydrogen bonding (usual in neat or concentrated solutions), the absorption is **broad** and occurs more to the right at 3500 to 3200 cm^{-1} (2.85–3.12 μ), sometimes overlapping C—H stretch absorptions.

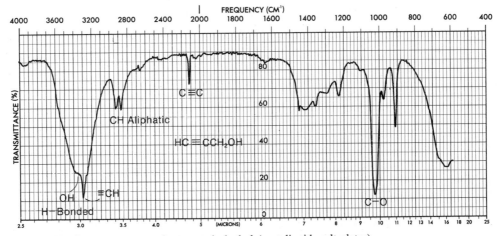

FIGURE IR–8.   Ir spectrum of propargyl alcohol (neat liquid, salt plates)

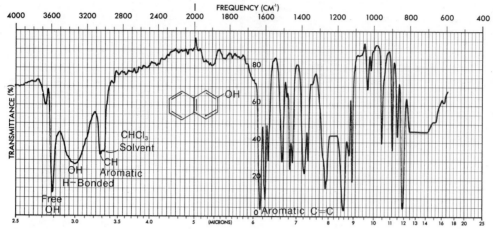

FIGURE IR–9. Ir spectrum of 2-naphthol, showing both free and hydrogen-bonded OH (CHCl$_3$ solution)

C—O   Stretch is usually in the range 1300 to 1000 cm^{-1} (7.7–10 μ).
Phenols are like alcohols. The 2-naphthol shown in Figure IR–9 has some molecules hydrogen-bonded and some free. The spectrum of cyclohexanol is given in Experiment 27. This alcohol, which was determined neat, would also have had a free OH spike to the left of its hydrogen-bonded band if it had been determined in dilute solution. The solution spectra of borneol and isoborneol are shown in Experiment 20.

Ethers

C—O   The most prominent band is due to C—O stretch at 1300 to 1000 cm^{-1} (7.7–10.0 μ). Absence of C=O and O—H bands is required to be sure C—O stretch is not due to an alcohol or ester. Phenyl and vinyl ethers are found in the left portion of the range, aliphatic ethers to the right. (Conjugation with the oxygen moves the absorption to the left.)

The spectrum of anisole is shown in Figure IR–10.

Amines

N—H   Stretch occurs in the range of 3500 to 3300 cm^{-1} (2.86–3.03 μ). Primary amines have **two** bands typically 30 cm^{-1} (0.03 μ) apart. Secondary amines have one band, often vanishingly weak. Tertiary amines have no NH stretch.

C—N   Stretch is weak and occurs in the range of 1350 to 1000 cm^{-1} (7.4–10 μ).

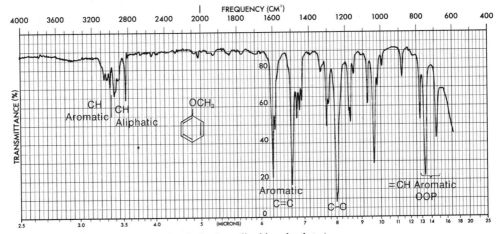

FIGURE IR–10. Ir spectrum of anisole (neat liquid, salt plates)

N—H   Scissoring mode occurs in the range of 1640 to 1560 cm^{-1} (6.1–6.4 μ) (broad).

An out-of-plane bending absorption can sometimes be observed at about 800 cm^{-1} (12.5 μ).

The spectrum of *n*-butylamine is shown in Figure IR–11.

Nitro Compounds

N=O   Stretch is usually two strong bands at 1600 to 1500 cm^{-1} (6.25–6.67 μ) and 1390 to 1300 cm^{-1} (7.2–7.7 μ).

The spectrum of nitrobenzene is shown in Figure IR–12.

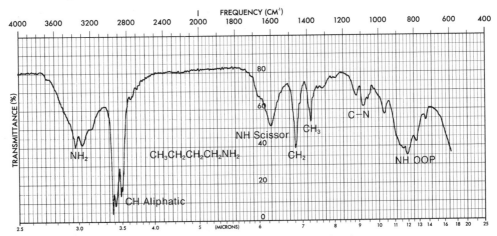

FIGURE IR–11. Ir spectrum of *n*-butylamine (neat liquid, salt plates)

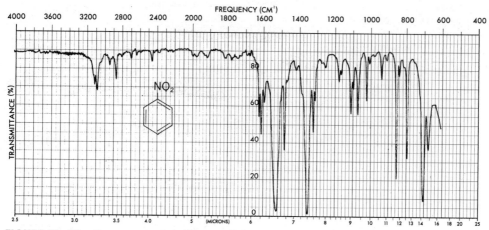

FIGURE IR–12.   Ir spectrum of nitrobenzene, neat

Nitriles

C≡N   Stretch is a sharp absorption near 2250 cm^{-1} (4.5 μ).
Conjugation with double bonds or aromatic rings moves the absorption
to the right.

The spectrum of benzonitrile is shown in Figure IR–13.

Carbonyl Compounds

The carbonyl group is one of the most strongly absorbing groups in the infrared region
of the spectrum. This is mainly due to its large dipole moment. It absorbs in a variety of
compounds (aldehydes, ketones, acids, esters, amides, anhydrides, and so on) in the

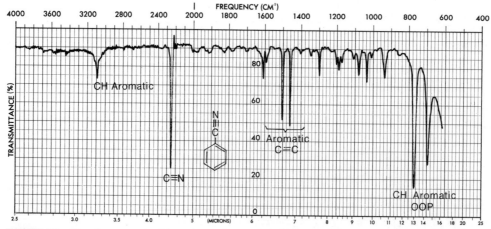

FIGURE IR–13.   Ir spectrum of benzonitrile (neat liquid, salt plates)

5.52	5.68	5.76	5.80	5.83	5.85	5.92	μ
1810	1760	1735	1725	1715	1710	1690	cm^{-1}
ANHYDRIDE (Band 1)		**ESTERS**		**KETONES**		**AMIDES**	
	ANHYDRIDE (Band 2)		**ALDEHYDES**		**CARBOXYLIC ACIDS**		

FIGURE IR–14. Normal values (± 10 cm^{-1}) for various types of carbonyl groups

range of 1850 to 1650 cm^{-1} (5.41–6.06 μ). In Figure IR–14, the normal values for the various types of carbonyl groups are compared. In the sections that follow, each type will be examined separately.

Aldehydes

C=O   Stretch at approximately 1725 cm^{-1} (5.80 μ) is normal.
Aldehydes **seldom** absorb to the left of this value.
Conjugation moves the absorption to the right.

C—H   Stretch, aldehyde hydrogen (—CHO), consists of **weak** bands at about 2750 cm^{-1} (3.65 μ) and 2850 cm^{-1} (3.50 μ). Note that the CH stretch in alkyl chains does not usually extend this far to the right.

The spectrum of nonanal is shown in Figure IR–15. In addition, the spectrum of benzaldehyde is shown in Experiment 38.

Ketones

C=O   Stretch at approximately 1715 cm^{-1} (5.83 μ) is normal.
Conjugation moves the absorption to the right.
Ring strain moves the absorption to the left in cyclic ketones.

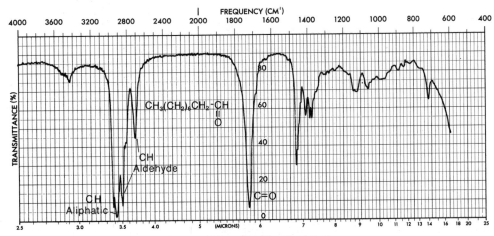

FIGURE IR–15. Ir spectrum of nonanal (neat liquid, salt plates)

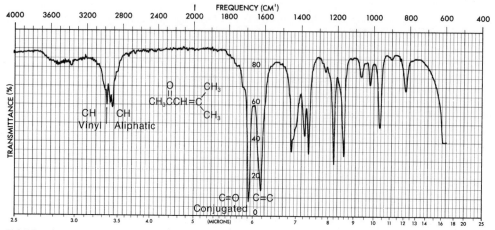

FIGURE IR–16. Ir spectrum of mesityl oxide (neat liquid, salt plates)

The spectra of methyl isopropyl ketone and mesityl oxide are shown in Figures IR–3 and IR–16. The spectrum of camphor is shown in Experiment 20.

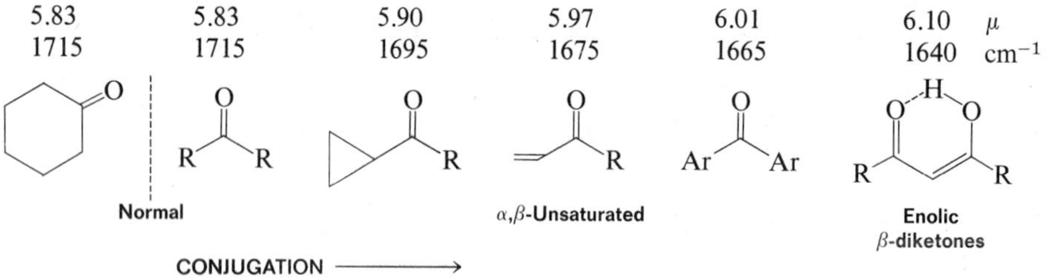

5.83	5.83	5.90	5.97	6.01	6.10	μ
1715	1715	1695	1675	1665	1640	cm^{-1}

Normal α,β-Unsaturated   Enolic
                                                     β-diketones

CONJUGATION ⟶

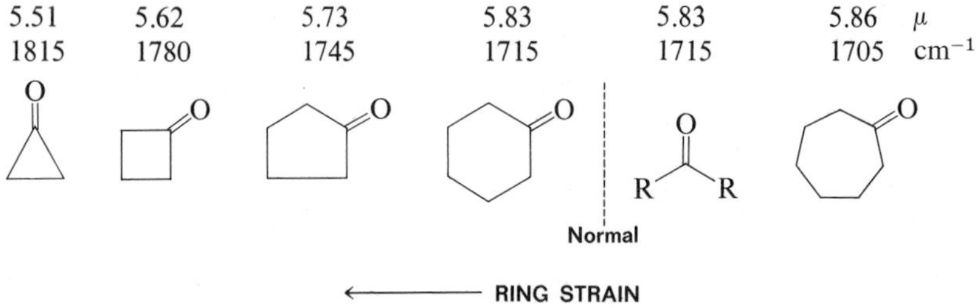

5.51	5.62	5.73	5.83	5.83	5.86	μ
1815	1780	1745	1715	1715	1705	cm^{-1}

Normal

⟵ RING STRAIN

Acids

O—H   Stretch, usually **very broad** (strongly hydrogen-bonded) at 3300 to 2500 cm^{-1} (3.0–4.0 μ), often interferes with C—H absorptions.

C=O   Stretch, broad, 1730 to 1700 cm^{-1} (5.8–5.9 μ)
            Conjugation moves the absorption to the right.

C—O   Stretch, in range of 1320 to 1210 cm^{-1} (7.6–8.3 μ), strong

The spectrum of benzoic acid is shown in Experiment 30B.

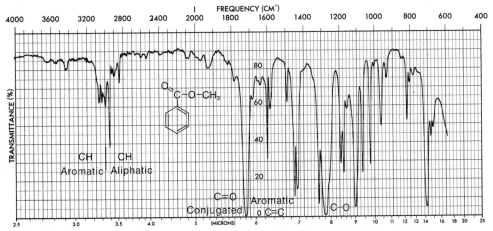

FIGURE IR–17. Ir spectrum of methyl benzoate (neat liquid, salt plates)

Esters (R—C(=O)—OR')

C=O   Stretch occurs at about 1735 cm^{-1} (5.76 μ) in normal esters.
      (a) Conjugation in the R part moves the absorption to the right.
      (b) Conjugation with the O in the R' part moves the absorption to the left.
      (c) Ring strain (lactones) moves the absorption to the left.
C—O   Stretch, two bands or more, one stronger than the others, is in the range of 1300 to 1000 cm^{-1} (7.69–10.0 μ).

The spectrum of methyl benzoate is shown in Figure IR–17. The spectra of isopentyl acetate and methyl salicylate are shown in Experiments 10 and 11.

Amides

C=O   Stretch is at approximately 1670 to 1640 cm^{-1} (6.0–6.1 μ).
      Conjugation and ring size (lactams) have the usual effects.
N—H   Stretch (if monosubstituted or unsubstituted) 3500 to 3100 cm^{-1} (2.85–3.25 μ)
      Unsubstituted amides have two bands (—NH$_2$) in this region.
N—H   Bending around 1640 to 1550 cm^{-1} (6.10–6.45 μ).

The spectrum of benzamide is shown in Figure IR–18.

Anhydrides

C=O   Stretch always has **two** bands: 1830 to 1800 cm^{-1} (5.46–5.56 μ) and 1775 to 1740 cm^{-1} (5.63–5.75 μ).

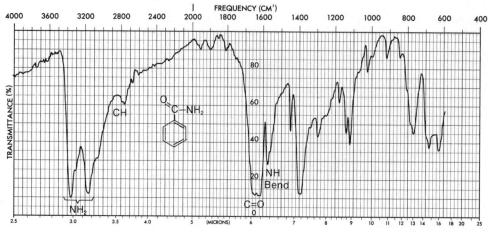

FIGURE IR–18.   Ir spectrum of benzamide (solid phase, KBr)

Unsaturation moves the absorptions to the right.
Ring strain (cyclic anhydrides) moves the absorptions to the left.
C—O   Stretch is at 1300 to 900 cm^{-1} (7.7–11 μ).

Halides

It is often difficult to determine either the presence or the absence of a halide in a compound by infrared spectroscopy. The absorption bands cannot be relied on, especially if the spectrum is being determined with the compound dissolved in CCl_4 or $CHCl_3$ solution.

    C—F   Stretch, 1350 to 960 cm^{-1} (7.41–10.4 μ)
    C—Cl   Stretch, 850 to 500 cm^{-1} (11.8–20.0 μ)
    C—Br   Stretch, to the right of 667 cm^{-1} (15.0 μ)
    C—I   Stretch, to the right of 667 cm^{-1} (15.0 μ)

The spectra of carbon tetrachloride and chloroform are shown in Technique 17, Figures 17–4 and 17–5, page 665.

REFERENCES

Bellamy, L. J. *The Infrared Spectra of Complex Molecules.* 3rd ed. New York: Methuen, 1975.
Dyer, J. R. *Applications of Absorption Spectroscopy of Organic Compounds.* Englewood Cliffs, N.J.: Prentice-Hall, 1965.
Nakanishi, K., and Soloman, P. H. *Infrared Absorption Spectroscopy.* 2nd ed. San Francisco: Holden-Day, 1977.
Pasto, D. J., and Johnson, C. R. *Laboratory Text for Organic Chemistry.* Englewood Cliffs, N.J.: Prentice-Hall, 1974.
Pavia, D. L., Lampman, G. M., and Kriz, G. S., Jr. *Introduction to Spectroscopy.* Philadelphia: W. B. Saunders, 1979.
Silverstein, R. M., Bassler, G. C., and Morrill, T. C. *Spectrophotometric Identification of Organic Compounds.* 4th ed. New York: Wiley, 1981.

Appendix 4
Nuclear Magnetic Resonance Spectroscopy

NMR.1 THE RESONANCE PHENOMENON

Nuclear magnetic resonance (**nmr**) spectroscopy is an instrumental technique that allows the number, type, and relative positions of certain atoms in a molecule to be determined. This type of spectroscopy applies only to those atoms that have nuclear magnetic moments because of their nuclear spin properties. Although many atoms meet this requirement, hydrogen atoms ($_1^1H$) are of the greatest interest to the organic chemist. Atoms of the ordinary isotopes of carbon ($_6^{12}C$) and oxygen ($_8^{16}O$) do not have nuclear magnetic moments, and ordinary nitrogen atoms ($_7^{14}N$), although they do have magnetic moments, generally fail to show typical nmr behavior for other reasons. The same is true of the halogen atoms, except for fluorine ($_9^{19}F$), which does show active nmr behavior. Atoms other than hydrogen will not be considered here.

Nuclei of nmr-active atoms placed in a magnetic field can be thought of as tiny bar magnets. In hydrogen, which has two allowed nuclear spin states ($+\frac{1}{2}$ and $-\frac{1}{2}$), either the nuclear magnets of individual atoms can be aligned with the magnetic field (spin $+\frac{1}{2}$), or they can be opposed to it (spin $-\frac{1}{2}$). A slight majority of the nuclei will be aligned with the field, as this spin orientation constitutes a slightly lower-energy spin state. If radiofrequency waves of the appropriate energy are supplied, nuclei aligned with the field can absorb this radiation and reverse their direction of spin, or become reoriented so that the nuclear magnet opposes the applied magnetic field (Figure NMR–1).

The frequency of radiation required to induce spin conversion is a direct function of the strength of the applied magnetic field. When a spinning hydrogen nucleus is placed in a magnetic field, the nucleus begins to precess with angular frequency ω, much like a child's toy top. This precessional motion is depicted in Figure NMR–2. The angular frequency of nuclear precession ω increases as the strength of the applied magnetic field is increased. The radiation that must be supplied to induce spin conversion in a hydrogen nucleus of spin $+\frac{1}{2}$ must have a frequency that just matches the angular precessional frequency ω. This is called the resonance condition, and spin conversion is said to be a resonance process.

$-1/2$ ———

$+1/2$ —|—

$\rightsquigarrow +h\nu \rightsquigarrow$

Magnetic Field Direction

FIGURE NMR–1. The nmr absorption process

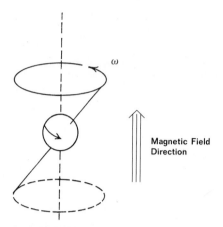

FIGURE NMR–2. Precessional motion of a spinning nucleus in an applied magnetic field

For the average proton (hydrogen atom), if a magnetic field of approximately 14,000 Gauss is applied, radiofrequency radiation of 60 MHz (60,000,000 cycles per second) is required to induce a spin transition. Fortunately, the magnetic-field strength required to induce the various protons in a molecule to absorb 60-MHz radiation varies from proton to proton within the molecule and is a sensitive function of the immediate **electronic** environment of each proton. The typical proton nuclear magnetic resonance spectrometer supplies a basic radiofrequency radiation of 60 MHz to the sample being measured and **increases** the strength of the applied magnetic field over a range of several parts per million from the basic field strength. As the field increases, various

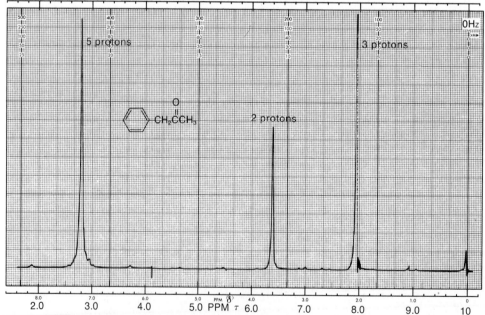

FIGURE NMR–3. Nuclear magnetic resonance spectrum of phenylacetone (the absorption peak at the far right is caused by the added reference substance TMS)

protons come into resonance (absorb 60-MHz energy), and a resonance signal is generated for each proton. An nmr spectrum is a plot of the strength of the magnetic field versus the intensity of the absorptions. A typical nmr spectrum is shown in Figure NMR–3.

NMR.2   THE CHEMICAL SHIFT

The differences in the applied field strengths at which the various protons in a molecule absorb 60-MHz radiation are extremely small. The different absorption positions amount to a difference of only a few parts per million (ppm) in the magnetic field strength. Since it is experimentally difficult to measure the precise field strength at which each proton absorbs to less than one part in a million, a technique has been developed whereby the **difference** between two absorption positions is measured directly. To achieve this measurement, a standard reference substance is used and the positions of the absorptions of all other protons are measured relative to the values for the reference substance. The reference substance that has been universally accepted is tetramethylsilane, $(CH_3)_4Si$, which is also called TMS. The proton resonances in this molecule appear at a higher field strength than the proton resonances in most all other molecules do, and all the protons of TMS have resonance at the same field strength.

To give the position of absorption of a proton a quantitative measurement, a parameter called the **chemical shift** (δ) has been defined. One δ unit corresponds to a 1-ppm change in the magnetic field strength. To determine the chemical shift value for the various protons in a molecule, the operator determines an nmr spectrum of the molecule with a small quantity of TMS added directly to the sample. That is, both spectra are determined **simultaneously.** The TMS absorption is adjusted to correspond to the $\delta = 0$ position on the recording chart, which is calibrated in δ units, and the δ values of the absorption peaks for all other protons can be read directly from the chart.

Since the nmr spectrometer increases the magnetic field as the pen moves from left to right on the chart, the TMS absorption appears at the extreme right edge of the spectrum ($\delta = 0$) or at the **upfield** end of the spectrum. The chart is calibrated in δ units (or ppm), and most other protons absorb at a lower field strength (or **downfield**) from TMS.

Because the frequency at which a proton precesses, and hence the frequency at which it absorbs radiation, is directly proportional to the strength of the applied magnetic field, a second method of measuring an nmr spectrum is possible. One could hold the magnetic field strength constant and vary the frequency of the radiofrequency radiation supplied. Thus, a given proton could be induced to absorb **either** by increasing the field strength, as described earlier, or alternatively, by decreasing the frequency of the radiofrequency oscillator. A 1-ppm decrease in the frequency of the oscillator would have the same effect as a 1-ppm increase in the magnetic field strength. For reasons of instrumental design, it is simpler to vary the strength of the magnetic field than to vary the frequency of the oscillator. Most instruments operate on the former principle. Nevertheless, the recording chart is calibrated not only in δ units but in Hz as well (1 ppm = 60 Hz when the frequency is 60 MHz), and the chemical shift is customarily defined and computed using Hertz rather than Gauss:

$$\delta = \text{chemical shift} = \frac{\text{(observed shift from TMS, in Hz)}}{\text{(60 MHz)}} = \frac{\text{Hz}}{\text{MHz}} = \text{ppm}$$

Although the equation defines the chemical shift for a spectrometer operating at 14,100 Gauss and 60 MHz, the chemical shift value that is calculated is **independent** of the field strength. For instance, at 23,500 Gauss the oscillator frequency would have to be 100 MHz. Although the observed shifts from TMS (in Hz) would be larger at this field strength, the divisor of the equation would be 100 MHz, instead of 60 MHz, and δ would turn out to be identical under either set of conditions.

NMR.3   CHEMICAL EQUIVALENCE—INTEGRALS

All the protons in a molecule that are in chemically identical environments often exhibit the same chemical shift. Thus, all the protons in tetramethylsilane (TMS), or all the protons in benzene, cyclopentane, or acetone, have their own respective resonance values all at the same δ value. Each compound gives rise to a single absorption peak in

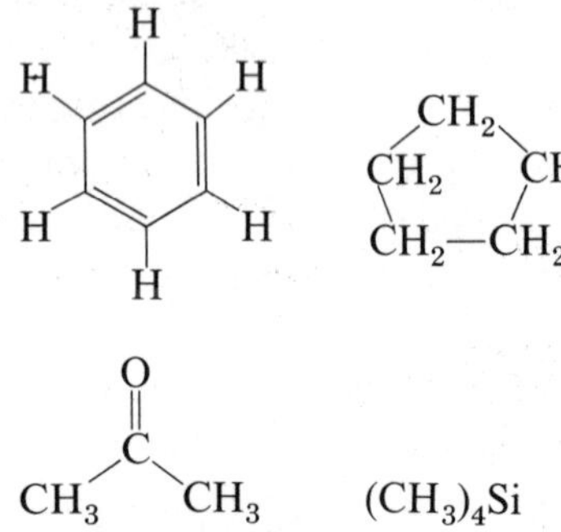

MOLECULES GIVING RISE TO ONE NMR ABSORPTION PEAK—ALL PROTONS CHEMICALLY EQUIVALENT

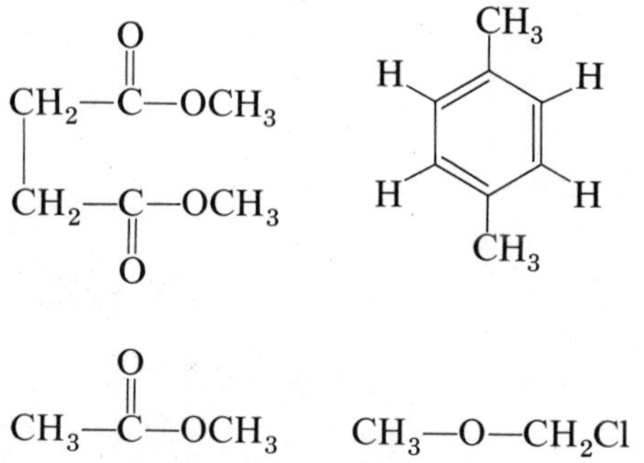

MOLECULES GIVING RISE TO TWO NMR ABSORPTION PEAKS—TWO DIFFERENT SETS OF CHEMICALLY EQUIVALENT PROTONS

its nmr spectrum. The protons are said to be **chemically equivalent.** On the other hand, molecules that have sets of protons that are chemically distinct from one another may give rise to an absorption peak from each set.

The nmr spectrum given in Figure NMR–3 is that of phenylacetone, a compound having **three** chemically distinct types of protons:

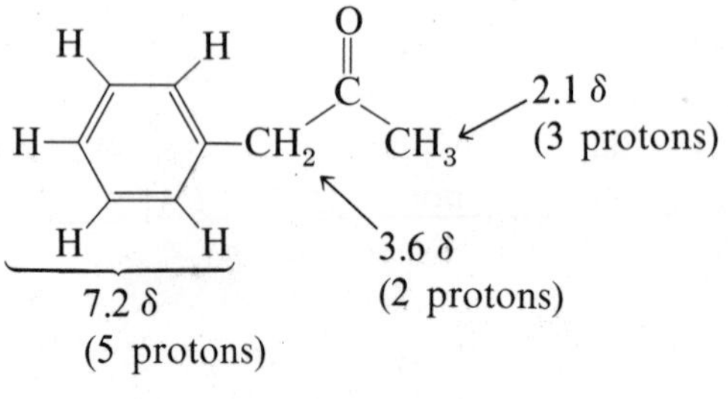

One can immediately see that the nmr spectrum furnishes a valuable type of information on this basis alone. In fact, the nmr spectrum can not only distinguish how many different types of protons a molecule has but also can reveal **how many** of each different type are contained within the molecule.

In the nmr spectrum, the area under each peak is proportional to the number of hydrogens generating that peak. Hence, in the above compound, the area ratio of the three peaks is $5:2:3$, the same as the ratio of the numbers of each type of hydrogen. The nmr spectrometer can electronically "integrate" the area under each peak. It does this by tracing over each peak a vertically rising line, which rises in height by an amount proportional to the area under the peak. Shown in Figure NMR–4 is an nmr spectrum of benzyl acetate, with each of the peaks integrated in this way.

It is important to note that the height of the integral line does not give the absolute number of hydrogens; it gives the **relative** numbers of each type of hydrogen. For a given integral to be of any use, there must be a second integral to which it is referred. The benzyl acetate case gives a good example of this. The first integral rises for 55.5 divisions on the chart paper, the second for 22.0 divisions, and the third for 32.5 divisions. These numbers are relative and give the **ratios** of the various types of protons. One can find these ratios by dividing each of the large numbers by the smallest number:

$$\frac{55.5 \text{ div}}{22.0 \text{ div}} = 2.52 \qquad \frac{22.0 \text{ div}}{22.0 \text{ div}} = 1.00 \qquad \frac{32.5 \text{ div}}{22.0 \text{ div}} = 1.48$$

Thus, the number ratio of the protons of each type is $2.52:1.00:1.48$. If we assume that the peak at 5.1δ is really caused by two hydrogens, and if we assume that the integrals are slightly in error (this can be as much as 10%), then we can arrive at the true ratios by multiplying each figure by two and rounding off; we then get $5:2:3$. Clearly the peak at 7.3δ, which integrates for 5, arises from the resonance of the aromatic ring protons, and the peak at 2.0δ, which integrates for 3, is caused by the methyl protons. The two-proton resonance at 5.1δ arises from the benzyl protons. Notice then that the integrals give the simplest ratios, but not necessarily the true ratios, of the number of protons of each type.

NMR.4 CHEMICAL ENVIRONMENT AND CHEMICAL SHIFT

If the resonance frequencies of all protons in a molecule were the same, nmr would be of little use to the organic chemist. However, not only do different types of protons have different chemical shifts, but they also have a value of chemical shift that characterizes the type of proton they represent. Every type of proton has only a limited range of δ values over which it gives resonance. Hence, the numerical value of the chemical shift for a proton indicates the **type of proton** originating the signal, just as the infrared frequency suggests the type of bond or functional group. Notice, for instance, that the aromatic protons of both phenylacetone (Figure NMR–3) and benzyl acetate (Figure NMR–4) have resonance near 7.3δ and that both methyl groups attached directly to a

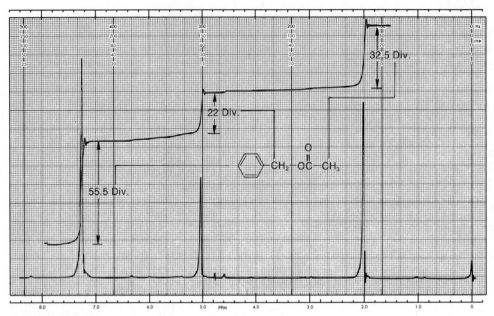

FIGURE NMR–4. Determination of the integral ratios for benzyl acetate

carbonyl group have a resonance of approximately 2.1δ. Aromatic protons characteristically have resonance near 7 to 8δ, and acetyl groups (the methyl protons) have their resonance near 2δ. These values of chemical shift are diagnostic. Notice also how the resonance of the benzyl ($-CH_2-$) protons comes at a higher value of chemical shift (5.1δ) in benzyl acetate than in phenylacetone (3.6δ). Being attached to the electronegative element, oxygen, these protons are more deshielded (see Section NMR.5) than the protons in phenylacetone. A trained chemist would have readily recognized the probable presence of the oxygen by the chemical shift shown by these protons.

It is important to learn the ranges of chemical shifts over which the most common types of protons have resonance. Figure NMR–5 is a correlation chart that contains the most essential and frequently encountered types of protons. For the beginner, it is often difficult to memorize a large body of numbers relating to chemical shifts and proton types. One need actually do this only crudely. It is more important to "get a feel" for the regions and the types of protons than to know a string of factual numbers.

The values of chemical shift given in Figure NMR–5 can be easily understood in terms of two factors: local diamagnetic shielding and anisotropy. These two factors are discussed in Sections NMR.5 and NMR.6.

NMR.5   LOCAL DIAMAGNETIC SHIELDING

The trend of chemical shifts that is easiest to explain is that involving electronegative elements substituted on the same carbon to which the protons of interest are attached. The chemical shift simply increases as the electronegativity of the attached element increases. This is illustrated in Table NMR–1 for several compounds of the type CH_3X.

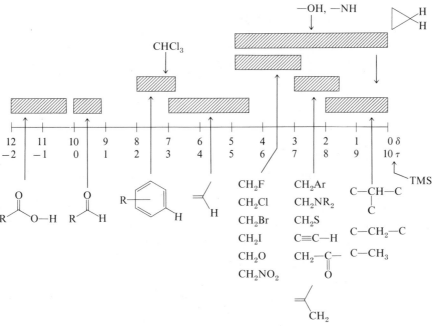

FIGURE NMR–5. Simplified correlation chart for proton chemical shift values

Multiple substituents have a stronger effect than a single substituent. The influence of the substituent, an electronegative element having little effect on protons that are more than three carbons away, drops off rapidly with distance. These effects are illustrated in Table NMR–2.

Electronegative substituents attached to a carbon atom, because of their electron-withdrawing effects, reduce the valence electron density around the protons attached to that carbon. These electrons **shield** the proton from the applied magnetic field. This effect, called **local diamagnetic shielding,** occurs because the applied magnetic field induces the valence electrons to circulate and thus to generate an induced magnetic field, which **opposes** the applied field. This is illustrated in Figure NMR–6. Electronegative substituents on carbon reduce the local diamagnetic shielding in the vicinity of the attached protons because they reduce the electron density around those protons. Substituents that produce this effect are said to **deshield** the proton. The greater the electronegativity of the substituent, the more the deshielding of the protons and, hence, the greater the chemical shift of those protons.

TABLE NMR–1.   Dependence of Chemical Shift of CH₃X on the Element X

Compound CH_3X	CH_3F	CH_3OH	CH_3Cl	CH_3Br	CH_3I	CH_4	$(CH_3)_4Si$
Element X	F	O	Cl	Br	I	H	Si
Electronegativity of X	4.0	3.5	3.1	2.8	2.5	2.1	1.8
Chemical shift (δ)	4.26	3.40	3.05	2.68	2.16	0.23	0

TABLE NMR–2.   Substitution Effects

	CHCl$_3$	CH$_2$Cl$_2$	CH$_3$Cl	—CH$_2$Br	—CH$_2$—CH$_2$Br	—CH$_2$—CH$_2$CH$_2$Br
δ	7.27	5.30	3.05	3.30	1.69	1.25

NMR.6   ANISOTROPY

Figure NMR–5 clearly shows that several types of protons have chemical shifts not easily explained by simple consideration of the electronegativity of the attached groups. Consider, for instance, the protons of benzene or other aromatic systems. Aryl protons generally have a chemical shift that is as large as that for the proton of chloroform! Alkenes, alkynes, and aldehydes also have protons whose resonance values are not in line with the expected magnitude of any electron-withdrawing effects. In each of these cases, the effect is due to the presence of an unsaturated system (π electrons) in the vicinity of the proton in question. In benzene, for example, when the π electrons in the aromatic ring system are placed in a magnetic field, they are induced to circulate around the ring. This circulation is called a **ring current.** Moving electrons (the ring current) generate a magnetic field much like that generated in a loop of wire through which a current is induced to flow. The magnetic field covers a spatial volume large enough to influence the shielding of the benzene hydrogens. This is illustrated in Figure NMR–7. The benzene hydrogens are deshielded by the **diamagnetic anisotropy** of the ring. An applied magnetic field is nonuniform (anisotropic) in the vicinity of a benzene molecule because of the labile electrons in the ring that interact with the applied field. Thus, a proton attached to a benzene ring is influenced by **three** magnetic fields: the strong magnetic field applied by the electromagnets of the nmr spectrometer and two

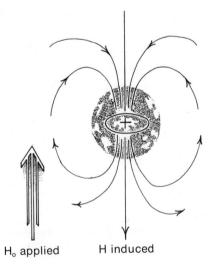

H$_o$ applied   H induced

FIGURE NMR–6.   Local diamagnetic shielding of a proton due to its valence electrons

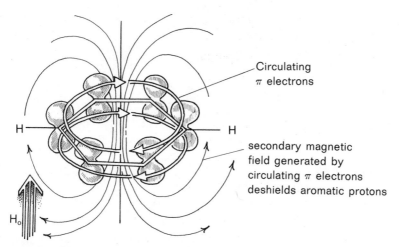

FIGURE NMR–7. Diamagnetic anisotropy in benzene

weaker fields, one due to the usual shielding by the valence electrons around the proton and the other due to the anisotropy generated by the ring system electrons. It is this anisotropic effect that gives the benzene protons a greater chemical shift than is expected. These protons just happen to lie in a **deshielding** region of this anisotropic field. If a proton were placed in the center of the ring rather than on its periphery, the proton would be shielded, since the field lines would have the opposite direction.

All groups in a molecule that have π electrons generate secondary anisotropic fields. In acetylene, the magnetic field generated by induced circulation of π electrons has a geometry such that the acetylene hydrogens are **shielded.** Hence, acetylenic hydrogens come at a higher field than expected. The shielding and deshielding regions due to the various π electron functional groups have characteristic shapes and directions; they are illustrated in Figure NMR–8. Protons falling within the cones are shielded, and those falling outside the conical areas are deshielded. Since the magnitude of the anisotropic field diminishes with distance, beyond a certain distance anisotropy has essentially no effect.

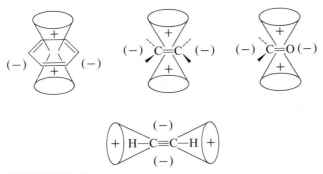

FIGURE NMR–8. Anisotropy caused by the presence of π electrons in some common multiple bond systems

NMR.7   SPIN-SPIN SPLITTING (*n* + 1 RULE)

We have already considered how the chemical shift and the integral (peak area) can give information about the number and type of hydrogens contained in a molecule. A third type of information available from the nmr spectrum is derived from spin-spin splitting. Even in simple molecules, one finds that each type of proton rarely gives a single resonance peak. For instance, in 1,1,2-trichloroethane there are two chemically distinct types of hydrogen:

$$Cl—\underset{\underset{Cl}{|}}{\overset{\overset{\textcircled{H}}{|}}{C}}—\textcircled{CH_2}—Cl$$

From information given thus far, one would predict **two** resonance peaks in the nmr spectrum of 1,1,2-trichloroethane with an area ratio (integral ratio) 2:1. In fact, the nmr spectrum of this compound has **five** peaks. A group of three peaks (called a triplet) exists at 5.77δ and a group of two peaks (called a doublet) at 3.95δ. The spectrum is shown in Figure NMR–9. The methine (CH) resonance (5.77δ) is split into a triplet, and the methylene resonance (3.95δ) is split into a doublet. The area under the three triplet peaks is **one,** relative to an area of **two** under the two doublet peaks.

This phenomenon is called spin-spin splitting. Empirically, spin-spin splitting can be explained by the ''*n* + 1 rule.'' Each type of proton ''senses'' the number of equivalent protons (*n*) on the carbon atom or atoms next to the one to which it is bonded, and its resonance peak is split into *n* + 1 components.

Let's examine the case at hand, 1,1,2-trichloroethane, using the so-called *n* + 1 rule. First, the lone methine hydrogen is situated next to a carbon bearing two methylene protons. According to the rule, it has two equivalent neighbors (*n* = 2) and is split

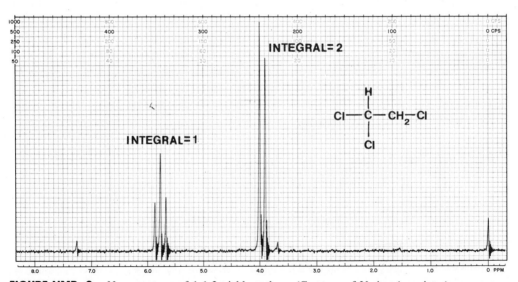

FIGURE NMR–9.   Nmr spectrum of 1,1,2-trichloroethane (Courtesy of Varian Associates)

into $n + 1 = 3$ peaks (a triplet). The methylene protons are situated next to a carbon bearing only one methine hydrogen. According to the rule, they have one neighbor ($n = 1$) and are split into $n + 1 = 2$ peaks (a doublet).

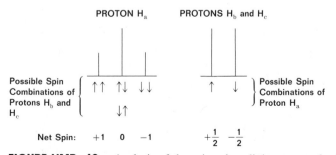

Two neighbors give | One neighbor gives
a triplet ($n + 1 = 3$) | a doublet ($n + 1 = 2$)
(area = 1) | (area = 2)

The spectrum of 1,1,2-trichloroethane can be explained easily by the interaction, or coupling, of the spins of protons on adjacent carbon atoms. The position of absorption of proton H_a may be affected by the spins of protons H_b and H_c attached to the neighboring (adjacent) carbon atom. If the spins of these protons are aligned with the applied magnetic field, the small magnetic field generated by their nuclear spin properties will augment the strength of the field experienced by the first-mentioned proton H_a. The proton H_a will thus be **deshielded.** If the spins of H_b and H_c are opposed to the applied field, they will decrease the field experienced by proton H_a. It will then be **shielded.** In each of these situations, the absorption position of H_a will be altered. Among the many molecules in the solution, one will find all the various possible spin combinations for H_b and H_c; hence, the nmr spectrum of the molecular solution will give **three** absorption peaks (a triplet) for H_a, since H_b and H_c have three different possible spin combinations (Figure NMR–10). By a similar analysis, it can be seen that protons H_b and H_c should appear as a doublet.

Some common splitting patterns that can be predicted by the $n + 1$ rule and that are frequently observed in a number of molecules are shown in Figure NMR–11. Notice particularly the last entry, where **both** methyl groups (6 protons in all) function as a unit and split the methine proton into a septet ($6 + 1 = 7$).

NMR.8 THE COUPLING CONSTANT

The quantitative amount of spin-spin interaction between two protons can be defined by the coupling constant. The spacing between the component peaks in a simple multiplet

PROTON H_a PROTONS H_b and H_c

Possible Spin Combinations of Protons H_b and H_c
$\uparrow\uparrow$ $\uparrow\downarrow$ $\downarrow\downarrow$
$\downarrow\uparrow$

Possible Spin Combinations of Proton H_a
$\uparrow$ $\downarrow$

Net Spin: $+1$ 0 -1 $+\frac{1}{2}$ $-\frac{1}{2}$

FIGURE NMR–10. Analysis of the spin-spin splitting pattern for 1,1,2-trichloroethane

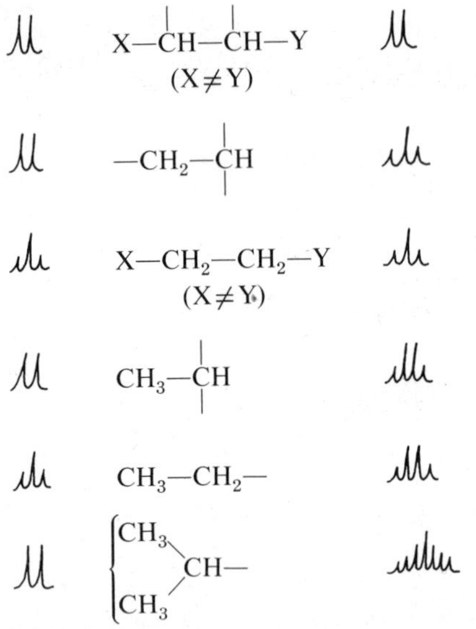

X—CH—CH—Y
              (X≠Y)

—CH₂—CH

X—CH₂—CH₂—Y
              (X≠Y)

CH₃—CH

CH₃—CH₂—

CH₃   CH—
          CH₃

FIGURE NMR–11.   Some common splitting patterns

is called the coupling constant J. This distance is measured on the same scale as the chemical shift and is expressed either in cycles per second (cps) or in Hertz (Hz).

For the interaction of most aliphatic protons in acyclic systems, the magnitudes of the coupling constants are always near 7.5 Hz. See, for instance, the nmr spectrum of 1,1,2-trichloroethane in Figure NMR–9, where the coupling constant is approximately 6 Hz. Different magnitudes of J are found for different types of protons. For instance, the *cis* and *trans* protons substituted on a double bond commonly have values where $J_{trans} \cong 17$ Hz and $J_{cis} \cong 10$ Hz are typical coupling constants. In ordinary compounds, coupling constants may range anywhere from 0 to 18 Hz. The magnitude of J often provides structural clues. One can usually distinguish, for example, between a *cis* olefin and a *trans* olefin on the basis of the observed coupling constants for the vinyl protons. The approximate values of some representative coupling constants are given in Table NMR–3.

NMR.9   MAGNETIC EQUIVALENCE

In the example of spin-spin splitting in 1,1,2-trichloroethane, one notices that the two protons H_b and H_c, which are attached to the same carbon atom, do not split one another. They behave as an integral group. Actually the two protons H_b and H_c **are** coupled to one another; however, for reasons we cannot explain fully here, protons that are attached to the same carbon and both of which have the **same chemical shift** do not show spin-spin splitting. Another way of stating this is that protons coupled to the same extent to **all** other protons in a molecule do not show spin-spin splitting. Protons that have the same chemical shift and are coupled equivalently to all other protons are **magnetically equivalent** and do not show spin-spin splitting. Thus, in 1,1,2-trichloro-

TABLE NMR–3. Representative Coupling Constants and Approximate Values (Hz)

 H H │ │ C—C │ H	6 to 8	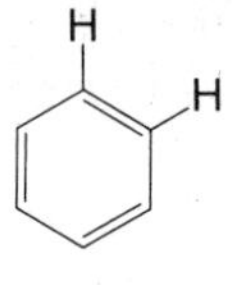 ortho 6 to 10	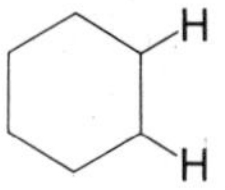 a,a 8 to 14 a,e 0 to 7 e,e, 0 to 5
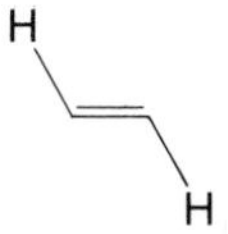 (H₂C=CH)	11 to 18	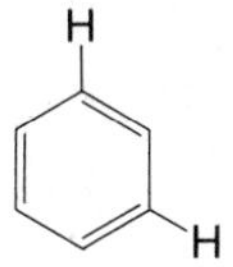 meta 1 to 4	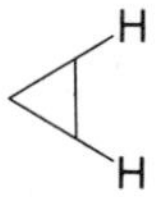 cis 6 to 12 trans 4 to 8
 (cis alkene)	6 to 15	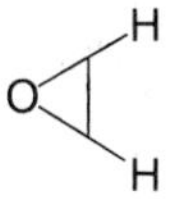 para 0 to 2	cis 2 to 5 trans 1 to 3
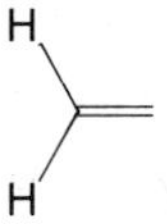 (gem alkene)	0 to 5		 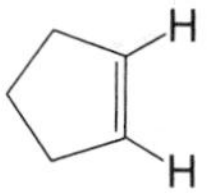 5 to 7
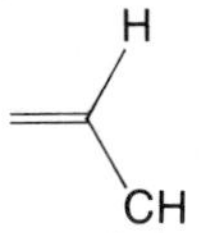 (allylic)	4 to 10	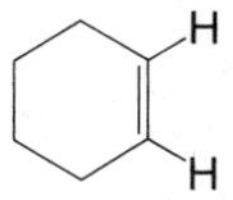 8 to 11	
H—C≡C—CH	0 to 3		

ethane, protons H_b and H_c have the same value δ and are coupled by the same value J to proton H_a. They are magnetically equivalent.

It is important to differentiate magnetic equivalence and chemical equivalence. Note the two compounds shown below.

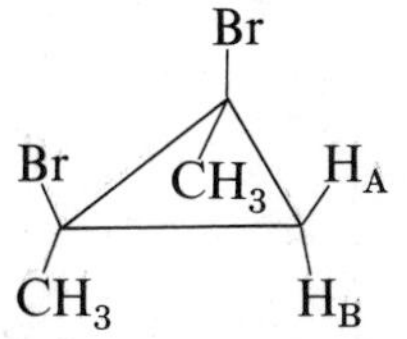

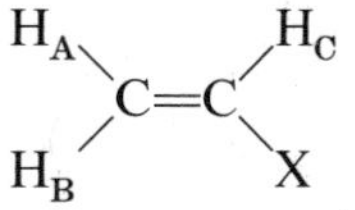

In the cyclopropane compound, the two geminal hydrogens are chemically equivalent; however, they are not magnetically equivalent. Proton H_A is on the same side of the ring as the two halogens. Proton H_B is on the same side of the ring as the two methyl groups. Protons H_A and H_B will have different chemical shifts, will couple to one another, and will show spin-spin splitting. Two doublets will be seen for H_A and H_B. For cyclopropane rings, $J_{geminal}$ is usually around 5 Hz.

Another situation in which protons are chemically equivalent but not magnetically equivalent exists in the vinyl compound. In this example, protons A and B are chemically equivalent but not magnetically equivalent. H_A and H_B have different chemical shifts. In addition, a second distinction can be made between H_A and H_B in this type of compound. Each has a different coupling constant with H_C. The constant J_{AC} is a *cis* coupling constant, and J_{BC} is a *trans* coupling constant. Whenever two protons have different coupling constants relative to a third proton, they are not mag-

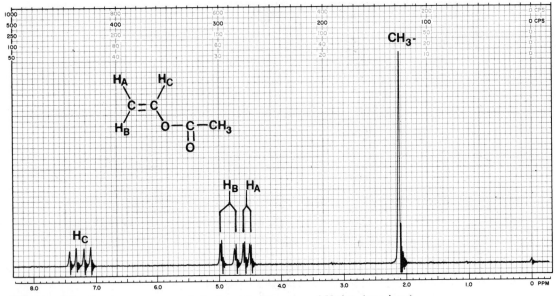

FIGURE NMR–12.   Nmr spectrum of vinyl acetate (Courtesy of Varian Associates)

netically equivalent. In the vinyl compound, H_A and H_B do not act as a group to split proton H_C. Each proton acts independently. Thus H_B splits H_C with coupling constant J_{BC} into a doublet, and then H_A splits each of the components of the doublet into doublets with coupling constant J_{AC}. In such a case, the nmr spectrum must be analyzed graphically, splitting by splitting. An nmr spectrum of a vinyl compound is shown in Figure NMR–12. The graphical analysis of the vinyl portion of the nmr spectrum is in Figure NMR–13.

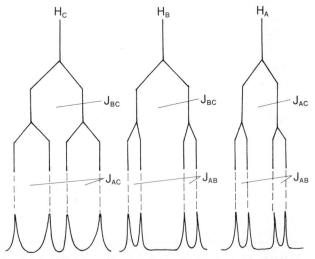

FIGURE NMR–13.   Analysis of the splittings in vinyl acetate

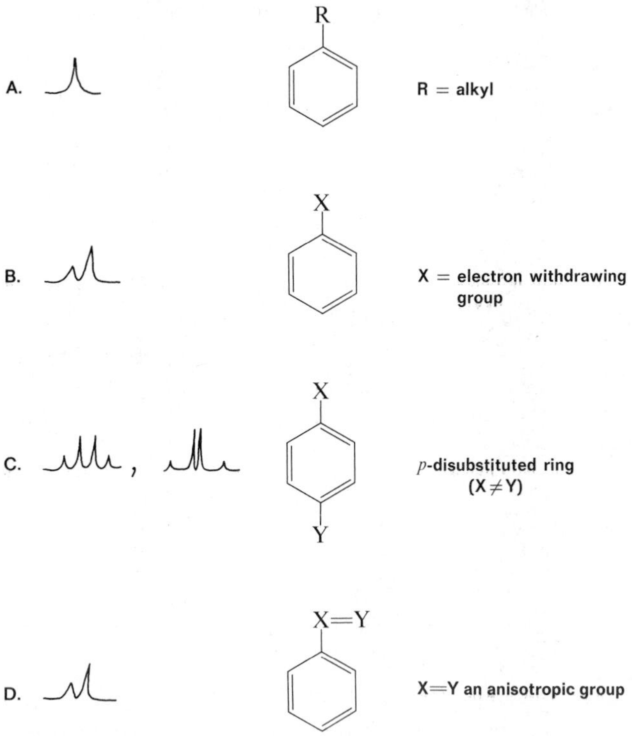

A. R = alkyl

B. X = electron withdrawing group

C. *p*-disubstituted ring (X ≠ Y)

D. X=Y an anisotropic group

FIGURE NMR–14. Some common aromatic patterns

NMR.10 AROMATIC COMPOUNDS

The nmr spectra of protons on aromatic rings are often too complex to explain by simple theories. However, some simple generalizations can be made that are useful in analyzing the aromatic region of the nmr spectrum. First of all, most aromatic protons have resonance near 7.0δ. In monosubstituted rings in which the ring substituent is an alkyl group, all the ring protons often have chemical shifts that are very nearly identical, and the five ring protons may appear as if they gave rise to an overly broad singlet (Figure NMR–14A). If an electronegative group is attached to the ring, all the ring protons are shifted downfield from where they would appear in benzene. However, often the *ortho* protons are shifted more than the others, as they are more affected by the group. This often gives rise to an absorption pattern like that in Figure NMR–14B. In a *para*-disubstituted ring with two substituents X and Y that are identical, all the protons in the ring are chemically and magnetically equivalent, and a singlet is observed. If X is different from Y in electronegativity, however, a pattern like that shown in the left side of Figure NMR–14C is often observed, clearly identifying a *p*-disubstituted ring. If X and Y are more nearly similar, a pattern more like the one on the right is observed. In monosubstituted rings that have a carbonyl group or a double bond attached directly to the ring, a pattern like that in Figure NMR–14D is not uncommon. In this case, the *ortho* protons of the ring are influenced by the anisotropy

TABLE NMR–4. Typical Ranges for Groups with Variable Chemical Shift

Acids	RCOOH	10.5–12.0 δ
Phenols	ArOH	4.0–7.0
Alcohols	ROH	0.5–5.0
Amines	RNH$_2$	0.5–5.0
Amides	RCONH$_2$	5.0–8.0
Enols	CH=CH—OH	≥ 15

of the π systems that make up the CO and CC double bonds and are deshielded by them. In other types of substitution, such as *ortho* or *meta,* or polysubstituted ring systems, the patterns may be much more complicated and require an advanced analysis.

NMR.11 PROTONS ATTACHED TO ATOMS OTHER THAN CARBON

Protons attached to atoms other than carbon often have a widely variable range of absorptions. Several of these groups are tabulated in Table NMR–4. In addition, under the usual conditions of determining an nmr spectrum, protons on heteroelements normally do not couple with protons on adjacent carbon atoms to give spin-spin splitting. This is primarily because such protons often exchange very rapidly with those of the solvent medium. The absorption position is variable because these groups also undergo varying degrees of hydrogen bonding in solutions of different concentrations. The amount of hydrogen bonding that occurs with a proton radically affects the valence electron density around that proton and produces correspondingly large changes in the chemical shift. The absorption peaks for protons that have hydrogen bonding or are undergoing exchange are frequently broad relative to other singlets and can often be recognized on that basis. For a different reason, called quadrupole broadening, protons attached to nitrogen atoms often show an extremely broad resonance peak, often almost indistinguishable from the baseline.

NMR.12 SPECTRA AT HIGHER FIELD STRENGTH

Occasionally the 60-MHz spectrum of an organic compound, or a portion of it, is almost undecipherable because the chemical shifts of several groups of protons are all very similar. In these cases all of the proton resonances occur in the same area of the spectrum, and often peaks overlap so extensively that individual peaks and splittings cannot be extracted. One of the ways in which such a situation can be simplified is by the use of a spectrometer that operates at a higher frequency. Although both 60- and 90-MHz instruments are quite common, it is not unusual at a large university or an industrial research center to find instruments with operating frequencies of 100, 220, 300 MHz, or even higher.

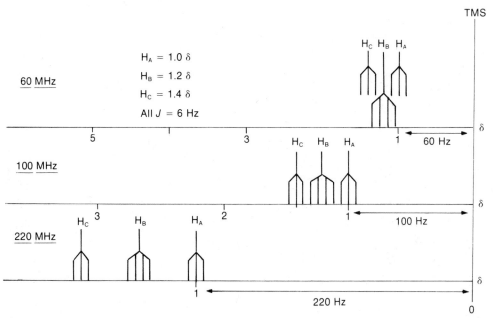

FIGURE NMR–15. A comparison of the spectrum of a compound with overlapping multiplets at 60 MHz, with spectra of the same compound also determined at 100 and 220 MHz. The drawing is to scale.

Although nmr coupling constants are not dependent on the frequency or the field strength of operation of the nmr spectrometer, chemical shifts in Hz are dependent on these parameters. This circumstance can often be used to simplify an otherwise undecipherable spectrum. Suppose, for instance, that a compound contained three multiplets: a quartet and two triplets derived from groups of protons with very similar chemical shifts. At 60 MHz these peaks might overlap, as illustrated in Figure NMR–15, and simply give an unresolved envelope of absorption.

Figure NMR–15 also shows the spectrum of the same compound at two higher field strengths (frequencies). In redetermining the spectrum at higher field strengths, the coupling constants do not change, but the chemical shifts in Hz (not δ) of the proton groups (H_A, H_B, H_C) responsible for the multiplets do increase. It should be noted that at 220 MHz the individual multiplets are cleanly separated and resolved.

NMR.13 CHEMICAL SHIFT REAGENTS

It has been known for some time that interactions between molecules and solvents, such as those due to hydrogen bonding, can cause large changes in the resonance positions of certain types of protons (e.g., hydroxyl and amino). It has also been known that the resonance positions of some groups of protons can be greatly affected by changing from the usual nmr solvents such as CCl_4 and $CDCl_3$ to solvents like benzene, which impose local anisotropic effects on surrounding molecules. In many cases it was found possible to resolve partially overlapping multiplets by such a solvent change. However, the use of chemical shift reagents for this purpose is a more recent innovation, dating from about 1969. Most of these chemical shift reagents are organic complexes of paramagnetic rare earth metals from the lanthanide series of elements. When these metal complexes are added to the compound whose spectrum is being determined, profound shifts in the resonance positions of the various groups of protons are observed. The direction of the shift (upfield or downfield) depends primarily on which metal is being used. Complexes of europium, erbium, thulium, and ytterbium shift resonances to lower field; complexes of cerium, praseodymium, neodymium, samarium, terbium, and holmium generally shift resonances to higher field. The advantage of using such reagents is that shifts similar to those observed at higher field can be induced without the purchase of an expensive higher field instrument.

Of the lanthanides, europium is probably the most commonly used metal. Two of its widely used complexes are *tris*-(dipivalomethanato)europium and *tris*-(6,6,7,7,8,8,8-heptafluoro-2,2-dimethyl-3,5-octanedionato)europium. These are frequently abbreviated Eu(dpm)$_3$ and Eu(fod)$_3$, respectively.

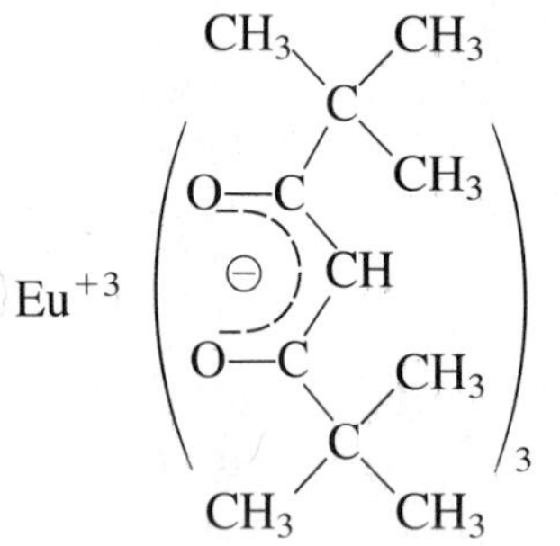

Eu(dpm)$_3$
or Eu(thd)$_3$

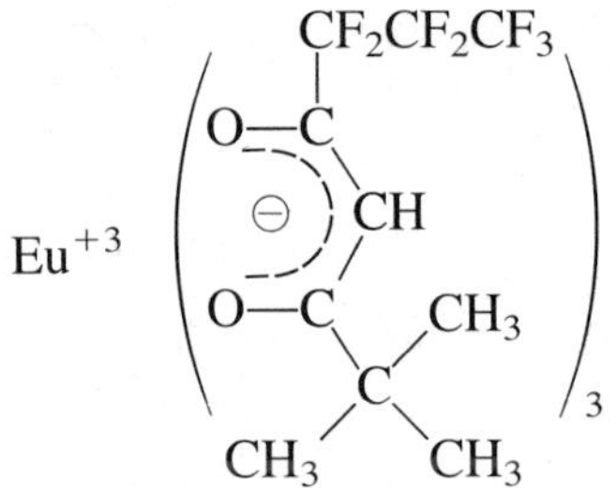

Eu(fod)$_3$

These lanthanide complexes produce spectral simplifications in the nmr spectrum of any compound that has a relatively basic pair of electrons (unshared pair) that can coordinate with Eu^{3+}. Typically, aldehydes, ketones, alcohols, thiols, ethers, and amines will all interact:

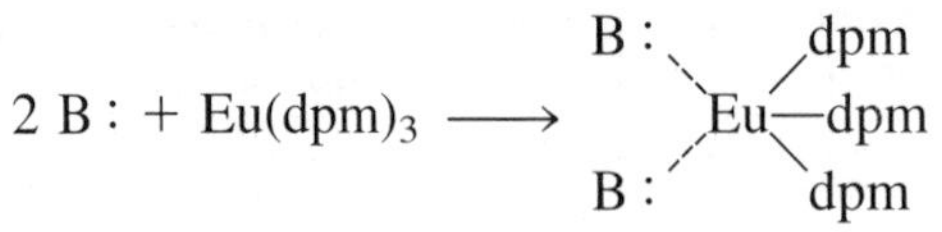

$$2 \text{ B}: + \text{Eu(dpm)}_3 \longrightarrow$$

The amount of shift that a given group of protons will experience depends (1) on the distance separating the metal (Eu^{3+}) and that group of protons and (2) on the concentration of the shift reagent in the solution. Because of the latter dependence, it is necessary when reporting a lanthanide-shifted spectrum to report the number of mole equivalents of shift reagent used or its molar concentration.

The distance factor is illustrated in the spectra of hexanol, which are given in Figures NMR−16 and NMR−17. In the absence of shift reagent, the normal spectrum

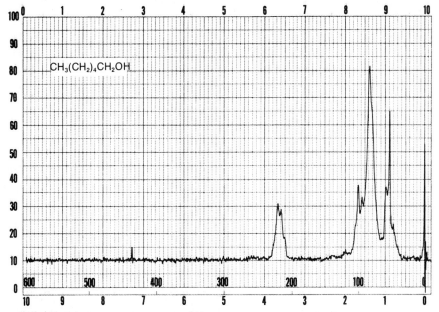

FIGURE NMR–16.   The normal 60-MHz nmr spectrum of hexanol

is obtained (Figure NMR–16). Only the triplet of the terminal methyl group and the triplet of the methylene group next to the hydroxyl are resolved in the spectrum. The other protons (aside from OH) are found together in a broad unresolved group. With shift reagent added (Figure NMR–17), each of the methylene groups is clearly separated and resolved into the proper multiplet structure. The spectrum is first order and simplified; all of the splittings are explained by the $n + 1$ rule.

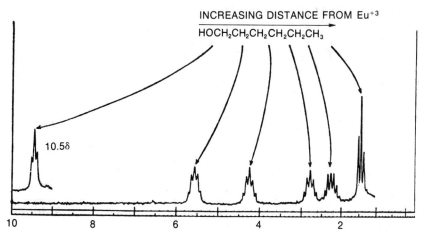

FIGURE NMR–17.   The 100-MHz nmr spectrum of hexanol with 0.29 mole equivalents of Eu(dpm)$_3$ added. (Reprinted with permission from J. K. M. Sanders and D. H. Williams, *Chem. Commun.*, 422 [1970])

One final consequence of the use of a shift reagent should be noted. Notice in Figure NMR–17 that the multiplets are not as nicely resolved into sharp peaks as one usually expects. This is due to the fact that shift reagents cause a small amount of peak broadening. At high shift reagent concentrations this problem becomes serious, but at most useful concentrations the amount of broadening experienced is tolerable.

REFERENCES

Dyer, J. R. *Applications of Absorption Spectroscopy of Organic Compounds.* Englewood Cliffs, N.J.: Prentice-Hall, 1965.

Jackman, L. M., and Sternhell, S. *Applications of Nuclear Magnetic Resonance in Organic Chemistry.* 2nd ed. New York: Pergamon Press, 1969.

Pasto, D. J., and Johnson, C. R. *Laboratory Text for Organic Chemistry.* Englewood Cliffs, N.J.: Prentice-Hall, 1979.

Paudler, W. W. *Nuclear Magnetic Resonance.* Boston: Allyn and Bacon, 1971.

Pavia, D. L., Lampman, G. M., and Kriz, G. S., Jr. *Introduction to Spectroscopy.* Philadelphia: W. B. Saunders, 1979.

Silverstein, R. M., Bassler, G. C., and Morrill, T. C. *Spectrometric Identification of Organic Compounds.* 4th ed. New York: Wiley, 1981.

Appendix 5
Carbon-13 Nuclear Magnetic Resonance Spectroscopy

CMR.1   CARBON-13 NUCLEAR MAGNETIC RESONANCE

Carbon-12, the most abundant isotope of carbon, does not possess spin ($I = 0$); it has both an even atomic number and an even atomic weight. The second principal isotope of carbon, ^{13}C, however, does have the nuclear spin property ($I = \frac{1}{2}$). ^{13}C atom resonances are not easy to observe, due to a combination of two factors. First, the natural abundance of ^{13}C is low; only 1.08% of all carbon atoms are ^{13}C. Second, the magnetic moment (μ) of ^{13}C is low. For these two reasons, the resonances of ^{13}C are about **6000 times weaker** than those of hydrogen. With special Fourier-Transform instrumental techniques, which will not be discussed here, it is possible to observe ^{13}C nuclear magnetic resonance (**cmr**) spectra on samples that contain only the natural abundance of ^{13}C.

The most useful parameter derived from cmr spectra is the chemical shift. Integrals are unreliable and are not necessarily related to the relative numbers of ^{13}C atoms present in the sample. Hydrogens that are attached to ^{13}C atoms cause spin-spin splitting, but spin-spin interaction between adjacent carbon atoms is rare. With its low natural abundance (0.0108), the probability of finding two ^{13}C atoms adjacent to one another is extremely low.

CMR.2   COMPLETELY COUPLED ^{13}C SPECTRA

Figure CMR–1 shows the cmr spectrum of ethyl phenylacetate. Consider first the upper trace shown in the figure. Chemical shifts, just as in proton nmr, are reported by the number of ppm (δ units) that the peak is shifted downfield from TMS. Keep in mind, however, that it is a ^{13}C atom of the methyl group of TMS that is being observed, not the 12 methyl hydrogens. Notice the extent of the scale. While the chemical shifts of protons encompass a range of only about 20 ppm, ^{13}C chemical shifts cover an extremely wide range of up to 200 ppm! Under these circumstances, even adjacent —CH_2— carbons in a long hydrocarbon chain generally have their own distinct resonance peaks, and these peaks are clearly resolved. It is unusual to find any two carbon atoms in a molecule having resonance at the same chemical shift unless these two carbon atoms are equivalent by symmetry.

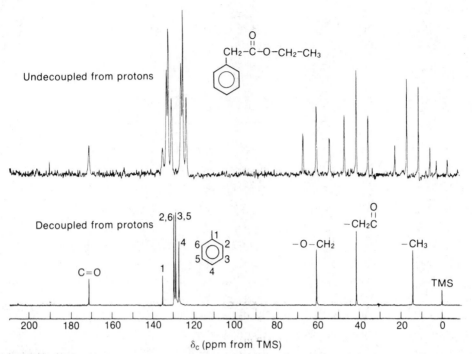

FIGURE CMR–1. Cmr spectra of ethyl phenylacetate. (From Moore, J. A., and Dalrymple, D. L., *Experimental Methods in Organic Chemistry,* copyright 1976 by W. B. Saunders Co., Philadelphia, PA. Reprinted by permission of the publisher.)

Returning to the upper spectrum in Figure CMR–1, the first quartet downfield from TMS (14.2δ) corresponds to the carbon of the methyl group. It is split into a quartet ($J = 127$ Hz) by the three attached hydrogen atoms. In addition, although it cannot be seen on the scale of this spectrum, each of the quartet lines is split into a **closely spaced** triplet ($J = $ ca 1 Hz). This additional fine splitting is caused by the two protons on the adjacent —CH_2— group. These are geminal couplings (H—C—^{13}C) of a type that commonly occur in cmr spectra, with coupling constants that are generally small ($J = 0$–2 Hz). The quartet is caused by **direct coupling** (^{13}C—H). Direct coupling constants are larger, usually about 100 to 200 Hz, and are more obvious on the scale in which the spectrum is presented.

There are two —CH_2— groups in ethyl phenylacetate. The one corresponding to the ethyl —CH_2— group is found further downfield (60.6δ), as this carbon is deshielded by the attached oxygen. It is a triplet because of the two attached hydrogens. Again, although it is not seen in this unexpanded spectrum, each of the triplet peaks is finely split into a quartet by the three hydrogens on the adjacent methyl group. The benzyl —CH_2— carbon is the intermediate triplet (41.4δ). Furthest downfield is the carbonyl group carbon (171.1δ). On the scale of presentation, it is a singlet (no directly attached hydrogens), but because of the adjacent benzyl —CH_2— group, actually it is split finely into a triplet. The aromatic ring carbons also appear in the spectrum, and they have resonances over the range from 127δ to 136δ.

CMR.3   BROAD-BAND DECOUPLED ^{13}C SPECTRA

Although the splittings in a simple molecule such as ethyl phenylacetate yield interesting structural information, namely the number of hydrogens attached to each carbon (as well as those adjacent if the spectrum is expanded), for large molecules the cmr spectrum becomes very complex due to these splittings, and the splitting patterns often overlap. It is customary, therefore, to decouple **all** the protons in the molecule by irradiating them all simultaneously with a broad spectrum of frequencies in the proper range. This type of spectrum is said to be **completely decoupled.** The completely decoupled spectrum is much simpler and, for larger molecules, much easier to interpret. The decoupled spectrum of ethyl phenylacetate is presented in the lower trace of Figure CMR–1.

In the completely decoupled cmr spectrum, each peak represents a different carbon atom. If two carbons are represented by a single peak, they must be equivalent by symmetry. Thus the carbons at positions 2 and 6 of the aromatic ring of ethyl phenylacetate give a single peak, and the carbons at positions 3 and 5 also give a single peak in the lower spectrum of Figure CMR–1.

CMR.4   CHEMICAL SHIFTS

Just as is the case for proton spectra, the chemical shift of each carbon indicates both its type and its structural environment. In fact, a correlation chart can be presented for ^{13}C chemical shift ranges, similar to the correlation chart for proton resonances shown in Figure NMR–5. Figure CMR–2 gives typical chemical shift ranges for the various types of carbon resonances.

Electronegativity, hybridization, and anisotropy effects all influence ^{13}C chemical shifts, just as they do for protons, but in a more complex fashion. These factors will not be discussed in any detail here, but note that the —CH_2— group carbon attached to oxygen in ethyl phenylacetate has a larger chemical shift than the —CH_2— carbon of the benzyl group. Note also that the carbonyl carbon appears relatively far downfield, probably due to an anisotropy effect.

CMR.5   SOME SAMPLE SPECTRA

The following spectra illustrate some of the effects that can be observed in cmr spectra. In Figure CMR–3, the spectrum of methylcyclopentane is presented. The methyl carbon appears at highest field (20.7 ppm), as expected. The carbon atoms of the ring appear over the range 25.5 to 34.9 ppm. The presence of an electronegative element should deshield a carbon atom closest to it, as is illustrated in the cases of bromocyclohexane (Figure CMR–4) and cyclohexanol (Figure CMR–5). The carbon bearing the bromine in bromocyclohexane appears at 53.0 ppm; the carbon bearing the hydroxyl group of cyclohexanol appears at 70.0 ppm. In each of these cases, note that as the ring

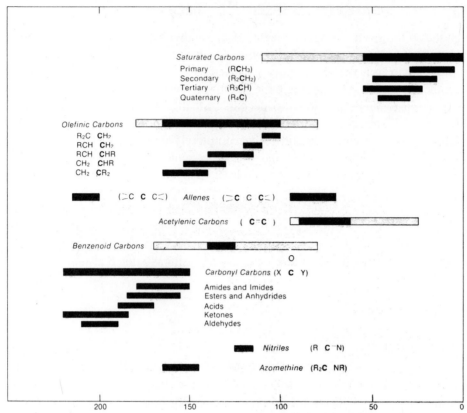

FIGURE CMR–2. ¹³C chemical shift ranges (ppm from TMS). Extended ranges sometimes occur when polar substituents are attached to the carbons. These extended ranges are indicated by the lightly shaded areas. (From Moore, J. A., and Dalrymple, D. L., *Experimental Methods in Organic Chemistry,* copyright 1976 by W. B. Saunders Co., Philadelphia, PA. Reprinted by permission of the publisher.)

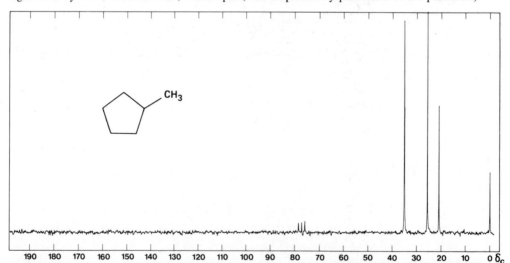

FIGURE CMR–3. Cmr spectrum of methylcyclopentane. (From Johnson, L. F., and Jankowski, W. C., *Carbon-13 NMR Spectra: A Collection of Assigned, Coded, and Indexed Spectra,* copyright 1972 by John Wiley and Sons, New York, NY. Reprinted by permission of the publisher.)

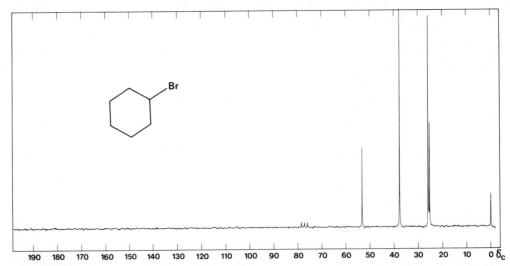

FIGURE CMR–4.   Cmr spectrum of bromocyclohexane. (From Johnson, L. F., and Jankowski, W. C., *Carbon-13 NMR Spectra: A Collection of Assigned, Coded, and Indexed Spectra,* copyright 1972 by John Wiley and Sons, New York, NY. Reprinted by permission of the publisher.)

carbons are located farther away from the electronegative element, their resonances appear at higher field. A carbon attached to a double bond appears deshielded, due to diamagnetic anisotropy. This effect can be seen in the spectrum of cyclohexene (Figure CMR–6). The carbon atoms of the double bond appear at 127.2 ppm. Again, it can be seen that as carbon atoms are located farther from the double bond, their resonances appear at higher field. The effect of diamagnetic anisotropy can be seen in the spectrum of toluene (Figure CMR–7), where the carbon atoms of the aromatic ring appear at low

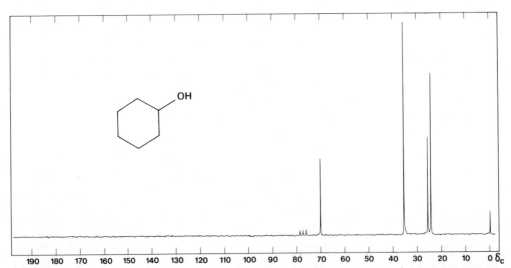

FIGURE CMR–5.   Cmr spectrum of cyclohexanol. (From Johnson, L. F., and Jankowski, W. C., *Carbon-13 NMR Spectra: A Collection of Assigned, Coded, and Indexed Spectra,* copyright 1972 by John Wiley and Sons, New York, NY. Reprinted by permission of the publisher.)

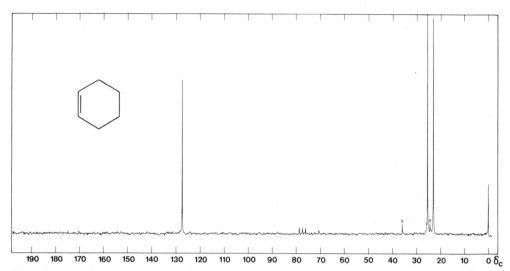

FIGURE CMR–6. Cmr spectrum of cyclohexene. (From Johnson, L. F., and Jankowski, W. C., *Carbon-13 NMR Spectra: A Collection of Assigned, Coded, and Indexed Spectra,* copyright 1972 by John Wiley and Sons, New York, NY. Reprinted by permission of the publisher.)

field (125.5 to 137.7 ppm). Finally, the strong deshielding experienced by the carbon atom of a carbonyl group can be seen in the cmr spectrum of cyclohexanone (Figure CMR–8). The carbon atom appears at a chemical shift of 211.3 ppm.

CMR.6   NUCLEAR OVERHAUSER EFFECT

As was mentioned above, integrals (areas under peaks) are not as reliable for carbon spectra as they are for hydrogen spectra. This is due in part to the **nuclear Overhauser**

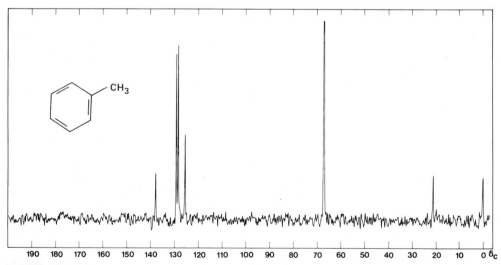

FIGURE CMR–7. Cmr spectrum of toluene. (NOTE: The peak at 67.4 ppm is due to the solvent, dioxane.) (From Johnson, L. F., and Jankowski, W. C., *Carbon-13 NMR Spectra: A Collection of Assigned, Coded, and Indexed Spectra,* copyright 1972 by John Wiley and Sons, New York, NY. Reprinted by permission of the publisher.)

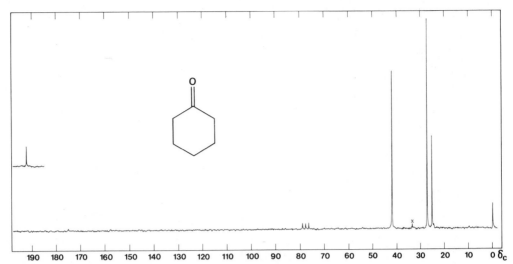

FIGURE CMR–8. Cmr spectrum of cyclohexanone. (From Johnson, L. F., and Jankowski, W. C., *Carbon-13 NMR Spectra: A Collection of Assigned, Coded, and Indexed Spectra*, copyright 1972 by John Wiley and Sons, New York, NY. Reprinted by permission of the publisher.)

effect. This effect operates when two dissimilar adjacent atoms (in this case carbon and hydrogen) both exhibit spins and are nmr active. The atoms can influence the nmr absorption intensities of each other. The effect can be either positive or negative, but in the case of carbon-13 interacting with hydrogen, the effect is positive. As a result, carbon-13 nmr absorptions vary in intensity with respect to the number of hydrogen atoms that are directly attached to the carbon atom being observed. In general, the more hydrogens that are attached to a given carbon, the stronger will be its nmr absorption. Other factors also influence the absorption intensities (they are related to molecular relaxation phenomena), so the number of attached hydrogens can only be taken as a single factor influencing absorption intensity; many times this is a very helpful factor in deciding which carbon to assign to a given absorption. In Figure CMR–1 note the low intensity of the carbonyl carbon (172δ), and in Figure CMR–7 note the low intensity of the ring carbon to which the methyl group is attached (138δ). The carbonyl peak in cyclohexanone (Figure CMR–8) is also weak. None of these carbons has attached hydrogens.

CMR.7 AN EXAMPLE OF SYMMETRY

As one example of the utility of cmr experiments, consider the cases of the isomers, 1,2- and 1,3-dichlorobenzene. Although these isomers could be difficult to distinguish from one another on the basis of their boiling points or their infrared spectra, each could be identified clearly by their cmr spectra. 1,2-Dichlorobenzene has a plane of symmetry that gives it only three different types of carbon atoms. 1,3-Dichlorobenzene has a plane of symmetry that gives it four different types of carbon atoms. The proton-decoupled cmr spectra of these two compounds are shown in Figures CMR–9 and CMR–10, respectively. It is easy to see the differences in the cmr spectra of these two isomers.

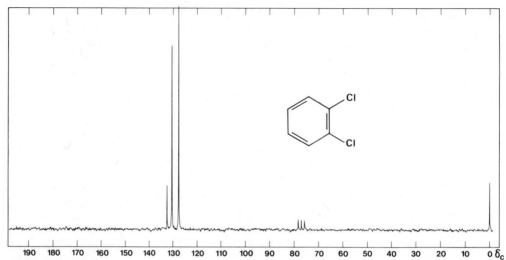

FIGURE CMR–9. Cmr spectrum of 1,2-dichlorobenzene. (From Johnson, L. F., and Jankowski, W. C., *Carbon-13 NMR Spectra: A Collection of Assigned, Coded, and Indexed Spectra,* copyright 1972 by John Wiley and Sons, New York, NY. Reprinted by permission of the publisher.)

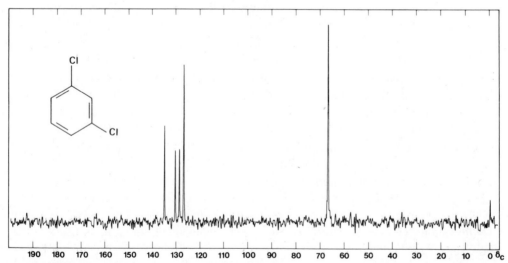

FIGURE CMR–10. Cmr spectrum of 1,3-dichlorobenzene. (NOTE: The peak at 67.4 ppm is due to the solvent, dioxane.) (From Johnson, L. F., and Jankowski, W. C., *Carbon-13 NMR Spectra: A Collection of Assigned, Coded, and Indexed Spectra,* copyright 1972 by John Wiley and Sons, New York, NY. Reprinted by permission of the publisher.)

APPENDIX 6
Index of Spectra

Infrared Spectra

NMR Spectra

¹³C NMR Spectra

Ultraviolet-Visible Spectra

Mixtures

Index